Ethnopharmacological Properties, Biological Activity and Phytochemical Attributes of Medicinal Plants, Volume 3

This book covers the morphological characteristics, ethnopharmacological properties, isolated and identified structurally diverse secondary metabolites, and biological and pharmacological activities of medicinal plants. Ethnopharmacology is the systematic study of folklore/traditional medicines, which continue to provide innovative drugs and lead molecules for the pharmaceutical industry. In fact, plant secondary metabolites, used as a single molecule or as a mixture, are medicines that can be effective and safe even when synthetic drugs fail. Therefore, the description of these secondary metabolites as well as methods for the targeted expression and/or purification is of high interest.

In addition to surveying the morphological features, ethnopharmacological properties, biological and pharmacological activities, and studies of clinical trials, this book offers a comprehensive treatment of 56 plant species. It also presents the cell culture conditions and various methods used for increasing the production of medicinally important secondary metabolites in plant cell cultures.

This volume:

- Provides the morphological features, habitat, and distribution of each species of 56 genera selected from different regions of the world.
- Presents ethnopharmacological applications of various species of the 56 genera included in this book. Different species of 56 genera are used for ethnomedicinal uses by the people of various countries of the world.
- Describes structures of various secondary metabolites identified in 56 plant species together with their biological and pharmacological activities.
- Discusses strategies of secondary metabolites production, such as organ culture, pH, elicitation, hairy root cultures, light, and mutagenesis.
- Provides a complete overview of each species of 56 genera and complete information up to 2022.

Ethnopharmacological Properties, Biological Activity and Phytochemical Attributes of Medicinal Plants is an important book for undergraduate and postgraduate students, pharmacologists, phytochemists, Ayurvedic practitioners, medical doctors, and biotechnologists interested in the ethnopharmacological properties, phytochemistry, and biological and pharmacological activities of plants.

Ethnopharmacological Properties, Biological Activity and Phytochemical Attributes of Medicinal Plants, Volume 3

Bharat Singh

CRC Press
Taylor & Francis Group
Boca Raton London New York

CRC Press is an imprint of the
Taylor & Francis Group, an **informa** business

First edition published 2024
by CRC Press
6000 Broken Sound Parkway NW, Suite 300, Boca Raton, FL 33487-2742

and by CRC Press
4 Park Square, Milton Park, Abingdon, Oxon, OX14 4RN

CRC Press is an imprint of Taylor & Francis Group, LLC

© 2024 Bharat Singh

ISBN: 9781032503363 (hbk)
ISBN: 9781032503370 (pbk)
ISBN: 9781003398035 (ebk)

DOI:10.1201/9781003398035

Typeset in Times
by Apex CoVantage, LLC

Dedicated to

---

my parents **Shri Mool Chand** and **Smt. Shanti Devi**, for encouraging and
motivating me toward learning
to my wife **Smt. Arti Singh** for her sacrifice and support extended to me during
my most difficult days of life and
my three lovely children **Gunjan Singh, Mansi Chaudhary**, and **Hemendra Singh**

Bharat Singh

Contents

About the Author x

Chapter 1   *Callistemon* Species 1

    1.1   Morphological Features, Distribution, Ethnopharmacological
            Properties, Phytochemistry, and Pharmacological Activities 1
    1.2   Phytochemistry 4
    1.3   Culture Conditions 17
    References 18

Chapter 2   *Cardamomum* or *Elettaria* Species 23

    2.1   Morphological Features, Distribution, Ethnopharmacological
            Properties, Phytochemistry, and Pharmacological Activities 23
    2.2   Phytochemistry 29
    2.3   Culture Conditions 31
    References 32

Chapter 3   *Carduus* Species 36

    3.1   Morphological Features, Distribution, Ethnopharmacological
            Properties, Phytochemistry, and Pharmacological Activities 36
    3.2   Phytochemistry 37
    3.3   Culture Conditions 44
    References 44

Chapter 4   *Centaurium* Species 47

    4.1   Morphological Features, Distribution, Ethnopharmacological
            Properties, Phytochemistry, and Pharmacological Activities 47
    4.2   Phytochemistry 50
    4.3   Culture Conditions 53
    References 55

Chapter 5   *Chenopodium* Species 60

    5.1   Morphological Features, Distribution, Ethnopharmacological
            Properties, Phytochemistry, and Pharmacological Activities 60
    5.2   Phytochemistry 63
    5.3   Culture conditions 73
    References 74

Chapter 6   *Chimaphila* Species 79

    6.1   Morphological Features, Distribution, Ethnopharmacological
            Properties, Phytochemistry, and Pharmacological Activities 79
    6.2   Phytochemistry 79

6.3   Culture Conditions 82
References 82

Chapter 7   *Cinnamomum* Species 84

7.1   Morphological Features, Distribution, Ethnopharmacological
       Properties, Phytochemistry, and Pharmacological Activities 84
7.2   Phytochemistry 91
7.3   Culture Conditions 111
References 111

Chapter 8   *Cissus* species 121

8.1   Morphological Features, Distribution, Ethnopharmacological
       Properties, Phytochemistry, and Pharmacological Activities 121
8.2   Phytochemistry 129
8.3   Culture Conditions 133
References 133

Chapter 9   *Citrus* Species 139

9.1   Morphological Features, Distribution, Ethnopharmacological
       Properties, Phytochemistry, and Pharmacological Activities 139
9.2   Phytochemistry 144
9.3   Culture Conditions 156
References 156

Chapter 10   *Clerodendrum* Species 163

10.1   Morphological Features, Distribution, Ethnopharmacological
       Properties, Phytochemistry, and Pharmacological Activities 163
10.2   Phytochemistry 169
10.3   Culture Conditions 186
References 187

Chapter 11   *Convolvulus* Species 194

11.1   Morphological Features, Distribution, Ethnopharmacological
       Properties, Phytochemistry, and Pharmacological Activities 194
11.2   Phytochemistry 198
11.3   Culture Conditions 204
References 205

Chapter 12   *Cordia* Species 209

12.1   Morphological Features, Distribution, Ethnopharmacological
       Properties, Phytochemistry, and Pharmacological Activities 209
12.2   Phytochemistry 214
12.3   Culture Conditions 229
References 230

Chapter 13 *Crassocephalum* Species 236

13.1 Morphological Features, Distribution, Ethnopharmacological
Properties, Phytochemistry, and Pharmacological Activities 236
13.2 Phytochemistry 239
13.3 Culture Conditions 242
References 242

Chapter 14 *Cyclopia* Species 246

14.1 Morphological Features, Distribution, Ethnopharmacological
Properties, Phytochemistry, and Pharmacological Activities 246
14.2 Phytochemistry 249
14.3 Culture Conditions 254
References 254

Chapter 15 *Cymbopogon* Species 258

15.1 Morphological Features, Distribution, Ethnopharmacological
Properties, Phytochemistry, and Pharmacological Activities 258
15.2 Phytochemistry 263
15.3 Culture Conditions 269
References 270

Chapter 16 *Cyperus* Species 276

16.1 Morphological Features, Distribution, Ethnopharmacological
Properties, Phytochemistry, and Pharmacological Activities 276
16.2 Phytochemistry 283
16.3 Culture Conditions 299
References 299

Index 305

About the Author

Bharat Singh, PhD, FBS, is a professor in the Amity Institute of Biotechnology, Amity University Rajasthan, Jaipur. His doctoral research was with Professor S. C. Jain and dealt with isolation and characterization of secondary metabolites as well as their pharmacological investigations from medicinal plants. He did his postdoctoral work in the Department of Botany, University of Rajasthan, Jaipur, India. His research at the Department dealt with the investigation of pharmacological potentials from Strobilanthes callosus. His current research focuses on ethnopharmacology, phytochemistry, plant cell culture engineering, and biotization strategies to increase the production of bioactive secondary metabolites in plants. By using plant cell culture engineering, the author has increased the production of secondary metabolites five- to eightfold higher than intact plants.

He has more than 20 years of teaching and research experience, teaching phytochemistry, plant cell culture engineering, plant biotechnology, genetics, molecular biology, and agricultural biotechnology in the last two decades. He regularly gives presentations to faculty and graduate students on ways to improve the quality and efficiency of teaching. Similarly, more than 50 research papers have been published by him in prominent national and international journal. The author has presented more than 20 papers in various National and International Conferences. He has written six books of phytochemistry and biotechnology and has guided several students for their PhD and postgraduate degrees.

1 *Callistemon* Species

1.1 MORPHOLOGICAL FEATURES, DISTRIBUTION, ETHNOPHARMACOLOGICAL PROPERTIES, PHYTOCHEMISTRY, AND PHARMACOLOGICAL ACTIVITIES

The Callistemon genus (Fam.—Myrtaceae) includes 34 species and is distinguished by its cylindrical stem, brush-like flowers, and shrubby nature (matching with traditional bottlebrushes). *Callistemon* species possess insecticidal as well as anticough and antibronchitis properties (Goyal et al. 2012). *Callistemon viminalis* (Sol. ex Gaertn.) G.Don is a perennial, woody, small tree or shrub, with long drooping branches, growing up to 15–20 feet high. Its leaves are simple, light green, entire, hairless, and long (1–9 cm). The leaf blade is linear (Wheeler 2005; Ahmad and Athar 2017). Inflorescence is a spike, with flowers arranged in clusters, looking like a bottle brush, inconspicuous, deciduous, and red in color. Fruit is a woody capsule containing seeds inside. It is native to Australia and widely distributed in *Tasmania* and New South Wales, Australia (Spencer and Lumley 1991; Wrigley and Fagg 1993). As per traditional Chinese medicine, the pills of *C. viminalis* are used for the treatment of hemorrhoids (Ji 2009; Islam et al. 2010). In the Jamaican medicine system, the hot water decoction (tea) of leaves is useful in gastroenteritis, diarrhea, and skin diseases (Cowan 1999; Elliot and Jones 1982; Salem et al. 2017). The aqueous, methanol and hexane extracts of *C. viminalis* leaves demonstrated antibacterial and antifungal activity against Gram-positive (*B. cereus*, *S. aureus*, and *S. pyogenes*) and Gram-negative bacterial species (*E. coli*, *P. aeruginosa*, *S. enteritidis*, and *S. sonnei*) and *C. albicans*. The hexane extract demonstrated significant antibacterial activity against Gram-positive bacteria and *C. albicans* (MBCs of 0.8–6.3 mg/ml) but showed moderate inhibition of Gram-negative bacteria (MBCs of 6.3–50 mg/ml). The aqueous and methanol extracts displayed lower MIC and MBC values than the hexane extract. These findings support the traditional claim that *C. viminalis* may be used as an effective agent in the treatment of bacterial diseases (Delahaye et al. 2009).

Tormentic acid congener (a pentacyclic triterpene saponin, isolated from *C. viminalis* leaves) displayed time-dependent antibacterial effects against *S. aureus* (MIC 25 μg/ml) and *P. aeruginosa* (12.5 μg/ml). Similarly, hydroethanol extract (MIC 25 μg/ml) and tormentic acid congener (MIC of 100 μg/ml) showed antibacterial activity against *S. pyogenes*. The hydroethanolic extract and tormentic acid congener produced disengagement of biofilms and reduced the discharge of extracellular DNA and capsular polysaccharides from biofilms of *P. aeruginosa* and *S. aureus*. Tormentic acid congener and hydroethanolic extract possess significant antibacterial and antibiofilm properties on *E. faecium*, *S. aureus*, *K. pneumoniae*, *A. baumannii*, *P. aeruginosa*, and Enterobacter species as well as can act as a source of lead compounds for the development of antibacterial agents (Chipenzi et al. 2020). The acetone and *n*-butanol extracts of *C. viminalis* flowers showed antibacterial activity against *Ralstonia solanacearum* (inhibition zone 18.5 mm), and *Agrobacterium tumefaciens* (inhibition zone 15.0 mm). The 5-hydroxymethylfurfural, palmitic acid, and pyrogallol from acetone extract and palmitic acid, 2-hydroxymyristic acid, 5-hydroxymethylfurfural, and shikimic acid have been identified from *n*-butanol extract of *C. viminalis* flowers (El-Hefny et al. 2017). The green-mediated iron (III) oxide nanoparticles were synthesized by using *C. viminalis* leaf and flower extracts. The average particle sizes were recorded as 17.91 nm (leaf) and 27.93 nm (flowers). Moreover, flower nanoparticles showed significant antimicrobial activity against *B. cereus* (inhibition zone 13 mm), *S. enteritidis* (inhibition zone 15 mm), and *V. cholerae* (inhibition zone 25 mm). The synthesized, green-mediated iron (III) oxide nanoparticles from *C. viminalis* leaf and flower extracts possess the efficacy of being used to check the pathogenic bacterial activity

DOI: 10.1201/9781003398035-1

and as an electrochemical sensor for biological and environmental analytes (Uwaya et al. 2020). *C. viminalis* essential oil showed significant antioxidant activity (IC_{50} 1.40 mg/ml and 1.77 mg/ml). Leaf oil displayed significant membrane stabilization property in hypotonic solution–stimulated hemolysis assay (IC_{50} 25.6 µg/ml) and results compared with indomethacin (17.02 µg/ml). However, leaf essential oils did not show any cytotoxicity against HCT-116 and HeLa human cancer cell lines (Gad et al. 2019).

C. citrinus (Curtis) Skeels (syn. *C. lanceolatus* DC.) is a shrub or small tree, with a slender stem, drooping branches, and 3–7 m tall. The stem bark is green or light gray. The leaves are evergreen, long (1–1.5 cm), aromatic, alternate, and lanceolate with entire margin. Inflorescence is spike type (bottle brush-like spikes). Flowers (20–50) are small, sessile, bracteate, bisexual, actinomorphic, complete, epigynous, and arrange in acropetal succession. Fruits are woody capsules and contain numerous seeds (Marzouk 2008; Arora et al. 2016; Spencer 1996; Sharanya et al. 2014). The plant species is distributed in India, Bangladesh, Ethiopia, Mexico, and Australia (Anonymous 1992; Netala et al. 2015; Chauhan et al. 2017). *C. citrinus* is used for its therapeutic potential in ethno-pharmacological applications (Tabuti et al. 2010). Methanol extract from flowers and leaves of *C. citrinus* inhibited the growth of *Listeria monocytogenes* significantly (Fayemi et al. 2017). The methanol extract of *C. citrinus* leaf and flower demonstrated strong antioxidant capacity (EC_{50} 0.474 ± 0.03 and 0.787 ± 0.15 ml sample/g) in 2,2-diphenyl-1-picrylhydeazyl assay. The leaf extract (IC50 3.69 ± 0.61 µg/ml) showed its inhibitory ability against α-glucosidase enzyme activity. The extract also demonstrated (200 µg/ml) maximum cytotoxicity against MCF-7 cell lines within 24 h of treatment (Fayemi et al. 2019). Methanolic and aqueous extracts of *C. lanceolatus* stem were screened for their antioxidant and antihyperglycemic activities. Methanol extract (100–400 µg/ml) demonstrated greater antioxidant (metal chelating) activity ($p < 0.001$) than the aqueous extract. Methanol extract (400 µg/ml) displayed 2.5-fold higher reducing power than the aqueous extract. Methanol extract (100 µg/ml) also displayed significant radical scavenging activity against DPPH (IC_{50} 31.31 µg/ml) radical. The methanol extract (400 mg/kg, p.o.) displayed more weight gain ($p < 0.05$) in treated rats than in diabetic control rats. Methanol extract (400 mg/kg, p.o.) treatment showed reduction in blood glucose and other metabolic markers accompanied by an improvement in body weight and high-density lipoprotein levels in hyperglycemic rats. The study findings suggest that methanol extract of stem possesses antioxidant and antihyperglycemic properties in diabetes (Kumar et al. 2020).

The water:ethanol (50:50) and the dichloromethane:methanol (50:50) extracts of *C. citrinus* flowers showed antibacterial effects against *S. aureus* (MIC 100 µg/ml and 50 µg/ml). Both extracts and tormentic acid suppressed the synthesis of extracellular proteases in *S. aureus*. The levels of proteolytic activity of tormentic acid were recorded as 1 while values of dichloromethane:methanol extract and the water:ethanol extract were 0.92 and 0.84, respectively. The tormentic acid suppressed extracellular protease production; hence, it is needed to explore antivirulence therapy to fight *S. aureus* infections (Mashezha et al. 2020). The crude methanol extract of *C. citrinus* leaves showed maximum free radical scavenging activity in 1,1 diphenyl-2-picrylhydrazyl and hydrogen peroxide radical assays. The petroleum ether fraction of the methanol extract showed stronger thrombolysis ($p = 0.024$) than other fractions of the extract. Similarly, the chloroform fraction of the extract showed membrane-stabilizing activity ($p = 1.000$) against hypotonic solution–induced hemolysis. The chloroform fraction of the extract was found more effective than that of the standard ($p = 0.0000$) in heat induced hemolysis. The different fractions of *C. citrinus* displayed free radical scavenging, thrombolysis and membrane-stabilizing anti-inflammatory activities (Ahmed and Rahman 2016).

The alkaloids-rich ethanol extract (1.67 mg/ml) of *C. citrinus* leaves showed antibacterial activity against *S. aureus* (MIC 0.0025 mg/ml and MBC of 0.835 mg/ml). The extract displayed bacteriostatic effect against *P. aeruginosa*. The alkaloids-rich extract demonstrated antibacterial activity as well as suppression in ATP-dependent transport of constituents across the cell membrane (Mabhiza et al. 2016). Methanolic extract of *C. lanceolatus* leaves demonstrated maximum inhibition against

S. aureus (20.0 ± 1.73 mm) and minimum inhibition against *C. albicans* (3.33 ± 1.52 mm). The extract did not show any activity against *K. pneumoniae* and *E. coli*. The extract showed maximum relative inhibition percentage against *S. aureus* (194.79%) and minimum against *C. albicans* (4.52%; Paluri et al. 2012). The methanol extract of *C. citrinus* leaves suppressed the growth of *A. hydrophilia*, *A. feacalis*, *B. cereus*, *B. subtilis*, *C. freundii*, *E. aerogenes*, *E. coli*, *K. pneumoniae*, *P. aeurogi-nosa*, *P. fluorescens*, *S. salford*, *S. marcescens*, *S. aureus* and *Y. enterocolitia* (43%; Cock 2012). The hydrodistilled extract of *C. citrinus* leaves showed the presence of eucalyptol and α-terpineol, whereas α-eudesmol, caryophyllene, (-)-bornyl-acetate, and eucalyptol were reported from the flower extract. Similarly, eucalyptol and α-pinene were obtained from the stem oil. The extract of essential oils of leaves and flowers demonstrated significant antioxidant activity in DPPH (IC_{50} 1.49 and 1.13 mg/ml) and 2,2′-azino-bis(3-ethylbenzothiazoline-6-sulfonate; IC_{50} 0.14 and 0.03 mg/ml) assays. The essential oils of leaves and flowers showed antibacterial activity against Gram-negative (*A. hydrophilia* ACC, *E. coli* ATCC 35150, *V. alginolyticus* DSM 2171, *S. typhi* ACC) and Gram-positive (*S. enteritis* ACC, *S. aureus* ACC, and *L. monocytogenes* ACC) bacterial strains (Larayetan et al. 2017). Ethyl acetate and methanolic extracts of *C. citrinus* leaves were evaluated for their *in vitro* antimalarial, antitrypanosomal, and cytotoxic properties against *T. brucei*, *P. falciparum* 3D7 strain, and human cervix adenocarcinoma cells (HeLa cells). Both extracts (methanol and ethyl acetate) demonstrated significant inhibitory effects against *E. coli* (ATCC 35150) and *V. algino-lyticus* (DSM 2171), *A. hydrophila* ACC, *S. typhi* (ACC), *P. aeruginosa* (ACC), *S. enteritis* (ACC), *S. aureus* (ACC), and *L. monocytogenes* (ACC). Ethyl acetate extract (MIC 0.025 ± 0.00, 0.025 ± 0.01, 0.025 ± 0.01, and 0.025 ± 0.00 mg/ml) displayed more potent activities than the methanol extract (MIC 0.15 ± 0.01, 0.100 ± 0.00, 0.100 ± 0.00, and 0.100 ± 0.01 mg/ml) against *A. hydrophila* (ACC), *V. alginolyticus* (DSM 2171), *S. enteritis* (ACC), and *L. monocytogenes* (ACC). Both crude extracts demonstrated antitrypanosomal (IC_{50} 6.6/9.7 μg/ml) and antiplasmodial (IC_{50} 8.4/13.0 μg/ml) activities (Larayetan et al. 2019). The essential oils (1,8-cineole and α-terpineol) of *C. citrinus* leaves (50 mg/ml) showed antibacterial activity against *S. typhi*, *S. aureus*, *E. coli* and *P. aerugi-nosa*. The maximum inhibition zone was recorded against *Salmonella typhi* (27.93 ± 2.10 mm), *S. aureus* (23.83 ± 2.75 mm). The activity was compared with gentamycin against both bacte-rial species *S. typhi* (15.00 ± 1.33 mm), *S. aureus* (13.25 ± 1.25 mm; Aweke and Yeshanew 2016). Microwave-assisted hydrodistillation-assisted extracted oil showed better antibacterial activity than hydrodistillation-assisted extracted oils. Similarly, both oil extracts displayed significant antioxidant activity in 2, 2-diphenyl-1-picrylhydrazyl radical scavenging and β-carotene/linoleic acid assays (Mande and Sekar 2020). Aqueous extract of *C. citrinus* leaves presented strongest free radicals scavenging effect (EC_{50} 43.88 ± 3.07 μg/ml) when compared to ascorbic acid (22.42 ± 0.74 μg/mL; $p ≤ 0.05$) on DPPHs assay. The half-highest efficacious ferric ion reducing power concentrations for ascorbic acid (EC_{50} 28.54 ± 2.3 μg/ml) was recorded to be one-third of the most efficacious aqueous extract (EC_{50} 88.74 ± 3.5 μg/ml). The extract contains the maximum total phenolics (308.2 ± 5.9 μg tannic acid equivalent/mg dry extract), flavonoids (516.7 ± 3.5 μg quercetin equivalent/mg dry extract), total antioxidant capacity (441.48 ± 12.8 μg ascorbic acid equivalent/mg dry extract; Ghosh et al. 2021).

1-(2,6- dihydroxy-4-methoxyphenyl)-3-methylbutan-1-one was isolated from *C. citrinus* flowers, showing significant anti-inflammatory and antinociceptive activities against experi-mental models (Radulović et al. 2015). The methanol extract (400 mg/kg) of *C. lanceola-tus* leaves showed significant ($p < 0.05$) anti-inflammatory activity against carrageenan-induced paw oedema model. The results were compared with diclofenac sodium (50 mg/kg; Kumar et al. 2011). *C. lanceolatus* leaf oil (25–100 mg/kg, p.o., for 3 days) demonstrated graded dose response, equivalent to 21.95%–89.90% protection, in the tail-flick latent test in rats. The oil (50 and 100 mg/kg, given orally for 3 days) was found effective in hot plate reaction time (64.05% and 112.97%, $p < 0.01$ and $p < 0.001$), analgesymeter induced mechanical pain (28.17% and 54.42%, $p < 0.01$ and $p < 0.001$), and acetic acid-induced writhing (26.68% and 51.79%, $p < 0.5$ and $p < 0.05$) assays. The analgesic activity of oil was compared with pentazocine (10 mg/kg, ip) and aspirin (25 mg/kg, ip). The oil (50 and 100 mg/kg, given orally for 3 days) reduced the carrageenan-induced oedema paw volume significantly (26.68% and 51.79%) in dose-dependent manner (1–3 h time interval). The results were compared with nimesulide (50 mg/kg, p.o.). As per findings, the leaf oils can be used as a significant antinociceptive and anti-inflammatory agents (Sudhakar et al. 2004).

Leaf extract of (250 mg/kg bw) *C. citrinus* was orally administered (daily) for 22 weeks to 1,2-dimethylhydrazine-induced colon carcinogenic rats. The extract significantly decreased the size of tumors, the number of aberrant crypt foci, and the crypt multiplicity. Moreover, *C. citrinus* leaf extract enhanced the activity of quinone reductase and glutathione S-transferase in the different tissues. The results were compared with animals of 1,2-dimethylhydrazine-treated group ($p > 0.05$). The results show that the leaf may be used as a chemopreventive agent against colon carcinogenesis (López-Mejia et al. 2019). The cytotoxic effects of essential oils of *C. citrinus* flowers (1,8-cineole and α-pinene) and leaves (α-pinene, limonene and α-terpineol) were evaluated against human lung carcinoma (A549), rat glioma (C-6), human colon cancer (Colo-205), and human cervical cancer (SiHa) cells. Essential oils of both leaf and flowers demonstrated maximum cytotoxicity on A549 cells (61.4% ± 5.0 and 66.7% ± 2.2, respectively), while flower oil (100 µg/ml) was found effective against C-6 cells (69.1% ± 3.1). Additionally, essential oils did not show any toxicity on normal cells. The synergistic actions of essential oils may possibly be responsible for cytotoxic activity against experimental cells. These essential oils may form a potential source of natural anticancer compounds and play a significant role in human health (Kumar et al. 2015). The *n*-hexane, chloroform, ethyl acetate, and *n*-butanol fractions of methanol extract of *C. citrinus* fruits were assessed for their antispasmodic and cytotoxic activities. Four fractions showed relaxations to the spontaneous and KCl-induced contractions. The ethyl acetate fraction showed effects on (EC_{50} 2.62 ± 0.78 mg/ml and EC_{50} 3.72 ± 0.86) spontaneous and KCl-induced contractions. Similarly, the *n*-butanol fraction displayed positive effects (EC_{50} 3.59 ± 0.2 and 5.57 ± 0.2 mg/ml) on spontaneous and KCl- induced contractions. The methanol extract (250 mg/ml) showed inhibitory effects in brine shrimp cytotoxicity assay (Ali et al. 2011).

C. rigidus R.Br. is an evergreen, stiff upright shrub, and grows up to 3–8 feet high. Its flowers are small, attractive, and bloom intermittently throughout the year in frost-free areas. Fruits are capsules, consisting of seeds. It is native to Australia, but widely distributed in the U.S. states of Florida, Texas, Arizona, and California, and in Cameroon, Australia, China, and Asia. Its leaves are used to treat cough, bronchitis, and other respiratory tract infections (Jirovetz et al. 1997). The methanol extract of *C. rigidus* leaves showed antibacterial activity against multidrug-resistant human pathogens (methicillin-resistant *S. aureus*, vancomycin-intermediate *S. aureus*, vancomycin-resistant *E. coli*, extended spectrum β-lactamase-producing *E. coli*, and multidrug-resistant Pseudomonas spp; Saxena and Gomber 2006). A major essential oil (1, 8-cineole, 89.9%) has been reported from *C. rigidus* leaves (Ji et al. 1991). *C. comboynensis* Cheel is a small shrub, slender, and 1–2 m tall. Leaves are narrow-oblanceolate, long (5–7 cm), wide (8–20 mm), apex obtuse to acute or usually mucronate, base tapering, thick, and leathery. The essential oils of *C. comboynensis* demonstrated antibacterial activity against Gram-positive (*B. subtilis* and *S. aureus*), Gram-negative (*P. vulgaris* and *P. aeruginosa*) bacteria and a pathogenic fungus *C. albicans*. Leaf oil (1000 µg/ml) extract demonstrated significant antioxidant activity (91.1 ± 0.3 %) in 1,1-diphenyl-2-picrylhydrazyl assay (Abdelhady and Aly 2012).

1.2 PHYTOCHEMISTRY

GC-MS analysis of *C. comboynensis* leaf oils showed the presence of essential oils (isobutyl acetate, α-pinene, camphene, β-pinene, β-myrcene, α-phellandrene, *p*-cymene, 1,8-cineole, limonene, γ-terpinene, terpinolene, linalool, α-fenchol, *cis*-limonene oxide, *trans*- limonene oxide, *trans*-pinocarveol, (*E*)-β-terpineol, n-dodecane, *trans*-carveol, citronellol, nerol, carvone, eugenol, β-caryophyllene, aromadendrene, *allo*-aromadendrene, terpinyl-n-butyrate, (*E*)-nerolidol, spathulenol, caryophyllene oxide, β-copaen-4-α-ol, globulol, viridiflorol, cubenol, τ-cadinol, tridecanoic acid). The extract showed antibacterial (*S. aureus, S. epidermidis, S. mutans, E. faecalis, K. pneumoniae, P. aeruginosa, E. aerogenes* and *E. coli*) and antifungal (*C. albicans, C. neoformans, T. rubrum, A. flavus, A. niger* and *S. schenckii*) activities (Gupta et al. 2008). Six polyphenolic compounds (gallic acid, ellagic acid, kaempferol 3-*O*-α-L-rhamnopyranoside, methoxy ellagic acid, quercetin, and kaempferol) have been reported from the ethanol extract of *C. comboynensis* leaves. The ethanol extract demonstrated moderate cytotoxicity against P388 cell lines. Moreover, the extract demonstrated significant antibacterial activity against Bacillus sp.HM03, Bacillus sp.HM07, and Exiguobacterium sp. HM04, Acinetobacter sp. HM01, Pseudomonas sp. HM05, and Pseudomonas sp. HM06 (Hassan et al. 2014).

The isolated compounds and *C. lanceolatus* extracts possess antimicrobial, antioxidant, antiinflammatory, antidiabetic, antiproliferative, and insecticidal properties (Sowndhararajan et al. 2021).

The presence of *n*-octane, 2-methyloctane, o-xylene, 3-methyloctane, *n*-nonane 1,3-cyclohexadiene, undecane, *n*-dodecane, 2-isopropyl-5-methylphenol, exo-2-hydroxy cineole, etyhl-5,9-dimethyl-2,4-decadienoate, 3-methyl-5-(2,6-dimethylheptyl)-1,5-pent-2-enolide, *n*-tetradecane, 4-Oxo-β-isodamascol, n-pentadecane, 2,4-di-tert-butylphenol, durohydroquinone, (10*E*,12*E*)-10,12-tetradecadienyl acetate, 2,5,5,6,8a-pentamethyl-*trans*-4a,5,6,7,8,8a-hexahydro-gamma-chromene, (−)-spathulenol, hexadecane, 10-methylicosane, heneicosane, origanene, n-nonadecane, eicosanoic acid, n-tetracosane, 1,54-dibromotetrapentacontane, *trans*-phytol, *cis*-9, *cis*-12-octadecadienoic acid, *Z*-7-tetradecenal, stearic acid, n-tetratricoaconate, ergost-7, 22-dien-9, 11-epoxy-3-ol, acetate(ester), furostan-12-one, mono(2-ethylhexyl)phthalate, *trans*-squalene, n-hexatriacontane, α-tocopherol-β-D-mannoside, γ-sitosterol has been determined by GC-MS from *n*-hexane extract of *C. viminalis* leaves. The extract showed antioxidant activity in DPPH radical scavenging (IC$_{50}$ 28.4–56.2 µg/ml; 40.1–70.2%) and linoleic acid peroxidation assays. The hemolytic activity of the extract was recorded in the range of 1.79–4.95 (Zubair et al. 2013).

A total of 34 essential oils (α-thujene, α-pinene, caphene, sabinene, β-pinene, myrcene, limonene, α-phellandrene, δ-3-carene, α-terpinene, α-terpinolene, β-cymene, limonene, *E*-ocimene, 1.8-cineol, β-ocimene-γ-terpinene, terpinolene, α-terpineol, linalool, fenchol, isopulegol, *cis*-verbenol, terpinen-4-ol, borneol, terpinen-4-ol, *p*-cymene-8-ol, α-terpineol, β-citronellol, geranial, δ-elemene, cis-citral, eugenol, geranyl acetate, methyl acetate, β-bourbonene, β-elemene, methyl eugenol, α-copaene, β-caryophyllense, α-gurjunene, α-bergamotene, funebrene, (*E*)-β-farnesene, aromadendrene, β-eudesmene, α-humulene, alloaromadendrene, δ-cadinene, germacrene D, β-cadinene, viridiflorene, β-bisabolene, δ-cadinene, globulol, β-sesquiphellandrene, α-bisabolene, calacorene, palustrol, spathulenol, caryophyllene epoxide, epiglobulol, ledol, rosifoliol, eudesm-7 (11)-en-4-ol, γ-eudesmol, α-cadinol, β-eudesmol, bulnesol, viridiflorol, bergamotol, *Z*-α-*trans*) has been determined from *C. viminalis* (Fall et al. 2017).

A new acylphloroglucinol (callistemenonone A) was isolated from *C. viminalis* leaves, and the compound showed potent bactericidal activity against methicillin resistant pathogenic microbes (Xiang et al. 2017). The acylphloroglucinols (callistenones F-K) have been identified from leaves of *C. viminalis*, and the isolated compounds showed inhibitory effects against four tumor cell lines (MCF-7, NCI-H460, SF-268 and HepG-2; Liu et al. 2016a). Five acylphloroglucinols (callistenones L-P) and three other compounds (watsonianone A, callistenones F and H) were obtained from *C. viminalis* fruits. Watsonianone A displayed antibacterial activity against *S. aureus* (MIC 20.3 µg/ml) and *E. coli* (MIC 15.6 µg/ml; Wu et al. 2017). Two unusual compounds, callistemonols A and B were separated and characterized from leaves of *C. viminalis*. Both compounds showed potent bactericidal activities against methicillin-resistant pathogenic microbes (Wu et al. 2019). Five phloroglucinols (callviminols A-E) were separated and identified from the leaves of *C. viminalis* (Liu et al. 2016b). Fifteen β-triketone with sesqui-or monoterpene adducts (callistiviminenes A-O) have been isolated and characterized from *C. viminalis* fruits. Callistiviminenes C and D showed inhibition against lipopolysaccharide-induced nitric oxide production in RAW 264.7 macrophages (IC$_{50}$ 20.3 *µ*M and 32.5 *µ*M; Wu et al. 2016).

The presence of α-thujene, α-pinene, camphene, β-pinene, myrcene, α-phellandrene, α-terpinene, 1,8-cineole, Z-(β)-ocimene, α-terpinolene, linalool, fenchol, *trans*-pinocarveol, pinocarvone, terpinen-4-ol, crypton, α-terpineol, *trans*-carveol, citronellol, carvone, geraniol, eugenol, geranyl acetate, spathulenol, caryophyllene oxide, and ledol has been determined by GC-MS from hydrodistilled extract of *C. citrinus* and *C. viminalis* leaves. The extract displayed antibacterial activity against *S. faecalis* (inhibition zone 20.3–24.0 mm), *S. aureus* (inhibition zone 23.0–26.3 mm), *B. cereus* (17.3–19.0 mm), and *S. macrcesens* (inhibition zone 11.3–23.7 mm). Results were compared with standard antibiotics gentamycin and tetracycline (Oyedeji et al. 2009).

Eight essential oils (eucalyptol, α-pinene, α-terpineol, *p*-cymene, limonene, 4-terpinenol, methyl eugenol, and methyl cinnamate) were isolated and identified from *Melaleuca viminalis* (syn. *C. viminalis*, red bottle brush) and *M. armillaris* (white Bottle brush) leaves. Essential oil of *M. viminalis* demonstrated potent *in vitro* cytotoxic effect against A549 (lung; IC_{50} 24.12 μg/ml), HCT-116 (colon; IC_{50} 21.5 μg/ml) and T47D (breast; IC_{50} 21.78 μg/ml) cell lines whereas essential oil from M. *armillaris* showed cytotoxicity against A549 (IC_{50} 10.2 μg/mL) lung cancer cell lines only (Bhagat et al. 2017). Six compounds {3-*O*-[α-L-arabinopyranosyl-(1→2)-α-L-arabinopyranosyl)]-3'-*O*-methylquercetin, 5,7,3',4' tetrahydroxy isoflavone-7-*O*-α- L-1C4-rhamnopyranosyl (1'''-6'')-*O*-β-D-4C1-glucopyranoside, 6-methyl-5,7-dihydroxy-4'- methoxyflavone, hyperoside, rutin, and isoquercitrin} were isolated and identified from the methanolic extract of *C. viminalis* leaves. The methanolic extract displayed strong cytotoxic effect against hepatocellular carcinoma cells (Hep G-2 cell lines) lines (Ahmed 2020).

Similarly, the presence of 1,8 cineole, α-pinene, α-phellandrene, limonene, and α-terpineol from leaf oil (Sharma et al. 2006), 1,8-cineole, α-pinene and α-terpineol from water-distilled extract (Chane-Ming et al. 1998; Srivastava et al. 2001), γ-butyrolactone, isobutyl isobutyrate, α-pinene, sabinene, β-pinene, myrcene, α-phellandrene, γ-terpinene, 1,8-cineole, linalool, *trans*-pinocarveol, borneol, terpinen-4-ol, α-terpineol, nerol, geranial, (Z)-β-farnesene, germacrene D, β-bisabolene, δ-cadinene, α-cadinene has been determined from *C. rigidus* and *C. citrinus* leaf essential oils. The α-terpineol and terpinen-4-ol rich fraction (1500 ppm) showed significant antifungal activity against *Phaeoramularia angolensis* (Jazet et al. 2009), 1,8-cineole, α-terpineol, linalool, 4-terpineol, spathulenol, β-gurjunene, and viridiflorol from leaves, and *p*-cymene, myrtanol, 1,8-cineole, linalool, spathulenol and rosifoliol were determined from flowers of *C. citrinus* (Petronilho et al. 2013). 3-methyltetradec-2-en-7-ol and the C-methylated flavones (5-hydroxy-7,4'-dimethoxy-6,8-dimethylflavone and 5-hydroxy-7,4'-dimethoxy-6-methylflavone) from leaves (Huq and Misra 1997), 5,7-dihydroxy-6,8-dimethyl- 4' -methoxy flavone and 8-(2-hydroxypropan-2-yl)-5-hydroxy-7-methoxy-6-methyl-4'-methoxy flavone along with the seven known phytoconstituents (α-amyrin, β-sitosterol, 3-epi-ursolic acid acetate, urs-12-en-3β-ol-β-D-glucopyranoside, betulinic acid, oleanolic acid, and kaempferol) have been characterized from aerial parts of *C. lanceolatus*. The isolated flavones reduced the levels of blood glucose in streptozotocin-induced diabetic rats (Nazreen et al. 2012). Five acylphloroglucinols (callistenones A–E) from the leaves of *C. lanceolatus* (Rattanaburi et al. 2013) and two flavonoids (eucalyptine and 8-demethyleucalyptine), two alcohols (blumenol A, *n*-tetratriacontanol), three benzoic acid derivatives (acid gallic, methyl gallate protocatechuic acid), one sterol (β-sitosterol), and one sesquiterpene (2,6,10- bisabolatriene) have been reported from *C. citrinus* stem (Khanh et al. 2016). The eucalyptol and α-terpineol from leaves, and α-eudesmol, caryophyllene, (–)-bornyl-acetate and eucalyptol have been determined from flowers of *C. citrinus*. The antioxidant potential of the leaves and flowers oils were assessed (IC_{50} 1.49 and 1.13) for DPPH and (0.14 and 0.03) for ABTS assays (Larayetan et al. 2017).

Three meroterpenoids (callistrilones L–N) were identified from *C. citrinus* leaf extract. The leaf extract suppressed the growth of PANC-1 human pancreatic cancer cells (PC_{50} 7.4 μg/ml). Callistrilone L showed potent cytotoxicity against PANC-1 cells (PC_{50} 10 to 65 nM) in nutrient-deprived medium. Callistrilone L also strongly inhibited the migration of PANC-1 cells in real time so, it may be considered as a promising lead for the development of anticancer drugs (Tawila et al. 2020a). A triterpene lactone (3β-hydroxy-13,28-epoxyurs-11-en-28-one; ursenolide) was obtained from *C. citrinus* extract. Ursenolide displayed strong cytotoxicity against PANC-1 cells under nutrient starvation (PC_{50} 0.4 μm). Ursenolide also suppressed PANC-1 cell migration significantly. In summary, it is suggested that ursenolide may be considered as a potential anticancer agent against pancreatic cancer (Tawila et al. 2020b). Two meroterpenoids (callistrilones O and P) have been reported from dichloromethane extract of *C. citrinus* leaves. Callistrilone O demonstrated strong cytotoxicity (PC_{50} 0.3 nM) against PANC-1 human pancreatic cancer cells. The activity was 2000 times stronger than the positive control arctigenin. Callistrilone O stimulated dramatic alterations in PANC-1 cell morphology leading to cell death under nutrient-deprived conditions. Callistrilone O also suppressed PANC-1 cell migration and -PANC-1 colony development under the nutrient-rich condition (Tawila et al. 2020c). Four anthocyanins [cyanidin-3,5-*O*-diglucoside (cyanin), peonidin-3,5-*O*-diglucoside (peonin), cyanidin-3-*O*-glucoside, and cyanidin-coumaroylglucoside-pyruvic acid] have been separated and identified from methanol extracts of *C. citrinus* flowers (Laganà et al. 2020).

Seven flavonoids (callistine A, 6,7- dimethyl-5,7-dihydroxy-4′-methoxy flavone, astragalin, quercetin, catechin, eucalyptin, 8-demethyleucalyptin) and five triterpenoids (3-β-acetylmorolic acid, 3β-hydroxy-urs-11-en-13(28)-olide, betulinic acid, diospyrolide and ursolic acid) were reported from the leaves and stems of *C. citrinus*. The isolated compounds [quercetin and 3β-hydroxy-urs-11-en-13(28)-olide] showed potent inhibitory effects against lipopolysaccharide-induced nitric oxide formation in macrophage RAW264.7 cells (Cuong et al. 2016). Callistrilone Q and epicallistrilone Q were isolated and identified from *C. citrinus* leaves. Both compounds presented cytotoxic effects against PANC-1 human pancreatic cancer cells at the nanomolar level (Tawila et al. 2022). The isolated compound (eucalyptol) showed radical scavenging activity (IC_{50} 16.71 μg/ml). Essential oils and the isolated compound showed (IC_{50} 21.19 μg/ml and 19.53μg/ml) protease inhibitory effects (Gogoi et al. 2021).

Ten essential oils (α-pinene, camphene, β-pinene, β-myrcene, β-pinene, eucalyptol, γ-terpinene, α-methyl-α-[4-methyl-3-pentenyl] oxiranemethanol, 2-carene, and linalool) were determined by GC-MS analysis from *C. citrinus*. Essential oils showed toxic effects against *Tribolium castaneum* by inhibiting acetylcholinesterase, α-carboxylesterase, β-carboxylesterase, glutathione-S-transferase, and alkaline phosphatase activities (Ramachandran et al. 2022). The presence of essential oils {*cis*-3-hexenyle, acetate isoamyle, isobutyrate isobutyle, R (-) α-phellandrene, L-α- pinene, camphene, β-pinene, myrcene, 3- octen-2-one, (*E*), R (-) α-phellandrene, isobutyrate isoamyle, ether, 3-butenyl pentyl, p- cymene, limonene, eucalyptol, *trans*-β-ocimene, 2-buten-1-ol, 3-methyl-, acetate, spiro [2.14] heptane, 1,5-dimethyl. . ., terpinolene, linalol, fenchol, *p*-menth-1(7)-en-9-ol, 3-oxatricyclo[4.1.1.20,4] octane . . . , pinocarveol, methanone, dicyclopropyl-bicyclo [2.2.1] heptan-2-ol, 2,3, 2-(1-Methylpropyl) pyrazine, borneol, pentyl 2 furyl cetone, terpinene-ol-4, cymenol-8, cycloheptane, 1,3,5-tris(methyl . . . , α- terpineol, *E*,*Z*-3-ethylidenecyclohexene, bicyclo [3.1.0] hexan-3-ol, 4-met . . . , 2-Cyclohexen-1-ol, 3-methyl-6-(. . . , (-) carveol, pyridine, 4-methyl-, 1-oxide, benzene, 1-ethenyl-4-methoxy-, bicyclo[3.1.0] hexan-3-ol, 4-met . . . , (-) carvone, gravenone, 2,4-pentanedione, 3- (2-propanyl)-, geraniol, ethanone, 1-cyclopentyl, 3-hexanone, 2-methyl, thymol, phenol, p-terp-butyl, 2-cyclopenten- 1-one, 3-methoxy, eugenol, 2-cyclohexen-1-one, 3- (3-hydrox, pyridine, 2-butyl-, acetate geranyle, centifolyl, β-caryophyllene, (+)-aromadendrene, copahu, patchouli, naphthalene, 1,2,3,4-tetrahydro, 1H-cyclopropa [a] naphthalene, La, exo-2-hydroxycineole, phenol, 3-methoxy-2,4,5 -trimethyl, phenanthrene, 3,6-dimethoxy9-m, furan, 2,5-dimethyl, endo-1,5,6,7-tetramethylbicyclo, caryophyllene oxide, eremophyllene, (1H)-pteridinone, 2-amino-, caryophyllene oxide, 1H-indene, 1-ethylideneoctahydr, 12-oxabicyclo [9.1.0] dodeca3,7-, 1H-indene, 1 -ethylideneoctahydr, benzaldehyde, 2,3,4-trimethoxy, 10,10-dimethyl-2,6-dimethyleneb, 10,10-dimethyl-2,6-dimethyleneb, 1H-cycloprop [e] azulen-7-ol, dec, dodecylcyclohexane, (3H)-isobenzofuranone, 3a,4,5, phenol, 2-methoxy-4- (1-propenyl)-, 1-methyl-2-methylene-transdecalin, (3H)-naphthalenone, 4,4a,5,6,7, (1H)-quinazolinone, 2,3-dihydr, acetic acid, bromo-, ethyl ester, methyl -4, 6-bis (isopropylamino)-, bicyclo [3.1.1] heptane, 6,6-dime, camphene} were determined by GC-MS analysis from *C. citrinus* leaves (Chachad et al. 2021).

Seven polyphenolic compounds (gallic acid, ellagic acid, isoquercetin, hyperin, 1,2:3,4-(bis(s)-hexahydroxy diphenoyl-β-D-glucopyranose, nilocitin and quercetin-3-*O*-α-L-glucuronopyranoside) have been reported from methanol extract of *C. viridiflorus* leaves. The methanol extract of leaves showed significant antimicrobial activity against *S. aureus*, *E. coli* and *C. albicans* (Abdelhady et al. 2012a).

Piceatannol and scirpusin B were isolated from the stem bark of *C. rigidus* and demonstrated inhibitory effects on α-amylase activity of plasma. Scirpusin B also suppressed α-amylase activity in mouse gastrointestinal tract (Kobayashi et al. 2006). Six new triketone-phloroglucinol-monoterpene hybrids (Callistrilones F–K) have been determined from twigs and leaves of *C. rigidus*. Callistrilone H and Callistrilone I demonstrated moderate inhibitory effects against herpes simplex virus (IC$_{50}$ 10.00 ± 2.50 and 12.50 ± 1.30 μm; Cao et al. 2018a). The triketone phloroglucinol monoterpene hybrids (callistrilones A and B) were isolated from the leaves of *C. rigidus* and callistrilone A demonstrated moderate inhibitory activity against Gram-positive multiresistant bacterial strains (Cao et al. 2016).

Two phloroglucinol-monoterpenoid adducts (callisretones A and B) along with callisalignene B, viminalin C, viminalin N, viminalin B, viminalin H, viminalin L, 2-methyl-1-[(5αR,8R,9αR)-5α,8,9,9α-tetrahydro-3-hydroxy-1-methoxy-5a-methyl8-(1-methylethyl)-4-dibenzofuranyl]-1-propanone were isolated and characterized from *C. rigidus*. Callisretones A and B displayed inhibitory activity on nitric oxide synthesis (IC$_{50}$ 15.3 ± 1.0 and 17.7 ± 1.1 μM; Cao et al. 2018b). Methanol, hexane, chloroform, ethyl acetate, and methanol fractions (125–1000 ppm) of methanol extract (250–2000 ppm) of *C. rigidus* leaves were screened for their antilarvicidal activity. The hexane fraction was found most effective against *A. gambiae* (LC$_{50}$ 17.11 ppm) and *A. aegypti* (LC$_{50}$ 56.25 ppm). The methanol extract and hexane fraction displayed significant activity against *C. quinquefasciatus* (LC$_{50}$ of 447.38 and 721.95 ppm; Pierre et al. 2014).

Three meroterpenoids of β-triketone and monoterpene (Callisalignenes G–I) together with (-)-callistenone F and Viminalin B were reported from *C. salignus*. Callisalignenes G, H and I showed cytotoxicity against HCT116 cells (IC$_{50}$ 8.51 ± 1.8, 9.12 ± 0.3, and 16.33 ± 3.3 μM; Qin et al. 2017a). Three acylphloroglucinol derivatives (callisalignones A–C), six meroterpenoids (callisalignenes A–F) and other compounds [2,6-dihydroxy-4-methoxyisobutyrophenone, 2,6-dihydroxy-4-methoxy-3-methylisopropiophenone, 2,6- dihydroxy-4-methoxyisovalerophenone, callistenone B, callistenone H, myrtucommulone D, myrtucommulone B, isomyrtucommulone B, Euglobal G$_1$, Euglobal G$_2$, Euglobal G$_3$, callistiviminene F, callistiviminene H, callistiviminene I, callistiviminene M, callistiviminene, callistiviminene O, and intermediate] were obtained from *C. salignus* twigs and leaves. Myrtucommulone D displayed significant antibacterial activity against *S. aureus* and three drug resistant *S. aureus* strains (MIC 1.953 and 0.975 μg/ml). Isomyrtucommulone B showed significant antibacterial activity against *E. coli* (MIC 0.122 μg/ml). Isomyrtucommulone B also demonstrated cytotoxic activity against HCT116 (IC$_{50}$ 2.09 ± 0.10 μM; Qin et al. 2017b).

A total of 15 polyphenols] quercetin-3-*O*-β-D-glucuronopyranoside, kaempferol-3-*O*-β-D-glucuronopyranoside, strictinin, quercetin-3-*O*-(2″-*O*-galloyl)-β-D-glucuronopyranoside, afzelin, digalloyl glucose, mono-galloyl glucose, acacetin, apigenin-6,7-dimethyl ether, kaempferol trimethyl ether, dimethoxy chrysin, quercetin, kaempferol, methyl gallate, and gallic acid] were isolated and identified from aqueous methanolic extract of *C. sabulatus* leaves. The aqueous methanolic extract showed greater significant antioxidant effect than that of Trolox or BHT. In addition, the extract showed elastase, tyrosinase, and collagenase suppressive effects. Moreover, significant peripheral and central analgesic effects were also displayed by the extract in a dose-dependent manner ($p < 0.0001$). The extract significantly increased the diarrhea onset but reduced the frequency of defecation and the weight of feces in a castor oil–induced diarrhea model. The extract also demonstrated a significant decrease in the gastrointestinal motility in a charcoal meal model ($p < 0.0001$)

as well as suppressive effect on gastrointestinal transit and peristaltic index ($p < 0.0001$; Mady et al. 2022). A total of 15 compounds [subulatone A (-), (-)-myrtucommulone A, (-)-myrtucommulone B′, (-)-myrtucommulone B′, (-)-myrtucommulone B, callistemenonone A, acacetin, callistiviminene F, calliviminone A, 8-demethyl eucalyptin, eucalyptin, isoguaiacin, uvaol, betulin, and betulinic acid] were isolated and identified from the *n*-hexane extract of *C. subulatus*. The isolated compounds were examined for their cytotoxic effects against PANC-1 human pancreatic cancer cell lines. The isolated compound (myrtucommulone A) presented potent cytotoxicity against PANC-1 cell lines (PC_{50} 0.28 μM; Maneenet et al. 2022). The presence of 1,8-cineole, α-pinene, α-terpineol, limonene, and α-phellandrene was reported from the leaves and α-pinene, 1,8-cineole, α-phellandrene, limonene and α-terpineol were recorded in the flowers of *C. macropunctatus* and *C. subulatus*. Essential oils showed cytotoxicity against the human liver cancer cell line HepG2 (IC_{50} 1.76 μg/mL to 13.41 μg/ml). Volatile oil demonstrated strong inhibition of topoisomerases I-β and II-β activities (IC_{50} 819.6 ng/ml; Ibrahim and Moussa 2020).

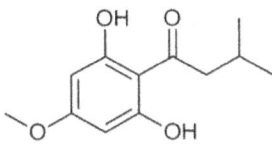

1-(2,6-Dihydroxy-4-methoxyphenyl)-3-methylbutan-1-one

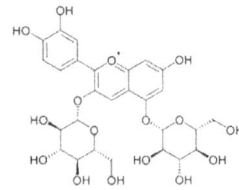

Cyanin

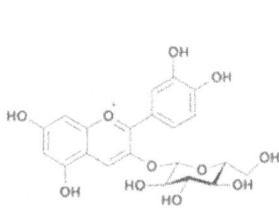

Cyanidin-3-*O*-glucoside

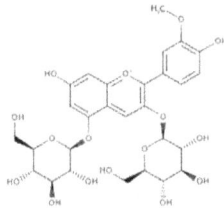

Peonidin-3,5-*O*-diglucoside

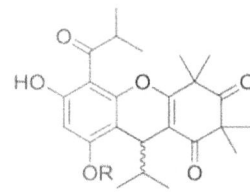

R = H (7*R*) - Myrtucommulone B

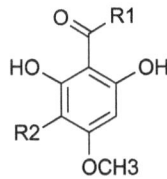

R^1 = iPr, R^2 = H - 2,6-Dihydroxy-4-methoxyisobutyrophenone; R^1 = iPr, R^2 = CH$_3$ - 2,6-Dihydroxy-4-methoxy-3-methylisopropiophenone; R^1 = iBu, R^2 = H - 2,6-Dihydroxy-4-methoxyisovalerophenone

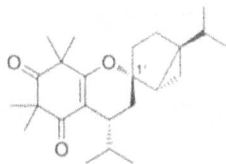

Callistenone F

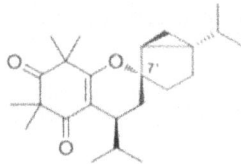

Callistiviminene F

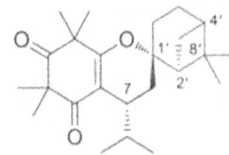

7*R*1′*S*2′*S*4′*R* - Callistiviminene I

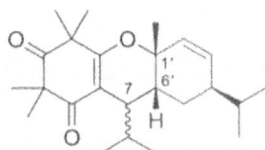

7*S* - Callistiviminene M; 7*R* - Callistiviminene N

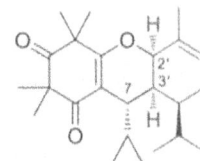

Callistiviminene O

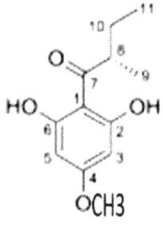

Callisalignone A

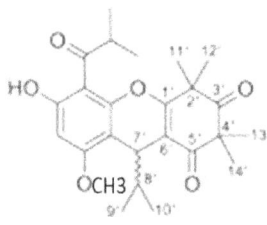

7*R* - Callisalignone B

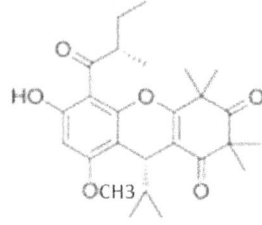

Callisalignone C

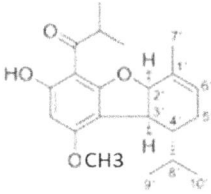

Callisalignene A

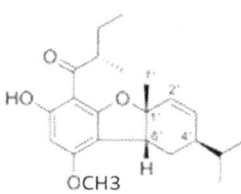

Callisalignene B

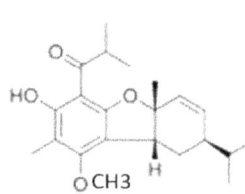

Callisalignene C

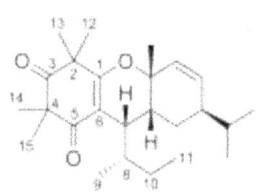

Callisalignene D

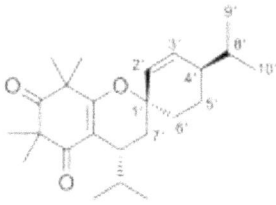

Callisalignene E

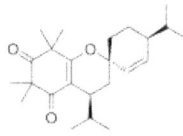

Callisalignene F

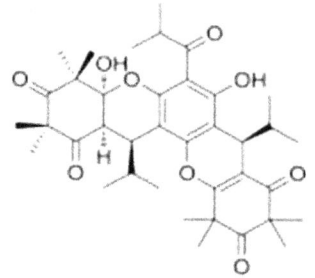

Myrtucommulone D

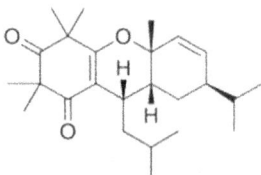

Callisalignene G

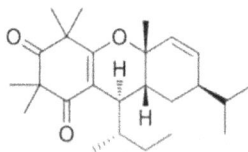

Callisalignene H

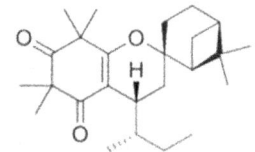

Callisalignene I

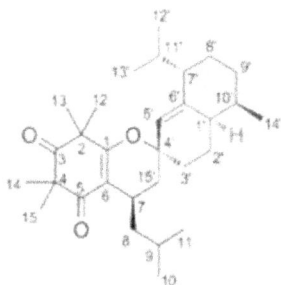

Callistiviminene A

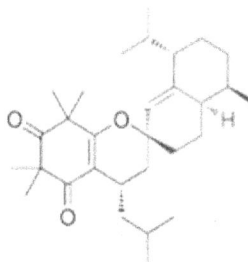

Callistiviminene B

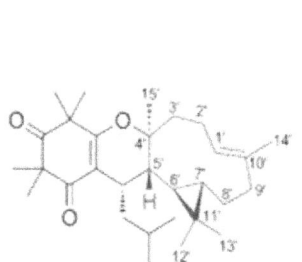

Callistiviminene C

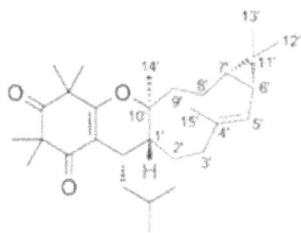

Callistiviminene D

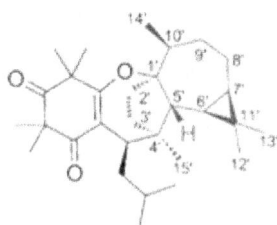

Callistiviminene E

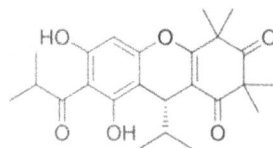

Isomyrtucommulone B

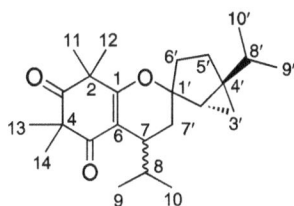

7*R*, 1'*S*,4'*R*, 2'*S* - Callistiviminene F; 7*R*, 1'*R*,4'*R*, 2'*S* - Callistiviminene G; 7S, 1'*R*,4'*R*, 2'*S* - Callistiviminene H

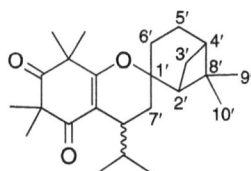

7*R*, 1'*S*,4'*R*, 2'*S* - Callistiviminene I; 7*S*, 1'*S*,4'*R*, 2'*S* - Callistiviminene J

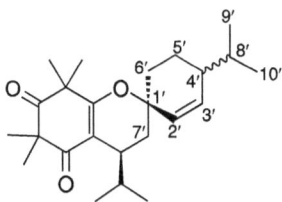

α-4'H - Callistiviminene K; β-4'H - Callistiviminene L

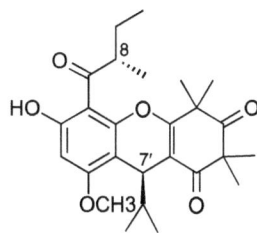

(−)-Callistenone F

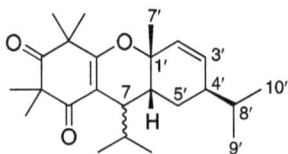

β-7H - Callistiviminene M; α-7H - Callistiviminene N

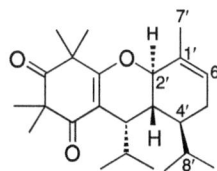

Callistiviminene O

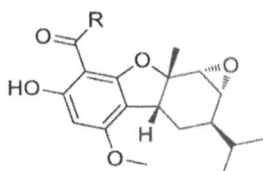

R = Isobutyl - Callviminol A; R = Isopropyl - Callviminol B

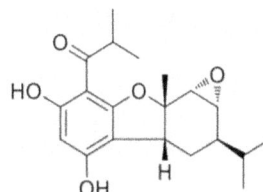

Viminalin B

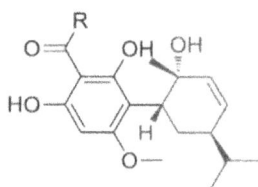

R = Isobutyl - Callviminol C; R = Isopropyl - Callviminol D

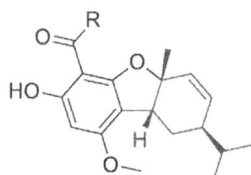

R = Isobutyl - Callviminol E

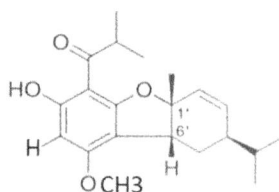

Intermediate

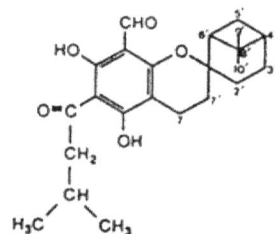

Euglobal G$_3$

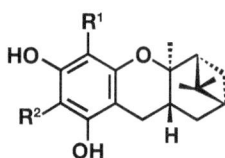

R^1 = COCH$_2$CH(CH$_3$)$_2$, R^2 = CHO – Euglobal G$_1$; R^1 = CHO, R^2 = COCH$_2$CH(CH$_3$)$_2$ - Euglobal G$_2$

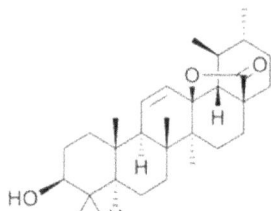

Ursenolide

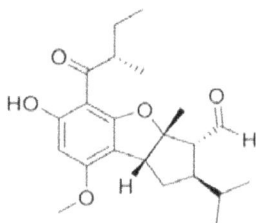

Callisretone A

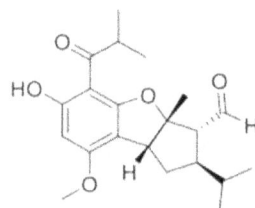

Callisretone B

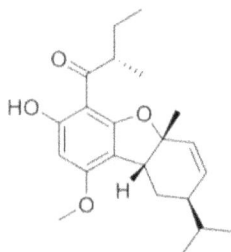

Callisalignene B

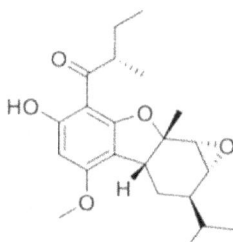

Viminalin C

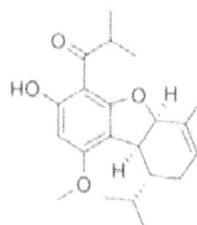

Viminalin N

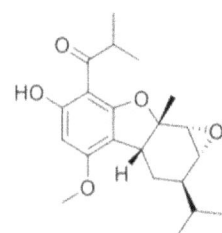

Viminalin B

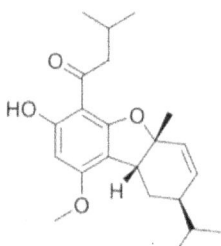

Viminalin H

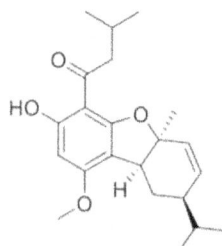

Viminalin L

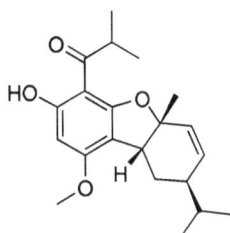

2-Methyl-1-[(5αR,8R,9αR)-5α,8,9,9α-tetrahydro-3-hydroxy-1-methoxy-5a-methyl8-(1-methylethyl)-4-dibenzofuranyl]-1-propanone

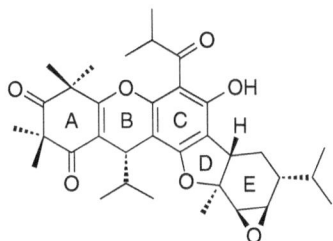

Callistrilone A

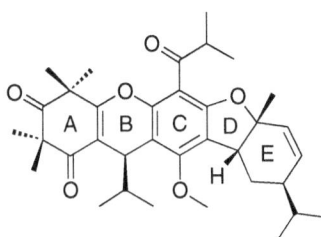

Callistrilone B

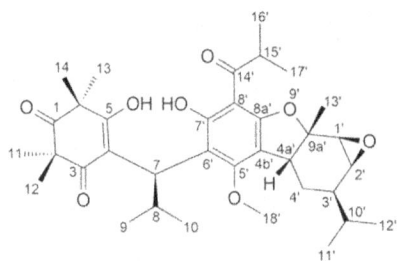

Callistrilone F

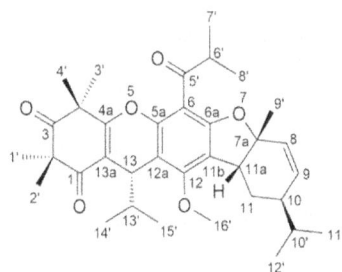

Callistrilone J

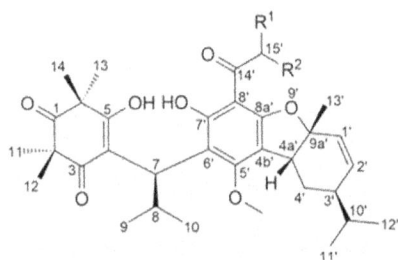

$R^1 = R^2 = CH_3$ - Callistrilone G; $R^1 = {}^iPr$, $R^2 = H_2$ - Callistrilone H; $R^1 = Et$, $R^2 = CH_3$ - Callistrilone I

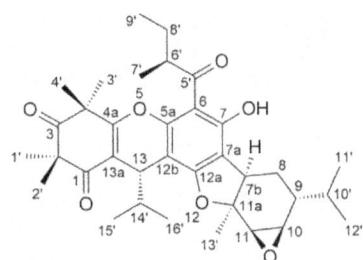

Callistrilone K

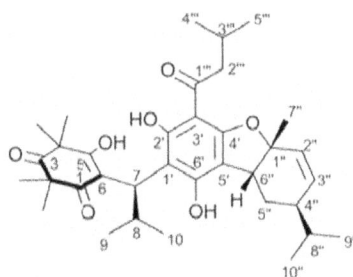

Callistrilone L

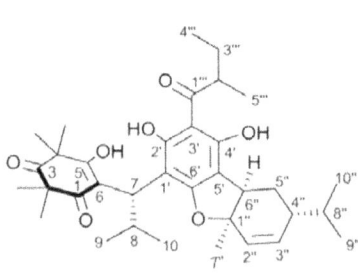

Callistrilone M

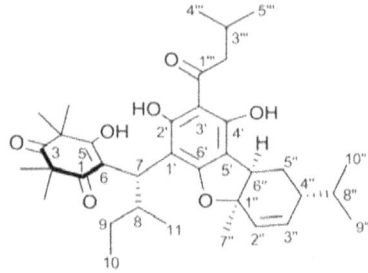

Callistrilone N

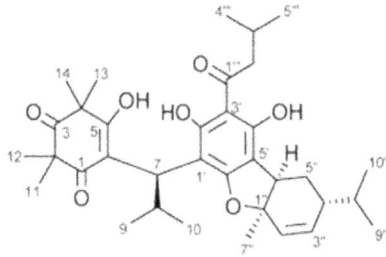

Callistrilone O

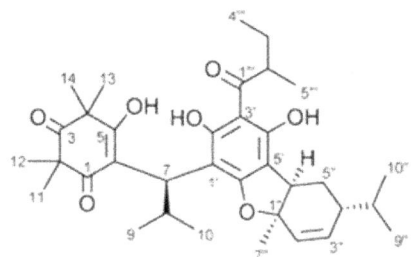

Callistrilone P

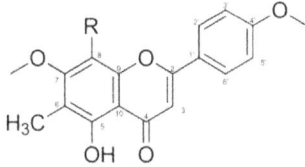

R = CH$_3$ - Eucalyptin; R = H - 8-Demethyleucalyptin

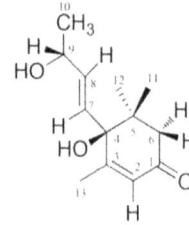

Blumenol A

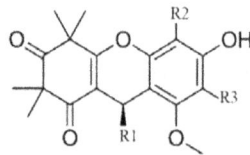

R^1 = Isopropyl, R^2 = 2-Methylbutanoyl, R^3 = H – Callistenone F; R^1 = Isopropyl, R^2 = H, R^3 = 2-Methylbutanoyl – Callistenone G; R^1 = Isopropyl, R^2 = Isovaleryl, R^3 = H – Callistenone H; R^1 = Isobutyl, R^2 = Isovaleryl, R^3 = H -Callistenone I

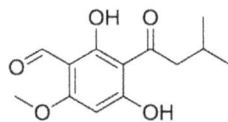

Callistenone J

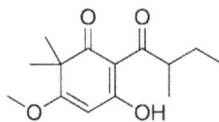

Callistenone K

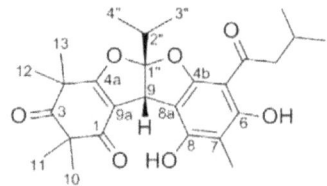

1″S* 9R* - Callistenone L

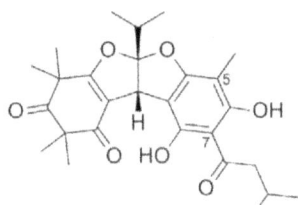

1″*S** 9*R** - Callistenone M

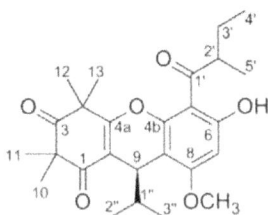

Callistenone N

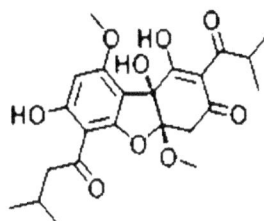

Callistemonol A

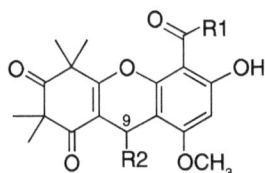

$R^1 = CH(CH_3)_2$, $R^2 = CH(CH_3)_2$ - (+)-Callistenone O (+)- or (−)-Callistenone O (−)-; $R^1 = CH_2CH(CH_3)_2$, $R^2 = CH(CH_2CH_3)CH_3$ - Callistenone P; $R^1 = CH_2CH(CH_3)_2$, $R^2 = CH(CH_3)_2$ - (+)-Callistenone H (+)- or (−)-Callistenone H (−)-

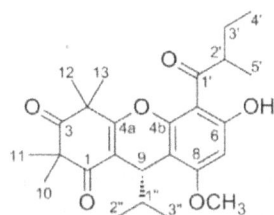

Callistenone F

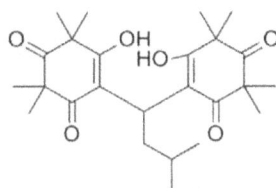

Watsonianone A

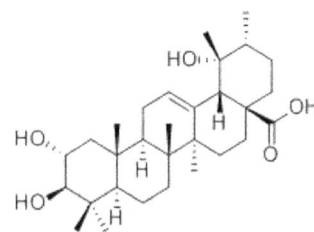

Tormentic acid

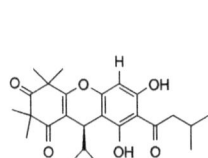

Callistenone A

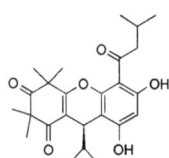

Callistenone B

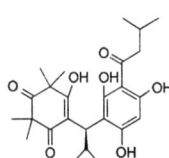

Callistenone C

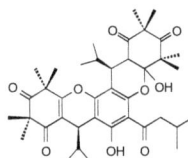

Callistenone D

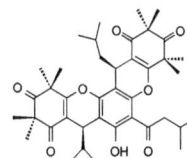

Callistenone E

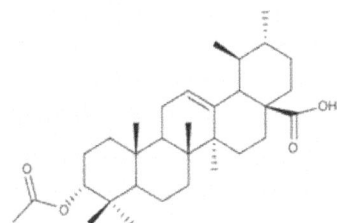

3-*epi*-Ursolic acid acetate

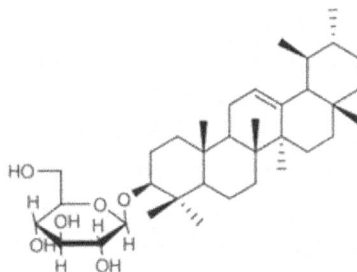

Urs-12-en-3β-ol-β-D-glucopyranoside

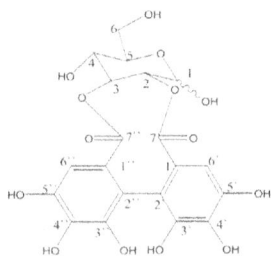

Nilocitin

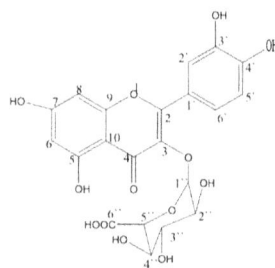

Quercetin-3-O-α-L-glucuronopyranoside

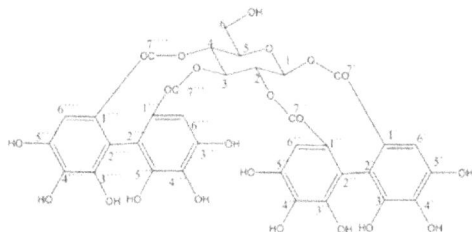

1,2:3,4-(Bis(s)- hexahydroxy diphenoyl-β-D-glucopyranose

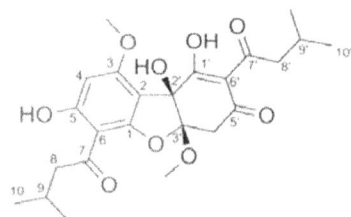

Callistemenonone A

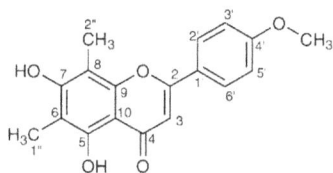

5,7-Dihydroxy-6,8-dimethyl-4′-methoxy flavone

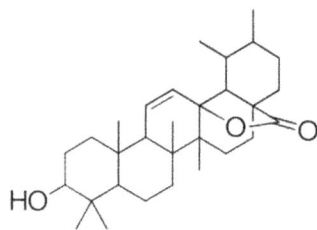

3β-Hydroxy-urs-11-en-13(28)-olide

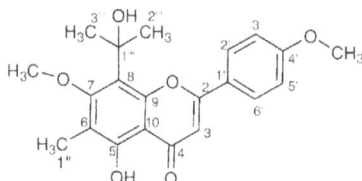

8-(2-hydroxypropan-2-yl)-5-hydroxy-7-methoxy-6-methyl-4′-methoxy flavone

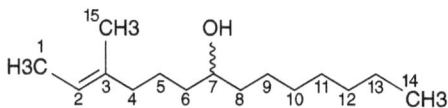

3-Methyltetradec-2-en-7-ol

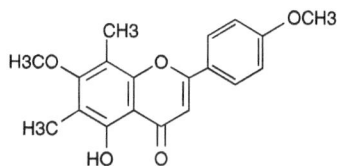

5-Hydroxy-7,4′-dimethoxy-6,8-dimethylflavone

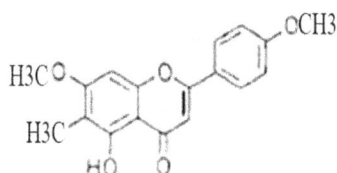

5-Hydroxy-7,4′-dimethoxy-6-methylflavone

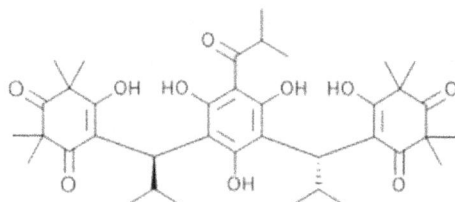

(–)-Myrtucommulone A

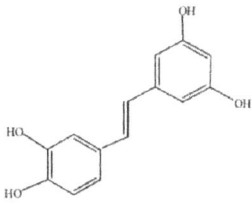

Piceatannol

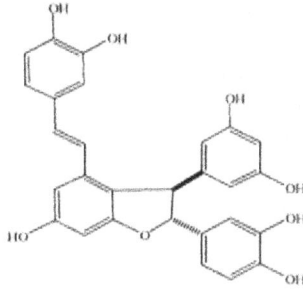

Scirpusin B

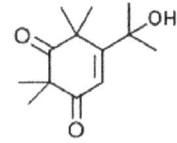

Subulatone A (−)

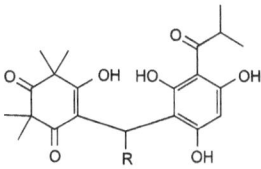

R = ... - (−)-Myrtucommulone B'; R = ... - (−)-Myrtucommulone B'

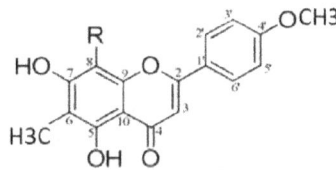

R = H - Callistine A; R = CH3 - 6,8-Dimethyl-5,7-dihydroxy-4'-methoxy flavone

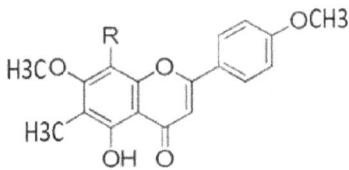

R = H - 8-Demethyleucalyptin

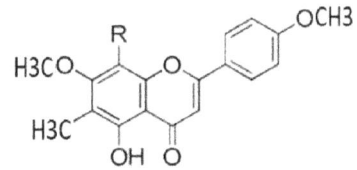

3β-Acetylmorolic acid

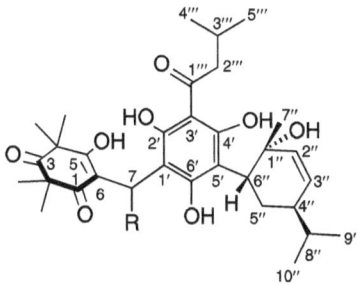

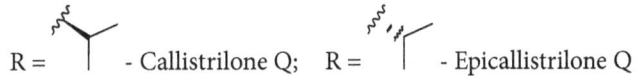

R = ... - Callistrilone Q; R = ... - Epicallistrilone Q

1.3 CULTURE CONDITIONS

Axillary buds and nodal parts of C. *viminalis* were cultured on modified de Fossard's (1978) medium for the induction and proliferation of callus. The medium was supplemented with sucrose (0.06 M), inositol (300 μM), nicotinic acid (20 μM), pyridoxine hydrochloride (3 μM), thiamine hydrochloride (2 µM), riboflavin (10 µM), cytokinins (5 µM), and auxins (0.1 µM). *p*-chlorophen-oxyacetic acid (0.1µM) promoted maximum growth of callus (Shipton 1982). Similarly, the nodal explants of *C. citrinus* were cultured on solid Woody Plant Medium (Lloyd and McCown 1980) containing agar (8 g/l), myo-inositol (100 mg/l), thiamine (1 mg/l), pyridoxine (0.5 mg/l), nicotinic acid (0.5 mg/l), and sucrose (30 g/l), and BA (1 mg/l) or NAA (1 mg/) and BA (0.25 mg/l). The composition enhanced maximum proliferation of callus (Papafotiou and Skylourakis 2010). *C. viridiflorous*, *C. comboynensis*, and *C. lanceolatus* explants were used for callus induction. The explants were cultured on MS (Murashige and Skoog 1962) medium with different combinations of growth hormones. For cell culture induction, the *C. viridiflorous*, *C. comboynensis*, and *C. lanceolatus* aliquotes of callus were acquired from the growth phase and transferred in a suspension MS medium. The medium was supplemented with kinetin (0.9 mg/l) and 1-naphthaleneacetic acid (1.1 mg/l; Abdelhady et al. 2012b). Semihard wood microcuttings of *C. viminalis*, pretreated with indole butyric acid (2000 mg/l), were inoculated on MS or half-strength MS medium containing indole butyric acid (0, 2, 4 and 6 mg/l) for induction and proliferation of callus (Shokri et al. 2012).

REFERENCES

Abdelhady MI, Aly HAH. 2012. Antioxidant and antimicrobial activities of *Callistemon comboynensis* essential oil. Free Radic Antioxid 2, 37–41.
Abdelhady MIS, Kamal AM, Tawfik NF, Abdelkhalik SM. 2012a. Polyphenolic constituents of the methanolic extract of *Callistemon viridiflorous* leaves and its antimicrobial activity. Pharmacogenomics J 4, 47–53.
Abdelhady MIS, Kamal AM, Tawfik NF, Abdelkhalik SM, Motaal AA, Beerhues L. 2012b. Induction of callus and cell suspension cultures of three *Callistemon* species. Biotechnology 6, 295–298.
Ahmad K, Athar F. 2017. Phytochemistry and pharmacology of *Callistemon viminalis* (Myrtaceae): A review. Nat Prod J 7, 166–175.
Ahmed AH. 2020. Phytochemical and cytotoxicity studies of *Callistemon viminalis* leaves extract growing in Egypt. Curr Pharm Biotechnol 21, 219–225.
Ahmed F, Rahman MS. 2016. Preliminary assessment of free radical scavenging, thrombolytic and membrane stabilizing capabilities of organic fractions of *Callistemon citrinus* (Curtis.) skeels leaves. BMC Complement Altern Med 16, 247.
Ali N, Ahmed G, Shah SWA, Shah I, Ghias M, Khan I. 2011. Acute toxicity, brine shrimp cytotoxicity and relaxant activity of fruits of *Callistemon citrinus* Curtis. BMC Complement Altern Med 11, 99.
Anonymous. 1992. The Wealth of India: A Dictionary of Indian Raw Materials and Industrial Products, Vol III. PID, CSIR, New Delhi.
Arora DS, Nim L, Kaur H. 2016. Antimicrobial potential of *Callistemon lanceolatus* seed extract and its statistical optimization. Appl Biochem Biotech 180, 289–305.
Aweke N, Yeshanew S. 2016. Chemical composition and antibacterial activity of essential oil of *Callistemon citrinus* from Ethiopia. Eur J Med Plants 17, 1–7.
Bhagat M, Sangral M, Pandita S, Vironica, Gupta S, Bindu K. 2017. Pleiotropic chemodiversity in extracts and essential oil of *Melaleuca viminalis* and *Melaleuca armillaris* of Myrtaceae family. J Explor Res Pharmacol 2, 113–120.
Cao J-Q, Huang X-J, Li Y-T, Wang Y, Wang L, Jiang R-W, Ye W-C. 2016. Callistrilones A and B, triketone—phloroglucinol—monoterpene hybrids with a new skeleton from *Callistemon rigidus*. Org Lett 18, 120–123.
Cao J-Q, Tian H-Y, Li M-M, Zhang W, Wang Y, Wang L, Ye W-C. 2018b. Rearranged phloroglucinol-monoterpenoid adducts from *Callistemon rigidus*. J Nat Prod 81, 57–62.
Cao J-Q, Wu Y, Zhong Y-L, Li N-P, Chen M, Li M-M, Ye W-C, Wang L. 2018a. Antiviral triketone-phloroglucinol-monoterpene adducts from *Callistemon rigidus*. Chem Biodivers 15, e1800172.
Chachad DP, Dias A, Uniyal K, Varma U, Jadhav P, Satvekar T, Ghag-Sawant M, Mondal M, Doshi N. 2021. Chemical characterization; antimicrobial and larvicidal activity of essential oil from *Callistemon citrinus* (Bottle brush) leaves. bioRxiv.
Chane-Ming J, Vera RR, Fraisse DJ. 1998. Chemical composition of essential oil of *Callistemon citrinus* (Curtis) skeel from reunion. J Essent Oil Res 10, 429–431.
Chauhan S, Chauhan SVS, Galetto L. 2017. Floral and pollination biology, breeding system and nectar traits of *Callistemon citrinus* (Myrtaceae) cultivated in India. S Afr J Bot 111, 319–325.

Chipenzi T, Baloyi G, Mudondo T, Sithole S, Chi GF, Mukanganyama S. 2020. An evaluation of the antibacterial properties of tormentic acid congener and extracts from *Callistemon viminalis* on Selected ESKAPE pathogens and effects on biofilm formation. Adv Pharmacol Pharm Sci 2020, 8848606.

Cock IE. 2012. Antimicrobial activity of *Callistemon citrinus* and *Callistemon salignus* methanolic extracts. Pharmacogn Commun 2, 50–57.

Cowan MM. 1999. Plant products as antimicrobial agents. Clin Microbiol Rev 12, 564–582.

Cuong NM, Khanh PN, Duc HV, Huong TT, Kim Y-C, Long PQ, Kim YH. 2016. Flavonoids and triterpenoids from *Callistemon citrinus* and their inhibitory effect on no production in LPS-stimulated RAW264.7 macrophages. Vietnam J Sci Technol 54, 214–223.

de Fossard RA. 1978. Tissue culture propagation of *Eucalyptus ficifolia*. In: Proceedings of a Symposium on Plant Tissue Culture, Muell F (ed.). Science Press, Peking, pp 425–438.

Delahaye C, Rainford L, Nicholson A. 2009. Antibacterial and antifungal analysis of crude extracts from the leaves of *Callistemon viminalis*. J Med Biol Sci 3, 1–7.

El-Hefny M, Ashmawy NA, Salem MZM, Salem AZM. 2017. Antibacterial activities of the phytochemicals-characterized extracts of *Callistemon viminalis*, *Eucalyptus camaldulensis* and *Conyza dioscoridis* against the growth of some phytopathogenic bacteria. Microb Pathog 113, 348–356.

Elliot WR, Jones DL. 1982. Encyclopedia of Australian Plants, Vol 2. Lothian Publishing Company, Melbourne.

Fall R, Ngom S, Sall D, Sembène M, Samb A. 2017. Chemical characterization of essential oil from the leaves of *Callistemon viminalis* (D.R.) and *Melaleuca leucadendron* (Linn.). Asian Pac J Trop Biomed 7, 347–351.

Fayemi PO, Ozturk I, Kaan D, Özcan S, Yerer MB, Dokumaci AH, Özcan C, Uwaya GE, Fayemi OE, Yetim H. 2019. Bioactivities of phytochemicals in *Callistemon citrinus* against multi-resistant foodborne pathogens, alpha glucosidase inhibition and MCF-7 cancer cell line. Biotechnol Biotechnol Equip 33, 764–778.

Fayemi PO, Ozturk I, Ozcan C. 2017. Antimicrobial activity of extracts from *Callistemon citrinus* flowers and leaves against *Listeria monocytogenes* in beef burgers. J Food Meas Charact 11, 924–929.

Gad HA, Ayoub IM, Wink M. 2019. Phytochemical profiling and seasonal variation of essential oils of three *Callistemon* species cultivated in Egypt. PLoS One 14, e0219571.

Ghosh P, Goswami S, Roy S, Das R, Chakraborty T, Ray S. 2021. Comparative *in vitro* antioxidant and antibacterial activities of leaf extract fractions of Crimson bottlebrush, *Callistemon citrinus* (Curtis.) Skeels. bioRxiv. https://doi.org/10.1101/2021.03.26.436274.

Gogoi R, Begum T, Sarma N, Pandey SK, Lal M. 2021. Chemical composition of *Callistemon citrinus* (Curtis) Skeels aerial part essential oil and its pharmacological applications, neurodegenerative inhibitory, and genotoxic efficiencies. J Food Biochem 45, e13767.

Goyal PK, Jain R, Jain S, Sharma AA. 2012. Review on biological and phytochemical investigation of plant genus *Callistimon*. Asian Pac J Trop Biomed 2, S1906–S1909.

Gupta S, Kumar A, Srivastava K, Srivastava SK, Luqman S, Maurya A, Darokar MP, Syamsundar KV, Khanuja SPS. 2008. Antimicrobial activity and chemical composition of *Callistemon comboynensis* and *C. citrinus* leaf essential oils from the Northern plains of India. Nat Prod Commun 3, 1931–1934.

Hassan HA, Kamal AM, Abdelhady MI, Ibrahim M. 2014. Investigation of polyphenolic compounds, cytotoxic and antimicrobial activities of *Callistemon comboynensis* leaves. Nat Prod 10, 13–16.

Huq F, Misra LN. 1997. An alkenol and C-methylated flavones from *Callistemon lanceolatus* leaves. Planta Med 63, 369–370.

Ibrahim N, Moussa AY. 2020. Comparative metabolite profiling of *Callistemon macropunctatus* and *Callistemon subulatus* volatiles from different geographical origins. Ind Crops Prod 147, 112222.

Islam MR, Ahamed R, Rahman MO, Akbar MA, Al-Amin M, Alam KD. 2010. *In vitro* antimicrobial activities of four medicinally important plants in Bangladesh. Eur J Sci Res 39, 199–206.

Jazet PM, Tatsadjieu LN, Ndongson BD, Kuate J, Zollo PHA, Menut C. 2009. Correlation between chemical composition and antifungal properties of essential oils of *Callistemon rigidus* and *Callistemon citrinus* of Cameroon against *Phaeoramularia angolensis*. J Med Plant Res 3, 9–15.

Ji T. 2009. Traditional Chinese medicine pills for treating hemorrhoid. Chinese Patent, CN101352524 A 20090128.

Ji X-D, Pu Q-L, Garraffo HF, Pannell LK. 1991. The essential oil of the leaves of *Callistemon rigidus* R. Br. J Essent Oil Res 3, 465–466.

Jirovetz L, Fleischhacker W, Buchbauer G, Ngassoum MB. 1997. Analysis of the essential oils of *Callistemon rigidus* (Myrtaceae) from Cameroun by GC=FID and GC=MS. Sci Pharm 65, 315–319.

Khanh PN, Duc HV, Huong TT, Ha VT, Van DT, Son NT, Kim YH, Viet DQ, Cuong NM. 2016. Phenolic compounds from *Callistemon citrinus* leaves and stems. Vietnam J Sci Technol 54, 190–197.

Kobayashi K, Ishihara T, Khono E, Miyase T, Yoshizaki F. 2006. Constituents of stem bark of *Callistemon rigidus* showing inhibitory effects on mouse α-amylase activity. Biol Pharm Bull 29, 1275–1277.

Kumar D, Sukapaka M, Kiran Babu GD, Padwad Y. 2015. Chemical composition and in vitro cytotoxicity of essential oils from leaves and flowers of *Callistemon citrinus* from western Himalayas. PLoS One 10, e0133823.

Kumar R, Gupta A, Singh AK, Bishayee A, Pandey AK. 2020. The antioxidant and antihyperglycemic activities of bottlebrush plant (*Callistemon lanceolatus*) stem extracts. Medicines 7, 11.

Kumar S, V Kumar V, Prakash OM. 2011. Pharmacognostic study and anti-inflammatory activity of *Callistemon lanceolatus* leaf. Asian Pac J Trop Biomed 1, 177–181.

Laganà G, Barreca D, Smeriglio A, Germanò MP, D'Angelo V, Calderaro A, Bellocco E, Trombetta D. 2020. Evaluation of anthocyanin profile, antioxidant, cytoprotective, and anti-angiogenic properties of *Callistemon citrinus* flowers. Plants 9, 1045.

Larayetan RA, Okoh OO, Sadimenko A, Okoh AI. 2017. Terpene constituents of the aerial parts, phenolic content, antibacterial potential, free radical scavenging and antioxidant activity of *Callistemon citrinus* (Curtis) Skeels (Myrtaceae) from Eastern Cape Province of South Africa. BMC Complement Altern Med 17, 292.

Larayetan RA, Ololade ZS, Ogunmola OO, Ladokun A. 2019. Phytochemical constituents, antioxidant, cyto-toxicity, antimicrobial, antitrypanosomal, and antimalarial potentials of the crude extracts of *Callistemon citrinus*. Evid Based Complement Altern Med 2019, 5410923.

Liu HX, Chen K, Liu Y, Li C, Wu JW, Xu ZF, Tan HB, Qiu SX. 2016b. Callviminols A-E, new terpenoid-conjugated phloroglucinols from the leaves of *Callistemon viminalis*. Fitoterapia 115, 142–147.

Liu HX, Chen YC, Liu Y, Zhang WM, Wu JW, Tan HB, Qiu SX. 2016a. Acylphloroglucinols from the leaves of *Callistemon viminalis*. Fitoterapia 114, 40–44.

Lloyd G, McCown BH. 1980. Commercially feasible micropropagation of mountain laurel, *Kalmia latifolia*, by use of shoot-tip culture. Proc Int Plant Propag Soc 30, 421–427.

López-Mejia A, Ortega-Pérez LG, Godinez-Hernández D, Nateras-Marin B, Meléndez-Herrera E, Rios-Chavez P. 2019. Chemopreventive effect of *Callistemon citrinus* (Curtis) Skeels against colon cancer induced by 1,2-dimethylhydrazine in rats. J Cancer Res Clin Oncol 145, 1417–1426.

Mabhiza D, Chitemerere T, Mukanganyama S. 2016. Antibacterial properties of alkaloid extracts from *Callistemon citrinus* and *Vernonia adoensis* against *Staphylococcus aureus* and *Pseudomonas aeruginosa*. Int J Med Chem 2016, Article ID 6304163.

Mady MS, Elsayed HE, El-Sayed EK, Hussein AA, Ebrahim HY, Moharram FA. 2022. Polyphenolic profile and ethnopharmacological activities of *Callistemon subulatus* (Cheel) Craven leaves cultivated in Egypt. J Ethnopharmacol 284, 114698.

Mande P, Sekar N. 2020. Comparison of chemical composition, antioxidant and antibacterial activity of *Callistemon citrinus* skeels (bottlebrush) essential oil obtained by conventional and microwave-assisted hydrodistillation. J Microw Power Electromagn Energy 54, 230–244.

Maneenet J, Tawila AM, Omar AM, Phan ND, Ojima C, Kuroda M, Sato M, Mizoguchi M, Takahashi I, Awale S. 2022. Chemical constituents of *Callistemon subulatus* and their anti-pancreatic cancer activity against human PANC-1 cell line. Plants 11, 2466.

Marzouk MSA. 2008. An acylated flavonol glycoside and hydrolysable tannins from *Callistemon lanceolatus* flowers and leaves. Phytochem Anal 19, 541–549.

Mashezha R, Mombeshora M, Mukanganyama S. 2020. Effects of tormentic acid and the extracts from *Callistemon citrinus* on the production of extracellular proteases by *Staphylococcus aureus*. Biochem Res Int 2020, 6926320.

Murashige T, Skoog F. 1962. A revised medium for rapid growth and bioassays with tobacco tissue cultures. Physiol Plant 15, 473–497.

Nazreen S, Kaur G, Alam MM, Shafi S, Hamid H, Ali M, Alam MS. 2012. New flavones with antidiabetic activity from *Callistemon lanceolatus* DC. Fitoterapia 83, 1623–1627.

Netala S, Penmetsa R, Nakka S, Polisetty BL. 2015. Pharmacognostic study of *Callistemon citrinus* L. bark. Int J Pharm Pharm Sci 7, 427–430.

Oyedeji OO, Lawal AO, Shode OF, Oyedeji OA. 2009. Chemical composition and antibacterial activity of the essential oils of *Callistemon citrinus* and *Callistemon viminalis* from South Africa. Molecules 14, 1990–1998.

Paluri V, Ravichandran S, Kumar G, Karthik L, Bhaskara Rao KV. 2012. Phytochemical composition and *in vitro* antimicrobial activity of methanolic extract of *Callistemon lanceolatus* D.C. Int J Pharm Pharm Sci 4, 699–702.

Papafotiou M, Skylourakis A. 2010. *In vitro* propagation of *Callistemon citrinus*. Acta Horticult 885, 267–270.

Petronilho S, Rocha SM, Ramírez-Chávez E, Molina-Torres J, Rios-Chavez P. 2013. Assessment of the terpenic profile of *Callistemon citrinus* (Curtis) Skeels from Mexico. Ind Crops Prod 46, 369–379.

Pierre DYS, Okechukwu EC, Nchiwan NE. 2014. Larvicidal and phytochemical properties of *Callistemon rigidus* R. Br. (Myrtaceae) leaf solvent extracts against three vector mosquitoes. J Vector Borne Dis 51, 216–223.

Qin X-J, Liu H, Yu Q, Yan H, Tang J-F, An L-K, Khan A, Chen Q-R, Hao X-J, Liu H-Y. 2017b. Acylphloroglucinol derivatives from the twigs and leaves of *Callistemon salignus*. Tetrahedron 73, 1803–1811.

Qin X-J, Shu T, Yu Q, Yan H, Ni W, An L-K. Li P-P, Zhi Y-E, Khan A, Liu H-Y. 2017a. Cytotoxic acylphloroglucinol derivatives from *Callistemon salignus*. Nat Prod Bioprospect 7, 315–321.

Radulović NS, Randjelović PJ, Stojanović NM, Cakić ND, Bogdanović GA, Živanović AV. 2015. Aboriginal bush foods: A major phloroglucinol from Crimson Bottlebrush flowers (*Callistemon citrinus*, Myrtaceae) displays strong antinociceptive and anti-inflammatory activity. Food Res Int 77, 280–289.

Ramachandran M, Baskar K, Jayakumar M. 2022. Essential oil composition of *Callistemon citrinus* (Curtis) and its protective efficacy against *Tribolium castaneum* (Herbst) (Coleoptera: Tenebrionidae). PLoS One 17, e0270084.

Rattanaburi S, Mahabusarakam W, Phongpaichit S, Carroll AR. 2013. Acylphloroglucinols from *Callistemon lanceolatus* DC. Tetrahedron 69, 6070–6075.

Salem MZM, EL-Hefny M, Nasser RA, Ali HM, El-Shanhorey NA, Elansary HO. 2017. Medicinal and biological values of *Callistemon viminalis* extracts: History, current situation and prospects. Asian Pac J Trop Med 10, 229–237.

Saxena S, Gomber C. 2006. Antimicrobial potential of *Callistemon rigidus*. Pharm Biol 44, 194–201.

Sharanya M, Aswani K, Sabu M. 2014. Pollination biology of *Callistemon citrinus* (Curtis) Skeels (Myrtaceae). Int J Plant Reprod Biol 6, 105–110.

Sharma RK, Kotoky R, Bhattacharyya PR. 2006. Volatile oil from the leaves of *Callistemon lanceolatus* DC grown in north-eastern India. Flavour Fragr J 21, 239–240.

Shipton WA. 1982. Clonal propagation of *Callistemon* by tissue culture techniques. Plant Cell Rep 1, 199–201.

Shokri S, Zarei H, Alizadeh M. 2012. Evaluation of rooting response of stem cuttings and *in vitro* micro-cuttings of bottlebrush tree (*Callistemon viminalis*) for commercial mass propagation. Wudpecker J Agric Res 1, 424–428.

Sowndhararajan K, Deepa P, Kim S. 2021. A review of the chemical composition and biological activities of *Callistemon lanceolatus* (Sm.) Sweet. J Appl Pharm Sci 11, 65–73.

Spencer R, Lumley P. 1991. Callistemon. In: Flora of New South Wale, Harden GJ (ed.). New South Wales University Press, Sydney, pp 168–173.

Spencer RD. 1996. *Callistemon*. In: Flora of Victoria: Dicotyledons Winteraceae to Myrtaceae, Vol 3, Walsh NG, Entwisle TJ (eds). Inkata Press, Melbourne.

Srivastava S, Ahmad A, Jain N, Aggarwal K, Syamasunder K. 2001. Essential oil composition of *Callistemon citrinus* leaves from the lower region of Himalayas. J Essent Oil Res 13, 359–361.

Sudhakar M, Rao CV, Rao AL, Ramesh A, Srinivas N, Raju D. 2004. Antinociceptive and anti-inflammatory effects of the standardized oil of Indian *Callistemon lanceolatus* leaves in experimental animals. East Cent Afr J Pharm Sci 7, 10–15.

Tabuti JRS, Kukunda CB, Waako PJ. 2010. Medicinal plants used by traditional medicine practitioners in the treatment of tuberculosis and related ailments in Uganda. J Ethnopharmacol 127, 130–136.

Tawila AM, Omar AM, Phan ND, Takahashi I, Maneenet J, Awale S. 2022. New callistrilone epimers from *Callistemon citrinus* and their antiausterity activity against the PANC-1 human pancreatic cancer cell line. Tetrahedron Lett 100, 153881.

Tawila AM, Sun S, Kim MJ, Omar AM, Dibwe DF, Awale S. 2020b. A triterpene lactone from *Callistemon citrinus* inhibits the PANC-1 human pancreatic cancer cells viability through suppression of unfolded protein response. Chem Biodivers 17, e2000495.

Tawila AM, Sun S, Kim MJ, Omar AM, Dibwe DF, Ueda J-Y, Toyooka N, Awale S. 2020a. Highly potent antiausterity agents from *Callistemon citrinus* and their mechanism of action against the PANC-1 human pancreatic cancer cell line. J Nat Prod 83, 2221–2232.

Tawila AM, Sun S, Kim MJ, Omar AM, Dibwe DF, Ueda J-Y, Toyooka N, Awale S. 2020c. Chemical constituents of *Callistemon citrinus* from Egypt and their antiausterity activity against PANC-1 human pancreatic cancer cell line. Bioorg Med Chem Lett 30, 127352.

Uwaya GE, Fayemi OE, Sherif E-SM, Junaedi H, Ebenso EE. 2020. Synthesis, electrochemical studies, and antimicrobial properties of Fe_3O_4 nanoparticles from *Callistemon viminalis* plant extracts. Materials (Basel) 13, 4894.

Wheeler GS. 2005. Maintenance of a narrow host range by *Oxyops vitiosa*; a biological control agent of *Melaleuca quinquenervia*. Biochem Syst Ecol 33, 365–383.

Wrigley JW, Fagg M. 1993. Bottlebrushes, Paperbarks and Tea Trees and All Other Plants in the Leptospermum Alliance. Angus & Robertson, Sydney.

Wu J-W, Li B-L, Tang C, Ke C-Q, Zhu N-L, Qiu S-X, Ye Y. 2019. Callistemonols A and B, potent antimicrobial acylphloroglucinol derivatives with unusual carbon skeletons from *Callistemon viminalis*. J Nat Prod 82, 1917–1922.

Wu L, Wang XB, Li RJ, Zhang YL, Yang MH, Luo J, Kong LY. 2016. Callistiviminenes A-O: Diverse adducts of β-triketone and sesqui- or monoterpene from the fruits of *Callistemon viminalis*. Phytochemistry 131, 140–149.

Wu L, Zhang Y, Wang X, Liu R, Yang M, Kong L, Luo J. 2017. Acylphloroglucinols from the fruits of *Callistemon viminalis*. Phytochem Lett 20, 61–65.

Xiang Y-Q, Liu H-X, Zhao L-Y, Xu Z-F, Tan H-B, Qiu S-X. 2017. Callistemenonone A, a novel dearomatic dibenzofurantype acylphloroglucinol with antimicrobial activity from *Callistemon viminalis*. Sci Rep 7, 2363.

Zubair M, Hassan S, Rizwan K, Rasool N, Riaz M, Zia-Ul-Haq M, De Feo V. 2013. Antioxidant potential and oil composition of *Callistemon viminalis* leaves. Sci World J 2013, Article ID 489071.

2 Cardamomum or Elettaria Species

2.1 MORPHOLOGICAL FEATURES, DISTRIBUTION, ETHNOPHARMACOLOGICAL PROPERTIES, PHYTOCHEMISTRY, AND PHARMACOLOGICAL ACTIVITIES

Cardamomum officinale Salisb. (syn. *Elettaria cardamomum* (L.) Maton; Fam.—Zingiberaceae) is a herbaceous, perennial, with underground rhizomes, aerial stems formed by encircling the leaf sheaths, and grows 2–5 m high. The leaves are long (30–35 cm) and wide (7–10 cm), lanceolate with acuminate tip and dark green (Murugan et al. 2016). Inflorescence arises from the rhizome (a anicle), containing a long cane-like peduncle possessing nodes and internodes (Telja et al. 2006; Husain and Ali 2014). It is distributed in rainforests of South India, Tanzania, Morocco, Guatemala, Sri Lanka, and other tropical countries (Kapoor 2000; Garg et al. 2016; Bacha et al. 2016; Merghni et al. 2018). *E. cardamomum* fruits have been used for the treatment and control of asthma, teeth and gum infections, cataracts, nausea, and diarrhea as well as cardiac, digestive, and kidney disorders (Menon et al. 1999; Ashokkumar et al. 2020a). The cardamom essential oils showed significant immunomodulatory activity (Vaidya and Rathod 2014). African cardamom (*Aframomum danielli* (Hook.f.) K. Schum) is native to southeast African countries (Tanzania, Cameroon, Madagascar, and Guinea), Nepal, and the Sikkim state of India (Govindarajan et al. 1982; Adegoke et al. 1998). Since long ago, cardamom capsules have been used for the treatment of asthma, teeth and gum infections, digestive and kidney disorders (Das et al. 2012), cataracts, nausea, diarrhea, and cardiac diseases (Gilani et al. 2008; Khan et al. 2011; Nirmala 2000). The capsules (containing essential oil and other bioactive metabolites) of cardamom are used as a functional food, pharmaceutical, and nutraceutical agents (Hamzaa and Osman 2012).

Ethanol extract of *E. cardamomum* seeds showed an inhibitory effect on glyceraldehyde 3-phosphate dehydrogenase, but it significantly increased the level of thiobarbituric acid reactive substances, succinate dehydrogenase, and catalase activities. The extract also induced (0.3 mg/g) toxicity and affected energy metabolism and oxidative stress in mice. After a 48 h treatment, a 3 mg/g dose induced mortality in the mice. In treated mice, the extract induced the levels of thiobarbituric acid reactive substances (at 0.3 mg/g concentration) and catalase activities (0.003 and 0.3 mg/g concentration), when compared with the control (Malti et al. 2007). The effects of aqueous extract of cardamom fruits on hemodynamic, biochemical, histopathological, and ultrastructural alterations were evaluated in isoproterenol-induced myocardial infarction in rats. Isoproterenol treatment showed a significant decrease in levels of endogenous antioxidants, superoxide dismutase, catalase, glutathione peroxidase, cardiomyocytes enzymes, and lactate dehydrogenase but an increase in lipid peroxidation in treated rats. Aqueous extract of cardamom fruits (100 and 200 mg/kg, p.o.) significantly checked the decrease in the superoxide dismutase, catalase, and glutathione peroxidase activities ($p < 0.05$), when compared with the control group. Moreover, the protective effects were enlarged by increased histopathological and ultrastructural alterations, which specifies the recovery of cardiomyocytes from the adverse effects of isoproterenol. The study results show that cardamom significantly saves the myocardium and triggers cardioprotective effects by increasing free radical scavenging and antioxidant properties (Goyal et al. 2015). The 1,8-cineole and an 1,8-cineole-rich supercritical carbon dioxide extracts of cardamom seeds were tested in prevention of oligomerization of amyloid β-peptide and suppressing iron-dependent oxyradical production in *in vitro*. Both pure 1,8-cineole and 1,8-cineole-rich extract of cardamom seeds (50 μM and 100 μM) inhibited the formation of reactive hydroxyl radicals from a mixture of Fe^{2+} and ascorbate. The 1,8-cineole-rich extract of cardamom seeds checked *in vitro* Aβ42 oligomerization more than synthetic 1,8-cineole. Both pure 1,8-cineole and 1,8-cineole-rich supercritical carbon dioxide extract prevented the iron-dependent cell death of SHSY5Y cells. Onset of Alzheimer's disease pathology is characterized by oxidative damage, Aβ42 aggregation, and loss of cell viability. The study results suggest a significant therapeutic role of 1,8-cineole-rich extract of cardamom seeds over pure 1,8-cineole in treating neurodegenerative disease (Paul et al. 2020b). The 1,8- cineole-rich supercritical carbon dioxide extract of cardamom seeds was tested for their antidiabetic activity. PEGylated nanoliposomes of extract demonstrated significant *in vitro* DPPH scavenging activity (59%) attributed to a synergistic group of antioxidants. Administration (p.o.) of nanoliposomes (550 mg/kg bw) in rats restored their normal fasting blood glucose levels and serum lipid profiles on day 35. Oral administration of the nanoliposomes (550 mg/kg bw) in rats restored the normal values of fasting blood glucose and serum lipid profiles on day 35. Therefore, the nanoliposome of 1,8-cineole rich extract may be considered as a therapeutic agent in managing type 2 diabetes and hypercholesterolemia (Paul et al. 2019). The hexane extract of *E. repens* showed DPPH and metal chelating activity (IC_{50} 464 ± 28.3μg/ml, and 199 ± 7.2μg/ml) while the reducing power and antioxidant activities were recorded as 289 ± 14.6 ascorbic acid equivalent/mg, 468 ± 22.7 gallic acid equivalents/mg. The hexane extract demonstrated protective effects against hydrogen peroxide–induced DNA injury and suppressed 2,2-azobis(2-amidinopropane) dihydrochloride-induced protein oxidation and lipid peroxidation. Additionally, extract administration (100 mg/kg) suppressed carrageenan-induced paw edema, and downregulated cytokines such as cyclooxygenase-2, interleukin-6, and tumor necrosis factor-α as well as inducible nitric oxide synthase mediated nitric oxide formation. In summary, the results show that *E. repens* may be used in managing inflammation and oxidative stress (Kandikattu et al. 2017). The protective effects of cardamom (250 mg/kg bw) extract were tested on uranyl acetate dehydrate (40 mg/kg bw)–induced toxicity in rats. The uranyl acetate dehydrate administration (p.o.) stimulated the enhancement of Fe^{3+} ions content in different areas of the brain and serum malondialdehyde whereas, significant reduction was found in serum glutathione and testosterone. The results showed that cardamom ameliorated the hazardous effects of uranyl acetate dehydrate on the central nervous system, antioxidant, and reproductive systems. Therefore, cardamom may have a protective effect against uranium hazards (Abdel Kader et al. 2015). *E. cardamomum* contains polyphenols, 1,8-cineole, limonene, α-terpinyl acetates and possesses antioxidant, antitumor, and anti-inflammatory properties (Qiblawi et al. 2020).

The effects of green cardamom were evaluated on overweight or obese patients with type 2 diabetes. In this clinical study, eighty overweight or obese patients were selected. These patients were divided into two groups: treated and placebo. The green cardamom (3 g/day) was given for 10 weeks to the treated patients. During treatment, the levels of blood glucose, lipid profile, oxidative stress biomarkers (in serum), and total antioxidant ability, malondialdehyde, glutathione peroxidase, and superoxide dismutase in red blood cells were monitored. The study results showed the potent efficacy of green cardamom in managing the type 2 diabetes mellitus in treated patients (Aghasi et al. 2018). The effects of green cardamom (*E. cardamomum*) on blood glucose indices, lipids, inflammatory profiles, and liver functions were investigated in obese patients with nonalcoholic fatty liver disease. A total of 80 obese patients were recruited in this clinical study and divided into two groups, cardamom, and placebo. Two capsules (500 mg each) were taken by patients (three times/day, taken with meals) for 3 months. General parameters such as dietary intake and physical activity were monitored during the case study. Simultaneously, the weight, height, and waist circumference, blood pressure, extent of fatty liver, and blood biomarkers, including fasting blood sugar and insulin of serum, homeostasis model assessment-insulin resistance, quantitative insulin sensitivity check index, triglyceride, low-density lipoprotein-cholesterol, high-density lipoprotein-cholesterol, total cholesterol, highly sensitive C-reactive protein, tumor necrosis factor-α, interleukin-6, alanine transaminase, aspartate transaminase, irisin, paraxonase-1, and sirtuin-1, were also determined in obese patients. The green cardamom treatment restored the altered parameters (several blood factors, including glucose indices, lipids, inflammatory markers, liver enzymes, irisin, parax-onase-1, and sirtuin-1, and blood pressure and anthropometry) in obese patients with nonalcoholic fatty liver diseases (Daneshi-Maskooni et al. 2017). The effects of green cardamom were evaluated on serum Sirtuin-1, inflammatory factors, and liver enzymes in overweight or obese nonalcoholic fatty liver disease patients. A total of 87 patients were selected in this study and randomly divided into two groups, as cardamom ($n = 43$) or placebo ($n = 44$). Two capsules (500 mg each/daily) were received with meal by patients for three months. General parameters such as dietary intake, physi-cal activity status, weight and height, tumor necrosis factor-α, highly sensitive c-reactive protein, interleukin-6, alanine transaminase, and aspartate transaminase were monitored in treated patients. The green cardamom significantly enhanced the levels of serum sirtuin-1 but reduced the levels of highly sensitive c-reactive protein, interleukin-6, alanine transaminase, and aspartate transaminase significantly ($p < 0.05$). The differences in weight, body mass index, and aspartate transaminase were not found significant ($p > 0.05$). Green cardamom administration improved some biomarkers related to inflammation, alanine transaminase, and sirt1 in overweight/obese nonalcoholic fatty liver disease patients (Daneshi-Maskooni et al. 2018). Similarly, the effects of green cardamom were evaluated on dyslipidemia, hepatomegaly, and fasting hyperglycemia in nonalcoholic fatty liver disease. A total of 87 patients were included in this study and divided into groups, cardamom (n = 43) or placebo (n = 44). Two capsules (500 mg each) were given 3 times a day with meals for 90 days. The levels of general parameters viz., serum irisin, fasting blood sugar, insulin, total cholesterol, triglyceride, low-density lipoprotein cholesterol, and high-density lipoprotein choles-terol were monitored in treated patients and placebo. The green cardamom significantly enhanced the levels of irisin, high-density lipoprotein cholesterol, and quantitative insulin sensitivity check index but reduced the fasting blood insulin, triglyceride, low-density lipoprotein cholesterol, and homeostasis model assessment-insulin resistance levels ($p < 0.05$). The differences in fasting blood sugar, total cholesterol, and body mass index of patients were not significant ($p > 0.05$). Green car-damom administration enhanced the grade of fatty liver, serum glucose indices, lipids, and irisin level among overweight or obese non-alcoholic fatty liver disease patients. The alterations in these biomarkers may produce some significant effects on patients of nonalcoholic fatty liver disease (Daneshi-Maskooni et al. 2019).

The mixture of essential oils (α-terpinyl acetate, 1,8-cineole, α-terpineol, linalool, α-terpinyl acetate, 5-hydroxymethylfurfural) of *E. cardamomum* seeds and pods, demonstrated strong inhibi-tory effects against bacterial (*S. aureus, B. cereus, E. coli* and *S. typhi*; at 3000 ppm concentration) and fungal (*A. terreus, P. purpurogenum, F. graminearum* and *P. madriti*) organisms (Singh et al. 2008). Cardamom extracts (fruit and seeds), rich in volatile compounds, exerted an antibacterial effect against *A. actinomycetemcomitans, F. nucleatum, P. gingivalis*, and *P. intermedia* (MIC 0.5%, v/v; 0.25%, 0.062%, 0.125%) and minimum bactericidal concentrations (MBC 1%, 0.25%, 0.062%, 0.25%). The volatile extract interrupted the cell membrane of *P. gingivalis*. The extract also suppressed biofilm development in bacterial species. Additionally, the extract significantly reduced the secretion of IL-1β, TNF-α, and IL-8 by lipopolysaccharide-stimulated macrophages. The results suggest that anti-inflammatory activity may be associated with the inhibition of the NF-κB sig-naling pathway. The study results indicate that the volatile extract of cardamom fruit and seeds may be used as a therapeutic agent against periodontal infections (Souissi et al. 2020). Aqueous extracts of cardamom seeds significantly increased splenocyte proliferation in a dose-dependent and synergistic manner. Cardamom extracts significantly suppressed the formation of T helper (Th)1 cytokine by splenocytes. The extract reduced the formation of nitric oxide by macrophages (Majdalawieh and Carr 2010). The aqueous and acidified (0.5 and 1N) methanol extracts of green cardamom (*E. cardamomum*) seeds and pods demonstrated strong antimutagenic activity against *S. typhimurium* TA98 (mutant strain) and *S. typhimurium* TA100. Therefore, both extracts from cardamom seeds and pods may be examined as a potential chemotherapeutic agent against different types of cancers (Saeed et al. 2014). The presence of monoterpenes (α-terpinyl acetate, 1,8-cineole, linalyl acetate, linalool, α-terpineol and limonene) was determined by GC–MS in green cardamom essential oils. Additionally, green cardamom oil showed greater antibacterial and antifungal activities against Gram-positive bacteria (mean diameter = 21.77 mm), cariogenic bacteria (mean diameter = 19.51 mm) and fungi (mean diameter = 39.5 mm). The cardamom oil demonstrated a diverse range of minimum inhibitory concentrations (MICs 0.023 to 0.046 mg/ml) against all screened bacterial and fungal strains. *E. cardamomum* oils demonstrated broad-spectrum antibacterial and antifungal activities. This oil may be used to check the contamination of food products and foodborne microbes (Mejdi et al. 2015). The crude extracts (aqueous, ethanol, methanol, ethyl acetate and hexane) of *E. cardamomum* dry fruits were screened for their antimicrobial activity. Aqueous extract showed potent inhibitory effect against *S. aureus* while methanol extract was found effective against *S. aureus* and *E. coli*. The ethanol and aqueous extracts demonstrated very mild inhibition against *S. typhi* and *S. pyogenes*. The extracts demonstrated significant overall MIC values (512 µg/ml) against bacterial species (Kaushik et al. 2010). The essential oils of cardamom showed significant antimicrobial activity against Gram-positive and Gram-negative bacterial (*B. cereus, B. subtilis* ATCC 6633, *E. faecalis* ATCC 29212, *L. monocytogenes* ATCC 19115, *M. luteus* NCIMB 8166, *S. aureus* MR (B2), *S. aureus* ATCC 6816, *S. epidermidis* ATCC 12228, *E. coli* ATCC 25922, *K. pneumoniae, P. aeruginosa* ATCC 27853, *P. mirabils, S. putrefaciens, S. typhimirium* ATCC 14028, *S. flexenerii* ATCC 12022, *V. alginolyticus* ATCC 17749, *V. alginolyticus* ATCC 33787, *V. cholerae* ATCC 9459, *V. parahemolyticus* ATCC 17802, *V. parahemolyticus* ATCC 43996, *V. vulnificus* ATCC 27562, *V. vulnificus* ATCC 33149, *Serratia marcescens*) and fungal strains (*Candida tropicalis* 06–085, *C. parapsilosis* ATCC 22019, *C. krusei* ATCC 6258, *C. glabrata* ATCC 900030, *C. guilliermondi* 06–018, *C. albicans* ATCC 2019 and *S. cerevisiae* 11–161). The Ethiopian cardamom was found most effective against fungal (inhibition zone 12.67 to 34.33 mm) and bacterial strains (MICs 0.048 to 0.19 mg/ml, and MBCs 0.19 to 1.75 mg/ml; Noumi et al. 2018).

The antiulcer effects of methanol extract (100–500 mg/kg), essential oil (12.5–50 mg/kg), petroleum ether soluble (12.5–150 mg/kg), and insoluble fractions of methanol extract (450 mg/kg) were evaluated on ethanol and aspirin-induced gastric lesions in rats. The extracts and fractions significantly suppressed gastric lesions caused by ethanol and aspirin but not those caused by pylorus ligation. Methanol extract (500 mg/kg) was found to be effective in reducing lesions (70%) in the ethanol-induced ulcer model. The petroleum ether fraction also (100 mg/kg) reduced the lesions (50%), and effects were similar to insoluble fraction (450 mg/kg). The petroleum ether fraction (12.5 mg/kg) showed 100% inhibition in ethanol-induced lesion of rats. In this study, the petroleum ether fraction (≥12.5mg/kg) were found more effective than ranitidine (50 mg/kg) against ethanol-induced lesions in rats (Jamal et al. 2006).

The effects of cardamom oil were evaluated in aluminum chloride-induced (100 mg/kg, *p.o.*) neurotoxicity in rats. Cardamom oil (200 mg/kg, p.o., for six weeks) demonstrated a significant increase in behavioral parameters, suppression of acetylcholinesterase activity ($p < 0.001$), and decrease in oxidative stress in the brain of rats. Essential oils also inhibited amyloid β expression and upregulated the brain-derived neurotrophic factor. The study results suggest that cardamom oil possesses neuroprotective effects in aluminum chloride-induced neurotoxicity; hence, cardamom oil may be useful in treatment of Alzheimer's disease (Auti and Kulkarni 2019).

The antihypertensive activity of cardamom powder was assessed on the cardiovascular risk factors in individuals with stage 1 hypertension. Cardamom powder (3 g) significantly ($p < 0.001$) reduced systolic, diastolic, and mean blood pressure but significantly ($p < 0.05$) enhanced fibrinolytic activity (at the end 12th week). Similarly, the total antioxidant potential was also significantly ($p < 0.05$) enhanced (by 90%) at the end of 90 days. The powder did not show any side effects in treated animals during the study. The study reports suggest that cardamom powder effectively decreases blood pressure and increases fibrinolysis activity and antioxidant potential, without significantly changing blood lipids and fibrinogen levels in stage 1 hypertensive individuals (Verma et al. 2009).

The antihypercholesterolemic effects of cardamom powder (50 g/kg bw), cardamom oil (3 g/kg, bw, equivalent to 50 g/kg cardamom), and de-oiled cardamom powder (50 g/kg) were assessed in hypercholesterolemia induced Wistar rats. Hypercholesterolemia was induced with a high-cholesterol diet (fed for 8 weeks) in Wistar rats. The administration of cardamom oil (p.o.) showed a significant decrease in levels of blood total cholesterol, serum triglycerides, and low-density lipoprotein cholesterol in hypercholesterolemic rats. Cardamom oil also decreased the profiles of cholesterol content of cardiac muscle in hypercholesterolemic rats. The de-oiled cardamom powder and cardamom oil countered the reduced the activity of catalase in hypercholesterolemic animals. Cardamom also increased the heart superoxide dismutase activity in hypercholesterolemic subjects. The profiles of ascorbic acid in serum were significantly enhanced by dietary cardamom or its fractions in hypercholesterolemic animals. In this study, the potential of cardamom oil in restoring the changes in lipid homeostasis in hypercholesterolemic conditions in rats was established. The significant decrease in atherogenicity index by dietary supplementation of cardamom powder and cardamom oil reveals the potent cardioprotective effects of cardamom in animals (Nagashree et al. 2017). To ascertain hypocholesterolemia activity, the melatonin-rich and 1,8-cineole-rich extracts of green cardamom seeds were co-administered (p.o.) with Triton X (at the pre-optimized dose 175 mg/kg bw) to Wistar albino rats (550, 175 and 55 mg/kg bw dose). The levels of total cholesterol of serum in the rats were monitored on the day 3, 7, 15, and 21. On day 21, the level of total cholesterol decreased appreciably (48.95%) in cardamom-treated rats. The extract showed inhibitory effects on hepatic 3-hydroxy-3-methyl-glutaryl-CoA reductase activity of treated rats. The extracts might be consumed per se as hypocholesterolaemic supplements or may be ingredients of new spice-based therapeutic food supplements (Chakraborty et al. 2019). The supercritical carbon dioxide extract of cardamom seeds was evaluated for antidiabetic effects in streptozotocin-induced Wistar albino rats. The levels of fasting blood glucose were monitored in the rats on days 8, 15, and 21. On day 15, the levels of fasting blood glucose were decreased appreciably by cardamom extract (31.49%) in treated rats. The results were compared to metformin (30.70%) and BGR-34 (a commercial polyherbal drug; 31.81%) in treated rats. Cardamom extract showed desirable efficacy on hepatic glucose-6-phosphatase, glucose-6-phosphate dehydrogenase, and catalase activities in treating diabetes. The findings showed that all bioactive constituents in the extract have binding harmony with the enzyme and presented to the antidiabetic activity of the extract as glucose-6-phosphate dehydrogenase proliferators. The extract administration showed an increase in both insulin sensitivity of the liver and glucose uptake in the gut. The results of supercritical carbon dioxide extract of cardamom were found to be a safe alternative to metformin and BGR-34 in controlling type 2 diabetes and may be used for clinical studies (Paul et al. 2020a).

Unbleached and dry capsules (seeds and pods) of cardamom were administered to thirty pregnant mice to test developmental, learning ability, and biochemical parameters of mice offspring. Cardamom capsules and Pillsbury's Diet were mixed in such ratio that it provided doses of 10% and 20% of cardamom (w/w). The mice were allocated into three groups. Groups II and III were administered Pillsbury's Diet containing 10% and 20% of cardamom (w/w) respectively, whereas Group I was used as control. Cardamom was given from the first day of pregnancy and was continued up to day 15, and afterwards, the mothers were administered plain Pillsbury's Diet. During the weaning period, cardamom perinatal treatment led to decreased body weight gain of treated offspring. The decrease in body weight was significant ($p < 0.001$) in 9, 11, and 13 days in high-dose-treated offspring. The effect of low dose of cardamom was significant ($p < 0.01$) in days 13 and 15, and was significant ($p < 0.05$) in day 17, as compared to the control group. The appearance of hairs was delayed significantly ($p < 0.05$) in perinatal cardamom-treated offspring. The opening of eyes was delayed ($p < 0.01$) also, as compared to offspring of control group. The cardamom treatment increased the learning and memory retention abilities significantly in treated offspring, as compared to control. Monoamines (dopamine and serotonin or 5-hydroxytryptamine) and glutathione levels were increased, while thiobarbituric acid-reactive substances formation was suppressed significantly in treated animals. In summary, perinatal cardamom treatment increased learning and memory abilities, as compared to control; therefore, more studies are needed on cardamom seeds to investigate its useful effects on other types of behaviors (Abu-Taweel 2018).

The inhibitory effect of aqueous extract of cardamom was evaluated on activity of human platelets. The activity of human platelets was induced by agonists including adenine diphosphate (2.5 mM), epinephrine (2.5 mM), collagen (10 mM), calcium ionophore A 23187 (6 μM), and ristocetin (1.25 μg/ml). Cardamom extract suppressed platelet aggregation with inducers such as ADP (IC_{50} 0.49 mg), epinephrine (IC_{50} 0.21 mg), collagen (IC_{50} 0.55 mg), and calcium ionophore A 23187 (IC_{50} 0.59 mg). Lipid peroxidation was stimulated by the iron-ascorbic acid system in platelet membranes. Higher concentration of cardamom extract decreased the levels of malondialdehyde development significantly. Therefore, the aqueous extract of cardamom may possess bioactive compounds, which protect human platelets from aggregation and lipid peroxidation (Suneetha and Krishnakantha 2005).

The effects of cardamom powder (supplement) were evaluated on inflammation and oxidative stress in hyperlipidaemic, overweight, and obese prediabetic women. A total of 80 prediabetic subjects were selected in this clinical trial study. Randomly, the subjects received the cardamom supplement (n = 40, 3 g/day) or placebo (n = 40) for 56 days. The supplementation of cardamom powder significantly reduced serum high-sensitivity C-reactive protein ($p = 0.02$), interleukin-6 ratio ($p = 0.008$), and malondialdehyde ($p = 0.009$) levels, compared with the placebo group. Cardamom powder increased inflammation and oxidative stress parameters in prediabetic subjects. Therefore, it may be used in decreasing complications attributed with inflammation and oxidative stress in some patients (Kazemi et al. 2017).

The effect of aqueous extract of cardamom was evaluated for their cardioprotective effect against myocardial damage stimulated by doxorubicin in rats. The doxorubicin showed notable myocardial injury (discontinuity and disorganization of cardiac muscle fibers, mononuclear cell infiltration, and enhancement in collagen fiber deposition), functional loss (increased levels of serum lactate dehydrogenase, creatine kinase, and cardiac troponin), induction of oxidative stress (higher levels of nitric oxide and malondialdehyde, and lower levels of superoxide dismutase, catalase, and glutathione peroxidase), apoptosis (high caspase 3 activity and immunostaining), and inflammation (high cardiac NFκB level) in treated rats. Though, extract administration not only reduced harmful effects of doxorubicin but also stimulated angiogenesis, as revealed by a significant enhancement in vascular endothelial growth factor immunoreactivity. The results reveal that the extract may relieve doxorubicin-induced cardiotoxicity, by decreasing oxidative stress, apoptosis, and inflammation and increasing tissue regeneration. So, cardamom extract can be considered as a significant cytoprotective agent against doxorubicin cardiotoxicity (Abu Gazia and El-Magd 2018).

The chemoprotective effects of cardamom (*E. cardamom*) oil (eucalyptol, α-pinene, β-pinene, δ-limonene and geraniol) was evaluated in *in vitro* and *in vivo* systems. Eucalyptol attaches to the caspase 3, which is well-suited for nucleophilic attacks by polar residues inside the caspase 3 catalytic site. Moreover, a handpicked set of anti-tumor targets for cardamom oil was developed, which may be further authenticated by *in vivo* and *in vitro* experiments (Bhattacharjee and Chatterjee 2013). The protective effects of the essential oil and methanolic extract of *E. cardamomum* fruits were examined against chemically (pentylentetrazole)- and electrically (maximal electroshock)-induced seizures in mice. The TLC analysis showed the presence of kaempferol, rutin, and quercetin in methanol extract of fruits. Similarly, 1,8-cineole, α-terpinyl acetate, sabinene, 4-terpinen-4-ol, and myrcene were determined from mixture of essential oils. The extract (1.5 g/kg) and essential oils (0.75 ml/kg) demonstrated significant neurotoxicity in the rotarod test. No mortalities were recorded (2 g/kg and 0.75 ml/kg) in extract (2 g/kg)- and essential oil (0.75 ml/kg)-treated rats. The essential oil was found to be effective in pentylentetrazole and maximal electroshock models while methanol extract showed effectivity in pentylentetrazole model only. The methanol extract did not show any significant lethality in treated mice. The methanol extract and essential oils displayed movement toxicity. Similarly, the anticonvulsant activity of *E. cardamomum* was found negligible against the seizures stimulated by pentylentetrazole and maximal electroshock (Masoumi-Ardakani et al. 2016).

The effects of cardamom powder (10-week green cardamom intake) were evaluated on blood pressure and concentrations of inflammatory and endothelial functional biomarkers in type 2 diabetes mellitus patients. Cardamom powder (3 g/day, 6 capsules) or placebo (rusk powder, 6 capsules) was administered to patients for a period of 10 weeks. The levels of systolic blood pressure and diastolic blood pressure, asymmetric dimethylarginine, and nitric oxide were monitored in treated patients. Serum inflammatory markers (interleukin 6, tumor necrosis factor-α, high-sensitivity C-reactive protein) and endothelial functions (intercellular adhesion molecule-1, vascular cell adhesion molecule 1, and CD62 antigen-like family member E) were also monitored in treated patients (Ghazi Zahedi et al. 2020). The protective effects of cardamonin were examined on delayed cerebral vasospasms after subarachnoid hemorrhage. Apoptosis and *p*-AKT and C-myc expressions in the subarachnoid hemorrhage + cardamonin group were significantly minimal than in the subarachnoid hemorrhage and subarachnoid hemorrhage + vehicle groups ($p < 0.05$), while expression of α-smooth muscle actin was greater than in the subarachnoid hemorrhage and subarachnoid hemorrhage + vehicle groups ($p < 0.01$). Cardamonin appears to alleviate cerebral vasospasms after subarachnoid hemorrhage. These actions can be associated with the suppression of *p*-AKT and C-myc expression and apoptosis, and the increase of α- smooth muscle actin expression (Ma et al. 2018).

The effect of supplementation of cardamom powder was assessed on levels of blood lipids and glycemic indices in type 2 diabetic patients. In this clinical study, a total of 83 overweight or obese type 2 diabetic patients were selected and divided into two groups (intervention, n = 41; or a control group, n = 42). Cardamom powder (3 g) was administered to patients of treated and the placebo (rusk powder) groups for 10 weeks. The levels of dietary intake, weight, height and waist circumference, glucose, insulin, hemoglobin A1c, triglyceride, total cholesterol, high-density lipoprotein cholesterol and low-density lipoprotein cholesterol, and serum sirtuin-1 were monitored in treated patients. A significant reduction in levels of serum hemoglobin A1c (−0.4%), insulin (−2.8 μIU/dl), homeostasis index of insulin resistance (−1.7) and triglycerides (−39.9 mg/dl), and an elevation in serum sirtuin-1 (2.3 ng/ml) was reported in treated group patients. No significant alterations were recorded in levels of serum total cholesterol, high-density lipoprotein cholesterol, and low-density lipoprotein cholesterol in treated patients. The results were compared with the patients of placebo. The reported results reveal that cardamom powder can reduce hemoglobin A1c, insulin level, homeostasis index of insulin resistance, and triglyceride levels by enhancing serum sirtuin-1 concentration in treated type 2 diabetes mellitus patients (Aghasi et al. 2019; Allen et al. 2013).

2.2　PHYTOCHEMISTRY

The effects of the extraction methods on the optimization of yield and composition of cardamom volatile oil were assessed by using two pressure values (9.0 and 11.0 MPa), two temperatures (40 and 50 °C), two flow rates (0.6 and 1.2 kg/h), and two particles' sizes (250–425 and >850 μM). The maximum yield (5.5%) was obtained with the following extraction conditions: pressure (9.0 MPa), temperature (40 °C), carbon dioxide flow (phi = 1.2 kg/h) and particles sizes (250–425 μM). α-terpinyl acetate, 1,8-cineole, linalyl acetate, limonene, and linalool were found as major essential oils (Marongiu et al. 2004). Supercritical fluid extraction methods improved to some extent fatty acid composition and profiles of volatile oils in the recovered extracts (Hamdan et al. 2008).

Monoterpenes (α-terpinyl acetate, 1,8-cineole, α-thujene, α- pinene, camphene, sabinene, β-pinene, dehydro-1,8-cineole, myrcene, α-terpinene, limonene, 1,8-cineole, γ-terpinene, *trans*-linalool oxide, terpinolene, *trans*-sabinene hydrate, linalool, limonene oxide, δ -terpineol, terpinen-4-ol, α-terpineol, nerol, carvone, neral, geraniol, linalyl acetate, undecanol, α-terpinyl acetate, neryl acetate, geranyl acetate, isoeugenol, β-caryophyllene, germacrene d, β-selinene, α-selinene, cis-nerolidol, caryophyllene oxide, octadecane) have been determined by GC-MS from *E. cardamomum* seeds and pods. The essential oils demonstrated significant antimicrobial activity against *S. aureus, B. subtilis, E. coli, K. pneumoniae, C. albicans*, and *A. niger* (Alam et al. 2019).

The profiles of monoterpenes (1,8-cineole, cis-ocimene, α-terpinene, linalool, thujyl alcohol, limonene-1,2-epoxide, citronellol, trans-pinocarveol, nerol and linalyl acetate and borneol, acetate, linalyl acetate and borneol, 1,4-cineole, pcymene-8-ol and isoborneol) were determined in essential oil mixture of fruits (Sereshti et al. 2012; Sultana et al. 2009; Bampidis et al. 2019). Similarly, the presence of 1, 8-cineole, α-pinene, sabinene, linalool, α-terpineol and nerol, ç-elemene and 1,6,10-dodecatrien-3-ol (Nerolidol), α-terpinyl acetate, ocimenyl acetate, and *E*5-dodecenyl acetate was determined by GC-MS in dry capsules of *E. cardamomum* (Ashokkumar et al. 2020b).

α-pinene, *cis-p*-mentha-2,8-dien-1-ol, α-thujene, neral, camphene,1,8-menthadien-4-ol, β-pinene, α-terpinyl acetate, sabinene, verbenone, myrecene, neryl acetate, α-terpinene, *trans-p*-menth-2-en-1,8-diol, dehydro-1,8-cineole, β-selinene, limonene, geranial, 1,8-cineole, carvone, *cis*-β-ocimene, *cis*-piperitol, γ-terpinene, geranyl acetate, *trans*-β-ocimene, γ-cadinene, *p*-cymene, *p*-mentha-1,5-dien-8-ol, terpinolene, cuminal, octanal, nerol, 6-methyl-5-hepten-2-one, *cis*-8-methylbicyclo[4.3.0]non-3-ene, 1,3,8-p-menthatriene, *trans*-carveol, *trans*-linalool oxide(furanoid), geraniol, *trans*-1,2-limonene epoxide, *p*-cymen-8-ol, *trans*-sabinene hydrate, *cis*-carveol, *cis*-linalool oxide(furanoid), caryophyllene oxide, *cis*-4-decenal, perilla alcohol, linalool, methyl eugenol, octanol, trans-nerolidol, linalyl acetate, cuminol, *trans-p*-menth-2-en-1-ol, eugenol, pinocarvone, thymol, bornyl acetate, carvacrol, terpinen-4-ol, carvone acetate, hotrienol, geranic acid, *cis*-dihydrocarvone, 2*E*,6*E*-farnesol, *cis-p*-2-menthen-1-ol, and hexadecanoic acid were characterized from hydro-distilled extract of *E. cardamomum* fruits (Kuyumcu Savan and Küçükbay 2013).

The antioxidant activity of hexane, dichloromethane, ethyl acetate, methanol, and water extracts of cardamom fruits was assessed in total reducing power and DPPH assays. Chemical analysis of volatile oils was carried out and showed the presence of α-pinene, sabinene, β-myrcene Δ³-carene, D-limonene, 1,8 cineole, γ-terpinene, sabinene hydrate, terpinolene, linalool, β-terpineol, terpene-4-ol Octyl acetate, α-terpineol, linalyl acetate, verbeneol, geraniol, geranial, ocimenol, terpenyl acetate, neryl acetate, 3-decen 2-ol cis-, dodecen 5 enol, cedrene, eudesmene, β-caryophyllene, nerolidol, *cis*, *trans* farnesol, *cis,* and *cis* farnesol. The ethyl acetate extract of fruits showed significant antioxidant activity (Padmakumari Amma et al. 2010). The essential oils viz., α-thujene, α-pinene, sabinene, β-pinene, 6-methyl-5-hepten-2-one, myrcene, α-phellandrene, α-terpinene, *p*-cymene, limonene, 1.8-cineole, γ-terpinene, *cis*-sabinene hydrate, *cis*-linalool oxide (furanoid), terpinolene, 2-hexyl furan, linalool, *cis-p*-menth-2-en-1-ol, *trans-p*-menth-2-en-1-ol, sabina ketone, δ-terpineol, 4-terpineol, *p*-cymen-8-ol, α-terpineol, *trans*-piperitol, neral, carvone, geraniol, linalyl acetate, (*E*)-2-decenal, geranial, 2-phenyl-2-butenal, thymol, *cis*-2.3-pinanediol, methyl geranate, myrtenyl acetate, exo-2-hydroxycineol acetate, α-terpinyl acetate, α-copaene, geranyl acetate, (*E*)-2-decenyl acetate, β-caryophyllene, γ-muurolene, δ-cadinene, elemol, (*E*)-nerolidol, and caryophyllene oxide have been determined from *E. cardamomum* fruits (Noumi et al. 2018).

The GC-MS analysis of the cardamom extract revealed the presence of a large number of monoterpenes viz., hydroxy-α-terpenyl acetate, λ -8(17),13(*E*)-diene-15, α-pinene, 1,8-cineole, sabinene, *trans*-sabinene hydrate, β-myrcene, linalool, 1,8-cineole, δ-terpineol, trans-sabinene hydrate, 3-cyclohexen-1-ol, linalool, linalyl propionate,δ-terpineol, 4-terpinenyl acetate, 3-cyclohexen-1-ol, linalyl propionate, linalyl acetate, *cis*-sabinene hydrate acetate, trans-geraniol, z-citral, linalyl acetate, δ-terpinyl acetate, *trans*-geraniol, α-terpinyl acetate, citral, eugenol, δ-terpinyl acetate, neryl acetate, 2,6-octadienoic acid, *trans*-caryophyllene, 1-*p*-menthen-8-yl acetate, 1-methyl-4-(1-acetoxy-1-methyl), eugenol, germacrene-D, neryl acetate, β-selinene, 2-octenyl acetate, (-)-α-selinene, *trans*-caryophyllene, 3-allyl-6-methoxyphenol, 1-methyl-4-(1-acetoxy-1-methyl), hydroxy-α-terpenyl acetate, β-selinene, δ-nerolidol, α-selinene, (*E*,*E*)-4,8,12-trimethyl-1,3,7,1 acetiisoeugenol, (±) 2-exo-hydroxycineole hydroxy-, α-terpenyl acetate, *cis*-farnesol, nerolidol, *trans*-farnesal, (*E*,*E*)-4,8,12-trimethyl-, farnesyl acetate, 1,3,7,1,1-tridecatetraene, geranyl linalool isomer B, *cis*-farnesol, 1-octadecanol, *trans*-caryophyllene, docosane, 11-decyl-, (+)-λ-8(17),13(*E*)-diene-15, hexadecanoic acid, farnesyl acetate, tritetracontane, chavicol, cyclohexane, α-copaene, cycloheptane, 3-allyl guaiacol, heptane, *trans*-caryophyllene, α-humulene, nonane, and germacrene-D (Mona 2020). The maximum yield of oil (74.83 mg oil/g seed) and β-sitosterol (4.73 mg/g seed) was recovered from cardamom seeds under optimized conditions (temperature of 40 °C, pressure of 200 bar and CO_2 flow rate of 4 l/min; İçen et al. 2017).

The monoterpenes (1, 8-cineole, α-pinene, sabinene, linalool, α-terpineol, ocimenyl acetate, E5-dodecenyl acetate and nerol) and two sesquiterpenes (ç-elemene and nerolidol) have been reported from cardamom seeds (Ashokkumar et al. 2019). Similarly, α-ionone, eucalyptol, santolina alcohol, 1,6-octadiene-3-ol,3.7-dimethyl-, 2,6- octadiene-1-ol,3-ol,3,7-dimethyl-,(Z)-, cinnamalde-hyde, (*E*)-, terpinen-4-ol, 1,6,10-dodecatrien-3-ol,3,7,11- trimethyl-, acetic acid, and 1-methyl-1-(4-methyl-5-oxy-cy) have been determined from cardamom fruits. The essential oils showed significant antibacterial activity against both Gram-negative and Gram-positive bacteria (Ahmed et al. 2019).

Essential oils {α-terpinyl acetate, 2-((1*R*,4*R*)-4-hydroxy-4-methylcyclohex-2-enyl) propan-2-yl acetate, 9-hexacosene, γ-sitoesterol, heneicosane, 8-acetoxycarvotanacetone, geranyl oleate, (Z)-3,7-dimethylocta-2,6-dien-1-yl palmitate, naphthalene, decahydro-4a-methyl-1-methylene-7-(1-methylethenyl)-, [4aR-(4aα,7α,8aα)]-, α-terpineol, and pentacosane} were isolated and identified from aqueous extract of *E. cardamomum* seeds. *E. cardamomum* showed significant cytotoxic activity against cancer cells {EC50 (µg/mL) of 473.84 (HeLa cells), 237.36 (J774A.1 cells), 257.51 (Vero E6 cells), and 431.16 (Balb/C peritoneal cells; Cárdenas Garza et al. 2021).

Monoterpenes [α-thujene, α-pinene, β-pinene, myrcene, α-phellandrene, α-terpinene, 1,8-cineole, β-ocimene, γ -terpinene, terpinolene, linalool, 4,8-dimethylnona-1,3,7-triene, *p*-mentha-1,5,8-triene, 1,3,8-*p*-menthatriene, *p*-mentha-1,5-dien-8-ol, terpinen-4-ol, α-terpineol, *trans-p*-menth-1(7),8-dien-2-ol, neral, linalyl acetate, citral, bornyl acetate, methyl geranate, α-terpinyl acetate, *cis*-geranyl acetate], sesquiterpenes (β-elemene, β-caryophyllene, aromandendrene, β-selinene, α-selinene, γ -cadinene, nerolidol, α-elemene), and diterpene (coronarin E) were reported from *E. cardamomum* (Alam et al. 2021). Twenty-four essential oils (α-thujene, α-pinene, sabinene, β-pinene, β-myrcene, 3-carene, α-terpinolene, limonene, 1,8-cineole, β-cymene, γ-terpinene, β-linalool, terpinen-4-ol, α-terpineol, β-terpineol, β-citral, nerol, linalyl acetate, α-citral, α-terpinyl acetate, geranyl acetate, p-cresol, γ-cadinene, nerolidol) were isolated and identified from *E. cardamomum* (Ashokkumar et al. 2021).

Thirty nine volatile oils [sabinene, myrcene, iso-sylvestrene, *p*-cymene, 1,8-cineole, Z-β-ocimene, E-β-ocimene, γ-terpinene, n-octanol, terpinolene, linalool, *cis-p*-menth-2-en-1-ol, *trans-p*-menth-2-en-1-ol, δ-terpineol, terpinen-4-ol, dihydrocarveol, *trans*-piperitol, *trans*-carveol, nerol, neral, Geraniol, geranial, geranyl formate, δ-terpinyl acetate, methyl-geranate, α-terpinyl acetate, eugenol, neryl acetate, Z-caryophyllene, α-humulene, germacrene D, *cis*-β-guaiene, viridiflorene, γ-cadinene, chavibetol acetate, *E*-nerolidol, Z-dihydro-apofarnesol, caryophyllene oxide, (2Z,6E)-farnesol] were obtained from *E. cardamomum* (Al-Zereini et al. 2022).

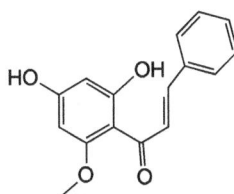

Cardamonin

2.3 CULTURE CONDITIONS

The inner core region of rhizome of *Elettaria cardamomum* was cultured on an MS (Murashige and Skoog 1962) medium for induction of callus. For the maximum amount of callus regeneration, the MS medium was supplemented with 6-benzylaminopurine (8.8 µM) and 1-naphthaleneacetic acid (0.5 µM). Similarly, the maximum frequency of embryogenic calli (68%) was reported in the MS medium containing 6-benzylaminopurine (4.4 µM) and 1-naphthaleneacetic acid (0.5 µM; Manohari et al. 2008). The nodal explants of cardamom were transferred on MS medium for callus development. The medium was supplemented 6-benzylaminopurine (5 µM). The slow growth of callus was achieved in half concentration of the MS medium (Tyagi et al. 2009).

REFERENCES

Abdel Kader SM, Bauomi AA, Abdel-Rahman M, Mohammaden TF, Rezk MM. 2015. Antioxidant potentials of (*Elletaria cardamomum*) cardamom against uranium hazards. Int J Basic Life Sci 3, 164–181.

Abu Gazia M, El-Magd MA. 2018. Ameliorative effect of Cardamom aqueous extract on doxorubicin-induced cardiotoxicity in rats. Cells Tissues Organs 206, 62–72.

Abu-Taweel GM. 2018. Cardamom (*Elettaria cardamomum*) perinatal exposure effects on the development, behavior and biochemical parameters in mice offspring. Saudi J Biol Sci 25, 186–193.

Adegoke GO, Rao LJM, Shankaracharya NB. 1998. A comparison of the essential oils of *Aframomum danielli* (Hook.f.) K. Schum. and *Amomum subulatum* Roxb. Flavour Fragr J 13, 349–352.

Aghasi M, Ghazi-Zahedi S, Koohdani F, Siassi F, Nasli-Esfahani E, Keshavarz A, Qorbani M, Khoshama H, Salari-Moghaddam A, Sotoudeh G. 2018. The effects of green cardamom supplementation on blood glucose, lipids profile, oxidative stress, sirtuin-1 and irisin in type 2 diabetic patients: A study protocol for a randomized placebo-controlled clinical trial. BMC Complement Altern Med 18, 18.

Aghasi M, Koohdani F, Qorbani M, Nasli-Esfahani E, Ghazi-Zahedi S, Khoshamal H, Keshavarz A, Sotoudeh G. 2019. Beneficial effects of green cardamom on serum SIRT1, glycemic indices and triglyceride levels in patients with type 2 diabetes mellitus: A randomized double-blind placebo controlled clinical trial. J Sci Food Agric 99, 3933–3940.

Ahmed HM, Ramadhani AM, Erwa IY. 2019. Phytochemical screening, chemical composition and antibacterial activity of essential oil of *Cardamom*. World J Pharm Res 8, 1166–1175.

Alam A, Jawaid T, Alam P. 2021. *In vitro* antioxidant and anti-inflammatory activities of green cardamom essential oil and *in silico* molecular docking of its major bioactives. J Taibah Univ Med Sci 15, 757–768.

Alam A, Majumdar RS, Alam P. 2019. Systematics evaluations of morphological traits, chemical composition, and antimicrobial properties of selected varieties of *Elettaria cardamomum* (L.) Maton. Nat Prod Commun 14, 1–7.

Allen RW, Schwartzman E, Baker WL, Coleman CI, Phung OJ. 2013. Cinnamon use in type 2 diabetes: An updated systematic review and meta-analysis. Ann Fam Med 11, 452–459.

Al-Zereini WA, Al-Trawneh IN, Al-Qudah MA, TumAllah HM, Al Rawashdeh HA, Abudayeh ZH. 2022. Essential oils from *Elettaria cardamomum* (L.) Maton grains and *Cinnamomum verum* J. Presl barks: Chemical examination and bioactivity studies. J Pharm Pharmacogn Res 10, 173–185.

Ashokkumar K, Murugan M, Dhanya MK, Raj S, Kamaraj D. 2019. Phytochemical variations among four distinct varieties of Indian cardamom *Elettaria cardamomum* (L.) Maton. Nat Prod Res 34, 1919–1922.

Ashokkumar K, Murugan M, Dhanya MK, Raj S, Kamaraj D. 2020b. Phytochemical variations among four distinct varieties of Indian cardamom *Elettaria cardamomum* (L.) Maton. Nat Prod Res 34, 1919–1922.

Ashokkumar K, Murugan M, Dhanya MK, Warkentin TD. 2020a. Botany, traditional uses, phytochemistry and biological activities of cardamom [*Elettaria cardamomum* (L.) Maton]—a critical review. J Ethnopharmacol 246, 112244.

Ashokkumar K, Vellaikumar S, Murugan M, Dhanya MK, Ariharasutharsan G, Aiswarya S, Akilan M, Warkentin TD, Karthikeyan A. 2021. Essential oil profile diversity in Cardamom accessions from Southern India. Front Sustain Food Syst 5, 639619.

Auti ST, Kulkarni YA. 2019. Neuroprotective effect of Cardamom oil against aluminum induced neurotoxicity in rats. Front Neurol 10, 399.

Bacha K, Tariku Y, Gebreyesus F, Zerihun S, Mohammed A, Weiland-Bräuer N, Schmitz RA, Mulat M. 2016. Antimicrobial and antiquorum sensing activities of selected medicinal plants of Ethiopia: Implication for development of potent antimicrobial agents. BMC Microbiol 16, 139–147.

Bampidis V, Azimonti G, Bastos MDL, Christensen H, Kouba M, Durjava MK, López-Alonso M, Puente SL, Marcon F, Mayo B, Pechová A, Petkova M, Ramos F, Sanz Y, Villa R, Woutersen R, Brantom P, Chesson A, Kolar B, Beelen PV, Westendorf J, Gregoretti L, Manini P, Dusemund B. 2019. Safety and efficacy of an essential oil from *Elettaria cardamomum* (L.) Maton when used as a sensory additive in feed for all animal species. EFSA J 17, e05721.

Bhattacharjee B, Chatterjee J. 2013. Identification of proapoptopic, anti-inflammatory, anti- proliferative, anti-invasive and anti-angiogenic targets of essential oils in cardamom by dual reverse virtual screening and binding pose analysis. Asian Pac J Cancer Prev 14, 3735–3742.

Cárdenas Garza GR, Elizondo Luévano JH, Bazaldúa Rodríguez AF, Chávez Montes A, Pérez Hernández RA, Martínez Delgado AJ, López Villarreal SM, Rodríguez Rodríguez J, Sánchez Casas RM, Castillo Velázquez U. 2021. Benefits of cardamom (*Elettaria cardamomum* (L.) Maton) and turmeric (*Curcuma longa* L.) extracts for their applications as natural anti-inflammatory adjuvants. Plants 10, 1908.

Chakraborty S, Paul K, Mallick P, Pradhan S, Das K, Chakrabarti S, Nandi DK, Bhattacharjee P. 2019. Consortia of bioactives in supercritical carbon dioxide extracts of mustard and small cardamom seeds lower serum cholesterol levels in rats: New leads for hypocholesterolaemic supplements from spices. J Nutr Sci 8, e32.

Daneshi-Maskooni M, Keshavarz SA, Mansouri S, Qorbani M, Alavian SM, Badri-Fariman M, Jazayeri-Tehrani SA, Sotoudeh G. 2017. The effects of green cardamom on blood glucose indices, lipids, inflammatory factors, paraxonase-1, sirtuin-1, and irisin in patients with nonalcoholic fatty liver disease and obesity: Study protocol for a randomized controlled trial. Trials 18, 260.

Daneshi-Maskooni M, Keshavarz SA, Qorbani M, Mansouri S, Alavian SM, Badri-Fariman M, Jazayeri-Tehrani SA, Sotoudeh G. 2018. Green cardamom increases Sirtuin-1 and reduces inflammation in overweight or obese patients with non-alcoholic fatty liver disease: A double-blind randomized placebo-controlled clinical trial. Nutr Metab (Lond) 15, 63.

Daneshi-Maskooni M, Keshavarz SA, Qorbani M, Mansouri S, Alavian SM, Badri-Fariman M, Jazayeri-Tehrani SA, Sotoudeh G. 2019. Green cardamom supplementation improves serum irisin, glucose indices, and lipid profiles in overweight or obese non-alcoholic fatty liver disease patients: A double-blind randomized placebo-controlled clinical trial. BMC Complement Altern Med 19, 59.

Das I, Acharya A, Berry DL, Sen S, Williams E, Permaul E, Sengupta A, Bhattacharya S. 2012. Antioxidative effects of the spice cardamom against nonmelanoma skin cancer by modulating nuclear factor erythroid-2-related factor 2 and NF-jB signalling pathways. Br J Nutr 108, 984–997.

Garg G, Sharma S, Dua A, Mahajan R. 2016. Antibacterial potential of polyphenol rich methanol extract of Cardamom (*Amomum subulatum*). J Innov Biol 3, 271–275.

Gilani AH, Jabeen Q, Khan A, Shah J. 2008. Gut modulatory, blood pressure lowering, diuretic and sedative activities of cardamom. J Ethnopharmacol 115, 463–472.

Govindarajan VS, Shanthi N, Raghuveer KG, Lewis YS, Stahl WH. 1982. Cardamom—production, technology, chemistry, and quality. Crit Rev Food Sci Nutr 16, 229–326.

Goyal SN, Sharma C, Mahajan UB, Patil CR, Agrawal YO, Kumari S, Arya DS, Ojha S. 2015. Protective effects of *Cardamom* in isoproterenol-induced myocardial infarction in rats. Int J Mol Sci 16, 27457–27469.

Hamdan S, Daood HG, Toth-Markus M, Illés V. 2008. Extraction of cardamom oil by supercritical carbon dioxide and sub-critical propane. J Supercrit Fluids 44, 25–30.

Hamzaa R, Osman N. 2012. Using of coffee and cardamom mixture to ameliorate oxidative stress induced in γ-irradiated rats. Biochem Anal Biochem 1, 113–119.

Husain SS, Ali M. 2014. Analysis of volatile oil of the fruits of *Elettaria cardamomum* (L.) Maton and its antimicrobial activity. World J Pharm Pharm Sci 3, 1798–1808.

İçen H, Çelik HT, Ekinci MS, Gürü M. 2017. Obtaining of β-sitosterol from *Cardamom* by supercritical CO_2 extraction. Preprints 2017, 110005.

Jamal A, Javed K, Aslam M, Jafri MA. 2006. Gastroprotective effect of cardamom, *Elettaria cardamomum* Maton. fruits in rats. J Ethnopharmacol 103, 149–153.

Kandikattu HK, Rachitha P, Jayashree GV, Krupashree K, Sukhith M, Majid A, Amruta N, Khanum F. 2017. Anti-inflammatory and antioxidant effects of Cardamom (*Elettaria repens* (Sonn.) Baill) and its phytochemical analysis by 4D GCXGC TOF-MS. Biomed Pharmacother 91, 191–201.

Kapoor LD. 2000. CRC Handbook of Ayurvedic Medicinal Plants. CRC Press Inc, Boca Raton.

Kaushik P, Goyal P, Chauhan A, Chauhan G. 2010. *In vitro* evaluation of antibacterial potential of dry fruit extracts of *Elettaria cardamomum* Maton (Chhoti Elaichi). Iran J Pharm Res 9, 287–292.

Kazemi S, Yaghooblou F, Siassi F, Foroushani AR, Ghavipour M, Koohdani F, Sotoudeh G. 2017. Cardamom supplementation improves inflammatory and oxidative stress biomarkers in hyperlipidemic, overweight, and obese pre-diabetic women: A randomized double-blind clinical trial. J Sci Food Agric 97, 5296–5301.

Khan AU, Khan QJ, Gilani AH. 2011. Pharmacological basis for the medicinal use of cardamom in asthma. Bangladesh J Pharmacol 6, 34–37.

Kuyumcu Savan E, Küçükbay ZF. 2013. Essential oil composition of *Elettaria cardamomum* Maton. J Appl Biol Sci 7, 42–45.

Ma Y, Yu T, Zhang Y, Yin Y, Zhao Z, Yu X, Yu Y. 2018. The protective effect of cardamonin on the factors involved in delayed cerebral vasospasm in a rat model of subarachnoid hemorrhage. Int J Clin Exp Pathol 11, 5955–5961.

Majdalawieh AF, Carr RI. 2010. In vitro investigation of the potential immunomodulatory and anti-cancer activities of black pepper (*Piper nigrum*) and cardamom (*Elettaria cardamomum*). J Med Food 13, 371–381.

Malti JE, Mountassif D, Amarouch H. 2007. Antimicrobial activity of *Elettaria cardamomum*: Toxicity, biochemical and histological studies. Food Chem 104, 1560–1568.

Manohari C, Backiyarani S, Jebasingh T, Somanath A, Usha R. 2008. Efficient plant regeneration in small cardamom (*Elletaria cardamomum* Maton.) through somatic embryogenesis. Indian J Biotechnol 7, 407–409.

Marongiu B, Piras A, Porcedda S. 2004. Comparative analysis of the oil and supercritical CO_2 extract of *Elettaria cardamomum* (L.) Maton. J Agric Food Chem 52, 6278–6282.

Masoumi-Ardakani Y, Mandegary A, Esmaeilpour K, Najafipour H, Sharififar F, Pakravanan M, Ghazvini H. 2016. Chemical composition, anticonvulsant activity, and toxicity of essential oil and methanolic extract of *Elettaria cardamomum*. Planta Med 82, 1482–1486.

Mejdi S, Emira N, Ameni D, Guido F, Mahjoub A, Madiha A, Al-Sieni Abdulbasit A-S. 2015. Chemical composition and antimicrobial activities of *Elettaria cardamomum* L. (Manton) essential oil: A high activity against a wide range of food borne and medically important bacteria and fungi. J Chem Biol Phys Sci 6, 248–259.

Menon AN, Chacko S, Narayanan CS. 1999. Free and glycosidically bound volatiles of cardamom (Elettaria cardamomum Maton var. miniscula Burkill). Flavour Fragr J 14, 65–68.

Merghni A, Noumi E, Hadded O, Dridi N, Panwar H, Ceylan O, Mastouri M, Snoussi M. 2018. Assessment of the antibiofilm and antiquorum sensing activities of *Eucalyptus globulus* essential oil and its main component 1,8-cineole against methicillin-resistant *Staphylococcus aureus* strains. Microb Pathog 118, 74–80.

Mona MAD. 2020. Insecticidal potential of cardamom and clove extracts on adult red palm weevil *Rhynchophorus ferrugineus*. Saudi J Biol Sci 27, 195–201.

Murashige T, Skoog F. 1962. A revised medium for rapid growth and bioassays with tobacco tissue cultures. Physiol Plant 15, 473–497.

Murugan M, Dhanya MK, Deepthy KB, Preethy TT, Aswathy TS, Sathyan T, Manoj VS. 2016. Compendium on Cardamom, 2nd ed. Kerala Agricultural University, Cardamom Research Station, Pampadumpara.

Nagashree S, Archana KK, Srinivas P, Srinivasan K, Sowbhagya HB. 2017. Anti-hypercholesterolemic influence of the spice cardamom (*Elettaria cardamomum*) in experimental rats. J Sci Food Agric 97, 3204–3210.

Nirmala MA 2000. Studies on the volatile of cardamom (*Elleteria cardamomum*). J Food Sci Technol 37, 406–408.

Noumi E, Snoussi M, Alreshidi MM, Rekha P-D, Saptami K, Caputo L, Martino LD, Souza LF, Msaada K, Mancini E, Flamini G, Al-Sieni A, De Feo V. 2018. Chemical and biological evaluation of essential oils from *Cardamom* species. Molecules 23, 2818.

Padmakumari Amma KPA, Rani MP, Sasidharan I, Nisha VNP. 2010. Chemical composition, flavonoid—phenolic contents and radical scavenging activity of four major varieties of cardamom. Int J Biol Med Res 1, 20–24.

Paul K, Bhattacharjee P, Chatterjee N, Pal TK. 2019. Nanoliposomes of supercritical carbon dioxide extract of small Cardamom seeds redresses type 2 diabetes and hypercholesterolemia. Recent Pat Biotechnol 13, 284–303.

Paul K, Chakraborty S, Mallick P, Bhattacharjee P, Pal TK, Chatterjee N, Chakrabarti S. 2020a. Supercritical carbon dioxide extracts of small cardamom and yellow mustard seeds have fasting hypoglycaemic effects: Diabetic rat, predictive iHOMA2 models and molecular docking study. Br J Nutr 1–12. https://doi.org/10.1017/S000711452000286X.

Paul K, Ganguly U, Chakrabarti S, Bhattacharjee P. 2020b. Is 1,8-cineole-rich extract of small Cardamom seeds more effective in preventing Alzheimer's disease than 1,8-cineole alone? Neuromolecular Med 22, 150–158.

Qiblawi S, Kausar MA, Shahid SMA, Saeed M, Alazzeh AY. 2020. Therapeutic interventions of *Cardamom* in cancer and other human diseases. J Pharm Res Int 32, 74–84.

Saeed A, Sultana B, Anwar F, Mushtaq M, Alkharfy KM, Anwarul-Hassan Gilani A-H. 2014. Antioxidant and antimutagenic potential of seeds and pods of green *Cardamom* (*Elettaria cardamomum*). Int J Pharmacol 10, 461–469.

Sereshti H, Rohanifar A, Bakhtiari S, Samadi S. 2012. Bifunctional ultrasound assisted extraction and determination of *Elettaria cardamomum* Maton essential oil. J Chromatogr A 1238, 46–53.

Singh G, Kiran S, Marimuthu P, Isidorov V, Vinogorova V. 2008. Antioxidant and antimicrobial activities of essential oil and various oleoresins of *Elettaria cardamomum* (seeds and pods). J Sci Food Agric 88, 280–289.

Souissi M, Azelmat J, Chaieb K, Grenier D. 2020. Antibacterial and anti-inflammatory activities of cardamom (*Elettaria cardamomum*) extracts: Potential therapeutic benefits for periodontal infections. Anaerobe 61, 102089.

Sultana S, Ali M, Ansari SH and Bagri P. 2009. Effect of physical factors on the volatile constituents of *Elettaria cardamomum* fruits. J Essent Oil Bear Plant 12, 287–292.

Suneetha WJ, Krishnakantha TP. 2005. Cardamom extract as inhibitor of human platelet aggregation. Phytother Res 19, 437–440.

Telja R, Olavl L, Roberto Q. 2006. Small cardamom-precious for people, harmful for mountain forests. Possibilities for sustainable cultivation in the east Usambaras, Tanzania. Mt Res Dev 26, 131–137.

Tyagi RK, Goswami R, Sanayaima R, Singh R, Tandon R, Agrawal A. 2009. Micropropagation and slow growth conservation of cardamom (*Elettaria cardamomum* Maton). *In Vitro* Cell Dev Biol Plant 45, 721–729.

Vaidya A, Rathod M. 2014. An in vitro study of the immunomodulatory effects of *Piper nigrum* (black pepper) and *Elettaria cardamomum* (cardamom) extracts using a murine macrophage cell line. Am Int J Res Form Appl Nat Sci 8, 18–27.

Verma K, Jain V, Katewa SS. 2009. Blood pressure lowering, fibrinolysis enhancing and antioxidant activities of cardamom (*Elettaria cardamomum*). Indian J Biochem Biophys 46, 503–506.

Zahedi SG, Koohdani F, Qorbani M, Siassi F, Keshavarz A, Nasli-Esfahani E, Aghasi M, Khoshamal H, Sotoudeh G. 2020. The effects of green cardamom supplementation on blood pressure and endothelium function in type 2 diabetic patients: A study protocol for a randomized controlled clinical trial. Medicine (Baltimore) 99, e11005.

3 *Carduus* Species

3.1 MORPHOLOGICAL FEATURES, DISTRIBUTION, ETHNOPHARMACOLOGICAL PROPERTIES, PHYTOCHEMISTRY, AND PHARMACOLOGICAL ACTIVITIES

The Carduus genus (Fam.—Asteraceae) comprises about 100 species and is widely distributed in the Mediterranean region (Mandaville 1990; Chaudhary 2000). As per Chinese traditional medicine, the Carduus genus is used in the treatment of cold, stomachache, and rheumatism (Esmaeili et al. 2005). Ethanol extract of *C. thoermeri, C. nutans*, and *C. candicans* ssp. Globifer possesses significant antioxidative activity (Zheleva-Dimitrova et al. 2011). *C. nutans* L. (syn. *C. thoermeri* Weinm) is an annual or biennial herb, with single or up to seven erect stems, much branched, contains spiny wings, and grows up to 20–200 cm high. It has a long fleshy taproot. The leaves are elliptic to lanceolate, long (15–30 cm), glabrous to densely pubescent, pinnately lobed, and each lobe ends in a spine. Cauline leaves are similar but smaller, simple, alternate, and decurrent. Its peduncle is naked or with a few bracts below the head (1.5–5 cm). The head is 1.5–4.5(7) cm in diameter, usually nodding. Fruits are achenes, long (3.5–4 mm), outer achenes curved, and turn brown at maturity (Desrochers et al. 1988). It is naturalized in Europe (including Britain, north to Norway), south and east to north Africa, Siberia, west Asia, Canada, China, India, Kazakhstan, Mongolia, New Zealand, North America, Russia, Siberia, Tasmania, and Turkey (Weber and Wittmann 2001). This is used in the treatment of atherosclerosis (Kindscher 1987). The inflorescence of *C. nutans* is macerated in cold water for ten days and is taken as a tea in the treatment of eczema (Mustafa et al. 2012). Methanol extract of *C. nutans* showed the strongest radical scavenging property (the EC_{50} 618 ± 10.03 µg/mL) on DPPH, and total phenolic contents were determined (TPC = 61.49 mg/g; Kozyra et al. 2022). In Turkish medicine, ethanol extract demonstrated hepatoprotective effects (Aktay et al. 2000). The fructan-rich extract of *C. thoermeri* Weinm. flower heads demonstrated antioxidant activity in 2,2-diphenyl-1-picrylhydrazy radical scavenging, 2,2'-azinobis-3 ethyl benxothiazoline-6-sulphonic acid cation decolorization, ferric-reducing/antioxidant power, and cupric ion-reducing antioxidant capacity assays (Petkova et al. 2015).

C. acanthoides L. is an annual or biennial herb, height 20–150 cm, and branched stem with spiny wings extending to flower heads. The leaves are narrowly elliptic or oblong, pinnate, lobes pointed (1–3), each point ends in a spine. Its flower heads are terminal, and solitary or clustered on young branches. Fruits are achenes, long (2.5–3 mm), turning to light brown at maturity (Desrochers et al. 1988). They are found in France, Italy, and western Turkey, Russia, and Kazakhstan to China (Stuckey and Forsyth 1971; Dunn 1976). It is (when used as the hormone-like herbicide 2,4-D) effective in controlling of growth of weeds (Desrochers et al. 1988; Palma-Bautista et al. 2020).

C. crispus L. is a biennial herb, stem branched, and grows up to 0.90 m high. The leaves are broad, palmate, and elliptic. It is found in Asia, Australia, Britain, Canada, China, Europe, Kazakhstan, Korea, Mongolia, North America, Russia, and the US. The root is alternative anodyne (Denev et al. 2018). Methanol extract of *C. crispus* leaves showed significant acetylcholinesterase inhibitory (60.9% at 1.000 µg/ml) activity. The chlorogenic acid, neochlorogenic acid, *p*-coumarylquinic acid, and apigenin derivatives have been isolated and characterized from the *n*-butanol fraction of methanol extract (Brantner et al. 2010). As per Mongolian traditional medicine, *C. crispus* is used in the treatment of cancer of the glands and as analgesics. The ethanol extract possesses cytotoxic activity (Qingyig et al. 2002). The silver nanoparticles (17 µg/ml) of *C. crispus* presented very low cytotoxicity on the HepG2 cell line but showed high antibacterial activity against *E. coli* (inhibition zone 5.5 ± 0.2 mm to 6.5 ± 0.3 mm) and *Micrococcus luteus* (inhibition zone 7 ± 0.4 mm to 7.7 ± 0.5 mm; Urnukhsaikhan et al. 2021).

DOI: 10.1201/9781003398035-3

C. lanuginosus Willd. is a biennial herb and contains a branched stem. The leaves are simple, alternate, and broadly elliptic. It is distributed in Turkey (Kazmi 1964; Davis 1975; Kabaktepe et al. 2014). The aqueous and hexane extracts demonstrated strong enzyme inhibitory effects against α-amylase and α-glucosidase. The maximum level of chlorogenic acid was recorded in methanol extract. The hexane and ethyl acetate extracts demonstrated significant antibacterial activity against *E. faecium*. Therefore, *C. lanuginosus* extracts might be used as a natural source for pharmaceutical drugs (Özcan 2021).

C. benedictus Auct. ex Steud. [Syn. *Cnicus benedictus* L. syn. *Centaurea benedicta* (L.) L.] is a biennial herb, erect and heavily branched, thistle-like, villous, and glutinous pubescent stem, 30 to 50 cm in height. The leaves are oblong, emarginated to pinnatifid, spiny-dentate, and reticulate. The flower is solitary, sessile, and pale-yellow. The outer bracts are terminate in a simple thorn while inner bracts end in a long rigid and pinnatifid thorn. The fruit has a tuft of hair (Fleming 1998). Leaves, stems, and flowers have been used in the stimulation of bile secretion, strengthening of liver, treatment of jaundice, and flatulence. As per European traditional and Indian Ayurvedic medicine systems, it is used in regulation of menstrual flow. It is useful in (as galactagogue) increasing the milk flow in breast feeding mothers (Grieve 2014; Rotblatt 2000; Hoffmann 1998). Similarly, in homeopathic medicine it is used for the treatment of nausea, left-sided stomach pain, gallstones, homesickness, intermittent fever, enlarged liver, and eye infections (Al-Snafi 2016). The tea is prepared from leaves and used to increase perspiration and in the reduction of congestion and fever (Coon 1979). Its mild infusion is used as an astringent and in relieving diarrhea. It is useful in heart ailments, cancers, as well as used as a contraceptive (Murray and Michael 1992; Rajendran et al. 2011). The water extract of *C. benedictus* flowers showed significant antimicrobial activity against *S. typhimurium* ATCC 14028, *S. enteritidis* ATCC 13076, *S. aureus* ssp. ATCC 25923, *S. aureus* ssp. ATCC 29213, *E. coli* ATCC 25922, *E. coli* ATCC 35218, *S. pyogenes* Gp ATCC 19615, *P. aeruginosa* ATCC 27853, *E. faecalis* ATCC 29212, and *S. sonnei* ATCC 25931 (Ildiko et al. 2009).

3.2 PHYTOCHEMISTRY

The isoquinoline alkaloid (crispine A N-oxide), along with quercetin and rutin, were isolated from ethanol extract of *C. crispus* aerial parts (Tunsag et al. 2011). A flavone glycoside (chrysoeriol 7-*O*-(2″-*O*-6‴-*O*-acetyl-β-D-glucopyranosyl-β-D-glucopyranoside), along with fourteen other compounds [sophoroside, linariifolioside, crispine B, crispine C, afzelin, astragaline, luteolin 3′ -*O*-α-lrhamnopyranoside, luteolin, apigenin, tricin, adenosine, adenine, syringin, and (*E*)-2-butenedioc acid] were separated and identified from the *C. crispus* whole plant. Crispine B demonstrated significant antitumor activity against HO-8901 (human ovarian neoplasm) cells (Xie et al. 2005). Five isoquinoline alkaloids (crispine A–E) have been isolated and characterized from *C. crispus*. All the isolated compounds showed cytotoxic activity against human-cancer lines *in vitro* (Zhang et al. 2002), two isoquinoline alkaloids (carcrisine A and B0 have been isolated and characterized from the *C. crispus* whole plant (Xie and Jia 2004).

3-*O*-acetyl-ursolic acid-28-ethyl ester, diosmetin-7-*O*-α-L-arabinopyransyl (1‴→4″)-β-D-glucopyranoside, bis (2-ethylhexyl) benzene-1,2-dicarboxylate, 3α, 24 dihydroxyolean-12-en-28, 30-dioic acid dimethyl ester, and kaempferol have been reported from alcoholic extract of *C. pycnocephalus*. The ethanol extract showed variable degrees of cytotoxic activity against MCF-7, A-549 and HepG-2 (IC$_{50}$ 17.9, 17.5 and 21.8 µg) cell lines. The ethanol extract demonstrated weak antioxidant activity in a DPPH scavenging assay (SC$_{50}$ 554.2 µg). Similarly, the extract displayed significant antibacterial (*P. aeruginosa* and *E. coli*) and antifungal (*Syncephalastrum racemosum*) activities (Hassan et al. 2015). The flavonoidal [apigenin, kaempferol-3-*O*-β-D-glucoside, kaempferol-3-*O*-α-L-rhamnoside, kaempferol-7-methoxy-3-*O*-α-L-rhamnoside, diosmetin-7- *O*-β-D-xylosyl-(1‴→6″)-β-D-glucopyranoside,diosmetin-7-*O*-α-L-arabinosyl(1‴→6″)-β-D-glucopyranoside, kaempferol-3-*O*-α-L-rhamnosyl-(1‴→2″)-α-L-rhamnoside] and steroidal (lupeol, β-sitosterol and β-sitosterol-3-*O*-β-D-glucoside)compounds from aerial parts (Al-Shammari et al. 2015), two flavonoids (diosmetin and kaempferol-3-*O*-α-L-rhamnoside) from ether extract of aerial parts (El-Lakany et al. 1995), chrysoeriol-7-*O*-β-D-(6-O-acetylglucopyranosyl) (1→3)-β-D-glucopyranoside, and diosmetin 7-*O*-α-L-arabinosyl (1→6)-β-D-glucoside have been isolated from ethyl acetate extract of *C. pycnocephalus* aerial parts (El-Lakany et al. 1997). Twenty methyl esters (butanedioic acid, dimethyl ester, benzoic acid methyl ester, pentanedioic, dimethyl ester, adipic acid dimethyl ester, pimelic acid dimethyl ester, 1,2-benzenedicarboxylic acid, dimethyl ester, 1,2-benzenedicarboxylic acid, 1-ethyl 2-methyl ester, azelaic acid dimethyl ester, decanedioic acid, dimethyl ester, tetradecanoic acid methyl ester, undecanedioc acid dimethyl ester, tetradecanoic acid, 12 methyl- methyl ester, pentadecanoic acid methyl ester, dodecanedioic acid dimethyl ester, palmitic acid methyl ester, 9-octadecenoic acid methyl ester, isostearic acid methyl ester, arachidic acid methyl ester, adipic acid di(2-ethylhexyl)ester, 1,2- benzenedicarboxylic acid diisononyl ester). Twenty-nine essential oils [(*E*)-β-Ionone, δ-cadenen, 1-butyl hexylbenzene, 1-propyl heptyl benzene, elemol, 1-ethyl octylbenzene, dodecanoic acid, caryophyllene oxide, hexadecane, 1-pentyl hexyl benzene, 1-butyl hexylbenzene, 1-propyl octylbenzene, β-eudesmol, α- cardinal, heptadecane, 1-methyl decylbenzene, 1-pentyl heptylbenzene, 1-butyl octylbenzene, 1-propyl nonyl benzene, tetradecanoic acid, octadecane, 1-pentyl octylbenzene, 1-butyl nonylbenzene, 6,10,14 trimethyl -2-pentadecanone, 1-propyl decylbenzene, nonadecane, dibutyl,1,2-benzene dicarboxylate, hexadecanoic acid, eicosane) have been identified from petroleum ether extract of *C. pycnocephalus* (Al-Shammari et al. 2012). The coumarin-hemiterpene ether (6-(3,3-dimethylallyloxy)-7-methoxycoumarin has been isolated and characterized from acetone extract of *C. tenuiflorus* Curtis (syn. *C. pycnocephalus*; Cardona et al. 1992).

Taraxasterol acetate, taraxasterol, and erythrodiol 3-acetate from petroleum extract of aerial parts (Abdel-Salam et al. 1983), hydroxytyrosol hexoside, 1-*O*-caffeoylquinic acid, 5-*O*-caffeoylquinic acid, *O*-caffeoylquinic acid isomer, luteolin-*O*-hexosyl-*O*-glucuronide, luteolin-*O*-glucuronide rhamnoside, luteolin-*O*-dihexoside, apigenin-*O*-hexosyl-*O*-pentoside, luteolin-*O*-hexosyl-*O*-pentoside, luteolin-7-*O*-rutinoside (scolymoside), luteolin-7-*O*-glucoside (cynaroside), apigenin-*O*-hexouronide-rhamnoside, tricin-*O*-malonyl hexoside, astragalin, luteolin-*O*-glucuronide, apigenin-*O*-hexouronide, luteolin-*O*-rhamnoside, chrysoeriol-*O*-hexoside, apigenin-*O*-glucoside, luteolin, apigenin, tricin, hydroxyoctadecatrienoic acid, hydroxyoctadecadienoic acid, hydroxyoctadecadienoic acid, linolenic acid, linoleic acid, palmitic acid, and oleic acid were reported from *C. getulus* aerial parts (Taha et al. 2019).

Apigenin-7-glucoside, luteolin-7-glucoside, kaempferol-3-rhamnoglucoside, kaempferol-3-glucoside, apigenin, luteolin and chlorogenic acid and *p*-coumaric acid were detected in the methanol extract of *C. acanthoides* flower heads (Kozyra et al. 2013). The flavones [luteolin 7-*O*-glucosyl-(1→2)-glucoside, luteolin 7-*O*-glucoside, chrysoeriol 7-*O*-glucosyl-(1→2)- glucoside, luteolin 7-*O*-6‴-acetyl-glucosyl-(1→2)- glucoside, apigenin 7-*O*-glucuronide, apigenin 7-*O*-glucoside, chrysoeriol 7-*O*-6‴-acetyl-glucosyl-(1→2)-glucoside, luteolin 3′ -*O*-rhamnoside], flavonols [kaempferol 3-*O*-rhamnoside-7-*O*-gulcuronide, kaempferol 3-*O*-glucoside, quercetin 3-*O*-α-L-rhamnoside, kaempferol 3-*O*-rhamnoside], quinic acids [5-*O*-*p*-coumaroylquinic acid, 4-*O*-*p*-coumaroylquinic acid, 4-*O*-caffeoylquinic acid (cryptochlorogenic acid), 3-*O*-caffeoylquinic acid (neochlorogenic acid), 5-*O*-caffeoylquinic acid (chlorogenic acid)], coumarins [3,3′-biisofraxidin, eleutheroside B1, isofraxidin], isoflavone (eurycarpin B), chalcone (licochalcone A), phenolic acid glycosides (2,6-dimethoxy-p-hydroquinone 4-*O*-glucoside, 2,6-dimethoxy-4-hydroxyphenol 1-*O*-glucoside, syringic acid 4-*O*-arabinoside, tachioside, vanillic acid 4-*O*-β-D-glucoside), prenylethanoid glycoside (3,5-dihydroxyphenethyl alcohol 3-*O*-glucoside, salidroside, syringin) have been reported from methanol extract of *C. acanthoides*. All analytes showed good linearity ($r^2 \geq 0.998$) with an LOD of 0.077–0.120 μg mL^{-1}. These nine compounds accounted for 2.1–3.5% of the herb (Li et al. 2014). The acanthoine and acanthoidine from *C. acanthoides* (Frydman and Deulofeu 1962), apigenin, and luteolin have been obtained from *C. acanthoides* (Bain and Desrochers 1988). Chlorogenic acid, protocatechuic acid, *p*-coumaric acid, caffeic acid, syringic acid, *p*-hydroxybenzoic acid, ferulic acid, vanillic acid, and gentisic and gallic acids were reported from methanolic extract

C. acanthoides flowering head. The methanol extract demonstrated antimicrobial activity against Gram-positive bacteria, Gram-negative bacteria, and fungal organisms (Kozyra et al. 2017).

Benzaldehyde, palmitic acid, methyl salicylate, heptacosane, tricosane, pentacosane, Z-12-pentacosene, and β-caryophyllene from *C. candicans* ssp. Globifer, and palmitic acid, methyl salicylate, benzaldehyde, trans-nerolidol, *p*-cymen-8-ol, and tricosane have been determined from *C. thoermeri* flower heads (Zhelev et al. 2012). Sinapic, chlorogenic, ferulic acids, luteolin, hesperidin, kaempferol, myrcetin, quercetin, and apigenin have been identified from ethanol extract of *C. rhodopaeus*. The ethanol extract showed antioxidant activity in 2,2-diphenyl-1-picrylhydrazyl, 2,2′-azino-bis (3-ethylbenzothiazoline-6-sulphonic acid) and ferric-reducing antioxidant power and copper-reduction antioxidant assays (Dimitrova-Dyulgerova et al. 2015).

Phenolic compounds (caffeic acid, cinnamic acid, chlorogenic acid, *p*-coumaric acid, 3,4-dihydroxybenzoic acid, ferulic acid, gallic acid, 2-hydroxybenzoic acid, sinapic acid, syringic acid, vanillic acid), flavonoid glycosides (hyperoside and rutin), and flavonoid aglycones (apigenin, hesperedin, kaempferol, luteolin, myricetin, quercetin) have been determined from *C. acicularis* flower heads. The methanol extract of *C. acicularis* flower heads showed scavenging activity on 2,2-diphenil-1-picrylhydrazyl radical, 2,2′-azinobis-(3-ethyl-benzothiazoline-6-sulfonate) radical cation decolorization assay, ferric reducing antioxidant power, and copper reduction antioxidant assays (Slavov et al. 2016). The profiles of flavonoids (1,8–3,2%) and total phenols (1,7–2,3%) were higher in comparison with this of phenolic acids (0,6–2,4%) and anthocyanins (0,5–1,5%). The maximum levels of total phenols and anthocyanins were determined in the *C. thracicus* while profiles of phenolic acids were found higher in *C. armatus* (Zhelev et al. 2013).

Nine flavonoids (isorhamnetin-3-glucoside, kaempferol-3-rhamnoglucoside, luteolin-7-glucoside, luteolin-7-glucuronide, apigenin-7-glucoside, kaempferol-3-rutinoside, kaempferol-3-rhamnoglucoside, apigenin, and luteolin) and five phenolic acids (neochlorogenic, chlorogenic, caffeic, protocatechuic, and *p*-hydroxybenzoic acids) have been obtained from methanol extract of *C. carlinoides, C. nigrescens, C. defloratus, C. nutans,* and *C. crispus* flower heads. The methanol extract of *C. nutans* (EC_{50} = 0.46 mg/ml), *C. carlinoides* (EC_{50} = 0.60 mg/ml), *C. crispus* (EC_{50} = 0.70 mg/ml), *C. nigrescens* (EC_{50} = 0.78 mg/ml), and *C. nigrescens* (EC_{50} = 0.80 mg/ml) showed antioxidant activity in DPPH assay (Kozyra et al. 2019). A lignan, (2-hydroxyolivil) along with hispidulin 7-glucoside and luteolin 3′-rhamnoside have been identified from *C assoi* aerial parts (Fernández et al. 1991).

Nineteen essential oils {3,5-dihydroxy-6-methyl-2,3-dihudro-4Hpyran-4-on, octanoic acid, octanoic acid, ethyl ester, hydroxymethylfurfurale 2-decenal (E)-, decanoic acid, ethyl ester, tridecane, hexadecanoic acid, ethyl ester, 4-((1E)-3-hydroxy-1-propenyl)-2-methoxyphenol, tetradecanoic acid, spiro[bicycle [3.3.0]octan-6one-3-cyclopropane], 9-hexadecenoic acid, 1,4-methyl ester, (Z)-, pentadecanoic acid, 1,4-methyl-,methyl ester, n-hexadecanoic acid, 13-docosenoic acid, methyl ester, (Z)-, 9,12-octadecadienoic acid, methyl ester, (E, E)-, 9-octadecenoic acid, methyl ester, (E)-, 11-octadecenoic acid, methyl ester, oleic acid} were obtained from the methanol extract of *C. edelbergii* aerial parts. Methanol extract and its synthesized nanoparticles displayed significant antioxidant and antimicrobial properties. The methanol extract (300 mg/Kg b.w.) and synthesized nanoparticles (5 and 10 mg/kg b.w.) showed a significant ($p < 0.05$) increase in body weight and HDL while a decrease in the levels of blood glucose, bilirubin, creatinine, urea, triglyceride, VLDL, LDL, ALP, ALT, and AST in treated animals (Jamil et al. 2022).

The anti-inflammatory and antinociceptive effects of aqueous extract of *C. schimperi* roots were evaluated. The extract (400 mg/kg *p.o.*) produced significant ($p < 0.05$–0.001) anti-inflammatory effects against carrageenan-induced acute inflammation and formalin-induced nociceptive pain stimulus in treated mice. The isolated compound (syringin, 100 mg/kg, p.o.) showed a significant anti-inflammatory and antinociceptive activities ($p < 0.001$). The present study suggested that *C. schimperi* exhibits promising anti-inflammatory and antinociceptive activities, lending pharmacological support to traditional use of this plant species in the treatment of painful inflammatory complaints (Wolde-Mariam et al. 2012).

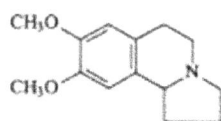

Crispine A

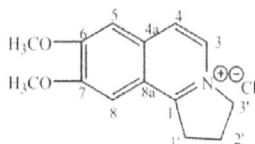

Carcrisine A

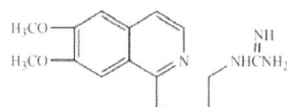

Carcrisine B

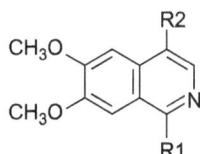

$R^1 = CH_2CH_2CH_2NHC(NH)NH_2$, $R^2 = CH_2CH_2CH_2CH_2NHC(NH)NH_2$ – Crispine D

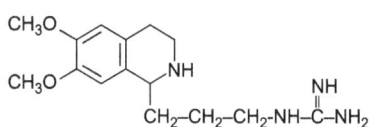

Crispine E

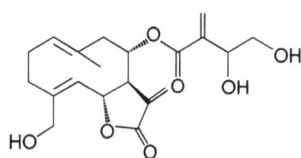

Cnicin

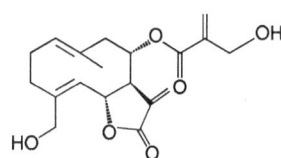

Onopordopicrin

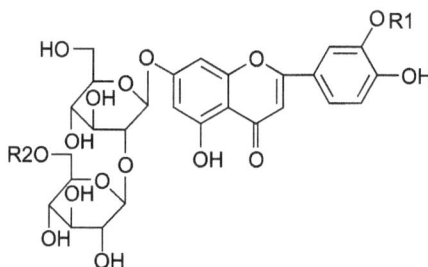

$R^1 = CH_3$, $R^2 = Ac$ - Chrysoeriol 7-O-(2''-O-6'''-O-acetyl-β-D-glucopyranosyl-β-D-glucopyranoside;
$R^1 = CH_3$, $R^2 = H$ - Chrysoeriol 7- sophoroside; $R^1 = H$, $R^2 = Ac$ - Linariifolioside

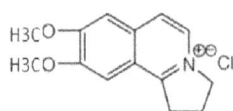

Crispine B

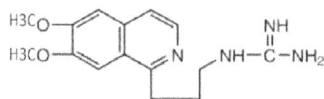

Crispine

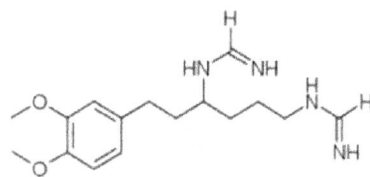

C Acanthoidine

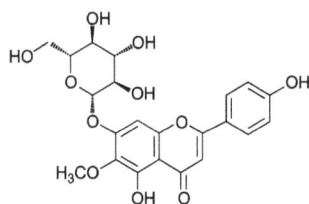

Hispidulin 7-glucoside

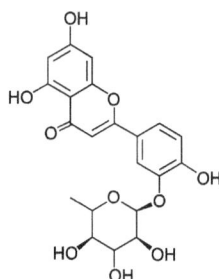

Luteolin 3'-rhamnoside

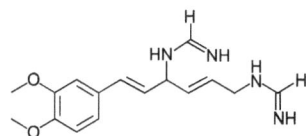

Acanthoine

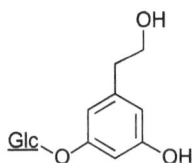

3,5-Dihydroxyphenethyl alcohol 3-*O*-glucoside

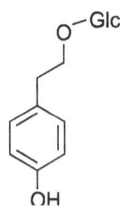

Salidroside

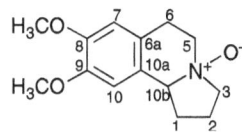

Crispine A N-oxide

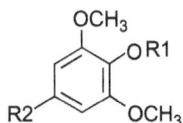

R^1 = Glc, R^2 = OH - 2,6-Dimethoxy-p-hydroquinone 4-*O*-glucoside; R^1 = H, R^2 = O-Glc - 2,6-Dimethoxy-4-hydroxyphenol 1-*O*-glucoside; R^1 = Ara, R^2 = COOH - Syringic acid 4-*O*-arabinoside

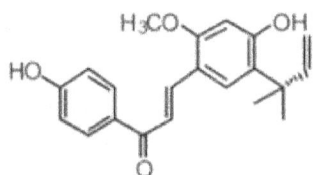

Licochalcone A

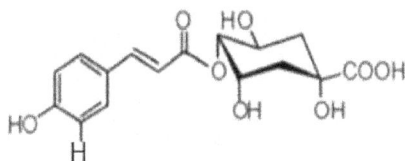

4-*O*-p-Coumaroylquinic acid

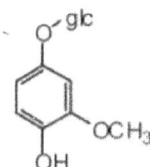

Tachioside

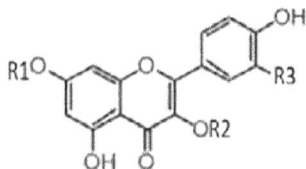

R^1 = GluA, R^2 = Rha, R^3 = H - Kaempferol *3*-*O*-rhamnoside-7-*O*-gulcuronide; R^1 = H, R^2 = Glc, R^3 = H - Kaempferol 3-*O*-glucoside; R^1 = H, R^2 = Rha, R^3 = OH - Quercetin 3-*O*-α-L-rhamnoside; R^1 = H, R^2 = Rha, R^3 = H - Kaempferol 3-*O*-rhamnoside

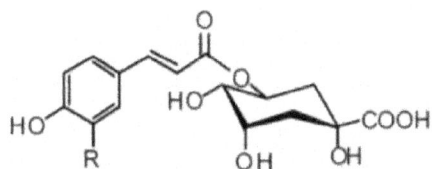

R = H - 5-*O*-p-Coumaroylquinic acid

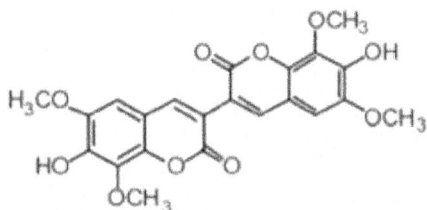

3,3′ -Biisofraxidin

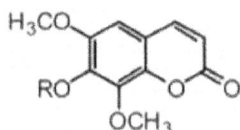

R = Glc - Eleutheroside B1; R = H – Isofraxidin

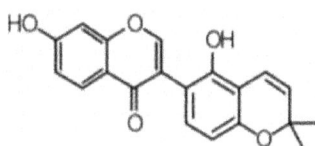

Eurycarpin B

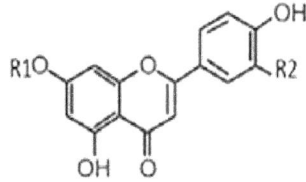

R^1 = Glc(2)glc, R^2 = OH - Luteolin 7-O-glucosyl-(1→2)-glucoside; R^1 = Glc, R^2 = OH - Luteolin 7-O-glucoside; R^1 = Glc(2)glc, R^2 = OCH$_3$ - Chrysoeriol 7-O-glucosyl-(1→2)- glucoside; R^1 = Glc(2)glc(6)Ac, R^2 = OH - Luteolin 7-O-6'''-acetyl-glucosyl-(1→2)- glucoside; R^1 = GluA, R^2 = H - Apigenin 7-O-glucuronide; R^1 = Glc, R^2 = H - Apigenin 7-O-glucoside; R^1 = Glc(2)glc(6)Ac, R^2 = OCH$_3$ - Chrysoeriol 7-O-6'''-acetyl-glucosyl-(1→2)-glucoside; R^1 = H, R^2 = O-Rha - Luteolin 3' -O-rhamnoside

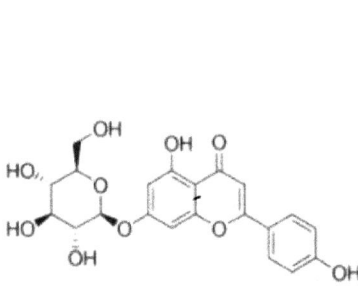

Apigenin-7-glucoside

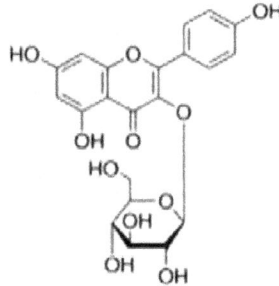

Kaempferol 3-glucoside

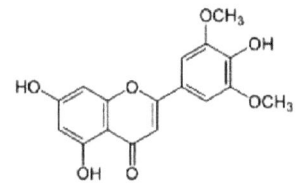

Tricin

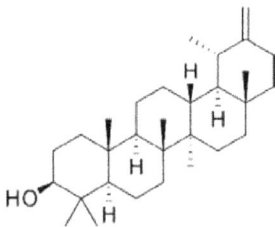

Taraxasterol

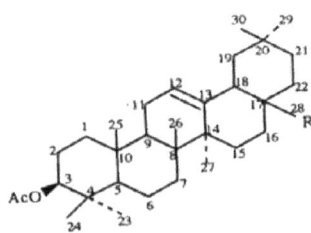

R = CH$_2$OH - Erythrodiol 3-acetate

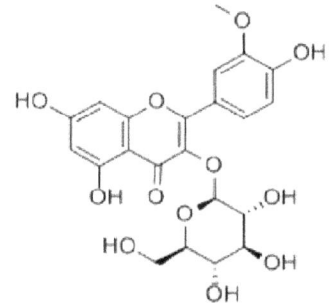

Isorhamnetin-3-glucoside

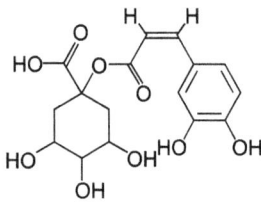

1-O-Caffeoylquinic acid

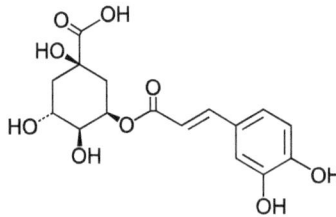

5-O-caffeoylquinic acid

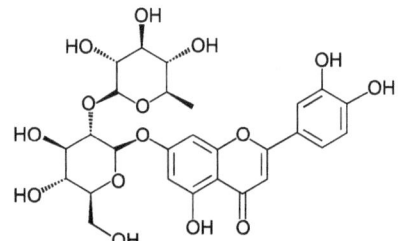

Scolymoside

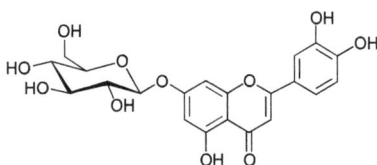

Cynaroside

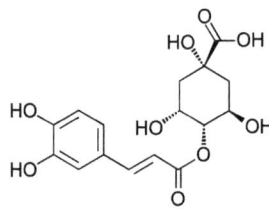

Cryptochlorogenic acid

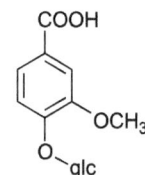

Vanillic acid 4-O-β-D-glucoside

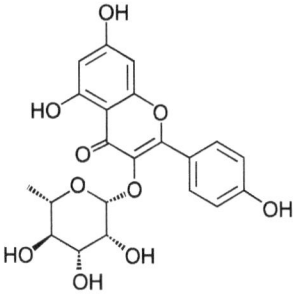

Kaempferol-3-*O*-α-L-rhamnoside

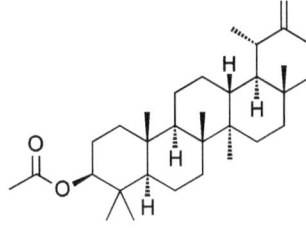

Taraxasterol acetate

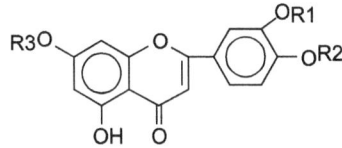

R^1 = CH$_3$, R^2 = H, R^3 = β-D-(6-*O*-acetylglucopyranosyl) (1→3)-β-D-glucopyranoside - Chrysoeriol-7-*O*-β-D-(6-O-acetylglucopyranosyl) (1→3)-β-D-glucopyranoside; R^1 = H, R^2 = CH$_3$, R^3 = α-L-arabinosyl(1→6)-β-D-glucopyranoside - Diosmetin 7-*O*-α-L-arabinosyl (1→6)-β-D-glucoside

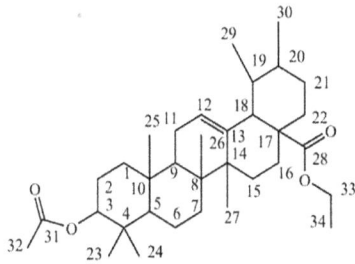

3-*O*-Acetyl-ursolic acid-28-ethyl ester

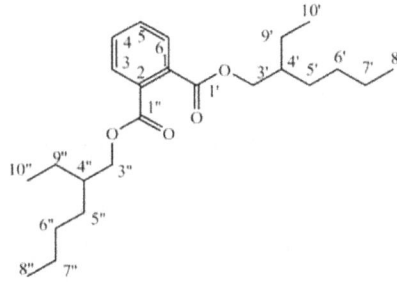

Bis (2-ethylhexyl) benzene-1,2-dicarboxylate

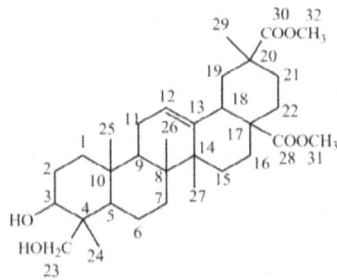

3α, 24-Dihydroxyolean-12-en-28, 30-dioic acid dimethyl ester

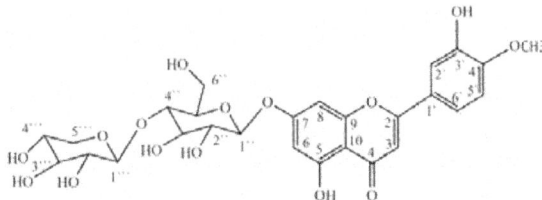

Diosmetin-7-*O*-α-L-arabinopyransyl (1‴→4″)-β-D- glucopyranoside

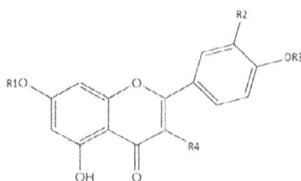

R^1 = H, R^2 = H, R^3 = H, R^4 = O-β-D-Glucose - Kaempferol-3-O-β-D-glucoside; R^1 = H, R^2 = H, R^3 = H, R^4 = O-α-L-Rhamnose - Kaempferol-3-O-α-L-rhamnoside; R^1 = CH$_3$, R^2 = H, R^3 = H, R^4 = O-α-L-Rhamnose - Kaempferol-7-methoxy-3-O-α-L-rhamnoside; R^1 = β-D-Glucose(6″→1‴)β-D-xylose, R^2 = H, R^3 = CH$_3$, R^4 = H - Diosmetin-7- O-β-D-xylosyl-(1‴→6″)-β-D-glucopyranoside; R^1 = β-D-Glucose(6″→1‴)α-L-arabinose, R^2 = H, R^3 = CH$_3$, R^4 = H - Diosmetin-7-O-α-L-arabinosyl (1‴→6″)-β-D-glucopyranoside; R^1 = H, R^2 = H, R^3 = CH$_3$, R^4 = O-α-L-Rhamnose(1‴→2″)α-L-rhamnose - Kaempferol-3-O-α-L-rhamnosyl-(1‴→2″)-α-L-rhamnoside

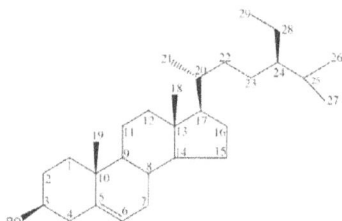

R = Glu - β-sitosterol-3-O-β-D-glucoside

3.3   CULTURE CONDITIONS

The leaf explants of *C. benedictus* were cultured on an MS medium (Murashige and Skoog 1962). Maximum callus growth was achieved in an MS medium containing 2.4-dichlorophenoxyacetic acid (1 mg/l) and 6-benzylaminopurine (0.2 mg/l). The levels of flavonoids (rutin) were higher in callus than the mother plant (Catană et al. 2020).

REFERENCES

Abdel-Salam NA, Morelli I, Catalano S. 1983. An unusual triterpene ester in a Compositaceous plant, *Carduus getulus* Pomel. Int J Crude Drug Res 21, 79–80.

Aktay G, Deliorman D, Ergun E, Ergun F, Yeşilada E, Cevik C. 2000. Hepatoprotective effects of Turkish folk remedies on experimental liver injury. J Ethnopharmacol 73, 121–129.

Al-Shammari LA, Hassan WHB, Al-Youssef HM. 2012. Chemical composition and antimicrobial activity of the essential oil and lipid content of *Carduus pycnocephalus* L. growing in Saudi Arabia. J Chem Pharm Res 4, 1281–1287.

Al-Shammari LA, Hassan WHB, Al-Youssef HM. 2015. Phytochemical and biological studies of *Carduus pycnocephalus* L. J Saudi Chem Soc 19, 410–416.

Al-Snafi AE. 2016. The constituents and pharmacology of *Cnicus benedictus*—a review. Pharm Chem J 3, 129–135.

Bain JF, Desrochers AM. 1988. Flavonoids of *Carduus nutans* and *C. acanthoides*. Biochem Syst Ecol 16, 265–268.

Brantner A, Baumberger J, Kunert O, Bucar F, Batkhuu J. 2010. Phytochemical and biological investigations on *Carduus crispus* L. Planta Med 76, 67.

Cardona L, Garcí B, Pedro JR, Pérez J. 1992. 6-Prenyloxy-7-methoxycoumarin, a coumarin-hemiterpene ether from *Carduus tenuiflorus*. Phytochemistry 31, 3989–3991.

Catană R, Helepciuc F-E, Zamfir M, Florescu L, Mitoi M. 2020. *In vivo* and *in vitro* antioxidant activity of *Cnicus benedictus*. AgroLife Sci J 9, 73–78.

Chaudhary SA. 2000. Flora of the Kingdom of Saudi Arabia, Vol II, Part 3. Ministry of Agriculture and Water, National Herbarium, National Agriculture and Water Research Center, Riyadh.

Coon N. 1979. An American Herbal-Using Plants for Healing. Rodale Press, Pennsylvania.

Davis PH. 1975. *Carduus* L. In: Flora of Turkey and the East Aegean Islands, Vol 5, Davis PH (ed.). Edinburgh University Press, Edinburgh, pp 420–438.

Denev I, Todorov K, Kirilova I, Mladenov R, Stoyanov P, Dimitrova-Dyulgerova I. 2018. Genetic diversity of Bulgarian representatives of genus *Carduus* L. (Asteraceae) as revealed by variability in sequences of internal transcribed spacers region. Biotechnol Biotechnol Equip 32, 387–396.

Desrochers AM, Bain JF, Warwick SI. 1988. The biology of Canadian weeds. 89. *Carduus nutans* L. and *Carduus acanthoides* L. Can J Plant Sci 68, 1053–1068.

Dimitrova-Dyulgerova I, Zhelev I, Mihaylova D. 2015. Phenolic profile and *in vitro* antioxidant activity of endemic Bulgarian *Carduus* species. Pharmacogn Mag 11, S575–S579.

Dunn PH. 1976. Distribution of *Carduus nutans, C. acanthoides, C. pycnocephalus*, and *C. crispus*, in the United States. Weed Sci 24, 518–524.

El-Lakany AM, Abdel-Kader MS, Hammoda HM, Ghazy NM, Mahmoud ZF. 1995. Flavonoids from *Carduus pycnocephalus* L. Alex J Pharm Sci 9, 41–43.

El-Lakany AM, Abdel-Kader MS, Hammoda HM, Ghazy NM, Mahmoud ZF. 1997. A new flavone glycoside with antimicrobial activity from *Carduus pycnocephalus* L. Pharmazie 52, 78.

Esmaeili F, Rustaiyan A, Nadimi M. 2005. Volatile constituents of *Centaurea depressa* and *Carduus pycno-cephalus* L. two compositae herbs growing wild in Iran. J Essent Oil Res 17, 539.

Fernández I, Garcia B, Pedro JR, Varea A. 1991. Lignans and flavonoids from *Carduus assoi*. Phytochemistry 30, 1030–1032.

Fleming T. 1998. PDR for Herbal Medicines. Medical Economics Company, Inc., Montvale.

Frydman B, Deulofeu V. 1962. Studies of Argentina plants—XIX: Alkaloids from *Carduus acanthoides* L. Structure of acanthoine and acanthoidine and synthesis of racemic acanthoidine. Tetrahedron 18, 1063–1072.

Grieve M. 2014. A Modern Herbal. Dover Publications Inc, New York.

Hassan WHB, Al-Youssef HM, AL-Shammari L, Abdallah RH. 2015. New pentacyclic triterpene ester and flavone glycoside from the biologically active extract of *Carduus pycnocephalus* L. J Pharmacogn Phytother 7, 45–55.

Hoffmann D. 1998. The Herbal Handbook: A User's Guide to Medical Herbalism. Inner Traditions/Bear & Co, Rochester.

Ildiko S, Pallag A, Bidar CF. 2009. The antimicrobial activity of the *Cnicus benedictus* L. extracts. Analele Univ Din Oradea Fasc Biol 1, 126–128.

Jamil S, Dastagir G, Foudah AI, Alqarni MH, Yusufoglu HS, Alkreathy HM, Ertürk Ö, Shah MAR, Khan RA. 2022. *Carduus edelbergii* Rech. f. mediated fabrication of gold nanoparticles; Characterization and evaluation of antimicrobial, antioxidant and antidiabetic potency of the synthesized AuNPs. Molecules 27, 6669.

Kabaktepe S, Köstekci S, Arabaci T. 2014. Host range and distrubiotion of rust fungi *Puccinia calcitrapae* DC. on *Carduus* L. (Asteraceae) species in Turkey. Biol Divers Conserv 7, 69–72.

Kazmi SMA. 1964. Revision der Gattung Carduus (Compositae). II. Mitteilungen der Botanischen Staatssammlung, München.

Kindscher K. 1987. Edible Wild Plants of the Prairie: An Ethnobotanical Guide. University Press of Kansas, Lawrence.

Kozyra M, Biernasiuk A, Malm A. 2017. Analysis of phenolic acids and antibacterial activity of extracts obtained from the flowering herbs of *Carduus acanthoides* L. Acta Pol Pharm Drug Res 74, 161–172.

Kozyra M, Główniak K, Boguszewska M. 2013. The analysis of flavonoids in the flowering herbs of *Carduus acanthoides* L. Curr Issues Pharm Med Sci 26, 10–15.

Kozyra M, Komsta L, Wojtanowski K. 2019. Analysis of phenolic compounds and antioxidant activity of methanolic extracts from inflorescences of *Carduus* sp. Phytochem Lett 31, 256–262.

Kozyra M, Kukula-Koch W, Szymański M. 2022. Phenolic composition of inflorescences of *Carduus nutans* L. Chem Biodivers 19, e202100827.

Li R, Liu S-K, Song W, Wang Y, Li Y-J, Qiao X, Liang H, Ye M. 2014. Chemical analysis of the Tibetan herbal medicine *Carduus acanthoides* by UPLC/DAD/qTOF-MS and simultaneous determination of nine major compounds. Anal Methods 6, 7181–7189.

Mandaville JP. 1990. Flora of Eastern Saudi Arabia. Kegan Paul International Ltd, London.

Murashige T, Skoog F. 1962. A revised medium for rapid growth and bioassays with tobacco tissue cultures. Physiol Plant 15, 473–497.

Murray T, Michael ND. 1992. The Healing Power of Herbs, 2nd ed. Prima Publishing, East Sussex.

Mustafa B, Hajdari A, Krasniqi F, Hoxha E, Ademi H, Quave CL, Pieroni A. 2012. Medical ethnobotany of the Albanian Alps in Kosovo. J Ethnobiol Ethnomed 8, 6.

Özcan K. 2021. Determination of biological activity of *Carduus lanuginosus*: An endemic plant in Turkey. Int J Environ Health Res 31, 45–53.

Palma-Bautista C, Belluccini P, Gentiletti V, Vázquez-García JG, Cruz-Hipolito HE, De Prado R. 2020. Multiple resistance to glyphosate and 2,4-D in *Carduus acanthoides* L. from Argentina and alternative control solutions. Agronomy 10, 1735.

Petkova N, Mihaylova D, Denev P, Krastanov A. 2015. Evaluation of biological active substances in flower heads of *Carduus thoermeri* Weinm. Rom Biotechnol Lett 20, 10592–10599.

Qingyig Z, Guangzhong T, Yuying Z. 2002. Novel bioactive isoquinoline alkaloids from *Carduus crispus*. Tetrahedron 58, 6795–6798.

Rajendran K, Awen BZS, Alogbi MIS, Alttawrghi NM, Elhri NE, Bharathi RV. 2011. An identity based pharmacognostical profile of *Carduus benedictus*. Res J Pharm Technol 4, 616–619.

Rotblatt M. 2000. Herbal Medicine: Expanded Commission E Monographs, Blumenthal M, Goldberg A, Brinckmann J (eds). American Botanical Council, Newton, p. 519.

Slavov IZ, Dimitrova-Dyulgerova IZ, Mladenov R. 2016. Phenolic profile and antioxidant activity of methanolic extract of *Carduus acicularis* Bertol. (Asteraceae). Ecol Balk 8, 41–46.

Stuckey RL, Forsyth JL. 1971. Distribution of naturalized *Carduus nutans* (Compositae) mapped in relation to geology in northwestern Ohio. Ohio J Sci 71, 1–14.

Taha HE, Farag MA, El Fishawy AM, El Toumy SA, Amer KF, Mansour AM. 2019. Hepatoprotective effect of *Carduus getulus* Pomel in relation to its metabolite fingerprint as analyzed via UPLC-MS technique. Biosci Res 16, 2459–2474.

Tunsag J, Davaakhuu G, Batsuren D. 2011. New isoquinoline alkaloid from *Carduus crispus* L. Mong J Chem 12, 85–87.

Urnukhsaikhan E, Bold B-E, Gunbileg A, Sukhbaatar N, Mishig-Ochir T. 2021. Antibacterial activity and characteristics of silver nanoparticles biosynthesized from *Carduus crispus*. Sci Rep 11, 21047.

Weber WA, Wittmann RC. 2001. Colorado Flora: Eastern Slope, 3rd ed. University Press of Colorado, Denver.

Wolde-Mariam M, Veeresham C, Asres K. 2012. Antiinflammatory and antinociceptive activities of extracts and syringin isolated from *Carduus schimperi* Sch. Bip. ex A. Rich. Phytopharmacology 3, 252–262.

Xie WD, Jia ZJ. 2004. Two new isoquinoline alkaloids from *Carduus crispus*. Chin Chem Lett 15, 1057–1059.

Xie W-D, Li P-L, Jia Z-J. 2005. A new flavone glycoside and other constituents from *Carduus crispus*. Pharmazie 60, 233–236.

Zhang Q, Tu G, Zhao Y. 2002. Novel bioactive isoquinoline alkaloids from *Carduus crispus*. Tetrahedron 58, 6795–6798.

Zhelev I, Dimitrova-Dyulgerova I, Belkinova D, Mladenov R. 2013. Content of phenolic compounds in the genus *Carduus* L. from Bulgaria. Ecol Balk 5, 13–21.

Zhelev I, Dimitrova-Dyulgerova I, Merdzhanov P, Stoyanova A. 2012. Chemical composition of *Carduus candicans* ssp. globifer and *Carduus thoermeri* essential oils. J Essent Oil Bear Plant 17, 196–202.

Zheleva-Dimitrova D, Zhelev I, Dimitrova-Dyulgerova I. 2011. Antioxidant activity of some *Carduus* species growing in Bulgaria. Free Radic Antioxid 1, 15–20.

4 *Centaurium* Species

4.1 MORPHOLOGICAL FEATURES, DISTRIBUTION, ETHNOPHARMACOLOGICAL PROPERTIES, PHYTOCHEMISTRY, AND PHARMACOLOGICAL ACTIVITIES

Centaurium erythraea Rafn. (syn. *C. umbellatum* Gilib., *C. minus* Moench, *Erythraea centaurium* Pers., common name centaury; Fam.—Gentianaceae) is a winter-annual or biennial and herb plant, an erect branched stem that is 2–50 cm in height. The leaves are obovate or elliptic (10–50 × 8–20 mm), and prominently three-to-seven veined. Cauline leaves are shorter, narrow, and acute, but never parallel-sided. Flowers are sessile or sub-sessile, pink, often clustered, forming a dense corymb-like cyme (Vágnerová 1992). It is naturalized in Europe, Southwest Asia, and North Africa (Mroueh et al. 2004; Subotić et al. 2006; Cunha et al. 2003). The tinctures, tonics, lotions, or tea, prepared from extract of aerial parts, have been used in the treatment of gastrointestinal diseases, dyspepsia, constipation, anorexia, hepatitis, jaundice, rheumatism, wounds, and sores (Kumarasamy et al. 2003a; Mroueh et al. 2004; Subotić et al. 2006). In traditional medicine, *C. erythraea* has been used in the treatment of fever, hypertension, anemia, scrofula, and gout. It is also used in homeopathic medicines (van der Sluis 1985; Šiler et al. 2012). As per Croatian medicine, it is useful in asthma, eczema, rheumatism, wounds, and sores (Tahraoui et al. 2010; Božunović et al. 2019). In European traditional medicine, *C. erythraea* aerial parts are used as stomachic, tonic, depurative, sedative, and as antipyretic agent (Van Hellemont 1986; Newall et al. 1996). In Moroccan traditional medicine, it is used in the treatment of diabetes. Methanol extract and pure secoiridoids were found effective against tested *Penicillium* species (Hill 1986; Božunović et al. 2018).

Water decoction of aerial parts showed inhibitory effects on 5-lipoxygenase, but no cytotoxicity displayed against RAW 264.7 macrophage cells line (Kachmar et al. 2019). The aqueous extract (1–15 g/kg) of *C. erythraea* was administered to IOPS OFA mice. General unpleasant behavior, mortality, and mortality latency were recorded for 14 days. In case of sub-chronic doses, the extract was given orally (100, 600, and 1200 mg/kg/daily) for 90 days to animals. In case of acute doses (15 g/kg), the extract did not show any signs of toxicity. The extract displayed progressive acute toxicity with increasing doses. The no-observed-adverse-effect level for the intraperitoneal dose was 6 g/kg while the lowest-observed-adverse-effect level was 8 g/kg (acute toxicity—LD_{50} 12.13 g/kg). Extract administration (100, 600, and 1200 mg/kg for 90 days) did not show any alterations in hematological and biochemical parameters, except a small decrease recorded in mean corpuscular volume, serum glucose, and triglyceride levels at the higher doses. Therefore, it is concluded that the aqueous extract is relatively non-toxic (Tahraoui et al. 2010). The aqueous (0.66 ml/100 g/bw) and butanol extracts (0.015 ml/100 g bw) of *E. centaurium* showed significant decrease in blood glucose levels. The results were compared with Glibenclamide Daonil® 5mg (0.25 mg/100 g bw), as a standard drug (Mansar-Benhamza et al. 2013). The antihyperglycemic and antilipidemic activities of methanol extract of *C. erythrea* aerial parts were evaluated in streptozotocin-induced hyperglycemic rats. Methanol extract showed significant effects on the hepatic carbohydrate metabolism by increasing the direct synthesis of glycogen. It normalized the phosphorylase activity and decreased glucose-6-phosphatase activity, which further produces a decrease in the production of blood glucose. During long-term (45 days) treatment, the level of hemoglobin A1c was lower in treated animals. The extract also reduced the levels of total cholesterol, triglycerides, high density lipoprotein, and low-density lipoprotein levels in treated animals (Stefkov et al. 2014). Similarly, the leaf extract (200 mg/kg bw/day, i.p; for 30 days) showed a significant decrease in levels of blood glucose and malondialdehyde in streptozotocin-induced diabetic rats. The extract displayed a significant increase

DOI: 10.1201/9781003398035-4

in both enzymatic and non-enzymatic antioxidants. In streptozotocin-induced diabetes rats, the extract treatment minimized the degenerative alterations in pancreatic β-cells. *C. erythrea* treatment demonstrated significant antidiabetic activity by reducing oxidative stress and pancreatic β-cells' damage that may be associated to its antioxidative potential (Sefi et al. 2011). The methanol extract of the *C. erythraea* aerial parts was evaluated for their antidiabetic activity against streptozotocin (40 mg/kg, for five consecutive days)-induced diabetic rats. Extract administration (100 mg/kg; daily; four weeks) and increased the profiles of serum insulin but decreased the levels of blood glucose and glycated hemoglobin in treated animals. The extract showed antioxidant activity in red blood cells of diabetic rats by decreasing lipid peroxidation. The extract showed ameliorative effect against oxidative damage due to the increase of superoxide dismutase, catalase, and glutathione reductase activities. The treatment also increased the levels of reduced/oxidized glutathione but decreased S-glutathionylated proteins levels. The results provide significant support for the application of *C. erythraea* used in traditional medicine for diabetes management (Đorđević et al. 2017).

The antioxidant and prosurvival effects of ethanol extract of *C. erythraea* were evaluated in β-cells and pancreatic islets of streptozotocin-induced diabetic rats. The extract (100 mg/kg) improved the levels of insulin and the structural and functional properties of pancreatic islets in streptozotocin-induced diabetic rats. The extract treatment also reduced the interruptions of islet morphology and islet cell contents in diabetic rats. These protective effects were associated with increases in levels of insulin, glucose transporter 2, and p-Akt in diabetic islets. The extract showed antioxidative effect by decreasing DNA damage, lipid peroxidation, and protein S-glutathionylation, and by increasing disruption in Mn superoxide dismutase, CuZn superoxide dismutase, and catalase enzymatic activities in diabetic rats. The extract treatment also improved the interruptions caused by oxidative stress in the transcriptional regulation of catalase, Mn superoxide dismutase, CuZn superoxide dismutase, glutathione peroxidase and glutathione reductase enzymes in β-cells in diabetic rats and readjusted for the presence and activities of redox-sensitive NFκB-p65, FOXO3A, Sp1 and Nrf-2 transcription factors. The extract-mediated stimulation of proliferative and pro-survival pathways and secretion of insulin after streptozotocin-induced oxidative stress in β-cells may be associated with well-defined modulation of the activities of pro-survival Akt, ERK, and p38 kinases and islet-enriched Pdx-1 and MafA regulatory factors. The study supports that the extract improves the structural and functional properties of pancreatic β-cells by stimulating the endogenous antioxidant regulatory mechanisms as well as by stimulating proliferative and pro-survival channels in β-cells (Đorđević et al. 2019). The effect of *C. erythraea* extract (2 g/kg/bw daily for 20 weeks) was evaluated on lipid tissue accumulation in high fat diet-induced diabetic rats. Liver steatosis was verified with high fat diet in animals. The extract increased liver steatosis by the end of the preventive (20 weeks) and curative periods (35 weeks) significantly ($p < 0.05$). No steatosis was reported in other tissues of treated animals. The results reveal that the extract possesses a significant hepatoprotective activity in high fat diet-induced diabetes in treated animals. Though it has antidiabetic potential, the extract may also be used as a promising agent to treat non-alcoholic liver steatosis (Hamza et al. 2015).

The aqueous extract of *E. centaurium* whole plant (500 mg/kg/day; for 12 days) demonstrated significant anti-inflammatory effects against polyarthritis model-induced by Freund's adjuvant in rats. Similarly, the extract also exhibited significant antipyretic activity (Berkan et al. 1991).

The antiulcer effects of aqueous-ethanolic extract (50:50%) of *C. erythraea* aerial parts were evaluated in an aspirin-induced acute gastric ulcer model. The extract (100 mg/kg dose) showed significant protection (77%) against aspirin-induced gastric lesions ($p < 0.05$). Aspirin administration (*p.o.*) reduced the levels of catalase and reduced glutathione but increased lipid peroxidation levels. However, myeloperoxidase activity was enhanced by aspirin, while it was lower in the aspirin-plus-extract-treated group animals; therefore, the extract protects animals from aspirin-induced injury (Tuluce et al. 2011). The methanol extract of *C. erythraea* leaves was assessed for their hepatoprotective effect against acetaminophen-induced liver toxicity in rats. The extract (300 mg/kg/day for 6 days or a single dose of 900 mg/kg for 1 day) demonstrated significant protective activity by reducing the levels of serum glutamate oxaloacetate transaminase, glutamate pyruvate transaminase. and lactate dehydrogenase in treated animals

(Mroueh et al. 2004). *C. erythraea* is distributed in the Morocco, Algeria, Italy, Spain, Portugal, and countries of the Balkan Peninsula. The essential oils and ethanol extracts of *C. erythraea* demonstrated antibacterial, antioxidant, antifungal, antileishmanial, anticancer, antidiabetic, anti-inflammatory, diuretic, gastroprotective, hepatoprotective, and neuroprotective properties (Menyiy et al. 2021). *C. erythraea* is used in the treatment of liver and kidney complaints and diabetes, as well as to enhance bile secretion and the contraction of uterine muscles (Budniak et al. 2021).

The methanol, ethanol, *n*-hexane, and ethyl acetate extracts of *C. erythraea* were tested for their antiradical, antibacterial, and antileishmanial activities. The hexane extract showed higher scavenging activity than methanol, ethanol, and ethyl acetate extracts in DPPH (IC_{50} = 49.54 ± 2.43 µg/ml) assay. The activity was compared with ascorbic acid (IC_{50} = 27.20 ± 0.17 µg/ml) and trolox (IC_{50} = 43.72 ± 0.31 µg/ml), used as standards. The methanol, *n*-hexane and ethyl acetate extracts showed remarkable antibacterial activity (inhibition zone ≥14 mm) against all tested strains (*S. aureus, E. coli, L. monocytogenes* and *P. aeruginosa*). The minimum inhibitory concentration and minimum bactericidal concentrations of extracts were recorded in the range of 0.25 to 8.00 mg/ml. The *n*-hexane extract showed significant inhibition against *Leishmania tropica* (IC_{50} = 37.20 ± 1.62 µg/ml) and *L. major* (IC_{50} = 64.52 ± 2.20 µg/ml). The results were compared to N-methyl glucamine antimoniate (Glucantime®; IC_{50}>500 µg/ml) used as a control. Therefore, the extract may be considered as a significant source of novel antioxidant, antibacterial, and antileishmanial agents (Bouyahya et al. 2017). The volatile compounds *C. erythraea* developmental stages (vegetative, flowering, and post-flowering) were evaluated for their antioxidant, antidiabetic, dermatoprotective, and antibacterial activities. The essential oils (flowering stage) showed antioxidant effects in 2,2-diphenyl-1-picrylhydrazyl (IC_{50} = 47.18 ± 3.62 µg/ml), ferric reducing antioxidant power (IC_{50} = 53.25 ± 2.19 µg/ml), and 2,2'-azino-bis(3-ethylbenzothiazoline-6-sulfonate) radical cation (IC_{50} = 65.34 ± 3.71 µg/ml) assays. Essential oils (vegetative stage) displayed a remarkable α-amylase (IC_{50} = 31.91 ± 0.336 µg/ml) and α-glucosidase (IC_{50} = 56.77 ± 1.02 µg/ml) suppressive effect. Similarly, the essential oils of flowering (IC_{50} 41.863 ± 0.031 µg/ml) and postflowering stages (IC_{50} = 49.183 ± 0.298 µg/ml) strongly suppressed tyrosinase enzyme activity. The essential oils (postflowering stage) demonstrated significant antibacterial activity against *S. aureus* (MIC = MBC = 0.125%, v/v), *L. monocytogenes* (MIC = MBC = 0.125%, v/v), and *P. mirabilis* (MIC = MBC = 0.125%, v/v; Bouyahya et al. 2019). The profiles of total polyphenols, total flavonoid contents, and total monomeric anthocyanin contents were evaluated in the aqueous extract of *C. erythraea*. The diverse range of compounds (total polyphenols 1.23 to 12.46 mg gallic acid equivalents/g dw; total flavonoids 1.18 to 3.35 mg quercetin/g dw, and total monomeric anthocyanins 1.70 to 6.15 mg/l) were obtained from the aqueous extract. Therefore, *C. erythraea* extract might be recommended as a good source of bioactive compounds and bio-antioxidative agent (Mihaylova et al. 2019).

The total polyphenolic and total flavonoid contents of the methanolic extract of *C. erythraea* were obtained as 35.45 ± 0.041 µg gallic acid equivalents/mg extract and 6.65 ± 0.060 µg quercetin/mg. The findings of the DPPH assay displayed a strong antioxidant effect of methanolic (IC_{50} 0.232 ± 0.002 mg/ml) and aqueous (IC_{50} 0.208 ± 0.002 mg/ml) extracts. The inhibitory effects of both extracts in the β-carotene/linoleic acid method was (86.781 ± 0.17% for methanolic; 77.816 ± 0.69% for the aqueous) found effective. Methanol extract showed greater reducing power (IC_{50} 0.35 ± 0.066 mg/ml) when compared to the aqueous extract (IC_{50} 1.31 ± 0.047 mg/ml; Merghem and Dahamna 2020). The methanol extract of *C. erythraea* suppressed the acetylcholinesterase activity (IC_{50} 51.33 ± 3.35%; Orhan et al. 2016).

Aqueous extract of *C. erythraea* leaves showed significant antioxidant activity in free radical 2,2-diphenyl-1-picrylhydrazyl reduction, lipoperoxidation, and NO radical scavenging capacity assays. The extract demonstrated significant acetylcholinesterase and 3-hydroxy-3-methylglutaryl coenzyme A reductase inhibitory activities (56% at 500 µg/ml and 48% at 10 µg/ml). As per molecular docking studies, the xanthones showed greater inhibition of acetylcholinesterase than gentiopicroside, whereas the compound demonstrates a greater shape complementarity with the 3-hydroxy-3-methylglutaryl coenzyme A reductase active site than xanthones (Guedes et al. 2019).

The phenolic compounds-rich lyophilized decoction of *C. erythraea* flower heads showed significant superoxide radical scavenging activity in enzymatic (xanthine/xanthine oxidase) and nonenzymatic (NADH/phenazine methosulfate) superoxide generating systems. The extract showed the presence of esters of hydroxycinnamic acids, namely, *p*-coumaric, ferulic, and sinapic acids (Valentão et al. 2001). *C. erythraea* infusion demonstrated significant antioxidant activity in its ability to scavenge hydroxyl radical and hypochlorous acid. The activity was lower than green tea (*Camellia sinensis*). Green tea demonstrated a dual effect in the hydroxyl radical scavenging assay, as well as stimulating deoxyribose degradation at lower dosages (Valentão et al. 2003a).

The effect of aqueous extracts of *C. erythraea* on urinary volume and the excretion of sodium, potassium, and chloride were evaluated in rats. Extract (10 ml/kg, 16%) did not significantly influence the excretion of water and electrolytes. The extract showed slight increase in urinary release of sodium and chloride (on day 7) and potassium (on day 6) significantly ($p < 0.05$) in treated animals. As per results, the extract exhibits diuretic effect on water and electrolyte excretion (Haloui et al. 2000).

The aqueous extract of *C. erythraea* aerial parts were evaluated for their myorelaxant and spasmolytic effects. Aqueous extract significantly decreased the jejunum's spontaneous contractions in treated animals *(p < 0.05)*. The extract also suppressed the recovery of spontaneous contraction as well as those stimulated by high K^+ during Ca^{2+} addition. The findings suggest that extract contains spasmolytic compounds mediating their effect at least via Ca^{2+} influx prevention and activation of NO-cGMP pathways (Chda et al. 2016).

C. pulchellum (Sw.) Druce is an annual herb, slender stem, and 5–20 cm tall. The leaves are oppositely arranged, stalkless, long (2–3 cm), and oblong. Flowers arrange in dichotomous cymes, 1–2 cm, star-shaped, and pink in color. Fruits are capsules (8–10 mm). It is distributed in Serbia and widely spread in Europe (Khafagy and Mnajed 1970; Zeltner 1970; Ubsdell 1976; Živković et al. 2007). In Egyptian medicine, they are used in the treatment of rheumatic pains, renal colic, and renal stones (Khafagy and Mnajed 1970). The methanol extract of *C. pulchellum* aerial parts and roots demonstrated significant antibacterial (0.05–0.2 mg/ml) and antifungal (0.1–2 mg/ml) activities. Similarly, pure secoiridoid glycosides showed potent antibacterial (0.01–0.04 mg/ml) and antifungal (0.001–0.1 mg/ml) activities (Šiler et al. 2010). Aqueous and ethanolic extracts (70% ethanol) of *C. umbellatum* showed significant acetylcholinesterase (72.24–94.24%; 3 mg/ml dose) and tyrosinase (66.96% and 94.03%; 3 mg/ml dose) suppressive activities, possibly associated with potent 2,2-diphenyl-1-picrylhydrazyl radical suppression (70.29–84.9%; 3 mg/ml dose; Neagu et al. 2018).

4.2　PHYTOCHEMISTRY

Swertiamarin and gentiopicroside were isolated and characterized from *Centaurium erythrea* and *C. turcicum* from various regions in Bulgaria (Nikolova-Damyanova and Handjieva 1996). The glucose (13.07 mg/g), saccharose (7.16 mg/g), and pinitol (3.71 mg/g) from aerial parts (Liliya et al. 2017), secoiridoid glycoside (gentiopicroside) have been identified from the methanol extract of *C. erythraea* aerial parts. The isolated compounds showed general toxicity in brine shrimp lethality bioassay (Kumarasamy et al. 2003b). Terpinene-4-ol, methone, *p*-cymene, γ-terpinene, and limonene, neophytadiene, thymol, carvacrol, and hexadecanoic acid have been determined by GC-MS from Croatian *C. erythraea*. The essential oil demonstrated antimicrobial activity against *E. coli*, *S. enteritidis*, *S. aureus* and *B. cereus* (Jerković et al. 2012).

Two secoiridoid glycosides (swertiamarin and sweroside) have been reported from *C. erythraea* aerial parts. Both compounds suppressed *B. cereus*, *B. subtilis*, and *C. freundii* and *E. coli* growth. The swertiamarin showed antibacterial activity against *P. mirabilis* and *S. marcescens*, while sweroside suppressed the growth of *S. epidermidis*. Swertiamarin (LD_{50} 8.0 µg/ml) and sweroside (LD_{50} 34 µg/ml) demonstrated significant general toxicity in brine shrimp lethality bioassay. The results were compared with podophyllotoxin (positive control; LD_{50} 2.79 µg/ml (Kumarasamy et al. 2003a).

Decussatin, loganic acid, methylbellidifolin, demethyleustomin amarogentin, 1,3,8-trihydroxy-5,6-dimethoxyxanthone, gentiopicroside and sweroside from *C. erythraea* (Aberham et al. 2011), secoiridoid glycosides (swertiamarin, sweroside, gentiopicroside, secologanol, secoxyloganin) were also isolated methanolic extract of *C. erythraea* (Mandova et al. 2017), xanthones (gentisin, isogentisin, 1-hydroxy-2,3,5-trimethoxyxanthone, 1-hydroxy-2,3,4,5- tetramethoxyxanthone, 1-hydroxy-2,3,4,7-tetramethoxyxanthone, swerchirin, methylswertianin, mesuaxanthone, bellidifolin, 1,6-dihydroxy-3,5-dimethoxyxanthone, 1,3,6-trihydroxy-2,5-dimethoxyxanthone, 1,6-dihydroxy-3,5,7,8- tetramethoxyxanthone, 1,6,8-trihydroxy-3,5,7-trimethoxyxanthone, tovopyrifolin C) were separated (Waltenberger et al. 2015), and sweroside, centapicrin, desacetylcentapicrin, decentapicrin A, decentapicrin B, and decentapicrin C were isolated and identified from *Centaurium* species (van der Sluis 1984).

Two methoxylated xanthone derivatives (eustomin and demethyleustomin) were obtained from the aerial parts of *C. erythraea*. Both compounds showed strong antimutagenic activity in *S. typhimurium* (strains TA98, TA100, and TA102) when screened against 2-nitrofluorene, 2-aminoanthracene, ethyl methanesulfonate, and nalidixic acid assays (Schimmer and Mauthner 1996). The presence of neophytadiene isomer III, erythrocentaurin, 5-formyl-2,3- dihydroisocoumarin, carvacrol, p-camphorene, hexadecanoic acid and thymol was determined by GC-MS from aerial parts (Jovanović et al. 2009) and isocoumarin (5-formyl-2,3-dihydroisocoumarin) was separated from chloroform extract of *C. erythraea* aerial parts (Valentão et al. 2003b).

Two isocoumarin derivatives (erythricin and erythrocentaurin) were isolated and characterized from *C. pulchellum* whole plant (Bibi et al. 2006). The ester of swertiamarin and secoxyloganic acid, swertiamarin, and secoxyloganic acid were identified from *C. spicatum*. The isolated compounds showed transaminases inhibitory effects (86%, 83%, 81%). The ester of swertiamarin and secoxyloganic acid demonstrated significant hepatoprotective effects (Allam et al. 2015). Three acetylated flavonol glycosides {quercetin 3-*O*-[(2,3,4-triacetyl-α-rhamnopyranosyl)-(1→6)]-β-galactopyranoside, quercetin 3-*O*-[(2,3,4-triacetyl-α-rhamnopyranosyl)-(1→6)]-3-acetyl-β-galactopyranoside, and quercetin 3-*O*-[(2,3,4- triacetyl-α-rhamnopyranosyl)-(1→6)]-4-acetyl-β-galactopyranoside} have been isolated from *C. spicatum* (Shahat et al. 2003).

Four secoiridoid glucosides (gentiopicrin, swertiamarin, eustomin, and demethyleustomin), and two xanthones (eustomin and demethyleustomin), were isolated and identified from the methanol extract of *C. erythraea*. The methanol extract showed greater antifungal activity than isolated compounds (Trifunović-Momčilov et al. 2019). The phytochemical analysis of *C. erythraea* infusion revealed the presence of secoiridoid glycosides (sweroside, gentiopicroside, secologanoside, swertiamarin), xanthones, and flavonoids (Nikolić et al. 2022). HPLC determination of etheric extract of *C. umbellatum* flowers showed the presence of six compounds (quercetin, orcinol, monohydrated gallic acid, β-resorcylic acid, *p*-anisic acid, and o-coumaric acid). The ether extract (150 mg/kg b.w.) significantly reduced (75.53%) the paw volume of Wistar strain rats when compared to indomethacin (62.12%). The aqueous extract (150 mg/kg) significantly reduced the paw oedema (60.86%) compared to piroxicam (62.17%; Berrak et al. 2017).

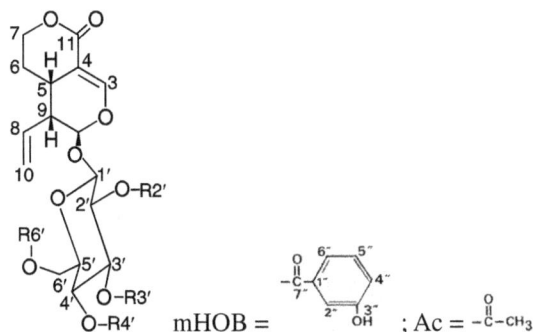

R$^{2'}$ = mOHB, R$^{3'}$ = Ac, R$^{4'}$ = H, R$^{6'}$ = H – Centapicrin; R$^{2'}$ = mOHB, R$^{3'}$ = H, R$^{4'}$ = H, R$^{6'}$ = H – Desacetylcentapicrin; R$^{2'}$ = H, R$^{3'}$ = mOHB, R$^{4'}$ = H, R$^{6'}$ = H – Decentapicrin A; R$^{2'}$ = H, R$^{3'}$ =H, R$^{4'}$ = mOHB, R$^{6'}$ = H – Decentapicrin B; R$^{2'}$ = H, R$^{3'}$ =H, R$^{4'}$ = H, R$^{6'}$ = mOHB – Decentapicrin C

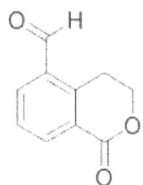

Erythrocentaurin

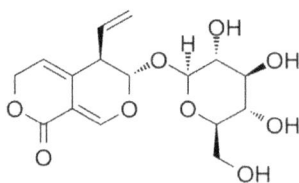

Gentiopicroside

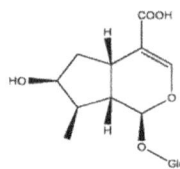

Loganic acid

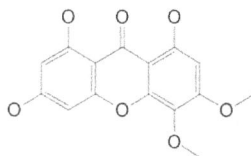

1,3,8-Trihydroxy-5,6-dimethoxy-xanthone

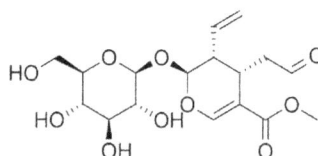

Secologanin

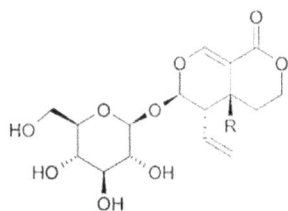

R = OH - Swertiamarin; R = H – Sweroside

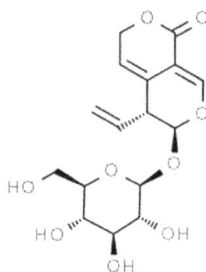

Gentiopicrin

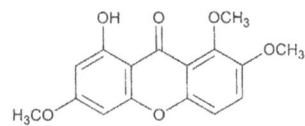

Decussatin

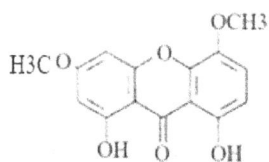

Methylbellidifolin

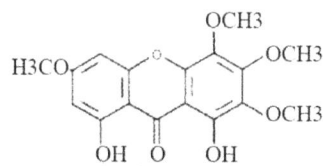

Demethyleustomin

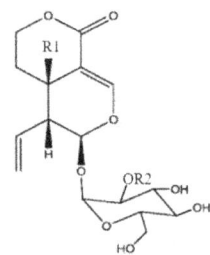

R^1 = H, R^2 =   - Amarogentin

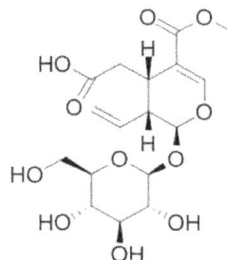

Secoxyloganin

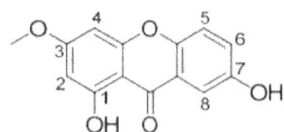

Gentisin

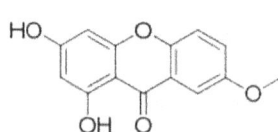

Isogentisin

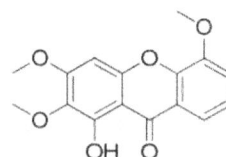

1-Hydroxy-2,3,5-trimethoxyxanthone

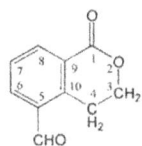
5-Formyl-2,3-dihydroisocoumarin

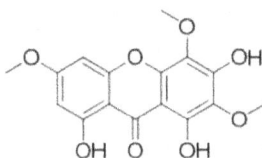
1,6,8-Trihydroxy-3,5,7-trimethoxyxanthone

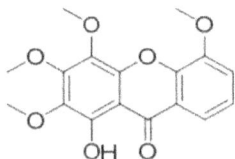
1-Hydroxy-2,3,4,5- tetramethoxyxanthone

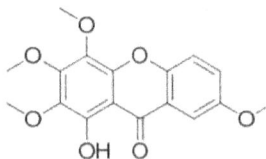
1-Hydroxy-2,3,4,7-tetramethoxyxanthone

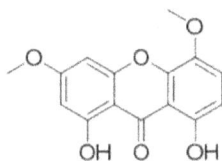
Swerchirin

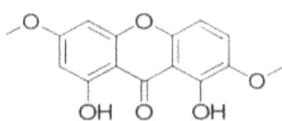
Methylswertianin

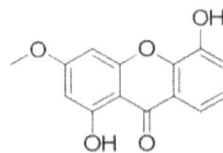
Mesuaxanthone A

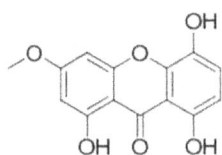
Bellidifolin

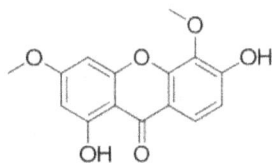
1,6-Dihydroxy-3,5-dimethoxyxanthone

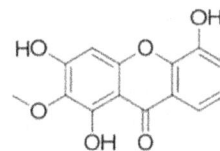
Tovopyrifolin C

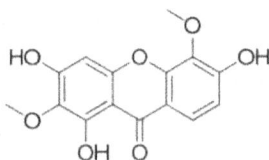
1,3,6-Trihydroxy-2,5-dimethoxyxanthone

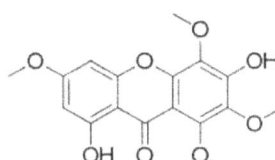
1,6-Dihydroxy-3,5,7,8-tetramethoxyxanthone

4.3 CULTURE CONDITIONS

The shoot tips of *C. erythraea* were cultured on an MS (Murashige and Skoog 1962) medium containing indole-3-acetic acid (0.1 mg/l) and 6-benzylaminopurine (1.0 mg/l). The higher levels of secoiridoid glucosides (gentiopicroside, swertiamarin and sweroside) were accumulated in micropropagated shoots of *C. erythraea* (10 weeks old) up to 149 mg/g dry weight. The profile was significantly greater than aerial parts of intact plants (Piatczak et al. 2005b). Similarly, the shoots were cultured in a 5 l mist trickling bioreactor (for 21 and 28 days) for enhancing their dry weight (0.54 g to 13.7 g and 18.3 g). Secoiridoids (gentiopicroside, sweroside and swertiamarin) accumulation in micropropagated shoots (after 21 days of culture) reached up to 303 mg/l (Piatczak et al. 2005a). Similarly, the 30-day old seedlings of *C. erythraea* were cultured on an MS medium, and maximum callusing was obtained with supplementation of indole-3-acetic acid (2.85 µM) and 6-benzylaminopurine (0.88 µM; Piątczak and Wysokińska 2003). The production of secoiridoids (gentiopicrin, swertiamarin, and sweroside) and xanthones (methylbellidifolin, demethyleustomin, and deccussatin) was monitored in the *in vitro* regenerated shoots and roots of *C. pulchellum*. Sweroside was reported as a major compound in the aerial parts of *in vivo* plants, whereas swertiamarin was found to be a major compound in callus. The demethyleustomin was reported as a major constituent in the shoots and roots of intact plant as well as in callus. The supplementation of sugars (glucose, fructose, and sucrose) also affected the production of secondary compounds (Krstić et al. 2003). The levels of phenolic compounds (1,2,3-trihydroxy-5-methoxyxanthone, 1-hydroxy-3,5,6,7, 8-pentamethoxyxanthone, and 1,8-dihydroxy-2,3,4,6-tetramethoxyxanthone) were increased in callus tissue of *C. erythraea* with increasing the age of callus cultures. In contrast, the levels of the phenolic acid derivatives (cinnamic, chlorogenic, and ferulic acids) were not changed with increasing the age of callus. The accumulation of phenolic acids was lower in suspension culture than callus, but low-molecular-weight phenolic compounds occur in higher concentration in the suspension culture than the callus tissue (Meravy 1987). The genes involved in somatic embryogenesis and shoot development may also be useful in biosynthesis of *C. erythraea* secondary metabolites (Ćuković et al. 2020). The addition of high salt concentrations to the cell cultures of C. *erythraea* induced the production of chlorophylls (a, b) and total chlorophylls (Šiler et al. 2007). Transverse sections of *C. erythraea* were cultured on an MS medium containing indole-3-acetic acid and sucrose. Lower concentration of indole-3-acetic acid (1.3×10^{-6} mol/l) induced the formation of callus (Barešová et al. 1985). Indole butyric acid (0.1 mg/l) supplementation in MS medium induced the synthesis of secoiridoids (150 mg/g dry weight) in cell cultures of *C. erythraea*. It also increased the total biomass of callii; hence, callus tissue makes them a better source of secondary metabolites (Piątczak et al. 2011). The roots or leaves-derived callus was obtained on an MS medium containing indole-3-acetic acid (2.85 μM) and 6-benzylaminopurine (0.88 μM; Piątczak and Wysokińska 2011).

The shoots of *C. erythraea* and *C. pulchellum* were infected with *Agrobacterium rhizogenes* (strain A4M70GUS) for establishment of hairy roots. Five clones of hairy roots were established. The hairy roots have a capacity for biosynthesis of xanthones. The accumulation of xanthones and secoiridoids (methylbellidifolin, demethyleustomin, decussatin, gentiopicrin, swertiamarin, and sweroside) was higher in hairy roots than naturally occurring plant roots (Janković et al. 2002; Subotić et al. 2003/2004). The profiles of secoiridoids were altered in different developmental stages of hairy roots. The total concentration of secoiridoids (gentiopicroside, sweroside, and swertiamarin) in 10-week-old hairy roots was eightfold (280 mg/g dry weight) higher than *C. erythraea* herb (Piatczak et al. 2006). The hairy roots of *C. erythraea* were cultured on half-strength MS medium containing kinetin, N^6-benzylaminopurine, 6-γ,γ-dimethylallylaminopurine, N-(2-chloro-4-pyridyl)-N'-phenylurea, 1-phenyl-3-(1,2,3-thiadiazol-5-yl)urea and 6-[4-hydroxy-3-methyl-but-2-enylamino]purine. The addition of *N*-(2-chloro-4-pyridyl)-N'-phenylurea (1.0 μM) induced the maximum number (25.6, 18.2, respectively) of adventitious hairy roots (Subotić et al. 2009b). Similarly, the half-strength MS medium with paclobutrazol showed better response in formation as well as growth of hairy roots (Subotić et al. 2009c). The addition of yariv phenylglycoside in half-strength MS medium induced the growth of hairy roots of *C. erythraea* (Trifunović- Momčilov et al. 2014). The *in vitro* grown AtCKX-transformed centaury lines of *C. erythraea* showed a significant increase in the production of eustomin and/or demethyleustomin contents. The increase was higher than *in vivo* plants. The xanthone eustomin demonstrated maximum cell growth inhibitory effects against the colorectal cancer cell line (DLD1) and its resistant counterpart (DLD1-TxR). Since xanthones possess pharmacological activities, AtCKX transgenic cell lines could be used as a useful source of plant material to produce novel drugs (Trifunović-Momčilov et al. 2015, 2016). The addition of NaCl concentrations (0, 50, 100, 150, 200 mM) in *AtCKX1* transgenic lines induced the accumulation of cytokinins (Trifunović-Momćilov et al. 2020). In hairy roots, better production of bitter secoiridoid glucosides and xanthones was obtained and may be considered as an alternative source for their production (Subotić et al. 2009a; Simonović et al. 2021).

The 3,5,6,7,8-pentamethoxy-1-*O*-primeverosylxanthone was isolated from cell cultures of *C. erythraea* and *C. littorale*. The addition of yeast extract in MS medium increased the production of 1,5-dihydroxy-3-methoxyxanthone (Beerhues 1994). Similarly, the accumulation of 1-hydroxy-3,5,6,7-tetramethoxyxanthone was increased by methyl jasmonate and yeast extract after 10 h addition in the cell suspension cultures of *C. erythraea*. Phenylalanine ammonia-lyase activity was not increased by elicitors. The addition of yeast extract enhanced the production of 1,5-dihydroxy-3-methoxyxanthone in cell suspension cultures. Both elicited xanthones were produced intracellularly in cells of *C. erythraea* and *C. littorale* (Beerhues and Berger 1995). The effects of methyl jasmonate and chitosan were investigated on the production of secoiridoid glycoside and xanthone in the cell suspension cultures of *C. erythraea*. The addition of methyl jasmonate and chitosan reduced the yield of sweroside and eustomin in cell cultures (Filipović et al. 2015; Boroduške et al. 2016). The production and accumulation of secoiridoid glucosides was reported for *in vitro* regenerated leaves. The secoiridoid glucosides biosynthesis in leaves of *C. erythraea* was increased by methyl jasmonate supplementation. The supplementation caused reprogramming of secoiridoid glucosides synthesis-related gene expression, leading to an enhanced production of valuable bioactive compounds in *in vitro* regenerated leaves (Matekalo et al. 2018). Somatic embryos were developed from leaf explants of *C. erythraea*. The arabinogalactan proteins were reported from somatic embryos (Simonović et al. 2015, 2021). Arabinogalactan proteins distribution with specific epitopes were detected in the indirect somatic embryogenesis and shoot organogenesis in *C. erythraea* leaf culture (Bogdanović et al. 2021). During the development of adventitious buds, the number of detected arabinogalactan proteins were also reduced (Paunović et al. 2021; Filipović et al. 2021).

REFERENCES

Aberham A, Pieri V, Croom Jr EM, Ellmerer E, Stuppner H. 2011. Analysis of iridoids, secoiridoids and xanthones in *Centaurium erythraea*, *Frasera caroliniensis* and *Gentiana lutea* using LC-MS and RP-HPLC. J Pharm Biomed Anal 54, 517–525.

Allam AE, Nafady AM, El-Shanawany MA, Takano F, Ohta T. 2015. New secoiridoid ester of swertiamarin and secoxyloganic acid with hepatoprotective activity from *Centaurium spicatum* L. J Pharm Pharmacogn Res 3, 69–76.

Barešová H, Herben T, Kamínek M, Krekule J. 1985. Hormonal control of morphogenesis in leaf segments of *Centaurium erythraea*. Biol Plant 27, 286–291.

Beerhues, L. 1994. Constitutive and elicitor-induced xanthones in cell cultures of *Centaurium erythraea* and *Centaurium littorale*. Acta Horticult 381, 250–257.

Beerhues L, Berger U. 1995. Differential accumulation of xanthones in methyl-jasmonate- and yeast-extract-treated cell cultures of *Centaurium erythraea* and *Centaurium littorale*. Planta 197, 608–612.

Berkan T, Ustünes L, Lermioglu F, Ozer A. 1991. Anti-inflammatory, analgesic, and antipyretic effects of an aqueous extract of *Erythraea centaurium*. Planta Med 57, 34–37.

Berrak M, Gheyouche-Siachi R, Allali H, Ouafi S, Bendifallah L, Lazeli N. 2017. Phenolic compounds and *in vivo* anti-inflammatory activity of aqueous extract of *Centaurium umbellatum* (Gibb.) beck. flowers in northern Algeria. J Med Plant Herb Ther Res 5, 18–26.

Bibi H, Ali I, Sadozai SK, Atta-ur-Rahman. 2006. Phytochemical studies and antibacterial activity of *Centaurium pulchellum* Druce. Nat Prod Res 20, 896–901.

Bogdanović MD, Ćuković KB, Subotić AR, Dragićević MB, Simonović AD, Filipović BK, Todorović SI. 2021. Secondary somatic embryogenesis in *Centaurium erythraea* Rafn. Plants (Basel) 10, 199.

Boroduške A, Nakurte I, Tomsone S, Lazdane M, Boroduskis M, Rostoks N. 2016. *In vitro* culture type and elicitation affects secoiridoid and xanthone LC-ESI-TOF MS profile and production in *Centaurium erythraea*. Plant Cell Tissue Organ Cult 126, 567–571.

Bouyahya A, Bakri Y, Belmehdi O, Et-Touys A, Abrini J, Dakka N. 2017. Phenolic extracts of *Centaurium erythraea* with novel antiradical, antibacterial and antileishmanial activities. Asian Pac J Trop Dis 7, 433–439.

Bouyahya A, Belmehdi O, Jemli ME, Marmouzi I, Bourais I, Abrini J, Abbes-Faouzi ME, Dakka N, Bakri Y. 2019. Chemical variability of *Centaurium erythraea* essential oils at three developmental stages and investigation of their *in vitro* antioxidant, antidiabetic, dermatoprotective and antibacterial activities. Ind Crops Prod 132, 111–117.

Božunović J, Skorić M, Matekalo D, Živković S, Dragićević M, Aničić N, Filipović B, Banjanac T, Šiler B, Mišić D. 2019. Secoiridoids metabolism response to wounding in common centaury (*Centaurium erythraea* Rafn) leaves. Plants 8, 589.

Božunović J, Živković S, Gašić U, Glamočlija J, Cirić A, Matekalo D, Šiler B, Soković M, Tešić Ž, Mišić D. 2018. *In Vitro* and *in vivo* transformations of *Centaurium erythraea* secoiridoid glucosides alternate their antioxidant and antimicrobial capacity. Ind Crops Prod 111, 705–721.

Budniak L, Slobodianiuk L, Marchyshyn S, Klepach P. 2021. Investigation of the influence of the thick extract of common centaury (*Centaurium erythraea* RAFN.) herb on the secretory function of the stomach. Pharmacologyonline 2, 352–360.

Chda A, Kabbaoui ME, Chokri A, Abida KE, Tazi A, Cheikh R. 2016. Spasmolytic action of *Centaurium erythraea* on rabbit jejunum is through calcium channel blockade and NO release. Eur J Med Plants 11, 1–13.

Ćuković K, Dragićević M, Bogdanović M, Paunović D, Giurato G, Filipović B, Subotić A, Todorović S, Simonović A. 2020. Plant regeneration in leaf culture of *Centaurium erythraea* Rafn. Part 3: *De novo* transcriptome assembly and validation of housekeeping genes for studies of *in vitro* morphogenesis. Plant Cell Tissue Organ Cult 141, 417–433.

Cunha AP, Silva AP, Roque O. 2003. Plantas e Produtos Vegetais em Fitoterapia. Fundação Calouste Gulbenkian, Lisbon.

Đorđević M, Grdović N, Mihailović M, Jovanović JA, Uskoković A, Jovana Rajić J, Sinadinović M, Tolić A, Mišić D, Šiler B, Poznanović G, Vidaković M, Dinić S. 2019. *Centaurium erythraea* extract improves survival and functionality of pancreatic β-cells in diabetes through multiple routes of action. J Ethnopharmacol 242, 112043.

Đorđević M, Mihailović M, Jovanović JA, Grdović N, Uskoković A, Tolić A, Sinadinović M, Rajić J, Mišić D, Šiler B, Poznanović G, Vidaković M, Dinić S. 2017. *Centaurium erythraea* methanol extract protects red blood cells from oxidative damage in streptozotocin-induced diabetic rats. J Ethnopharmacol 202, 172–183.

Filipović BK, Simonović AD, Trifunović MM, Dmitrović SS, Savić JM, Jevremović SB, Subotić AR. 2015. Plant regeneration in leaf culture of *Centaurium erythraea* Rafn. Part 1: The role of antioxidant enzymes. Plant Cell Tissue Organ Cult 121, 703–719.

Filipović BK, Trifunović-Momčilov MM, Simonović AD, Jevremović SB, Milošević SM, Subotić AR. 2021. Immunolocalization of some arabinogalactan protein epitopes during indirect somatic embryogenesis and shoot organogenesis in leaf culture of centaury (*Centaurium erythraea* Rafn). *In Vitro* Cell Dev Biol Plant 57, 470–480.

Guedes L, Reis PBPS, Machuqueiro M, Ressaissi A, Pacheco R, Serralheiro ML. 2019. Bioactivities of *Centaurium erythraea* (Gentianaceae) decoctions: Antioxidant activity, enzyme inhibition and docking studies. Molecules 24, 3795.

Haloui M, Louedec L, Michel JB, Lyoussi B. 2000. Experimental diuretic effects of *Rosmarinus officinalis* and *Centaurium erythraea*. J Ethnopharmacol 71, 465–472.

Hamza N, Berke B, Cheze C, Marais S, Lorrain S, Abdoulfath A. 2015. Effect of *Centaurium erythraea* Rafn, *Artemisia herba-alba* Asso and *Trigonella foenum-graecum* L. on liver fat accumulation in C57BL/6J mice with high-fat diet-induced type 2 diabetes. J Ethnopharmacol 171, 4–11.

Hill RA. 1986. Naturally occurring isocoumarins. In: Progress in the Chemistry of Organic Natural Products, Vol 49, Herz W, Grisebach H, Kirby GW, Tamm C (eds). Springer-Verlag, New York, pp 1–78.

Janković T, Krstić D, Savikin-Fodulović K, Menković N, Grubisić D. 2002. Xanthones and secoiridoids from hairy root cultures of *Centaurium erythraea* and *C. pulchellum*. Planta Med 68, 944–946.

Jerković I, Gašo-Sokač D, Pavlović H, Marijanović Z, Gugić M, Petrović I, Kovač S. 2012. Volatile organic compounds from *Centaurium erythraea* Rafn (Croatia) and the antimicrobial potential of its essential oil. Molecules 17, 2058–2072.

Jovanović O, Radulović N, Stojanović G, Palić R, Zlatković B, Gudžić B. 2009. Chemical composition of the essential oil of *Centaurium erythraea* Rafn (Gentianaceae) from Serbia. J Essent Oil Res 21, 317–322.

Kachmar MR, Oliveira AP, Valentão P, Gil-Izquierdo A, Domínguez-Perles R, Ouahbi A, Badaoui KE, Andrade PB, Ferreres F. 2019. HPLC-DAD-ESI/MS phenolic profile and *in vitro* biological potential of *Centaurium erythraea* Rafn aqueous extract. Food Chem 278, 424–433.

Khafagy SM, Mnajed HK. 1970. Phytochemical investigation of *Centaurium pulchellum* (Sw.) Druce. Acta Pharm Suec 7, 667–672.

Krstić D, Janković T, Šavikin-Fodulović K, Menković N, Grubišićs D. 2003. Secoiridoids and xanthones in the shoots and roots of *Centaurium pulchellum* cultured *in vitro*. *In Vitro* Cell Dev Biol Plant 39, 203–207.

Kumarasamy Y, Nahar L, Cox PJ, Jaspars M, Sarker SD. 2003a. Bioactivity of secoiridoid glycosides from *Centaurium erythraea*. Phytomedicine 10, 344–347.

Kumarasamy Y, Nahar L, Sarker SD. 2003b. Bioactivity of gentiopicroside from the aerial parts of *Centaurium erythraea*. Fitoterapia 74, 151–154.

Liliya S, Iryna D, Olena P, Svitlana M. 2017. Polysaccharides in *Centaurium erythraea* Rafn. Int J Res Ayurveda Pharm 8, 252–255.

Mandova T, Audo G, Michel S, Grougnet R. 2017. Off-line coupling of new generation centrifugal partition chromatography device with preparative high pressure liquid chromatography-mass spectrometry triggering fraction collection applied to the recovery of secoiridoid glycosides from *Centaurium erythraea* Rafn. (Gentianaceae). J Chromatogr A 1513, 149–156.

Mansar-Benhamza L, Djerrou Z, Hamdi Pacha Y. 2013. Evaluation of anti-hyperglycemic activity and side effects of *Erythraea centaurium* (L.) Pers. in rats. Afr J Biotechnol 12, 6980–6985.

Matekalo D, Skorić M, Nikolić T, Božunović J, Aničić N, Filipović B, Mišić D. 2018. Organ-specific and genotype-dependent constitutive biosynthesis of secoiridoid glucosides in *Centaurium erythraea* Rafn, and its elicitation with methyl jasmonate. Phytochemistry 155, 69–82.

Menyiy NE, Guaouguaou F-E, Baaboua AE, Omari NE, Taha D, Salhi N, Shariati MA, Aanniz T, Benali T, Zengin G, El-Shazly M, Chamkhi I, Bouyahya A. 2021. Phytochemical properties, biological activities and medicinal use of *Centaurium erythraea* Rafn. J Ethnopharmacol 276, 114171.

Meravy L. 1987. Phenolic substances in tissue cultures of *Centaurium erythraea*. Biol Plant 29, 81–87.

Merghem M, Dahamna S. 2020. Antioxidant activity of *Centaurium erythraea* extracts. J Drug Deliv Ther 10, 171–174.

Mihaylova D, Vrancheva R, Popova A. 2019. Phytochemical profile and *in vitro* antioxidant activity of *Centaurium erythraea* Rafn. Bulg Chem Commun 51, 95–100.

Mroueh M, Saab Y, Rizkallah R. 2004. Hepatoprotective activity of *Centaurium erythraea* on acetaminophen-induced hepatotoxicity in rats. Phytother Res 18, 431–433.

Murashige T, Skoog F. 1962. A revised medium for rapid growth and bio-assays with tobacco tissue cultures. Physiol Plant 15, 473–497.

Neagu E, Radu GL, Albu C, Paun G. 2018. Antioxidant activity, acetylcholinesterase and tyrosinase inhibitory potential of *Pulmonaria officinalis* and *Centaurium umbellatum* extracts. Saudi J Biol Sci 25, 578–585.

Newall CA, Anderson LA, Phillipson JD. 1996. Herbal Medicines—A Guide for Health-care Professionals. The Pharmaceutical Press, London.

Nikolić VG, Zvezdanović JB, Konstantinović SS. 2022. UHPLC-DAD-ESI-MS analysis of the *Centaurium erythraea* infusion. Adv Technol 11, 13–21.

Nikolova-Damyanova B, Handjieva N. 1996. Quantitative determination of swertiamarin and gentiopicroside in *Centarium erythrea* and *C. turcicum* by densitometry. Phytochem Anal 7, 140–142.

Orhan IE, Senol FS, Haznedaroglu MZ, Koyu H, Erdem SA, Yılmaz G, Cicek M, Yaprak AE, Ari E, Kucukboyaci N, Toker G. 2016. Neurobiological evaluation of thirty-one medicinal plant extracts using microtiter enzyme assays. Clin Phytosci 2, 9.

Paunović DM, Ćuković KB, Bogdanović MD, Todorović SI, Trifunović-Momčilov MM, Subotić AR, Simonović AD, Dragićević MB. 2021. The arabinogalactan protein family of *Centaurium erythraea* Rafn. Plants (Basel) 10, 1870.

Piatczak E, Chmiel A, Wysokinska H. 2005a. Mist trickling bioreactor for *Centaurium erythraea* Rafn growth of shoots and production of secoiridoids. Biotechnol Lett 27, 721–724.

Piatczak E, Krolicka A, Wysokinska H. 2006. Genetic transformation of *Centaurium erythraea* Rafn by *Agrobacterium rhizogenes* and the production of secoiridoids. Plant Cell Rep 25, 1308–1315.

Piątczak E, Królicka A, Wysokińska H. 2011. Morphology, secoiridoid content and RAPD analysis of plants regenerated from callus of *Centaurium erythraea* RAFN. Acta Biol Crac Ser Bot 53, 79–86.

Piatczak E, Wielanek M, Wysokinska H. 2005b. Liquid culture system for shoot multiplication and secoiridoid production in micropropagated plants of *Centaurium erythraea* Rafn. Plant Sci 168, 431–437.

Piątczak E, Wysokińska H. 2003. *In vitro* regeneration of *Centaurium erythraea* Rafn from shoot tips and other seedling explants. Acta Soc Bot Pol 72, 283–288.

Schimmer O, Mauthner H. 1996. Polymethoxylated xanthones from the herb of *Centaurium erythraea* with strong antimutagenic properties in *Salmonella typhimurium*. Planta Med 62, 561–564.

Sefi M, Fetoui H, Lachkar N, Tahraoui A, Lyoussi B, Boudawara T, Zeghal N. 2011. *Centaurium erythrea* (Gentianaceae) leaf extract alleviates streptozotocin-induced oxidative stress and β-cell damage in rat pancreas. J Ethnopharmacol 135, 243–250.

Shahat AA, Cos P, Hermans N, Apers S, Bruyne TD, Pieters L, Berghe DV, Vlietinck AJ. 2003. Anticomplement and antioxidant activities of new acetylated flavonoid glycosides from *Centaurium spicatum*. Planta Med 69, 1153–1156.

Šiler B, Avramov S, Banjanac T, Cvetković J, Zivković JN, Patenković A, Mišić D. 2012. Secoiridoid glycosides as a marker system in chemical variability estimation and chemotype assignment of *Centaurium erythraea* Rafn from the Balkan Peninsula. Ind Crops Prod 40, 336–344.

Šiler B, Mišić D, Filipović B, Popović Z, Cvetić T, Mijović A. 2007. Effects of salinity on *in vitro* growth and photosynthesis of common centaury (*Centaurium erythraea* Rafn). Arch Biol Sci 59, 129–134.

Šiler B, Mišić D, Nestorović J, Banjanac T, Glamočlija J, Soković M, Ćirić A. 2010. Antibacterial and antifungal screening of *Centaurium pulchellum* crude extracts and main secoiridoid compounds. Nat Prod Commun 5, 1525–1530.

Simonović AD, Filipović BK, Trifunović MM, Malkov SN, Milinković VP, Jevremović SB, Subotić AR. 2015. Plant regeneration in leaf culture of *Centaurium erythraea* Rafn. Part 2: The role of arabinogalactan proteins. Plant Cell Tissue Organ Cult 21, 721–739.

Simonović AD, Trifunović-Momčilov MM, Filipović BK, Marković MP, Bogdanović MD, Subotić AR. 2021. Somatic embryogenesis in *Centaurium erythraea* Rafn—current status and perspectives: A review. Plants 10, 70.

Stefkov G, Miova B, Dinevska-Kjovkarovska S, Stanoeva J.P, Stefova M, Petrusevska G, Kulevanova S. 2014. Chemical characterization of *Centaurium erythrea* L. and its effects on carbohydrate and lipid metabolism in experimental diabetes. J Ethnopharmacol 152, 71–77.

Subotić A, Budimir S, Grubisić D, Momoćilović I. 2003/2004. Direct regeneration of shoots from hairy root cultures of *Centaurium erythraea* inoculated with *Agrobacterium rhizogenes*. Biol Plant 47, 617–619.

Subotić A, Jankovic T, Jevremović S, Grubišic D. 2006. Plant tissue culture and secondary metabolites productions of *Centaurium erythraea* Rafn, a medical plant. In: Floriculture Ornamental and Plant Biotechnology: Advances and Topical Issues, Vol 2, 1st ed., Teixeira da Silva JA (ed.). Global Science Books, London, pp 564–570.

Subotić A, Jevremović S, Grubišić D. 2009b. Influence of cytokinins on *in vitro* morphogenesis in root cultures of *Centaurium erythraea*—valuable medicinal plant. Sci Hortic 120, 386–390.

Subotić A, Jevremović S, Grubisić D, Janković T. 2009a. Spontaneous plant regeneration and production of secondary metabolites from hairy root cultures of *Centaurium erythraea* Rafn. Methods Mol Biol 547, 205–215.

Subotić A, Jevremović S, Trifunović M, Petrić M, Milošević S, Grubišić D. 2009c. The influence of gibberellic acid and paclobutrazol on induction of somatic embryogenesis in wild type and hairy root cultures of *Centaurium erythraea* Gillib. Afr J Biotechnol 8, 3223–3228.

Tahraoui A, Israili ZH, Lyoussi B. 2010. Acute and sub-chronic toxicity of a lyophilised aqueous extract of *Centaurium erythraea* in rodents. J Ethnopharmacol 132, 48–55.

Trifunović-Momčilov M, Krstić-Milošević D, Trifunović S, Podolski-Renić A, Pešić M, Subotić A. 2016. Secondary metabolite profile of transgenic centaury (*Centaurium erythraea* Rafn.) plants, potential producers of anticancer compounds. In: Transgenesis and Secondary Metabolism. Reference Series in Phytochemistry, Jha S (ed.). Springer, Cham, pp 1–26.

Trifunović-Momčilov M, Krstić-Milošević D, Trifunovic SS, Ciric A. 2019. Antimicrobial activity, antioxidant potential and total phenolic content of transgenic AtCKX1 centaury (*Centaurium erythraea* Rafn.) plants grown *in vitro*. Environ Eng Manag J 18, 2063–2072.

Trifunović-Momčilov M, Motyka V, Dragićević IČ, Petrić M, Jevremović S, Malbeck J, Holík J, Dobrev PI, Subotić A. 2015. Endogenous phytohormones in spontaneously regenerated *Centaurium erythraea* Rafn. plants grown *in vitro*. J Plant Growth Regul 35, 543–552.

Trifunović-Momčilov M, Paunović D, Milošević S, Marković M, Jevremović S, Dragićević IC, Subotić A. 2020. Salinity stress response of non-transformed and AtCKX transgenic centaury (*Centaurium erythraea* Rafn) shoots and roots grown *in vitro*. Ann Appl Biol 177, 74–89.

Trifunović-Momčilov M, Tadić V, Petrić M, Jontulović D, Jevremović S, Subotić A. 2014. Quantification of arabinogalactan proteins during *in vitro* morphogenesis induced by β-D-glucosyl Yariv reagent in *Centaurium erythraea* root culture. Acta Physiol Plant 36, 1187–1195.

Tuluce Y, Ozkol H, Koyuncu I, Ine H. 2011. Gastroprotective effect of small centaury (*Centaurium erythraea* L.) on aspirin-induced gastric damage in rats. Toxicol Ind Health 27, 760–768.

Ubsdell RAE. 1976. Studies on variation and evolution in *Centaurium erythraea* Rafn., and *C. littorale*. In: Gilmour in the British Isles, Turner D (ed.). Watsonia 11, 7–31.

Vágnerová H. 1992. Micropropagation of common centaury (*Centaurium erythraea* Rafn.). In: Biotechnology in Agriculture and Forestry, Vol 19, Bajaj YPS (ed.). Springer-Verlag, Berlin, Heidelberg, pp 388–396.

Valentão P, Andrade PB, Silva AMS, M. Moreira M, Seabra RM. 2003b. Isolation and structural elucidation of 5-formyl-2,3-dihydroisocoumarin from *Centaurium erythraea* aerial parts. Nat Prod Res 17, 361–364.

Valentão P, Fernandes E, Carvalho F, Andrade PB, Seabra RM, Bastos ML. 2001. Antioxidant activity of *Centaurium erythraea* infusion evidenced by its superoxide radical scavenging and xanthine oxidase inhibitory activity. J Agric Food Chem 49, 3476–3479.

Valentão P, Fernandes E, Carvalho F, Andrade PB, Seabra RM, Bastos ML. 2003a. Hydroxyl radical and hypochlorous acid scavenging activity of small centaury (*Centaurium erythraea*) infusion. A comparative study with green tea (*Camellia sinensis*). Phytomedicine 10, 517–522.

van der Sluis WG. 1984. Chemotaxonomieal investigations of the genera *Blackstonia* and *Centaurium* (Gentianaceae). Plant Syst Evol 149, 253–286.

van der Sluis WG. 1985. Secoiridoids and xanthones in the genus *Centaurium* Hill (Gentianaceae)—a pharmacognostical study. PhD Thesis, Rijksuniv Utrecht-Utrecht University, Utrecht.

Van Hellemont J. 1986. Compendium de Phytotherapie. Association Pharmaceutique Belge, Bruxelles.

Waltenberger B, Liu R, Atanasov AG, Schwaiger S, Heiss EH, Dirsch VM, Stuppner H. 2015. Nonprenylated xanthones from *Gentiana lutea, Frasera caroliniensis*, and *Centaurium erythraea* as novel inhibitors of vascular smooth muscle cell proliferation. Molecules 20, 20381–20390.

Zeltner L. 1970. Resherches de biosystematique sur les genres *Blackstonia* et *Centaurium*. Soc Neuchatel Sci Nat 93, 57–85.

Živković S, Dević M, Filipović B, Giba Z, Grubišić D. 2007. Effect of NaCl on seed germination in some *Centaurium* Hill. species (Gentianaceae). Arch Biol Sci Belgrade 59, 227–231.

5 *Chenopodium* Species

5.1 MORPHOLOGICAL FEATURES, DISTRIBUTION, ETHNOPHARMACOLOGICAL PROPERTIES, PHYTOCHEMISTRY, AND PHARMACOLOGICAL ACTIVITIES

Chenopodium album Linn. (syn. *C. quinoa* Willd., *C. nuttalliae* Saff.; Fam.—Chenopodiaceae) is an erect, annual weed plant, branched, and reaches up to a height of 10–150 cm. The leaves are simple, alternate, ovate, lanceolate to rhomboid, and irregularly dentated. The leaves have long petioles while stipules are absent. The inflorescence is panicle in leaf axils and flowers arrange in clusters. The flowers are green, small, and sessile in irregular spikes. The fruit is a glabrous utricle, containing a single seed (Nishimura et al. 2010). It is mainly grown as a weed in the fields of wheat, barley, mustard, gram, and other crops in the Indian sub-continent (Khurana et al. 1986; Bhattacharjee 2001; Yadav et al. 2007). It is distributed in Western Asia, Europe, and North American (Bailey 1977; Risi and Galwey 1984), the Andean region of South America (Del Castillo et al. 2008; Bazile et al. 2014), and the Lake Titicaca region of Bolivia and Peru (Wilson 1990; Schlick and Bubenheim 1996; Bhargava et al. 2007; Maliro et al. 2017). The roasted seeds are added to soups and even fermented into beer (traditional drink of the Andes; Jancurová et al. 2009; Vega-Gálvez et al. 2010; Cooper 2015). In India, it is naturalized in western Rajasthan, the Kulu valley, and Shimla (Kirtikar and Basu 1964–65). It possesses sperm immobilizing activity (Kumar et al. 2007b). As per Atharva Veda, Charak Samhita, and Sushruta Samhita medicine, it is used for anthelmintic, cardiotonic, carminative, digestive, diuretic, and laxative properties (Bakshi et al. 1999; Agrawal et al. 2014). In Ayurvedic medicine, the *Chenopodium* species is well-known for uses in the treatment of pectoral disease, cough, abdominal pain, pulmonary blocks, and in nervous affections (Yadav et al. 2007). It is also useful in peptic ulcer, dyspepsia, flatulence, strangury, pharyngopathy, splenopathy, opthalmopathy, and general debility. The leaf juice is used to treat burns. The water decoction of aerial parts is mixed with alcohol and rubbed on arthritis and rheumatism-affected body parts (Prajapati et al. 2003; Pal et al. 2011). Its seeds are naturally gluten free and have high nutrient profiles (Kilinc et al. 2016). The soft shoots are eaten raw in salad or cooked as a vegetable, or the cooked shoots are mixed with curd and eaten. The leaves are rich in vitamin C and potassium (Murphy and Matanguihan 2015; Pérez et al. 2011). The aerial parts are used for the treatment of hepatic diseases, spleen enlargement, and intestinal ulcers (Sarma et al. 2008; Poonia and Upadhayay 2015). In Lebanese medicine, it possesses antirheumatic properties (Nelly et al. 2008). In Italy, it is used in boiled or in salads, soups, and stews (Bianco et al. 1996). The decoction of leaves possesses anthelmentic, antiphlogistic, antirheumatic, mildly laxative, and odontalgic activities. The leaf poultice is applied to bug bites, sunstroke, treatment of rheumatic joints, and swollen feet (Kokanova-Nedialkova et al. 2009). Four phenolic compounds (protocatechuic acid, catechin hydrate, *trans*-ferulic acid, and benzoic acid) were determined with the HPLC from the methanol and ethanol extracts *C. botrys*. The ethanolic extract is more efficacious than the methanolic extract in all parameters (total phenolic contents, antioxidant capacity, and cytotoxicity; Şimşek Sezer and Uysal 2021).

C. album is widely cultivated in Europe, North America, Iran, South Africa, Australia, South America, and Asia (Chamkhi et al. 2022). *C. album* is traditionally used as an herbal medicine to treat abdominal pains, eye infection, throat troubles, piles, heart, and spleen disorders. Its leaves are used in the treatment of digestive problems, peptic ulcer, and hepatic disorder (Verza et al. 2012; Das and Borthakur 2022). The plant contains phenolic amide, saponin, cinnamic acid amide, chinoalbicin, apocortinoid, xyloside, phenols and lignans, protein, carbohydrates, suberin, and glucosides. The extracts of plant parts possess antioxidant potential, antimicrobial, antinociceptive, anthelmintic properties (Bhatia et al. 2020; Kasali et al. 2021).

DOI: 10.1201/9781003398035-5

C. album possess allelopathic effects and inhibit the germination and growth of native vegetation and/or crop plants. Both species may be used as an alternative herbicide in the integration of chemicals (Bajwa et al. 2019). The white, yellow, red, or black seeds contain significant levels of protein (12–21%; García-Parra et al. 2020). It has been established that digestibility of the quinoa seed proteins is comparable to that of other high quality food proteins. One study results showed that the saponins do not trigger any adverse effect on the nutritional quality of the protein (Ruales and Nair 1992). The quinoa seed, rich in dietary fiber, can reduce the levels of cholesterol in the blood and improve digestion, which may increase interest in including quinoa into one's supplementary diet (Repo-Carrasco et al. 2003). The seed possesses proteins, lipids, fiber, vitamins, and minerals, and has an extraordinary balance of essential amino acids (Navruz-Varli and Sanlier 2016). The quinoa consists of a high amount of essential fatty acids, minerals, vitamins, dietary fibers, and carbohydrates, and possesses significant hypoglycemic effects (gluten-free; Angeli et al. 2020).

The acetone extract (200 mg/kg, p.o.) of *C. album* aerial parts showed significant decrease in rat paw edema (80.13%) within 21 days. The extract significantly reduced the expression of NFκB protein in paraventricular nucleus of hypothalamus. The results were compared with standard indomethacin in complete Freund's adjuvant-induced arthritis model in rats. The polyphenolic and flavonoid contents of ethyl acetate, acetone, methanol, and 50% methanol extracts were in the range of $14.56 \pm 0.21 - 42.00 \pm 0.2$ mg (gallic acid equivalent/g extract) and $2.20 \pm 0.003 - 7.33 \pm 0.5$ mg (rutin equivalent/g extract) respectively. The antiarthritic activity can be associated with its antioxidant ability, high-flavonoidal contents, and its ability to suppress the expression of NFκB protein (Arora et al. 2014).

The antilithiatic effects of methanol and aqueous extracts of the *C. album* leaves were evaluated in ethylene glycol-induced urolithiasis in rats. Both extracts (200 and 400 mg/kg/daily, p.o.) significantly reduced the ethylene glycol-induced increase in the urine and plasma levels of calcium, phosphorus, urea, uric acid, and creatinine along with reduction in urine volume, pH, and oxalates in treated rats. The extracts also reduced the levels of renal tissue oxalate and deposition of oxalate crystals in the kidneys of rats. The results were compared with the antilithiatic drug, cystone in treated animals. The results reveal suppressive effects of extracts on crystallization and stone dissolution. These effects were associated to the presence of flavonoids and saponins in both extracts. The leaves of *C. album* demonstrated antilithiatic effect and justifies its ethnomedicinal use in treatment of urinary diseases and kidney stones (Sikarwar et al. 2017).

The ethanolic extract (100–400 mg/kg) of *C. album* fruits suppressed scratching behavior stimulated by 5-hydroxytryptamine (10 μg/mouse, s.c.) or compound 48/80 (50 μg/mouse, s.c.) in mice. The extract (100 and 200 mg/kg) did not show any effect on hind-paw swelling caused by 5-hydroxytryptamine or compound 48/80 in mice but showed weak suppression on the swelling at higher dose (400 mg/kg). Moreover, ethanol extract (200 and 400 mg/kg) significantly reduced the writhing responses stimulated by an acetic acid (i.p.) and the inflammatory pain response stimulated by formalin (i.p. injection) in mice. The extract (400 mg/kg) also suppressed the neurogenic pain response induced by formalin injection. The extract possesses antipruritic and antinociceptive properties and the results support evidence for the clinical use of extract to treat cutaneous pruritus (Dai et al. 2002).

Methanolic extract of *C. album* minimized the levels of alanine transaminase (68.75 ± 8.38 U/L, 200 mg/kg), aspartate transaminase (140.75 ± 13.35 U/L, 200 mg/kg) and alkaline phosphatase (248.25 ± 4.03 U/L, 300 mg/kg) at different concentrations. Extract increased the levels of triglycerides (64.75 ± 12.66 mg/dl at 200 mg/kg) when compared to carbon tetrachloride treated group (33.25 ± 1.26 mg/dl). Carbon tetrachloride increased the levels of urea (43.25 ± 6.6) but reduced by high dose of extract (18 ± 8.17, 300 mg/kg). In addition, extract ameliorated the levels of WBCs (10.59×10^3/Cu.mr, 200 mg/kg), RBCs (6.97×10^3/Cu.mr, 200 mg/kg) and hemoglobin (14.05 G/dL at 300 mg/kg; Hussain et al. 2022). The aqueous extract (400 mg/kg) of *C. album* leaves significantly checked the carbon tetrachloride-stimulated increase of the levels of serum glutmate oxaloacatate transaminase, pyruvate oxaloacatate transaminase, alkaline phosphatase, bilirubin, lactate dehydrogenase, and triglycerides, whereas it reduced the total protein. Aqueous extract reduced the carbon tetrachloride-stimulated fibrosis in liver tissue and increased levels of lipid peroxidation (Wanjari et al. 2016).

The antioxidant, anticancer, bioaccessibility, and bioavailability effects of *C. quinoa* leaves were evaluated. The substantial yields of ferulic, sinapinic and gallic acids, kaempferol, isorhamnetin, and rutin were reported in the leaf extract. The extract was associated with its suppressive activity on prostate cancer cell growth, motility, and cellular competence for gap junctional communication. Leaf extract showed an inhibitory activity on lipoxygenase, chelating, antioxidative, antiradical, and reducing power activities. The extract may trigger chemoprotective and anticarcinogenic activity on oxidative stress and reactive oxygen species-dependent intracellular signaling *via* synergic effects (Gawlik-Dziki et al. 2013; Fischer et al. 2013). It is considered as a rich source of protein (9.1–15.7 g), total fat (4.0–7.6 g), and dietary fiber (8.8–14.1 g), mineral, and vitamins (Nowak et al. 2016). The petroleum ether extract of *C. album* aerial parts demonstrated significant growth suppressive effects (IC_{50} 33.31 ± 2.79 μg/ml) on human non-small cell lung cancer A549 cells. Moreover, extract-treated A549 cells displayed dose-dependent cell growth arrest at the G1 phase of the cell cycle and cell apoptosis. Therefore, the extract may be used in developing lung cancer therapeutic agents (Zhao et al. 2016).

A quinoa protein isolate (purity 40.73 ± 0.90%) was hydrolyzed at 50 °C for 3 h with two enzyme formulations: papain and a microbial papain-like enzyme to form quinoa protein hydrolysates. Both hydrolysates were assessed for their dipeptidyl peptidase-IV suppressive and oxygen radical absorbance ability activities. The hydrolysis of protein was recorded in the quinoa protein isolate control, possibly due to quinoa endogenous proteinase activities. The quinoa protein isolate control displayed greater dipeptidyl peptidase-IV half maximal inhibitory concentrations (IC_{50}) and lower oxygen radical absorbance capacity values than quinoa protein hydrolysate-papain and quinoa protein hydrolysate-papain-like enzyme ($p < 0.05$). The quinoa protein hydrolysate -papain showed significant dipeptidyl peptidase -IV (IC_{50} 0.88 ± 0.05 mg/ml) and an oxygen radical absorbance capacity activity (501.60 ± 77.34 μmol trolox equivalent/g). The obtained results support the dipeptidyl peptidase-IV inhibitory activities of quinoa protein hydrolysates. Therefore, quinoa protein hydrolysate may have capacity as functional ingredients with serum glucose lowering activities (Nongonierma et al. 2015). The hydroalcoholic extract of quinoa seed coats showed the presence of thiol compounds, polyphenols, and antioxidants. The extract suppressed microsomal lipid peroxidation and the deficit of microsomal thiol contents (oxidative processes induced by Cu^{2+}/ascorbate). Microsomal glutathione *S*-transferase is suppressed by decreasing the content of catalytically active disulfide-linked dimers. Due to the presence of thiol compounds, the extract may be considered as a potent antioxidant agent (Letelier et al. 2011). Due to color, hardness, and specific volume profiles in gluten-free cookies, the optimized formulation contains 30% quinoa flour, 25% quinoa flakes, and 45% corn starch. Quinoa-based cookies are rich in dietary fiber and essential amino acids, linolenic acid, and minerals (Brito et al. 2015).

The profiles of total phenolics and total flavonoids were estimated and showed the presence of 6.02 to 43.47 mg/gallic acid equivalents/g and 1.30 and 12.26 mg/quercetin equivalents/g of plant material. The seeds and leaves showed antioxidant activity in 2,2-diphenyl-1-picrylhydrazyl free radical scavenging and ferric-reducing antioxidant power assays (Buitrago et al. 2019). The methanol extract of *C. nuttalliae* Saff (250 and 500 mg/kg) showed antioxidant potential in DPPH assay. The extract (500 mg/kg) also showed a significant decrease in levels of glucose in alloxan-induced diabetic rats (Rodríguez-Magaña et al. 2019). Quinoa contains high levels of lysine-rich proteins (Coulter and Lorenz 1990), polyunsaturated fatty acids, micronutrients, and vitamins C and E (Chauhan et al. 1992; Ranhotra et al. 1993; Gutzmán-Maldonado and Paredes-Lopez 1999). Bread is made from a blend of quinoa flour (5–10%) and wheat flour (Lorenz and Coulter 1991). In quinoa flours, different levels of prolamin fraction (0.8%), albumins (31%), and globulins (37%) have been reported (Fairbanks et al. 1990; Prakash and Pal 1998). Gliadin-like proteins were also reported in quinoa flours (Berti et al. 2004).

C. ambrosioides L. [syn. *Dysphania ambrosioides* (L.) Mosyakin & Clemants] is an annual herb, highly branched, and 1 m in height. The leaves are alternate, elongated, acute apex, glabrous, sessile, and have different sizes. It has short petioles. The plants have strong and specific flavors. The inflorescence is racemose type, with green and small flowers. The seeds are many, spherical, and black in color (Cruz 1995; Lorenzi and Matos 2002; Matos 2007; Lorenzi 2008). It is distributed in Mexico, Africa, Mediterranean to Central Europe, and Central and South America (Kliks 1985; Kismann 1991; Albuquerque et al. 2009). It is also naturalized in all the territories of Brazil (Sousa et al. 2004). It is used as a remedy for the removal of worms, increasing menstrual flow, and useful in abortions (Comway and Slocumb 1979). It possesses diuretic and anthelmintic properties. It is used in the treatment of wounds, respiratory complaints, inflammations, bronchitis, tuberculosis, and rheumatism (Kumar et al. 2007a). It is used as a poultice to treat snake bites, hemorrhoids, and wounds (Chevallier 1996).

The essential oils of *C. ambrosioides* leaves showed antifungal activity against *A. niger, A. fumigatus, B. theobromae, F. oxysporum, S. rolfsii, M. phaseolina, C. cladosporioides, H. oryzae* and *P. debaryanum* (MIC 100 µg/ml). The essential oils also displayed significant effectiveness in suppressing the aflatoxin B1 formation by *A. flavus* (Kumar et al. 2007a; Sá et al. 2016). The methanol extract of Mexican *C. ambrosioides* fruits showed inhibitory effects against *E. faecalis* (MIC 4375 µg/ml), *E. coli* (MIC 1094 µg/ml), and *S. typhimurium* (MIC 137 µg/ml). The fruit extract also displayed cytotoxic activity against Caco-2 cell lines (CC_{50} 45 µg/ml; Knauth et al. 2018). *C. ambrosioides* extract demonstrated significant antifungal activity against *C. albicans* (MIC and MFC 0.25 mg/ml). The minimum inhibitory concentration is sufficient to decrease the counts and activity of the biofilm cells ($p < 0.0001$), while fivefold minimum inhibitory concentration resulted in almost complete destruction of fungal infections (Zago et al. 2019).

5.2 PHYTOCHEMISTRY

Fatty acids (caprylic acid, pelargonic acid, lauric acid, myristic acid, 4,8,12-trimethyltridecanoic acid, pentadecanoic acid, tetradecanoic acid, 5,9,13-trimethyl-, palmitelaidic acid, palmitic acid, oleic acid, margaric acid, isooleic acid, stearic acid, arachidic acid, linoleic acid, behenic acid, tricosylic acid), terpenes (dihydroactinidiolide, and neophytadiene), triterpenes (β-sitosterol) from *n*-hexane fraction, and phenolic acids (benzoic acid, methylethylmaleimide, 2,4-di-tert-butylphenol, loliolide, chlorogenic acid, rutin, coniferyl alcohol, ferulic acid) have been identified as a dichloromethane fraction of *C. album* (Amodeo et al. 2019).

The genistein and daidzein have been estimated from *C. quinoa* seeds (Lutz et al. 2013). The quinoa flour contains improved levels of protein, total phenolic content, and antioxidant activity from 8.1%, 0.7 mg gallic acid equivalent/g, and 13.4%, up to 12.7%, 1.5 mg gallic acid equivalent/g, and 28.8%, respectively (Demir and Bilgiçli 2017). Similarly, quinoa seed estimation showed the presence of higher levels of ash (37%), protein (111%), total phenolic content (123%), and antioxidant activity (17%), but lower level of phytic acid (decreased by 77%) was reported (Demir and Bilgiçli 2021). Quinoa starch possesses significant physicochemical properties (such as viscosity, freeze stability); hence, it may be used as a novel functional food (James 2009).

The quinoa seeds consist of higher levels of D-xylose (120.0 mg in 100 g sample) and maltose (101.0 mg in 100 g sample), and lower levels of glucose (19.0 mg in 100 g sample) and fructose (19.6 mg in 100 g sample). The quinoa flour has high-water absorption capacity (147.0%) and low foaming ability and stability (9.0%, 2.0%; Ogungbenle 2003). Mean levels for isoleucine, leucine, lysine, tryptophan, valine, sulfur, and aromatic amino acids were greater in Washington-grown samples of quinoa. A total of 23 amino acids have been estimated from quinoa seeds (Craine and Murphy 2020).

The effects of dehulling, boiling, extrusion, heating under pressure, and baking on the content of amino acids and fatty acids and the release of glucose from quinoa seeds were evaluated. The retention rate of essential amino acids and fatty acids of dehulled and boiled quinoa was 100%. The profiles of essential oils of the extruded quinoa samples of two cultivars were 47.71% and 39.75%.

The dehulling, boiling, extrusion, heating under pressure, and baking processes influence essential amino acid content, composition of fatty acids, glucose release, and nutritional quality of quinoa seeds (Wu et al. 2020).

Quercetin-3-*O*-(2",6"-di-*O*-R-L-rhamnopyranosyl)-β-D-glucopyranoside, kaempferol-3-*O*-(2", 6"-di-*O*-R-L-rhamnopyranosyl)-β-D-glucopyranoside, quercetin-3-*O*-β-D-glucopyranosyl-(1'''→6")-β-D-glucopyranoside, rutin, quercetin-3-*O*-β-D-glucopyranoside, kaempferol-3-*O*-β-D-glucopyranoside were identified from methanol extract of *C. album*. Methanol extract demonstrated significant anti-oxidant activity (Chludil et al. 2008). A flavonoid [2-(3, 4-dihydroxyphenyl)-3, 5, 7- trihydroxy-4H-chromen-4-one] has been identified from acetone extract of *C. album* aerial parts. The findings showed that maximum yield of the flavonoid (7.335 mg/g) was reported from acetone extract (Arora and Itankar 2018).

Seven cinnamic acid amides (*N-trans*-feruloyl 4' -*O*-methyldopamine, *N-trans*-feruloyl 3'-*O*-methyldopamine, *N-trans*-feruloyltyramine, *N-trans*-4-*O*-methylferuloyl 3',4'-*O*-dimethyldopamine, *N-trans*-4-*O*-methylcaffeoyl 3'-*O*-methyldopamine, *N-trans*-4-*O*-methylferuloyl 4'-*O*-methyldopamine, and *N-trans*-feruloyl tryptamine) have been isolated from *C. album* (Cutillo et al. 2003). Quercetin-3-ramnoglucoside (rutin) and kaempferol-3-galactoside (trifolin) from *C. hircinum* and *C. album* leaves (Gonzalez et al. 1998), and a phenolic glycoside (chenoalbuside) has been identified from methanol extract of *C. album* seeds. The isolated compound showed antioxidant potential in DPPH assay (RC_{50} 1.4 × 10^{-4} mg/ml; Nahar and Sarker 2005).

Cinnamic acid amide (chenoalbicin) has been reported from ethyl acetate fraction of methanol extract of *C. album* roots (Cutillo et al. 2004). Similarly, 3β,14α-dihydroxy-5β-pregn-7-ene-2,6,20-trione has been identified from acetone fraction of water/methanol infusion of *C. album* leaves (DellaGreca et al. 2005). β-sitosterol, lupeol and 3 hydroxy nonadecyl henicosanoate were identified from petroleum ether extract of *C. album* leaves (Jhade et al. 2009). Two new compounds [(3*R*,6*R*,7*E*,9*E*,11*E*)-3-hydroxy-13-apo-α-caroten-13- one and (6*S*,7*E*,9*E*,11*E*)-3-oxo-13-apo-α-caroten-13-one] along with 16 apocarotenoids [(+)-abscisic alcohol, grasshopper ketone, zeaxantine, (3*R*,6*R*,7*E*)-3-hydroxy-4,7- megastigmadien-9-one, (6*R*,7*E*,9*R*)-9-hydroxy-4,7-megastigmadien-3-one, (3*R*,6*R*,7*E*,9*R*)-3,9-Dihydroxy-4,7-megastigmadiene, (6*R*,7*E*)-4,7-megastigmadien-3,9-dione, blumenol A, (+)-dehydrovomifoliol, *S*-(+)-3-hydroxy-β-ionone, 3,9-dihydroxy-4-megastigmene, 4-megastigmen-3,9-dione, 3,6,9-trihydroxy-4-megastigmene, (6*Z*,9*S*)-9-hydroxy-4,6-megastigmadien-3-one C-13 nor-terpenes, (6*R*,9*R*)-9-hydroxy-4-megastigmen-3-one] were isolated and characterized from *C. album* (DellaGreca et al. 2004). GC-MS analysis of hydrodistilled leaves of *C. album* showed the presence of *p*-cymene, ascaridole, pinane-2-ol, α-pinene, β-pinene, and α-terpineol. The essential oils showed strong anti-inflammatory activity against 12-*O*-tetradecanoylphorbol-13-acetate-induced ear edema in mice (Usman et al. 2010).

Four new compounds {3β-[(*O*-β-D-glucopyranosyl-(1→3)-α-L-arabinopyranosyl)oxy]-23-oxo-olean-12-en-28-oic acid β-D-glucopyranoside, 3β-[(*O*-β-D-glucopyranosyl-(1→3)-α-L-arabinopyranosyl)oxy]-27-oxo-olean-12-en-28-oic acid β-D-glucopyranoside, 3-*O*-α-L-arabinopyranosyl serjanic acid 28-*O*-β-D-glucopyranosyl ester, and 3-*O*-β-D-glucuronopyranosyl serjanic acid 28-*O*-β-D-glucopyranosyl ester} together with two bidesmosides of serjanic acid (3-*O*-β-D-glucopyranosyl-(1→2)-β-D-glucopyranosyl-(1→3)-α-L-arabinopyranosyl serjanic acid 28-*O*-β-D-glucopyranosyl ester, 3-*O*-β-D-glucopyranosyl-(1→3)-α-L-arabinopyranosyl serjanic acid 28-*O*-β-D-glucopyranosyl ester), four bidesmosides of oleanolic acid (quinoside D, quinoa saponin 7, 3-*O*-β-D-glucopyranosyl-(1→3)-α-L-arabinopyranosyl oleanolic acid 28-*O*-β-D-glucopyranosyl ester, chikusetsusaponin IVa), five bidesmosides of phytolaccagenic acid (quinoa saponin 4, quinoa saponin 3, quinoa saponin 5, quinoa saponin 8, 3-*O*-β-D-glucopyranosyl-(1→4)-*O*-β-Dglucopyranosyl-(1→4)-*O*-β-D-glucopyranosyl phytolaccagenic acid 28-*O*-β-D-glucopyranosyl ester), and four bidesmosides of hederagenin (quinoa saponin 2, *Hedera nepalensis* saponin F, quinoa saponin 9, quinoa saponin 1), and one bidesmoside of 3β,23,30-trihydroxy olean-12-en-28-oic acid) have been reported from the flowers, fruits, seed coats, and seeds of *C. quinoa*. 3β-[(*O*-β-D-glucopyranosyl-(1→3)-α-L-arabinopyranosyl)oxy]-23-oxo-olean-12-en-28-oic acid β-D-glucopyranoside, 3β-[(*O*-β-D-glucopyranosyl-(1→3)-α-L-arabinopyranosyl) oxy]-27-oxo-olean-12-en-28-oic acid β-D-glucopyranoside, 3-*O*-α-L-arabinopyranosyl serjanic acid 28-*O*-β-D-glucopyranosyl ester, and 3-*O*-β-D-glucuronopyranosyl serjanic acid 28-*O*-β-D-glucopyranosyl ester showed cytotoxicity against HeLa cells (Kuljanabhagavad et al. 2008).

Two saponins {3-*O*-(β-D-Glucopyranosyl)-oleanolic acid, and 3-*O*-[(β-D-xylopyranosyl) (1→3)]-β-D-glucuronopyranosyl-6-O -methyl ester]-oleanolic acid} from warm-aqueous extract of quinoa seeds from Mexico (Ma et al. 1989), 3-*O*-β-D-glucuropyranosyl oleanolic acid, 3-*O*-β-D-glucopyranosyl-(1→3)-α-L-arabinopyranosyl hederagenin, and 3-*O*-β-D-glucopyranosyl-(1→3)-α-L-arabinopyranosyl-3'-*O*-methyl spergulagenate 28-*O*-β-D-glucopyranosyl ester were identified from quinoa seeds (Zhu et al. 2002). The major saponins [3-*O*-β-D-glucuronopyranosyl oleanolic acid 28-*O*-β-D-glucopyranosyl ester, 3-*O*-α-L-arabinopyranosyl hederagenin 28-*O*-β-D-glucopyranosyl ester, 3-*O*-β-D-glucopyranosyl-(1→3)-α-L-arabinopyranosyl hederagenin 28-*O*-β-D-glucopyranosyl ester, 3-*O*-α-L-arabinopyranosyl phytolaccagenic acid 28-*O*-β-D-glucopyranosyl ester, 3-*O*-β-D-glucopyranosyl-(1→3)-α-L-arabinopyranosyl phytolaccagenic acid 28-*O*-β-D-glucopyranosyl ester, and 3-*O*-β-D-glucopyranosyl-(1→3)-α-L-arabinopyranosyl phytolaccagenic acid] were isolated and characterized from *C. quinoa* seeds. The isolated compounds showed anti-fungal activity against *C. albicans* (Woldemichael and Wink 2001).

1-*O*-galloyl-β-D-glucoside, acacetin, protocatechuic acid 4-*O*-glucoside, penstebioside, ethyl-m-digallate, (epi)-gallocatechin, and canthoside C have been tentatively identified from *C. quinoa* flour (Gómez-Caravaca et al. 2011). The oleanolic acid, hederagenin, phytolaccagenic acid, and serjanic acid were reported from *C. quinoa* (Madl et al. 2006). Oleanolic acid, chikusetsusaponin Iva, quinoside D, quinoa-saponin-1, quinoa-saponin-2, quinoa-saponin-3, quinoa-saponin-4, quinoa-saponin-5, quinoa-saponin-6, 30-*O*-methyl spergulagenate, prosapogenin, quinoa-saponin-7, quinoa-saponin-8, chikusetsusaponin Iva, quinoa-saponin-9, quinoside D, quinoa-saponin-10, 3-*O*-[β-glucopyranosyl(1→2)-β-glucopyranosyl(1→3)-α-arabinopyranoside]-28-*O*-β-glucopyranoside of 3'-*O*-methyl spergulagenate, oleanolic acid and phytolaccagenic acid, hederagenin 3-*O*-β-glucuronide-28-*O*-β-glucopyranoside, and hederagenin 3-*O*-β-xylopyranosyl(1→3)-β-glucuronide-28-*O*-β-glucopyranoside were identified from *C. quinoa* seeds (Mizui et al. 1990).

Five saponins [28-*O*-β-glucopyranosyl esters of hederagenin 3-*O*-β-glucopyranosyl-(1→3)-α-arabinopyranoside, 3-*O*-β-glucopyranosyl-(1→3)-β-galactopyranoside, 28-*O*-β-glucopyranosyl esters of phytolaccagenic acid 3-*O*-α-arabinopyranoside, 3-*O*-β-glucopyranosyl-(1→3)-α-arabinopyranoside, and 3-O-β-glucopyranosyl-(1→3)-β-galactopyranoside were separated and identified from *C. quinoa* seeds (Mizui et al. 1988). Six triterpenoid saponins {phytolaccagenic acid 3-*O*-[α-L-arabinopyranosyl-(1''→3')-β-D-glucuronopyranosyl]-28-*O*-β-D-glucopyranoside, spergulagenic acid 3-*O*-[β-D-glucopyranosyl-(1→2)-β-D-glucopyranosyl-(1→3)-α-L-arabinopyranosyl-28-*O*-β-D-glucopyranoside, hederagenin 3-*O*-[β-D-glucopyranosyl-(1→3)-α-L-arabinopyranosyl]-28-*O*-β-D-glucopyranoside, phytolaccagenic acid 3-*O*-[β-D-glucopyranosyl-(1→4)-β-D-glucopyranosyl-(1→4)-β-D-glucopyranosyl]-28-*O*-β-D-glucopyranoside, hederagenin 3-*O*-[β-D-glucopyranosyl-(1→4)-β-D-glucopyranosyl-(1→4)-β-D-gluco-pyranosyl]-28-*O*-β-D-glucopyranoside, and spergulagenic acid 3-*O*-[α-L-arabinopyranosyl-(1''→3')-β-D-glucuronopyranosyl]-28-*O*-β-D-glucopyranoside were obtained from seeds of *C. quinoa* (Dini et al. 2001b).

A quinoside A [(olean-12-ene-28-oic acid, 3,23-bis(*O*-β-D-glucopyranosyloxy)-*O*-β-D-glucopyranosyl-(1 leads to 3)-*O*-α-L-arabinopyranosyl ester (3 β,4 α)], saponin A (β-D-glucopyranosyl-[β-D-glucopyranosyl-(1→3)-α-l-arabino-pyranosyl-(1→3)]-3-β-23-dihydroxy-12-en-28-oate methyl ester) and saponin B (β-D-glucopyranosyl-[β-d-glucopyranosyl-(1 → 3)-α-l-arabino-pyranosyl-(1→3)]-3-β-23-dihydroxyolcan-12-en-28-oate have been reported from quinoa seeds (Ruales and Nair 1993). Six triterpenoid saponins (phytolaccagenic acid 3-*O*-[α-L-arabinopyranosyl-(1''→3')-β-D-glucuronopyranosyl]-28-*O*-β-D-glucopyranoside, phytolacca-genic acid 3-*O*-[β-D-glucopyranosyl-(1''→3')-α-L-arabinopyranosyl]-28-*O*-β-D-glucopyranoside, phytolaccagenic acid 3-*O*-[β-D-glucopyranosyl-(1'''→3'')-β-D-xylopyranosyl-(1''→2')-β-D-glucopyranosyl]-28-*O*-β-D-glucopyranoside, phytolaccagenic acid 3-*O*-[β-D-glucopyranosyl-(1'''→2'')-β-d-glucopyranosyl-(1''→3')-α-L-arabinopyranosyl]-28-*O*-β-D-glucopyranoside, oleanolic acid 3-O-[α-L-arabinopyranosyl-(1″→3′)-β-D-glucuronopyranosyl]-28-O-β-D-glucopy-ranoside, and oleanolic acid 3-O-[β-D-glucopyranosyl-(1″→3′)-α-l-arabinopyranosyl]-28-O-β-D-glucopyranoside were isolated from the edible seeds of *C. quinoa* (Dini et al. 2001a).

The phytolaccagenic, oleanolic, and serjanic acids, hederagenin, 3β,23,30 trihydroxy olean-12-en-28-oic acid, 3β-hydroxy-27-oxo-olean-12en-28-oic acid, and 3β,23,30 trihydroxy olean-12-en-28-oic acid were obtained from quinoa seeds. These metabolites demonstrate molluscicidal, antifungal, anti-inflammatory, hemolytic, and cytotoxic activities (Hazzam et al. 2020). Four flavonol glycosides [quercetin 3-O-(2″,6″-di-O-α-rhamnopyranosyl)-β-galactopyranoside, and kaempferol 3-O-(2″,6″-di-O-α-rhamnopyranosyl)-β-galactopyranoside, quercetin 3-O-(2″,6″-di-O-α-rhamnopyranosyl)-β-glucopyranoside, and quercetin 3-O-(2″-O-β-apiofuranosyl-6″-O-α-rhamnopyranosyl)-β-galactopyranoside] were determined in Japanese quinoa seeds. The Japanese quinoa seeds are considered as the most effective functional foodstuff—in terms of being a source of antioxidative and bioactive flavonoids—among cereals and pseudo-cereals (Hirose et al. 2010).

The presence of betacyanins (betanin and isobetanin) was confirmed from the red and black quinoa seeds. Dark quinoa seeds showed higher levels of phenolic compounds and antioxidant ability (Tang et al. 2015). Oleanolic acid, hederagenin, serjanic acid, and phytolaccagenic acid were identified in 22 quinoa varieties and 6 original breeding lines grown in North America (Medina-Meza et al. 2016). The bioactive compounds—total phenolic (1.23–3.24 mg gallic acid equivalents/g) and flavonol contents (0.47–2.55 mg quercetin equivalents/g)—were highly matched ($r = 0.910$) to *C. quinoa*. Betalains content (0.15–6.10 mg/100 g) was matched with L color parameter ($r = -0.569$), total phenolics ($r = 0.703$) and flavonols content ($r = 0.718$). The betaxanthins to betacyanins ratios (0.0–1.41) were negatively matched with L value ($r = -0.744$). Although, high total antioxidant capacity values (119.8–335.9 mmol trolox equivalents/kg) were negatively matched with L value ($r = -0.779$), but positively with betalains ($r = 0.730$), as well as with free ($r = 0.639$), bound ($r = 0.558$), and total phenolic compounds ($r = 0.676$). Therefore, colored quinoa seeds may be considered as a valuable source of phenolics and betalains with significant antioxidant property (Abderrahim et al. 2015).

Kaempferol 3-O-α-L-^{1}C$_4$-rhamnosyl-(1‴→2″)-β-D-^{4}C$_1$-xylopyranoside, afzelin, kaempferol 7-O-α-L-^{1}C$_4$-rhamnopyranoside, caffeic acid, 1,2-benzopyrone, and kaempferol were obtained from the *n*-butanol fraction of *C. ambrosioides* leaves. Kaempferol 3-O-α-L-^{1}C$_4$-rhamnosyl-(1‴→2″)-β-D-^{4}C$_1$-xylopyranoside showed *in vitro* antioxidant activity (SC$_{50}$ 12.45 μg/ml) when compared to ascorbic acid (SC$_{50}$ 7.50 μg/ml; Ghareeb et al. 2016). Two flavone glycosides [scutellarein-7-O-α-rhamnopyranosyl-(1→2)-α-rhamnopyranosyl-(1→2)-α-rhamnopyranoside and 5,6,7,4′-tetrahydroxyflavone-7-O-α-rhamnopyranosyl-(1→2)-α-rhamnopyranosyl-(1→2)-αrhamnopyranoside] were isolated and identified from the ethyl acetate extract of *C. ambrosioides* (Hammoda et al. 2015). Seven compounds {4-hydroxy-4(α or β)-isopropyl-2-methyl-2-cyclohexen-1-one, 1-methyl-4β- isopropyl-1-cyclohexene-4α,5α,6α-triol, (1S,2S,3R,4S)-1-methyl-4-(propan-2-yl)cyclohexane-1,2,3,4-tetrol, (1R,2S,3S,4S)- 1,2,3,4-tetrahydroxy-p-menthane, (1R,2S)-3-p-menthen-1,2-diol, (1R,4S)-*p*-menth-2-en-1-ol and 1,4-dihydroxy-p-menth-2-ene} were reported from the ethyl alcohol extract (95%) of *C. ambrosioides* stems. The isolated compound {4-hydroxy-4(α or β)-isopropyl-2-methyl-2-cyclohexen-1-one} showed moderate capacity to suppress nitric oxide formation of lipopolysaccahride-induced RAW 264.7 macrophages (IC$_{50}$ 16.83 μM; Hou et al. 2017).

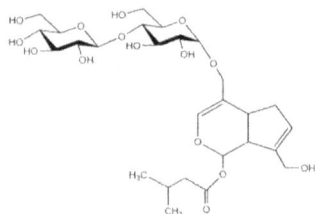

Penstebioside   Canthoside C

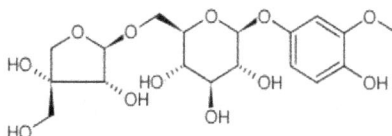

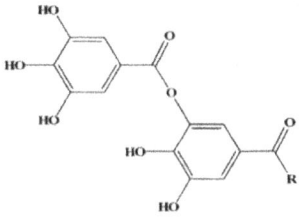

R = OC$_2$H$_5$ - Ethyl-*m*-digallate

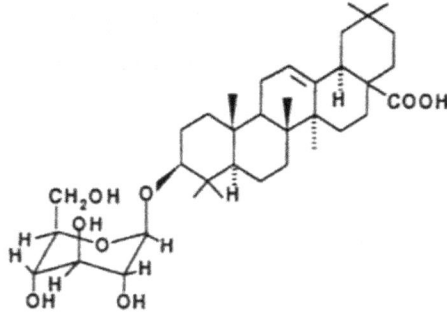

3-*O*-(β-D-Glucopyranosyl)-oleanolic acid

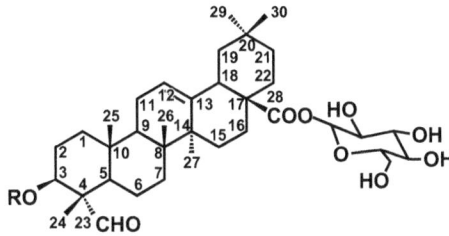

R = β-D-Glc(1→3)-α-L-ara - 3β-[(O-β-D-glucopyranosyl-(1→3)-α-L-arabinopyranosyl)oxy]-23-oxo-olean-12-en-28-oic acid β-D-glucopyranoside

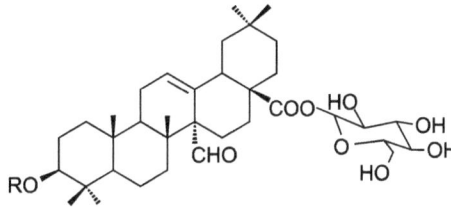

R = β-D-Glc(1→3)-α-L-ara - 3β-[(O-β-D-glucopyranosyl-(1→3)-α-L-arabinopyranosyl)oxy]-27-oxo-olean-12-en-28-oic acid β-D-glucopyranoside

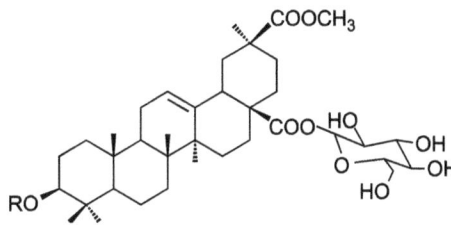

R = α-L-Ara - 3-*O*-α-L-Arabinopyranosyl serjanic acid 28-*O*-β-D-glucopyranosyl ester; R = β-D-GlcA - 3-*O*-β-D-Glucuronopyranosyl serjanic acid 28-*O*-β-D-glucopyranosyl ester

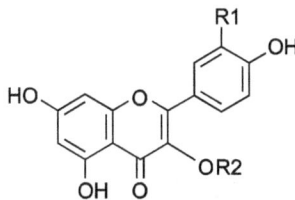

R^1 = OH, R^2 = 2,6-Di-*O*-α-rhamnopyranosyl-β-galactopyranosyl - Quercetin 3-*O*-(2″,6″-di-*O*-α-rhamnopyranosyl)-β-galactopyranosides; R^1 = H, R^2 = 2,6-Di-*O*-α-rhamnopyranosyl-β-galactopyranosyl- Kaempferol3-*O*-(2″,6″-di-*O*-α-rhamnopyranosyl)-β-galactopyranosides; R^1=OH, R^2

= 2,6-Di-*O*-α-rhamnopyranosyl-β-glucopyranosyl - Quercetin 3-*O*-(2″,6″-di-*O*-α-rhamnopyranosyl)-β-glucopyranoside; R¹ = OH, R² = 2-*O*-β-Apiofuranosyl-6-*O*-α-rhamnopyranosyl -β-galactopyranosyl - Quercetin 3-*O*-(2″-O-β-apiofuranosyl-6″-*O*-α-rhamnopyranosyl)-β-galactopyranoside

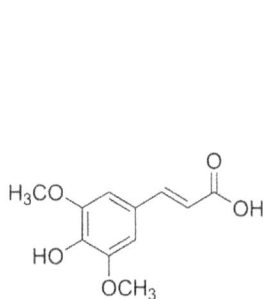

Sinapinic acid

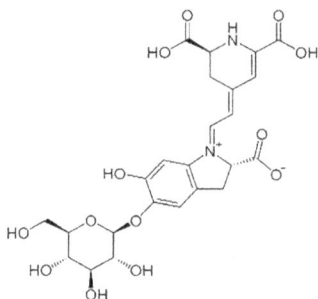

Betanin

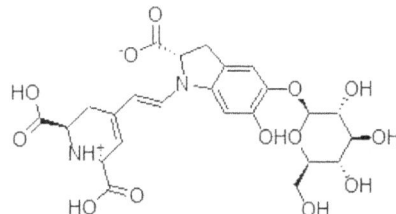

Isobetanin

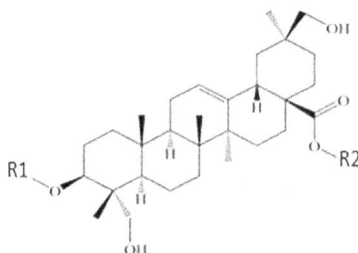

R¹ = Glc(1-3)ara, R² = Glc - 3,23,30-trihydroxyolean-12-en-28-oic acid 3-*O*-β-D-Glucopyranosyl-(1,3)-α-L-arabinopyranosyl-28 -*O*-β-D-glucopyranoside

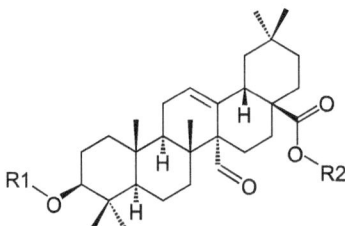

R¹ = Glc(1-3)ara, R² = Glc - 3β-*O*-β-D-glucopyranosyl-(1,3)-α- L -arabinopyranosyl-oxy-27-oxo-olean-12-en-28-oic acid β-D-glucopyranoside

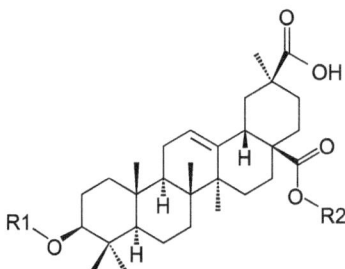

R¹ = Glc(1-2)glc(1-3)ara, R² = H - Spergulagenic acid 3-*O*-β-D-glucopyranosyl-(1,2)-β-D-glucopyranosyl-(1,3) -α-L-arabinopyranoside; R¹ = Glc(1-2)glc(1-3)ara, R² = Glc - Spergulagenic acid 3-*O*-β-D-glucopyranosyl-(1,2)-β-D-glucopyranosyl-(1,3) -α-L-arabinopyranosyl-28-*O*-β-D-glucopyranoside; R¹ = Ara(1-3)glcUA, R² = Glc - Spergulagenic acid 3-*O*-α-L-arabinopyranosyl-(1,3)-β-D-glucuronopyranosyl -28-*O*-β-D-glucopyranoside

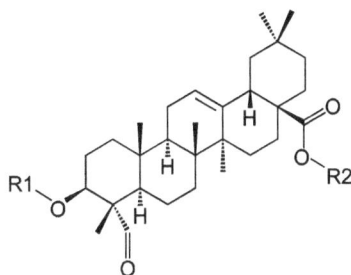

R^1 = Glc(1-3)ara, R^2 = Glc - 3β-*O*-β-D-glucopyranosyl-(1,3)- α -L-arabinopyranosyl-oxy-23-oxo-olean-12-en-28-oic acid β-D-glucopyranoside

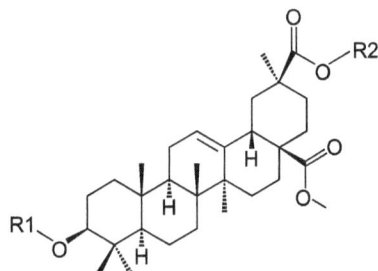

R^1 = Glc(1-3)ara, R^2 = Glc - Serjanic acid 3-*O*-[β-D-glucopyranosyl-(1,3)- α -L-arabinopyranosyl]-28-*O*-β-D-glucopyranoside = 30-*O*-methyl spergulagenate 3-*O*-β-D-glucopyranosyl-(1,3)-α-L-arabinopyranosyl-28 -*O*-β-D-glucopyranoside; R^1 = Glc(1-2)glc(1-3)ara, R^2 = Glc - Serjanic acid 3-*O*-β-D-glucopyranosyl-(1,2)-β-D-glucopyranosyl-(1,3) -α-L-arabinopyranosyl-28-*O*-β-D-glucopyranoside=30-*O*-methyl spergulagenate 3-*O*-β-D-glucopyranosyl-(1,2)-β-D-glucopyranosyl-(1,3) -α-L-arabinopyranosyl-28-*O*-β-D-glucopyranoside; R^1 = Ara, R^2 = Glc - Serjanic acid 3-*O*-α-L-arabinopyranosyl-28-*O*-β-D-glucopyranoside; R^1 = GlcUA, R^2 = Glc - Serjanic acid 3-*O*-β-D-glucuronopyranosyl-28-*O*-β-D-glucopyranoside; R^1 = Ara(1-3)glc, R^2 = Glc - Serjanic acid 3-*O*-α-L-arabinopyranosyl-(1,3)-β-D-glucuronopyranosyl -28-*O*-β-D-glucopyranosyl ester

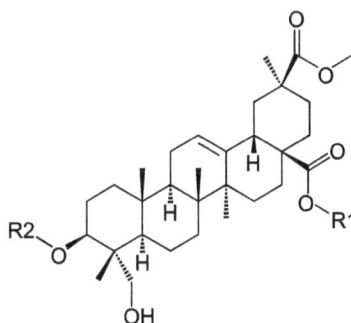

R^1 = -Glc(1-3)ara, R^2 = -H - Phytolaccagenic acid 3-*O*-β-D-glucopyranosyl (1,3)-α-L-arabinopyranoside; R^1 = Ara(1-3)glcUA, R^2 = -Glc - Phytolaccagenic acid 3-*O*- [α -L-arabinopyranosyl-(1,3)-β-D-glucuronopyranosyl]-28 -*O*-β-D-glucopyranoside; R^1 = Glc(1-3)ara, R^2 = -Glc - Phytolaccagenic acid 3-*O*-[β-D-glucopyranosyl-(1,3)-α-L-arabinopyranosyl]-28 -*O*-β-D-glucopyranoside; R^1 = Glc(1-2) xyl(1-3)glc, R^2 = -Glc - Phytolaccagenic acid 3-*O*-[β-D-glucopyranosyl-(1,3)-β-D-xylopyranosyl-(1,2)-β-D-glucopyranosyl]-28-*O*-β-D-glucopyranoside; R^1 = Ara, R^2 = -Glc - Phytolaccagenic acid 3-*O*-α -L-arabinopyranosyl 28-*O*-β-D-glucopyranoside; R^1 = Gal(1-3)glc, R^2 = -Glc - Phytolaccagenic acid 3-*O*-β-D-galactopyranosyl-(1,3)-β-D-glucopyranosyl 28-*O*-β-D-glucopyranoside; R^1 = Glc(1-3)gal, R^2 = - Glc - Phytolaccagenic acid 3-*O*-β-D-Glucopyranosyl-(1,3)-β-D-galactopyranosyl 28-*O*-β-Dglucopyranoside; R^1 = Glc(1-2)glc(1-3)ara, R^2 = -Glc - Phytolaccagenic acid 3-*O*-β-Dglucopyranosyl-(1,2)-β-D-glucopyranosyl-(1,3) -α-L-arabinopyranosyl 28-*O*-β-Dglucopyranoside; R^1 = Glc(1-4)

glc(1-4)glc, R^2 = -Glc - Phytolaccagenic acid 3-*O*-β-Dglucopyranosyl-(1,4)-β-D-glucopyranosyl-(1,4)-β-D-glucopyranosyl 28-*O*-β-D-glucopyranoside

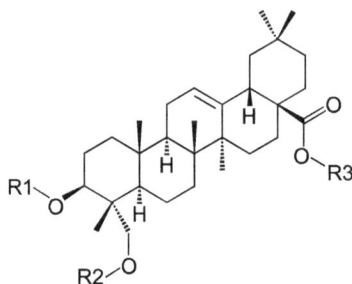

R^1 = Glc(1-3)ara, R^2 = -H, R^3 = H - Hederagenin 3-*O*-β-D-glucopyranosyl-(1,3)- α -Larabinopyranoside; R^1 = Ara, R^2 = -H, R^3 = H - Hederagenin 3-*O*-α-L-arabinopyranoside; R^1 = Glc(1-3)ara, R^2 = -Glc, R^3 = H - Hederagenin 3-*O*-[β-D-glucopyranosyl-(1,3)-α-Larabinopyranosyl]- 28 -*O*-β-D-glucopyranoside; R^1 = Ara, R^2 = -Glc, R^3 = H - Hederagenin 3-*O*- α-L-arabinopyranosyl-28-*O*-β-D-glucopyranoside; R^1 = Glc(1-3)gal, R^2 = -Glc, R^3 = H - Hederagenin 3-*O*-β-D-Glucopyranosyl-(1,3)-α-L-galactopyranosyl 28-*O*-β-D-glucopyranoside; R^1 = GlcUA, R^2 = -Glc, R^3 = H - Hederagenin 3-*O*- β- D-Glucuronopyranosyl 28-*O*-β-Dglucopyranoside; R^1 = Xyl(1-3)glcUA, R^2 = -Glc, R^3 = H - Hederagenin 3-*O*-β-D-Xylopyranosyl-(1,3)-β-D-glucuronopyranosyl -28-*O*-β-D-glucopyranoside; R^1 = Glc(1-4)glc(1-4)glc, R^2 = -Glc, R^3 = H - Hederagenin 3-*O*-β-D-G-lucopyranosyl-(1,4)-β-D-glucopyranosyl-(1,4) -β-D-glucopyranosyl-28-*O*-β-D-glucopyranoside; R^1 = Glc, R^2 = -Glc-glc, R^3 = Glc - 3,23-bis(*O*-β-D-Glucopyranosyloxy) olean-12-en28-oicacid28-*O*-α-L-arabinopyanosyl-(1,3) -β-D-glucopyranoside

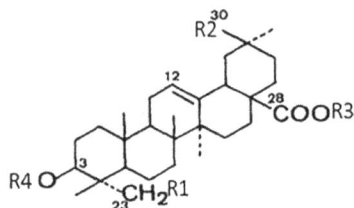

Ara = α-L-Arabinopyranosyl; GlcA = β-D-glucuronopyranosyl; Glc = β-D-Glucopyranosyl; Gal = β-D-Galactopyranosyl; Xyl = β-D-Xylopyranosyl

R^1 = -OH, R^2 = - CH_3, R^3 = Glc, R^4 = Ara^3 - Glc - Quinoa-saponin-1; R^1 = -OH, R^2 = - CH_3, R^3 = Glc, R^4 = Gal^3 - Glc - Quinoa-saponin-2; R^1 = -OH, R^2 = - $COOCH_3$, R^3 = Glc, R^4 = Ara - Quinoa-saponin-3; R^1 = -OH, R^2 = - $COOCH_3$, R^3 = Glc, R^4 = Ara^3 - Glc - Quinoa-saponin- 4; R^1 = -OH, R^2 = - $COOCH_3$, R^3 = Glc, R^4 = Gal^3 - Glc Quinoa-saponin-5; R^1 = -H, R^2 = - $COOCH_3$, R^3 = Glc, R^4 = Ara^3 -Glc^3-Glc - Quinoa-saponin-6; R^1 = -H, R^2 = - $COOCH_3$, R^3 = H, R^4 = - H – 30-*O*-Methyl spergulagenate; R^1 = -H, R^2 = - COOH, R^3 = H, R^4 = Ara^3 - Glc^2 - Glc - Prosapogenin; R^1 = -H, R^2 = -CH_3, R^3 = Glc, R^4 = Ara^3 - Glc^2 - Glc - Quinoa-saponin-7; R^1 = - OH, R^2 = - $COOCH_3$, R^3 = -Glc, R^4 = Ara^3 - Glc^2 - Glc - Quinoa-saponin-8; R^1 = -H, R^2 = -CH_3, R^3 = -Glc, R^4 = GlcA – Chikusetsusaponin IVa; R^1 = -OH, R^2 = -CH_3, R^3 = -Glc, R^4 = -GlcA – Quinoa-saponin-9; R^1 = -H, R^2 = -CH_3, R^3 = -Glc, R^4 = $GlcA^3$ - Xyl - Quinoside D; R^1 = -OH, R^2 = -CH_3, R^3 = -Glc, R^4 = $GlcA^3$ - Xyl- Quinoa-saponin-10

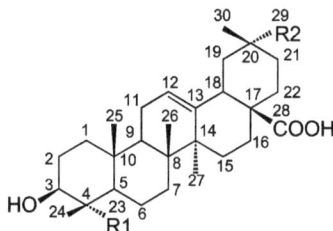

R^1 = CH_3, R2 = CH_3 – Oleanolic acid; R^1 = CH_2OH, R^2 = CH_3 – Hederagenin; R^1 = CH_2OH, R^2 = $COOCH_3$ – Phytolaccagenic acid; R^1 = CH_3, R^2 = $COOCH_3$ – Serjanic acid

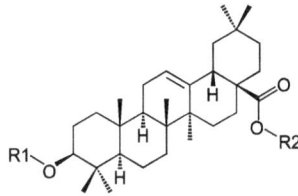

R^1 = Glc, R^2 = H - Oleanolic acid 3-*O*-β -D-glucopyranoside; R^1 = GlcUA, R^2 = H - Oleanolic acid 3-*O*-β-D-glucuropyranoside; R^1 = -Xyl(1-3)GlcUA, R^2 = H - Oleanolic acid 3-*O*-β-D-xylopyranosyl-(1,3)-β-D-glucuronopyranoside; R^1 = -Xyl(1-3)-6-*O*-CH$_3$-GlcUA, R^2 = H - Oleanolic acid 3-*O*-β -D-xylopyranosyl(l,3)-β-D-glucuronopyranosyl-methyl-ester; R^1 = -Glc(1-3)ara, R^2 = -Glc - Oleanolic acid 3-*O*- [β -D- glucopyranosyl-(1,3)- α-L-arabinopyranosyl]-28-*O*-β-D-glucopyranoside; R^1 = -ara(1-3)GlcUA, R^2 = -Glc - Oleanolic acid 3-*O*-α-L-arabinopyranosyl-(1,3)-β-D-glucuronopyranosyl-28-*O*-β-D -glucopyranoside; R^1 = -GlcUA, R^2 = -Glc - Oleanolic acid 3-*O*-[β-D-glucuronopyranosyl]-28-*O*-β-D-glucopyranoside; R^1 = -Xyl(1-3)GlcUA, R^2 = -Glc - Oleanolic acid 3-*O*-β-D-xylopyranosyl-(1,3)-β-D-glucuronopyranosyl-28 -*O*-β-D-glucopyranoside; R^1 = -Glc(1-2)Glc(1-3)ara, R^2 = -Glc - Oleanolic acid 3-*O*-β-D-glucopyranosyl-(1,2)-β-D-glucopyranosyl-(1,3) -α-L-arabinopyranosyl 28-*O*-β-D-glucopyranoside

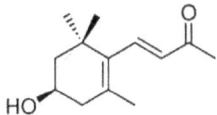

S-(+)-3-Hydroxy-β-ionone

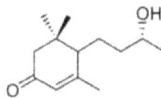

(6*R*,9*R*)-9-Hydroxy-4-megastigmen-3-one

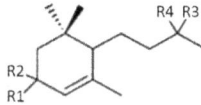

R^1 = H, R^2 = OH, R^3 = H, R^4 = OH - 3,9-Dihydroxy-4-megastigmene; R^1 = R^2 = O, R^3 = R^4 = O - 4-Megastigmen-3,9-dione

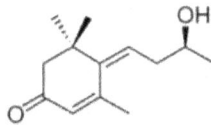

(6*Z*,9*S*)-9-Hydroxy-4,6-megastigmadien-3-one C-13 nor-terpenes

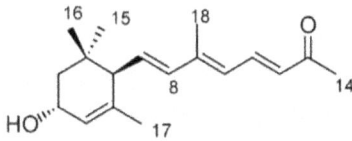

(3*R*,6*R*,7*E*,9*E*,11*E*)-3-Hydroxy-13-apo-α-caroten-13- one

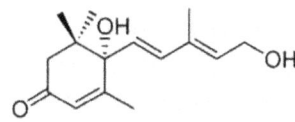

(+)-Abscisic alcohol

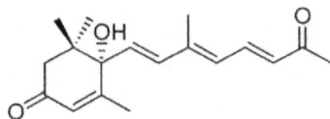

(6*S*,7*E*,9*E*,11*E*)-3-Oxo-13-apo-α-caroten-13-one

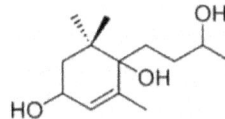

3,6,9-Trihydroxy-4-megastigmene

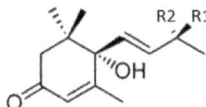

R^1 = H, R^2 = OH - Blumenol A; R^1 = R^2 = O - (+)-Dehydrovomifoliol

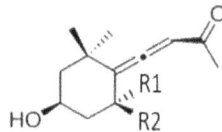

R^1 = OH, R^2 = CH$_3$ - Grasshopper ketone; R^1 = CH$_3$, R^2 = OH - Zeaxantine

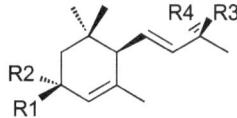

R^1 = H, R^2 = OH, R^3 = R^4 = O - (3R,6R,7E)-3-Hydroxy-4,7- megastigmadien-9-one; R^1 = R^2 = O, R^3 = H, R^4 = OH - (6R,7E,9R)-9-Hydroxy-4,7-megastigmadien-3-one; R^1 = H, R^2 = OH, R^3 = H, R^4 = OH - (3R,6R,7E,9R)-3,9-Dihydroxy-4,7-megastigmadiene; R^1 = R^2 = O, R^3 = R^4 = O - (6R,7E)-4,7-Megastigmadien-3,9-dione

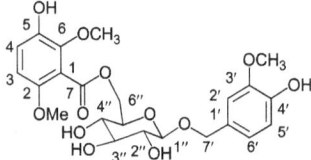

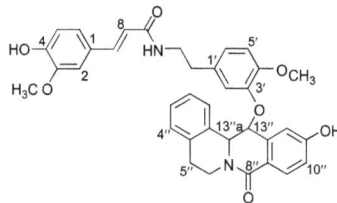

Chenoalbuside   Chenoalbicin   N-trans-Feruloyl tryptamine

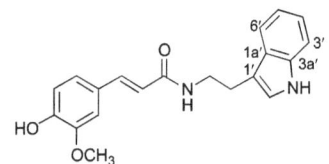

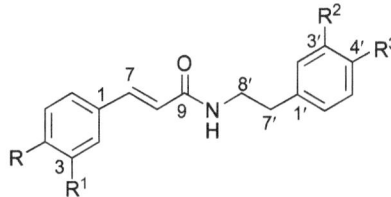

R = R^2 = OH, R^1 = R^3 = OCH$_3$ - N-trans-Feruloyl 4' -O-methyldopamine; R = R^3 = OH, R^1 = R^2 = OCH$_3$ - N-trans-Feruloyl 3' -O-methyldopamine; R = R^3 = OH, R^1 = OCH$_3$, R^2 = H - N-trans-Feruloyl tyramine; R = R^1 = R^2 = R^3 = OCH$_3$ - N-trans-4-O-Methylferuloyl 3',4'-O-dimethyldopamine; R = R^2 = OCH$_3$, R^1 = R^3 = OH - N-trans-4-O-Methylcaffeoyl 3' -O-methyldopamine; R = R^1 = R^3 = OCH$_3$, R^2 = OH - N-trans-4-O-Methylferuloyl 4' -O-methyldopamine.

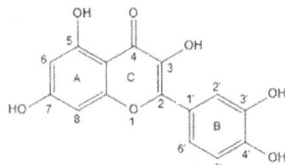

2-(3, 4-Dihydroxyphenyl)-3, 5, 7- trihydroxy-4H-chromen-4-one   Trifolin

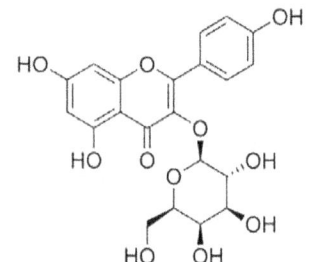

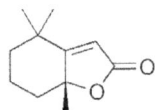

Dihydroactinidiolide   Neophytadiene   Coniferyl alcohol

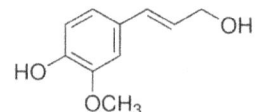

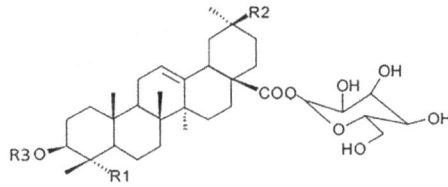

$R^1 = R^2 = R^3 =$ Hexose-Hexose-Pentose - AG487; R1 = CH_2OH, $R^2 = COOCH_3$, $R^3 =$ Hexose-Hexose-Pentose - Phytolaccagenic acid; $R^1 = CH_3$, $R^2 = COOCH_3$, $R^3 =$ Hexose-Hexose-Pentose - Serjanic acid; $R^1 = CH_2OH$, $R^2 = CH_3$, $R^3 =$ Pentose - Hederagenin

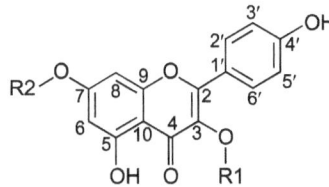

$R^1 =$ Rha-Xyl, $R^2 = H$ - Kaempferol 3-O-α-L-$^1C_4$-rhamnosyl-(1‴→2″)-β-D-$^4C_1$-xylopyranoside; $R^1 = H$, $R^2 =$ Rha - Kaempferol 7-O-α-L-1C_4 –rhamnopyranoside

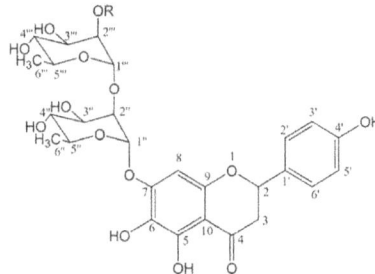

R = Rhamnopyranose - Scutellarein-7-O-α-rhamnopyranosyl-(1→2)-α-rhamnopyranosyl-(1→2)-α-rhamnopyranoside: 5,6,7,4′-tetrahydroxyflavone-7-O-α-rhamnopyranosyl-(1→2)-α-rhamnopyranosyl-(1→2)-αrhamnopyranoside; R=H - Scutellarein-7-O-α-rhamnopyranosyl-(1→2)-α-rhamnopyranoside: 5,6,7,4′-tetrahydroxyflavone -7-O-α- rhamnopyranosyl-(1→2)-α-rhamnopyranoside

5.3 CULTURE CONDITIONS

The hypocotyl explants of *C. quinoa* were cultured on an MS (Murashige and Skoog 1962) medium containing 2,4-D (0.45 μM). The maximum growth callus was achieved from hypocotyl explants within 12–14 days of culture (Eisa et al. 2005). Similarly, shoot tips of seedlings and adult plants were inoculated on modified B5 medium. The medium was modified with reducing levels of sucrose (10 g/l), enhancing levels of nitrate (2,700 mg/l) and phosphate (315 mg/l) salts, and glycine (4 mg/l). The callus was achieved with supplementation of benzyladenine (0.22 mg/l) and naphthaleneacetic acid (0.018 mg/l; Burnouf-Radosevich and Paupardin 1985). The callus induction of hypocotyl of *C. quinoa* was obtained in different concentrations of plant growth regulators. Maximum callus growth (93.33%) was recorded on an MS medium containing 2,4-dichlorophenoxyacetic acid (0.5 mg/l) + 6-benzymaminopurine (0.05 mg/l; Hesami and Daneshvar 2016).

The petioles of *in vitro* grown *C. rubrum* were infected with *A. rhizogenes* (strain A4M70GUS) for hairy roots development. The maximum growth of hairy roots was achieved on plant growth hormone-free solid or liquid half-strength of MS medium. Fresh weight of hairy roots was increased 30- to 90-fold higher in liquid medium, than solid medium (Dmitrović et al. 2010). Similarly, the roots, cotyledons, leaves, and internodes of *C. murale* seedlings were used for the development of transgenic hairy roots (*A. rhizogenes* A4M70GUS). The levels of caffeic, ferulic, and *p*-coumaric acids (0.07–2.85 mol/l) were determined by HPLC analysis in hairy roots of quinoa (Mitić et al. 2012).

REFERENCES

Abderrahim F, Huanatico E, Segura R, Arribas S, Gonzalez MC, Condezo-Hoyos L. 2015. Physical features, phenolic compounds, betalains and total antioxidant capacity of coloured quinoa seeds (*Chenopodium quinoa* Willd.) from Peruvian Altiplano. Food Chem 183, 83–90.

Agrawal MY, Agrawal YP, Shamkuwar PB. 2014. Phytochemical and biological activities of *Chenopodium album*. Int J Pharmtech Res 6, 383–391.

Albuquerque UP, Araújo TAS, Ramos MA, Nascimento VT, Lucena RFP, Monteiro JM, Alencar NL, Araújo EL. 2009. How ethnobotany can aid biodiversity conservation: Reflections on investigations in the semi-arid region of NE Brazil. Biodivers Conserv 18, 127–150.

Amodeo V, Marrelli M, Pontieri V, Cassano R, Trombino S, Conforti F, Statti G. 2019. *Chenopodium album* L. and *Sisymbrium officinale* (L.) Scop.: Phytochemical content and in vitro antioxidant and anti-inflammatory potential. Plants 8, 505.

Angeli V, Silva PM, Massuela DC, Khan MW, Hamar A, Khajehei F, Graeff-Hönninger S, Piatti C. 2020. Quinoa (*Chenopodium quinoa* Willd.): An overview of the potentials of the "golden grain" and socio-economic and environmental aspects of its cultivation and marketization. Foods 9, 216.

Arora S, Itankar P. 2018. Extraction, isolation and identification of flavonoid from *Chenopodium album* aerial parts. J Tradit Complement Med 8, 476–482.

Arora SK, Itankar PR, Verma PR, Bharne AP, Kokare DM. 2014. Involvement of NFκB in the antirheumatic potential of *Chenopodium album* L., aerial parts extracts. J Ethnopharmacol 155, 222–229.

Bailey LH. 1977. Manual of Cultivated Plants. Macmillan Publishing, New York.

Bajwa AA, Zulfiqar U, Sadia S, Bhowmik P, Chauhan BS. 2019. A global perspective on the biology, impact and management of *Chenopodium album* and *Chenopodium murale*: Two troublesome agricultural and environmental weeds. Environ Sci Pollut Res Int 26, 5357–5371.

Bakshi DNG, Sen Sarma P, Pal DC. 1999. A Lexicon of Medicinal Plants in India. Naya Prakash Publisher, Calcutta.

Bazile D, Bertero D, Nieto C. 2014. Estado del arte de la quinua en el mundo en 2013. In: Organización de las Naciones Unidas para la Alimentación y la Agricultura, Santiago, Chile, y Centre de Coopération Internationale en Recherche Agronomique pour le Développement, Montpellier.

Berti C, Ballabio C, Restani P, Porrini M, Bonomi F, Iametti S. 2004. Immunochemical and molecular properties of proteins in *Chenopodium quinoa*. Cereal Chem J 81, 275–277.

Bhargava A, Shukla S, Rajan S, Ohri D. 2007. Genetic diversity for morphological and quality traits in quinoa (*Chenopodium quinoa* Willd.) germplasm. Genet Resour Crop Evol 54, 167–173.

Bhatia M, Singh S, Pagare S, Kumar B. 2020. A pharmacological comprehensive review on *Chenopodium album* L. Int J Curr Res 12, 15360–15368.

Bhattacharjee SK. 2001. Handbook of Medicinal Plant. Pointer Publishers, Jaipur.

Bianco VV, Santamaria P, Elia A. 1996. Nutritional value and nitrate content in edible wild species used in southern Italy. Acta Hortic 467, 71–90.

Brito IL, Leite de Souza E, Felex SSS, Madruga MS, Yamashita F, Magnani M. 2015. Nutritional and sensory characteristics of gluten-free quinoa (*Chenopodium quinoa* Willd)-based cookies development using an experimental mixture design. J Food Sci Technol 52, 5866–5873.

Buitrago D, Buitrago-Villanueva I, Barbosa-Cornelio R, Coy-Barrera E. 2019. Comparative examination of antioxidant capacity and fingerprinting of unfractionated extracts from different plant parts of quinoa (*Chenopodium quinoa*) grown under greenhouse conditions. Antioxidants 8, 238.

Burnouf-Radosevich M, Paupardin C. 1985. Vegetative propagation of *Chenopodium quinoa* by shoot tip culture. Am J Bot 72, 278–283.

Chamkhi I, Charfi S, Hachlafi NE, Mechchate H, Guaouguaou F-E, Omari NE, Bakrim S, Balahbib A, Zengin G, Bouyahya A. 2022. Genetic diversity, antimicrobial, nutritional, and phytochemical properties of *Chenopodium album*: A comprehensive review. Food Res Int 154, 110979.

Chauhan GS, Eskin NAM, Tkachuk R. 1992. Nutrients and antinutrients in quinoa seed. Cereal Chem 69, 85–88.

Chevallier A. 1996. The Encyclopedia of Medicinal Plants. Dorling Kindersley, London.

Chludil HD, Corbino GB, Leicach SR. 2008. Soil quality effects *Chenopodium album* flavonoid content and antioxidant potential. J Agric Food Chem 56, 5050–5056.

Comway GA, Slocumb JC. 1979. Plants used as abortifacients and emmenagogues by Spanish New Mexicans. J Ethnopharmacol 1, 241–261.

Cooper R. 2015. Re-discovering ancient wheat varieties as functional foods. J Tradit Complement Med 5, 138–143.

Coulter L, Lorenz K. 1990. Quinoa – composition, nutritional value, food applications. Food Sci Technol 23, 203–207.

Craine EB, Murphy KM. 2020. Seed composition and amino acid profiles for quinoa grown in Washington State. Front Nutr 7, 126.

Cruz GL. 1995. Dicionário das plantas úteis do Brasil, 5th ed. Editora Bertrand Brasil, Rio de Janeiro.

Cutillo F, D'Abrosca B, DellaGreca M, Marino CD, Golino A, Previtera L, Zarrelli A. 2003. Cinnamic acid amides from *Chenopodium album*: Effects on seeds germination and plant growth. Phytochemistry 64, 1381–1387.

Cutillo F, D'Abrosca B, Greca MD, Zarrelli A. 2004. Chenoalbicin, a novel cinnamic acid amide alkaloid from *Chenopodium album*. Chem Biodivers 1, 1579–1583.

Dai Y, Ye WC, Wang ZT, Matsuda H, Kubo M, But PPH. 2002. Antipruritic and antinociceptive effects of *Chenopodium album* L. in mice. J Ethnopharmacol 81, 245–250.

Das A, Borthakur MK. 2022. *Chenopodium album* Linn.: A review on various pharmacological activities. Uttar Pradesh J Zool 43, 9–14.

Del Castillo C, Mahy G, Winkel T. 2008. La quinoa en Bolivie: Une culture ancestrale devenue culture de rente bioéquitable. Biotechnol Agron Soc Environ 12, 445–454.

DellaGreca M, D'Abrosca B, Fiorentino A, Previtera L, Zarrelli A. 2005. Structure elucidation and phytotoxicity of ecdysteroids from *Chenopodium album*. Chem Biodivers 2, 457–462.

DellaGreca M, Di Marino C, Zarrelli A, D'Abrosca B. 2004. Isolation and phytotoxicity of apocarotenoids from *Chenopodium album*. J Nat Prod 67, 1492–1495.

Demir B, Bilgiçli N. 2017. Utilization of quinoa flour in cookie production. Int Food Res J 24, 2394–2401.

Demir B, Bilgiçli N. 2021. Utilization of quinoa flour (*Chenopodium quinoa* Willd.) in gluten-free pasta formulation: Effects on nutritional and sensory properties. Food Sci Technol Int 27, 242–250. https://doi.org/10.1177/1082013220940092.

Dini I, Schettino O, Simioli T, Dini A. 2001a. Studies on the constituents of *Chenopodium quinoa* seeds: Isolation and characterization of new triterpene saponins. J Agric Food Chem 49, 741–746.

Dini I, Tenore GC, Schettino O, Dini A. 2001b. New oleanane saponins in *Chenopodium quinoa*. J Agric Food Chem 49, 3976–3981.

Dmitrović S, Mitić N, Zdravković-Korać S, Vinterhalter B, Ninković S, Ćulafić LJ. 2010. Hairy roots formation in recalcitrant-to-transform plant *Chenopodium rubrum*. Biol Plant 54, 566–570.

Eisa S, Koyro HW, Kogel KH, Imani J. 2005. Induction of somatic embryogenesis in cultured cells of *Chenopodium quinoa*. Plant Cell Tissue Organ Cult 81, 243–246.

Fairbanks DJ, Burgener KW, Robinson LR, Andersen WR, Ballon E. 1990. Electrophoretic characterization of quinoa seed proteins. Plant Breed 104, 190–195.

Fischer S, Wilckens R, Jara J, Aranda M. 2013. Variation in antioxidant capacity of quinoa (*Chenopodium quinoa* Will) subjected to drought stress. Ind Crops Prod 46, 341–349.

García-Parra M, Zurita-Silva A, Stechauner-Rohringer R, Roa-Acosta D, Jacobsen S-E. 2020. Quinoa (*Chenopodium quinoa* Willd.) and its relationship with agroclimatic characteristics: A Colombian perspective. Chil J Agric Res 80, 290–302.

Gawlik-Dziki U, Świeca M, Sułkowski M, Dziki D, Baraniak B, Czyż J. 2013. Antioxidant and anticancer activities of *Chenopodium quinoa* leaves extracts—*In vitro* study. Food Chem Toxicol 57, 154–160.

Ghareeb MA, Saad AM, Abdou AM, Refahy LA-G, Ahmed WS. 2016. A new kaempferol glycoside with antioxidant activity from *Chenopodium ambrosioides* growing in Egypt. Orient J Chem 32, 3053–3061.

Gómez-Caravaca AM, Segura-Carretero A, Fernández-Gutiérrez A, Caboni MF. 2011. Simultaneous determination of phenolic compounds and saponins in quinoa (*Chenopodium quinoa* Willd) by a liquid chromatography—diode array detection—electrospray ionization—time-of-flight mass spectrometry methodology. J Agric Food Chem 59, 10815–10825.

Gonzalez JA, Gallardo M, de Israilev LA. 1998. Leaf flavonoids in *Chenopodium hircinum* Schrad and *Chenopodium album* L. (Chenopodiaceae). Phyton 63, 279–281.

Gutzmán-Maldonado SH, Paredes-Lopez O. 1999. Functional products of plants indigenous to Latin America: Amaranth, quinoa, common beans, and botanicals. In: Functional Foods: Biochemical and Processing Aspects, Mazza G (ed.). Technomic Publishing, Lancaster, PA, pp 293–328.

Hammoda HM, Harraz FM, El Ghazouly MG, Radwan MM, ElSohly MA, Wanas AS, Bassam SM. 2015. Two new flavone glycosides from *Chenopodium ambrosioides* growing wildly in Egypt. Rec Nat Prod 9, 609–613.

Hazzam KE, Hafsa J, Sobeh M, Mhada M, Taourirte M, Kacimi KE, Yasri A. 2020. An insight into saponins from quinoa (*Chenopodium quinoa* Willd): A review. Molecules 25, 1059.

Hesami M, Daneshvar MH. 2016. Development of a regeneration protocol through indirect organogenesis in *Chenopodium quinoa* Willd. Indo Am J Agric Vet Sci 4, 25–32.

Hirose Y, Fujita T, Ishii T, Ueno N. 2010. Antioxidative properties and flavonoid composition of *Chenopodium quinoa* seeds cultivated in Japan. Food Chem 119, 1300–1306.

Hou S-Q, Li Y-H, Huang X-Z, Li R, Lu H, Tian K, Ruan R-S, Li S-K. 2017. Polyol monoterpenes isolated from *Chenopodium ambrosioides*. Nat Prod Res 31, 2467–2472.

Hussain S, Asrar M, Rasul A, Sultana S, Saleem U. 2022. *Chenopodium album* extract ameliorates carbon tetrachloride induced hepatotoxicity in rat model. Saudi J Biol Sci 29, 3408–3413.

James LEA. 2009. Quinoa (*Chenopodium quinoa* Willd.): Composition, chemistry, nutritional, and functional properties. Adv Food Nutr Res 58, 1–31.

Jancurová M, Minarovicova L, Dandar A. 2009. Quinoa—a review. Czech J Food Sci 27, 71–79.

Jhade D, Paarakh PM, Gavani U. 2009. Isolation of phytoconstituents from the leaves of *Chenopodium album* Linn. J Pharm Res 2,1192–1193.

Kasali FM, Tusiimire J, Kadima JN, Agaba AG. 2021. Ethnomedical uses, chemical constituents, and evidence-based pharmacological properties of *Chenopodium ambrosioides* L.: Extensive overview. Future J Pharm Sci 7, 153.

Khurana SC, Malik YS, Pandita ML. 1986. Herbicidal control of weeds in potato C.V. kufribadshah. Pesticides 20, 55–56.

Kilinc OK, Ozgen S, Selamoglu Z. 2016. Bioactivity of triterpene saponins from quinoa (*Chenopodium quinoa* Willd.). Res Rev Res J Biol 4, 25–28.

Kirtikar KR, Basu BD. 1964–65. Indian Medicinal Plants, Vol III. International Book Distributor, Dehradun.

Kismann KG. 1991. Plantas Infestantes e Nocivas. BASF Brasileira, São Paulo.

Kliks MM. 1985. Studies on the traditional herbal anthelmintic *Chenopodium ambrosioides* L.: Ethnopharmacological evaluation and clinical field trials. Soc Sci Med 21, 879–886.

Knauth P, Acevedo-Hernández GJ, Cano ME, Gutiérrez-Lomelí M, López Z. 2018. In vitro bioactivity of methanolic extracts from *Amphipterygium adstringens* (Schltdl.) Schiede ex Standl., *Chenopodium ambrosioides* L., *Cirsium mexicanum* DC., *Eryngium carlinae* F. Delaroche, and *Pithecellobium dulce* (Roxb.) Benth. used in traditional medicine in Mexico. Evid Based Complement Alternat Med 2018, Article ID 3610364.

Kokanova-Nedialkova Z, Nedialkov P, Nikolov S. 2009. The genus *Chenopodium*: Phytochemistry, ethnopharmacology and pharmacology. Pharmacogn Rev 3, 280–306.

Kuljanabhagavad T, Thongphasuk P, Chamulitrat W, Wink M. 2008. Triterpene saponins from *Chenopodium quinoa* Willd. Phytochemistry 69, 1919–1926.

Kumar R, Kumar MA, Dubey NK, Tripathi YB. 2007a. Evaluation of *Chenopodium ambroisioides* oil as a potential source of antifungal, anti-aflatoxigenic and antioxidant activity. Int J Food Microbiol 115, 159–164.

Kumar S, Biswas S, Mandal D, Roy HN, Chakraborty S, Kabir SN, Banerjee S, Mondal NB. 2007b. *Chenopodium album* seed extract: A potent sperm immobilizing agent both *in vitro* and *in vivo*. Contraception 75, 71–78.

Letelier ME, Rodríguez-Rojas C, Sánchez-Jofré S, Aracena-Parks P. 2011. Surfactant and antioxidant properties of an extract from *Chenopodium quinoa* Willd seed coats. J Cereal Sci 53, 239–243.

Lorenz K, Coulter L. 1991. Quinoa flour in baked products. Plant Foods Human Nutr 41, 213–223.

Lorenzi H. 2008. Plantas Daninhas do Brasil: Terrestres, Aquáticas, Parasitas e Tóxicas, 4th ed. Editora Instituto Plantarum, Nova Odessa.

Lorenzi H, Matos FJA. 2002. Plantas Medicinais no Brasil: Nativas e Exóticas. Editora Instituto Plantarum, Nova Odessa.

Lutz M, Martínez A, Martínez EA. 2013. Daidzein and Genistein contents in seeds of quinoa (*Chenopodium quinoa* Willd.) from local ecotypes grown in arid Chile. Ind Crops Prod 49, 117–121.

Ma WW, Heinstein PF, McLaughlin JL. 1989. Additional toxic, bitter saponins from the seeds of *Chenopodium quinoa*. J Nat Prod 52, 1132–1135.

Madl T, Sterk H, Mittelbach M, Rechberger GN. 2006. Tandem mass spectrometric analysis of a complex triterpene saponin mixture of *Chenopodium quinoa*. J Am Soc Mass Spectrom 17, 795–806.

Maliro MFA, Guwela VF, Nyaika J, Murphy KM. 2017. Preliminary studies of the performance of quinoa (*Chenopodium quinoa* Willd.) genotypes under irrigated and rainfed conditions of Central Malawi. Front Plant Sci 8, 227.

Matos FJA. 2007. Plantas Medicinais: Guia de selec̨ão e Emprego das Plantas Usadas em Fitoterapia no Nordeste do Brasil, 3rd ed. Imprensa Universitária, Fortaleza.

Medina-Meza IG, Aluwi NA, Saunders SR, Ganjyal GM. 2016. GC-MS Profiling of triterpenoid saponins from 28 quinoa varieties (*Chenopodium quinoa* Willd.) grown in Washington state. J Agric Food Chem 64, 45, 8583–8591.

Mitić N, Dmitrović S, Djordjević M, Zdravkovic-Korać S, Nikolić R, Raspor M, Djordjević T, Maksimović V, Zivković S, Krstic-Milošević D, Stanišić M, Ninkovic S. 2012. Use of *Chenopodium murale* L. transgenic hairy root *in vitro* culture system as a new tool for allelopathic assays. J Plant Physiol 169, 1203–1211.

Mizui F, Kasai R, Ohtani K, Tanaka O. 1988. Saponins from brans of quinoa, *Chenopodium quinoa* Willd. I. Chem Pharm Bull 36, 145–1418.

Mizui F, Kasai R, Ohtani K, Tanaka O. 1990. Saponins from Bran of quinoa, *Chenopodium quinoa* Willd. II. Chem Pharm Bull 38, 375–377.

Murashige T, Skoog F. 1962. A revised medium for rapid growth and bioassays with tobacco tissue cultures. Physiol Planta 15, 473–497.

Murphy K, Matanguihan J. 2015. Quinoa Improvement and Sustainable Production. John Wiley and Sons, Hoboken.

Nahar L, Sarker SD. 2005. Chenoalbuside: An antioxidant phenolic glycoside from the seeds of *Chenopodium album* L. (Chenopodiaceae). Braz J Pharmacogn 15, 279–282.

Navruz-Varli S, Sanlier N. 2016. Nutritional and health benefits of quinoa (*Chenopodium quinoa* Willd.). J Cereal Sci 69, 371–376.

Nelly A, Annick DD, Frederic D. 2008. Plants used as remedies antirheumatic and antineuralgic in the traditional medicine of Lebanon. J Ethnopharmacol 120, 315–334.

Nishimura E, Suzaki E, Irie M, Nagashima H, Hirose T. 2010. Architecture and growth of an annual plant *Chenopodium album* in different light climates. Ecol Res 25, 383–393.

Nongonierma AB, Maux SL, Dubrulle C, Barre C, FitzGerald RJ. 2015. Quinoa (*Chenopodium quinoa* Willd.) protein hydrolysates with *in vitro* dipeptidyl peptidase IV (DPP-IV) inhibitory and antioxidant properties. J Cereal Sci 65, 112–118.

Nowak V, Du J, Charrondière UR. 2016. Assessment of the nutritional composition of quinoa (*Chenopodium quinoa* Willd.). Food Chem 193, 47–54.

Ogungbenle HN. 2003. Nutritional evaluation and functional properties of quinoa (*Chenopodium quinoa*) flour. Int J Food Sci Nutr 54, 153–158.

Pal A, Banerjee B, Banerjee T, Masih M, Pal K. 2011. Hepatoprotective activity of *Chenopodium album* Linn. plant against paracetamolinduced hepatic injury in rats. Int J Pharm Pharm Sci 3, 3.

Pérez SG, Zavala MS, Arias LG, Ramos ML. 2011. Anti-inflammatory activity of some essential oils. J Essent Oil Res 2011, 23, 38–44.

Poonia A, Upadhayay A. 2015. *Chenopodium album* Linn: Review of nutritive value and biological properties. J Food Sci Technol 52, 3977–3985.

Prajapati ND, Purohit SS, Sharma AK, Kumar TA. 2003. Handbook of Medicinal Plants: A Complete Source Book. Agrobios, Jodhpur.

Prakash D, Pal M. 1998. Chenopodium: Seed protein, fractionation and amino acid composition. Int J Food Sci Nutr 49, 271–275.

Ranhotra GS, Gelroth JA, Glaser BK, Lorenz KJ, Johnson DL. 1993. Composition and protein nutritional quality of quinoa. Cereal Chem 70, 303–305.

Repo-Carrasco R, Espinoza C, Jacobsen S-E. 2003. Nutritional value and use of the Andean crops quinoa (*Chenopodium quinoa*) and Kañiwa (*Chenopodium pallidicaule*). Food Rev Int 19, 179–189.

Risi J, Galwey NW. 1984. The *Chenopodium* grains of the Andes: Inca crops for modern agriculture. In: Advances in Applied Biololgy, Coaker TH (ed.). Academic, London, pp 145–216.

Rodríguez-Magaña MP, Cordero-Pérez P, Rivas-Morales C, Oranday-Cárdenas MA, Moreno-Peña DP, García-Hernández DG, Leos-Rivas C. 2019. Hypoglycemic activity of *Tilia americana, Borago officinalis*, *Chenopodium nuttalliae*, and *Piper sanctum* on Wistar rats. J Diabetes Res 2019, Article ID 7836820.

Ruales J, Nair BM. 1992. Nutritional quality of the protein in quinoa (*Chenopodium quinoa*, Willd) seeds. Plant Foods Hum Nutr 42, 1–11.

Ruales J, Nair, BM. 1993. Saponins, phytic acid, tannins, and protease inhibitors in quinoa (*Chenopodium quinoa* Willd) seeds. Food Chem 48, 137–143.

Sá RD, Santana ASCO, Silva FCL, Soares LAL, Randau KP. 2016. Anatomical and histochemical analysis of *Dysphania ambrosioides* supported by light and electron microscopy. Braz J Pharmacogn 26, 533–543.

Sarma H, Sarma A, Sarma CM. 2008. Traditional knowledge of weeds: A study of herbal medicines and vegetables used by the Assamese people (India). Herba Pol 54, 80–88.

Schlick G, Bubenheim DL. 1996. Quinoa: Candidate crop for NASA's controlled ecological life support systems. In: Progress in New Crops, Janick J (ed.). ASHS Press, Arlington, pp 632–640.

Sikarwar I, Dey YN, Wanjari MM, Sharma A, Gaidhani SN, Jadhav AD. 2017. *Chenopodium album* Linn. leaves prevent ethylene glycol-induced urolithiasis in rats. J Ethnopharmacol 195, 275–282.

Şimşek Sezer EN, Uysal T. 2021. Phenolic screening and biological activities of *Chenopodium botrys* L. extracts. Anatol J Bot 5, 78–83.

Sousa MP, Matos MEO, Matos FJA, Machado MIL, Craveiro AA. 2004. Constituintes químicos ativos e propriedades biológicas de plantas medicinais brasileiras. Editora UFC, Fortaleza.

Tang Y, Li X, Zhang B, Chen PX, Liu R, Tsao R. 2015. Characterisation of phenolics, betanins and antioxidant activities in seeds of three *Chenopodium quinoa* Willd. genotypes. Food Chem 166, 380–388.

Usman LA, Hamid AA, Muhammad NO, Olawore NO, Edewor TI, Saliu BK. 2010. Chemical constituents and anti-inflammatory activity of leaf essential oil of Nigerian grown *Chenopodium album* L. EXCLI J 9, 181–186.

Vega-Gálvez A, Miranda M, Vergara J, Uribe E, Puente L. 2010. Nutrition facts and functional potential of quinoa (*Chenopodium quinoa* Willd) and ancient Andean grain: A review. J Sci Food Agric 90, 2541–2547.

Verza SG, Silveira F, Cibulski S, Kaiser S, Ferreira F, Gosmann G, Roehe PM, Ortega GG. 2012. Immunoadjuvant activity, toxicity assays, and determination by UPLC/Q-TOF-MS of triterpenic saponins from *Chenopodium quinoa* seeds. J Agric Food Chem 60, 3113–3118.

Wanjari MM, Bajpai V, Dey YN, Gaidhani SN, Babu G. 2016. *Chenopodium album* Linn. leaves prevent carbon tetrachloride-induced liver fibrosis in rats. ASIO J Exp Pharmacol Clin Res 1, 16–22.

Wilson HD. 1990. Quinoa and relatives (*Chenopodium* sect. *Chenopodium* subsect. cellulata). Econ Bot 44(Suppl. 3), 92–110.

Woldemichael GM, Wink M. 2001. Identification and biological activities of triterpenoid saponins from *Chenopodium quinoa*. J Agric Food Chem 49, 2327–2332.

Wu L, Wang A, Shen R, Qu L. 2020. Effect of processing on the contents of amino acids and fatty acids, and glucose release from the starch of quinoa. Food Sci Nutr 8, 4877–4887.

Yadav N, Vasudeva N, Singh HS, Sharma SK. 2007. Medicinal properties of genus *Chenopodium* Linn. Nat Prod Rad 6, 131–134.

Zago PMM, Branco SJDSC, Fecury LAB, Carvalho LT, Rocha CQ, Madeira PLB, de Sousa EM, de Siqueira FSF, Paschoal MAB, Diniz RS, Gonçalves LM. 2019. Anti-biofilm action of *Chenopodium ambrosioides* extract, cytotoxic potential and effects on acrylic denture surface. Front Microbiol 10, 1724.

Zhao T, Pan H, Feng Y, Li H, Zhao Y. 2016. Petroleum ether extract of *Chenopodium album* L. prevents cell growth and induces apoptosis of human lung cancer cells. Exp Ther Med 12, 3301–3307.

Zhu N, Sheng S, Sang S, Jhoo J-W, Bai N, Karwe MV, Rosen RT, Ho C-T. 2002. Triterpene saponins from debittered quinoa (*Chenopodium quinoa*) seeds. J Agric Food Chem 50, 865–867.

6 *Chimaphila* Species

6.1 MORPHOLOGICAL FEATURES, DISTRIBUTION, ETHNOPHARMACOLOGICAL PROPERTIES, PHYTOCHEMISTRY, AND PHARMACOLOGICAL ACTIVITIES

Chimaphila umbellata (L.) W. Bart (Fam.—Pyrolaceae) is an evergreen, perennial plant, and grows up to 10–35 cm height. It has bright green, toothed margin leaves, arranged in opposite pairs or whorls of 3–4 along the stem. The flowers are white or pink, produced in a small umbel of 4–8 together. It is distributed in Canada, Korea, northeast China, Japan, Russia, and Europe. In Canadian medicine, it is used in the treatment of infections and inflammations, kidney stones, gonorrhea, stomachache, backache, and coughs. It has also been used as a blood purifier and astringent agent (Arnason et al. 1981; Chevalier 1996; Marles et al. 2000). *C. umbellata* flowers possesses diuretic property and used as an expectorant and stimulant (Alok et al. 2013). In traditional medicine, *C. umbellata* leaves and stems are used as an antibiotic, diuretic, and astringent (Oka et al. 2007). The white or pink flowers and thick waxy leaves of *C. maculata* have been used in the treatment of urinary tract infections as well as in removal of kidney stones (Boughman and Oxendine 2004). The ethanol extract of *C. umbellata* demonstrates great potential in preventing breast cancer MCF-7 cell lines by increasing caspase-independent necroptosis through participating in RIP1/RIP3 kinases and MLKL proteins (Das et al. 2022).

The ethanol extract of *C. umbellata* showed significant antibacterial [*S. aureus* (ATCC 29213), *E. coli* (ATCC 25922), *P. aeruginosa* (ATCC 27853), *M. phlei* (ATCC 11758), and *S. lactis* (ATCC 19435)] and antifungal activities [*C. albicans* (ATCC 10231) and *S. octosporus*]. Similarly, six compounds (gallic acid, ethyl gallate, caffeic acid, sinapic acid, gentisic acid, and chlorogenic acid) have also been isolated and characterized from ethanol extract. The isolated compounds showed MIC ranges from 62.5 to 1000 mg/ml against experimental bacterial strains (Vandal et al. 2015; Boyko et al. 2014). Water extracts of leaves, stems, and roots of *C. maculata* were tested for their antimicrobial activity against *S. aureus, B. subtilis, P. mirabilis, P. vulgaris, C. xerosis* and *E. facecalis*. The extracts showed 28% to 80% inhibition against tested bacterial strains (McLean 2017).

The water-alcohol extract of the aerial part of *C. umbellata* was tested for their immunocorrecting activity in an azathioprine immunosuppression model (F1 line laboratory mice (C57Bl/6 × CBA). During a 14-day treatment, the extract restored the immune status (by the delayed-type hypersensitivity reaction) to 85.6% (in intact control 100%) and 69.9% in the positive control (azathioprine). The extract affected the number of antibody-forming cells by enhancing their absolute (2.36) and the relative values (1.84 times in positive control-azathioprine). The immunocorrecting potential of *C. umbellata* extract was found consistent in this study (Kusheev et al. 2020).

6.2 PHYTOCHEMISTRY

Chimaphilin, isohomoarbutin, β-amyrin, ursolic acid, flavonoids, and phenolic glycosides have been reported from *C. umbellata* (Veitch and Welton 1951; Zellnig et al. 1996; Raja et al. 2014). A new naphthalene glycoside (2,7-dimethyl-1,4-dihydroxynaphthalene-1-*O*-β-D-glucopyranoside) has been isolated and characterized from *C. umbellata* leaves and stems. The isolated compound significantly suppressed the receptor activator of nuclear factor-κB ligand-induced tartrate-resistant acid phosphatase activity and the formation of multinucleated osteoclasts in a

DOI: 10.1201/9781003398035-6

dose-dependent manner. Moreover, the isolated compound suppressed the receptor activator of nuclear factor-κB ligand-induced mRNA expression of osteoclast-associated genes that encode tartrate-resistant acid phosphatase, cathepsin K, and another transcription factor-nuclear factor of activated T-cells c1 so, the compound may be used for therapeutic uses (Shin et al. 2015). Chimaphilin (2,7- dimethyl-1,4-naphthoquinone) was identified from ethanol extract of *C. umbellata*. The isolated compound showed antifungal activity against *Saccharomyces cerevisiae* (0.05 mg/ml), *Malassezia globosa* (0.39 mg/ml) and *M. restricta* (0.55 mg/ml). The ethanol extract displayed significant antioxidant activity in DPPH (2,2-diphenyl-1-picrylhydrazyl) assay (Galván et al. 2008). Chimaphilin, 3-chloro-chimaphilin, and 3-hydroxy-chimaphilin were isolated and characterized from *C. umbellata*, and isolated compounds showed significant effect on deficiency of mitochondrial complex I. Therefore, these isolated compounds may be useful in complex I disease biology (Vafai et al. 2016). The different compounds viz., chimaphilin (Duke 1992), toluquinol, renifolin (Pedersen 2002; Gruenwald 2000), arbutin, isohomoarbutin, methyl salicylate and salicylic acid methyl ester, epicatechin gallate (Galván et al. 2008), quercetin, hyperoside, kaempferol, avicularin, and lignan (Pedersen 2002) have been reported from *C. umbellata* (Walewska and Thieme 1969; Trubachev and Batyuk 1968, 1969).

8-chloro-2,7-dimethyl-l,4-naphthoquinone (8-chlorochimaphilin), together with chimaphilin and 3-hydroxychimaphilin, 2,7- Dimethyl-l,3-dihydroxynaphthyl 4-*O*-α-L-rhamnopyranoside, and 2,7-dimethoxy-1,4,8-trihydroxynaphthalene were also isolated and characterized from *Moneses uniflora* (syn. *C. rhombifolia* Hayata). 8-chloro-2,7-dimethyl-l,4-naphthoquinone (8-chlorochimaphilin), chimaphilin, and 3-hydroxychimaphilin showed antimicrobial activity against selected microbes (Saxena et al. 1996). A 1,4-naphthoquinone derivative (5,8-dihydro-3-hydroxychimaphilin) and five other compounds (chimaphilin, umbelliferone, isofraxetin, 3-hydroxychimaphilin, and 4-hydroxy-2,7-dimethylnaphthylene-1-*O*-β-D-glucopyranoside) were reported from *M. uniflora* (syn. *C. rhombifolia* Hayata). 5,8-dihydro-3-hydroxychimaphilin demonstrated potent antimycobacterial activity (*Mycobacterium tuberculosis* H37Ra; IC_{50} 5.4 μM) and moderate cytotoxicity (mammalian HEK 293 cell line; IC_{50} 30 μM). Similarly, 3-hydroxychimaphilin exhibited moderate effect in both assays (IC_{50} 44 and 55 μM, respectively; Li et al. 2018). A total of 23 compounds, including triterpenoids, flavonoids, sterols, quinonoids, saccharide derivative, phenolic glycoside, and megastigmane glycoside were also identified from *C. japonica* (Yu et al. 2021).

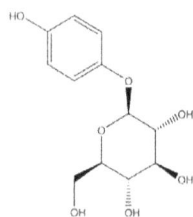

Arbutin

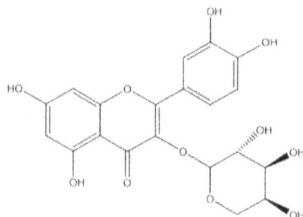

Avicularin

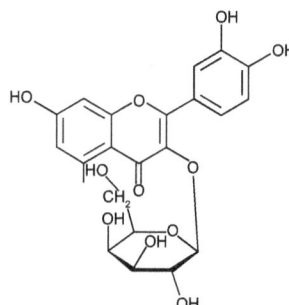

Hyperoside

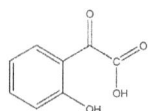

Methylsalicylate

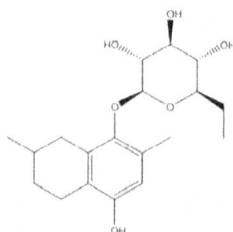

Renifolin

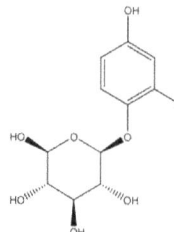

Isohomoarbutin

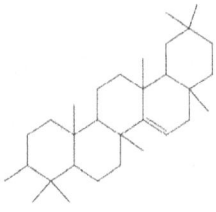

Taraxerol

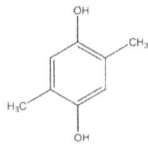

Toluquinol

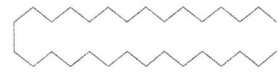

Triacontane

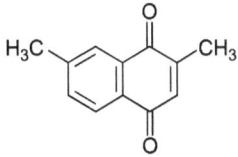

Chimaphilin

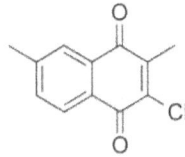

3-Chloro-chimaphilin

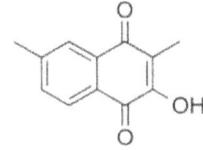

3-Hydroxy-chimaphilin

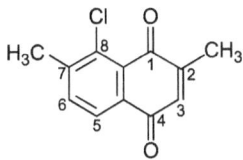

8-Chlorochimaphilin

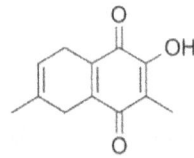

5,8-Dihydro-3-hydroxychimaphilin

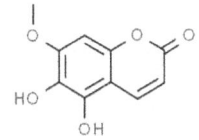

Isofraxetin

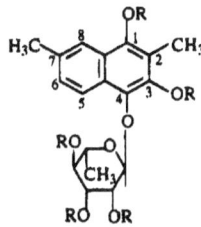

R = H - 1,3-Dihydroxy-2,7-dimethylnaphthyl 4-*O*-α-L-rhamnopyranoside

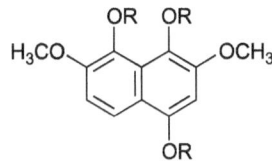

R = H - 2,7-Dimethoxy-1,4,8-trihydroxynaphthalene

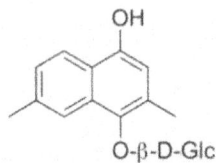

4-Hydroxy-2,7-dimethylnaphthylene-1-*O*-β-D-glucopyranoside

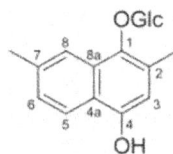

Glc = β-D-Glucopyranosyl - 2,7-Dimethyl-1,4-dihydroxynaphthalene-1-*O*-β-D-glucopyranoside

6.3   CULTURE CONDITIONS

The *Chimaphilla umbellata* was germinated on Knudson C medium containing activated charcoal and sugar (1–3%). The medium showed positive effects on seedling germination (Figura et al. 2018; Thomas 2008).

REFERENCES

Alok S, Jain SK, Verma A, Kumar M, Sabharwal M. 2013. Pathophysiology of kidney, gallbladder, and urinary stones treatment with herbal and allopathic medicine: A review. Asian Pac J Trop Dis 3, 496–504.

Arnason T, Hebda RJ, Johns T. 1981. Use of plants for food and medicine by native peoples of eastern Canada. Can J Bot 59, 2189–2325.

Boughman AL, Oxendine LO. 2004. Herbal Remedies of the Lumbee Indians. McFarland & Co, Jefferson.

Boyko N, Zaytsev A, Osolodchenko T. 2014. Screening of antimicrobial properties of ethanolic extracts from some kinds of raw materials with quinone derivatives. Ann Mech Inst 4, 67–72.

Chevalier A. 1996. The Encyclopedia of Medicinal Plants. Dorling Kindersley Publishing, London.

Das N, Samantaray S, Ghosh C, Kushwaha K, Sircar D, Roy P. 2022. *Chimaphila umbellata* extract exerts anti-proliferative effect on human breast cancer cells via RIP1K/RIP3K-mediated necroptosis. Phytomed Plus 2, 100159.

Duke JA. 1992. Handbook of Phytochemical Constituents of GRAS Herbs and Other Economic Plants. CRC Press, Boca Raton, FL.

Figura T, Tylová E, Šoch J, Selosse MA, Ponert A. 2018. In vitro axenic germination and cultivation of mixotrophic pyroloideae (Ericaceae) and their post-germination ontogenetic development. Ann Bot 123, 625–639.

Galván IJ, Mir-Rashed N, Jessulat M, Atanya M, Golshani A, Durst T, Petit P, Amiguet VT, Boekhout T, Summerbell R, Cruz I, Arnason JT, Smith ML. 2008. Antifungal and antioxidant activities of the phytomedicine pipsissewa, *Chimaphila umbellata*. Phytochemistry 69, 738–746.

Gruenwald J. 2000. PDR for Herbal Medicines. 2nd ed. Medical Economics, Montrose, NJ.

Kusheev CB, Kutaev EM, Lomboeva SS, Khobrakova VB, Pavlov SA. 2020. The impact of the *Chimaphila umbellata* (L.) W.P.C.Barton extract on the immune response in animals. IOP Conf Ser Earth Environ Sci 548, 072018.

Li H, Bos A, Jean S, Webster D, Robichaud GA, Johnson JA, Gray CA. 2018. Antimycobacterial 1,4-napthoquinone natural products from *Moneses uniflora*. Phytochem Lett 27, 229–233.

Marles RJ, Clavelle C, Monteleone L, Tays N, Burns D. 2000. Aboriginal Plant Use in Canada's Northwest Boreal Forest. University of British Columbia Press, Vancouver.

McLean KY. 2017. Characterizing the antibacterial properties of *C. maculata*. Project Report, Esther G. Maynor Honors College, University North Carolina, Chapel Hill.

Oka M, Tachibana M, Noda K, Inoue N, Tanaka M, Kuwabara K. 2007. Relevance of anti-reactive oxygen species activity to anti-inflammatory activity of components of Eviprostat, a phytotherapeutic agent for benign prostatic hyperplasia. Phytomedicine 14, 465–472.

Pedersen JA. 2002. On the application of electron paramagnetic resonance in the study of naturally occurring quinones and quinols. Spectrochim Acta A Mol Biomol Spectrosc 58, 1257–1270.

Raja S, Ravindranadh K, Keerthi T. 2014. A complete profile on *Chimphila umbellata*-traditional uses, pharmacological activites and phytoconstituents. Int J Phytomed 6, 464–470.

Saxena G, Farmer SW, Hamock REW, Towers GHN. 1996. Chlorochimaphilin: A new antibiotic from *Moneses uniflora*. J Nat Prod 59, 62–65.

Shin B-K, Kim J, Kang KS, Piao H-S, Park JH, Hwang GS. 2015. A new naphthalene glycoside from *Chimaphila umbellata* inhibits the RANKL-stimulated osteoclast differentiation. Arch Pharm Res 38, 2059–2065.

Thomas TD. 2008. The role of activated charcoal in plant tissue culture. Biotechnol Adv 26, 618–631.

Trubachev AA, Batyuk VS. 1968. Flavonoids of *Chimaphila umbellata*. Chem Nat Comp 4, 271–272.

Trubachev AA, Batyuk VS. 1969. Phytochemical study of *Chimaphila umbellata* (L.) Nutt. Farmatsiia 18, 48–51.

Vafai SB, Mevers E, Higgins KW, Fomina Y, Zhang J, Mandinova A. 2016. Natural product screening reveals naphthoquinone complex I bypass factors. PLoS One 11, e0162686.

Vandal J, Abou-Zaid MM, Ferroni G, Leo G, Leduc LG. 2015. Antimicrobial activity of natural products from the flora of Northern Ontario, Canada. Pharm Biol 53, 800–806.

Veitch FP, Welton PA. 1951. β-Amyrin from *Chimaphila umbellata*. J Am Chem Soc 73, 3530.

Walewska E, Thieme H. 1969. Isolation of isohomoarbutin from *Chimaphila umbellata* (L.) Barton. Pharmazie 24, 423.

Yu Y, Elshafei A, Zheng X, Cheng S, Wang Y, Piao M, Wang Y, Jin M, Li G, Zheng M. 2021. Chemical constituents of *Chimaphila japonica* Miq. Biochem Syst Ecol 95, 104219.

Zellnig K, Michelitsch A, Likussar W, Schubert-Zsilavecz M, Baumesiter A, Salama ZB. 1996. Polarographic assay of chimaphilin in *Chimaphila umbellata* primary tincture. Eur J Pharm Sci 4(Suppl 1), S127.

7 *Cinnamomum* Species

7.1 MORPHOLOGICAL FEATURES, DISTRIBUTION, ETHNOPHARMACOLOGICAL PROPERTIES, PHYTOCHEMISTRY, AND PHARMACOLOGICAL ACTIVITIES

Cinnamomum verum J.Presl (syn. *C. iners* Reinw. ex Blume; *C. zeylanicum* Blume; Fam.—Lauraceae) is an evergreen and erect tree, with thick, smooth, reddish-brown bark and reaches a height of about 18–24 feet. The leaves are opposite, ovate, or ovate-lanceolate, hard, and coriaceous, contain 3–5 nerves, and glabrous. Many small flowers are arranged in axillary or sub-terminal cymes. The fruits are 1.5–2 cm long, ovate, dark purple in color. The fruit contains a single seed. It is widely distributed in Sri Lanka, China, Sumatra, Eastern Islands, Brazil, Mauritius, India, and Jamaica (Rawat et al. 2019; Mehrpouri et al. 2020). It possesses stomachic and carminative properties. It is also a useful lineament in rheumatic pains. The paste is prepared from bark, mixed with lemon juice, and applied for the treatment of pimples (Aluyor and Oboh 2014; Premakumara and Abeysekera 2021; Abeysinghe et al. 2021). In the Ayurvedic system of medicine, the leaves and bark of *C. iners* possess aromatic, astringent, stimulant, and carminative properties. It is used in the treatment of rheumatism, colic, diarrhea, nausea, and vomiting. The leaf essential oils are used as carminative, antiflatulent, diuretic, and cardiac tonic agents (Swanston-Flatt et al. 1989; Devi et al. 2007; Rao et al. 2008; Ranasinghe et al. 2013). In traditional medicine, cinnamon species are used for respiratory, cardiovascular, and digestive disorders (Błaszczyk et al. 2021). It contains antimicrobial, wound healing, antidiabetic, anti-HIV, anti-anxiety, and anti-Parkinson's properties (Thakur et al. 2021). Cinnamomum species contains lignans, butanolides, flavonoids, phenylpropanoids, and alkaloids (Wu et al. 2022). The acetone and ethanol extracts of *C. iners* stem bark showed significant anticancer activity against DAL and EAC cell lines (Ramalingam and Balasubramanian 2015; Sharifi-Rad et al. 2021). About 250 species of Cinnamomum genus are distributed in Asia, South and Central America, China, and Australia (Wuu-Kuang 2011; Cardoso-Ugarte et al. 2016). The *C. zeylanicum* dried stem bark is (Jakhetia et al. 2010) used in preparation of chocolates, beverages, spicy candies, and liquors (Krishnamoorthy and Rema 2004). *C. verum* bark is also considered as a spice (Samy et al. 2005). *C. burmanii* bark is also useful in curing diarrhea and malaria (Burkill 1966; Leela 2008). *C. iners* roots are boiled, and its decoction is given to mothers after childbirth. *C. iners* mucilage is used in the preparation of mosquito coils, as well as fragrant joss sticks (Pereira and Hastie 2014; Kumar et al. 2019). It possesses wound-healing properties (Sulaiman 2013; Wang et al. 2020; Pathak and Sharma 2021). The cinnamaldehyde, eugenol, caryophyllene, cinnamyl acetate, and cinnamic acid rich extracts of *C. verum* possess antioxidant, antimicrobial, anti-inflammatory, anticancer, antidiabetic, wound healing, anti-HIV, anti-anxiety, and antidepressant activities (Singh et al. 2021). *C. walaiwarense*, and *C. travancoricum* bark extracts significantly reduced liver weight and liver volume in carbon tetrachloride-induced in hepatic injury in albino rats (Maridass 2009).

Microwave digestion, pressurized water extraction, steam distillation, solvent extraction, decoction water extraction, and infusion water extraction methods were used for cinnamon quill extract preparations for investigation of antidiabetic actions. α-amylase and α-glucosidase inhibitory potential and phytochemical profiling were investigated in water extracts of two new *C. zeylanicum* accessions (Sri Wijaya and Sri Gemunu). Minimum (IC_{50}) inhibitory values were recorded in pressurized water extraction and decoction water extraction methods of Sri Wijaya. Maximum levels of proanthocyanidin and total phenolic contents were found in pressurized water extraction and decoction water extraction methods of Sri Wijaya. Pressurized water and cold decoctions were used for the extraction of antidiabetic constituents from cinnamon. Benzoic acid, cinnamyl alcohol, benzyl alcohol, and 4-allyl-2,6-dimethoxyphenol were identified from Sri Wijaya extracts. Due

DOI: 10.1201/9781003398035-7

to the presence of these phytochemicals, the extracts showed strong α-amylase and α-glucosidase inhibitory activities. Therefore, *C. zeylanicum* Sri Wijaya may be used in the management of diabetes (Wariyapperuma et al. 2020a). Similarly, both accessions (Sri Wijaya and Sri Gemunu) of *C. zeylanicum* were pretreated with *Trichoderma harzianum* (MH298760). *T. harzianum* (MH298760) pretreatment showed the maximum cell lysis ability; hence, it was used for the microbial pretreatment process. Sri Wijaya extract treated with *T. harzianum* species afforded the maximum yield total phenolic content (2.24 ± 0.02 mg gallic acid equivalent/g), and proanthocyanidin content (48.2 ± 0.4 mg of catechin equivalent/g), significant inhibition ($p < 0.05$) of α-amylase and α-glucosidase activities (IC_{50} 57 ± 8 and 36 ± 8 µg/ml). The obtained results were compared with nontreated samples. *T. harzianum* treatment increased the hypoglycemic activities, proanthocyanidin, and total phenolic contents in cinnamon extracts and offer new understanding into the recovery of phytoconstituents (Wariyapperuma et al. 2020b). The cinnamon bark extract showed inhibitory effect against intestinal α-glucosidase and pancreatic α-amylase activities. Thai cinnamon extract showed significant inhibition against intestinal maltase (IC_{50} 0.58 ± 0.01 mg/ml) activity. The cinnamon extract also displayed significant suppression of intestinal sucrase and pancreatic α-amylase activities (IC_{50} 0.42 ± 0.02 and 1.23 ± 0.02 mg/ml, respectively). Moreover, cinnamon extracts exhibited significant inhibition against intestinal α-glucosidase and pancreatic α-amylase activities. The results were compared with acarbose (reference drug). The findings suggest that bark extracts may be used for the control of postprandial glucose in diabetic patients by suppressing intestinal α-glucosidase and pancreatic α-amylase activities (Adisakwattana et al. 2011). *C. zeylanicum* essential oils demonstrates antioxidant activity on DPPH radical assays (Ashfaq et al. 2021).

The dichloromethane:methanol (1:1, v/v) extract of Ceylon cinnamon leaves showed significant differences ($p < 0.05$) among different maturity stages (immature, partly mature, and mature) of leaf against antioxidant and glycemic regulatory activities. The antioxidant and glycemic regulatory effects of immature, partly mature, and mature leaves were recorded as total polyphenolic content: 0.68 ± 0.02–22.35 ± 0.21 mg gallic acid equivalents/g of gallic acid equivalents; total flavonoid content: 0.85 ± 0.01–4.68 ± 0.06 mg quercetin equivalents/gallic acid equivalents; 1, 1-diphenyl-2-picryl-hydrazyl: 0.42 ± 0.01–27.09 ± 0.65 mg trolox equivalents/gallic acid equivalents; 2-azino-bis (3-ethylbenzothiazoline-6-sulfonic acid: 3.57 ± 0.10–43.91 ± 1.46 trolox equivalents/gallic acid equivalents; oxygen radical absorbance capacity: 0.71 ± 0.01–18.70 ± 0.26 trolox equivalents/gallic acid equivalents, ferric reducing antioxidant power: 0.31 ± 0.02–69.16 ± 0.52 trolox equivalents/gallic acid equivalents; and antiamylase: 18.05 ± 0.24–36.62 ± 4.00% inhibition at 2.5 mg/ml. Mature leaf extract displayed maximum antioxidant and antiamylase effects; therefore, the extract may be used in the management of oxidative stress-associated chronic diseases (Abeysekera et al. 2019). The ethanol and dichloromethane:methanol (1:1, v/v) extracts of Ceylon cinnamon bark and leaf were tested for the reversing activities of antiamylase, antiglucosidase, anticholinesterases, antiglycation, and glycation in bovine serum albumin-glucose and bovine serum albumin-methylglyoxal *in vitro* models. The bark and leaf extracts displayed significant ($p < 0.05$) biological activities (except antiglucosidase) and total proanthocyanidins. Bark extract displayed higher ($p < 0.05$) effects than leaf extract for antiamylase (IC_{50} 214 ± 2 – 215 ± 10 µg/ml), antibutyrylcholinesterase (IC_{50} 26.62 ± 1.66–36.09 ± 0.83 µg/ml), and glycation-reversing in a bovine serum albumin-glucose model (EC_{50} 94.33 ± 1.81–107.16 ± 3.95 µg/ml). The total proanthocyanidins were significantly greater ($p < 0.05$) in bark extract than leaf extract (bark and leaf extracts: 1097.90 ± 73.01 – 1381.53 ± 45.93 and 309.52 ± 2.81–434.24 ± 14.12 mg cyanidin equivalents/g extract). In summary, both bark and leaf extracts of Ceylon cinnamon possess antidiabetic activities and therefore may be used in the management of diabetes (Arachchige et al. 2017). The ethanol (95%) and dichloromethane:methanol (1: 1) extracts of Ceylon cinnamon bark were tested for their HMG-CoA reductase, lipase, cholesterol esterase, and cholesterol micellization inhibitory and bile acids–binding activities (*in vitro*). Bark extract displayed significant HMG-CoA reductase, lipase, cholesterol esterase, and cholesterol micellization inhibitory and bile acids–binding activities (IC_{50} 153.07 ± 8.38 – 277.13 ± 32.18, 297.57 ± 11.78–301.09 ± 4.05, 30.61 ± 0.79–34.05 ± 0.41, and 231.96 ± 9.22–478.89 ± 9.27 µg/ml, for

HMG-CoA reductase, lipase, cholesterol esterase, and cholesterol micellization inhibitory effects). The bile acids–binding activity (3 mg/ml) for taurocholate, glycodeoxycholate, and chenodeoxycholate were recorded in ranges of $19.74 \pm 0.31 - 20.22 \pm 0.31$, $21.97 \pm 2.21 - 26.97 \pm 1.61$, and $16.11 \pm 1.42 - 19.11 \pm 1.52\%$. The ethanol extract showed a greater quantity of individual compounds as well as antilipidemic activities. Results reveal scientific support for the traditional claim that Ceylon cinnamon possesses antilipidemic activities (Abeysekera et al. 2017).

Procyanidin-B2 (active component from *C. zeylanicum*) suppressed advanced glycation end products formation *in vitro*. The ability of procyanidin-B2-enriched fraction of cinnamon to stop *in vivo* accumulation of advanced glycation end products and to reduce renal changes was investigated in diabetic rats. Administration of procyanidin-B2 rich-fraction prohibited glycation mediated RBC-IgG cross-links and HbA1c accumulation in diabetic rats. Procyanidin-B2 rich-fraction also inhibited the accumulation of *N*-carboxy methyl lysine, a prominent advanced glycation product, in kidneys of diabetic rats. In summary, procyanidin-B2 rich-fraction from cinnamon suppressed advanced glycation end products accumulation in diabetic rat kidneys and reduced advanced glycation end products mediated pathogenesis of diabetic nephropathy (Muthenna et al. 2014). The methanol extract, ethyl acetate, *n*-butanol, chloroform, and aqueous fractions of *C. iners* leaves were evaluated for their antidiabetic and antihyperlipidemic effects on streptozotocin-induced diabetic rats. The results showed significant enhancement in the levels of body weight and high-density lipoproteins but reduction in the levels of blood glucose, total cholesterol, triglycerides, low-density lipoproteins, and very low-density lipoproteins. *C. iners* leaves showed significant antidiabetic and hypolipidemic effects (Mustaffa et al. 2014). The chloroform fraction also showed significant glucose tolerance effect in STZ-induced diabetic rats (Mustaffa et al. 2016).

Methanol extract of *C. iners* showed antioxidant (in 2,2-diphenyl-1-picrylhydrazyl assay) and antibacterial activity against *B. cereus* (ATCC 11778), *E. coli* (ATCC 29214), *S. aureus* (ATCC 13150), *P. vulgaris* (TISTR 100), *S. cremoris* (TISTR 058), *S. typhi* (ATCC 43579), and *C. krusei* (TISTR 5256; Butkhup and Samappito 2011). The ethanol, acetone, and water extracts (15 and 30 μg) of *C. iners* leaf and bark were screened for their antimicrobial activity. The antimicrobial activity was tested by using *in vitro* disc diffusion assay. Bark extracts demonstrated a higher inhibition zone against bacterial (*E. coli, S. aureus, S. marcescens, K. pneumoniae* and *P. aeruginosa*) and fungal strains *Trichophyton rubrum, A. fumigatus* and *C. albicans* (Vigila et al. 2018). The cinnamon bark essential oil in combination with piperacillin showed significant antimicrobial and synergistic activity against β-lactamase TEM-1 plasmid-conferred *Escherichia coli* J53 R1. GC-MS analysis of *C. verum* showed the presence of *trans*-cinnamaldehyde, benzyl alcohol, and eugenol. Bark essential oils can reverse *E. coli* J53 R1 resistance to piperacillin through two pathways; modification in the permeability of the outer membrane or bacterial quorum sensing inhibition (Yap et al. 2015).

Anticancer activities of *C. zeylanicum* powder was investigated in rat, mouse, and cell line breast carcinoma models. *C. zeylanicum* bark powder (0.1% and 1% w/w) was administered to chemically-induced rat mammary carcinomas and a syngeneic 4T1 mouse model. Powder (1% dose) significantly reduced the tumor volumes by 44% (as compared with control). Moreover, treated tumors displayed a significant dose-dependent reduction in mitotic activity index by 45.5% (as compared with control group). In treated rat carcinomas, an increase in caspase-3 and Bax expression and reduction in Bcl-2, Ki67, VEGF, and CD24 expressions and MDA levels were observed. A significant reduction in lysine methylation of H3K4m3 and H3K9m3 levels and enhancement in H4K16ac levels were also observed in treated groups. The essential oil of *C. zeylanicum* showed significant anticancer activity in MCF-7 and MDA-MB-231 cells. In summary, *C. zeylanicum* powder displayed chemopreventive and therapeutic effects in animal breast carcinoma models (Kubatka et al. 2020). The ethanolic and acetone extract of *C. cassia* and *C. zeylanicum* displayed potent inhibition of *S. pyogenes, S. aureus, B. cereus, E. faecalis, P. aeruginosa* and *S. bongori* (Singh et al. 2020). Ethanol and procyanidin-B2-enriched F2 fraction *C. zeylanicum* bark suppressed the catalytic activities of the proteasome in cancer cells. Both extracts did not show any effect on normal cells. Ethanol extract and procyanidin-B2-enriched F2 fraction significantly reduced the proliferation of human prostate cancer cells and the expression of anti-apoptotic and angiogenic markers in prostate cancer cell lysates. The findings reveal that cinnamon extract and its PCB2-enriched fraction act as proteasome suppressors and hence have prospects for development as anticancer agents (Gopalakrishnan et al. 2018). The cytotoxic effects of chloroform and alcoholic extracts of *C. iners* leaves were tested against human colorectal tumor cells. The chloroform extract demonstrated significant cytotoxicity (IC_{50} 31 μg/ml; $p < 0.01$). Ethanol extract showed moderate cytotoxicity ($IC_{50} > 200$ μg/ml). Caryophyllene, eicosanoic acid ethyl ester, 2,2′-bithiazolidine, 3-methyl-2-(3,7,11-trimethyldodecyl) furan, hexadecanoic acid, ethyl ester, 3-(4,8,12-trimethyltridecyl) furan, 3,7,11,15-tetramethyl-2-hexadecen-1-ol, 1,4-*S*,*S*-2,5-bis[carbethoxy]phenylene bis[*N*,*N* dimethyldithiocarbamate], and (3á,5á)-2-methylene-cholestan-3-ol were determined in chloroform extract (Ghalib et al. 2012). Methanol and acetone extracts of *C. iners* leaves (6 mg) exhibited significant antikinase activity against MKK1 in the signal transduction pathway. Both extracts contain polyphenol and flavonoid contents with potent antioxidation activity in DPPH free radicals (IC_{50} 0.2 and 0.3 mg/ml respectively). The results reveal that extracts may act as potent MKK1 inhibitors that can be used for development as anticancer agents (Pang et al. 2009). Methanol extract of *C. iners* leaves (200 and 500 mg/kg) displayed significant analgesic activity ($p < 0.05$) in the formalin-induced pain model (late phase) on the rats. The extract did not show any toxicity in treated animals at tested doses throughout the 14-day period (Mustaffa et al. 2010). *C. zeylanicum* bark essential oil showed positive effects on human skin cells. Essential oils displayed strong antiproliferative activity on skin cells and significantly suppressed the formation of several inflammatory biomarkers (vascular cell adhesion molecule-1, intercellular cell adhesion molecule-1, monocyte chemoattractant protein-1, interferon γ-induced protein 10, interferon-inducible T-cell alpha chemoattractant, and monokine induced by γ-interferon). Essential oils significantly suppressed the formation of several tissue remodeling molecules (epidermal growth factor receptor, matrix metalloproteinase-1, and plasminogen activator inhibitor-1). Essential oils also inhibited the production of macrophage colony-stimulating factor. Moreover, essential oils significantly regulated global gene expression and changed signaling pathways. The results show that bark essential oils may be considered as a promising anti-inflammatory agent (Han and Parker 2017). The *C. verum* leaf oils showed significant acaricide activity against *Rhipicephalus microplus*. The *C. verum* essential oil was found to be more effective (3.3 times) on the *R. microplus* larvae than benzyl benzoate (Monteiro et al. 2017).

C. bejolghota (Buch.-Ham.) Sweet (syn. *C. obtusifolium* (Roxb.) Nees) is an evergreen tree with erect stems, brownish-white bark, 4–8 mm thick and grows up to 6–8 m high. Leaves are alternate, sub-opposite or opposite, narrowly oblong to oblong-elliptic-lanceolate or ovate-lanceolate, base cuneat to the decurrently acute, glossy above, glabrous, and pale beneath. Inflorescence is panicle, pseudo-terminal, solitary axillary, slender, pale yellowish green, minutely pubescent, usually equal to the leaves, ovate-elliptic-lanceolate, and silky on both surfaces. Flowers contain 3 + 3+3, 1.5 – 1.75 mm long, pale yellowish green, introrse, silky pubescent to villous stamens (Gogoi et al. 2016). It is distributed in India (Gogoi et al. 2014), Bangladesh, Bhutan, Laos, Myanmar, Nepal, Thailand, and Vietnam (Anonymous 1992; Wu and Raven 1996), China, Sri Lanka, Madagascar, and East of Thailand (Li et al. 2013). Its bark is used as a spice in curry and leaves for the preparation of a special kind of rice-beer (Gogoi et al. 2014; Baruah and Nath 2001). *C. bejolghota* is useful in cough, cold, toothache, and liver complaints (Rao 1979; Choudhury et al. 1998). The bark is taken as a remedy in curing bone fractures as well as in healing of wounds by smashing (Wu and Raven 1996). The root paste is applied on the forehead to treat headache and dizziness (Ghorbani 2005). Water decoction of *C. obtusifolium* fresh bark is taken for the treatment of stomach diseases (Hossan et al. 2009). The slurry of woody parts is applied as lotion for the treatment of muscle stiffness and pain, tingling and numbness, arthritis, skin rashes, and diseases (DeFilipps and Krupnick 2018). The plant leaves are used in the treatment of diarrhea (Chopra et al. 1956) and the bark is useful in the treatment of fever (Sajem et al. 2008). The Nyshi tribe of Arunachal Pradesh (India) use this plant species in liver diseases (Doley et al. 2010). In Manipur, its bark is used in the treatment of urinary stone pains (Baruah et al. 1997; Lokendrajit et al. 2011).

The methanol extract of *C. bejolghota* was tested for its antihyperglycemic activity on strep-tozotocin-induced type-2 diabetic rats and showed significant enhancement in body weights and rapid reduction in hyperglycemic peak in treated animals. After treatment (15 days), levels of total cholesterol, triglycerides, low density lipoprotein were decreased but high-density cholesterol level increased significantly. Methanol extract significantly decreased the elevated levels of aspartate transaminase, alanine transaminase, and alkaline phosphatase. The extract also decreased the levels of lipid peroxidation but increased catalase and glutathione levels of liver. The results reveal that *C. bejolghota* possesses potent antihyperglycemic activity and supports the *in vivo* antioxidative status (Gogoi et al. 2014).

C. tamala (Buch. -Ham.) T.Nees & Eberm. is an evergreen, perennial tree, rough stem with gray-brown, soft wrinkled bark and attaining 24–35 feet height with a girth of 150 cm. The leaves are large, long (12–20 cm) and broad (5–8 cm), ovate-lanceolate, thick and leathery, acuminate, glabrous, opposite, reticulate venation, sub-opposite or alternate, and short-stalked. Petioles are slender and long (0.8–1.8 cm). The flowers are small, bisexual, white, numerous, axillary cymes, and terminal pubescent panicles. The fruit is an ellipsoidal drupe and seeds require approxi-mately one year for attaining maturity. Ripe fruits are dark purple in color and contain a single seed (Singh and Singh 1992; Sharma et al. 2009; Sharma and Nautiyal 2011; Kumar et al. 2012a). The tree is distributed in tropical and subtropical regions of Australia, the Pacific region, South America, and the Himalayan region of Asia (Ahmed et al. 2000), India, Nepal, Bhutan, and China (Rema et al. 2005). The plant species possesses antidiarrheal, hepatoprotective, gastroprotective, antibacterial, and immunomodulatory properties (Kumar et al. 2012b). In Nepalese traditional medicine, *C. tamala* bark is used in the treatment of intestinal diseases, nausea, and diarrhea (Kunwar and Adhikari 2005). The *C. tamala* leaves are useful in repelling stored grain pests in India (Bhattacharjee and Ray 2010). The leaves are used as a spice, carminative, anti-flatulent, diuretic, and cardiotonic agent, (Showkat et al. 2004) used to treat bladder diseases, mouth dry-ness, coryza, diarrhea, nausea, and spermatorhea (Kapoor 2000). As per Ayurvedic medicine, dried leaves and bark have been used for the treatment of fever, anemia, and body odor. The crushed seeds are mixed with honey and given to children to treat dysentery or cough (Dhar et al. 2002; Shah and Panchal 2010). *C. tamala* oil (100 mg/kg and 200 mg/kg), cinnamaldehyde (20 mg/kg) and glibenclamide (0.6 mg/kg) reduced the levels of blood glucose in streptozotocin-induced diabetic rats. There was significant enhancement in body weight, liver glycogen content, plasma insulin level and reduction in the levels of blood glucose, glycosylated hemoglobin, and total plasma cholesterol recorded in treated animals. Both C. *tamala* oil and cinnamaldehyde showed significant antidiabetic effects and results compared with standard drug glibenclamide (Kumar et al. 2012c). Crude methanolic extract of *C. tamala* leaves possesses antimicrobial effects against *S. typhi* (Hassan et al. 2016). Hydroalcoholic extract of *C. tamala* leaves showed the presence of 48.1 mg gallic acid/g, 22.1 mg quercetin/g, 59.9 mg/g and 1.75 mg rutin)/g on quantification of phenolic compounds (Raksha et al. 2021).

C. kanehirai (Hayata) Hayata (syn. *C. micranthum* (Hayata) Hayata; *C. xanthophyllum* H.W.Li) is a perennial tree, with rough stem. The leaves are alternate, ovate-lanceolate, thick leathery, and acuminate. It is native to Taiwan (Wu et al. 2017) and widely distributed in tropical and subtropical regions of eastern Asia, Australia, and the Pacific islands (Liao et al. 2010). In Chinese traditional medicine, it is used in the treatment of infection of lungs and nervous depressions. Essential oils of leaves are useful in skin infections. Essential oils also possess antimicrobial (Yeh et al. 2009), anticancer (Straub et al. 2002; Chen et al. 2009; Liu et al. 2013), and hepatoprotective properties (Zisman et al. 2002). The ethanol extract of *C. kanehirai* leaves suppressed the cellular viability of both HepG2 and HA22T/VGH human hepatoma cell lines in a dose- and time-dependent manner. The extract caused cleavage of caspase-3 and enhanced enzyme activities of caspase-8 and cas-pase-9. In summary, ethanol extracts stimulated apoptosis intrinsic pathways via caspase-3 cascade in human hepatoma HA22T/VGH and HepG2 cells and therefore may be used in hepatoma therapy (Liu et al. 2015).

C. insularimontanum Hayata is a perennial, broad-leaved evergreen tree with bird-dispersed seeds (Chung et al. 2003). The plant species is distributed in the different regions of Taiwan. It is used for the treatment of inflammations, gastric ulcers, and rheumatic diseases. *C. insularimontanum* exhibited significant anti-HSV-2 activity (IC_{50} 180 ± 30 µg/ml) and showed selectivity index in the range of 1.18–12.28 (Lin et al. 2003). The presence of α-pinene, camphene, β-pinene, limonene, citronellal, citronellol, and citral was determined by GC-MS analysis in *C. insularimontanum* fruits. The crude essential oil and citral showed significant nitric oxide production inhibitory activity (IC_{50} 18.68 and 13.18 µg/ml) in nitric oxide inhibitory activity assay. Additionally, citral decreased the expression of IKK, iNOS, and nuclear NF-κB pathways but increased IκBα pathway in a dose-dependent manner. The compound did not show any inhibitory effect on COX-2 production pathway. Moreover, citral (0.3 mg per ear) demonstrated potent anti-inflammatory effect in the croton oil-induced mice ear edema assay (83% reduction). Therefore, fruit essential oil of *C. insularimontanum* and/or citral may be used in the development of anti-inflammatory drugs in the future (Lin et al. 2008a). Actinodaphnine isolated and identified *C. insularimontanum* and stimulated apoptosis in Mahlavu cells. The activity is associated with enhanced levels of intracellular nitric oxide and reactive oxygen species, disruptive mitochondrial transmembrane potential, and activation of caspase 3/7. The actinodaphnine downregulated the activity of the nuclear factor κB (NF-κB) pathway. The actinodaphnine-stimulated apoptosis was mediated by increasing the nitric oxide and/or reactive oxygen species and caspases-dependent pathway (Hsieh et al. 2006a).

C. glaucescens (Nees) Hand.-Mazz. is a perennial tree with rough bark. The leaves are alternate, long (7–10 cm) with elliptic shape, a well-marked midrib, and found in Himalayan range of Nepal and India (Baruah and Nath 2006). It has been traditionally used as a demulcent and stimulant and possesses analgesic, antiseptic, astringent, and carminative activities. Its seeds are used for the treatment of cold, cough, toothache, and taenias (Chopra et al. 1956; Sthapit and Tuladhar 1993; Ravindran et al. 2004). Seed paste is externally applied to treat muscular pain and swellings. Crude oil of seeds is used for the treatment of muscular spasm, joint pain, and body aches (Prakash et al. 2013). In Manipuri medicine (India), bark powder is used to treat kidney complaints (Mikawlrawng and Kuma 2014). Essential oils of *C. glaucescens* demonstrated antibacterial activity against *E. coli*, *S. aureus* and *K. pneumoniae* (20,000 µg/20 µl). Maximum activity was demonstrated against *E. coli* (MIC 125 µg/20 µl; Gyawali et al. 2013).

C. mercadoi S. Vidal is a small perennial tree with a thick and aromatic bark and grows up to 6 to 10 m. The plant is widely distributed in the Philippines (Joshi et al. 2009; Shahwar et al. 2010; Sultana et al. 2010). In Philippines, it is used as remedy for the treatment of stomach diseases, flatulence, headache, rheumatism, and diarrhea (Palis 1995; Langerberger et al. 2009; Fuentes et al. 2010). The methanol extract of *C. mercadoi* (500 mg/kg) showed 84% protection in plantar test method while aspirin (300 mg/kg) displayed 72.07% protection in treated mice (Torres et al. 2003). The methanol extracts of *C. mercadoi* leaves and bark were tested for their antidiarrheal effect in a castor oil-induced diarrhea model. The bark extract (500 mg/kg bw) displayed a significant decrease in diarrheal feces (78.1%) in treated mice. The antidiarrheal effect of the bark extract was compared with loperamide (reference drug, 3 mg/kg; Gorgonio and Fuentes 2011). The methanol extract of leaves showed higher DPPH radical scavenging activity (IC_{50} 91.7 ± 15.7 µg/ml) than bark extract (IC_{50} 12.7 ± 0.2 µg/ml). The bark extract also consists of more total phenolic content (1331 ± mg gallic acid equivalents/g) than leaf extract (216 ± 7 mg gallic acid equivalents/g). Moreover, bark extract showed significant antibacterial activity against *Bacillus subtilis* and *Staphylococcus aureus* (Fuentes et al. 2010).

C. altissimum Kosterm is a medium-sized plant species, growing up to 39 m high. The stem girth is about 1.5 m. The leaves are opposite or sub-opposite, with long stalks (1–1.5 cm). The leaf blade is thick, leathery, slightly glaucous, elliptic, or oblong, apex pointed or blunt, base cuneate, or rounded with midrib (Kochummen 1989). It is native to Peninsular Malaysia as well as in lowland and hill forests of Sumatra (Kochummen 1989). The leaves, stem bark, and stem wood are used in treating wound infections (Salleh et al. 2015a). *C. altissimum* bark extract showed the presence of phenolic content (130.1 mg gallic acid/g) and demonstrated significant free radical scavenging activity in DPPH (IC_{50} 126.2 μg/ml) and ferric reducing antioxidant power assays (341.2 mg ascorbic acid/g; Salleh et al. 2015a).

C. javanicum Blume is a medium-sized tree and grows up to 30–35 m high. The stem bark is smooth and white/gray in color with yellowish white sapwood. Twigs are stout, terete to subangular, 2–5 mm diameter, and glabrescent. Leaves are opposite or sub-opposite, thinly coriaceous to coriaceous, and sometimes glabrescent. It is distributed in southern China to Peninsular Malaysia, Sumatra, Java, and Borneo (Wuu-Kuang 2011). Stem crude extract of *C. javanicum* showed significant antimicrobial activity against *Listeria monocytogenes* (MIC 0.13 mg/ml). The total phenolic content of the extract was 78.3 mg gallic acid equivalents/g and demonstrated significant antioxidant activity (57.2–326.5 mg trolox equivalents/g extract). Stem crude extract and eucalyptol caused significant membrane damage in treated *L. monocytogenes* (Yuan et al. 2017).

C. osmophloeum Kaneh is a perennial tree and is distributed in Taiwan and South and Southeast Asia. Its essential oils are used in the treatment of infections (Kurniawati et al. 2017). The essential oils of *C. osmophloeum* leaves strongly suppressed nitric oxide production (IC_{50} 9.7–15.5 μg/ml). Moreover, *trans*-cinnamaldehyde also demonstrated significant anti-inflammatory activity (Tung et al. 2010). *C. osmophloeum* leaf essential oils (60 μg/ml) inhibited proIL-1β protein expression induced by LPS-treated J774A.1 murine macrophage. Moreover, essential oils (60 μg/ml) also inhibited IL-1β and IL-6 production but did not show any effects on TNF-α (Chao et al. 2005). The essential oils of *C. osmophloeum* leaves showed larvicidal activity against *Anopheles gambiae*. The cinnamaldehyde and caryophyllene oxide rich crude oil displayed significant larvicidal activity (LC_{90} 57.71 to 91.54 μg/ml) in *in vitro* as well as in semi-field environments (52.07 to 173.77 μg/ml; Mdoe et al. 2014). The essential oils of *C. osmophloeum* leaves possess antiviral, antifungal, and antibacterial activities (Reichling et al. 2009). Similarly, the essential oil of *C. osmophloeum* leaves showed strongest mosquito larvicidal activities (Cheng et al. 2004, 2009). The *n*-butanol soluble fraction (rich in proanthocyanidins and condensed tannins) of *C. osmophloeum* twigs demonstrated significant antihyperglycemic and PTP1B inhibitory activities. The fraction also demonstrated significant DPPH free-radical scavenging and ferrous ion-chelating activities with no toxicity on 3T3-L1 preadiocytes (Lin et al. 2016).

Anti-inflammatory effects of crude essential oils, cinnamaldehyde, and linalool (isolated from *C. osmophloeum*) were investigated in endotoxin-induced inflammation (i.p. 10 mg/ml/kg bw) in C57BL/6 mice. Cinnamaldehyde (0.45 or 0.9 mg/kg bw), linalool (2.6 or 5.2 mg/kg bw), and positive control group animals were administered with leaf essential oil (13 mg/kg body weight). Cinnamaldehyde and linalool restored the altered endotoxin-induced body weight loss and lymphoid organ enlargement in treated animals significantly ($p < 0.05$). Both compounds significantly reduced the levels of endotoxin-induced peripheral nitrate/nitrite, interleukin (IL)-1β, IL-18, tumor necrosis factor-α, interferon-γ, and high-mobility group box 1 protein, nitrate/nitrite, IL-1β, TNF-α, and IFN-γ in spleen and mesenteric lymph nodes ($p < 0.05$). Both isolated compounds also suppressed the endotoxin-stimulated expression of toll-like receptor 4, myeloid differentiation primary response gene 88, myeloid differentiation protein 2, NOD-like receptor family, pyrin domain containing 3, and apoptosis-associated speck-like protein consisting of a caspase-recruitment domain, and caspase-1 in spleen and mesenteric lymph nodes significantly ($p < 0.05$). Moreover, both compounds also inhibited the stimulation of the nuclear factor (NF)-κB and the activity of caspase-1 in spleen and mesenteric lymph nodes significantly ($p < 0.05$). The obtained results confirm prophylactic uses of both compounds in health complaints associated with inflammations that being associated to over-activated toll-like receptor 4 and/or NOD-like receptor family, pyrin domain containing 3 signaling pathways (Lee et al. 2018). Similarly, *C. osmophloeum* leaf essential oils significantly reduced the levels of peripheral tumor necrosis factor-α, interleukin-1β, IL-18, interferon-γ, and nitric oxide and suppressed the expression of toll-like receptor 4, myeloid differentiation primary response gene (88), myeloid differentiation factor 2, apoptosis-associated speck-like protein containing a caspase-recruitment domain (ASC), caspase-1, and NOD-like receptor family, pyrin domain containing 3 in treated animals. Essential oils also suppressed the activation of nuclear factor-κB, activity of caspase-1 in the small intestine, and reduced intestinal edema; therefore the oils may be used in the inhibition of the toll-like receptor 4 and NOD-like receptor family, pyrin domain-containing 3 signaling pathways in the intestine (Lee et al. 2015).

C. subavenium Miq is a perennial tree, 18–27 m in height, 13–50 cm diameter, bark smooth, grayish, inner bark fibrous, pinkish brown, and with whitish sapwood. The twigs are slender, 2–3 mm diameter, apically subangular, and dark brown. Leaves are opposite, sub-opposite or rarely alternate, subcoriaceous, hairs straight to curly, elliptic to narrowly elliptic, and apex acuminate (Wuu-Kuang 2011). The tree is distributed in China, Malaysia, Cambodia, Indonesia, and Burma (Lin et al. 2008b). Its peel, fruit, and leaves are used in the treatment of abdominal pain, chest pain, hernia, rheumatism, vomiting, nausea, and diarrhea (Liu et al. 2011). Different parts of this species possess antitumor activity against urothelial carcinoma cells, colorectal cancer cells, skin cancer melanoma cells, human bladder cancer cells, human lung cancer cell, and human prostate cancer cell lines (Chen et al. 2007; Kuo et al. 2008; Shen et al. 2011; Wang et al. 2011). Moreover, the leaf oil of *C. subavenium* possesses potent antioxidant and antimicrobial activities (Ho et al. 2008).

C. camphora (L.) Siebold is an evergreen tree, with a broad, dense, and symmetrical crown, and tall with wide branching. The leaves are green and broad. The flowers are white in color. The plant species is distributed in China, Korea, Japan, and Vietnam. In Ayurvedic medicine, it is used for the treatment of cough, cold, toothache, congestion, diarrhea, dysentery, gum complaints, skin infections, and vomiting (Nishida et al. 2006; Hsieh et al. 2006b). Moreover, in the Yunani medicine system, it is used as cephalic tonic, cardiac tonic, and as an expectorant agent (Singh and Jawaid 2012). The dried bark of *C. cassia* is used to treat diabetes, breast cancer, leukemia and other diseases (Liu et al. 2021).

7.2 PHYTOCHEMISTRY

Bicyclo[3.1.1]hept-2-ene, 2,6,6-trimethyl, camphene, benzaldehyde, β-pinene, eucalyptol, 1,6-octadien-3-ol,3,7-dimethyl, benzene propanal, borneol, (−)-α-terpineol (*p*-menth-1-en-8-ol), cinnamaldehyde, phenol, 2-methoxy-4-(2-propenyl)-, acetate (eugenol), and 1-phenyl-propane-2,2-diol diethanoate were determined in *C. pubescens*. The appreciable profiles of total phenolic (50.6 and 33.41 g/kg as gallic acid equivalent) and total flavonoid contents (205.6 and 244.8 g/kg as rutin equivalent) were estimated in *C. pubescens*. *C. pubescens* showed significant DPPH free radical scavenging activity (IC_{50} 77.2 µg/ml). The essential oils of *C. pubescens* displayed significant antibacterial activity against methicillin-resistant *S. aureus*, *B. subtilis*, *P. aeruginosa* and *S. choleraesuis* (Abdelwahab et al. 2010).

The presence of (*E*)-cinnamaldehyde, benzyl benzoate, linalool, (*E*)-caryophyllene, caryolan-8-ol, and borneol was determined in *C. verum*. *C. verum* leaf oils displayed significant antifungal, and antioxidant activities (Farias et al. 2020). Twenty-one essential oils [α-pinene, camphene, benzaldehyde, α-phellandrene, p-cymene, β-phellandrene, eucalyptol, linalool, benzenepropanal, *cis*-cinnamaldehyde, saffrole, *trans*-cinnamaldehyde, eugenol, α-cubebene, caryophyllene, coumarin (2H-1-benzopyran2-one), cinnamyl acetate (*E*), α-muurolene, trans-cadina-1(6),4-diene, eugenyl acetate benzyl benzoate] were determined in powder of whole parts of *C. zeylanicum* (Gotmare and Tambe 2019). Neral, geranial, *trans*-cinnamaldehyde, hydrocinnamaldehyde, and salicylaldehyde were reported from C. verum leaf oils. *trans*-cinnamaldehyde and geranial showed significant antifungal activity against *Raffaelea quercus-mongolicae* and *Rhizoctonia solani* (Lee et al. 2020a). *Trans*-cinnamaldehyde, salicylaldehyde, and hydrocinnamaldehyde from *C. verum* bark showed antibacterial activity against *Agrobacterium tumefaciens* (Lee et al. 2020b). *C. zeylanicum* bark essential oils and cinnamaldehyde showed antibacterial activity against *Porphyromonas gingivalis* (MICs 6.25 µg/ml and 2.5 µM (Wang et al. 2018). Thirty five essential oils [α-thujene, α-pinene, camphene, sabinene, myrcene, α-phellandrene, α-terpinene, p-cymene, β-phellandrene, γ-terpinene, isoterpinolene, linalool, borneol, terpinen-4-ol, α-terpineol, (*Z*)-cinnamaldehyde, (*E*)-cinnamaldehyde,

(*E*)-anethole, (*E*)-cinnamyl alcohol, eugenol, α-copaene, (*Z*)-caryophyllene, (E)-β-caryophyllene, (*E*)-cinnamyl acetate, α-humulene, δ-cadinene, (*E*)-*o*-methoxycinnamaldehye cinnamaldehyde, caryophyllene alcohol, caryophyllene oxide] were identified in bark of *C. zeylanicum*. Volatile oils demonstrated antityrosinase activity (Marongiu et al. 2007). The presence of (*E*)-cinnamyl acetate and (*E*)-caryophyllene were determined by GC-MS in *C. zeylanicum*. Volatile oils (200 ppm concentration) displayed 66.9% antioxidant activity (Jayaprakasha et al. 2003). Eugenol, β- caryophyllene, camphor and linalool from leaves, and linalool, α- pinene, (*E*) -cinamyl acetate, α-phelandreno, (*E*)-cinamaldehyde, limonene, and β-cariophylene were determined in branches of *C. zeylanicum* (da Paz Lima et al. 2005). Various compounds [β-caryophyllene, α-humulene, α-copaene, δ- and γ-cadinene, germacrence-B, τ-and α-cadinol, α-bergamotene, α-copaene, α-humulene, and δ-cadinene, cinnamic acid, benzoic acid *p*-coumaric acid, p-hydroxy benzoic acid, caffeic acid, protocatecheuic acid, ferulic acid, vanillic acid, sinapic acid, syringic acid, 3,4,5-trihydroxy cinnamic acid, gallic acid, *o*-coumaric acid, salicylic acid, 2,5-dihydroxy cinnamic acid, gentisic acid, 2,4-dihydroxy cinnamic acid, eugenol, (*E*)-cinnamaldehyde, and linalool] have been determined by GC-MS from *C. zeylanicum* leaves, bark, fruits, root bark, flowers, and buds (Jayaprakasha and Rao 2011). A total of six compounds {4-piperidineacetic acid, 1-acetyl-5-ethyl-2-[3-(2-hydroxyethyl)-1*H*-indol-2-yl]-a-methyl-methyl ester, pentadecanoic acid, 14-methyl-, methyl ester, 10-octadecenoic acid, methyl ester, cyclopropanebutanoic acid, 2-[[2-[[2-[(pentylcyclopropyl) methyl]cyclopropyl]methyl]cyclopropyl]methyl]-, methyl ester, cyclopentaneundecanoic acid, methyl ester, and oxiraneundecanoic acid, 3-pentyl-, methyl ester, *cis-*} were determined by GC–MS analysis in methanol extract of *C. iners* leaves. The total phenolic (7.434 ± 0.04 mg tannic acid equivalents/g dry weight), total flavonoid content (3.2 ± 0.12 µg quercetin equivalents/g dry weight), and total antioxidant content (119.412 ± 0.39 mg tannic acid equivalents/g dry weight) were estimated in methanol extract *C. iners* leaves. The methanol extract demonstrated antioxidant assay in DPPH free radical scavenging assay (IC$_{50}$ 15 µg/ml; Udayaprakash et al. 2015). The ethyl acetate fraction of methanol extract of *C. iners* leaves demonstrated significant antibacterial activity against methicillin resistant *S. aureus* and *E. coli* (MIC 100 and 200 µg/ml). The xanthorrhizol was isolated and characterized from the active fraction. The identified compound was also found active against methicillin resistant *S. aureus* (Mustaffa et al. 2011). The presence of geraniol, linalool, (*E*)-caryophyllene, geranyl propanoate, (*E*)-phytol acetate, dill apiole, β-selinene, α-selinene, β-pinene, (*E*)-nerolidol, α-pinene, (*E*)-β-ocimene, and α-humulene was determined by GC-FID in leaves and stems (Suhaimi et al. 2017); 5,7-dimethoxy-3′,4′-methylenedioxyflavan-3-ol and β-sitosterol, 4-(4-hydroxy-3-methoxyphenyl)but-3-en-2-one, cinnamaldehyde, linoleic acid, and vanillin from dichloromethane extract of the twigs, eugenol, linoleic acid, and β-sitosterol were identified from leaves of *C. iners* (Espineli et al. 2013). *Cis*-linalool oxide, *trans*-linaool oxide, linalool, borneol L, terpinen-4-ol, α-terpineol, 2,6-octadien-1-ol, geraniol, propanoic acid, (-)-bornyl acetate, cyclohexene, 3-allyl-6-methoxyphenol, α-copaene, β-elemene, dodecanal, *cis*-α-bergamotene, caryophyllene, (+)-aromadendrene, 2-propen-1-ol, α-humulene, aromadendrene, alloaromadendrene, α-amorphene, germacrene-D, β-selinene, α-muurolene, β-bisabolene, α-cubebene, δ-cadinene, epiglobulol, nerolidol, palustrol, spathulenol, caryophyllene oxide, viridiflorol, tetradecanal, calarene, naphthalene, 1,2,3,4,4a,7- hexahydro, isospathulenol, and α-cadinol were reported from leaves of *C. iners* (Phutdhawong et al. 2007).

Catechin, epicatechin, procyanidin B2, and phenol polymers were identified from the subfractions of aqueous *C. verum* bark extract. The isolated compounds demonstrated significant inhibitory activity on the formation of advanced glycation end products (Peng et al. 2008). *C. zeylanicum* bark oils showed significant inhibition of reducing power, linoleic acid peroxidation, and DPPH radical-scavenging activities. The essential oils also demonstrated significant antibacterial (against *B. subtilis, E. coli, P. multocida* and *S. aureus*) and antifungal (*Aspergillus niger* and *A. flavus*) activities. Cinnamaldehyde, limonene, copaene, naphthalene, heptane, bicyclo[4.2.0]octa-1,3,5-triene, and 2-propenal were separated from crude essential oils of *C. zeylanicum* bark (Saleem et al. 2015). The bark volatile oil and oleoresin of *C. zeylanicum* demonstrated antifungal activity against *A. flavus, A. ochraceus, A. niger, A. terreus, P. citrinum* and *P. viridicatum* (6 µl concentration;

Singh et al. 2007). The presence of tricylene, α-thujene, α-pinene, camphene, sabinene, β-myrcene, α-phellandrene, α-terpinene, *o*-cymene, limonene, (*Z*)-β-ocimene, €-β-ocimene, γ-terpinene, *trans*-sabinene hydrate, linalool oxide, α-terpinolene, linalool, *allo*-ocimenelinanenen-1-ol, *trans*-pinocarveol, borneoelinanenen-4-ol, α-terpineol, verbenone, *trans*-piperitol, *trans*-carveol, (*Z*)-2-decenal, geraniol, bornyl acetate, (*Z*)-citral, bicycloelemene, eugenol, α-copaene, geranyl acetate, β-elemene, α-gurjunene, methyl eugenol, β-caryophyllene, α-santalene, γ-elemene, α-humulene, germacrene D, α-amorphene, β-selinene, eudesma-4,11-diene, bicyclogermacrene, α-muurolene, β-bisabolene, (*E,E*)-α-farnesene, γ-cadinene, β-himachalene, δ-cadinene, selina-4(15),7(11)-diene, calacorene, elemol, (*E*)-nerolidol, spathulenol, caryophyllene oxide, longiborneol, α-guaiol, β-oplopenone, caryophyllenol, β-eudesmol, α-cadinol, *cis*-α-santalol, valerenol, β-santalol, and farnesol was determined by GC-MS in *C. galucescens* and *C. verum* leaves (Chinh et al. 2017).

(*Z*)-3-phenylacrylaldehyde, (*E*)-cinnamaldehyde, (*E*)-3-phenyl-2-propen-1-ol, cinnamic acid, (+)-cyclosativene, α-copaene, (+)-sativene, chromen-2-one, acetic acid cinnamyl ester, γ-muurolene, cadina-1(10),4-diene, cinnamaldehyde, *o*-methoxy-, calacorene, 1-phenylbicyclo(4.1.0)heptane, epi-cubenol, T-muurolol, 6-phenyl-3,5-hexadien-2-one, 2-methylbenzofuran, (*Z*)-3-phenylacrylaldehyde, (*E*)-cinnamaldehyde, cinnamyl alcohol, (+)-cyclosativene, α-copaene, (+)-sativene, isosativene, caryophyllene, coumarin, acetic acid, cinnamyl ester, α-caryophyllene, eremophila-1(10),11-diene, γ-muurolene, α-bisabolene, cadina-1(10),4-diene, *p*-methoxy-cinnamaldehyde, α-calacorene, caryo-phyllenyl alcohol, caryophyllene oxide, γ-elemene, globulol, carotol, cubenol, α-cadinol, and pal-mitic acid vinyl ester were determined by GC-MS from acetone extract of *C. verum* bark. Acetone and ethyl acetate extracts of *C. verum* bark suppressed the growth of *Babesia bovis, B. bigemina, B. divergens, B. caballi*, and *Theileria equi* (IC$_{50}$ 23.1 ± 1.4, 56.6 ± 9.1, 33.4 ± 2.1, 40.3 ± 7.5, 18.8 ± 1.6 μg/ml, and 40.1 ± 8.5, 55.6 ± 1.1, 45.7 ± 1.9, 50.2 ± 6.2, and 61.5 ± 5.2 μg/ml). Similarly, both extracts (acetone and ethyl acetate) showed significant cytotoxic activity against MDBK, NIH/3T3, and HFF cells (EC$_{50}$ 440 ± 10.6, 816 ± 12.7 and 914 ± 12.2 μg/ml and 376 ± 11.2, 610 ± 7.7 and 790 ± 12.4 μg/ml; Batiha et al. 2020).

α-thujene, α-pinene, sabinene, β-pinene, β-myrcene, α-phellandrene, α-terpinene, o-cymene, limonene, 1,8-cineole, (*E*)-β-ocimene, γ-terpinene, cis-sabinene hydrate, α-terpinolene, linalool, *trans-p*-menth-2-en-1-ol, *trans*-sabinol, *trans*-pinocarveol, borneol, linalool oxide (pyranoid), ter-pinen-4-ol, α-terpineol, nerol, (*E*)-cinnamaldehyde, bornyl acetate, carvacrol, (*Z*)-dimethoxy citral, bicycloelemene, δ-elemene, α-cubebene, eugenol, cyclosativene, α-ylangene, α-copaene, β-cubebene, β-elemene, dodecanal, β-caryophyllene, γ-elemene, aromadendrene, α-humulene, α-amorphene, β-selinene, δ-selinene, *cis*-cadina-1,4-diene, α-muurolene, bicyclogermacrene, γ-cadinene, *cis-Z-*α-bisabolene epoxide, δ-cadinene, α-calacorene, spathulenol, caryophyllene oxide, viridiflorol, *trans*-β-elemenone, methoxy eugenol, tetradecanal, fonenol, 5-*epi*-neointermedeol, τ-muurolol (*epi*-αmuurolol), α-cadinol, valerianol, phenylethyl benzoate, 1,2-benzenedicarboxylic acid, hexadeca-noic acid methyl ester, hexadecanoic acid, and (*Z*)-13-docosenamide are essential oils from the stem barks of the *Cinnamomum* species (Son et al. 2015). 2-hexenal, α-thujene, α-pinene, benzaldehyde, sabinene, β-pinene, myrcene, α-phellandrene, α-terpinene, *p*-cymene, limonene, β-phellandrene, 1,8-cineole, δ-3-carene, benzyl alcohol, (*Z*)-β-ocimene, (*E*)-β-ocimene, γ-terpinene, trans-lin-alool oxide, *cis*-linalool oxide, terpinolene, linalool, camphor, isopulegol, citronellal, terpinen-4-ol, α-terpineol, cis-piperitol, nerol, geraniol, safrole, eugenol, methyl cinnamate, β-bourbonene, β-elemene, methyl eugenol, cis-α-bergamotene, β-caryophyllene, trans-α-bergamotene, γ-elemene, aromadendrene, α-humulene, β-(*E*)-farnesene, ar-curcumene, γ-muurolene, β-selinene, *cis*-β-guaiene, β-bisabolene, γ-cadinene, δ-cadinene, (*Z*)-nerolidol, elemol, spathulenol caryophyllne oxide, globulol, viridiflorol, β-eudesmol, α-eudesmol, (*E,Z*)-farnesol, (*E*)-asarone, benzyl benzoate, and benzyl salicylate were determined by GC-MS from *Cinnamomum* species (*C. rhyncophyllum, C. cordatum, C. microphyllum, C. scortechinii, C. pubescens, C. impressicostatum, C. mollissi-mum*, and *C. sintoc*). These essential oils demonstrated strong larvicidal effects against mosquitoes after 3 h exposure (LC$_{50}$ 133.0 to 243.0 μg/ml against *Aedes aegypti* and from 118.0 to 194.0 μg/ml against *A. albopictus* (Jantan et al. 2005a).

α-thujene, α-pinene, camphene, sabinene, β-pinene, myrcene, α-phellandrene, δ-3-carene, α-terpinene, p-cymene, limonene, β-phellandrene, sylvestrene, 1,8-cineole, (*E*)-β-ocimene, γ-terpinene, *cis*-linalool oxide (furanoid), terpinolene, linalool, trans-pinocarveol, *cis*-verbenol, camphor, sabina ketone, pinocarvone, terpinen-4-ol, cryptone, α-terpineol, verbenone, citronellol, nerol, cumin aldehyde, geraniol, safrole, bornyl acetate, thymol, ρ-cymen-7-ol, carvacrol, δ-elemene, α-cubebene, α-terpinyl acetate, α-ylangene, α-copaene, β-cubebene, β-elemene, methyl eugenol, (*E*)-caryophyllene, α-humulene, *allo*-aromadendrene, α-amorphene, germacrene D, β-selinene, bicyclogermacrene, α-muurolene, δ-amorphene, γ-cadinene, δ-cadinene, elemol, elemicin, caryophyllene alcohol, spathulenol, caryophyllene oxide, humulene epoxide II, 1-*epi*-cubenol, α-muurolol, cubenol, β-eudesmol, α-eudesmol, α-cadinol, shyobunol, benzyl benzoate were determined by GC-MS from *C. wightii* bark and leaves. The essential oil showed better hypoglycemic activity (IC_{50} 1.617 ± 0.02 and 1.146 ± 0.02 mg/ml for α-amylase and α-glucosidase suppressive effects). The essential oils also demonstrated stronger antioxidant activity (IC_{50} 2.552 ± 0.13 and 3.485 ± 0.09 mg/ml in ABTS and DPPH assays; Sriramavaratharajan and Murugan 2018).

α-thujene, α-pinene, α-fenchene, camphene, sabinene, β-pinene, myrcene dehydroxy-*trans*-linalool oxide, dehydroxy-*cis*-linalool oxide, α-phellandrene, α-terpinene, *p*-cymene, limonene, β-phellandrene, 1,8-cineole, (*E*)-β-ocimene, γ-terpinene, terpinolene, rosefuran, perillene, linalool, hotrienol, isocitral, camphor, nerol oxide, isoneral, *cis*-linalool oxide (pyranoid), *trans*-linalool oxide (pyranoid), borneol, isogeranial, terpinen-4-ol, α-terpineol, decanal, citronellol, nerol, cuminal, neral, geraniol, geranial, (*E*)-cinnamaldehyde, bornyl acetate, safrole, δ-elemene, eugenol, α-ylangene, α-copaene, β-elemene, methyl eugenol, β-caryophyllene, *trans*-α-bergamotene, *allo*-aromadendrene, α-humulene, α-amorphene, β-selinene, *trans*-β-bergamotene, α-selinene, germacrene D, bicyclogermacrene, β-bisabolene, eugenyl acetate, δ-cadinene, (*E*)-α-bisabolene, (*E*)-nerolidol, germacrene B, spathulenol, caryophyllene oxide, intermedeol isomer, selin-11-en-4-one, selina-3,11-dien-6α-ol, α-cadinol, selin-11-en-4α-ol, germacra-4(15),5,10(14)-trien-1α-ol, aromadendrane-4,10-diol, oplopanone, α-cyperone, cyclocolorenone, monoterpene hydrocarbons, oxygenated monoterpenoids, sesquiterpene hydrocarbons, oxygenated sesquiterpenoids, and phenylpropanoids were determined in *Cinnamomum* species (*C. tonkinense* and *C. polyadelphum*) from north central Vietnam. The essential oils of *C. tonkinense* leaves showed excellent antimicrobial activity against *E. faecalis* (MIC 32 μg/ml) and *Candida albicans* (MIC 32 μg/ml). Similarly, essential oils also displayed larvicidal activity against *Aedes aegypti* (24 h LC_{50} of 17.4 μg/ml) and *Culex quinquefasciatus* (14.1 μg/ml). *C. polyadelphum* leaf essential oil also demonstrated significant antimicrobial activity against Gram-positive bacteria and mosquito larvicidal activity (Dai et al. 2020).

Monoterpene hydrocarbons (camphene, β-pinene, p-cymene, limonene, α-terpinene), oxygenated monoterpenes (linalool oxide, linalool, camphor, borneol, terpinene-4-ol, α-terpineol, geraniol, citral, bornyl acetate, geranyl acetate), sesquiterpene hydrocarbons (cyclosativene, α-cubebene, α-fenchene, humulene, β-caryophyllene, α-caryophyllene, aromadendrene, T-muurolene, valencene, α-muurolene, δ-cadinene, γ-muurolene, β-cadinene γ-elemene, isoledene, α-guaiene, copaene, azunol), oxygenated sesquiterpenes (caryophyllene oxide, guaiol, T-cadinol, α-cadinol, β-cadinol), diterpene hydrocarbons (rimuen, labda-8(20),12,14,-triene, kaur-16-ene), oxygenated diterpenes (verticiol), and others (benzaldehyde, salicylaldehyde, benzyl alcohol, 2-methylbenzofuran, benzenepropanal, *p*-allylanisole, anethole, *cis*-cinnamaldehyde, 4-allyphenol, *trans*-cinnamaldehyde, cinnamyl alcohol, eugenol, coumarin, cinnamyl acetate) were determined by GC-MS in *C. osmophloeum* leaves. Among identified essential oils, cinnamaldehyde showed maximum antifungal activity against *Trametes versicolor*, *Lenzites betulina* and *Laetiporus sulphureus* (IC50 73, 74 and 73 μg/ml; Cheng et al. 2006).

1-methyl-cyclopentanol, α-pinene, camphene, β-thujene, sabinene, β-pinene, β-myrcene, α-phellandrene, δ-2-carene, δ-3-carene, *p*-cymene, limonene, 1,8-cineole, (*E*)-β-ocimene, γ-terpinene, α-terpinolene, verbenol, (*Z*)-*p*-Menth-2-en-1-ol, camphor, terpinen-4-ol, α-terpineol, *cis*-geraniol, (*E*)-citral, (*E*)-cinnamaldehyde, bornyl acetate, α-cubebene, α-copaene, α-gurjunene, (*E*)-cinnamyl acetate, β-caryophyllene, aromadendrene, (*Z*)-β-farnesene, γ-muurolene, germacrene D, β-vatirenene, β-bisabolene, γ-cadinene, β-curcumene, δ-cadinene, *trans*-calamenene, β-calacorene, nerolidol, ledol, spathulenol, caryophyllene oxide, *allo*-aromadendrene oxide, α-guaiol, *trans*-longipinocarveol, isoaromadendrene epoxide, longiverbenone, β-acorenol, τ-cadinol, cubenol, α-cadinol, 7(11)-selinen-4α-ol, and α-santalol were determined in *C. curvifolium* and *C. mairei* (Dai et al. 2019).

Various essential oils (α-thujene, α-pinene, camphene, sabinene, β-pinene, 6 *cis-m*-mentha-2,8-diene, myrcene,8 α-phellandrene, δ-3-carene, 1,4-cineole, α-terpinene, *p*-cymene, β-phellandrene, 1,8-cineole, (Z)-β-ocimene, (E)-β-ocimene, γ-terpinene, *cis*-sabinene hydrate, *cis*-linalool oxide, *trans*-linalool oxide, terpinolene, linalool, α- pinene oxide, *trans*-sabinene hydrate, myrcenol, camphor, citronellal, camphene hydrate, (1Z)-2-propenyl phenol, terpinen-4-ol, α-terpineol, 4(E)-decenal, γ-terpineol, (Z)-cinnamaldehyde, citronellol, piperitone, *trans*-myrtanol, safrole, (Z)-methyl cinnamate, δ-elemene, α-terpinyl acetate, eugenol, hydrocinnamyl acetate, α-copaene, β-cubebene, β-elemene, methyl eugenol, (E)-caryophyllene, γ-elemene, (E)-cinnamyl acetate, α-humulene, allo-aromadendrene, *cis*-cadina-1(6),4-diene, *trans*-cadina-1(6)4-diene, γ-muurolene, α-amorphene, germacrene D, β-selinene, δ-selinene, α-selinene, α-muurolene, bicyclogermacrene, δ-amorphene, γ-cadinene, 7-*epi*-α-selinene, δ-cadinene, cubebol, (E)-γ-bisabolene, 10-*epi*-cubebol, α-cadinene, α-calacorene, *cis*-murrol-5-en-4-α-ol, germacrene B, maaliol, spathulenol, caryophyllene oxide, globulol, viridiflorol, rosifoliol, guaiol, 5-*epi*-7-epi-α-eudesmol, humulene epoxide (II), tetradecanal, 1,10-di-*epi*-cubenol, 1-*epi*-cubenol, eremoligenol, muurola-4,10(14)-dien-1-β-ol, γ-eudesmol, caryophyllene-4-(12),8(13)-dien-5α-ol, *cis*-cadin-4-en-7-ol, *epi*-α-cadinol, *epi*-α-muurolol, α-muurolol, α-eudesmol, α-cadinol, selin-11-en-4-α-ol, 7-*epi*-α-eudesmol, intermedeol, 14-hydroxy-Z-caryophyllene, khusinol, germacra-4(15), 5, 10(14)-trien-1-α-ol, α-bisabolol, shyobunol, 2(E),6(Z)-farnesol, 2(Z),6(E)-farnesol, 6R,7R-bisabolene, 6S,7R-bisabolene, benzyl benzoate, γ-curcumen-15-al, (Z)-nerolidylisobutyrate, (E)-nerolidylisobutyrate, benzyl cinnamate) have been determined in *Cinnamomum* species (*C. dubium* Nees, *C. litseifolium* Thwaites, *C. mohanense*, *C. palghatensis* Gangop., *C. riparium* Gamble, *C. travancoricum* Gamble, *C. walaiwarense* Kosterm. and *C. wightii* Meissn.) and two chemotypes (*C. agasthyamalayanum* Robi et al. and *C. keralaense* Kosterm.) from the Western Ghats, south India (Ananthakrishnan et al. 2018).

The presence of α-thujene, α-pinene, camphene, sabinene, β-pinene, myrcene, α-phellandrene, α-terpinene, *p*-cymene, limonene, 1,8-cineole, *cis*-β-ocimene, *trans*-β-ocimene, γ-terpinene, *cis*-sabinene hydrate, *trans*-linalool oxide (furanoid), terpinolene, linalool, fenchol, *trans*-sabinene hydrate, *trans-p*-mentha-2-en-1-ol, borneol, terpinen-4-ol, α-terpineol, *trans*-piperitol, nerol, geraniol, δ-elemene, eugenol, α-copaene, methyl eugenol, β-caryophyllene, *trans*-α-bergamotene, α-humulene, alloaromadendrene, γ-muurolene, germacrene D, β-selinene, α-selinene, *trans, trans*- α-farnesene, γ-cadinene, δ-cadinene, *trans*-nerolidol, caryophyllenyl alcohol, viridiflorol, α-muurolol, α-cadinol, and α-bisabolol was determined by GC-MS in methanol extract of *C. altissimum* Kosterm bark. Methanol extract displayed antioxidant activities, with an IC50 value of 38.5 ± 4.72 lg/ml, using 1,1-diphenyl-2-picrylhydrazyl (DPPH) assay, and 345.2 ± 14.8 lM Fe (II)/g dry mass using ferric reducing/antioxidant power (FRAP) assay (Abdelwahab et al. 2017).

α-amorphene, α-copaene, dcadinene, (E)-nerolidol, camphor, perillene, cterpinene, α-fenchol, 1-linalool, aphellandrene, citronellol, β-pinene, bcaryophyllene, β-phellandrene, linalool, (Z)-bocimene (Kumar et al. 2019), α-muurolene, spathulenol, 1,8- cineole, caryophyllene oxide, *p*-cymene; terpinen-4-ol, α-humulene, α-terpineol, limonene, bpinene, caryophyllene · alcohol, germacrene-D, α-cadinol, α-muurolol, α-selinene, β-bisabolene, β-elemene, β-selinene, borneol, *cis*-linalool oxide (furanoid), citronellal, geranial, geranyl acetate, myrcene, neral, terpinolene, *trans*-linalool oxide (furanoid), camphene, geraniol, nerol, α- pinene, geranyl formate, piperitone, α-fenchol (E)-cinnamaldehyde, (Z)- cinnamaldehyde, δ-cadinene, α-copaene, camphor, caryophyllene oxide, bphellandrene, β-pinene, limonene, linalool, β-pinene, and spathulenol (Rameshkumar and George 2006b), 1,8 cineole, borneol, camphene, geranial, p-cymene, terpinen-4- ol, β-bisabolene, β-caryophyllene, α-fenchol, α- humulene, α-muurolene, α-pinene, and α-terpineol were determined by GC-MS in *C. sulphuratum* leaves and bark (Baruah et al. 1999).

Three coumarins (cinnakanin A-C) and two cyclobutyl aliphatic alcohols (cinnakanol A-B) were isolated and characterized from *C. kanehirae*. The ethyl acetate extract of *C. kanehirae* leaves showed significant cytotoxic effects against human esophageal adenocarcinoma (BE-3) and human esophageal squamous cell carcinoma (CE 81T/VGH) cell lines (Leu et al. 2014). The presence of (-)-nerolidol, (-)-terpinen-4-ol, 1,8-cineole, 10-*epi*-cubebol, 3-carene, caryophyllene oxide, *cis*-linalool oxide, *cis*-β-ocimene, citronellol, citronellol acetate, *E*-citral, epi-cubenol, geraniol, germacrene-D, guaiol, limonene, linalool, nerol, *p*-cymene, sabinene, spathulenol, T-cadinol, T-muurolol, *trans*-linalool oxide, *trans*-β-caryophyllene, *trans*-bocimene, Z-citral, α-cadinol, α-humulene, amuurolene, α-phellandrene, α-pinene, α-terpinene, α-terpineol, α-terpinyl acetate, α-thujene, α-ylangene, β-elemene, β-myrcene, β-pinene, β-selinene, γ-murolene, γ -terpinene, δ-cadinol, δ-selinene, and δ-terpineol was determined by GC-MS in *C. kanehirae* (Cheng et al. 2015).

The essential oils {3-*O*-[β-D-Xylopyranosyl-(1→2)-α-L-arabinofuranoside], 7-*O*-α-L-rhamnopyranoside, sagittatin A, L-borneol, α-terpineol, *p*-allylanisole, *trans*-cinnamaldehyde, L-bornyl acetate, eugenol, acopaene, β-caryophyllene, cinnamyl acetate, acaryophyllene, curcumene, δ-cadinene, α-calacorene, elemicin, γ-nerolidol, spathulenol, caryophyllene oxide, *trans*-β-elemenone, γ-eudesmol, caryophylla-4 (14), 8(15)-dien-5-α-ol, δ-cadinol, T-cadinol, cadalene, guaiol acetate} from twigs (Tung et al. 2008); kaempferol glycosides (kaempferol 3-*O*-β-Dxylopyranosyl-(1→2)-α-L-arabinofuranosyl-7-*O*-α-L-rhamnopyranoside, kaempferol 3-*O*-β-D-xylopyranosyl-(1→2)-α-L-rhamnopyranosyl-7-*O*-α-L-rhamnopyranoside, kaempferol 3-*O*-β-D-glucopyranosyl-(1→2)-α-L-arabinofuranosyl-7-*O*-α-L-rhamnopyranoside, kaempferol 3-*O*-α-L-rhamnopyranosyl-(1→2)-α-L-arabinofuranosyl-7-*O*-α-L-rhamnopyranoside, kaempferol 3-*O*-β-D-apiofuranosyl-(1→2)-α-L-arabinofuranosyl-7-*O*-α-L-rhamnopyranoside, kaempferol 3-*O*-β-Dglucopyranosyl-(1→2)-α-L-rhamnopyranosyl-7-*O*-α-L-rhamnopyranoside, kaempferol 3-*O*-β-D-glucopyranosyl-(1→4)-α-L-rhamnopyranosyl-7-*O*-α-L-rhamnopyranoside, kaempferitrin, and kaempferol 7-*O*-α-L-rhamnopyranoside were determined from twigs of *C. osmophloeum*. Kaempferol 7-*O*-α-L-rhamnopyranoside showed inhibitory effect against LPS-induced production of nitric oxide in RAW 264.7 macrophages (IC$_{50}$ 41.2 μM; Lin and Chang 2012). (*E*)-cinnamaldehyde, (*Z*)-cinnamaldehyde, γ-muurolene, 1,8-cineole, 2-hydroxy-1,8-cineol, 4- terpineol, camphene, caryophyllene, caryophyllene oxide, cedrol, citronellol, copacamphene, δ-carvone, geranial, geraniol, geranyl acetate, isoborneol, limonene, linalool, menthone, myrcene, neral, nerol, ocimene, *p*-cymene, phytol, piperitone, spathulenol, T-cadinol, terpinolene, *trans*-linalool oxide, β-pinene, γ-cadinene, γ-humulene, α-pinene, α-terpinene, α- terpineol, α-terpinyl acetate from bark (Fang et al. 1989); *cis*-cinnamaldehyde, *trans*-cinnamaldehyde, cinnamaldehyde, γ-cadinene, δ-cadinene, α-copaene, copaene, camphor, γ-muurolene, limonene, linalool, camphene, caryophyllene oxide, geraniol, α-cadinol, α-humulene, β-pinene, α-terpineol, geranyl acetate, α-caryophyllene, α-muurolene, guaiol, β-caryophyllene, β-phellandrene, γ-elemene, T-cadinol, caryophyllene, β-bourbonene, 1,8-cineol, cedrol, spathulenol, T-muurolol, *trans*-bcaryophyllene, δ-cadinene, δ-cadinol, *p*-cymene, apinene from leaves (Huang et al. 2007); (+)-4-carene, (*E*)-ocimene, 3-carene, *cis*-β-terpineol, D-limonene, eucalyptol, germacrene-D, neral, nerol, sabinene, terpin-4-ol, terpinolene, bhumulene, β-myrcene, γ-terpinene, α-farnesene, α-phellandrene, α-thujene, kaempferol 3-*O*-α-L-rhamnopyranoside, kaempferol 7-*O*-α-L-rhamnopyranoside, kaempferol 3-*O*-α-Lrhamnopyranoside-7-*O*-α-L-rhamnopyranoside, kaempferol 3-*O*-α-L-rhamnopyranosyl-(1–2)- α-L-rhamnopyranoside from stem (Chen et al. 2004); 9,9′-di-*O*-feruloyl-5,5′ -dimethoxysecoisolariciresinol, (7′ *S*,8′ *R*,8*R*)-lyoniresinol-9-*O*-(*E*)-feruloyl ester, (7′ *S*,8′ *R*,8*R*)-lyoniresinol-9,9′-di-*O*-(*E*)-feruloyl ester, secoisolariciresinol, and (-)-lyoniresinol were determined by GC-MS from heartwood and roots of *C. osmophloeum* (Chen et al. 2010a).

The hot water extract of *C. osmophloeum* leaves (100 mg/kg body weight for 5 and 10 weeks; p.o.) administration significantly decreased the levels of total cholesterol, triglyceride and low-density lipoproteins. The levels of plasma low-density lipoprotein-cholesterol were also decreased to 27.77% after 10 weeks feeding. Similarly, kaempferol 3-*O*-β-D-apiofuranosyl-(1→2)-α-L-arabinofuranosyl-7-*O*-α-L-rhamnopyranoside and kaempferitrin were identified from hot aqueous extract of *C. osmophloeum* leaves (Lin et al. 2011a). The essential oils (cinnamaldehyde, benzaldehyde and 3-phenylpropionaldehyde) of *C. osmophloeum* leaves (1 mg/kg bw) did not show any cytokine-modulatory effects in ovalbumin (OVA)-primed Balb/C mice; though, the serum IL-2, IL-4 and IL-10 concentrations significantly were enhanced in treated animals (Lin et al. 2011b, 2011c). The antifungal effect of leaf essential oils of *C. osmophloeum* leaves was assessed against *Coriolus versicolor* and *Laetiporus sulphureus*. The cinnamaldehyde showed significant antifungal activity against *C. versicolor* and *L. sulphureus* (MIC 50 and 75 ppm; Wang et al. 2005; Bakar et al. 2020).

β-pinene, α-terpinene, limonene, *p*-cymene, camphor, borneol, α-terpineol, safrole, α-cubebene, α-ylangene, α-copaene, β-patchoulene, methyl eugenol, longifolene, γ-selinene, aromadendrene, dehydroaromadendrene, β-cadinene, γ-gurjunene, germacrene D, β-selinene, δ -selinene, valencene, α-selinene, epizonarene, δ-cadinene, cis-calamenene, α-cadinene, α-calacorene, selina-3,7(11)-diene, elemicin, germacrene B, spathulenol, and caryophyllene oxide were reported from *C. griffthii* and *C. macrocarpum* (Salleh et al. 2015b). Benzaldehyde, hydrocinnamaldehyde, benzoic acid, butanoic acid, propanoic acid, isovaleric acid, cinnamaldehyde, copaene, cyclohexene, naphthalene, γ-cadinene, (+)-α-terpineol, 1,3-cyclohexadiene, 1,4- cyclohexadiene, 1,6-cyclodecadiene, 1,6-octadien-3- Ol, 1,6-octadiene, 2-norbornanol, 3-carene, 4- terpineol, bicyclo hept-3-ene,-3-care, camphene, camphor, caryophyllene, cineole, cubenol, cymene, cymol, eucalyptol, linalool, nerolidol A, sabinene, spathulenol, β-myrcene, β-pinene, γ-elemene, cmuurolene, δ-cadinene, α-caryophyllene, α-cubebene, α-fenchol, α-humulene, α-phellandrene, α-pinene, α- selinene, α-terpinene, α- terpinolene from bark and cinnamaldehyde, naphthalenol, copaene, naphthalene, β-cadinene, δ-cadinene, (+)-bornanone, 1,3-cyclohexadiene, 1,4- cyclohexadiene, 1,6-cyclodecadiene, 1,6-octadien-3- ol, 3-carene, 4-carvomenthenol, 4-terpineol, bicyclo α-thujene, caryophyllene, cedr-8-ene, cineole, cis-α- bisabolene, cymene, cymol, eucalyptol, germacrene-D, linalool, sabinene, spathulenol, thujene, β-myrcene, bpinene, γ-terpinene, γ-ocimene, α-(+)-pinene, α- caryophyllene, α-cubebene, α-humulene, α- phellandrene, α-terpinene, α-terpinolene, camphor, cis-calamenene, and germacrene B were determined by GC-MS from leaves of *C. macrocarpum* (Hrideek et al. 2016).

Kaempferol 3-*O*-α-L-[2-(*Z*)-*p*-coumaroyl-4-(*E*)-*p*-coumaroyl]rhamnopyranoside and kaempferol 3-*O*-α-L-[2,4-di-(*E*)-*p*-coumaroyl]rhamnopyranoside have been isolated from the methanolic extract of *C. kotoense* leaves. Both compounds inhibited human peripheral blood mononuclear cells proliferation stimulated by phytohemagglutinin (IC$_{50}$ 5.0 ± 1.3 and 6.0 ± 1.5 μM (Kuo et al. 2005). Similarly, C$_{19}$ γ-lactone (cinnakotolactone and isolinderanolide B) were isolated and characterized from the *n*-hexane extract of *C. kotoense* leaves. Both identified compounds showed anti-proliferation activities against human HT29 and MCF-7 cancer cell lines (IC$_{50}$ 3.3±0.3 to 25.8±5.3 μM; Yang et al. 2006). Kaempferol 3-*O*-α-L-[2-(*Z*)-*p*-coumaroy-4-(*E*)-pcoumaryl]rhamnopyranoside (Kuo et al. 2005), three butanolides (kotomolides A and B and isokotomolide A), and one secobutanolide (secokotomolide A; Chen et al. 2006) from leaves, three butanolides (kotolactones A and B and secokotomolide), one long chain alcohol (kotodiol), and one furan (2-acetyl-5-dodecylfuran) and others (isoobtusilactone A and lincomolide B) from stem wood (Chen et al. 2005); and one butanolide (kotomolide) were obtained from the stem of *C. kotoense* (Chen 2006). Obtusilactone A was identified from *C. kotoense* and demonstrated effective anticancer activity. The compound also stimulated osteogenesis of bone marrow-derived mesenchymal stem cells. The compound induced the expression of osteogenesis markers *BMP2*, *Runx2*, *Collagen I*, and *Osteocalcin* in compound-treated bone marrow-derived mesenchymal stem cells (Lin et al. 2017). Obtusilactone A and (-)-sesamin were identified from *C. kotoense* and investigated for their anticancer activity. Obtusilactone A and (-)-sesamin connect with Ser855 and Lys898 residues in the active site of the Lon protease. Both compounds induced p53-independent DNA injury responses in non-small-cell lung cancer cell lines, as supported by phosphorylation of checkpoint proteins, caspase-3 cleavage, and sub-G1 deposition (Wang et al. 2010). Five compounds (isoobtusilactone A, obtusilactone

A, (+)-syringaresinol, β-sitosterol, and stigmasterol) were isolated and characterized from methanol extract of *C. kotoense* fruits (Chen and Hong 2011). Seven compounds (kotolactone A, kotolactone B, secokotomolide, kotodiol, and 2-acetyl-5-dodecylfuran, isoobtusilactone A and lincomolide B) have been isolated and characterized from *C. kotoense* stem wood. Isoobtusilactone A and lincomolide B demonstrated *in vitro* antitubercular activities (MICs 22.48 and 10.16 µM) against *Mycobacterium tuberculosis* 90–221387 (Chen et al. 2005).

Methyl eugenol, bicyclogermacrene, safrole, methyl eugenol elemicin, bicyclogermacrene, camphor, safrole and methyl eugenol, p-cymene, β-eudesmol, α-pinene, (*E*)-nerolidol, bicyclogermacrene, β-selinene, spathulenol, and β-eudesmol were determined by GC-MS in Australian *Cinnamomum (C. baileyi, C. oliveri* and one chemotype of *C. laubatii*) species (Brophy et al. 2001). EBC-23 from C. *laubatii* showed strong inhibition of the growth of an androgen-independent prostate tumor cell line DU145 in a mouse model (Dong et al. 2009).

The crude aqueous extract of *C. loureiroi* stem bark (100 µg/10 ml) displayed counter irritant when compared with a reference drug (dexamethasone). Extract demonstrated the counter-irritant effect (maximum inhibition 91.97% and lowest inhibition 41.39 % inhibition) in the ear of treated rabbits (Khan et al. 2014). Crude hexane extract of *C. loureiroi* stem bark demonstrated strong HIV-1 reverse transcriptase inhibitory activity (Silprasit et al. 2011). Cinnamic aldehyde, 3-methoxycinnamaldehyde, cinnamaldehyde, copaene, α-amorphene, bcadinene, caryophyllene, phellandrene, pinene, cadinadiene-4,9, cubenol, limonene, α-cedrene oxide, α-guaiene, α-myrcene, and β-pinene were determined from the bark of young branches C. *loureiroi* grown in China (Jiang et al. 2008). *C. cassia* and *C. loureirii* exhibited strong inhibition of cyclooxygenase-2 activity (>80% inhibition at the test concentration of 10 µg/ml; Hong et al. 2002).

Cinnamtannin B1 (20 mg/kg), a condensed tannin (from *C. validinerve*), reduced local redness, inflammation, and neutrophil formation in the mouse model of rosacea *in vivo*. Cinnamtannin B1 inhibited myeloperoxidase and macrophage inflammatory protein 2 formation at the lesions. Cinnamtannin B1 reversed the formation of LL-37-induced IL-8 in human keratinocytes HaCaT and monocyte THP-1 cells. Cinnamtannin B1 suppressed IL-8 formation through downregulating the extracellular signal-regulated kinase phosphorylation in the mitogen-activated protein kinase pathway (Kan et al. 2020). *Erythro*-guaiacylglycerol-β-*O*-4′-(5′)-methoxylariciresinol, caryolane-1,9β-diol, cinnamtannin B1, validinol, reticuol, burmanol, validinolide, lincolomide A, linderanolide B, isolinderanolide B, (-)-5,7-dimethoxy-3′,4′-methylenedioxy-flavan-3-ol, taxifolin, isophilippinolide A, secosubamolide, (-)-yangambin, (-)-pinoresinol, (+)-monomethylpinoresinol, and (+)-syringaresinol were identified from ethyl acetate fraction of methanol extract of *C. validinerve* stem. Among the isolated compounds, lincomolide A, secosubamolide, and cinnamtannin B1 demonstrated potent inhibitory effects on both superoxide anion generation (IC$_{50}$ 2.98 ± 0.3 µM, 4.37 ± 0.38 µM, and 2.20 ± 0.3 µM) and elastase formation (IC$_{50}$ 3.96 ± 0.31 µM, 3.04 ± 0.23 µM, and 4.64 ± 0.71 µM) by human neutrophils. Moreover, isophilippinolide A, secosubamolide, and cinnamtannin B1 demonstrated bacteriostatic activity against *Propionibacterium acnes* (MIC 16 µg/ml, 16 µg/ml, and 500 µg/ml; Yang et al. 2020).

Styrene, α-thujene, α-pinene, benzaldehyde, sabinene, β-pinene, myrcene, α-phellandrene, α-terpinene, *p*-cymene, limonene, β-phellandrene, 1,8-cineole, δ-3-carene, benzyl alcohol, (Z)-β-ocimene, (*E*)-β-ocimene, γ−terpinene, trans-linalool oxide, cis-linalool oxide, terpinolene, linalool, camphor, isopulegol, citronellal, borneol, terpinen-4-ol, α-terpineol, cinnamaldehyde, *cis*-piperitol, nerol, geraniol, safrole, methyl(Z)-cinnamate, eugenol, methyl(*E*)-cinnamate, β-bourbonene, neryl acetate, α-copaene, β-elemene, methyl eugenol, *cis*-α-bergamotene, β−caryophyllene, *trans*-α-bergamotene, aromadendrene, β-(Z)-farnesene, α-humulene, β-(*E*)-farnesene, curcumene, γ-muurolene, β-selinene, *cis*-β-guaiene, α-(*E,E*)-farnesene, β-bisabolene, γ -cadinene, δ-cadinene, (Z)-nerolidol, (*E*)-nerolidol, spathulenol, caryophyllne oxide, globulol, viridiflorol, β-eudesmol, α-eudesmol, (*E,Z*)-farnesol, (*E*)-asarone, benzyl benzoate, and benzyl salicylate identified from *Cinnamomum rhyncophyllum, C. cordatum, C. microphyllum, C. scortechinii, C. pubescens, C. impressicostatum, C. mollissimum*, and *C. zeylanicum* (Jantan et al. 2008). 2-(4-hydroxy-3-methoxyphenyl)ethyl hexacosanoate,

2-(4-hydroxy-3-methoxyphenyl)ethyl octacosanoate (Kuo and Shue 1991), isoreticulide (Lin et al. 2011c; Chen and Yeh 2011), and reticuol (Cheng et al. 2010b; Chia et al. 2011) from leaves, and reticuone (Cheng et al. 2010a), a mixture of 4-hydroxy-3-methoxyphenethyl pentadecyrate, 4-hydroxy-3-methoxyphenethyl stearate, 4- hydroxy-3-methoxyphenethyl heneicosyrate (Lin et al. 2010), and cinnaretamine were reported from the stem of *C. reticulatum* (Chen et al. 2011). Eight compounds (subamolides A-C, secosubamolide from stem; Chen et al. 2007), subamolides (D and E, and secosubamolide A from leaves; Kuo et al. 2008), and one sesquiterpeneoid (subamol from roots; Chen et al. 2010b) were separated from whole plant of *C. subavenium*. Four butanolides (tenuifolide A, isotenuifolide A, tenuifolide B, and secotenuifolide A) and one sesquiterpenoid, tenuifolin from stem (Lin et al. 2009) and two (ethyl 3,5-dihydroxy-4-nitrobenzoate (Cheng et al. 2011) and the benzodioxocinone, 2,3-dihydro-6,6- dimethylbenzo[b][1,5]-dioxocin-4(6H)-one were determined from leaves of *C. tenuifolium* (Chen et al. 2012).

Ethyl 3,5-dihydroxy-4-nitrobenzoate, 2,3-dihydro6,6-dimethylbenzo[b][1,5]dioxocin-4(6H)-one, (+) spathulenol, 1,8-cineole, 1-phellandrene, 2 borneol L, benzaldehyde, bicyclo, bornyl acetate, calarene, camphene, carvone, caryophyllene oxide, elemol, limonene, linalool, *p*-cymen-8-ol, sabinene, terpinen4-ol, α-pipene, α-terpineol, β-eudesmol, β-myrcene, β-pinene, γ-gurjunene, δ-selinene, δ3-carene, l-kaurene, δ-δ-cadinene, l-copaene, epsilon-cadinene, 1,8-cineole, camphene, camphor, *cis*-linalool oxide, citronellol, δ-*cis*-yabunikkeol, geraniol, l-caryophyllene, l-linalool, l-*trans*-yabunikkeol, l-α- phellandrene, l-α-terpineol, limonene, nerol, *p*-cymene, terpinen-4-ol, *trans*-linalool oxide, β-elemene, β-myrcene, β-pinene, α-cadinol, α- humulene, α-pinene from leaves; δ-δ-cadinene, l-copaene, ecadinene, 3-hexen-1-ol, calamenene, elemol, eugenol, lcarvone, methyl eugenol, safrole, β-calacorene, α-calacorene, α- terpinyl acetate, 1,8-cineole, camphene, camphor, *cis*-linalool oxide, citronellol, δ-*cis*-yabunikkeol, geraniol, l-caryophyllene, l-linalool, l-*trans*-yabunikkeol, l-α-phellandrene, l-α-terpineol, limonene, nerol, *p*-cymene, terpinen-4-ol, *trans*-linalool oxide, β-elemene, β-myrcene, β-pinene, α-cadinol, α- humulene, α-pinene from twigs; δ-δ-cadinene, l-copaene, ecadinene, 1,8-cineole, camphene, camphor, *cis*-linalool oxide, citronellol, δ-*cis*-yabunikkeol, geraniol, l-caryophyllene, l-linalool, l-*trans*-yabunikkeol, l-α- phellandrene, l-α-terpineol, limonene, nerol, pcymene, terpinen-4-ol, *trans*-linalool oxide, β-elemene, β-myrcene, β-pinene, α-cadinol, α- humulene, and α-pinene were identified by GC-MS from *C. tenuifolium* (syn. *C. japonicum* Siebold ex Nakai; Fujita et al. 1971).

Monoterpene hydrocarbons (α-thujene, α-pinene, camphene, β-pinene, myrcene, α-phellandrene, δ-3-carene, *p*-cymene, *o*-cymene, limonene, *cis*-ocimene, *trans*-ocimene, terpinolene), oxygenated monoterpenes (1,8-cineole, linalool, *cis-p*-menth-2-en-1-ol, terpinen-4-ol, α-terpineol, verbenone, bornyl acetate), sesquiterpene hydrocarbons (δ-elemene, α-copaene, β-elemene, β-caryophyllene, aromadendrene, α-humulene, *allo*-aromadendrene, γ-muurolene, germacrene D, β-selinene, bicyclogermacrene, δ-cadinene, germacrene B), oxygenated sesquiterpenes (spathulenol, globulol, viridiflorol, *epi*-α-cadinol, T-muurolol, α-cadinol), phenyl propanoids (eugenol, and eugenyl acetate) and other compounds (3-hexen-1-ol, and ethyl hexanoate) were determined from *C. tamala* leaves (Rana et al. 2012). Eugenol, eugenyl acetate, α-phellandrene, β-phellandrene, α-pinene, elixene, *cis*-caryophyllene, myrcene, and limonene were determined from *C. tamala* leaves (Sankaran et al. 2015). Eugenol demonstrated significant radical scavenging activity against DPPH and superoxide radicals with a potent metal chelating activity (Padmakumari Amma et al. 2013; Sharma and Rao 2014).

α-thujene, α-pinene, camphene, sabinene, β-pinene, myrcene, α- phellandrene, δ-3-carene, α-terpinene, ρ-cymene, limonene, β-phellandrene, (*E*)-β-ocimene, γ-terpinene, *cis*-sabinene hydrate, terpinolene, linalool, *trans*-sabinene hydrate, *cis*-ρ-mentha-2-en-1-ol, citronellal, borneol, terpinen-4-ol, cryptone, α-terpineol, citronellol, piperitone, safrole, *n*-tridecane, δ-elemene, α-cubebene, eugenol, α-ylangene, α-copaene, β-cubebene, β-elemene, methyl eugenol, dodecanal, (*E*)-caryophyllene, α-*trans*-bergamotene, α-humulene, *allo*-aromadendrene, γ-muurolene, α-amorphene, germacrene D, bicyclogermacrene, α-muurolene, δ-amorphene, γ-cadinene, δ-cadinene, α-cadinene, hedycaryol, elemol, (*E*)-nerolidol, germacrene D-4-ol, spathulenol, viridiflorol, humulene epoxide

II, tetradecanal, *cis*-cadin-4-en-7-ol, caryophyllene (II), *epi*-α-muurolol, α-muurolol, cubenol, α-eudesmol, α-cadinol, *trans*-calamenen-10-ol, shyobunol, and benzyl benzoate were determined by GC-FID and GC-MS in *C. keralaense* leaves (Sriramavaratharajan et al. 2017).

Santolina triene, tricyclene, α-thujene, α-pinene, camphene, thuja-2,4(10)-diene, verbenene, sabinene, β-pinene, β-myrcene, α-phellandrene, δ-3-carene, α-terpinene, *p*-cymene, *o*-cymene, limonene, 1,8-cineole, (*E*)-β-ocimene, γ-terpinene, *cis*-sabinene hydrate, α-terpinolene, *p*-mentha-1,3,8-triene, linalool, *p*-cymenene, α-campholenal, *trans*-sabinol, *trans*-pinocarveol, neo-*allo*-ocimene, *trans*-verbenol, camphor, pinocarvone, borneol, linalool oxide (pyranoid), terpinen-4-ol, cryptone, α-terpineol, myrtenal, verbenone, fenchyl acetate, *trans*-carveol, *m*-cumenol, nerol, cis-ocimenone, carvone, methyl-3-phenyl-propanal, piperitone, geraniol, (*E*)-cinnamaldehyde, (*E*)-citral, p-menth-1-en-7-al, teresantalol, safrole, bornyl acetate, carvacrol, (Z)-dimethoxy citral, linalyl propanoate, bicycloelemene, α-cubebene, eugenol, neryl acetate, α-ylangene, α-copaene, β-bourbonene, β-cubebene, β-elemene, methyl eugenol, α-santalene, β-caryophyllene, calarene (=β-gurjunene), γ-elemene, aromadendrene, undecanoic acid, α-humulene, β-santalene, γ-gurjunene, 2,6-di-t-butyl-4-methylene2,5-cyclohexadiene-1-one, γ-humulene, germacrene D, α-amorphene, bicyclosesquiphellandrene, β-selinene, α-selinene, cis-cadina-1,4-diene, bicyclogermacrene, β-bisabolene, (*E,E*)-α-farnesene, *cis*-α-bisabolene, γ-cadinene, *cis*-*Z*-α-bisabolene epoxide, eugenol acetate, δ-cadinene, *cis*-calamenene, germacrene B, (*E*)-nerolidol, 1,5-epoxysalvial-4(14)-ene, 3-hexenyl benzoate, dodecanoic acid, palustrol, spathulenol, caryophyllene oxide, globulol, viridiflorol, *trans*-β-elemenone, ledol, β-oplopenone, methoxy eugenol, isospathulenol, 5-*epi*-neointermedeol, α-cadinol, tridecanoic acid, tetradecanol, cis-α-santalol, *epi*-α-bisabolol, juniper camphor, α-cyperotundone, *n*-heptadecene, (*E*, *E*)-farnesol, (*E*)-β-santalol, benzyl benzoate, aristolone, tetradecanoic acid, 7-isopropyl-1,4-dimethyl- azulene, guaiazulene, phenylethyl benzoate, 1,2-benzenedicarboxylic acid, hexadecanoic acid methyl ester, hexadecanoic acid, benzyl cinnamate, hexadecanamide, octadecanoic acid, (Z)-9-octadecenamide, and (Z)-13-docosenamide were determined by GC-MS and GC-FID from leaves of Vietnamese *Cinnamomum* species (*C. kunstleri, C. damhaensis, C. cambodianum, C. curvifolium, C. mairei, C. rigidifolium* and *C. caryophyllus*; Son et al. 2014). Linalool, α-pinene, verbenone, *cis*-verbenol, spathulenol, caryophyllene oxide, α-pinene, β-pinene, spathulenol, guaia-6,10(14)-diene-40-ol, (*E*)-β-caryophyllene, caryophyllene oxide, 1,8-cineole, α-pinene, ledol, and caryophyllene oxide were identified from *C. rigidifolium* (Wanner et al. 2016). The α-pinene, *tau*-cadinol and α-cadinol were determined in *C. perrottetii* leaves (Sriramavaratharajan et al. 2016).

Linalyl acetate, cinnamaldehyde, 1,8-cineole, a-terpinol linalool, nerolidol, terpinen-4-ol, β-caryophyllene, β-phellandrene, δ-3- carene, α-farnesene, α-phellandrene, α-pinene from leaves (Barua and Nath 2006), α-thujene, α-pinene, camphene, sabinene, β-pinene, myrcene, δ-3-carene, α-terpinene, 1,8-cineole, *E*-β-ocimene, γ-terpinene, cis-sabinene hydrate, ρ-mentha-2,4(8)-diene, ρ-cymenene, linalool, endo-fenchol, cis-ρ-mentha-2-en-l-ol, camphor,camphene hydrate, isoborneol, borneol, terpinen-4-ol, γ-terpineol, verbenone, Z-cinnamaldehyde, thymol methyl ester, chavicol, trans-piperitone epoxide, geraial, dihydro-linalool acetate, isobornyl acetate, safrole, Geranyl formate, dihydro-carveol acetate, limonene aldehyde, δ-elemene, α-cubebene, eugenol, α-ylangene, β-elemene, α-chamipinene, α-gurjunene, α-trans-bergamotene, prezizaene, α-humulene, α-zingiberene, gernacrene A, 7-epi-α-selinene, δ-cadinene, zonarene, α-cadinene, selina-3,7(11)-diene, trans-cadinene ether, himachalene epoxide, neryl isovalerate, guaiol, β-cedrene epoxide, epi-α-cadinol, isoamyl geranate, 5-iso-cedranol, α-amyl cinnamyl alcohol, *Z*-α-*trans*-bergamotol, eumesm-7(11)-en-4-ol, longifolol, iso-longifolol, *E*-isoamyl cinnamate, *E*-2-hexyl cinnamaldehyde, benzyl benzoate, epi-cyclocolorenone, *Z*-α-*trans*-bergamotol acetate, α-bisabolol acetate, β-chenopodiol, acorone, 7-hydroxy coumarin, Z,Z-farnesyl acetone, Z-spiroether, 8*S*,14-cedranediol, cataponone, 11,12-dihydroxy valencene, carissone, isohibaene, phytol, columellarin, 3Z-cembrene A, 4-methoxy stilbene, Z,E-geranyl linalool, 13-*epi*-manool oxide, juvibione, bergaptene, sclareolide, benzyl cinnamate, laurenan-2-one, *E*-isoeugenyl benzyl ether, abienol, phenethyl cinnamate from bark (Atiphasaworn et al. 2017); α-campholenal, *cis*-pinocarvyl acetate, α-caryophyllene alcohol, (*Z*)-β-ocimene, (*E*)-nerolidol, (*E*)-β-ocimene, (*E,E*)-farnesol, 1,8-cineole, borneol, camphene, carvone, *endo*-fenchol (α-fenchol), linalool, limonene, *p*-cymene, myrcene, terpinen-4-ol, *trans*-carveol, *trans*-verbenol, tricyclene, verbenone, β-caryophyllene, β-elemene, β-pinene, β-selinene, δ-cadinol, δ-selinene, α-humulene, α-panasinsene, α-pinene, α- selinene, α-terpineol, and α-thujene were determined by GC-MS from flowers of *C. bejolghota* (Choudhury et al. 1998).

α-thujene, α-pinene, camphene, sabinene, β-pinene, myrcene, δ-3-carene, α-terpinene, 1,8-cineole, e-β-ocimene, γ-terpinene, *cis*-sabinene hydrate, ρ-mentha-2,4(8)-diene, ρ-cymenene, linalool, *endo*-fenchol, *cis*-ρ-mentha-2-en-l-ol, camphor, camphene hydrate, isoborneol, borneol, terpinen-4-ol, γ-terpineol, verbenone, Z-cinnamaldehyde, thymol methyl ester, chavicol, *trans*-piperitone epoxide, geraial, dihydro-linalool acetate, isobornyl acetate, safrole, geranyl formate, dihydro-carveol acetate, limonene aldehyde, δ-elemene, α-cubebene, eugenol, α-ylangene, β-elemene, α-chamipinene, α-gurjunene, α-*trans*-bergamotene, prezizaene, α-humulene, α-zingiberene, gernacrene a, 7-*epi*-α-selinene, δ-cadinene, zonarene, α-cadinene, selina-3,7(11)-diene, trans-cadinene ether, himachalene epoxide, neryl isovalerate, guaiol, β-cedrene epoxide, *epi*-α-cadinol, isoamyl geranate, 5-iso-cedranol, amyl cinnamyl alcohol, Z-α-*trans*-bergamotol, eumesm-7(11)-en-4-ol, longifolol, iso-longifolol, *E*-isoamyl cinnamate, *E*-2-hexyl cinnamaldehyde, benzyl benzoate, epi-cyclocolorenone, Z-α-*trans*-bergamotol acetate, α-bisabolol acetate, β-chenopodiol, acorone, 7-hydroxy coumarin, Z,Z-farnesyl acetone, Z-spiroether, 8S,14-cedranediol, cataponone, 11,12-dihydroxy valencene, carissone, isohibaene, phytol, columellarin, 3Z-cembrene A, 4-methoxy stilbene, Z,E-geranyl linalool, 13-*epi*-manool oxide, juvibione, bergaptene, sclareolide, benzyl cinnamate, laurenan-2-one, *E*-isoeugenyl benzyl ether, abienol, and phenethyl cinnamate have been determined *C. bejolghota* bark oil (Atiphasaworn et al. 2017).

Benzyl benzoate, benzyl alcohol, linalool, (Z,Z)-farnesol, 1,8-cineole, borneol, α-guaiol, myrcene, terpinen-4-ol, β-caryophyllene, γ-terpinene, α-guaiene, α-terpineol from leaves whereas cinnamaldehyde, linalool, and eugenol were identified the bark oil of *C. aureofulvum*. The oils were tested for their antibacterial (against *S. aureus, S. epidermidis, P. cepacia* and *P. aeruginosa*) and antifungal (*C. albicans, C. glabrata, Microsporum canis,* and *Trichophyton mentagrophytes*) activities. The bark oils demonstrated significant activity against *P. aeruginosa* (MIC 1.87 µg/µl) but displayed moderate to strong activities toward all the tested fungi (MIC 0.63 to 2.50 µg/µl). As per results, bark oil exhibited more potent activity than the leaf oil (Ali et al. 2002).

Tricyclene, methyl hexanoate, α-thujene, α-pinene, camphene, thuja-2,4(10)-diene, benzaldehyde, sabinene, α-pinene, myrcene, δ-2-carene, α-phellandrene, δ-3-carene α-terpinene, *p*-cymene, *o*-cymene, limonene, 1,8-cineole, salicylaldehyde, γ-terpinene, *cis*-sabinene hydrate, *cis*-linalool oxide (furanoid), *trans*-linalool oxide (furanoid), *trans*-dihydrorose oxide, *p*-cresol, terpinolene, fenchone, 6,7-epoxymyrcene, α-pinene oxide, perillene, linalool, *endo*-fenchol, *cis-p*-menth-2-en-1-ol, methyl octanoate, *trans*-pinocarveol, *trans*-sabinol, *trans*-verbenol, camphor, sabina ketone, pinocarvone, isoborneol, borneol, pinocarvone, δ-terpineol, *cis*-linalool oxide (pyranoid), terpinen-4-ol, *p*-methylacetophenone, thuj-3-en-10-al, cryptone, *p*-cymen-8-ol, α-terpineol, myrtenal, estragole (= methyl chavicol), myrtenol, *cis*-piperitol, *trans*-piperitol, verbenone, *trans*-carveol, *cis*-carveol, nerol, ascaridole, neral, cuminal, *o*-anisaldehyde, carvone, piperitone, geraniol, (2*E*)-decenal, (*E*)-cinnamaldehyde, geranial, neryl formate, safrole, neoiso-3-thujanol acetate, bornyl acetate, *p*-cymen-7-ol, terpinen-4-ol acetate, carvacrol, methyl (*Z*)-cinnamate, (*E*)-cinnamyl alcohol, methyl geranate, 8-hydroxy-neo-menthol, δ-elemene, α-cubebene, neryl acetate, hydrocinnamyl acetate, α-copaene, trans-soberol, *trans-p*-menth-6-ene-2,8-diol, geranyl acetate, methyl (*E*)-cinnamate, β-cubebene, β-elemene, (*E*)-caryophyllene, carvone hydrate, β-copaene, coumarin, aromadendrene, (*E*)-cinnamyl acetate, α-humulene, alloaromadendrene, 9-*epi*-(*E*)-caryophyllene, γ-muurolene, amorpha-4,7(11)-diene, widdra-2,4(14)-diene, β-selinene, *trans*-muurola-4(14),5-diene, *epi*-cubebol, bicyclogermacrene, α-muurolene, germacrene, γ-cadinene, cubebol, trans-calamenene, δ-cadinene, (*E*)-o-methoxycinnamaldehyde, α-cadinene, α-calacorene, elemol, (*Z*)-caryophyllene oxide, germacrene B, β-calacorene, spathulenol, caryophyllene oxide, viridiflorol, cubeban-11-ol, ledol, humulene epoxide II, α-corocalene, 1-*epi*-cubenol, caryophylla-4(12),8(13)-dien-5-ol, τ-muurolol, α-muurolol (=torreyol), β-eudesmol, α-cadinol, selin-11-en-4α-ol, *cis*-calamenen-10-ol, *trans*-calamenen-10-ol, 14-hydroxy-9-epi-(*E*)-caryophyllene, cadalene, mustakone, (2*Z*,6*E*)-farnesol, (2*E*,6*E*)-farnesal, and kaur-16-ene have been determined from *C. camphora* leaf, *C. glaucescens* fruit essential oil, and *C. tamala* root essential oil from Nepal (Satyal et al. 2013). Dotriaconyl-transcoumarate (Kuo et al. 1984) and (+)-diasesamin were reported from *C. camphora* leaves (Hsieh et al. 2006b) leaves.

The presence of essential oils {tricyclene, α-thujene, α-pinene, α-fenchene, camphene, sabinene, β-pinene, myrcene, octanal, δ-2-carene, α-phellandrene, δ-3-carene, α-terpinene, *p*-cymene, limonene, 1,8 cineole, (*Z*)-β-ocimene, (*E*)-β-ocimene, γ-terpinene, cis-sabinene hydrate, fenchone, terpinolene, *p*-cymenene, 6-camphenone, *trans*-sabinene hydrate, nopinone, *endo*-fenchol, *cis-p*-menth-2-en-1-ol, α-campholenal, *trans*-pinocarveol, *trans-p*-menth-2-en-1, camphor, citronellal, *trans*-β-terpineol, sabina ketone, pinocarvone, δ-terpineol, borneol, p-1,8-menthadien-4-ol, terpinen-4-ol, p-cymen-8-ol, cryptone + cymenol, α-terpineol, methyl chavicol, *trans*-piperitol, verbenone, trans-carveol, nerol, citronellol, neral, carvone, geraniol, piperitone, geranial, bornyl acetate, safrole, δ-elemene, α-santalal, α-cubebene, eugenol, neryl acetate, citronellyl acetate, α-ylangene, α-copaene, geranyl acetate, *trans*-β-elemene, methyl eugenol, dodecanal, *cis*-α-bergamotene, α-santalene, β-caryophyllene, γ-elemene, *trans*-α-bergamotene, 6,9-guaiadiene, *trans*-isoeugenol, *epi*-β-santalene, α-humulene, β-santalene, selina-4,11-diene, *trans*-β-bergamotene, germacrene D, β-selinene, α-selinene, β-bisabolene, ridecanal, γ-cadinene, δ-cadinene, myristicin, α-elemol, germacrene B, (*E*)-nerolidol, germacrene D-4-ol, caryophyllene oxide, globulol, methoxy eugenol, guaiol, *cis*-bisabol-11-ol, humuleneepoxide II, tetradecanal, selina-6-en-4-ol, 1-*epi*-cubenol, *epi*-α-cadinol, valerianol, α-cadinol, and selin-11-en-4α-ol} was determined in aerial parts of *C. camphora* (Poudel et al. 2021).

Cyclohexane methanol, *trans*-cinnamaldehyde, *trans*-cinnamyl acetate, bornyl acetate, (-)-spathulenol, caryophyllene, D-borneol, eucalyptol, guaiol from leaves (Chen et al. 2011); anthocyanins, proanthocyanidins, camphene, caryophyllene, citral, elemene, fenchol, guaiene, linalool, myrcene, nerolidol, pinene, sylvestrene, terpineol from fruits; camphene, caryophyllene, citral, elemene, fenchol, guaiene, linalool, myrcene, nerolidol, pinene, sylvestrene, and terpineol were determined from shoots *C. burmannii* (Al-Dhubiab 2012). 1R-α-pinene, 1,4-pentadiene, 5,5-dimethyl-7-oxabicyclo, 7,7-dimethyl-2-methylene-1,4-cyclohexadiene, limonene cyclohexene, eucalyptol, 2-aminobenzoate, 1,7,7-trimethyl-, D-borneol, 3-cyclohexen-1-o, oxime, bicyclo[2.2.1]hept-2-ene, bornyl acetate, ethylene oxide, caryophyllene, 1-phenyl-2-nitro-1-propene, 3-carene, 2,3-dimethylamphetamine, benzene ethanamine, benzyl isocyanate, γ-elemene, 8-quinolinol, cyclohexanol, 3,7,11-trimethyldodeca-2,6,10-trienylacetat, 2(5H)-furanone, acetamide, (-)-spathulenol, caryophyllene oxide, guaiol, cyclohexane, 3-heptyne, cyclohexane methanol, docosahexaenoic acid methyl ester, cyclohexane methanol, patchoulene, caryophyllene-(II), cinnamaldehyde, and phenyl 2-propynyl ether were determined from leaves of *C. burmannii*. D-borneol showed significant DPPH and hydroxyl radicals scavenging activities and reducing power (Chen et al. 2011). Metileugenol, verbanone, terpinol, sinamaldehid, coumarin, spathulenol massoyalakton, butanoat acid, and pentadekanoat acid were determined by GC-MS from *C. burmanii*, and *C. culilaban* (Hapsari and Simanjuntak 2010).

(*E*)-β-ocimene, (*E*)-β-santalol, camphene, *cis*-α-santalol; *epi*-α-bisabolol, spathulenol, germacrene B, isospathulenol, limonene, linalool oxide (pyranoid), *neo-allo*-ocimene, *o*-cymene, phytol, sabinene, terpinen-4-ol, *trans*-nerolidol, verbenone, α-bisabolol, α-pinene, α-terpinene, α-terpineol, athujene, β-myrcene, β-spathulenol, β-vetivenene, cterpinene, δ-3-carene, 1,8-cineole, carvone, geraniol, guaiol, teresantalol, α-amorphene, α-phellandrene, asantalene, γ-eudesmol from leaves and 4-terpineol, cadalene, isospathulenol, viridiflorol, α-cadinol, α-epi-cadinol, α-terpinene, α-terpinol, α-terpinolene, β-spathulenol, β-terpineol, and γ-terpinene from the bark of *C. cambodianum*. *C. cambodianum* showed inhibitory effects against *Hemophilus influenzae*, *Streptococcus pneumoniae*, and *Staphylococcus aureus* (Houdkova et al. 2018; Chhouk et al. 2018). Methyl eugenol, terpinen-4-ol, and 1,8-cineole were identified from *C. kunstleri*; eugenol, 1, 8-cineole, neryl acetate and eugenol acetate were identified from *C. mairei*; linalool, α-pinene, β-pinene, and 1,8-cineole were identified from *C. damhaensis*; linalool and terpinen-4-ol from *C. cambodianum*; 1,8-cineole, α-pinene, and camphene were identified from *C. caryophyllus* and α-selinene, β-caryophyllene and α-copaene were identified from *C. rigidifolium* (Son et al. 2014).

Cryptone, *p*-cymene, *p*-cymene, limonene, linalool, terpinene-1-ol, veratrole, cryptone, α-terpineol, trans-carveole, carvone, piperotine, safrole, germacrene B, longiborneol, viridiflorol, cuminaldehyde, and limonene were identified by GC-MS analysis from *C. filipedicellatum* (Rameshkumar and George 2006a).

The presence of (*E*)-nerolidol and elemicin was determined *C. glaucescens* and *C. glanduliferum* leaf oil (Baruah and Nath 2006). Twenty-six essential oils (α-thujene, α-pinene, camphene, sabinene, β-pinene, β-myrcene, δ-2-carene, α-terpinene, *p*-cymene, limonene, eucalyptol, γ-terpinene, *cis*-sabinene hydrate, terpinolene, *trans*-sabinene hydrate, terpinene-4-ol, α-terpineol, β-elemene, *trans*-caryophyllene, α-humulene, germacrene D, germacrene B, spathulenol, caryophyllene oxide, and globulol) were identified from leaf and branches of *C. glanduliferum*. Eucalyptol, sabinene, α-terpineol, α-pinene were reported as major compounds in leaves and branches. The leaf essential oils (250, 500, and 1000 mg/kg) significantly decreased the paw volume (94, 82 and 69%) in carrageenan-induced rat oedema model. The same doses of oils significantly decreased cyclooxygenase-2 activity (73.8, 50.7 and 21.4 nmol/min/ml). A significant decrease in prostaglandin E2 concentration was also recorded at the same doses of essential oils (2.95 ± 0.2, 2.45 ± 0.15 and 1.75 ± 0.015 pg/ml). The bark essential oils demonstrated a significant modulatory activity on ethanol-induced gastritis in rats. The levels of nitric oxide were significantly decreased to 32, 37, and 41 μM nitrate/g. The oils also showed significant suppression in values of lipid peroxidation by reducing the levels of malondialdehyde (1.15, 1.11, and 1.04 nmol/g). The essential oils demonstrated significant anti-inflammatory activity and showed protection against ethanol-induced non-ulcerative gastritis (Azab et al. 2017). Three major essential oils (eucalyptol, terpinen-4-ol, α-terpineol) were determined by GC and GC/MS from *C. glanduliferum* bark. The crude essential oils demonstrated strong antibacterial activity against *Escherichia coli* (activity index 1.0 and MIC 0.49 μg/ml). Similarly, the crude essential oils exhibited good antimicrobial activity against methicillin-resistant *Staphylococcus aureus, Geotrichum candidum, Pseudomonas aeruginosa, Bacillus subtilis, Helicobacter pylori, Aspergillus fumigatus* (MIC 7.81, 1.95, 7.81, 0.98, 31.25, and 32.5 μg/ml). The crude essential oils also displayed significant activity against *Staphylococcus aureus* and *Mycobacterium tuberculosis* (MIC 32.5 and 31.25 μg/ml). The essential oils exhibited significant cytotoxicity to colon (HCT-116), liver (HepG2), and breast (MCF-7) carcinoma cell lines (IC$_{50}$ 9.1, 42.4, and 57.3 μg/ml; Taha and Eldahshan 2017). The essential oils of (concentration 0.85–440.0 μg/ml in a 96-microtiter plate) *C. glanduliferum* showed potent antibacterial activity against *A. salmonicida* (MIC 1.72 μg/ml), *Escherichia coli* (MIC 3.43 μg/ml), and *Pseudomonas aeruginosa* (MIC 3.43 μg/ml). The results were compared with standards gentamicin and kanamycin (Singh et al. 2013). The ethanol extracts of *C. travancoricum, C. walaiwarense, C. wightii, C. verum, C. sulphuratum, C. riparium,* and *C. perrottetii* barks showed cytotoxic activity against brine shrimp lethality of *Artemia salina* (Maridass 2008).

α-cadinene, γ-cadinene, δ-cadinene, 1,8-cineole, caryophyllene oxide, germacrene-D, limonene, linalool, myrcene, p-cymene, terpinen-4-ol, *trans*-linalool oxide, α-cadinol, α-humulene, α-terpineol, β-caryophyllene, β-elemene, β-pinene, γ-elemene, γ-muurolene, γ-terpinene from leaves; 1,8-cineol, 4-terpineol, aromadendrene, benzyl benzoate, borneol, bornyl acetate, camphor, caryophyllene oxide, derivative eugenol, eugenic acid, eugenol, germacrene, globulol, hexadecanoic acid, isomyristicin, isopulegol, juniper camphor, L-limonene, L-linalool, methyl eugenol, myristicin, safrole, spathulenol, trans-caryophyllene, thymol, viridiflorol, a-cadinol, α-calacorene, α-curcumene, α-muurolene, α-terpineol, α-copaene, β-caryophyllene, γ-muurolene, δ-cadinene, δ-cadinol, dodecanal, tetradecanal, undecanal, tetradecanoic acid, dodecane, γ-cadinene, δ-cadinene, borneol, camphene, cislinalool oxide, cubenol, *epi*-α-cadinol, geraniol, germacrene B, limonene, linalool, myrcene, nerol, *p*-cymene, terpinen-4-ol, *trans*-linalool oxide, ahumulene, α-muurolene, α-pinene, α-selinene, aterpineol, α-thujene, α-ylangene, β-elemene, bpinene, β-selinene, γ-muurolene, γ-terpinene from bark; benzaldehyde, decanal, dodecanal, hexanal, octadecanal, tetradecanal, undecanal, hexadecanoic acid, octadecanoic acid, pentadecanoic acid, tetradecanoic acid, dodecane, α-copaene, γ-cadinene, 1,8-cineole, cis-linalool oxide, epi-acadinol, linalool, p-cymene, terpinen-4-ol, translinalool oxide, α-humulene, α-terpineol, aylangene from wood; benzaldehyde, tetradecanal, α-copaene, γ-cadinene, dcadinene, (E)-β-farnesene, (E,E)-α-farnesene, 1,8- cineole, cis-linalool oxide, linalool, p-cymene, terpinen-4-ol, trans-linalool oxide, α-terpineol, β-caryophyllene, and β-elemene were identified by GC-MS from twigs of C. sintoc (Jantan et al. 2005b). L-limonene, 1,8-cineol, L-linalool, isopulegol, camphor, borneol, 4-terpineol, α-terpineol, bornyl acetate, safrol, timol, eugenol, α-copaene, metylceugenol, trans-caryophyllen, β-caryophyllen, aromadendrene, γ-murolene, α-curcumene, α-murolene, gernacrene, Δ-cadinene, myristicin, α-calocorene, spatulenol, caryophyllene oxide, globulol, viridiflorol, isomyristicin, Δ-cadinol, α-cadinol, junifer camphor, eugenic acid, benzyl benzoate, eugenol derivative, hexadecanoic acid from sintok from Jogyakarta; L-linalool, t-sabinen hidrat, camphor, borneol, 4-terpineol, α-terpineol, bornil asetat, safrol, timol, eugennol, α-copaene, methyl eugeunol, isoeugenol, trans-caryophyllen, β-caryophyllen, aromadendrene, γ-murolene, α-curcumene, α-murolene, gernacrene, Δ-cadinene, myristicin, α-calocorena, spatulenol, kariofilen oksida, globulol, viridiflorol, isomyristicin, Δ-cadinol, α-cadinol, junifer camphor, eugeunic acid, methyl octadecenoate, benzyl benzoate, eugenol derivate, and hexadecanoic acid were determined by GC-MS from C. sintoc of Jember districts of Java (Muchtaridi et al. 2017).

Thirty essential oils (α-pinene, camphene, β-pinene, 1,8-cineole, linalool oxide, linalool oxide, l-linalool, fenchyl alcohol, camphor, lis-sabinene hydrate, endo-borneol, terpinene-4-ol, α-terpineol, α-fenchyl acetate, vinyl benzaldehyde, bornyl acetate, cinnamaldehyde, α-copaene, cinnamyl acetate, α-muurolene, calamenene, cinnamic acid, β-cedrene, caryophyllene oxide, isoaromadendrene epoxide, cubenol, α-cadinol, δ-cadinol, veridiflorol, globulol) were determined from C. zeylanicum (Elgammal et al. 2020).

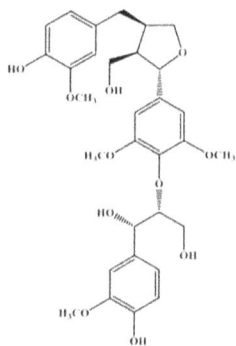

erythro-Guaiacylglycerol-β-O-4′-(5′)-methoxylariciresinol

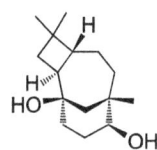

Caryolane-1,9β-diol

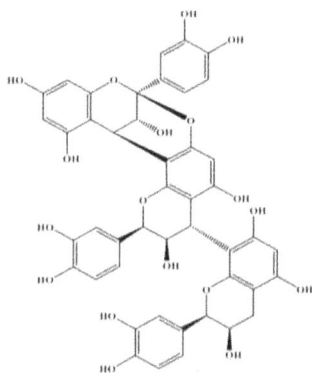

Cinnamtannin B1

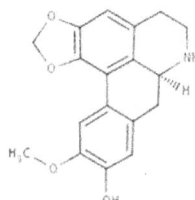

Actinodaphnine

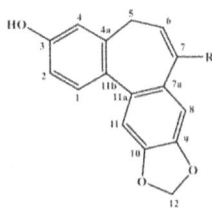

R = CH$_2$OCH$_3$ - Validinol; R = CH$_2$OH – Reticuol

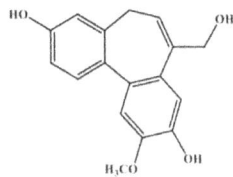

Burmanol

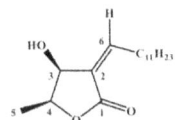

Validinolide

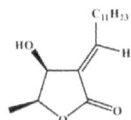

Lincolomide A

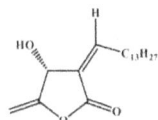

Linderanolide B

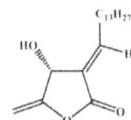

Isolinderanolide B

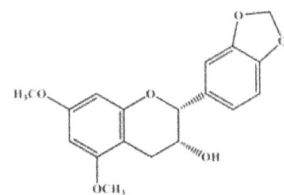

(-)-5,7-Dimethoxy-3′,4′-methylenedioxy-flavan-3-ol

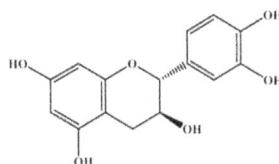

Taxifolin

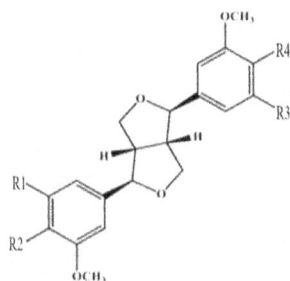

R^1 = R^2 = R^3 = R^4 = OCH$_3$ – (-)-Yangambin; R^1 = R^3 = H, R^2 = R^4 = OH – (-)-Pinoresinol

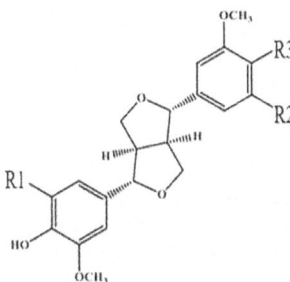

R^1 = R^2 = H, R^3 = OCH$_3$ – (+)-Monomethylpinoresinol; R^1 = R^2 = OCH$_3$, R^3 = OH – (+)-Syringaresinol

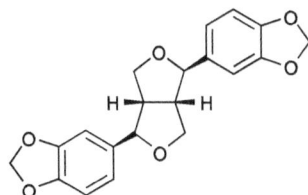

(-)-Sesamin

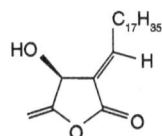

Isophilippinolide A

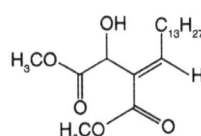

Secosubamolide

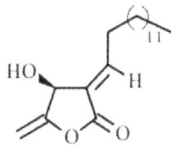

Isoobtusilactone A

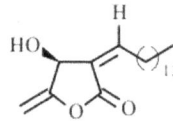

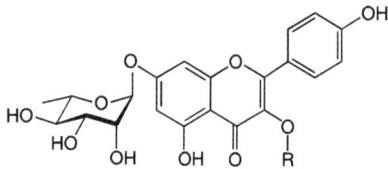

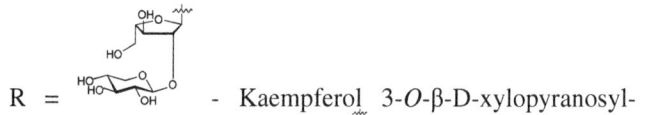

Obtusilactone A

R = [structure] - Kaempferol 3-O-β-D-xylopyranosyl-

(1→2)-α-L-arabinofuranosyl-7-O-α-L-rhamnopyranoside; R = [structure] - Kaempferol 3-O-β-D-xylopyranosyl-(1→2)-α-L-rhamnopyranosyl-7-O-α-L-rhamnopyranoside

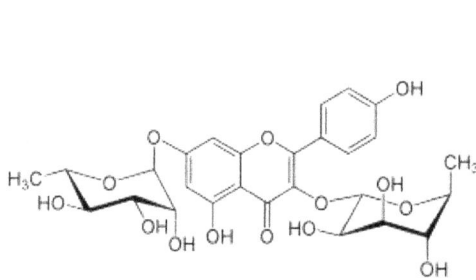

Kaempferitrin

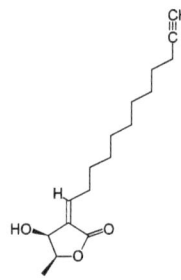

Lincomolide B

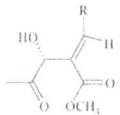

R = $(CH_2)_{23}CH_3$ - Secotenuifolide A

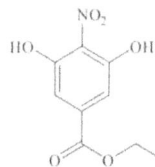

Ethyl 3,5-dihydroxy-4-nitrobenzoate

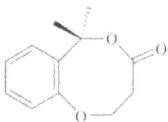

2,3-Dihydro-6,6- dimethylbenzo[b][1,5]-dioxocin-4(6H)-one

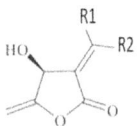

$R^1 = (CH_2)_{25}CH_3$, R^2 = H - Tenuifolide A; R^1 = H, $R^2 = (CH_2)_{25}CH_3$ - Isotenuifolide A

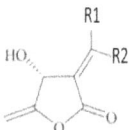

R^1 = H, $R^2 = (CH_2)_9CH_3$ - Subamolide D; $R^1 = (CH_2)_9CH_3$, R^2 = H - Subamolide E

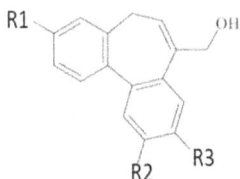

$R^1 = OCH_3$, $R^2 = R^3 = OH$ – Subamol

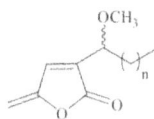

n = 18 - Tenuifolide B

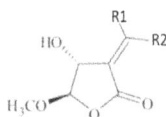

$R^1 = H$, $R^2 = (CH_2)_{12}CH_3$ - Subamolide A; $R^1 = (CH_2)_{12}CH_3$, $R^2 = H$ - Subamolide B

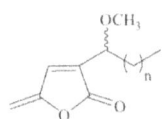

n = 13 – Subamolide C

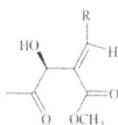

$R = (CH_2)_{12}CH_3$ – Secosubamolide; $R = (CH_2)_{14}CH_3$ - Secosubamolide A

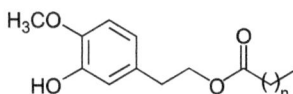

n=24-2-(4-Hydroxy-3-methoxyphenyl)ethyl hexacosanoate; n=26-2-(4-hydroxy-3-methoxyphenyl) ethyl octacosanoate; n = 13 - 4-Hydroxy-3-methoxyphenethyl pentadecyrate; n = 16 - 4-Hydroxy-3-methoxyphenethyl stearate; n= 19 - 4-Hydroxy-3-methoxyphenethyl heneicosyrate

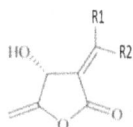

$R^1 = (CH_2)_{16}CH_3$, $R^2 = H$ – Isoreticulide

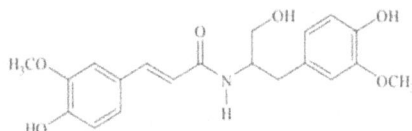

Cinnaretamine

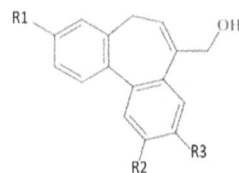

$R^1 = OH$, $R^2+R^3 = OCH_2O$ – Reticuol; $R^1 = OCH_3$, $R^2+R^3 = OCH_2O$ - Tenuifolin

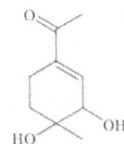

Reticuone

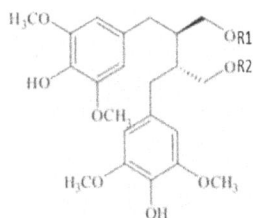

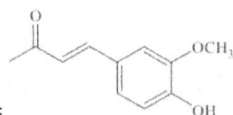

Feruloyl =

$R^1 = R^2 = Feruloyl$ - 9,9′-Di-*O*-feruloyl-(+)-5,5′-dimethoxy secoisolariciresinol

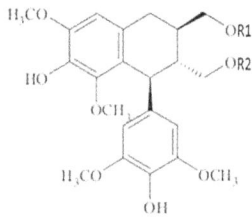

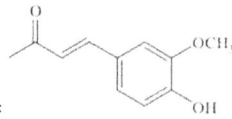

Feruloyl =

R^1 = Feruloyl, R^2 = H - (7′S,8′R,8R)-lyoniresinol-9-O-(E)-feruloyl ester; R^1 = R^2 = Feruloyl -(7′S,8′R,8R)-lyoniresinol-9,9′ -di-O-(E)-feruloyl ester

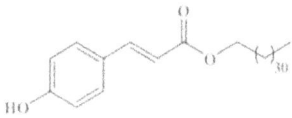

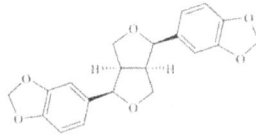

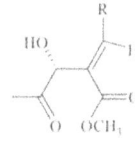

Dotriaconyl-*trans*-coumarate　　　(+)-Diasesamin　　　R = $(CH_2)_{12}CH_3$ - Secokotomolide A

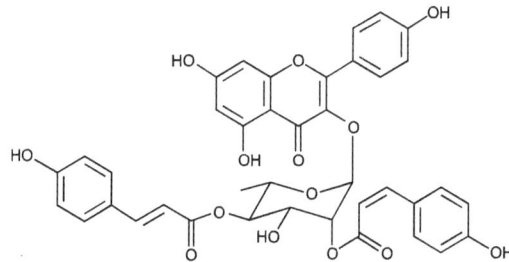

Kaempferol 3-O-α-L-[2-(Z)-p-coumaroy-4-(E)-pcoumaryl]rhamnopyranoside

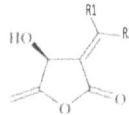

R^1 = H, R^2 = $(CH_2)_6CH_3$ - Kotomolide A; R^1 = $(CH_2)_6CH_3$, R^2 = H - Isokotomolide A

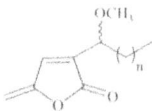

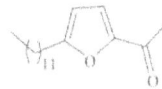

n = 17 - Kotomolide B; n = 14 – Kotomolide　　　2-acetyl-5-dodecylfuran

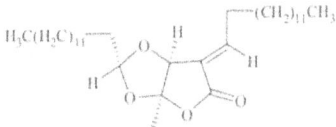

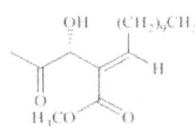

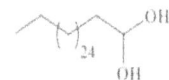

Kotolactone A　　　　　　Secokotomolide　　　Kotodiol

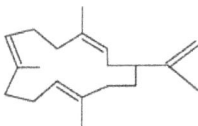

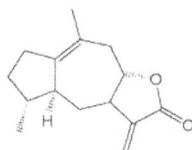

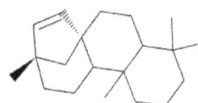

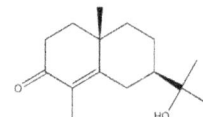

3Z-Cembrene A　　　　Columellarin　　Isohibaene　　　　Carissone

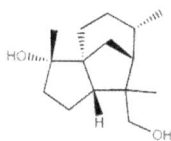

8S,14-Cedranediol

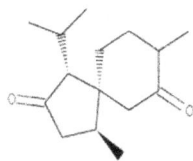

Acorone

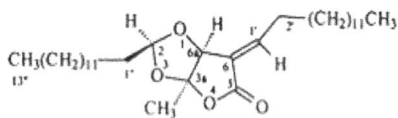

Kotolactone

Z-α-trans-Bergamotol acetate

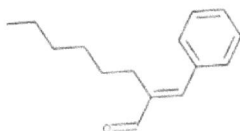

E-2-Hexyl cinnamaldehyde

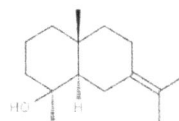

Eumesm-7(11)-en-4-ol

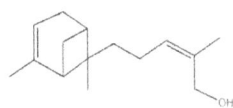

Z-α-*trans*-Bergamotol

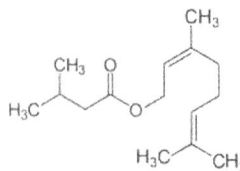

Neryl isovalerate

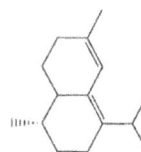

Zonarene

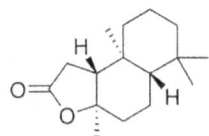

Sclareolide

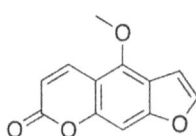

Bergaptene

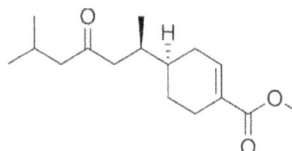

Juvibione

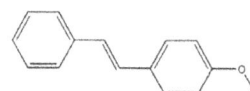

4-Methoxy stilbene

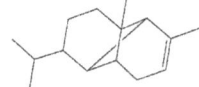

α-Ylangene

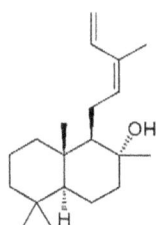

Abienol

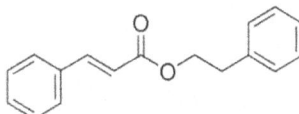

Phenethyl cinnamate

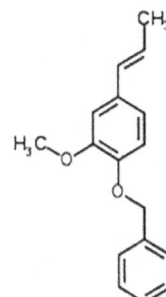

E-Isoeugenyl benzyl ether

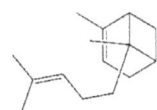

α-*trans*-Bergamotene

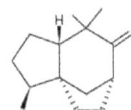

Prezizaene

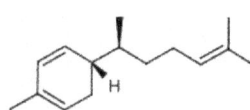

α-Zingiberene

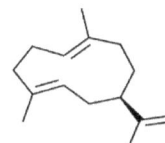

Gernacrene A

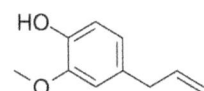

Eugenol

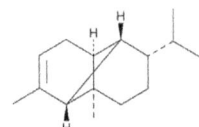

α-Copaene

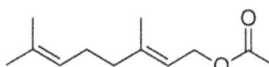

Geranyl acetate

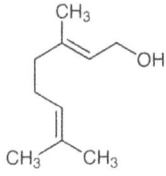

Geraniol

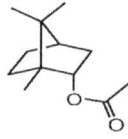

Bornyl acetate

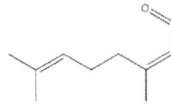

(*Z*)-citral

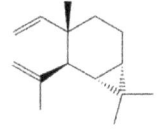

Bicycloelemene

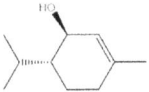

trans-Piperitol

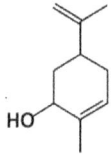

trans-Carveol

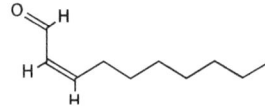

(*Z*)-2-Decenal

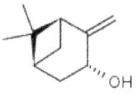

trans-Pinocarveol

Terpinen-4-ol

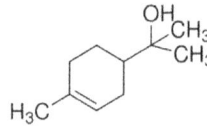

α-Terpineol

Verbenone

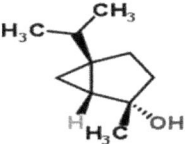

trans-Sabinene hydrate

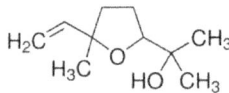

Linalool oxide

α-Terpinolene

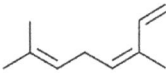

(*Z*)-β-Ocimene

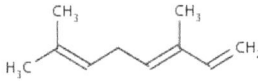

(*E*)-β-Ocimene

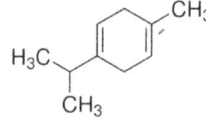

γ-Terpinene

Tricycline

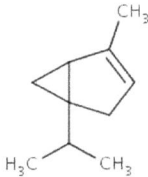

α-Thujene

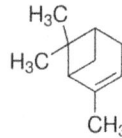

α-Pinene

Camphene

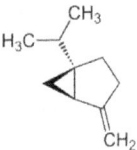

Sabinene

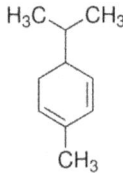

α-Phellandrene

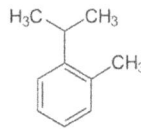

o-Cymene

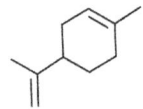

Limonene

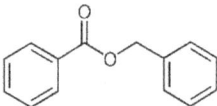

Benzyl benzoate

Benzyl alcohol

Linalool

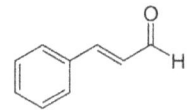

Cinnamaldehyde

Eugenol   1,8-Cineole   (Z,Z)-Farnesol   Borneol

Guaiol   Myrcene   Terpinen-4-ol   β-caryophyllene

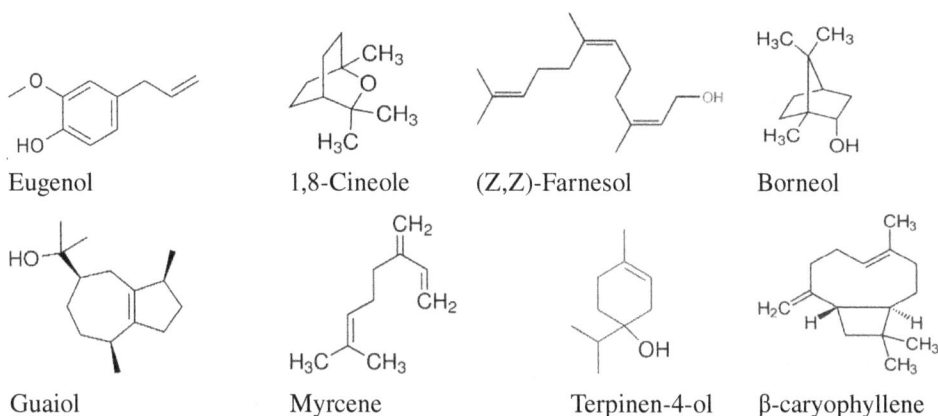

7.3   CULTURE CONDITIONS

Nodal segments of *C. camphora* tips were cultured on Murashige and Skoog medium (MS; Murashige and Skoog 1962) containing thidiazuron. The supplementation of thidiazuron induced the formation and growth of callus in this study (Huang et al. 1998). Similarly, the shoot tips and nodal segments of *C. camphora* were transferred on MS basal medium and Woody plant medium (McCown and Amos 1979). *C. camphora* leaf explants responded only to MS medium containing BAP (1.0 mg/l) and thidiazuron (0.005–5.0 mg L^{-1}) for the formation of compact callus (Soulange et al. 2007). For regeneration of callus, the explants were grown on kinetin (2.32 mM) containing MS medium. The explants were also transferred on the WPM medium. The medium was supplemented with 6-benzyl aminopurine, kinetin and zeatin (Babu et al. 2003). The nodal regions of seedlings of *C. zeylanicum* responded in the formation of callus on WMP medium. The medium was supplemented with 2,4-D (0.5 mg/l) + kinetin (0.5 mg/l; Rai and Chandra 1987).

REFERENCES

Abdelwahab SI, Mariod AA, Taha MME, Zaman FQ, Abdelmageed AHA, Khamis S, Sivasothy Y, Awang K. 2017. Chemical composition and antioxidant properties of the essential oil of *Cinnamomum altissimum* Kosterm. (Lauraceae). Arab J Chem 10, 131–135.

Abdelwahab SI, Zaman FQ, Mariod AA, Yaacob M, Abdelmageed AHA, Khamis S. 2010. Chemical composition, antioxidant and antibacterial properties of the essential oils of *Etlingera elatior* and *Cinnamomum pubescens* Kochummen. J Sci Food Agric 90, 2682–2688.

Abeysekera WPKM, Arachchige SPG, Abeysekera WKSM, Ratnasooriya WD, Medawatta HMUI. 2019. Antioxidant and glycemic regulatory properties potential of different maturity stages of leaf of Ceylon cinnamon (*Cinnamomum zeylanicum* Blume) *in vitro*. Evid Based Complement Altern Med 2019, 2693795.

Abeysekera WPKM, Arachchige SPG, Ratnasooriya WD. 2017. Bark extracts of Ceylon cinnamon possess antilipidemic activities and bind bile acids *in vitro*. Evid Based Complement Altern Med 2017, 7347219.

Abeysinghe PD, Bandaranayake PCG, Pathirana R. 2021. Botany of endemic *Cinnamomum* species of Sri Lanka. In: Cinnamon: Botany, Agronomy, Chemistry and Industrial Applications, Senaratne R, Pathirana R (eds). Springer International Publishing, Cham, pp 85–118.

Adisakwattana S, Lerdsuwankij O, Poputtachai U, Minipun A, Suparpprom C. 2011. Inhibitory activity of cinnamon bark species and their combination effect with acarbose against intestinal α-glucosidase and pancreatic α-amylase. Plant Foods Hum Nutr 66, 143–148.

Ahmed A, Choudhary MI, Farooq A, Demirci B, Demirci F, Başer KHC. 2000. Essential oil constituents of the spice *Cinnamomum tamala* (Ham.) Nees & Eberm. Flavour Fragr J 15, 388–390.

Al-Dhubiab BE. 2012. Pharmaceutical applications and phytochemical profile of *Cinnamomum burmannii*. Pharmacogn Rev 6, 125.

Ali NAM, Mohtar M, Shaari K, Rahmanii M. 2002. Chemical composition and antimicrobial activities of the essential oils of *Cinnamomum aureofulvum* Gamb. J Essent Oil Res 14, 135–138.

Aluyor EO, Oboh IO. 2014. Traditional preservatives—vegetable oils. In: Encyclopedia of Food Microbiology, Batt CA, Tortorello ML (eds). Academic Press, London, pp 137–140.

Ananthakrishnan R, Santhosh Kumar FS, Rameshkumar KB. 2018. Comparative chemical profiles of essential oil constituents of eight wild *Cinnamomum* species from the Western Ghats of India. Nat Prod Commun 13, 621–625.

Anonymous. 1992. The Wealth of India–A Dictionary of Indian Raw Materials and Industrial Products, Vol 3. Ca-Ci, CSIR, New Delhi.

Arachchige SPG, Abeysekera WPKM, Ratnasooriya WD. 2017. Antiamylase, anticholinesterases, antiglycation, and glycation reversing potential of bark and leaf of Ceylon cinnamon (*Cinnamomum zeylanicum* Blume) *in vitro*. Evid Based Complement Altern Med 2017, 5076029.

Ashfaq MH, Siddique A, Shahid S. 2021. Antioxidant activity of *Cinnamon zeylanicum*: (A review). Asian J Pharm Res 11, 106–116.

Atiphasaworn P, Monggoot S, Pripdeevech P. 2017. Chemical composition, antibacterial and antifungal activities of *Cinnamomum bejolghota* bark oil from Thailand. J Appl Pharm Sci 7, 69–73.

Azab SS, Jaleel GAA, Eldahshan OA. 2017. Anti-inflammatory and gastroprotective potential of leaf essential oil of *Cinnamomum glanduliferum* in ethanol-induced rat experimental gastritis. Pharm Biol 55, 1654–1661.

Babu KN, Sajina A, Minoo D, John CZ, Mini PM, Tushar KV, Rema J, Ravindran PN. 2003. Micropropagation of camphor tree (*Cinnamomum camphora*). Plant Cell Tissue Organ Cult 74, 179–183.

Bakar A, Yao P-C, Ningrum V, Liu C-T, Lee S-C. 2020. Beneficial biological activities of *Cinnamomum osmophloeum* and its potential use in the alleviation of oral mucositis: A systematic review. Biomedicines 8, 3.

Baruah A, Nath SC, Hazarika AK, Sarma TC. 1997. Essential oils of the leaf, stem bark and panicle of *Cinnamomum bejolghota* (Buch.-Ham.) Sweet. J Essent Oil Res 9, 243–245.

Baruah A, Nath SC, Leclercq PA. 1999. Leaf and stem bark oils of *Cinnamomum sulphuratum* Nees from Northeast India. J Essent Oil Res 11, 194–196.

Baruah A, Nath SC. 2001. Taxonomic status and composition of stem bark oil of a variant of *Cinnamomum bejolghota* (Lauraceae) from Northeast India. Nord J Bot 21, 571–576.

Baruah A, Nath SC. 2006. Leaf essential oils of *Cinnamomum glanduliferum* (Wall) Meissn and *Cinnamomum glaucescens* (Nees) Meissn. J Essent Oil Res 18, 200–202.

Batiha GE-S, Beshbishy AM, Guswanto A, Nugraha A, Munkhjargal T, Abdel-Daim MM, Mosqueda J, Igarashi I. 2020. Phytochemical characterization and chemotherapeutic potential of *Cinnamomum verum* extracts on the multiplication of protozoan parasites *in vitro* and *in vivo*. Molecules 25, 996.

Bhattacharjee PP, Ray DC. 2010. Pest management beliefs and practices of Manipuri rice farmers in Barak Valley, Assam. Indian J Tradit Knowl, 9, 673–676.

Błaszczyk N, Rosiak A, Kałuzna-Czaplińska J. 2021. The potential role of Cinnamon in human health. Forests 12, 648.

Brophy JJ, Goldsack RJ, Forster PI. 2001. The leaf oils of the Australian species of *Cinnamomum* (Lauraceae). J Essent Oil Res 13, 332–335.

Burkill IH. 1966. A Dictionary of the Economic Products of the Malay Peninsula. Ministry of Agriculture and Co-operatives, Kuala Lumpur.

Butkhup L, Samappito S. 2011. *In vitro* free radical scavenging and antimicrobial activity of some selected Thai medicinal plants. Res J Med Plant 5, 254–265.

Cardoso-Ugarte GA, López-Malo A, Sosa-Morales ME. 2016. Cinnamon (*Cinnamomum zeylanicum*) essential oils. In: Essential Oils in Food Preservation, Flavor and Safety, Preedy VR (ed.). Academic Press, San Diego, pp 339–347.

Chao LK, Hua K-F, Hsu H-Y, Cheng S-S, Liu J-Y, Chang S-T. 2005. Study on the antiinflammatory activity of essential oil from leaves of *Cinnamomum osmophloeum*. J Agric Food Chem 53, 7274–7278.

Chen CH, Lo WL, Liu YC, Chen CY. 2006. Chemical and cytotoxic constituents from the leaves of *Cinnamomum kotoense*. J Nat Prod 69, 927–933.

Chen CY. 2006. Butanolides from the stem of *Cinnamomum kotoense*. Nat Prod Commun 1, 453–455.

Chen CY, Chen CH, Wong CH, Liu YW, Lin YS, Wang YD, Hsui YR. 2007. Cytotoxic constituents of the stems of *Cinnamomum subavenium*. J Nat Prod 70, 103–106.

Chen CY, Hong Z-L. 2011. Chemical constituents from the fruits of *Cinnamomum kotoense*. Chem Nat Comp 47, 450–451.

Chen CY, Hsieh S-L, Hsieh M-M, Hsieh S-F. 2004. Substituent chemical shift of rhamnosides from the stems of *Cinnamomum osmophleum*. Chin Pharm J 56, 141–146.

Chen CY, Yang WL, Hsui YR. 2010b. A novel sesquiterpenoid from the roots of *Cinnamomum subavenium*. Nat Prod Res 24, 423–427.

Chen CY, Yeh HC. 2011. A new amide from the stems of *Cinnamomum reticulatum* Hay. Nat Prod Res 25, 26–30.

Chen FC, Peng CF, Tsai IL, Chen IS. 2005. Antitubercular constituents from the stem wood of *Cinnamomum kotoense*. J Nat Prod 68, 1318–1323.

Chen HL, Kuo SY, Li YP, Kang YF, Yeh YT, Huang JC, Chen CY. 2012. A new benzodioxocinone from the leaves of *Cinnamomum tenuifolium*. Nat Prod Res 26, 1881–1886.

Chen L, Su J, Li L, Li B, Li W. 2011. A new source of natural D-borneol and its characteristic. J Med Plant Res 5, 3440–3447.

Chen TH, Huang YH, Lin JJ, Liau BC, Wang SY, Wu YC, Jong TT. 2010a. Cytotoxic lignan esters from *Cinnamomum osmophloeum*. Planta Med 76, 613–619.

Chen YJ, Chou CJ, Chang TT. Compound MMH. 2009. Compound MMH01 possesses toxicity against human leukemia and pancreatic cancer cells. Toxicol In Vitro 23, 418–424.

Cheng KC, Hsueh MC, Chang HC, Lee AY, Wang HM, Chen CY. 2010a. Antioxidants from the leaves of *Cinnamomum kotoense*. Nat Prod Commun 5, 911–912.

Cheng MJ, Lo WL, Tseng WS, Yeh HC, Chen CY. 2010b. A novel normonoterpenoid from the stems of *Cinnamomum reticulatum* Hay. Nat Prod Res 24, 732–736.

Cheng MJ, Yeh YT, Wang CJ, Chen CY. 2011. Isolation of a nitrobenzoate from the leaves of *Cinnamomum tenuifolium*. Nat Prod Res 25, 118–122.

Cheng SS, Lin C-Y, Yang C-K, Chen Y-J, Chung M-J, Chang S-T. 2015. Chemical polymorphism and composition of leaf essential oils of *Cinnamomum kanehirae* using gas chromatography/mass spectrometry, cluster analysis, and principal component analysis. J Wood Chem Technol 35, 207–219.

Cheng S-S, Liu J-Y, Hsui Y-R, Chang S-T. 2006. Chemical polymorphism and antifungal activity of essential oils from leaves of different provenances of indigenous cinnamon (*Cinnamomum osmophloeum*). Bioresour Technol 97, 306–312.

Cheng S-S, Liu J-Y, Huang C-G, Hsui Y-R, Chen W-J, Chang S-T. 2009. Insecticidal activities of leaf essential oils from *Cinnamomum osmophloeum* against three mosquito species. Biores Technol 100, 457–464.

Cheng S-S, Liu J-Y, Tsai K-H, Chen W-J, Chang S-T. 2004. Chemical composition and mosquito larvicidal activity of essential oils from leaves of different *Cinnamomum osmophloeum* provenances. J Agric Food Chem 52, 4395–4400.

Chhouk K, Wahyudiono, Kanda H, Goto M. 2018. Efficacy of supercritical carbon dioxide integrated hydrothermal extraction of Khmer medicinal plants with potential pharmaceutical activity. J Environ Chem Eng 6, 2944–2956.

Chia YC, Yeh HC, Yeh YT, Chen CY. 2011. Chemical constituents from the leaves of *Cinnamomum reticulatum*. Chem Nat Comp 47, 220–222.

Chinh HV, Luong NX, Thin DB, Dai DN, Hoi TM, Ogunwande IA. 2017. Essential oils leaf of *Cinnamomum glaucescens* and *Cinnamomum verum* from Vietnam. Am J Plant Sci 8, 2712–2721.

Chopra R, Nayar S, Chopra I. 1956. Glossary of Indian Medicinal Plants. CSIR, New Delhi.

Choudhury S, Ahmed R, Barthel A, Leclercq PA. 1998. Composition of the bark and flower oils of *Cinnamomum bejolghota* (Buch.-Ham.) sweet from two locations of Assam, India. J Essent Oil Res 10, 245–250.

Chung MY, Nason JD, Epperson BK, Chung MG. 2003. Temporal aspects of the fine-scale genetic structure in a population of *Cinnamomum insularimontanum* (Lauraceae). Heredity 90, 98–106.

da Paz Lima M, das Graças B. Zoghbi M, Andrade EHA, Silva TMD, Fernandes CS. 2005. Volatile constituents of the leaves and branches of *Cinnamomum zeylanicum* Blume (Lauraceae). Acta Amazon 35, 363–366.

Dai DN, Chung NT, Huong LT, Hung NH, Chau DTM, Yen NT, Setzer WN. 2020. Chemical compositions, mosquito larvicidal and antimicrobial activities of essential oils from five species of *Cinnamomum* growing wild in North Central Vietnam. Molecules 25, 1303.

Dai DN, Lam NTT, Chuong NT, Ngan TQ, Truong NS, Ogunwande IA. 2019. Essential oils of *Cinnamomum curvifolium* (Lour.) Nees and *Cinnamomum mairei* H. Lev. Am J Essent Oil Nat Prod 7, 11–14.

DeFilipps RA, Krupnick GA. 2018. The medicinal plants of Myanmar. PhytoKeys 102, 1–341.

Devi SL, Kannappan S, Anuradha CV. 2007. Evaluation of *in vitro* antioxidant activity of Indian bay leaf, *Cinnamomum tamala* (Buch.-Ham.) T. Nees & Eberm using rat brain synaptosomes as model system. Indian J Exp Biol 45, 778–784.

Dhar U, Manjkhola S, Joshi M, Bhatt A, Bisht AK. 2002. Current status and future strategy for development of medicinal plant sector in Uttaranchal, India. Curr Sci 83, 956–963.

Doley B, Gajurel PR, Rethy P, Saikia B. 2010. A check list of commonly used species by the Nyshi tribes of Papumpare District, Arunachal Pradesh. J Biosci Res 1, 9–12.

Dong L, Schill H, Grange RL, Porzelle A, Johns JP, Parsons PG, Gordon VA, Reddell PW, Williams CM. 2009. Anticancer agents from the Australian tropical rainforest: Spiroacetals EBC-23, 24, 25, 72, 73, 75 and 76. Chem Eur J 15, 11307–11318.

Elgammal EW, El Gendy AENG, Elgamal AEBA. 2020. Mechanism of action and bioactivities of *Cinnamomum zeylanicum* essential oil against some pathogenic microbes. Egypt Pharm J 19, 162–171.

Espineli DL, Agoo EMG, Shen C-C, Ragasa CY. 2013. Chemical constituents of *Cinnamomum iners*. Chem Nat Comp 49, 932–933.

Fang JM, Chen SA, Cheng YS. 1989. Quantitative analysis of the essential oil of *Cinnamomum osmophloeum* Kanehira. J Agric Food Chem 37, 744–746.

Farias APP, Monteiro ODS, da Silva JKR, Figueiredo PLB, Rodrigues AAC, Monteiro IN, Maia JGS. 2020. Chemical composition and biological activities of two chemotype-oils from *Cinnamomum verum* J. Presl growing in North Brazil. J Food Sci Technol 57, 3176–3183.

Fuentes RG, Diloy FN, Tan IL, Balanquit BJR. 2010. Antioxidant and antibacterial properties of crude methanolic extracts of *Cinnamomum mercadoi* Vidal. Philipp J Nat Sci 15, 9–15.

Fujita Y Fujita S-I, Yoshikawa H. 1971. Biogenesis of the essential oils in camphor trees. XXVIII. On the components of the essential oil of *Cinnamomum japonicum* Sieb. Bull Chem Soc Jpn 44, 784–786.

Ghalib RM, Hashim R, Sulaiman O, Mehdi SH, Anis Z, Rahman SZ, Ahamed BMK, Majid AMSA. 2012. Phytochemical analysis, cytotoxic activity and constituents-activity relationships of the leaves of *Cinnamomum iners* (Reinw. ex Blume-Lauraceae). Nat Prod Res 26, 2155–2158.

Ghorbani A. 2005. Studies on pharmaceutical ethnobotany in the region of Turkmen Sahra, North of Iran (Part 1): General results. J Ethnopharmacol 102, 58–68.

Gogoi B, Kakoti BB, Borah S, Borah NS. 2014. Antihyperglycemic and in vivo antioxidative activity evaluation of *Cinnamomum bejolghota* (Buch.-Ham.) in streptozotocin induced diabetic rats: An ethnomedicinal plant in Assam. Asian Pac J Trop Med 7, S427–S434.

Gogoi B, Kakoti BB, Sharma N, Borah S. 2016. Pharmacognostic and preliminary phytochemical evaluation of *Cinnamomum bejolghota* (Buch.-Ham.) sweet bark. Indian J Nat Prod Resour 7, 59–64.

Gopalakrishnan S, Ediga HH, Reddy SS, Reddy GB, Ismail A. 2018. Procyanidin-B2 enriched fraction of cinnamon acts as a proteasome inhibitor and anti-proliferative agent in human prostate cancer cells. IUBMB Life 70, 445–457.

Gorgonio SRP, Fuentes RG. 2011. Antidiarrheal activity of *Cinnamomum mercadoi* methanolic leaf and bark extracts. Philipp J Nat Sci 16, 43–47.

Gotmare S, Tambe E. 2019. Identification of chemical constituents of cinnamon bark oil by GCMS and comparative study garnered from five different countries. Glob J Sci Front Res XIX, 35–42.

Gyawali R, Bhandari J, Amatya S, Piya E, Pradhan UL, Paudyal R, Shrestha R, Shrestha TM. 2013. Antibacterial and cytotoxic activities of high-altitude essential oils from Nepalese Himalaya. J Med Plant Res 7, 738–743.

Han X, Parker TL. 2017. Anti-inflammatory activity of cinnamon (*Cinnamomum zeylanicum*) bark essential oil in a human skin disease model. Phytother Res 31, 1034–1038.

Hapsari Y, Simanjuntak P. 2010. Study senyawa kimia dalam fase ekstrak etil asetat simplisia *Cinnamomum* spp. Secara KCKT dan KG-SM. J Kimia Mulawarman 8, 23–27.

Hassan W, Kazmi SNZ, Noreen H, Riaz A, Zaman B. 2016. Antimicrobial activity of *Cinnamomum tamala* leaves. J Nutr Disord Ther 6, 2.

Ho C-L, Wang EI-C, Wei X-T, Lu S-Y, Su Y-C. 2008. Composition and bioactivities of the leaf essential oils of *Cinnamomum subavenium* Miq. From Taiwan. J Essent Oil Res 20, 328–334.

Hong CH, Hur SK, Oh O-J, Kim SS, Nam KA, Lee SK. 2002. Evaluation of natural products on inhibition of inducible cyclooxygenase (COX-2) and nitric oxide synthase (iNOS) in cultured mouse macrophage cells. J Ethnopharmacol 83, 153–159.

Hossan MS, Hanif A, Khan M, Bari S, Jahan R, Rahmatullah M. 2009. Ethnobotanical survey of the Tripura tribe of Bangladesh. Am Eurasian J Sustain Agric 3, 253–261.

Houdkova M, Urbanova K, Doskocil I, Rondevaldova J, Novy P, Nguon S, Chrun R, Kokoska L. 2018. *In vitro* growth-inhibitory effect of Cambodian essential oils against pneumonia causing bacteria in liquid and vapour phase and their toxicity to lung fibroblasts. S Afr J Bot 118, 85–97.

Hrideek TK, Ginu J, Raghu AV, Jijeesh CM. 2016. Phytochemical profiling of bark and leafvolatile oil of two wild *Cinnamomum* species from evergreen forests of Western Ghats. Plant Arch 16, 266–274.

Hsieh TJ, Chen CH, Lo WL, Chen CY. 2006b. Lignans from the stem of *Cinnamomum camphora*. Nat Prod Commun 1, 21–25.

Hsieh T-J, Liu T-Z, Lu F-J, Hsieh P-Y, Chen C-H. 2006a. Actinodaphine induces apoptosis through increased nitric oxide, reactive oxygen species and down-regulation of NF-kappaB signaling in human hepatoma Mahlavu cells. Food Chem Toxicol 44, 344–354.

Huang L-C, Huang B-L, Murashige T. 1998. A micropropagation protocol for *Cinnamomum camphora*. *In Vitro* Cell Dev Biol Plant 34, 141–146.

Huang TC, Fu H-Y, Ho C-T, Tan D, Huang Y-T, Pan M-H. 2007. Induction of apoptosis by cinnamaldehyde from indigenous cinnamon *Cinnamomum osmophloeum* Kaneh through reactive oxygen species production, glutathione depletion, and caspase activation in human leukemia K562 cells. Food Chem 103, 434–443.

Jakhetia V, Patel R, Khatri P, Pahuja N, Pandey A, Gyan S. 2010. *Cinnamon*: A pharmacological review. J Adv Sci Res 1, 19–23.

Jantan I, Moharam BAK, Santhanam J, Jamal JA. 2008. Correlation between chemical composition and antifungal activity of the essential oils of eight *Cinnamomum* species. Pharm Biol 46, 406–412.

Jantan IB, Yalvema MF, Ahmad NW, Jamal JA. 2005a. Insecticidal activities of the leaf oils of eight *Cinnamomum* species against *Aedes aegypti* and *Aedes albopictus*. Pharm Biol 43, 526–532.

Jantan IB, Yalvema MF, Ayop N, Ahmad SA. 2005b. Constituents of the essential oils of *Cinnamomum sintoc* Blume from a mountain forest of Peninsular Malaysia. Flavour Fragr J 20, 601–604.

Jayaprakasha GK, Rao LJM. 2011. Chemistry, biogenesis, and biological activities of *Cinnamomum zeylanicum*. Crit Rev Food Sci Nutr 51, 547–562.

Jayaprakasha GK, Rao LJM, Sakariah KK. 2003. Volatile constituents from *Cinnamomum zeylanicum* fruit stalks and their antioxidant activities. J Agric Food Chem 51, 4344–4348.

Jiang ZT, Li R, Wang Y. 2008. Essential oil composition of *Cinnamomum loureiroi* grown in China extracted by supercritical fluid extraction. J Essent Oil Bear Plant 11, 267–270.

Joshi B, Lekhak S, Sharma A. 2009. Antibacterial property of different medicinal plants: *Ocimum sanctum*, *Cinnamomum zeylanicum*, *Xanthoxylum armatum*, and *Origanum majorna*. Kathmandu Univ J Sci Eng Technol 5, 143–150.

Kan H-L, Wang C-C, Cheng Y-H, Yang C-L, Chang H-S, Chen I-S, Lin Y-C. 2020. Cinnamtannin B1 attenuates rosacea-like signs via inhibition of pro-inflammatory cytokine production and down-regulation of the MAPK pathway. Peer J 8, e10548.

Kapoor LD. 2000. CRC Handbook of Ayurvedic Medicinal Plants. CRC Press Inc, Boca Raton.

Khan IA, Aziz A, Munawar SH, Manzoor Z, Sattar M, Afzal A. 2014. Evaluation of counter irritant potential of aqueos bark extract of *Cinnamon loureiroi*. Int J Pharm Res Allied Sci 3, 30–35.

Kochummen KM. 1989. Family Lauraceae. In: Tree Flora of Malaya: A Manual for Foresters, Vol 4, Whitmore T (ed.). Longmans, Kuala Lumpur, pp 124–132.

Krishnamoorthy B, Rema J. 2004. End uses of cinnamon and cassia. In: *Cinnamon* and *Cassia*: The Genus *Cinnamomum*, Ravindran PN, Babu KN (eds). CRC Press, Boca Raton, pp 311.

Kubatka P, Kello M, Kajo K, Samec M, Jasek K, Vybohova D, Uramova S, Liskova A, Sadlonova V, Koklesova L, Murin R, Adamkov M, Smejkal K, Svajdlenka E, Solar P, Samuel SM, Kassayova M, Kwon TK, Zubor P, Pec M, Danko J, Büsselberg D, Mojzis J. 2020. Chemopreventive and therapeutic efficacy of *Cinnamomum zeylanicum* L. bark in experimental breast carcinoma: Mechanistic *in vivo* and *in vitro* analyses. Molecules 25, 1399.

Kumar S, Kumari R, Mishra S. 2019. Pharmacological properties and their medicinal uses of *Cinnamomum*: A review. J Pharm Pharmacol 71, 1735–1761.

Kumar S, Sharma S, Vasudeva N. 2012a. Chemical compositions of *Cinnamomum tamala* oil from two different regions of India. Asian Pac J Trop Dis 2, S761–S764.

Kumar S, Vasudeva N, Sharma S. 2012b. GC-MS analysis and screening of antidiabetic, antioxidant and hypolipidemic potential of *Cinnamomum tamala* oil in streptozotocin induced diabetes mellitus in rats. Cardiovasc Diabetol 11, 95.

Kumar S, Vasudeva N, Sharma S. 2012c. Pharmacological and pharmacognostical aspects of *Cinnamomum tamala* Nees and Eberm. J Pharm Res 5, 480–484.

Kunwar RM, Adhikari N. 2005. Ethnomedicine of Dolpa district, Nepal: The plants, their vernacular names and uses. Lyonia 8, 43–49.

Kuo S-Y, Hsieh T-J, Wang Y-D, Lo W-L, Hsui Y-R, Chen C-Y. 2008. Cytotoxic constituents from the leaves of *Cinnamomum subavenium*. Chem Pharm Bull 56, 97–101.

Kuo YC, Lu C-K, Huang L-W, Kuo Y-H, Chang C, Hsu F-L, Lee T-H. 2005. Inhibitory effects of acylated kaempferol glycosides from the leaves of *Cinnamomum kotoense* on the proliferation of human peripheral blood mononuclear cells. Planta Med 71, 412–415.

Kuo YH, Chen WC, Lin YT. 1984. Chemistry of leave extractives from cineole tree. J Chin Chem Soc 31, 159–163.

Kuo YH, Shue MJ. 1991. New esters, 2-(4-hydroxy-3-methoxyphenyl)ethyl hexa- and octacosanoates from the leaves of *Cinnamomum reticulatum* Hay. J Chin Chem Soc 38, 65–69.

Kurniawati AD, Huang TC, Kusnadi J. 2017. Effect of fermentation on compositional changes of *Cinnamomum osmophloeum* Kaneh leaves. IOP Conf Ser Mater Sci Eng 193, 012013.

Langerberger G, Prigge V, Martin K, Belonias B, Sauerborn J. 2009. Ethnobotanical knowledge of Philippine lowland farmers and its application in agroforestry. Agrofor Syst 76, 173–194.

Lee J-E, Jung M, Lee S-C, Huh M-J, Seo S-M, Park I-K. 2020b. Antibacterial mode of action of trans-cinnamaldehyde derived from cinnamon bark (*Cinnamomum verum*) essential oil against *Agrobacterium tumefaciens*. Pestic Biochem Physiol 165, 104546.

Lee J-E, Seo S-M, Huh M-J, Lee S-C, Park I-K. 2020a. Reactive oxygen species mediated-antifungal activity of cinnamon bark (*Cinnamomum verum*) and lemongrass (*Cymbopogon citratus*) essential oils and their constituents against two phytopathogenic fungi. Pestic Biochem Physiol 168, 104644.

Lee S-C, Hsu J-S, Li C-C, Chen K-M, Liu C-T. 2015. Protective effect of leaf essential oil from *Cinnamomum osmophloeum* Kanehira on endotoxin-induced intestinal injury in mice associated with suppressed local expression of molecules in the signaling pathways of TLR4 and NLRP3. PLoS One 10, e0120700.

Lee S-C, Wang S-Y, Li C-C, Liu C-T. 2018. Anti-inflammatory effect of cinnamaldehyde and linalool from the leaf essential oil of *Cinnamomum osmophloeum* Kanehira in endotoxin-induced mice. J Food Drug Anal 26, 211–220.

Leela J. 2008. *Cinnamon* and *Cassia*. In: Chemistry of Spices, Parthasarathy V, Chempakam B, Zachariah T (eds). CABI, Cambridge.

Leu YL, Chung YM, Lai JY. 2014. The chemical principles of the leaves of *Cinnamomum kanehirae* Hayata. Planta Med 80, P1L125.

Li YQ, Kong Dx, Wu H. 2013. Analysis and evaluation of essential oil components of cinnamon barks using GC-MS and FTIR spectroscopy. Ind Crops Prod 41, 269–278.

Liao P-C, Kuo D-C, Lin C-C, Ho K-C, Lin T-P, Hwang S-Y. 2010. Research article historical spatial range expansion and a very recent bottleneck of *Cinnamomum kanehirae* Hay. (Lauraceae) in Taiwan inferred from nuclear genes. BMC Evol Biol 10, 124.

Lin CC, Cheng H-Y, Fang B-J. 2003. Anti-herpes virus type 2 activity of herbal medicines from Taiwan. Pharm Biol 41, 259–262.

Lin C-T, Chen C-J, Lin T-Y, Tung JC, Wang S-Y. 2008a. Anti-inflammation activity of fruit essential oil from *Cinnamomum insularimontanum* Hayata. Biores Technol 99, 8783–8787.

Lin G-M, Chen Y-H, Yen P-L, Chang S-T. 2016. Antihyperglycemic and antioxidant activities of twig extract from *Cinnamomum osmophloeum*. J Tradit Complement Med 6, 281–288.

Lin HY, Chang ST. 2012. Kaempferol glycosides from the twigs of *Cinnamomum osmophloeum* and their nitric oxide production inhibitory activities. Carbohydr Res 364, 49–53.

Lin IJ, Lo WL, Chia YC, Huang LY, Cham TM, Tseng WS, Yeh YT, Yeh HC, Wang YD, Chen CY. 2010. Isolation of new esters from the stems of *Cinnamomum reticulatum* Hay. Nat Prod Res 24, 775–780.

Lin IJ, Yeh HC, Cham TM, Chen CY. 2011c. A new butanolide from the leaves of *Cinnamomum reticulatum*. Chem Nat Comp 47, 43–45.

Lin RJ, Cheng MJ, Huang JC, Lo WL, Yeh YT, Yen CM, Lu CM, Chen CY. 2009. Cytotoxic compounds from the stems of *Cinnamomum tenuifolium*. J Nat Prod 72, 1816–1824.

Lin R-J, Lo W-L, Wang Y-D, Chen C-Y. 2008b. A novel cytotoxic monoterpenoid from the leaves of *Cinnamomum subavenium*. Nat Prod Res 22, 1055–1059.

Lin S-SC, Lu T-M, Chao P-C, Lai Y-Y, Tsai H-T, Chen C-S, Lee Y-P, Chen S-C, Chou M-C, Yang C-C. 2011b. In vivo cytokine modulatory effects of cinnamaldehyde, the major constituent of leaf essential oil from *Cinnamomum osmophloeum* Kaneh. Phytother Res 25, 1511–1518.

Lin T-Y, Liao J-W, Chang S-T, Wang S-Y. 2011a. Antidyslipidemic activity of hot-water extracts from leaves of *Cinnamomum osmophloeum* Kaneh. Phytother Res 25, 1317–1322.

Lin Y-H, Chen C-Y, Chou L-Y, Chen C-H, Kang L, Wang C-Z. 2017. Enhancement of bone marrow-derived mesenchymal stem cell osteogenesis and new bone formation in rats by obtusilactone A. Int J Mol Sci 18, 2422.

Liu C-H, Chen C-Y, Huang A-M, Li J-H. 2011. Subamolide A, a component isolated from *Cinnamomum subavenium*, induces apoptosis mediated by mitochondria-dependent, p53 and ERK1/2 pathways in human urothelial carcinoma cell line NTUB1. J Ethnopharmacol 137, 503–511.

Liu S, Yang L, Zheng S, Hou A, Man W, Zhang J, Wang S, Wang X, Yu H, Jiang H. 2021. A review: The botany, ethnopharmacology, phytochemistry, pharmacology of *Cinnamomi cortex*. RSC Adv 11, 27461–27497.

Liu Y, Chen K, Leu Y, Way T, Wang L, Chen Y, Liu Y. 2015. Ethanol extracts of *Cinnamomum kanehirai* Hayata leaves induce apoptosis in human hepatoma cell through caspase-3 cascade. Onco Targets Ther 8, 99–109.

Liu YM, Liu YK, Lan KL, Lee YW, Tsai TH, Chen YJ. 2013. Medicinal fungus *Antrodia cinnamomea* inhibits growth and cancer stem cell characteristics of hepatocellular carcinoma. Evid Based Complement Altern Med 2013, 569737.

Lokendrajit N, Swapana N, Singh CD, Singh CB. 2011. Herbal folk medicines used for urinary and calculi/ stone cases complaints in Manipur. NeBIO 2, 1–5.

Maridass M. 2008. Evaluation of brine shrimp lethality of *Cinnamomum* species. Ethnobot Leafl 12, 772–775.

Maridass M. 2009. Hepatoprotective activity of barks extract of six *Cinnamomum* species on carbon tetrachloride-induced in albino rats. Folia Med Indones 45, 204–207.

Marongiu B, Piras A, Porcedda S, Tuveri E, Sanjust E, Meli M, Sollai F, Zucca P, Rescigno A. 2007. Supercritical CO2 extract of *Cinnamomum zeylanicum*: Chemical characterization and antityrosinase activity. J Agric Food Chem 55, 10022–10027.

McCown BH, Amos R. 1979. Initial trials of commercial micropropagation with birch. Proc Int Plant Propag Soc 29, 387–393.

Mdoe FP, Sen-Sung Cheng S-S, Msangi S, Nkwengulila G, Chang S-T, Kweka EJ. 2014. Activity of *Cinnamomum osmophloeum* leaf essential oil against *Anopheles gambiae* s.s. Parasit Vectors 7, 209.

Mehrpouri M, Hamidpour R, Hamidpour M. 2020. Cinnamon inhibits platelet function and improves cardiovascular system. J Med Plants 19, 1–11.

Mikawlrawng K, Kumar S. 2014. Vandana Current scenario of urolithiasis and the use of medicinal plants as antiurolithiatic agents in Manipur (north east India): A review. Int J Herb Med 2, 1–12.

Monteiro IN, Monteiro ODS, Costa-Junior LM, da Silva Lima A, de Aguiar Andrade EH, Maia JGS, Filho VEM. 2017. Chemical composition and acaricide activity of an essential oil from a rare chemotype of *Cinnamomum verum* Presl on *Rhipicephalus microplus* (Acari: Ixodidae). Vet Parasitol 238, 54–57.

Muchtaridi M, Sumiwi SI, Nuwarda RF. 2017. Chemical composition of essential oils and its locomotor activity from the barks of *Cinnamomum sintoc* Bl. of two districts in middle Java. Asian J Pharm Clin Res 10, 84–87.

Murashige T, Skoog F. 1962. A revised medium for rapid growth and bioassays with tobacco tissue culture. Physiol Plant 15, 473–493.

Mustaffa F, Hassan Z, Asmawi MZ. 2016. *Cinnamomum iners* leaves as an alternative therapy for diabetes. Asian J Biochem 11, 44–52.

Mustaffa F, Hassan Z, Yusof NA, Razak KNA, Asmawi MZ. 2014. Antidiabetic and antihyperlipidemic potential of standardized extract, fraction and subfraction of *Cinnamomum iners* leaves. Int J Pharm Pharm Sci 6, 220–225.

Mustaffa F, Indurkar J, Ismail S, Mordi MN, Ramanathan S, Mansor SM. 2010. Analgesic activity, toxicity study and phytochemical screening of standardized *Cinnomomum iners* leaves methanolic extract. Pharmacogn Res 2, 76–81.

Mustaffa F, Indurkar J, Ismail S, Shah M, Mansor SM. 2011. An antimicrobial compound isolated from *Cinnamomum iners* leaves with activity against methicillin-resistant *Staphylococcus* aureus. Molecules 16, 3037–3047.

Muthenna P, Raghu G, Anil Kumar P, Surekha MV, Reddy GB. 2014. Effect of cinnamon and its procyanidin-B2 enriched fraction on diabetic nephropathy in rats. Chem Biol Interact 222, 68–76.

Nishida S, Tsukaya S, Nagamasu H, Nozaki M. 2006. A Comparative study on the anatomy and development of different shapes of domatia in *Cinnamomum camphora* (Lauraceae). Ann Bot 97, 601–610.

Padmakumari Amma KP, Priya Rani M, Sasidharan I, Sreekumar MM. 2013. Comparative chemical composition and in vitro antioxidant activities of essential oil isolated from the leaves of *Cinnamomum tamala* and *Pimenta dioica*. Nat Prod Res 27, 290–294.

Palis HG. 1995. Non-timber Forest Products in Manupali Watershed, Bukidnon, Philippines. ERDB-DENR Terminal Report. Los Baños, Laguna.

Pang K-L, Thong W-L, How S-E. 2009. *Cinnamomum iners* as mitogen-activated protein kinase kinase (MKK1) inhibitor. Int J Eng Technol 1, 310–313.

Pathak R, Sharma H. 2021. A review on medicinal uses of *Cinnamomum verum* (Cinnamon). J Drug Deliv Ther 11(6-S), 161–166.

Peng X, Cheng K-W, Ma J, Chen B, Ho C-T, Lo C, Chen F, Wang M. 2008. Cinnamon bark proanthocyanidins as reactive carbonyl scavengers to prevent the formation of advanced glycation endproducts. J Agric Food Chem 56, 1907–1911.

Pereira JT, Hastie AYL. 2014. The Cinnamon Trees, *Cinnamomum* Schaeff. (Lauraceae) in Sabah. Annual Report. Sabah Forestry Department, Malaysia.

Phutdhawong W, Kawaree R, Sanjaiya S, Sengpracha W, Buddhasukh D. 2007. Microwaveassisted isolation of essential oil of *Cinnamomum iners* Reinw. ex Bl.: Comparison with conventional hydrodistillation. Molecules 12, 868–877.

Poudel DK, Rokaya A, Ojha PK, Timsina S, Satyal R, Dosoky NS, Satyal P, Setzer WN. The 2021. Chemical profiling of essential oils from different tissues of *Cinnamomum camphora* L. and their antimicrobial activities. Molecules 26, 5132.

Prakash B, Singh P, Yadav S, Singh SC, Dubey NK. 2013. Safety profile assessment and efficacy of chemically characterized *Cinnamomum glaucescens* essential oil against storage fungi, insect, aflatoxin secretion and as antioxidant. Food Chem Toxicol 53, 160–167.

Premakumara GAS, Abeysekera WPKM. 2021. Pharmacological properties of Ceylon Cinnamon. In: Cinnamon: Botany, Agronomy, Chemistry and Industrial Applications, Senaratne R, Pathirana R (eds). Springer International Publishing, Cham, pp 307–325.

Rai VRS, Chandra KSJ. 1987. Clonal propagation of *Cinnamomum zeylanicum* Breyn. by tissue culture. Plant Cell Tissue Organ Cult 9, 81–88.

Raksha R, Kumar R, Sharma P, Ahmad HY, Rai S. 2021. Phytochemical screening and free radical scavenging activity of *Cinnamomum tamala* leaf extract. Int J Zool Invest 7, 376–386.

Ramalingam K, Balasubramanian A. 2015. *In-vitro* anticancer activity of *Cinnamomum iners* Reinw. against DAL and EAC cell lines. Ind J Appl Res 5, 546–547.

Rameshkumar KB, George V. 2006a. Chemical constituents and antimicrobial activity of the leaf oil of *Cinnamomum filipedicellatum* Kosterm. J Essent Oil Res 18, 234–236.

Rameshkumar KB, George V. 2006b. *Cinnamomum sulphuratum* Nees—a benzyl benzoate-rich new chemotype from Southern Western Ghats, India. J Essent Oil Res 18, 521–522.

Rana VS, Langoljam RD, Verdeguer M, Blázquez MA. 2012. Chemical variability in the essential oil of *Cinnamomum tamala* L. leaves from India. Nat Prod Res 26, 1355–1357.

Ranasinghe P, Pigera S, Sirimal Premakumara GA, Galappaththy P, Constantine GR, Katulanda P. 2013. Medicinal properties of 'true' cinnamon (*Cinnamomum zeylanicum*): A systematic review. BMC Complement Altern Med 13, 275.

Rao CV, Vijayakumar M, Sairam K, Kumar V. 2008. Antidiarrhoeal activity of the standardised extract of *Cinnamomum tamala* in experimental rats. Nat Med (Tokyo) 62, 396–402.

Rao R. 1979. Ethnobotanical studies on the flora of Meghalaya—some interesting reports of herbal medicines. In: Glimpses of Indian Ethnobotany, Jain SK (ed.). Oxford and IBH Publishing, New Delhi, pp 137–148.

Ravindran PN, Babu K, Shylaja M. 2004. Cinnamon and Cassia. The Genus *Cinnamomum*. CRC Press, Boca Raton.

Rawat I, Verma N, Joshi K. 2019. Cinnamon (*Cinnamomum zeylanicum*). In: Medicinal Plant in India: Importance & Cultivation, Ghosh SN (ed.). Narendra Publishing House, New Delhi, pp 164–177.

Reichling J, Schnitzler P, Suschke U, Saller R. 2009. Essential oils of aromatic plants with antibacterial, antifungal, antiviral, and cytotoxic properties—an overview. Forsch Komplementmed 16, 79–90.

Rema J, Leela NK, Krishnamoorthy B, Mathew PA. 2005. Chemical composition of *Cinnamomum tamala* essential oil—a review. J Med Aromat Plant Sci 27, 515–519.

Sajem AL, Rout J, Nath M. 2008. Traditional tribal knowledge and status of some rare and endemic medicinal plants of North Cachar Hills District of Assam, Northeast India. Ethnobot Leafl 12, 261–267.

Saleem M, Bhatti HN, Jilani MI, Hanif MA. 2015. Bioanalytical evaluation of *Cinnamomum zeylanicum* essential oil. Nat Prod Res 29, 1857–1859.

Salleh WMNHW, Ahmad F, Yen KH. 2015a. Evaluation of antioxidant, anticholinesterase and antityrosinase activities of Malaysian *Cinnamomum* species. Dhaka Univ J Pharm Sci 14, 125–132.

Salleh WMNHW, Ahmad F, Yen KH. 2015b. Antioxidant and anticholinesterase activities of essential oils of *Cinnamomum griffithii* and *C. macrocarpum*. Nat Prod Commun 10, 1465–1468.

Samy J, Sugumaran M, Lee KLW. 2005. Herbs of Malaysia: An Introduction to Medicinal, Culinary, Aromatic and Cosmetic Use of Herbs. Marshall Cavendish Publ Sdn Bhd, Selangor.

Sankaran V, Chakraborty A, Jeyaprakash K, Ramar M, Chellappan DR. 2015. Chemical analysis of leaf essential oil of *Cinnamomum tamala* from Arunachal Pradesh, India. J Chem Pharm Sci 8, 246–248.

Satyal P, Paudel P, Poudel A, Dosoky NS, Pokharelc KK, Setzer WN. 2013. Bioactivities and compositional analyses of cinnamomum essential oils from Nepal: *C. camphora, C. tamala*, and *C. glaucescens*. Nat Prod Commun 8, 1777–1784.

Shah M, Panchal M. 2010. Ethnopharmacological properties of *Cinnamomum tamala*—a review. Int J Pharm Sci Rev Res 5, 141–144.

Shahwar D, Rehman S, Ahmad N, Ullah S, Raza MA. 2010. Antioxidant activities of the selected plants from the family Euphorbiaceae, Malvaceae and Balsaminaceae. Afr J Biotechnol 9, 1086–1096.

Sharifi-Rad J, Dey A, Koirala N, Shaheen S, El Omari N, Salehi B, Goloshvili T, Cirone Silva NC, Bouyahya A, Vitalini S, Varoni EM, Martorell M, Abdolshahi A, Docea AO, Iriti M, Calina D, Les F, López V, Caruntu C. 2021. *Cinnamomum* species: Bridging phytochemistry knowledge, pharmacological properties and toxicological safety for health benefits. Front Pharmacol 12, 600139.

Sharma G, Nautiyal A. 2011. *Cinnamomum tamala*: A valuable tree from Himalayas. Int J Med Aromat Plant 1, 1–4.

Sharma G, Nautiyal BP, Nautiyal AR. 2009. Seedling emergence and survival in *Cinnamomum tamala* under varying micro-habitat conditions: Conservation implications. Trop Ecol 50, 201–209.

Sharma V, Rao LJM. 2014. An overview on chemical composition, bioactivity and processing of leaves of *Cinnamomum tamala*. Crit Rev Food Sci Nutr 54, 433–448.

Shen K-H, Lin E-S, Kuo P-L, Chen C-Y, Hsu Y-L. 2011. Isolinderanolide B, a butanolide extracted from the stems of Cinnamomum subavenium, inhibits proliferation of T24 human bladder cancer cells by blocking cell cycle progression and inducing apoptosis. Integr Cancer Ther 10, 350–358.

Showkat RM, Mohammed A, Kapoor R. 2004. Chemical composition of essential oil of *Cinnamomum tamala* Nees and Eberm. leaves. Flavour Fragr J 19, 112–114.

Silprasit K, Seetaha S, Pongsanarakul P, Hannongbua S, Choowongkomon K. 2011. Anti-HIV-1 reverse transcriptase activities of hexane extracts from some Asian medicinal plants. J Med Plant Res 5, 4899–4906.

Singh C, Singh S, Pande C, Tewari G, Pande V, Sharma P. 2013. Exploration of antimicrobial potential of essential oils of *Cinnamomum glanduliferum, Feronia elephantum, Bupleurum hamiltonii* and *Cyclospermum leptophyllum* against foodborne pathogens. Pharm Biol 51, 1607–1610.

Singh G, Maurya S, DeLampasona MP, Catalan CAN. 2007. A comparison of chemical, antioxidant and antimicrobial studies of cinnamon leaf and bark volatile oils, oleoresins and their constituents. Food Chem Toxicol 45, 1650–1661.

Singh J, Singh R, Parasuraman S, Kathiresan S. 2020. Antimicrobial activity of extracts of bark of *Cinnamomum cassia* and *Cinnamomum zeylanicum*. Int J Pharm Invest 10, 141–145.

Singh JS, Singh SP. 1992. Forest of Himalaya: Structure, Functioning and Impact of Man. Gyanodaya Prakashan, Nainital.

Singh N, Rao AS, Nandal A, Kumar S, Yadav SS, Ganaie SA, Narasimhan B. 2021. Phytochemical and pharmacological review of *Cinnamomum verum* J. Presl-a versatile spice used in food and nutrition. Food Chem 338, 127773.

Singh R, Jawaid T. 2012. *Cinnamomum camphora* (Kapur): Review. Pharmacogenomics J 4, 1–5.

Son LC, Dai DN, Thang TD, Huyen DD, Ogunwande IA. 2014. Study on *Cinnamomum* oils: Compositional pattern of seven species grown in Vietnam. J Oleo Sci 63, 1035–1043.

Son LC, Dai DN, Thang TD, Huyen DD, Olayiwola TO, Ogunmoye AR, Ogunwande IA. 2015. Chemical composition of essential oils from the stem barks of three *Cinnamomum* species. Br J Appl Sci Technol 11, 1–7.

Soulange JG, Ranghoo-Sanmukhiya VM, Seeburrun SD. 2007. Tissue culture and RAPD analysis of *Cinnamomum camphora* and *Cinnamomum verum*. Biotechnology 6, 239–244.

Sriramavaratharajan V, Murugan R. 2018. Screening of chemical composition, in vitro antioxidant, α-amylase and α-glucosidase inhibitory activities of the leaf essential oils of *Cinnamomum wightii* from different populations. Nat Prod Commun 13, 1539–1542.

Sriramavaratharajan V, Stephan J, Sudha V, Murugan R. 2017. Variation in volatile constituents of *Cinnamomum keralaense*, endemic to the Western Ghats, India. Nat Prod Res 31, 840–843.

Sriramavaratharajan V, Sudha V, Murugan R. 2016. Characterization of the leaf essential oils of an endemic species *Cinnamomum perrottetii* from Western Ghats, India. Nat Prod Res 30, 1085–1087.

Sthapit VM, Tuladhar PM. 1993. Sugandha kokila oil: A gift to perfumers from the Himalayan Kingdom of Nepal. J Herbs Spices Med Plants 1, 31–35.

Straub A, Benet-Buckholz J, Frode R. 2002. Metabolites of orally active NO-independent pyrazolopyridine stimulators of soluble guanylate cyclase. Bioorg Med Chem 10, 1711–1717.

Suhaimi AT, Ariffin ZZ, Ali NAM, Halim MIA, Mahat MM, Safian MF. 2017. Essential oil chemical constituent analysis of *Cinnamomum iners*. J Eng Appl Sci 12, 5369–5372.

Sulaiman SAB. 2013. Extraction of essential oil from *Cinnamomum zeylanicum* by various methods as a perfume oil. Bachelor Thesis, Universiti Malaysia Pahang, Gambang, Pahang.

Sultana S, Ripa FA, Hamid K. 2010. Comparative antioxidant activity study of some commonly used spices in Bangladesh. Pak J Biol Sci 13, 340–343.

Swanston-Flatt SK, Day C, Bailey CJ, Flatt PR. 1989. Evaluation of traditional plant treatments for diabetes: Studies in streptozotocin diabetic mice. Acta Diabetol Lat 26, 51–55.

Taha AM, Eldahshan OA. 2017. Chemical characteristics, antimicrobial, and cytotoxic activities of the essential oil of Egyptian *Cinnamomum glanduliferum* bark. Chem Biodivers 14, e1600443.

Thakur S, Walia B, Chaudhary G. 2021. Dalchini (*Cinnamomum zeylanicum*): A versatile spice with significant therapeutic potential. Int J Pharm Drug Ana 9, 126–136.

Torres RC, Sison FM, Ysrael MC. 2003. Phytochemical screening and biological studies on the crude methanol extract of *Cinnamomum mercadoi* Vidal. Philipp J Sci 132, 27–32.

Tung Y-T, Chua M-T, Wang S-Y, Chang S-T. 2008. Anti-inflammation activities of essential oil and its constituents from indigenous cinnamon (*Cinnamomum osmophloeum*) twigs. Bioresour Technol 99, 3908–3913.

Tung Y-T, Yen P-L, Lin C-Y, Chang S-T. 2010. Anti-inflammatory activities of essential oils and their constituents from different provenances of indigenous cinnamon (*Cinnamomum osmophloeum*) leaves. Pharm Biol 48, 1130–1136.

Udayaprakash NK, Ranjithkumar M, Deepa S, Sripriya N, Al-Arfaj AA, Bhuvaneswari S. 2015. Antioxidant, free radical scavenging and GC-MS composition of *Cinnamomum iners* Reinw. ex Blume. Ind Crops Prod 69, 175–179.

Vigila AG, Sahayaraj K, Baskaran X. 2018. *In vitro* antimicrobial activities of *Cinnamomum iners* leaf and bark extracts against pathogens of food borne diseases. Approaches Poult Dairy Vet Sci 3, 1–5.

Wang H-M, Cheng K-C, Lin C-J, Hsu S-W, Fang W-C, Hsu T-F, Chiu C-C, Chang H-W, Hsu C-H, Lee A-Y-L. 2010. Obtusilactone A and (-)-sesamin induce apoptosis in human lung cancer cells by inhibiting mitochondrial Lon protease and activating DNA damage checkpoints. Cancer Sci 101, 2612–2620.

Wang H-M, Chiu C-C, Wu P-F, Chen C-Y. 2011. Subamolide E from Cinnamomum subavenium induces sub-G1 cell-cycle arrest and caspase-dependent apoptosis and reduces the migration ability of human melanoma cells. J Agric Food Chem 59, 8187–8192.

Wang J, Su B, Jiang H, Cui N, Yu Z, Yang Y, Sun Y. 2020. Traditional uses, phytochemistry and pharmacological activities of the genus *Cinnamomum* (Lauraceae): A review. Fitoterapia 146, 104675.

Wang S-Y, Chen P-F, Chang S-T. 2005. Antifungal activities of essential oils and their constituents from indigenous cinnamon (*Cinnamomum osmophloeum*) leaves against wood decay fungi. Bioresour Technol 96, 813–818.

Wang Y, Zhang Y, Shi Y-Q, Pan X-H, Lu Y-H, Cao P. 2018. Antibacterial effects of cinnamon (*Cinnamomum zeylanicum*) bark essential oil on *Porphyromonas gingivalis*. Microb Pathol 116, 26–32.

Wanner JKR, Dai DN, Huong LT, Hung NV, Schmidt E, Jirovetz L. 2016. Chemical composition of Vietnamese essential oils of *Cinnamomum rigidifolium, Dasymaschalon longiusculum, Fissistigma maclurei* and *Goniothalamus albiflorus*. Nat Prod Commun 11, 1701–1703.

Wariyapperuma WANM, Kannangara S, Wijayasinghe YS, Subramanium S, Jayawardena B. 2020a. *In vitro* anti-diabetic effects and phytochemical profiling of novel varieties of *Cinnamomum zeylanicum* (L.) extracts. Peer J 8, e10070.

Wariyapperuma WANM, Kannangara S, Wijayasinghe YS, Subramanium S, Jayawardena B. 2020b. Fungal pretreatment to enhance the yield of phytochemicals and evaluation of α-amylase and α-glucosidase inhibition using *Cinnamomum zeylanicum* (L.) quills pressurized water extracts. Lett Appl Microbiol. https://doi.org/10.1111/lam.13410.

Wu C-C, Chu F-H, Ho C-K, Sung C-H, Chang S-H. 2017. Comparative analysis of the complete chloroplast genomic sequence and chemical components of *Cinnamomum micranthum* and *Cinnamomum kanehirae*. Holzforschung 71, 189–197.

Wu M, Lin Z, Huang B, Xu K, Zou S, Ni L, Chen Y. 2022. An update on phytochemistry and biological activities of *Cinnamomum*. Rec Nat Prod 16, 1–26.

Wu ZY, Raven PH. 1996. Flora of China: Myrsinaceae through Loganiaceae, Vol 15. Science Press, Beijing, and Missouri Botanical Garden Press, St. Louis.

Wuu-Kuang S. 2011. Taxonomic revision of *Cinnamomum* (Lauraceae) in Borneo. Blumea Biodivers Evol Biogeogr Plant 56, 241–264.

Yang C-L, Wu H-C, Hwang T-L, Lin C-H, Cheng Y-H, Wang C-C, Kan H-L, Kuo Y-H, Chen I-S, Chang H-S, Lin Y-C. 2020. Anti-inflammatory and antibacterial activity constituents from the stem of *Cinnamomum validinerve*. Molecules 25, 3382.

Yang S-S, Hou W-C, Huang LW, Lee T-H. 2006. A new γ-lactone from the leaves of *Cinnamomum kotoense*. Nat Prod Res 20, 1246–1250.

Yap PSX, Krishnan T, Chan K-G, Lim SHE. 2015. Antibacterial mode of action of *Cinnamomum verum* bark essential oil, alone and in combination with piperacillin, against a multi-drug-resistant *Escherichia coli* strain. J Microbiol Biotechnol 25, 1299–1306.

Yeh RY, Shiu YL, Shei SC. 2009. Evaluation of the antibacterial activity of leaf and twig extracts of stout camphor tree, Cinnamomum kanehirae, and the effects on immunity and disease resistance of white shrimp, Litopenaeus vannamei. Fish Shellfish Immunol 27, 26–32.

Yuan W, Lee HW, Yuk HG. 2017. Antimicrobial efficacy of *Cinnamomum javanicum* plant extract against *Listeria monocytogenes* and its application potential with smoked salmon. Int J Food Microbiol 260, 42–50.

Zisman A, Pantuck AJ, Chao DH. 2002. Renal cell carcinoma with tumor thrombus: Is cytoreductive nephrectomy for advanced disease associated with an increased complication rate? J Urol 168, 962–967.

8 *Cissus* species

8.1 MORPHOLOGICAL FEATURES, DISTRIBUTION, ETHNOPHARMACOLOGICAL PROPERTIES, PHYTOCHEMISTRY, AND PHARMACOLOGICAL ACTIVITIES

Cissus quadrangularis L. (Fam.—Vitaceae) is a tendril-climbing shrub with a thick quadrangular fleshy stem, internodes long (4–15 cm) and thick (1–2cm). The stem surface is smooth, pilos, and buff-colored with greenish shade (Mallika and Shyamala Devi 2005). The leaves are long, broadly ovate, denticulate, glabrous, cordate, rounded, and cuneate at the base (Rajpal 2005; Deka et al. 1994). Flowers arrange in short peduncle cymes with spreading umbellate branches. Calyx is cup-shaped, truncate or very obscurely lobed. There are four petals, ovate-oblong, short, and stout. Fruit is berry, obovoid or globose, long apiculate, red when ripe, one-seeded (Longman 1994; Rajpal 2005). Flowers bloom from June to December (Guhabakshi et al. 2001). They are naturalized in tropical and subtropical regions of India, Sri Lanka, South Africa, Thailand, Brazil, the southern United States, Java, and the Philippines (Anonymous 1992; Frank et al. 1995), Malaysia, West Africa, and Ceylon (Udupa et al. 1970; Sivarajan and Balachandran 1994). *C. quadrangularis* is used for the treatment of piles, anorexia, indigestion, chronic ulcers, asthma, otorrhea, and wound healing (Anonymous 1949; Viswanatha Swamy et al. 2006; Shah 2011). Its stem is useful in bone fracture, piles, chronic ulcers, asthma, scurvy, irregular menstruation, constipation, and blindness (Zaki et al. 2020). In Ayurvedic medicine, it is used in the treatment of gout, syphilis, venereal disease, piles, leucorrhea, and as an aphrodisiac. The stem paste is taken in asthma, burns, and wounds, bites of poisonous insects and for saddle sores of horses and camels (Ghouse 2015). Young shoots and stems of *C. quadrangularis* are used in the treatment of fractured bones, dyspepsia, indigestion, and asthma (Chopra et al. 1986). As per Siddha medicine, it is useful in piles, diarrhea, and dysentery (Shirwaikar et al. 2003). Gond tribals of Raisen district (India) use the stem of plant species in the treatment of scurvy, irregular menstruation, otorrhea, and epistaxis. The roots are used in bone fractures (Kumbhojkar et al. 1991; Khan et al. 1991). The plant extracts possess fracture-healing properties Udupa and Prasad 1964a, 1964b. The stem juice is used in the treatment of gastritis, constipation, eye diseases, piles, and anemia (Potu et al. 2011; Phimarn et al. 2014). The whole plant is used in the treatment of dyspepsia, colic, flatulence, tumors, convulsions, asthma, epistaxis, irregular menstruations, inflammations, pain, and syphilitic infections (Chatterjee and Chandraprakash 1997). *C. quadrangularis* has been used in dyspepsia, anorexia, seizures, tumors, epistaxis, asthma, antibacterial infections, and obesity (Nash et al. 2019; Sawangjit et al. 2017; Lee et al. 2018). In the Unani medicine system, the stems are alterative, anthelmintic, dyspeptic, digestive, tonic, analgesic in eye and ear diseases, used for irregular menstruation, piles, tumors, fractures of bones, wounds, and scurvy (Kritikar and Basu 2000; Stohs and Ray 2013).

The ethanolic and methanol extracts of *C. quadrangularis* aerial parts were tested for their antioxidant activities in DPPH, nitric oxide, superoxide, metal chelation, and hydrogen peroxide radical scavenging assays. The ethanol extract demonstrated significant dose-dependent free radical scavenging activity in different assays. Both extracts produced significant anticancer activity against leukemic cells HL-60 (IC_{50} 36 µg/ml and 40 µg/ml). Antioxidant and anticancer properties of both extracts of aerial parts of *C. quadrangularis* offers safe chemoprotective agents in fighting cancers (Dhanasekaran 2020). Methanol extract of *C. quadrangularis* was evaluated for their antioxidant activity in DPPH free radical, superoxide radical, hydroxyl radical production, and lipid peroxide production in erythrocytes. The effects of liver markers and antioxidant defense enzymes in liver homogenate were assessed in carbon tetrachloride and methanol extract-treated animals.

DOI: 10.1201/9781003398035-8

Carbon tetrachloride produced a significant enhancement in levels of aspartate aminotransferase, alanine aminotransferase, alkaline phosphatase, malondialdehyde, and a reduction in activities of superoxide dismutase, catalase, glutathione peroxidase, glutathione-S-transferase, and glutathione. The methanol extract reverted all these parameters. Methanol extract significantly displayed antilipid peroxidative, free-radical scavenging effects, and treated liver damage by an enhancing the activities of antioxidant enzymes; therefore, the plant extract may be used for therapeutic purposes against tissue damage (Jainu and Shyamala Devi 2005a, 2005b). The methanol extract of *C. quadrangularis* of root, stem, leaf, tendril, and standard (ascorbic acid) showed significant antioxidant activity (38.70 ± 0.04, 86.64 ± 0.04, 59.53 ± 0.04, 28.21 ± 0.04 and 99.96 ± 0.02 µg/ml) at 100 µg/ml concentration. The methanol extract produced significant inhibitory effects (IC_{50} 303 µg/ml (root), 267 µg/ml (stem), 325 µg/ml (leaves), 456 mg/ml (tendrils), and 204 µg/ml (ascorbic acid). Maximum hydrogen peroxide scavenging ability of methanol extract followed by using ethyl acetate, acetone, chloroform, and petroleum ether extracts. Stem extract showed higher antioxidant potential than leaves, roots, and tendrils. Hydrogen peroxide scavenging effect of methanolic extracts of root, stem, leaves, tendrils, and control were 20.23 ± 0.01, 31.15 ± 0.01, 24.61 ± 0.01, 19.28 ± 0.01, and 38.53 ± 0.05 µg/ml at 100 mg/ml concentration. Methanol extract of different parts showed significant hydrogen peroxide scavenging effect (IC_{50} values of root, stem, leaves, and tendrils were recorded as 430, 289, 329, and 489 µg/ml). Ascorbic acid also displayed significant inhibition (254 µg/ml). The inhibitory values of methanolic extract of root, stem, leaves, tendril, and ascorbic acid in ferric reducing antioxidant power assay were recorded as follows: root, stem, leaves, tendril, and ascorbic acid were 12.45 ± 0.04, 17.52 ± 0.02, 14.23 ± 0.02, 9.71 ± 0.02 and 20.92 ± 0.03 µg/ml at 100 mg/ml concentration. Methanol extract of different parts (root, stem, leaves, and tendril) and ascorbic acid displayed significant effect in ferric-reducing antioxidant power assay (IC_{50} 690, 589, 683, 725 µg/ml and 503 µg/ml; Sasi and Tamizhiniyan 2018). The total phenolic and total flavonoid contents in the ethanol extract of *C. quadrangularis* aerial parts were recorded in significant amounts (28.6 mg/g dry weight/gallic acid equivalents, and 15.8 mg/g/quercetin equivalents). The ethanol extract (IC_{50} 98 µg/ml, 125 µg/ml, and 96 µg/ml) and flavonoid-rich fraction (10 µg/ml, 12 µg/ml, and 10 µg/ml) showed significant free radical scavenging activity in all assays (nitric oxide, DPPH and hydroxyl radical scavenging assays). The flavonoid rich fraction demonstrated significant anticancer activity against breast cancer cells (MCF7; IC_{50} 40 µg/ml). The methanol extract of *C. quadrangularis* aerial parts possess potent antioxidant and anticancer effects (Vijayalakshmi et al. 2013). The ethanol extract of *C. quadrangularis* aerial parts showed maximum DPPH radical scavenging activity (51.17% at 300 µg/ml concentration). Ethanol extract also displayed significant phosphomolybednum reduction and Fe3+ reduction (84.21% and 58.57% at 300 µg/ml concentration; Sasi Rekha and Devika 2019).

The anti-obesity effect of *C. quadrangularis* formula (Cylaris™, contains a minimum 2.5% phytosterols and a minimum of 15% soluble plant fiber) were investigated in this clinical study. A total 123 overweight and obese persons (92 obese persons were randomized to a placebo or one of two treatment groups; 47.2% males and 52.8% females; ages 19–50 years) were recruited in this randomized, double-blind, placebo-controlled designed study. In 123 eligible persons, body mass index ranged from 25.5 to 45.6; waist circumferences from 85.5 cm to 125 cm; and weight from 62.6 kg to 142 kg. All persons received two daily doses (514 mg each) of the formulation or placebo and remained on a normal or calorie-controlled diet for 8 weeks. At the end of study, statistically significant net decrease in weight and central obesity, as well as in fasting blood glucose, total cholesterol, LDL-cholesterol, triglycerides, and C-reactive protein were recorded in persons who received the formulation, regardless of diet. *C. quadrangularis* formulation was found useful in the management of weight loss and metabolic diseases (Oben et al. 2006). Similarly, *C. quadrangularis*-only and a *C. quadrangularis/Irvingia gabonensis* combination were tested for their weight loss in overweight and obese human subjects. A total of 72 obese or overweight people were selected for the 10-week clinical study. Among selected 72 persons, 33 (45.8%) males and 39 (54.2%) females were recruited. The mean body mass index was recorded as >26 kg/m whereas the age range was

21–44 (mean age = 29.3). The persons were randomly grouped into three equal groups (n = 24): placebo; extract of *C. quadrangularis*-only, and *C. quadrangularis*/*Irvingia gabonensis* combination. The placebo (250 mg) or active formulations (150 mg *C. quadrangularis*-only and 250 mg *C. quadrangularis*/*Irvingia gabonensis* combination) were administered twice a day before meals with 8–10 oz. In comparison to the placebo group, the two active groups displayed a statistically significant difference in all six variables (body weight, body fat, waist size; total plasma cholesterol, low-density lipoproteins-cholesterol, fasting blood glucose level) by week 10. The magnitude of the differences was promising by week 4 and continued to increase during the period of study. The combination showed higher reductions in the studied variable than *C. quadrangularis*-only. The synergistic combination was found helpful in the management of obesity and its related complaints (Oben et al. 2008). Similarly, the anti-obesity effect of proprietary extract of *C. quadrangularis* and CQR-300 (CQR-300, a standardized extract of *C. quadrangularis* containing 2.5% keto-steroids and 15% soluble plant fiber) on weight, blood lipids, and oxidative stress in overweight and obese people were investigated. A total of 168 persons (38.7% males; 61.3% females; ages 19–54 years) were selected in this study. The persons (given normal diet or on an energy-restricted 2100 Kcal/ day diet) received two daily doses in the form of capsules of a proprietary formulation containing CQR-300 for 8 weeks and proprietary extract of *C. quadrangularis* or placebo for 6 weeks. Both CQR-300 and proprietary extract of *C. quadrangularis* demonstrated antioxidant activities *in vitro*. Both CQR 300 and proprietary extract of *C. quadrangularis* displayed significant ($p < 0.001$) decrease in plasma thiobarbituric acid reactive substances and carbonyls. The CQR-300 and proprietary extract of *C. quadrangularis* also showed significant decrease in levels of weight, body fat, total cholesterol, low-density lipoproteins-cholesterol, triglycerides, and fasting blood glucose during this study. CQR-300 (300 mg daily) and proprietary extract of *C. quadrangularis* (1028 mg daily) showed significant decrease in levels of body weight, blood glucose, and serum lipids, thus improving cardiovascular risk factors. The enhancement in levels of plasma 5-hydroxytryptamine and creatinine theorizes a mechanism of regulating appetite and stimulating the enhancement of lean muscle mass by *C. quadrangularis* hence, supporting the clinical study for weight loss and ameliorating cardiovascular health (Oben et al. 2007). A CQR-300 (preparation of *C. quadrangularis*) was used for anti-obesity effects. CQR-300 (300 mg/daily) administration was found effective in reducing weight, enhancing blood parameters attributed with metabolic disorders, as well as levels of serotonin in obese and overweight persons. Persons were grouped into four different groups: placebo overweight, placebo obese, overweight received 300 mg CQR-300 (A), and obese received 300 mg CQR-300 (B). A significant weight loss (difference, $p < 0.05$) in obese placebo group, the overweight group A, and the obese group B was observed in treated persons (3.1 ± 1.2%, 8.4 ± 1.1% and 11.3 ± 0.8). The decrease was significant at $p < 0.05$ at week 4 and at the end of the trial (week 8; $p < 0.01$). The significant reduction ($p < 0.05$) in waist circumference was observed in all the groups (placebo and treated) after 8 weeks (Kuate et al. 2015). The aqueous extract of *C. quandrangularis* leaf and stem and CQR-300 formulation (300 mg each) were tested for reducing body weight fat. A total of 67 individuals—placebo (32 persons) and the CQR-300 group (35 participants received 300 mg of corn starch or CQR-300 daily)—were selected in this study. After 8 weeks of treatment, persons of the placebo group displayed a significant decrease (1.05%) in body fat. CQR-300 treatment showed significant decrease (8.9% and 12.8%) in the body fat. The reduced values were significantly ($p < 0.05$) lessened than the placebo. The CQR-300 exhibited significant ($p < 0.05$) reductions in the circumferences of waist and hip, systolic and diastolic blood pressures, total cholesterol, triglycerides, fasting blood glucose, as well as in levels of leptin. Conversely, the CQR 300 increased the levels of HDL-cholesterol and adiponectin significantly in treated persons ($p < 0.05$). CQR-300 administration (single 300 mg dose daily) was found effective in decreasing body fat as well as increasing blood parameters attributed with metabolic disorders (Nash et al. 2019). Effects of *C. quadrangularis* on hemorrhoids and bone pain were studied. *C. quadrangularis* combination (CQR-300) showed significant decrease in levels of body weight, low-density lipoprotein, triglyceride, total cholesterol, and fasting blood sugar. The combination did not show any serious adverse effects during treatment. *C. quadrangularis* combination showed significant effects on bone fractures (Sawangjit et al. 2017). Effects of *C. quadrangularis* extract on obesity, lipolysis, and browning of human white adipocytes were examined in obese humans. Extract treatment significantly reduced the waist circumference ($p < 0.05$ all) and significantly reduced the hip circumference ($p < 0.05$ all) when compared with baseline levels. In human white adipocytes, glycerol secretion was decreased in the extract-treated persons. Similarly, ethanol extract (10, 100, and 1000 ng/ml) significantly increased the expression of UCP1 in a dose-dependent manner ($p < 0.001$ all). In summary, *C. quadrangularis* extract reduced the waist and hip circumferences in obese persons but increased UCP1 mRNA expression in human white adipocytes (Chatree et al. 2021).

C. quadrangularis hot water extract (CQR-300) was examined for its anti-obesity effects in 3T3-L1 adipocytes. CQR-300 suppressed lipid accumulation without displaying cytotoxicity to 3T3-L1 adipocytes. Moreover, CQR-300 reduced the adipogenesis/lipogenesis-related mRNA expression levels of fatty acid binding protein, fatty acid synthase, lipoprotein lipase, stearoyl-CoA desaturase-1, and acetyl-CoA carboxylase. Hot extract also downregulated the expression profiles of adipogenesis/lipogenesis-associated proteins (peroxisome proliferator-activated receptor γ, CCAAT/enhancer-binding protein α, sterol regulatory element binding protein-1c, and fatty acid synthase). The findings reveal that CQR-300 might have an anti-obesity activity by its potential to reduce expression levels of adipogenesis/lipogenesis-related genes and proteins (Lee et al. 2018). The anti-obesity activity of methanol extracts of *C. quadrangularis* and *Achyranthes aspera* (mixture) were assessed in high fat diet fed mice. Results revealed that consumption of high fat diet induced 9.17% body weight gain. The extract mixture (*C. quandrangualaris* and *A. aspera* extracts) displayed a significant decrease (3.02 gm) in body weight of treated mice. Similarly, anti-obesity effects of *A. aspera* mixed biscuits and *C. quadrangularis* capsules were administered to obese persons. Overweight and obese people (aged between 19 and 60 years) of both sexes were recruited in this clinical study. *A. aspera* biscuits (1g/day/patient) were taken by the patients. Similarly, *C. quadrangularis* capsules were also given (500 mg/day/patient) to the patients. These doses of biscuits and capsules were administered daily before meals in the morning to the patients. The *C. quadrangularis* capsules and *A. aspera* biscuits showed significant weight loss in treated obese patients than placebo (Talreja et al. 2017).

The petroleum ether extract of *C. quadrangularis* (100, 200, 300 µg/ml) increased the differentiation of marrow mesenchymal stem cells into alkaline phosphatase-positive osteoblasts and enhanced extracellular matrix calcification. Extract (300 µg/ml) also increased the growth of marrow mesenchymal stem cells. Cells cultured in osteogenic media containing methanol extract demonstrated greater proliferation, differentiation, and calcification rates than the cells of control. *C. quadrangularis* extract induces osteoblastogenesis and may be used as preventive/alternative natural medicine for the treatment of osteoporosis (Potu et al. 2009a). Petroleum ether extract (500 mg/kg) of *C. quadrangularis* decreased bone loss, as supported by the weight gain in the femur, and decreased the osteoclastic activity in ovariectomized rats. The extract showed positive effects on both enzymes (tartrate resistant acid phosphatase and alkaline phosphatase), and it might be considered as a potential subject for prevention and treatment of postmenopausal osteoporosis (Potu et al. 2009b). The effects of the petroleum ether fraction of *C. quadrangularis* were evaluated on the development of osteopenia in a type 1 diabetic rat model. The cortical and trabecular bone thickness and bone strength were significantly reduced in diabetic rats. The extract did not change the glycemic levels in these diabetic rats. In treated animals, the increased levels of serum alkaline phosphatase, tartrate-resistant acid phosphatase, and hydroxyproline were recorded. Extract also improved bone health in hyperglycemic conditions by increasing the cortical and trabecular bone growth; therefore, it may be used as an effective therapy against diabetes-associated bone disorders (Sirasanagandla et al. 2014). Petroleum ether extract of *C. quadrangularis* (500 mg/kg bw; daily for 90 days) was evaluated for anti-osteoporotic effects. The extract significantly enhanced the force necessitated to break the femur ($p < 0.001$) and significantly enhanced the thickness of both cortical ($p < 0.001$) and trabecular bones ($p < 0.001$). This activity was compared with the activity of raloxifene. The petroleum ether extract appears to demonstrate anti-osteoporotic effect in rats (Potu et al. 2010). The petroleum ether fraction of ethanol extract (95%) of *C. quadrangularis* fleshy stem was used to assess their actions in the formation of long bones during the intra-uterine developmental stage in pregnant rats. Pregnant rats were administered with petroleum ether extract (500 mg/kg bw) from gestation day 9 until delivery. Morphometric studies showed that the total length percentage of ossified cartilage (bone) in pups born to treated dams was significantly greater ($p < 0.001$–0.0001) than the control group. The extract stimulated the development of fetal bone growth significantly during the intra-uterine developmental period (Potu et al. 2008). *C. quadrangularis* extract enhances alkaline phosphatase in treated rats. Besides this, the extract is found useful in the treatment of obesity, anorexia, colic, dyspepsia, flatulence, helminthiasis, hemorrhage skin diseases, and many others (Verma 2018). Phytoestrogen-rich ethanol fraction of *C. quadrangularis* aerial parts (100 mg/kg) was evaluated for their anti-osteoporotic effects in ovariectomized rats. Ethanol fraction and estrogen exposure displayed significant enhancement in bone thickness, bone density, and bone hardness. The fraction also ($p < 0.05$) enhanced the levels of serum vitamin D3 and serum calcium. The extract significantly decreased the levels of alkaline phosphatase in treated animals and prevented bone loss. The study results confirm that phytoestrogen-rich fraction possess significant antiosteoporotic effects (Aswar et al. 2012).

The ethanol extract of *C. quadrangularis* did not show any toxic effects on osteoblastic cells. Extract enhanced the DNA synthesis of human osteoblastic SaOS-2 cells. Extract also enhanced the matrix mineralization of human osteoblastic SaOS-2 cells. Ethanolic extract's anabolic actions in SaOS-2 cells are mediated via enhanced mRNA and protein expression of Runx2. The study findings reveal positive regulation of ethanol extract on the growth, differentiation, and matrix mineralization of SaOS-2 cells (Muthusami et al. 2011a). The effects of ethanol extract of *C. quadrangularis* on the regulation of insulin-like growth factor system components in SaOS-2 cells were investigated. The extract increased the mRNA expression of insulin-like growth factor-I receptor and insulin-like growth factor-II receptor in treated cells. No significant difference in mRNA expression of insulin-like growth factor binding protein-3 was reported in between control and extract treated cells. As per findings, positive regulation of the extract on the insulin-like growth factor system components of SaOS-2 cells was reported (Muthusami et al. 2011b).

As per molecular docking data, quercetin and β-sitosterol showed maximum and minimum potential to bind to estrogen receptors. The hexane fraction of ethanol extract was enriched with rutin (65.36 ± 0.75 mg/g) and quercetin (1.06 ± 0.12 mg/g). Alkaline phosphatase activity was significantly increased in osteoblasts treated with hexane fraction. Hydroxyproline concentration was slightly enhanced during ethanol treatment but, synthesis of osteocalcin was suppressed. In addition, hexane fraction significantly activated osteoprotegerin synthesis but suppressed receptor activator of nuclear factor κ ligand expression. The rutin and quercetin-rich hexane fraction triggered the synthesis molecules associated with the formation of bones but suppressed bone resorption (Ruangsuriya et al. 2020).

The effects of hexane and dichloromethane fractions of *C. quadrangularis* on the differentiation and mineralization of mouse preosteoblast cell line MC3T3-E1 were studied. In growth, viability, and proliferation assays, the lower concentrations (0.1, 1, and 100 ng/ml) of both fractions (hexane and dichloromethane fractions) were found nontoxic, while higher concentrations were recorded as toxic to the cells. The nontoxic concentration of dichloromethane fraction delayed significant differentiation and mineralization of MC3T3-E1 cells. Similarly, nontoxic concentration of hexane fraction increased mineralization in MC3T3-E1 cells as well as osteoblast marker genes (Runx2, alkaline phosphatase, collagen, integrin-related bone sialoprotein, osteopontin, and osteocalcin). The results reveal that hexane fraction was found most efficacious. Further investigation is needed to isolate and identify the active constituents from hexane fraction (Toor et al. 2019a). The butanol and ethyl acetate extracts of *C. quadrangularis* were investigated for their osteoblast differentiation and mineralization potential on mouse pre-osteoblast cell line MC3T3-E1 (sub-clone 4). Both extracts were found nontoxic in (100 μg/ml) the treated animals. Both extracts increased the cell growth at minimal concentrations (0.1 µg/ml and 1 µg/ml) and upregulated the osteoblast formation as well as the process of mineralization in MC3T3-E1 cells. In this study, ethyl acetate extract showed higher potential than butanol extract in early differentiation and mineralization in MC3T3-E1 cells (Toor et al. 2019b). The effectiveness of ethanol extract of *C. quadrangularis* was investigated in promoting osteoblast formation of the murine pre-osteoblast cell line, MC3T3-E1. Ethanol extract showed significant effects on the growth kinetics of MC3T3-E1 cells in a dose-dependent manner. Higher concentrations (more than 10 µg/ml) of extract showed adverse effects, while lower concentrations (0.1 and 1 µg/ml) were found non-toxic in MC3T3-E1 cells. Lower concentration increased cell proliferation of MC3T3-E1 cells significantly. Ethanol extract treatment also enlarged osteoblast formation as well as extracellular matrix mineralization. The results reveal dose-dependent effects of extract with lower doses on anabolic and osteogenic activities (Tasadduq et al. 2017). The effects of ethanol extract of *C. quadrangularis* on osteoblast growth and function were assessed using murine osteoblastic cells. The expression of mRNA of osteoblast-related genes were not affected by the extract treatment. The extract treatment also increased the alkaline phosphatase activity and the formation of mineralized nodules. The findings reveal that the extract may control osteoblastic activity by increasing alkaline phosphatase activity and processes of mineralization. The enhanced alkaline phosphatase activity of the extract is likely to be mediated by MAP kinase-dependent pathway (Parisuthiman et al. 2009). The ethyl acetate and aqueous extracts of *C. quadrangularis* were prepared and standardized for ketosteroid contents. The aqueous extract was rich in 3-ketosteroids (α-amyrin, friedelin, and β-sitosterol). Aqueous extract increased the mean bone density, strength, and calcium contents in treated an orchidectomized rat model (Jadhav et al. 2016). The hexane extract of *C. quadrangularis* stem was tested for their anti-osteoporotic activity. The extract was rich in triterpenes, fatty acid methyl esters, glycerolipids, steroids, phytols, and cerebrosides. The glycerolipids and squalene induced alkaline phosphatase activity significantly (10 µg/ml). In addition, the hexane extract and glycerolipids and squalene showed significant synergistic effects on formation of bones; therefore, the plant species may be used in the treatment of osteoporosis (Pathomwichaiwat et al. 2015). The effects of *C. quadrangularis* extract loading on the mechanical, chemical, and degeneration properties as well as *in vitro* cytotoxicity, cell growth, and differentiation of the mixtures were examined. Extract internalization increased the compression modulus of scaffolds from 76 to 654 kPa. Extract scaffolds gave a favorable substrate for osteoblast adhesion, growth, and mineralization. On the 7th and 14th days of incubation, extract enhanced the alkaline phosphatase activity of the SaOS-2 cells on scaffolds, so *C. quadrangularis*/chitosan/Na-carboxymethyl cellulose scaffold may be used as a biomaterial for bone tissue engineering applications (Tamburaci et al. 2018). *C. quadrangularis* extract significantly enhanced the levels of alkaline phosphatase in mandibular fractures. The radiographic observations (increase in density and rate of change of density), clinical data (mobility, swelling, decrease in pain) and biochemical parameters (serum calcium, serum phosphorous) were not found statistically significant between the treated and control groups (Nayak and Keerthi 2020). Pain, swelling, tenderness, and mobility reduction were greater in the osteoseal group while lower in placebo. Extract increased the levels of serum calcium and phosphorus at different doses in the treated group but there was a reduction in the placebo group (Singh et al. 2011). *C. quadrangularis* capsules effects were evaluated in the healing process of maxillofacial fracture. In this clinical study, the patients were divided into two groups. First group patients were administered with one capsule of *C. quadrangularis* (500 mg) thrice a day for 6 weeks. In the second group patients, no supplementary prescription drug was given. Several parameters (pain, swelling, fragment mobility, serum calcium, and serum phosphorus levels) were examined pre- and post-operatively on days 1, 21, and 45 in patients. The levels of serum calcium and serum phosphorus were higher, and bone healing was clearly visible in first group patients on day 21 when compared with the control group. *C. quadrangularis* capsules assist in decreasing pain, swelling, and fracture mobility, and increasing the healing of fractured jaw bones (Brahmkshatriya et al. 2015). *C. quadrangularis* L. is used in bone regeneration, ulceration, and pain relief (Sundaran et al. 2020). *C. quadrangularis* extract (capsules) was used for osteogenic activity. In this study, a total of 60 patients (age, 20–35 years) of mandible fracture were selected in this study. The patients were divided into two groups. The first group patients were administered capsules and healing of fractures was examined. The second group was the control group. In treated patients, the capsules showed significant increase in profiles of osteopontin protein expression as well as in CD4+ T cells expressing osteopontin protein. Hot extract (capsules) increases the healing process of bone fractures and produces early remodeling of fracture callus (Singh et al. 2013).

Ethanol extract of *C. quadrangularis* (500 and 750 mg/kg/day) was investigated for their anti-osteoporotic activity in ovariectomized rat model of osteoporosis. Ethanol extract (5000 mg/kg bw) produced neither any mortality nor any signs of clinical abnormality in treated animals. The LD_{50} of extract was found as 5000 mg/kg bw. Extract (750 mg/kg bw) decreased the levels of tartrate resistant acid phosphatase in the treated animals. The extract (750 mg/kg bw) showed almost complete recovery with essential features of normal bone (Shirwaikar et al. 2003). The effects of sequential extracts (hexane, dichloromethane, ethanol, and water) and freeze-dried *C. quadrangularis* juice were examined to assess their effects on bone metabolism in an ovariectomized mouse model. Sequential extracts (equal to crude powder 5 g/kg/day for 8 weeks; p.o.) were administered to the experimental animals. The hexane extract increased the levels of bone mineral densities of the femur and tibia in treated animals and compared with those of sham-operated group. Bone mineral densities were attributed with the restoration of the trabecular bone of the lumbar spine in hexane extract treated animals. The hexane extract did not show any abnormality or pathological alterations in the internal organs of treated animals. The hexane extract displayed antiosteoporotic effects in ovariectomized mice without developing toxicity throughout the experiment. So, the hexane extract is most significant in further bioassay-guided identification of pharmacologically active constituents (Pathomwichaiwat et al. 2012). Bone healing effects of hexane fraction of ethanol extract of *C. quadrangularis* were examined. In the molecular docking study, the quercetin and β-sitosterol showed maximum and minimum potential to bind to estrogen receptors. Rutin and quercetin-rich ethanol fraction significantly enhanced the alkaline phosphatase activity in osteoblasts. In addition, the extract significantly stimulated the osteoprotegerin formation and suppressed receptor activator of nuclear factor κ ligand expression. Rutin and quercetin-rich ethanol fraction were involved in the formation of bones (Ruangsuriya et al. 2020). *C. quadrangularis* extract was tested for its bone protective properties in ovariectomized female Sprague Dawley rats. *C. quadrangularis* ovariectomized rats displayed protected bone mass and microarchitecture of trabecular bone in the distal femoral metaphysis and the proximal tibial metaphysis. The extract increased serum protein expression levels of P1NP and Trap5b but reduced CTX levels. The extract significantly increased the levels of osteocalcin but decreased Wnt/β catenin antagonist DCAT1 levels. In summary, the extract showed significant protection of the microarchitecture of the long bones in ovariectomy-induced bone loss (Guerra et al. 2019). Similarly, the extract effects on bone loss after ovariectomy in C57BL/6 mice were examined. The ethanol extract displayed no significant disagreement in the trabecular number, thickness, and connectivity density between sham and ovariectomized mice. There were no alterations recorded in the bone at the tibio-fibular junction between groups. It is suggested that the extract significantly suppressed bone loss in the cancellous and cortical bones of the femur and proximal tibia in these mice (Banu et al. 2012).

Aqueous extract of *C. quadrangularis* was evaluated for their antiepileptic effect in the maximal electroshock and isonicotinic hydrazide acid models and analgesic effect in the hot plate procedure, and for its smooth muscle relaxant effect in the rotarod method. The extract (500 mg/kg bw) showed significant protection in mice against maximal electroshock seizure and slowed down the onset period of seizures as stimulated by isonicotinic hydrazide acid. A significant analgesic effect was recorded using the hot plate method. In this method, the paw licking time was also slowed down significantly. The extract also showed significant smooth muscle relaxant activity in treated models (Kumar et al. 2010). Similarly, the chloroform extract (300 mg/kg) of *C. quadrangularis* displayed significant ($p < 0.05$) analgesic activity in hot plate and tail flick tests. Analgesic effect was not obstructed ($p < 0.05$) in the group pretreated with phentolamine but obstructed in the groups pretreated with naloxone and *p*-chlorophenylalanine methyl ester. The chloroform extract was found to be more active than methanol extract in stimulating analgesic activity in mice, and the analgesic effect may be moderated through opioidergic and serotonergic pathways (Nie et al. 2015). The effects of micronized purified flavonoid fraction of *C. quadrangularis* were determined on hemorrhoids pain (bleeding, pain, discharge, pruritis, and erythema). In this study, total 570 patients (299 females, 271 males) were registered, from which 80 patients received the flavonoid mixture. On second day, acute bleeding was stopped in treated patients. All the symptoms were improved in treated patients. No adverse effects and no blood chemistry changes were recorded in treated patients (Panpimanmas et al. 2010). The venotonic effect and analgesic and anti-inflammatory activities of the methanol extract of *C. quadrangularis* were compared with reference drugs in different models. Methanol extract showed a significant decrease in the number of writhes in acetic acid-induced writhing responses in mice. Extract also significantly decreased the licking time in both phases of the formalin test. In acute inflammation, the extract showed significant inhibitory effects on the formation of edema of rats' ear stimulated by ethyl phenylpropiolate and paw edema formation in rats stimulated by both carrageenin and arachidonic acid. Moreover, the extract showed significant venotonic effect on isolated human umbilical veins. *C. quadrangularis* extract is used for the treatment of pain and inflammation attributed with hemorrhoid as well as decreasing the size of hemorrhoids (Panthong et al. 2007).

Ethanol extract of *C. quandrangularis* increased the reduction of ulcer area in a dose-dependent manner. Additionally, extract significantly enhanced the (3)H-thymidine incorporation and the profiles of polyamines (putrescine, spermine and spermidine) in ulcerated rats. Moreover, the extract shows gastroprotection in the ulcerated area by enhanced expression of TGF-α. It also restored the alterations in the gastric mucosa of ulcerated rats with significant increase in the levels of mitochondrial tricarboxylic acid cycle enzymes. In summary, the healing activity of extract on acetic acid-induced gastric mucosal damage in rats may be associated to its growth promoting and cytoprotective roles (Jainu et al. 2010).

The ethyl acetate fraction of methanol extract of *C. quadrangularis* stem (100 ppm) displayed significant antioxidant activity in β-carotene linoleic acid (64.8%) and 1,1-diphenyl-2-picrylhydrazyl (DPPH) assays (61.6%). The ethyl acetate fraction and methanol extract of both fresh and dry stems demonstrated antimicrobial activity against *Bacillus subtilis*, *Bacillus cereus*, *S. aureus*, and *Streptococcus* species (Chidambara Murthy et al. 2003). The ethanol extract of *C. quadrangularis* was evaluated for its antimicrobial, antioxidant, and anti-cancer activities. Ethanol extract displayed significant antibacterial activity against *V. parahemolyticus*, *V. cholera*, enterotoxigenic *E. coli*, and *S. typhi* at 5–25 mg/well concentration. The extract also demonstrated dose-dependent scavenging of 2,2-diphenyl-1-picrylhydrazyl and 2,2′-azino-bis(3-ethylbenzothiazoline-6-sulfonic acid) radicals. The ethanol extract showed significant anticancer activity against MCF-7 cancer cells (IC_{50} 40 μg/ml). Over 24 h, the ethanol extract effectively suppressed the growth of MCF-7 cells in a dose-dependent manner (Subramani 2018).

The genotoxic and apoptotic activities of ethanol extract of *C. quadrangularis* were evaluated in A431 cells. Ethanol extract (100 μg/ml) stimulated the formation of intracellular reactive oxygen species in A431 cells. The extract also caused DNA injury in the A431 cells as well as cell death in a concentration-dependent manner (Arshad et al. 2016). *C. quadrangularis* extract (CQR-300) was used in sub-chronic toxicity and gentotoxicity studies. In the sub-chronic study, CQR-300 was administered to Sprague Dawley rats (0, 100, 1000, and 2500 mg/kg bw/day) for 90 days. During this study, no toxicity, mortality or changes in body weights, body weight gain or food consumption were recorded. Based on the sub-chronic study, the no-observed-adverse-effect level for CQR-300 was also assessed in this study (2500 mg/kg bw/day; Kothari et al. 2011).

The aqueous extract of *C. quadrangularis* leaves (372.21 mg/kg, p.o.) significantly and dose-dependently enhanced the latency to clonic and generalized tonic-clonic seizures and reduced the number and time of seizures. The extract also enhanced the number of crossings, and the time spent in the center of the open field paradigms. The findings suggest that the anticonvulsant effects of extract are accompanied by its anxiolytic effects (Moto et al. 2018). *C. quadrangularis* showed the presence of flavonoids, triterpenoids, alkaloids, saponins, iridoids, stilbenes, vitamins, steroids, and glycosides (Ayyanar and Ignacimuthu 2008; Bafna et al. 2021; Pandey et al. 2022). The methanolic extract of *C. quadrangularis* leaves showed significant anticancer activity against MG63 cells (IC$_{50}$ 100 µg/ml; Suresh et al. 2019; Packialakshmi and Makesh 2020). Ethyl acetate extract of C. quadrangularis displays the lowest percentage of viability which has significant activity against cancerous cells (Shabi Ruskin et al. 2014). Ethyl acetate extract presented a significant decrease in inflammatory conditions compared to control and standard treated groups (Tiwari et al. 2021). The ethanol extract possesses antioxidant activity as well as flavonoids and tannins (Arunagiri and Srinivasan 2021).

C. assamica (M.A.Lawson) Craib is a woody climber with reddish spots bearing angular stems. Leaves are simple, roundly cordate, orbicular, and cuspidate. Flowers are minute and arranged in slender cymes. Fruits are turbinate, the size of a pea, and black (Yusuf et al. 2009). It is naturalized in mainland of China, Vietnam, India, Thailand, Indonesia, the southern part of Taiwan, and Lanyu Island (Yang 1998). As per Chinese medicine, *C. assamica* stems are useful in activation of blood circulation to expel blood stasis and in the treatment of bruises, fractures, and rheumatoid arthritis (Zhang et al. 2006). The methanol extract of stems, leaves, and roots of *C. assamica* were investigated for their *in vivo* antinociceptive and antipyretic activities. The 200 and 400 mg/kg (bw) doses were administered to Swiss albino mice. In case of peripheral antinociceptive activity, the methanol extract of stem and its fractions (dichloromethane, chloroform, petroleum ether) and methanol extract of roots (200 and 400 mg/kg/bw) displayed significant antinociceptive effects as compared with diclofenac sodium (60.49% suppression). In the case of central antinociceptive activity, the methanol extracts of stems and roots displayed significant inhibitory effects. The results were compared with standard drug (morphine). In antipyretic activity, the methanol extract of leaves significantly decreased pyrexia level (400 and 200 mg/kg bw) after two, three, and four hours of administration (Dutta et al. 2020).

C. javana DC. is an herbaceous climber with distinct leaves, the upper surface containing white patches, while the lower surface is purple. Leaves are ovate with a heart-shaped base and an apiculate apex. It is distributed in Myanmar and India. In Myanmar, its roots are used in the treatment of ovarian cancer (Myint 2000), whereas in India, the leaves are useful in management of diabetes (Ali et al. 2017). The methanol extract of *C. javana* tuberous roots was evaluated for its glucose uptake stimulatory activity in L6 rat muscle cells and inhibitory effects against α-glucosidase. Methanol extract (100 µg/ml) demonstrated a significant increase in glucose uptake stimulatory activity (70.9%) and potent inhibition of α-glucosidase enzyme activity (100% inhibition). Three compounds (bergenin, stigmast-4-en-3-one and β-sitosterol) were isolated and identified from the methanol extract. Bergenin (100 µg/ml or 0.3046 mM) enhanced the glucose uptake stimulation by L6 myotubes (50.5%) without toxicity. At the same dose, bergenin displayed no inhibition on α-glucosidase enzyme, whereas, stigmast-4-en-3-one and β-sitosterol demonstrated 98.6 and 40.6% inhibition (San et al. 2020).

8.2 PHYTOCHEMISTRY

Forty bioactive compounds {5,9-undecadien-2-one, 6,10-dimethyl-, (Z), phenol, 2,4-bis(1,1-dimethylethyl), benzoic acid, 4-ethoxy-, ethyl ester, dodecanoic acid, ethyl ester, estra-1,3,510-trien-17β-ol, 2,6,10-trimethyl,14-ethylene-14-pentadecne, pentadecanoic acid, 1,2-benzenedicarboxylic acid, bis(2-methylpropyl) ester, 2-hexadecen-1-ol, 3,7,11,15-tetramethyl-, [R-[R*,R*-(E)]]-, pentadecanoic acid, ethyl ester, n-hexadecanoic acid, hexadecanoic acid, ethyl ester, heptadecanoic acid, phytol, 6-octadecenoic acid, (Z), n-propyl 9,12-octadecadienoate, ethyl oleate, octadecanoic acid, ethyl ester, 4,8,12,16-tetramethylheptadecan-4-olide, icosanoic acid, docosanoic acid, ethyl ester, L-(+)-ascorbic acid 2,6-dihexadecanoate, heptadecanoic acid, ethyl ester, vitamin E, ergost-5-en-3-ol, (3β, 24R), stigmasterol, β-sitosterol, cholest-4-en-3-one, lup-20(29)-en-3-one, 4,22-stigmastadiene-3-one, lupeol,

lup-20(29)-en-3-yl acetate, stigmast-4-en-3-one, cholesta-4,6-dien-3-one, friedelin, cyclopropa[5,6]
stigmast-22-en-3-one, 3′,6-dihydro-, (5β,6α,22E)-, 4,4a,6b,8a,11,11,12b,14a-octamethyl-eicosahydro-
picen-3-one, 3α,7β-dihydroxy-5β,6β-epoxycholestane, stigmastane-3,6-dione, (5α), solanesol} were
identified from ethanol extract of *C. quadrangularis*. Osteoblasts treated with different concentrations
of ethanol extract demonstrated biphasic variation in cell proliferation and differentiation. In addition,
lower concentrations (10–50 µg/ml) of extract slightly decreased the intensity of reactive oxygen species, although they increased matrix mineralization, DNA contents in the S phase of the cell cycle, and
Runx2 expression levels. Although, higher concentrations (75–100 µg/ml) considerably stimulated the
reactive oxygen species intensity and nuclear condensation in osteoblasts, while it decreased mineralization levels and proportion of cells in the S phase and Runx2 level of the osteogenic gene. The
ethanol extract showed concentration-dependent biphasic effects, which would contribute notably to
future assessment of preclinical efficacy and safety studies (Siddiqui et al. 2015).

Flavonoids and stilbenes [quercitrin, 3-*O*-α-L-rhamnopyranosyl kaempferol, rhamnitrin, 3-*O*-α-L-
rhamnopyranosylrhamnocitrin, quercitrin-3-*O″*-acetate, 3-(4-hydroxybenzylidene)-2-(2,5-dihydroxy-
phenyl)-1-(4-hydroxyphenyl)indane-4,6-diol), (3-*O*-α-L-rhamnopyranosylkaempferol)] from aqueous
extract of stems and leaves (Sharp et al. 2007), δ-amyrin acetate, aliphatic acid hexadecanoic acid,
trans-resveratrol-3-*O*-glucoside, δ- amyrone, δ-amyrin, β-sitosterol, kaempferol, quercetin, and resveratrol were isolated and characterized from *C. quadrangularis* stem extract (Thakur et al. 2009).
A tetracyclic triterpenoid (7-oxoonocer-8-ene-3β,21α-diol) from *C. quadrangularis* (Gupta and Verma
1990), two unsymmetric tetracyclic triterpenoids (onocer-7-ene-3α,21β-diol and onocer-7-ene-3β,21α-
diol) along with β-sitosterol, δ-amyrin and δ-amyrone from *C. quadrangularis* (Bhutani et al. 1984)
subsequently, another unsymmetric tetracyclic triterpenoid (7-oxoonocer-8-ene-3β,21α-diol) and
3,3′,4,4′-tetrahydroxybiphenyl has been reported from *C. quadrangularis* (Mehta et al. 2001).

Fifty five compounds [β-sitosterol, stigmasterol, β-sitosteryl glucoside, a mixture of 3β-hydroxyl
stigmast-5-en-7-one, 3β-hydroxystigmast-5, 22-dien-7-one, a mixture of β-sitostenone, stigmasta-
4,22-dien-3-one, 6β-hydroxy-β-sitostenone, ergosterol peroxide, 3,5,7,4′-tetramethoxyflavone, 3′,4′,
3,6,7-pentamethoxyflavone, 3′,4′,5,6,7-pentamethoxyflavone, 4′,5,6,7-tetramethoxy-flavone, a mixture of oleanolic acid and ursolic acid, betulinic acid, friedelin, epi-glut-5(6)-en-ol, taraxerol, *epi*-
friedelinol, glutinone, lup-28-al-20(29)-en-3-ol, a mixture of α-amyrin and β-amyrin, bergenin,
p-hydroxybenzaldehyde, vanillin, methyl gallate, gallic acid, 4-methoxybenzoic acid, vanillic acid,
a mixture of 4-hydroxy-*trans*-cinnamic acid methyl ester and 4-hydroxy-*cis*-cinnamic acid methyl
ester, a mixture of octadecyl-*trans*-ferulate and octadecyl-*cis*-ferulate, 1-(4-methoxy-phenyl)undecan-
1-one, 3-hydroxy-4-methoxybenzoic acid, hexadecyl ferulate, 2-hydroxybenzoquinone, 2,6-dimeth-
yoxybenzoquinone, 3,3′,4-tri-*O*-methyl-ellagic acid, 3,3′,4,4′ -tetra-*O*-methylellagic acid, methyl
pheophorbide-a, a mixture of methyl-21-hydroxy-(21*S*)-pheophorbide-a, methyl-21-hydroxy-(21*R*)-
pheophorbide-a, methyl-21-hydroxyl-(21*S*)-pheophorbide-b, pheophytin-a, α-tocopherol, tocopherol
trimer IVa, tocopherol trimer IVb, 1,2-bis-(5-γ-tocopheryl)ethane, α-tocospirol B, 5,6-dimethoxy-
3-methyl-2-cyclohexa-2,5-dien-1,4-dione, 3-methyl-8-hydroxy-3,4-dihydroisocoumarin, and methyl
linoleate] have been separated and identified from fresh stems of *C. assamica*. Betulinic acid demonstrated strong suppression of superoxide anion generation (IC$_{50}$ 0.2 ± 0.1 µM). Similarly, betulinic acid
and pheophytin-a also suppressed elastase formation significantly (IC$_{50}$ 2.7 ± 0.3 and 5.3 ± 1.0 µM).
Moreover, betulinic acid and epi-glut-5(6)-en-ol demonstrated significant cytotoxicity to non-small-
cell lung carcinoma (NCI-H226) and colon cancer (HCT-116) cell lines (IC$_{50}$ 1.6 to 9.1 µM; Chan et al.
2018). One phenolic glycoside (cissusic acid) and two new lignan glycosides (cissuside and cissusol)
have been reported from *C. quadrangularis* stems (Kumar et al. 2019).

The presence of 2-furancarboxaldehyde, 5-(hydroxymethyl)-, 3,7,11,15-tetramethyl-2-hexadecen-
1-ol, E-10- pentadecenol, cyclopentaneundecanoic acid, methyl ester, 9- octadecenoic acid (*Z*)-,
methyl ester, docosanoic acid, ethyl ester, phytol, squalene, urs-12-en-24-oic acid, 3-oxo-, methyl
ester, (+)-,4,8,13-cyclotetradecatriene-1,3-diol, 1,5,9-trimethyl-12-(1- methylethyl)-, 2H-pyran, and
2-(7-heptadecynyloxy) tetra hydro- was determined by GC-MS analysis in the *C. quadrangularis*
(Chenniappan et al. 2020).

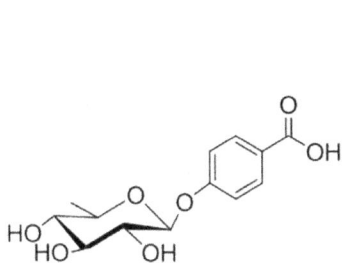

Cissusic acid

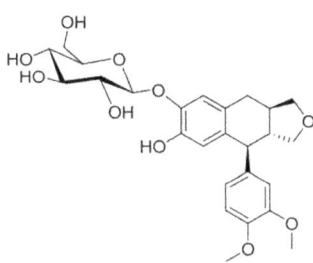

Cissuside

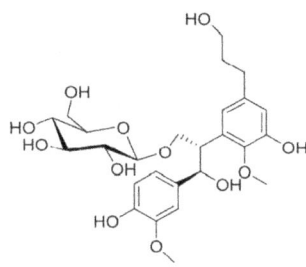

Cissusol

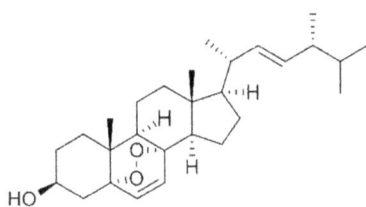

Ergosterol peroxide

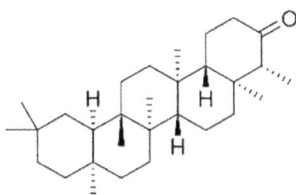

Friedelin

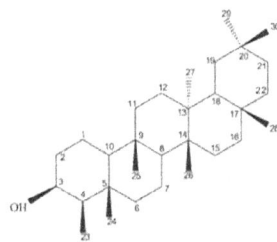

epi-Friedelinol

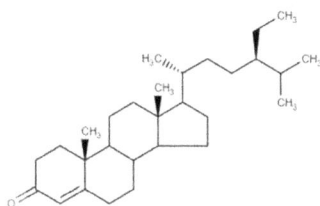

β-Sitostenone

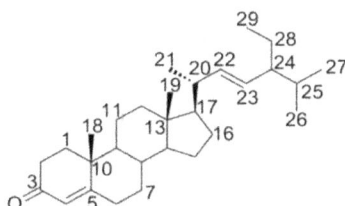

Stigmasta-4,22-dien-3-one

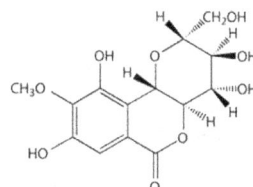

Bergenin

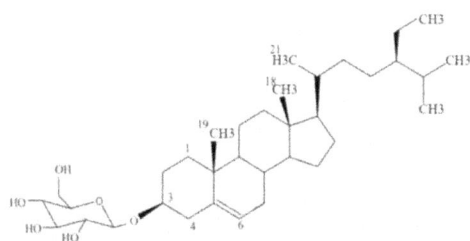

β-sitosteryl glucoside

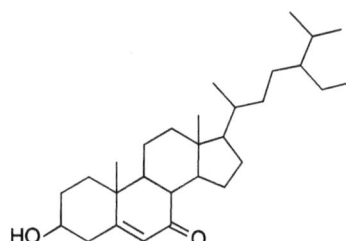

3β-Hydroxyl stigmast-5-en-7-one

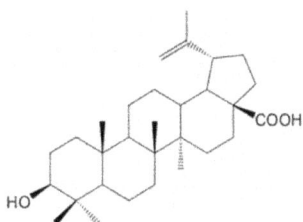

Betulinic acid

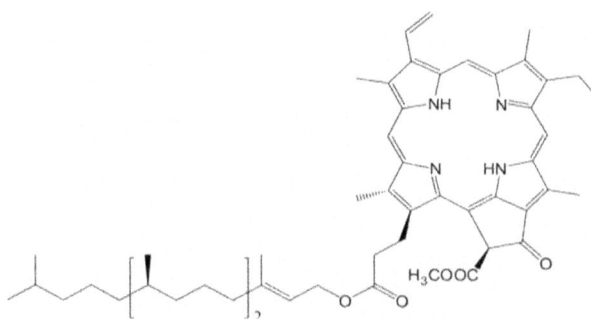

Pheophytin-a

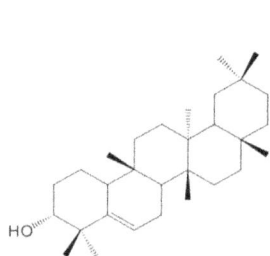

epi-Glut-5(6)-en-ol

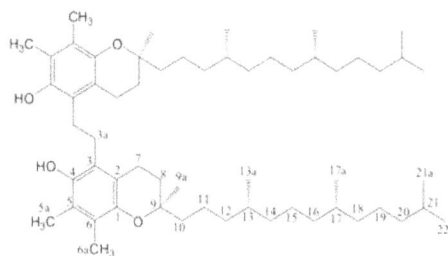

1,2-bis-(5-γ-Tocopheryl)ethane

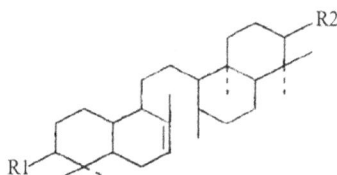

R^1 = α-OH, R^2 = β-OH - Onocer-7-ene-3α,21β-diol; R^1 = β-OH, R^2 = α-OH - Onocer-7-ene-3β,21α-diol

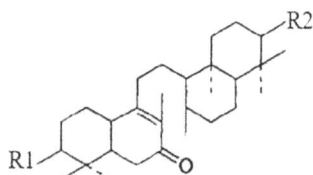

R^1 = β-OH, R^2 = α-OH - 7-Oxoonocer-8-ene-3β,21α-diol

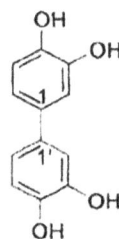

3,3′,4,4′-Tetrahydroxybiphenyl

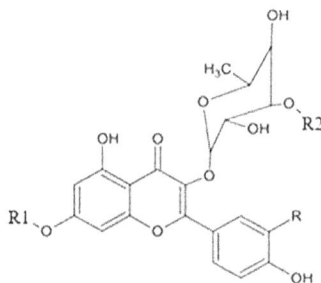

R = OH, R^1 = H, R^2 = H - Quercitrin; R = H, R^1 = H, R^2 = H - 3-*O*-α-L-Rhamnopyranosyl kaempferol; R = OH, R^1 = CH$_3$, R^2 = H - Rhamnitrin; R = H, R^1 = CH$_3$, R^2 = H - 3-*O*-α-L-Rhamnopyranosylrhamnocitrin; R = OH, R^1 = H, R^2 = CH$_3$ - Quercitrin-3-*O*″-acetate

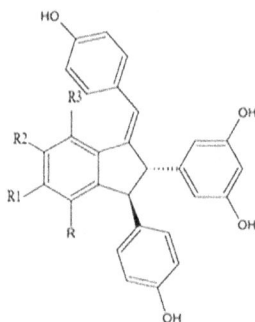

R = OH, R^1 = H, R^2 = OH, R^3 = H - Parthenocissin A; R = H, R^1 = OH, R^2 = H, R^3 = OH - 3-(4-Hydroxybenzylidene)-2-(2,5-dihydroxyphenyl)-1-(4-hydroxyphenyl)indane-4,6-diol

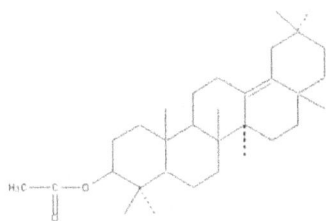
δ-Amyrin acetate

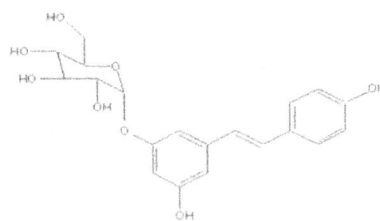
Trans-resveratrol-3-*O*-glucoside

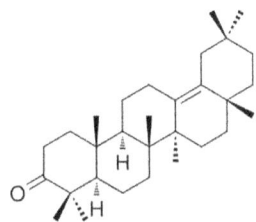
δ- Amyrone

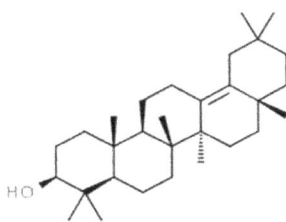
δ- Amyrin

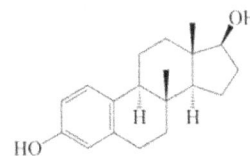
Estradiol

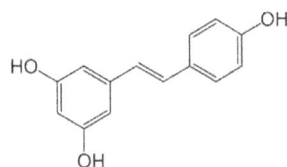
Resveratrol

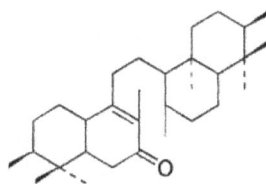
7-Oxoonocer-8-ene-3β,21α-diol

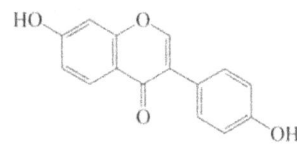
Diadzein

8.3   CULTURE CONDITIONS

Mature nodal segments of *C. quadrangularis* were cultured onto an MS (Murashige and Skoog 1962) medium for callus regeneration. The medium was supplemented with different concentrations of auxins and cytokinins. The accumulation of steroidal substances (β-sitosterol and stigmasterol) was increased by 3% sucrose, 1-naphthaleneacetic acid (2.0 mg/l) and 6-benzylaminopurine (0.5 mg/l) supplementation in MS medium. Increased production of steroidal substances up to ten- to twelvefold was obtained with IAA (5 ppm) and six- to eightfold with 2, 4-D (5 ppm) in callus tissue. Therefore, the callus tissue and increased steroidal contents may be used for scale-up procedure (Sharma et al. 2011). Similarly, the combination of 2, 4-dichlorophenoxyacetic acid (1 mg/L) and 6-benzylaminopurine (6 mg/l) with kinetin (1 mg L-1) also increased the growth of callus (Garg and Malik 2012). The shoot apices of *C. quadrangularis* were transferred on MS medium containing 6-benzylaminopurine (4.4–8.8 µM) and kinetin (4.65 µM) or adenine sulphate (1.35 µM) for obtaining callusing (Anand and Kaur 2012). Similarly, the apical bud and meristem of *C. qudrangularis* were also cultured on an MS medium. The successful callusing was induced by addition of 6-benzylaminopurine (2.5 mg/l) + 1-naphthaleneacetic acid (1.5 mg/l) in MS medium (Teware 2016). Unorganized callus tissue of *C. quadrangularis* was obtained on MS medium containing 6-benzylaminopurine. Total sterol contents (β-sitosterol and stigmasterol) were also found higher in concentration (0.0439 %) in six-week-old tissue of *C. quadrangularis* (Sharma and Patni 2007).

REFERENCES

Ali A, Manikandan S, Lakshmanan GM. 2017. Review on phytochemical and pharmacological activities of the genus *Cissus* Linn. J Pharm Res Int 8, 1–7.

Anand M, Kaur T. 2012. A micropropagation system for the cloning of *Cissus quadrangularis*–a valuable medicinal plant. Med Plant Int J Phytomed Relat Ind 10, 125.

Anonymous. 1949. The wealth of India: A dictionary of Indian raw materials and industrial products. Indian Med Gaz 84, 476–477.

Anonymous. 1992. Indian Medicinal Plants. Orient Longman Ltd., Hyderabad.

Arshad M, Siddiquia S, Alib D. 2016. *In vitro* anti-proliferative and apoptotic effects of ethanolic extract of *Cissus quadrangularis*. Caryol Int J Cytol Cytosystem Cytogenet 69, 128–132.

Arunagiri S, Srinivasan KS. 2021. Premilinary phytochemical study of different extracts of one medicinal plant *Cissus quadrangularis*. J Pharm Res Int 33, 42–48.

Aswar UM, MohanV, Bodhankar SL. 2012. Antiosteoporotic activity of phytoestrogen-rich fraction separated from ethanol extract of aerial parts of *Cissus quadrangularis* in ovariectomized rats. Indian J Pharmacol 44, 345–350.

Ayyanar M, Ignacimuthu S. 2008. Pharmacological actions of *Cassia auriculata* L. and *Cissus quadrangularis* Wall.: A short review. J Pharmacol Toxicol 3, 213–221.

Bafna PS, Patil PH, Maru SK, Mutha RE. 2021. *Cissus quadrangularis* L: A comprehensive multidisciplinary review. J Ethnopharmacol 279, 114355.

Banu J, Varela E, Bahadur AN, Soomro R, Kazi N, Fernandes G. 2012. Inhibition of bone loss by *Cissus quadrangularis* in mice: A preliminary report. J Osteoporos 2012, 101206.

Bhutani KK, Kapoor R, Atal CK. 1984. Two unsymmetric tetracyclic triterpenoids from *Cissus quadrangularis*. Phytochemistry 23, 407–410.

Brahmkshatriya HR, Shah KA, Ananthkumar GB, Brahmkshatriya MH. 2015. Clinical evaluation of *Cissus quadrangularis* as osteogenic agent in maxillofacial fracture: A pilot study. Ayu 36, 169–173.

Chan Y-Y, Wang C-Y, Hwang T-L, Juang S-H, Hung H-Y, Kuo P-C, Chen P-J, Wu T-S. 2018. The constituents of the stems of *Cissus assamica* and their bioactivities. Molecules 23, 2799.

Chatree S, Sitticharoon C, Maikaew P, Pongwattanapakin K, Keadkraichaiwat I, Churintaraphan M, Sripong C, Sririwichitchai R, Tapechum S. 2021. *Cissus quadrangularis* enhances UCP1 mRNA, indicative of white adipocyte browning and decreases central obesity in humans in a randomized trial. Sci Rep 11, 2008.

Chatterjee A, Chandraprakash S. 1997. The Treatise of Indian Medicinal Plants, Vol 3. PID, CSIR, New Delhi.

Chenniappan J, Sankaranarayanan A, Arjunan S. 2020. Evaluation of antimicrobial activity of *Cissus quadrangularis* L. stem extracts against avian pathogens and determination of its bioactive constituents using GC-MS. J Sci Res 64, 90–96.

Chidambara Murthy KN, Vanitha A, Mahadeva Swamy M, Ravishankar GA. 2003. Antioxidant and antimicrobial activity of *Cissus quadrangularis* L. J Med Food 6, 99–105.

Chopra RN, Nayar SL, Chopra IC. 1986. Glossary of Indian Medicinal Plants (Including the Supplement). CSIR, New Delhi.

Deka DK, Lahon LC, Saikia J, Mukit A. 1994. Effect of *Cissus quadrangularis* in accelerating healing process of experimentally fractured RadiusUlna of dog: A preliminary study. Indian J Pharmacol 26, 44–48.

Dhanasekaran S. 2020. Phytochemical characteristics of aerial part of *Cissus quadrangularis* (L) and its *in-vitro* inhibitory activity against leukemic cells and antioxidant properties. Saudi J Biol Sci 27, 1302–1309.

Dutta T, Paul A, Majumder M, Sultan RA, Emran TB. 2020. Pharmacological evidence for the use of *Cissus assamica* as a medicinal plant in the management of pain and pyrexia. Biochem Biophys Rep 21, 100715.

Frank S, Hubner G, Breier G, Longaker MT, Greenhalgh DG, Werner S. 1995. Regulation of vascular endothelial growth factor expression in cultured keratinocytes. J Biol Chem 270, 2607–2613.

Garg P, Malik CP. 2012. Efficient micropropagation protocol for *Cissus quadrangularis*. Med Plants 4, 1–6.

Ghouse MS. 2015. A pharmacognostical review on *Cissus quadrangularis* Linn. Int J Res Pharm Biosci 2, 28–35.

Guerra JM, Hanes MA, Rasa C, Loganathan N, Innis-Whitehouse W, Gutierrez E, Nair S, Banu J. 2019. Modulation of bone turnover by *Cissus quadrangularis* after ovariectomy in rats. J Bone Miner Metab 37, 780–795.

Guhabakshi DN, Pal DC, Sersuma P. 2001. A Lexicon of Medicinal Plants in India, Vol 1. Naya Prokash, New Delhi.

Gupta MM, Verma RK. 1990. Unsymmetric tetracyclic triterpenoid from *Cissus quadrangularis*. Phytochemistry 29, 336–337.

Jadhav AN, Rafiq M, Devanathan R, Azeemuddin M, Anturlikar SD, Ahmed A, Sundaram R, Babu UV, Paramesh R. 2016. Ketosteroid standardized *Cissus quadrangularis* L. extract and its anabolic activity: Time to look beyond ketosteroid? Pharmacogn Mag 12(Suppl 2), S213–S217.

Jainu M, Shyamala Devi CS. 2005a. *In vitro* and *in vivo* evaluation of free-radical scavenging potential of *Cissus quadrangularis*. Pharma Biol 43, 773–779.

Jainu M, Shyamala Devi CS. 2005b. *In vitro* and *in vivo* evaluation of free radical scavenging potential of *Cissus quadrangularis*. Afr J Biomed Res 8, 95–99.

Jainu M, Vijaimohan K, Kannan K. 2010. *Cissus quadrangularis* L. extract attenuates chronic ulcer by possible involvement of polyamines and proliferating cell nuclear antigen. Pharmacogn Mag 6, 225–233.

Khan SS, Singh MP, Chaghtai SA. 1991. Ethnomedicobotany of *Cissus quadrangularis* Linn. Orient J Chem 7, 170–172.

Kothari SC, Shivarudraiah P, Venkataramaiah SB, Koppolu KP, Gavara S, Jairam R, Krishna S, Chandrappa RK, Soni MG. 2011. Safety assessment of *Cissus quadrangularis* extract (CQR-300): Subchronic toxicity and mutagenicity studies. Food Chem Toxicol 49, 3343–3357.

Kritikar KR, Basu BD. 2000. Illustrated Indian Medicinal Plants III, 3rd rev. and Enlarged ed. International Book Distribution, Dehradun.

Kuate D, Nash RJ, Bartholomew B, Penkova Y. 2015. The use of *Cissus quadrangularis* (CQR-300) in the management of components of metabolic syndrome in overweight and obese participants. Nat Prod Commun 10, 1281–1286.

Kumar P, Dev K, Sharma K, Sahai M, Maurya R. 2019. New lignan glycosides from *Cissus quadrangularis* stems. Nat Prod Res 33, 233–238.

Kumar R, Sharma AK, Saraf SA, Gupta R. 2010. CNS activity of aqueous extract of root of *Cissus quadrangularis* Linn. (Vitaceae). J Diet Suppl 7, 1–8.

Kumbhojkar M, Kulkarni D, Upadhye A. 1991. Ethnobotany of *Cissus quadrangularis* {L}. from India. Barkhuis.

Lee HJ, Le B, Lee D-R, Choi B-K, Yang SH. 2018. *Cissus quadrangularis* extract (CQR-300) inhibits lipid accumulation by downregulating adipogenesis and lipogenesis in 3T3-L1 cells. Toxicol Rep 5, 608–614.

Longman O. 1994. Indian Medicinal Plants, Vol 2. South Asia Distributors & Publishers, New Delhi.

Mallika J, Shyamala Devi CS. 2005. *In vitro* and *in vivo* evaluation of free radical scavenging potential of *Cissus quadrangularis*. Afr J Biomed Res 8, 95–99.

Mehta M, Kaur N, Bhutani KK. 2001. Determination of marker constituents from *Cissus quadrangularis* Linn. and their quantitation by HPTLC and HPLC. Phytochem Anal 12, 91–95.

Moto FCO, Arsàa A, Ngoupaye GT, Taiwe GS, Njapdounke JSK, Kandeda AK, Nkantchoua GCN, Omam Omam JP, Pale S, Kouemou NE, Ayissi Mbomo ER, Pahaye DB, Ojong L, Mairara V, Ngo Bum E. 2018. Anxiolytic and antiepileptic properties of the aqueous extract of *Cissus quadrangularis* (Vitaceae) in mice pilocarpine model of epilepsy. Front Pharmacol 9, 751.

Murashige T, Skoog F. 1962. A revised medium for rapid growth and bioassays with tobacco tissue cultures. Physiol Plant 15, 473–497.

Muthusami S, Ramachandran I, Krishnamoorthy S, Govindan R, Narasimhan S. 2011b. *Cissus quadrangularis* augments IGF system components in human osteoblast like SaOS-2 cells. Growth Horm IGF Res 21, 343–348.

Muthusami S, Senthilkumar K, Vignesh C, Ilangovan R, Stanley J, Selvamurugan N, Srinivasan N. 2011a. Effects of *Cissus quadrangularis* on the proliferation, differentiation and matrix mineralization of human osteoblast like SaOS-2 cells. J Cell Biochem 112, 1035–1045.

Myint, KW. 2000. Survey on the Lesser-Known Species of Medicinal Plants of East Yoma and the Socio-Economic Status of the Local Communities. Forest Research Institute, Forest Department, Ministry of Forestry, Myanmar.

Nash R, Azantsa B, Kuate D, Singh H, Julius Oben J. 2019. The use of a stem and leaf aqueous extract of *Cissus quadrangularis* (CQR-300) to reduce body fat and other components of metabolic syndrome in overweight participants. J Altern Complement Med 25, 98–106.

Nayak T, Keerthi R. 2020. An assessment of the osteogenic potential of *Cissus quadrangularis* in mandibular fractures: A pilot study. J Maxillofac Oral Surg 19, 106–112.

Nie TW, Shukkoor MSA, Nair RS, Amiruddin FKB, Ramasamy S. 2015. Involvement of opioidergic and serotonergic systems in the analgesic activity of *Cissus quadrangularis* L. stem extract in mice. J Basic Clin Physiol Pharmacol 26, 35–41.

Oben J, Kuate D, Agbor G, Momo C, Talla X. 2006. The use of a *Cissus quadrangularis* formulation in the management of weight loss and metabolic syndrome. Lipids Health Dis 5, 24.

Oben JE, Enyegue DM, Fomekong GI, Soukontoua YB, Agbor GA. 2007. The effect of *Cissus quadrangularis* (CQR-300) and a *Cissus* formulation (CORE) on obesity and obesity-induced oxidative stress. Lipids Health Dis 6, 4.

Oben JE, Ngondi JL, Momo CN, Agbor1 GA, Sobgui CSM. 2008. The use of a *Cissus quadrangularis/Irvingia gabonensis* combination in the management of weight loss: A double-blind placebo-controlled study. Lipids Health Dis 7, 12.

Packialakshmi B, Makesh S. 2020. *In vitro* antioxidant, anti-arthritic, and anti-cancer activities of *Cissus quadrangularis* stem extract. Indo Glob J Pharm Sci 10, 50–56.

Pandey S, Parmar S, Shukla M, Sharma V, Dwivedi A, Pandey A, Mishra M. 2022. Phytochemical and pharmacological investigation of *Cissus quadrangularis* L. Herb Med J 7, 1–7.

Panpimanmas S, Sithipongsri S, Sukdanon C, Manmee C. 2010. Experimental comparative study of the efficacy and side effects of *Cissus quadrangularis* L. (Vitaceae) to Daflon (Servier) and placebo in the treatment of acute hemorrhoids. J Med Assoc Thai 93, 1360–1367.

Panthong A, Supraditaporn W, Kanjanapothi D, Taesotikul T, Reutrakul V. 2007. Analgesic, anti-inflammatory and venotonic effects of *Cissus quadrangularis* Linn. J Ethnopharmacol 110, 264–270.

Parisuthiman D, Singhatanadgit W, Dechatiwongse T, Koontongkaew S. 2009. *Cissus quadrangularis* extract enhances biomineralization through up-regulation of MAPK-dependent alkaline phosphatase activity in osteoblasts. *In Vitro* Cell Dev Biol Anim 45, 194–200.

Pathomwichaiwat T, Ochareon P, Soonthornchareonnon N, Ali Z, Khan IA, Prathanturarug S. 2015. Alkaline phosphatase activity-guided isolation of active compounds and new dammarane-type triterpenes from *Cissus quadrangularis* hexane extract. J Ethnopharmacol 160, 52–60.

Pathomwichaiwat T, Suvitayavat W, Sailasuta A, Piyachaturawat P, Soonthornchareonnon N, Prathanturarug S. 2012. Antiosteoporotic effect of sequential extracts and freezedried juice of *Cissus quadrangularis* L. in ovariectomized mice. Asian Biomed 6, 377–384.

Phimarn W, Caichompoo W, Sungthong B, Saramunee K. 2014. A systematic review and meta-analysis on effectiveness of *Cissus quadrangularis* (Linn.) in hemorrhoid treatment. Isan J Pharm Sci 10, 403–418.

Potu BK, Bhat KMR, Rao MS, Nampurath GK, Chamallamudi MR, Nayak SR, Muttigi MS. 2009a. Petroleum ether extract of *Cissus quadrangularis* (Linn.) enhances bone marrow mesenchymal stem cell proliferation and facilitates osteoblastogenesis. Clinics (Sao Paulo) 64, 993–998.

Potu BK, Nampurath GK, Rao MS, Bhat KM. 2011. Effect of *Cissus quadrangularis* Linn on the development of osteopenia induced by ovariectomy in rats. Clin Ther 162, 307–312.

Potu BK, Rao MS, Gopalan Kutty Nampurath GK, Chamallamudi MR, Nayak SR, Thomas H. 2010. Antiosteoporotic activity of the petroleum ether extract of *Cissus quadrangularis* Linn. in ovariectomized Wistar rats. Chang Gung Med J 33, 252–257.

Potu BK, Rao MS, Kutty NG, Bhat KMR, Chamallamudi MR, Nayak SR. 2008. Petroleum ether extract of *Cissus quadrangularis* (LINN) stimulates the growth of fetal bone during intra uterine developmental period: A morphometric analysis. Clinics (Sao Paulo) 63, 815–820.

Potu BK, Rao MS, Nampurath GK, Chamallamudi MR, Prasad K, Nayak SR, Dharmavarapu PK, Kedage V, Bhat KMR. 2009b. Evidence-based assessment of antiosteoporotic activity of petroleum-ether extract of *Cissus quadrangularis* Linn. on ovariectomy-induced osteoporosis. Upsala J Med Sci 114, 140–148.

Rajpal V. 2005. Standardization of Botanicals, Vol 1. Eastern Books Publishers, New Delhi.

Ruangsuriya J, Charumanee S, Jiranusornkul S, Sirisa-ard P, Sirithunyalug B, Sirithunyalug J, Pattananandecha T, Saenjum C. 2020. Depletion of β-sitosterol and enrichment of quercetin and rutin in *Cissus quadrangularis* Linn fraction enhanced osteogenic but reduced osteoclastogenic marker expression. BMC Complement Med Ther 20, 105.

San HT, Boonsnongcheep P, Putalun W, Sritularak B, Likhitwitayawuid K. 2020. Bergenin from *Cissus javana* DC. (Vitaceae) root extract enhances glucose uptake by rat L6 myotubes. Tropical J Pharm Res 19, 1081–1086.

Sasi B, Tamizhiniyan P. 2018. Antioxidant properties of various parts *Cissus quadrangularis* L. in different solvents. J Med Bot 2, 27–33.

Sasi Rekha GV, Devika PT. 2019. Antioxidant activities and GC-MS analysis of ethanol extract of creeper stems of *Cissus quadrangularis* L. J Pharmacogn Phytochem 8, 760–765.

Sawangjit R, Puttarak P, Saokaew S, Chaiyakunapruk N. 2017. Efficacy and safety of *Cissus quadrangularis* L. in clinical use: A systematic review and metaanalysis of randomized controlled trials. Phytother Res 31, 555–567.

Shabi Ruskin R, Priya Kumari VM, Gopukumar ST, Praseetha PK. 2014. Evaluation of phytochemical, antibacterial and anti-cancerous activity of *Cissus quadrangularis* from South-Western Ghats regions of India. Int J Pharm Sci Rev Res 28, 12–15.

Shah U. 2011. *Cissus quadrangularis* L.: Phytochemicals, traditional uses and pharmacological activities—a review. Int J Pharm Pharm Sci 3, 41–44.

Sharma N, Nathawat RS, Gour K, Patni V. 2011. Establishment of callus tissue and effect of growth regulators on enhanced sterol production in *Cissus quadrangularis* L. Int J Pharmacol 7, 653–658.

Sharma N, Patni V. 2007. Isolation of phytosterols from static tissue cultures of *Cissus quadrangularis* L. (Vitaceae)-a potent bone healer. Plant Cell Biotechnol Mol Biol 8, 93–96.

Sharp H, Hollinsheada J, Bartholomewa BB, Obenb J, Watsona A, Nash RJ. 2007. Inhibitory effects of *Cissus quadrangularis* L. derived components on lipase, amylase and α-glucosidase activity *in vitro*. Nat Prod Commun 2, 817–822.

Shirwaikar A, Khan S, Malini S. 2003. Antiosteoporotic effect of ethanol extract of *Cissus quadrangularis* Linn. on an ovariectomized rat. J Ethnopharmacol 89, 245–250.

Siddiqui S, Ahmad E, Gupta M, Rawat V, Shivnath N, Banerjee M, Khan MS, Arshad M. 2015. *Cissus quadrangularis* Linn exerts dose-dependent biphasic effects: Osteogenic and anti-proliferative, through modulating ROS, cell cycle and Runx2 gene expression in primary rat osteoblasts. Cell Prolif 48, 443–454.

Singh N, Singh V, Singh RK, Pant AB, Pal US, Malkunje LR, Mehta G. 2013. Osteogenic potential of *Cissus qudrangularis* assessed with osteopontin expression. Nat J Maxillofac Surg 4, 52–56.

Singh V, Singh N, Pal US, Dhasmana S, Mohammad S, Singh N. 2011. Clinical evaluation of *Cissus quadrangularis* and *Moringa oleifera* and osteoseal as osteogenic agents in mandibular fracture. Natl J Maxillofac Surg 2, 132–136.

Sirasanagandla SR, Karkala SRP, Potu BK, Bhat KMR. 2014. Beneficial effect of *Cissus quadrangularis* Linn. on osteopenia associated with streptozotocin-induced type 1 diabetes mellitus in male Wistar rats. Adv Pharmacol Sci 2014, Article ID 483051.

Sivarajan VV, Balachandran I. 1994. Ayurvedic Drugs and Their Plant Sources. Oxford India Book House, New Delhi.

Stohs SJ, Ray SD. 2013. A review and evaluation of the efficacy and safety of *Cissus quadrangularis* extracts. Phytother Res 27, 1107–1114.

Subramani B. 2018. Analysis of potential toxicological, phytochemical and anticancer properties from *Cissus quandrangularis*. Res J Life Sci Bioinform Pharm Chem Sci 4, 94–105.

Sundaran J, Begum R, Vasanthi M, Kamalapathy M, Bupesh G, Sahoo U. 2020. A short review on pharmacological activity of *Cissus quadrangularis*. Bioinformation 16, 579–585.

Suresh P, Xavier AS, Karthik VP, Punnagai K. 2019. Anticancer activity of *Cissus quadrangularis* L. methanolic extract against MG63 human osteosarcoma cells—an *in-vitro* evaluation using cytotoxicity assay. Biomed Pharmacol J 12, 975–980.

Talreja T, Kumar M, Sirohi P, Sharma T. 2017. Preparation and use of *Cissus quadrangularis* and *Achyranthes aspera* formulation in the management of weight loss. Pharm Innov J 6, 143–151.

Tamburaci S, Kimna C, Tihminlioglu F. 2018. Novel phytochemical *Cissus quadrangularis* extract—loaded chitosan/Na-carboxymethyl cellulose—based scaffolds for bone regeneration. J Bioact Compat Polym 33, 088391151879391.

Tasadduq R, Gordon J, Al-Ghanim KA, Lian JB, Wijnen AJV, Stein JL, Stein GS, Shakoori AR. 2017. Ethanol extract of *Cissus quadrangularis* enhances osteoblast differentiation and mineralization of murine preosteoblastic MC3T3-E1 cells. J Cell Physiol 232, 540–547.

Teware T. 2016. *In vitro* shoot induction and multiple shoot regeneration in *Cissus qudrangularis* L. (Hadjod) a medicinally important plant. World J Pharm Pharm Sci 5, 1201–1207.

Thakur A, Jain V, Hingorani L, Laddha KS. 2009. Phytochemical studies on *Cissus quadrangularis* Linn. Pharmacogn Res 1, 213–215.

Tiwari M, Gupta PS, Sharma N. 2021. A preliminary study on *in vitro* antioxidant and *in vivo* anti-inflammatory activity of *Cissus quadrangularis* Linn. Res J Pharm Technol 14, 2619–2624.

Toor RH, Malik S, Qamar H, Batool F, Tariq M, Nasir Z, Tassaduq R, Lian JB, Stein JL, Stein GS, Shakoori AR. 2019a. Osteogenic potential of hexane and dichloromethane fraction of *Cissus quadrangularis* on murine preosteoblast cell line MC3T3-E1 (subclone 4). J Cell Physiol 234, 23082–23096.

Toor RH, Tasadduq R, Adhikari A, Chaudhary MI, Lian JB, Stein JL, Stein GS, Shakoori AR. 2019b. Ethyl acetate and n-butanol fraction of *Cissus quadrangularis* promotes the mineralization potential of murine pre-osteoblast cell line MC3T3-E1 (sub-clone 4). J Cell Physiol 234, 10300–10314.

Udupa KN, Chaturvedi GN, Tripathi SN. 1970. Advances in Research in Indian Medicine, Vol 12. Banaras Hindu University, Varanasi.

Udupa KN, Prasad G. 1964a. Biomechanical and calcium-45 studies on the effect of *Cissus quadrangularis* in fracture repair. Indian J Med Res 52, 480–487.

Udupa KN, Prasad GC. 1964b. Further studies on the effect of *Cissus quadrangularis* in accelerating fracture healing. Indian J Med Res 52, 26–35.

Verma S. 2018. Role of *Cissus quadrangularis* in the treatment of osteoporosis: A review. Acute Med Res 1, 1.

Vijayalakshmi A, Kumar PR, Sakthi Priyadarsini S, Meenaxshi C. 2013. *In vitro* antioxidant and anticancer activity of flavonoid fraction from the aerial parts of *Cissus quadrangularis* Linn. against human breast carcinoma cell lines. J Chem 2013, Article ID 150675.

Viswanatha Swamy AHM, Thippeswamy AHM, Manjula DV, Mahendra Kumar CB. 2006. Some neuropharmacological effects of the methanolic root extract of *Cissus quadrangularis* in mice. Afr J Biomed Res 9, 69–75.

Yang TY. 1998. Flora of Taiwan, 2nd ed. Editorial Committee of the Flora of Taiwan, Taipei.

Yusuf M, Begum J, Hoque MN, Chowdhury JU. 2009. Medicinal Plants of Bangladesh (Revised and Enlarged). Bangladesh CSIR Laboratories, Chittagong.

Zaki S, Malathi R, Latha V, Sibi G. 2020. A review on effi cacy of *Cissus quadrangularis* in pharmacological mechanisms. Int J Clin Microbiol Biochem Technol 3, 49–53.

Zhang YQ, Xie YH, Huang LP. 2006. Studies on the chemical constituents and biological activities from *Cissus* L. Lishizhen Med Materia Med Res 17, 107–114.

9 *Citrus* Species

9.1 MORPHOLOGICAL FEATURES, DISTRIBUTION, ETHNOPHARMACOLOGICAL PROPERTIES, PHYTOCHEMISTRY, AND PHARMACOLOGICAL ACTIVITIES

Citrus limon (L.) Osbeck (syn. *C. hystrix* DC., syn. *C. medica* var. limon L., *C. limonum* Risso, syn. *C. bergamia* Risso & Poit., syn. *C. aurantium* var. bergamia (Risso) Brandis; Fam.—Rutaceae) is a shrub or evergreen tree, spiny branches, and 5 m tall. The leaves are alternate, and 1-foliolate, winged petiole, axillary and solitary flowers. Petals are linear, oblong, thick, imbricate, and 4–8 in number. Fruit is berry, large, oblong, or globose, cells few seeded, and filled with horizontal fusiform cells. Seed is pendulous and membranous, embryo sometimes two or more in one seed (Hooker 1872). It is distributed in India, China, Brazil, the United States, Mexico, and Spain (Lv et al. 2015). Its fruit is used as toothpowder in maintaining the health of teeth (Khalid et al. 2010; Goetz 2014). The fruits and leaves are useful in anorexia, asthma, constipation, diarrhea, dysentery, dysmenorrhea, fever, halitosis, headaches, hemorrhoids, intestinal disorders, jaundice, piles, pulmonary, skin diseases, and vomiting (Jitin 2013). As per Chinese medicine, it is used as a stimulant and appetite suppressant and in the treatment of indigestion, nausea, constipation, and cardiovascular diseases (Suryawanshi 2011). Fruit juice has been used as a remedy for the treatment of scurvy (Mabberley 2004), high blood pressure, the common cold, and irregular menstruation and coughs (Papp et al. 2011; Clement et al. 2015; Bhatia et al. 2015). As per Romanian medicine, essential oils have been used in the treatment of cold and coughs (Papp et al. 2011), while the fruit juice is useful in sore throats, fevers, rheumatism, and chest pain (Balogun and Ashafa 2019). In Trinidadian medicine, the lemon juice and alcohol or coconut oil mixture is used in the treatment of fever, coughs, and high blood pressure. The mixture of juice and olive oil is useful in womb infections and in removal of kidney stones (Clement et al. 2015). In Indian folk medicine, its fruit juice (two teaspoons taken twice a day) induces menstruation (Bhatia et al. 2015). In Italian medicine, the sweetened fruit juice is administered to relieve gingivitis, stomatitis, and tongue inflammations. Lemon juice mixed in hot water possesses laxative property and is used in prevention of cold. Honey and lemon juice, or lemon juice with salt or ginger, is taken as a remedy in the treatment of common cold (Joy et al. 1998). *C. medica* ripens fruits are used in the treatment of sore throat, cough, asthma, hiccough, earache, nausea, vomiting. The aqueous decoction of fruits possesses antiscorbutic, stomachic, and analgesic properties. Seeds are cardiotonic, sedative, and are used in palpitation. In ancient medicine, citron is taken as an antidote in various types of poisons (Khare 2007; Nadkarni 1996; Peter et al. 2008; Singh and Ali 1998). The fruits and leaves of *C. medica* var limetta have been used in common colds, regulation of blood lipids, inflammations, the treatment of fever, and digestion-related complaints (Beatriz and Luis 2005; Clement et al. 2007). In German homeopathic medicine, *C. limon* fresh fruits are used for the treatment of gingival bleeding and debilitating complaints (Klimek-Szczykutowicz et al. 2020; Shakour et al. 2020). In Ayurvedic medicine, *C. medica* is used as appetizer, cardiac stimulant, and antiemetic agent. *C. Jambhiri* is used as an antidiarrheal agent (Nagy 1980; Chaudhari et al. 2016). The citrons possess organoleptic properties (García-Salas et al. 2013). Moreover, *Citrus* fruits are used in the food, beverage, cosmetic, and pharmaceutical industries (He et al. 2011; Kelebek and Selli 2011; Cirmi et al. 2016). *C. hystrix* DC. is useful in the treatment of flu, fever, hypertension, abdominal pains, and diarrhea in infants (Fortin et al. 2002). Its fruits are used as a digestion stimulant, blood purifier, and high blood pressure regulator (Dasuki 2011; Norkaew et al. 2013). Moreover, fruits are useful in the production of shampoo for washing heads (Dassanayake 1985; Koh and Ong 1999). Citrus species possess antioxidant, anticancer, anti-inflammation, anti-aging, and cardiovascular protection activities (Addi et al. 2022; Liu et al. 2022).

Essential oils (limonene and β-pinene) of *C. limon* showed significant antibacterial activity against bacterial species, demonstrating weaker cytotoxic activity in comparison to the positive control in a dose-dependent manner (Nikolić et al. 2017). Ethanol extract and isolated flavonoids (neohesperidin, hesperetin, neoeriocitrin, eriodictyol, naringin, and naringenin) from *C. bergamia* displayed significant antimicrobial effect against Gram-negative (*E. coli, P. putida*, and *S. enterica*), Gram-positive bacteria (*L. innocua, B. subtilis, S. aureus*, and *Lactococcus lactis*), and yeast (*S. cerevisiae*). Similarly, neohesperidin, hesperetin, neoeriocitrin, eriodictyol, naringin, and naringenin showed significant minimum inhibitory concentration (MIC 200–800 µg/ml; Mandalari et al. 2006, 2007). Ethanol extract of *C. limon* normalized the levels of aspartate aminotransferase, alanine aminotransferase, alkaline phosphatase, and total and direct bilirubin, which had changed in carbon tetrachloride–intoxicated rats. In liver tissue, the extract significantly decreased the malondialdehyde levels but increased the profiles of superoxide dismutase and catalase. It increased the decreased levels of glutathione in treated rats. The extract significantly reduced the viability of human liver derived HepG2 cell lines in carbon tetrachloride exposed rats; hence, it may be considered as a hepatoprotective agent (Bhavsar et al. 2007). Ethyl acetate, acetone, methanol, Methanol:water (80:20) extracts of *C. limon* were investigated for their radical scavenging and apoptotic effects in human breast adenocarcinoma (MCF-7) and nonmalignant breast (MCF-12F) cell lines. Maximum radical scavenging effect (62.2% and 91.3%) was displayed by methanol:water (80:20) extract (833 µg/ml) in 1,1-diphenyl-2-picryl hydrazyl and 2,2′-azino-di-(3-ethylbenzothiazoline)-6-sulfonic acid assays. Moreover, the methanol:water (80:20) extract displayed maximum (29.1%, $p < 0.01$) suppression of the growth of MCF-7 cells. Similarly, methanol:water (80:20) extract also stimulated DNA fragmentation and poly(ADP-ribose) polymerase cleavage. Methanol:water (80:20) extract enhanced the levels of Bax and cytosolic cytochrome C but reduced the levels of Bcl2 in treated MCF-7 cells. In summary, methanol:water (80:20) extract possesses significant antioxidant activity and stimulates apoptosis in MCF-7 cells, leading to growth suppression. Therefore, the methanol:water (80:20) extract may serve as a promising chemoprotective agent for the treatment of breast cancer cells (Kim et al. 2012c).

Acetone and ethanol extracts of *C. limon* fruits were screened for their antioxidant and antimicrobial activities against several microbes associated in skin infections. Acetone extract exhibited maximum antibacterial activity against *E. faecalis, S. typhimurium, S. sonnei*, and *B. subtilis*. Nitric oxide and DPPH scavenging effects of both extracts (acetone and ethanol) were lower than vitamin C and rutin (positive control). Both extracts demonstrated significant antibacterial and antioxidant properties comparable to reference drugs and validated the ethnopharmacological claims of this plant species (Otang and Afolayan 2016). Methanol extract of *C. limon* fruit seeds demonstrated antioxidant activity in free radical 1,1-diphenyl2-picrylhydrazyl scavenging and superoxide (O^{2-}) generated by xanthine/xanthine oxidase assays. The results were compared with tert-butyl-4-hydroxytoluene, vitamin C, quercetin, and gallic acid (Orhan et al. 2003). Essential oils of *C. limon* showed significant minimum inhibitory concentration (4.5 mg/ml) against *S. mutans*. Essential oils also decreased Gtf activity and gtfs transcription in a dose-dependent manner ($p < 0.05$; Liu et al. 2013).

A moderate dose of *C. limon* fruit juice (0.2ml/kg, 0.4ml/kg and 0.6ml/kg) displayed significant increase in latency for test animals to enter the black compartment (3 and 24 h) than control, whereas high doses demonstrated significant enhancement in latency only after 3 h. Therefore, fruit juice may be used as a memory booster (Riaz et al. 2014). In an open field assay, a moderate dose (0.2, 0.4, and 0.6 ml/kg; low, moderate, and high doses) of *C. limon* fruits showed a significant increase in distance traveled, number of central entries, and number of rearings in treated animals. The moderate dose showed a significant increase in the elevated plus-maze, number of open arm entries methods. In the forced swimming test, the moderate dose also displayed a decrease in duration of immobility and enhancement in climbing duration. Therefore, study results suggest that *C. limon* fruit juice (at moderate dose) possess promising anxiolytic activity (Khan and Riaz 2015). Different flavones (naringin, neoeriocitrin, and neohesperidin) were identified from *C. limon*. Ethanol extract integrated in phospholipid vesicles (glycerosomes, hyalurosomes, and glycerol-containing hyalurosomes; prepared with extract/phospholipid (ratio 1/3.5, w/w). Hyalurosomes (containing glycerol) were able to block oxidative injury and death of both keratinocytes and fibroblasts (Manconi et al. 2016).

C. *bergamia* displayed significant hypocholesterolemic and antioxidant/radical scavenging activities (Cappello et al. 2016). *C. bergamia* fruit juice (1 ml/day, 30 days) showed significant protection against hypercholesterolemic diet-stimulated renal damage in rats. Fruit juice showed a significant decrease in the levels of plasma cholesterol, triglycerides, and low-density lipoproteins but induced a significant increase in the levels of high-density lipoproteins ($p < 0.05$). Fruit juice significantly reduced the elevated levels of malonaldehyde in hyperlipidemic controls (4.10 ± 0.10 nmol/ mg protein and 4.78 ± 0.15 nmol/mg protein). Fruit juice showed significant antioxidant potential in DPPH scavenging (IC_{50} = 25.01 ± 0.70 ml) and reducing power assays (IC_{50} 1.44 ± 0.01 ascorbic acid equivalents/ml). The results suggest that fruit juice has a protective effect in hypercholesterolemic diet-induced renal injury, which may be associated to its antioxidant activities (Trovato et al. 2010; Nauman and Johnson 2019). Chloroform extract of *C. bergamia* fruits displayed significant inhibitory effect on IL-8. β-Pinene, sabinene, α-pinene, β-myrcene, α-terpinene, *p*-cymene, limonene, *trans-E*-ocimene, γ-terpinene, linalool, α-terpineol, nerol, linalyl acetate, geranial, γ-terpinyl acetate, neryl acetate, geranyl acetate, caryophyllene, trans-α-bergamotene, bergapten, citropten, β-bisbolene, and coumarins + psoralens were reported from the chloroform extract of *C. bergamia* fruit pericarp. The isolated compounds (bergapten and citropten) also displayed significant inhibition of IL-8 expression, and hence may be used in the reduction of lung inflammation in patients of cystic fibrosis (Borgatti et al. 2011). Hypolipidemic and hepatoprotective effects of *C. bergamia* fruit juice were investigated on hyperlipidemic rats. Long-term administration of *C. bergamia* (1 ml/rat/day) produced a significant decrease in the levels of serum cholesterol, triglycerides, and low-density lipoproteins with an enhancement in the levels of high-density lipoproteins. Moreover, excretion of fecal neutral sterols and fecal bile acids were increased after *C. bergamia* treatment in rats. Besides hypolipidemic effects, the fruit juice demonstrated significant radical scavenging activity in DPPH assay. The observed results suggest that the positive intake of *C. bergamia* fruit juice may reduce the risk of some cardiovascular diseases via its radical scavenging and hypocholesterolemic functions (Miceli et al. 2007).

Naringenin, nobiletin, and hesperetin were isolated from *C. aurantium* dried fruits and demonstrated anti-inflammatory activity on lipopolysaccharide-induced RAW cells. Naringenin and nobiletin displayed significant inhibition on isolated jejunum contraction. The naringenin action is slightly associated to synthesis of nitric oxide synthase and cyclooxygenase, inositol triphosphate. *C. aurantium* dried fruits and its flavonoids showed regulatory effects on inflammatory bowel disease via anti-inflammation and suppression of intestine muscle contraction. This action may offer basic guidance on formulating new medicinal agents against inflammatory bowel disease (He et al. 2018; Musumeci et al. 2020). The effects of *C. aurantium* flavonoids (contains naringin, hesperidin, poncirin, isosiennsetin, hexamethoxyflavone, sineesytin, hexamethoxyflavone, tetramrthnl-o-isoscutellaeein, nobiletin, heptamethoxyflavone, 3-hydroxynobiletin, tangeretin, hydroxypentamethoxyflavone, and hexamethoxyflavone) were investigated on the suppression of adipogenesis and adipocyte formation in 3T3-L1 cells. The insulin-stimulated expression of C/ EBPb and PPARg mRNA and proteins were significantly downregulated in a dose-dependent manner by flavonoids-rich extract. The extract reduced the C/EBPa expression, which is required for the acquirement of insulin sensitivity by adipocytes. In addition, aP2 and FAS gene expressions were reduced significantly after extract treatment. The extract also reduced the insulin-induced serine phosphorylation of Akt and GSK3b. Moreover, the extract not only suppressed triglyceride accumulation during adipogenesis but also presented to the lipolysis of adipocytes. The anti-adipogenesis of the extract was mediated by the suppression of Akt activation and GSK3b phosphorylation, which finally suppressed the formation of adipocytes (Kim et al. 2012a). Flavonoids-rich extract of *C. aurantium* fruits reduced the formation of inducible nitric oxide synthase, cyclooxygenase-2, interleukin-6, and tumor necrosis factor-alpha by inhibiting the NF-κB and MAPKs signal pathways in LPS-induced L6 skeletal muscle cells. The results suggest that the flavonoids-rich extract of *C. aurantium* possesses anti-inflammatory potential that regulates the expression of inflammatory mediators in L6 skeletal muscle cells (Kim et al. 2012b). *C. aurantium* essential oils (1.0%; 2.5% and 5.0%, w/w) were tested for their antianxiety activity in elevated plus-maze and open-field methods. Essential oils (2.5% concentration) enhanced the duration of the animals in the open arms of the elevated plus-maze test and the time of active social interaction in the open-field test. The presence of limonene and myrcene was determined in *C. aurantium* fruits, and essential oils possessed antidepressant activity on the central nervous system (Leite et al. 2008).

Protective effects of naringenin, nobiletin, and hesperetin from *C. aurantium* were examined on ulcerative colitis induced by trinitrobenzenesulfonic acid in a mice model. Naringenin, nobiletin, and hesperetin reduced levels of body weight loss and colon shortness, diseased activity index score, and upregulated claudin-2, occluding, and zona occludens-1 expression significantly. Flavonoids also increased trans epithelial electric resistance but reduced the permeability and upregulated the expression of occludin and ZO-1 in LPS-injured epithelial monolayers system. Naringnien, nobiletin, and hesperetin demonstrate regulatory action on ulcerative colitis by preserving colonic mucosa layer integrity (He et al. 2019). Nobiletin significantly ($p < 0.05$) suppressed the release of nitric oxide and expression levels of inducible nitric oxide synthase and cyclooxygenase-2 in a dose-dependent manner. In addition, nobiletin significantly ($p < 0.05$) induced autophagy, up- and down-regulations of LC3II and p62 proteins, and the expression of autophagy-related genes. Nobiletin showed activation of the IL-6/STAT3/FOXO3a signal pathway via the downregulation of IL-6 and STAT3 phosphorylation and the upregulation of FOXO3a phosphorylation in the cell nucleus. Nobiletin inhibits inflammatory responses via increasing the autophagy through inducing the IL-6/STAT3/FOXO3a pathway in macrophages (Rong et al. 2021).

C. aurantifolia (Christm.) is an evergreen shrub or tree with spiny stem and 3–5 m height. The leaves are ovate, long (5–9 cm), and thick (3–5 cm). Flowers are white to yellow white. The fruits are globose to ovoid berry, green, and turn yellow after ripening (Saidan 2013). It is distributed in Spain, the United States, Israel, Morocco, South Africa, Japan, India, Brazil, Nigeria, Turkey, and Cuba (Enejoh et al. 2015). They are used to improve digestion and in the reduction of the levels of sugar, fat, and cholesterol in blood (Samah 2009). Fruit essential oils can be used in the treatment of cold, asthma, arthritis, and bronchitis (Kunow 2003). The fruit juice is useful in prevention of pimples, stamina enhancement, dysfunctional uterine bleeding treatment, and act as an antidote for poison (Aibinu et al. 2007; Khare 2007; Cheong et al. 2011). Fruit juice mixed with honey is considered as a cough reliever, antifever, and meat softener (Effiom et al. 2012; Lyle 2006). *C. aurantifolia* has been used as a laxative, appetizer, stomachic, and anthelmintic agent. It is helpful in rheumatic arthritis, obesity, and in the treatment of herpes, cuts, and insect bites (Scora 1975; Srinivasan et al. 2008).

C. sinensis (L.) Osbeck is an evergreen flowering tree, with large spines and 9–10 m in height. Leaves are alternate, elliptical, oblong to oval, bluntly toothed, with winged petioles (Goldhamer et al. 2012). Flowers are axillary, single, or in whorls of six, with five white petals. The fruits are globose to oval, turn yellow to orange or yellow at ripening (Orwa et al. 2009; Han 2008). The pericarp is largely made of parenchymatous cells and cuticles (Goudeau et al. 2008; Sharon-Asa et al. 2003). Its fruit consists of sweet pulp and numerous seeds (Rao et al. 2011). Perennial fruits are adapted to a variety of climates (Ulloa et al. 2012). It is native to China, Ecuador, and distributed in India and Malaysia. It is used in the treatment of constipation, cramps, colic, diarrhea, bronchitis, tuberculosis, cough, cold, obesity, menstrual disorder, angina, hypertension, anxiety, depression, and stress disorder (Etebu and Nwauzoma 2014). In China, aqueous decoction of husked orange seeds is useful in urinary tract infections while dried leaves are taken as a carminative or emmenagogue or applied on sores and ulcers. Seed extract prescribed as a medicine in malaria treatment (Milind and Dev 2012; González-Mas et al. 2019).

C. grandis (L.) Osbeck (syn. *C. maxima* (Burm.) Merr. is a perennial shrub with a crooked trunk and grows up to the height of 5–15 m. Leaves are large, evergreen, oblong to elliptic, long, and pubescent (Kharjul et al. 2012). It is distributed in India, Myanmar, and China (Kritikar and Basu 2008). Leaves are used in the treatment of epilepsy, chorea, cough, and hemorrhage. Fruits are prescribed as a remedy in the treatment of leprosy, asthma, cough, hiccough, and mental aberration (Vijaylakshmi and Radha 2015). *C. grandis* is useful in curing fever, gout, arthritis, kidney disorders, and ulcers (Orwa et al. 2009). The pulp and peels of fruits are used as an appetizer and stomach tonic (Thavanapong et al. 2010; Md Othman et al. 2016). Ethanol extract of *C. grandis* showed significant scavenging activity in hydroxyl, 1,1-diphenyl-2-picrylhydrazyl and 2,2′-azino-bis(3-ethylbenzthiazoline-6-sulphonic acid radicals assays (He et al. 2012).

C.s reticulata Blanco is a spiny, evergreen, bushy shrub with slender branches, and 2–8 m tall. The leaves are dark green and lance-shaped with a prominent midrib and narrowly winged petioles. Flowers are oval flattened and sweet-fleshed. Berry fruit is 4–8 cm in diameter with easily separable segments (Nishikawa 2013; Putnik et al. 2017; Jhade et al. 2018). It is found in different regions of China and Cochin-China, South China, Vietnam, India, and Japan (Gmitter and Hu 1990; Liu et al. 2012; Mahato et al. 2019). Its fruits are used as a laxative, aphrodisiac, antiemetic, astringent, and tonic agent (Chopra et al. 1986; Anonymous 2000), whereas the fruit peel is useful in the regulation of skin moisture and softening of hard and rough skin (Khan et al. 2010; Musara et al. 2020). In traditional medicine, it is also used as a stomachic and carminative. Both the pericarp and endocarp possess anticholesterolemic, analgesic, antiseptic, and antiasthmatic properties (Sultana et al. 2012; Yabesh et al. 2014; Apraj and Pandita 2016). Poncirin, isolated and characterized from *C. reticulata* fruits, demonstrated significant *in vitro* inhibitory activity on the growth of human gastric cancer cells, SGC-7901 (Zhu et al. 2013). The physiologic effect of naringenin, and hesperetin on mRNA levels of enzymes involved in hepatic fatty acid oxidation and serum and liver lipid levels were evaluated in male ICR mice. Naringenin showed a significant increase (56.1%) in the rate of hepatic cyanide-insensitive palmitoyl-coenzyme A oxidation. Naringenin displayed a significant increase (9.8% to 55.6%) in the activity levels of several enzymes associated with hepatic fatty acid oxidation. Naringenin also significantly enhanced (more than fivefold higher) the levels of mRNA of microsomal cytochrome P-450 IV A1 connected in fatty acids′ omega-oxidation. Naringenin significantly reduced the levels of serum triacylglycerol, cholesterol, phospholipids, and free fatty acids in treated animals (Huong et al. 2006). Antihyperglycemic, antihyperlipidemic, and antioxidant effects of aqueous-ethanol extract, hesperidin, and quercetin (100 mg/kg bw/day for four week) from *C. reticulata* fruit peel were evaluated in nicotinamide/streptozotocin-induced type 2 diabetic rats. Aqueous-ethanol extract, hesperidin, and quercetin significantly improved the impaired oral glucose tolerance and serum fructosamine levels but reduced the levels of serum insulin and C-peptides, the changed homeostatic model assessment-insulin resistance, homeostatic model assessment of insulin-sensitivity, and homeostatic model assessment-β cell function. The extract also reduced the levels of liver glycogen, the elevated liver glucose-6-phosphatase and glycogen phosphorylase activities, the increased serum aspartate transaminase and alanine aminotransferase activities, and the increased levels of serum creatinine and urea in the diabetic rats. The extract and isolated compounds showed significant improvement in the antioxidant defense system as indicated by a reduction in increased liver lipid peroxidation and an elevation in reduced glutathione and glutathione, glutathione-S-transferase, and superoxide dismutase activities. In addition, test samples also increased the mRNA expression of glucose transporter type 4 and the insulin receptor β-subunit in diabetic rats. In summary, the aqueous-ethanol extract, hesperidin, and quercetin of *C. reticulata* fruit peel possesses potent antidiabetic activities that is mediated via their insulinotropic effects and insulin-sensitizing actions. Moreover, the alleviation of the antioxidant defense system by the peel extract, hesperidin, and naringin may have an important role in increasing antidiabetic activities and in increasing liver and kidney functions in nicotinamide/streptozotocin-induced type 2 diabetic rats (Ali et al. 2020).

The presence of essential oils (limonene, β-caryophyllene, β-pinene, geranial edulinine, ribalinine, isoplatydesmine) was determined in *C. macroptera* fruits. Different parts (fruits, leaves, and stems) of *C. macroptera* possess antioxidant, cytotoxic, antimicrobial, thrombolytic, hypogly-cemic, anxiolytic, antidepressant, cardioprotective, and hepatoprotective properties (Tripoli et al. 2007; Aktar and Foyzun 2017). Methanol extract of *C. macroptera* fruits showed antimicrobial activity against *B. sublitis, S. aureus*, and *E. coli* (Uddin et al. 2014). The lupeol and stigmas-terol were isolated and identified from the methanol extract of *C. macroptera* stem bark. The hot methanol extract displayed potent antioxidant effect (IC_{50} 178.96 µg/ml) on DPPH assays while the cold methanol and dichloromethane extracts presented moderate effect (IC_{50} 242.78 µg/ml and 255.78 µg/ml, respectively). The hexane extract showed mild antioxidant activity (IC_{50} 422.94 µg/ml; Chowdhury et al. 2008). The ethanol extract (1000 mg/kg, p.o.) of *C. macroptera* fruits sig-nificantly ameliorated the levels of transaminase, alkaline phosphatase, lactate dehydrogenase, γ-glutamyl transferase activities and total bilirubin, total cholesterol, triglyceride and creatinine, urea, uric acid, sodium, potassium, and chloride ions, and TBARS levels in acetaminophen-induced hepatotoxicity and nephrotoxicity. The results suggested that ethanol extract displays protection against acetaminophen-induced hepatonephrotoxicity, which might be via the suppression of lipid peroxidation (Paul et al. 2016).

9.2 PHYTOCHEMISTRY

C. aurantium flavonoids (naringin, hesperidin, neohesperidin, narirutin, naringenin, poncirin, isosinesetin, hexamethoxyflavone, quercetin, sinesetin, hexamethoxyflavone, kaempferol, tetramethyl-*O*-isoscutellarein, nobiletin, heptamethoxyflavone, 3-hydroxynobiletin, tangeretin, hydroxypentamethoxyflavone, hexamethoxyflavone, didymin) were identified from *C. aurantium* and possess antioxidant, antimicrobial, anti-inflammatory, anticancer, and antidiabetic activities (Nagappan et al. 2014). The levels of flavanones were increased (twofold higher) by fermentation and showed significant antioxidant activity in trolox equivalent antioxidant capacity and ferric reducing antioxidant power assays (Escudero-López et al. 2013).

Essential oils (meranzin hydrate, herniarin, citropten, meranzin, isomeranzin, epoxyaurap-ten, osthol, 5-isopentenyloxy-7-methoxy-coumarin, aurapten, 5-geranyloxy-7methoxy-coumarin, byakangelicin, 8-methoxy-psoralen, psoralen, oxypeucedanin hydrate, isopimpinellin, bergapten, isobergapten, byakangelicol, oxypeucedanin, epoxybergamottin hydrate, isoimperatorin, impera-torin, cnidilin, phellopterin, epoxybergamottin, cnidicin, 8-geranyloxy-psoralen, bergamottin, sinensetin, hexamethoxyflavone, nobiletin, tetra-*O*-methyl-scutellarein, heptamethoxyflavone, tangeretin, 5-demethyl-tangeretin 5-demethyl-nobiletin) from *C. limon* fruits (Russo et al. 2015), monoterpene hydrocarbons (α-pinene, β-pinene, myrcene, δ-limonene, and γ-terpinene), oxygen-ated monoterpenes (α-terpineol, nerol, and geraniol, α-pinene, β-pinene, myrcene, δ-limonene, and γ-terpinene) from *C. limon* (Simeone et al. 2020), α-thujene, α-pinene, camphene, β-thujene, sabi-nene, β-pinene, β-myrcene, octanal, α-phellandrene, δ-3-carene, α-terpinene, p-cimene, limonene, 1,8-cineole, (*Z*)-β-ocimene, (*E*)-β-ocimene, γ-terpinene, *cis*-sabinene hydrate, terpinolene *p*-mentha-2,4(8)-diene, linalool, (3*Z*)-heptyl acetate, dehydro-sabina ketone, allo-ocimene, *cis*-limonene, oxide *trans*-limonene oxide, citronellal, isoborneol, (2*E*)-nonen-1-al, neoiso-isopulegol, terpinen-4-ol, α-terpineol, decanal, n-octyl acetate, citronellol, neral, carvone, linalyl acetate, geranial, peril-laldehyde bornyl acetate, nonyl acetate, citronellyl acetate, neryl acetate, neryl propionate, neryl butyrate, (*Z*)-β-caryophyllene, α-*cis*-bergamotene, (*Z*)-β-farnesene, γ-muurolene, germacrene D, epi-cubebol, 1,2 bicyclogermacrene, β-bisabolene, (*Z*)-α-bisabolene, (*Z*)-nerolidol, and heneicosane have been determined by GC-MS analysis from *C. bergamia, C. myrtifolia* and *C. limon* (Caputo et al. 2020). Eight flavanones (didymin, eriocitrin, hesperidin, naringin, narirutin, neoeriocitrin, neohesperidin, poncirin) have been identified from *C. limon* (Peterson et al. 2006). Flavonoid con-tents (hesperetin, naringenin, eriodictyol, kaempferol, myricetin, quercetin, apigenin, luteolin), amino acids (tryptophan, threonine, isoleucine, leucine, lysine, methionine, cystine, phenylalanine, tyrosine, valine, arginine, histidine, alanine, aspartic acid, glutamic acid, glycine, proline, serine, hydroxyproline) from fruits and juice (Liu et al. 2012), and two compounds (limonin and nomilin) have been separated and identified from *C. limon* fruits (Gualdani et al. 2016). The maximum amounts of these compounds were recorded in young fruits (Huang et al. 2019). Nomilin significantly reduced the numbers of tartrate-resistant acid phosphatase-positive multinucleated cells and demonstrated no cytotoxicity. Nomilin reduced bone resorption activity. Nomilin downregulated the expression of osteoclast-specific genes and levels of NFATc1 and TRAP mRNA. In addition, nomilin inhibited the MAPK signaling pathways. The results reveal that nomilin-containing herbal formulations have the potential ability to prevent bone metabolic diseases (Kimira et al. 2015).

Flavonoids (naringin, hesperidin, poncirin, isosinensetin, hexamethoxyflavone, sinensetin, hexamethoxyflavone, tetramethyl-*O*-isoscutellarein, nobiletin, heptamethoxyflavone, 3-hydroxynobiletin, tangeretin, hydroxypentamethoxyflavone, hexamethoxyflavone) were reported from *C. aurantium* and screened for their anticancer activity. Flavonoids-rich extract showed significant anticancer effect against human gastric adenocarcinoma cell-line (IC_{50} 99 µg/ml) in dose dependent manner. Flavonoids-rich extract suppressed cell cycle progression in the G2/M phase and reduced the levels of cyclin B1, cdc 2, cdc 25c expressions. Extract promoted apoptosis via activating caspase and deactivating the poly(ADP-ribose) polymerase activities. The study suggests that the isolated flavonoids from *C. aurantium* may be useful in the treatment of gastric cancer (Lee et al. 2012). 8-prenylnaringenin, cosmosiin, didymin, diosmin, hesperetin, hesperidin, isosiennsetin, naringenin, naringin, neohesperidin, nobiletin, poncirin, quercetin, rhoifolin, rutin, sineesytin, sudachitin, tangeretin, and xanthohumol have been reported from *Citrus* species (*C. grandis*, and *C. aurantium*). The isolated flavonoids regulate the levels of lipid profiles, renal function, hepatic, and antioxidant enzymes; therefore, these flavonoids may be considered as promising antidiabetic agents for human clinical studies (Gandhi et al. 2020). Neoeriocitrin, naringin, neohesperidin, apigenin 6,8-di-C-glucoside, diometin 6,8-di-C-glucoside, lucenin-2, vicenin-2, stellarin-2, lucenin-2-4'- methyl ether, scoparin, orientin 4'-methyl ether, brutieridin, melitidin, rhoifolin 4'-*O*-glucoside, chrysoeriol 7-*O*-neohesperidoside-4'-*O*- glucoside, diosmin, rhoifolin, chrysoeriol 7-*O*-neohesperidoside, narirutin, and neodiosmin from fruits (Tranchida et al. 2006), brutieridin, and melitidin were isolated and identified from *C. bergamia* fruits (Donna et al. 2009). Meliditin (from *C. grandis*) displayed antitussive activity on cough induced by citric acid in Guinea pig (Zou et al. 2013).

Essential oils [heptanol, tricycline, α-thujene, α- pinene, α-fenchene, camphene, verbenene, β-pinene, sabinene, β-myrcene, *n*-octanal, n-decane, δ-2-carene, iso-sylvestrene, α-terpinene, ρ-cymene, limonene, (*Z*)-β-ocimene, (*E*)-β-ocimene, γ-terpinene, *cis*-sabinene hydrate, ρ-mentha-3,8-diene, terpinolene, ρ- cymenene, linalool, n- nonanal, *cis*-ρ-menth-2-en-1-ol, allo-ocimene, *trans*-ρ-menth-2-en-1-ol, (*E*) tagetone, *trans*-verbenol, citronellal, terpinen-4-ol, ρ-cymen-8-ol, α-terpineol, myrtenol, *cis*-piperitol, *trans*-dihydrocarvone, *trans*-piperitol, *trans*- carveol, citronellol, *cis*-carveol, thymol, methyl ether, carvone, isogeijerene C, geraniol, (2*E*)-decenal, geranial, perilla aldehyde, α-terpinen-7-al, carvacrol, n-tridecane, undecanal, ρ-vinyl- guaiacol, methyl geranate, δ-elemene, α-terpinyl acetate, citronellyl acetate, neryl acetate, carvacrol acetate, geranyl acetate, tetradecene, β-elemene, n-tetradecene, α-funebrene, dodecanal, α-*cis*-bergamotene, γ-elemene, aromadendrene, (*Z*)-β-farnesene, *cis*-prenyllimponene, α-humulene, (*E*)-β-farnesene, α-gurjunene, γ-muurolene, α-amorphene, germacrene D, α-selinene, α-muurolene, γ-patchoulene, β-bisabolene, germacrene A, δ-amorphene, α-alaskene, δ-cadinene, citronellyl butanoate, *cis*-sesquisabinene hydrate, elemol, germacrene B, (*E*)-nerolidol, spathulenol, *trans*-sesquisabinene hydrate, gleenol, globulol, rosifoliol, n-hexadecane, humulene epoxide II, *cis*-solongifolanone, isolonfifolan,α-ol, muurola-4,10(14)dien-1-β-ol, *allo*-aromadendrene epoxide, *epi*-α-muurolol, α-muurolol, β-eudesmol, α-cadinol, (*Z*)-α-santalol, β-sinensal, n-heptadecane, (*E*)-apritone, (2*E*,6*Z*)-farnesal, (2*Z*,6*E*)-farnesal, (2*Z*,6*E*)-farnesol, α-sinensal, (*Z*)-α-santalol acetate, (2*Z*,6*E*)-farnesyl acetate, (2*E*,6*E*)-farnesyl acetate, n-nonadecane, phytol, (6*Z*,10*E*)pseudo phytol, n-heneicosane, methyl octadecenoate, oleic acid, incensole, n-tricosane, n-tetracosane, labd-(13*E*)-8,15-diol, n-pentacosane, hexacosane, heptacosane, octacosane, nonacosane] from leaf and fruit peel (Hamdan et al. 2016), α-thuyene, α-pinene, sabinene, myrcene, α-terpinene, limonene, β -phellandrene, γ-terpinene, *p-* cymene, octanal, decanal, linalool, terpinene-4-ol, l-caryophyllene, α-terpinenol, n-n-butylpyrrole, dimethylanthranilate, germacrene-D, thymol, δ-muurolen, β-cubebene, copaen, 2-isopropyl-5-metylphenol from *C. reticulata* fruit peel (Boughendjioua and Boughendjioua 2017), α-phellandrene, caryophyllene, α-pinene, γ-elemene, β-pinene, (*E*)-geranylacetone, β-myrcene, (6*E*)-7,11-dimethyl-3-methylene-1,6,10-dodecatriene, p-cymene, (-)-α-cubebene, δ-limonene, β-selinene, β-ocimene, δ-selinene, 1-methyl-1,4-cyclohexadiene, α-caryophyllene, terpinolene, α-muurolene, 1,3-cyclohexadiene, 2,6-di-tert-butyl-4-methylphenol, 2-cyclopenten-1-one 0.05 38 cadinene, *p*-mentha-1,3,8-triene, germacrene D, 1-(1,4-dimethyl-3-cyclohexen-1-yl)-ethanone, anti- (+)-nerolidol, terpene ketones, lauric acid, α-terpineol, spathulenol, decaldehyde, δ- cadinol, thymol methyl ether, β-orange aldehyde, terpineol, (*E,E,E*)-2,6,10-trimethyl2,6,9,11-dodecanetetraen1-al, *cis*-1-methyl-4-isopropyl2-cyclohexen-1-ol, myristic acid, perillaldehyde, diisobutyl phthalate, carvacrol, methyl palmitate, undecanal, palmitic acid, citronellone acetate, ethyl palmitate, nerol acetate, methyl linoleate, decanoic acid, trihexadecane, geranyl butyrate, eicosane) have been identified *Citrus reticulata* (Hou et al. 2019). The flavone C-glycosides (vicenin-2, apigenin-8-C-glucoside, diosmetin-6-C-glucoside), flavanone *O*-glycosides (eriocitrin, neoeriocitrin, narirutin, naringin, hesperidin, neohesperidin, didymin, poncirin, melitidin), polymethoxyflavonoids (monohydroxy-trimethoxyflavone, gardenin B, monohydroxytrimethoxyflavone, trihydroxydimethoxyflavone, monohydroxy-pentamethoxyflavone, isosinensetin, monohydroxytetramethoxyflavone, monohy-droxypentamethoxyflavone, hexamethoxyflavone, sinensetin, tetramethyl-*O*-isoscutellarein, dihydroxytrimethoxyflavone, hexa-*O*-methylgossypetin, 5,7,3′,4′,5′-pentamethoxyflavone), and flavone *O*-glycoside (rhoifolin) were reported from *C. reticulata* (Wang et al. 2017).

α-pinene, *p*-cymene, (*E*)-β-ocimene, and sabinene from *C. macroptera* leaves; terpinen-4-ol, α-terpineol, 1,8-cineole, and citronellol have been determined from *C. hystrix* leaves. *C. macroptera* leaf oil demonstrated antifungal activity against *Trichophyton mentagrophytes* var. interdigitale (MIC 12.5 μg/ml; Waikedre et al. 2010). *Citrus* essential oils (α-/β-pinene, sabinene, β-myrcene, δ-limonene, linalool, α-humulene, and α-terpineol), monoterpene aldehyde/alcohol, and sesquiterpenes possess antioxidant, anti-inflammatory, anticancer properties (Bora et al. 2020).

The flavonoids (naringin, hesperidin, hydroxypentamethoxyflavone, hydroxypentamethoxyflavone, sinensetin, pectolinarigenin, dihydroxytetramethoxyflavone, nobiletin, heptamethoxyflavone, tetramethyl-*O*-isoscutellarein, hydroxypentamethoxyflavone, hydroxyhexatamethoxyflavone, hydroxypentamethoxyflavone) from fruits (Lee et al. 2015), neoeriocitrin, naringin, hesperidin, isosinensetin, sinensetin, tetramethyl-*O*-isoscutellarein, nobiletin, tetramethoxyflavone, heptamethoxyflavone, tangeretin and hydroxypentamethoxyflavone have been reported from Korean *C. platymamma* fruits. *C. platymamma* flavonoids stimulated caspase-3 activation and subsequent poly (adenosine diphosphate-ribose) polymerase cleavage and increased the B-cell lymphoma (Bcl)-2-associated X protein/Bcl-extra-large ratio in A549 cells (Nagappan et al. 2016).

Nobiletin reduced the formation of rat epithelial proinflammatory cytokines and mediators. Nobiletin normalized the functions of impaired barriers in colitic rats and Caco-2 monolayer. Nobiletin also minimized the protein expressions of Akt, nuclear factor-κB, and myosin light chain kinase in intestinal epithelial tissues of treated rats. Nobiletin showed anti-inflammatory activity in 2,4,6-trinitrobenzene sulfonic acid-induced colitis via downregulating the expression of inducible nitric oxide synthase and cyclooxygenase 2. Nobiletin re-established the barrier roles, which were disturbed after 2,4,6-trinitrobenzene sulfonic acid administration, via the suppression of the Akt-NF-κB-MLCK pathway (Xiong et al. 2015).

Hedycaryol, β-sesquiphellandrene, and α-eudesmol from *C. macrocarpa*; geranial, phytol, β-caryophyllene, limonene, and neral were determined by GC-MS from *C. aurantifolia* fruit peel (Jantan et al. 1996). Ethanol extract of *C. aurantifolia* lime peels (6 μg/ml) stimulated apoptosis and cell accumulation at the G1 phase, whereas the extract (15 μg/ml) produced apoptosis and cell accumulation at the G2/M phase. The doxorubicin (200 nM) and extract (6 μg/ml) combination enhanced apoptosis induction higher than their single treatment and cell accumulation at the G2/M phase.

The results of apoptosis and protein expression of p53 and Bcl-2 revealed that both single uses and the combination of extract and doxorubicin are able to enhance apoptotic bodies of MCF-7 cells by enhancing the proteins expression; therefore, the extract may be used as a co-chemotherapeutic agent in the treatment of breast cancer cells (Adina et al. 2014).

The profiles of sakuranetin (0.5 to 100 μg/ml) were estimated in *C. sinensis* (Takemoto et al. 2008). The polymethoxylated flavones, flavanone-*O*-trisaccharides, flavanone- and flavone-*O*-disaccharides, and flavone-C-glycosides (sinensetin, quercetagetin hexamethyl ether, nobiletin, tetramethylscutellarein, heptamethoxyflavone, tangeretin, isosakuranetin rutinoside, hesperidin, narirutin, hesperetin trisaccharide, narirutin 4'-glucoside, 6,8-di-C-glucosylapigenin coniferin, limocitrin, limocitrol, and chrysoeriol) have been determined from *C. sinensis* molasses (Manthey 2004). Tetracosane, ethyl pentacosanoate, tetratriacontanoic acid, tangertin, β-sitosteryl-β-D-glucoside, and 3,5,4'-trihydroxy-7,3'-dimethoxy flavanone 3-*O*-β-glucoside have been identified from *C. sinensis* (Rani et al. 2009). Similarly, 5, 8-dihydroxy-6, 7, 4'-trimethoxyflavone has been obtained from ethyl acetate extract of *C. sinensis* roots (Intekhab and Aslam 2009). Isoflavonoids (daidzin, genistin, daidzein, genistein, formononetin, biochanin A, and prunetin) were characterized from *Citrus* species (*C. aurantium* L, *C. grandis* Osbeck, *C. limon* Osbeck., and *C. sinensis* Osbeck; Lapčík et al. 2004). The flavonone glycoside and two steroids [(*E*)-*N*-(1,3,4,5-tetrahydroxyhexadecan-2-yl)dec-4-enamide, atripliside B, β-sitosterol, and β-sitosterol-3-*O*-β-D-glucopyranoside] from leaves (Saleem et al. 2010), narirutin/hesperidin from fruit juice (Leuzzi et al. 2000), cyanidin 3-glucoside and cyanidin 3-(6"-malonylglucoside) along with cyanidin 3,5-diglucoside, delphinidin 3-glucoside, cyanidin 3-sophoroside, delphinidin 3-(6"-malonylglucoside), peonidin 3-(6"-malonylglucoside), 4-vinylphenol, 4-vinylcatechol, 4-vinylguaiacol, 4-vinylsyringol adducts of cyanidin 3-glucoside, hesperidin (4'-methoxy-3',5,7-trihydroxyflavanone-7-rutinoside), C-glycosylflavone vicenin 2 (apigenin, 6,8-di-C-glucoside), and cyanidin 3-(6"-dioxalylglucoside) have been reported from *C. sinensis* fruit juice (Gil-Izquierdo et al. 2001; Hillebrand et al. 2004). Nobiletin, tangeretin, 3,5,6,7,8,3',4'-heptamethoxyflavone, and 5,6,7,4'-tetramethoxyflavone from fruit peel (Li et al. 2007), polymethoxyflavone (naringin, hesperetin, nobiletin, tangeretin, 5-demethyltangeretin, sinensetin, naringenin, hesperidin, 5HTMF, 5HPMF, 5HHMF) have been reported from *C. sinensis* (Wang et al. 2014). The disaccharides [neohesperidose (rhamnosyl-α-1,2-glucose) or rutinose (rhamnosyl-α-1,6-glucose)] and flavonoids (hesperetin, naringenin, diosmetin, quercetin, and kaempferol, quercetin 7-*O*-rhamnoside, quercetin 7-*O*-glucoside, and kaempferol 7-*O*-glucoside) from *C. sinensis* fruits (Liu et al. 2018), nobiletin, sinensetin, and tangeretin from fruit peel (Ortuño et al. 2006), and flavanones (hesperetin, naringenin, eriodictyol, isosakuranetin, and their respective glycosides) were estimated from *C. sinensis* fruit and juices (range ~180 to 740 mg/l; Barreca et al. 2017).

Polymethoxyflavanone, hydroxylated polymethoxyflavanone, and hydroxylated polymethoxychalcones (5,7,8,3',4'-pentamethoxyflavone, 5,7,8,4'-tetramethoxyflavone, 5,6,7,4'-Tetramethoxyflavone, 3,5,6,7,4'-pentamethoxyflavone, 3,5,6,7,8,3'4'-heptamethoxyflavone, 3,5,7,8,3',4'-hexamethoxyflavone, 5,7,3',4'-tetramethoxyflavone, 3,5,6,7,3',4'-hexamethoxyflavone, 5,7,4'-trimethoxyflavone, 5-hydroxy-7, 8,3',4'-tetramethoxyflavone, 5-hydroxy-3,6,7,8,3',4'-hexamethoxyflavone, 5-hydroxy-6,7,8,4'-tetramethoxyflavone, 7-hydroxy-3,5,6,8,3',4'-hexamethoxyflavone, 7-hydroxy-3,5,6,3',4'-pentamethoxyflavone, 5,7-dihydroxy-3,6,8,3',4'-pentamethoxyflavone, 5-hydroxy-6,7,4'-trimethoxyflavone, 3-hydroxy-5,6,7, 4'-tetramethoxyflavone, 3-hydroxy-5,6,7,8,4'-pentamethoxyflavone, 5-hydroxy-3,7,3',4'-tetramethoxyflavone, 5-hydroxy-3,7,8,3',4'-pentamethoxyflavone, 5-hydroxy-6,7,3',4'-tetramethoxyflavone, 5,6,7,4'-tetramethoxyflavanone, 5-hydroxy-6,7,8,3',4'-pentamethoxyflavanone, 2'-hydroxy-3,4,4',5',6'-pentamethoxychalcone, 2'-hydroxy-3,4,3',4',5',6'-hexamethoxychalcone, 5-hydroxy-6,7,8,3',4'-pentamethoxyflavone) were isolated and characterized from *C. sinensis* peel (Li et al. 2006). Six citrus flavonoids (hesperetin, neohesperidin, tangeretin, nobiletin, naringin, and naringenin) were obtained from *C. sinensis*. Naringin and naringenin significantly reduced the tumor number [5.00 (control group), 2.53 (naringin group), and 3.25 (naringenin group)] in a hamster cheek pouch model, so, they may be used as an antitumor agent (Miller et al. 2008). Four cyclic peptides (citrusins I, II, III, and IV) were isolated from *C. unshiu* Marcov., *C. sinensis* Osbeck. and *C. natsudaidai* peelings (Matsubara et al. 1991). β-pinene, octanal, limonene, γ-terpinene, linalool, terpinen-4-ol, α-terpineol, *cis*-dihydrocarvone, decanal, *cis*-carveol, carvone, thymol, valencene, and eudesm-7(11)-en-4-ol, nootkatone were determined by GC-MS from *C. sinensis* fruit peels (Zouaghi et al. 2019).

Nobiletin and 3,5,6,7,8,3′,4′-heptamethoxyflavone (HMF) and two major monodemethylated PMFs [5-hydroxy-3,7,8,3′,4′-pentamethoxyflavone (5HPMF), and 5-hydroxy-3,6,7,8,3′,4′-hexamethoxyflavone (5HHMF) identified from *C. sinensis*] showed significant inhibition on the growth of human lung cancer H1299, H441, and H460 cells. The monodemethylated polymethoxyflavone produced significant enhancement in the sub-G0/G1 phase of H1299 cells (Xiao et al. 2009). A mixture of nonhydroxylated PMFs and hydroxylated PMFs and hydroxylated PMFs-rich fractions of *C. sinensis* fruits induced apoptosis in MCF-7 breast cancer cells. These fractions suppressed the growth and stimulated apoptosis attributed to an increase in the basal level of intracellular Ca$^{(2+)}$ of MCF-7 breast cancer cells. The bioactive PMFs fractions showed proapoptotic activity in human breast cancer cells (Sergeev et al. 2007).

Antimutagenic activity of naringin, hesperidin, nobiletin, and tangeretin were tested against benzo[a]pyrene, 2-aminofluorene, quercetin, and nitroquinoline N-oxide mutagens. Naringin and hesperidin displayed minimum antimutagenic effect against benzo[a]pyrene mutagens. Tangeretin showed antimutagenic activity against all mutagens (benzo[a]pyrene, 2-aminofluorene, quercetin, and nitroquinoline N-oxide). The tangeretin and nobiletin may be used in the chemoprevention of cancer (Calomme et al. 1996; Assini et al. 2013). Hesperidin and naringin significantly enhanced the levels of glucokinase mRNA, whereas naringin also minimized the mRNA expression of phosphoenolpyruvate carboxykinase and glucose-6-phosphatase in the liver. The hesperidin and naringin effectively minimized the levels of plasma free fatty acid and plasma and hepatic triglyceride, and simultaneously decreased the hepatic fatty acid oxidation and carnitine palmitoyl transferase activity. Both compounds also decreased the levels of plasma, and hepatic cholesterol may have been possibly associated with reduced activities of hepatic 3-hydroxy-3-methylglutaryl-coenzyme reductase and cholesterol acyltransferase (Jung et al. 2006).

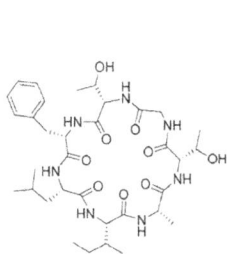 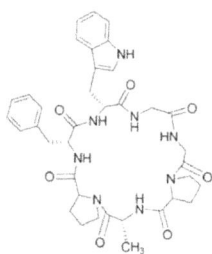 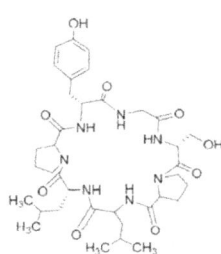 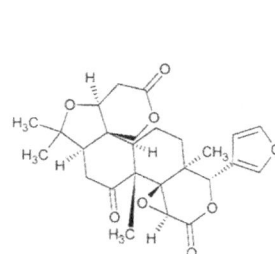

Citrusin I   Citrusin II   Citrusin III   Limonin

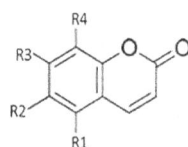

R^1 = H, R^2 = OCH$_3$, R^3 = OCH$_3$, R^4 = H – Scoparone; R^1 = H, R^2 = H, R^3 = OCH$_3$, R^4 = OCH$_3$ – Limettin; R^1 = H, R^2 = H, R^3 = OCH$_3$, R^4 = CH$_2$CHC(CH$_3$)$_2$ – Osthol

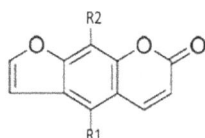

R^1 = H, R^2 = OCH$_3$ – Xanthotoxin; R^1 = OCH$_3$, R^2 = H – Bergapten; R^1 = OCH$_3$, R^2 = OCH$_3$ – Isopimpinellin; R^1 = OH, R^2 = H - Bergaptol

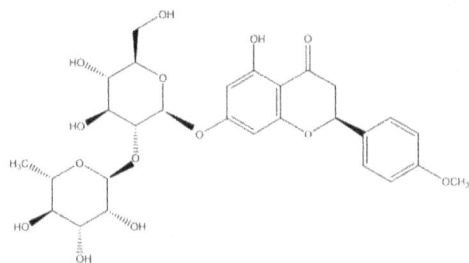

Poncirin

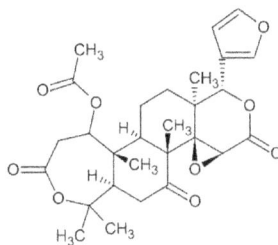

Nomilin

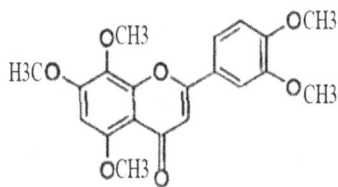

5,7,8,3′,4′-Pentamethoxyflavone

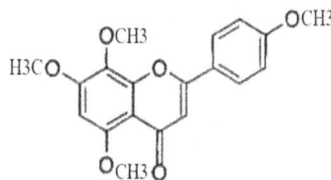

5,7,8,4′-Tetramethoxyflavone

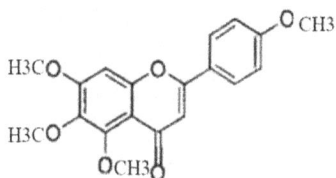

5,6,7,4′-Tetramethoxyflavone

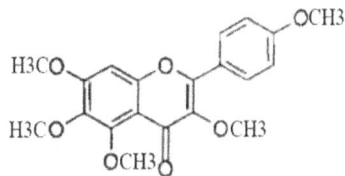

3,5,6,7,4′-Pentamethoxyflavone

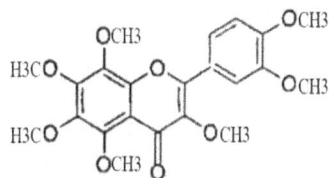

3,5,6,7,8,3′4′-Heptamethoxyflavone

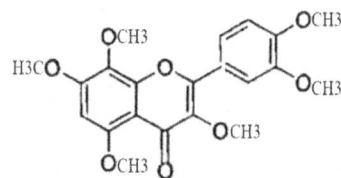

3,5,7,8,3′,4′-Hexamethoxyflavone

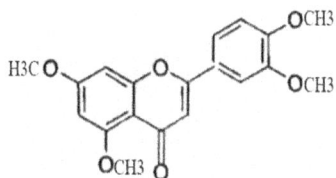

5,7,3′,4′-Tetramethoxyflavone

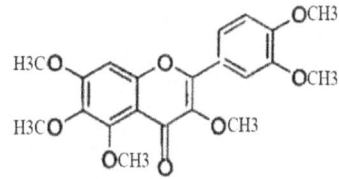

3,5,6,7,3′,4′-Hexamethoxyflavone

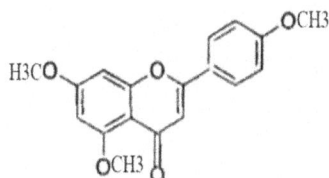

5,7,4′-Trimethoxyflavone

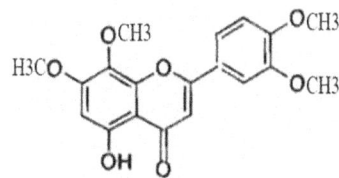

5-Hydroxy-7,8,3′,4′-tetramethoxyflavone

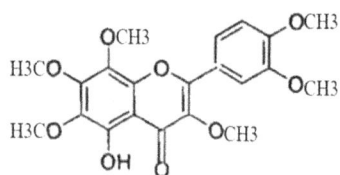

5-Hydroxy-3,6,7,8,3',4'-hexamethoxyflavone

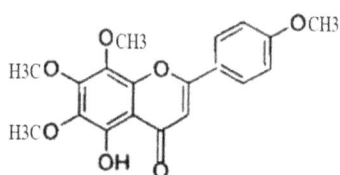

5-Hydroxy-6,7,8,4'-tetramethoxyflavone

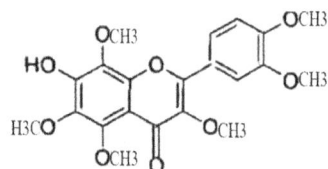

7-Hydroxy-3,5,6,8,3',4'-hexamethoxyflavone

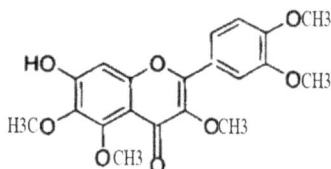

7-Hydroxy-3,5,6,3',4'-pentamethoxyflavone

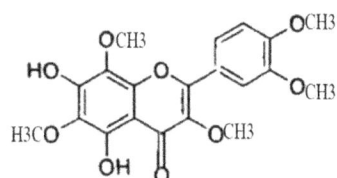

5,7-Dihydroxy-3,6,8,3',4'-pentamethoxyflavone

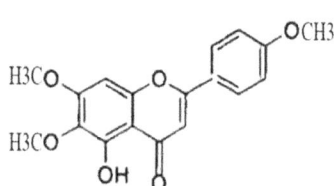

5-Hydroxy-6,7,4'-trimethoxyflavone

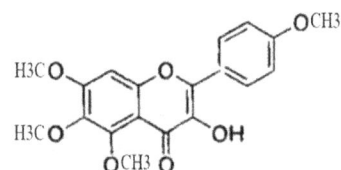

3-Hydroxy-5,6,7,4'-tetramethoxyflavone

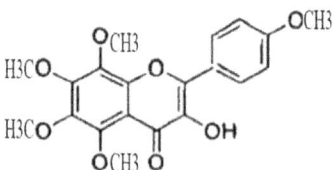

3-Hydroxy-5,6,7,8,4'-pentamethoxyflavone

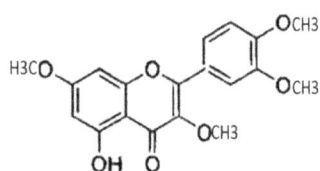

5-Hydroxy-3,7,3',4'-tetramethoxyflavone

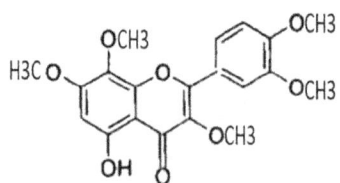

5-Hydroxy-3,7,8,3',4'-pentamethoxyflavone

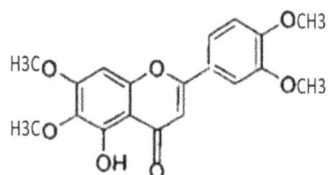

5-Hydroxy-6,7,3',4'-tetramethoxyflavone

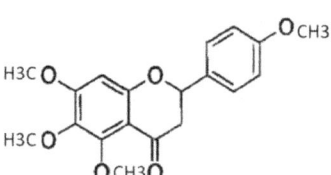

5,6,7,4'-Tetramethoxyflavanone

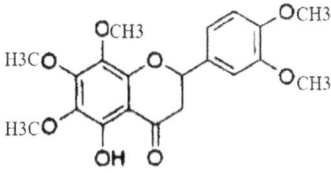

5-Hydroxy-6,7,8,3′,4′-pentamethoxyflavanone

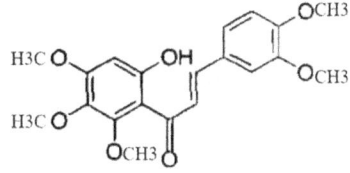

2′-Hydroxy-3,4,4′,5′,6′-pentamethoxychalcone

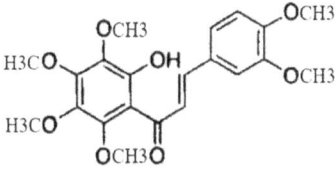

2′-Hydroxy-3,4,3′,4′,5′,6′-hexamethoxychalcone

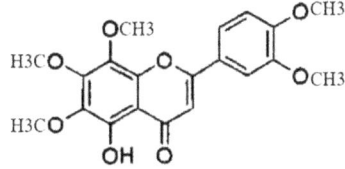

5-Hydroxy-6,7,8,3′,4′-pentamethoxyflavone

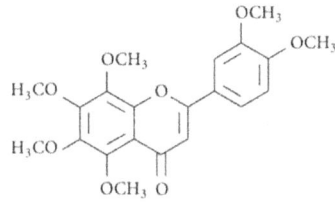

Nobiletin

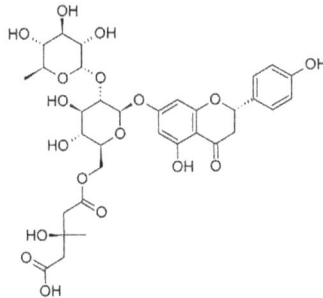

Melitidin

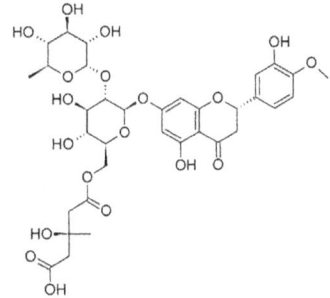

Brutieridin

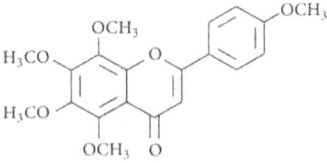

Tangeretin

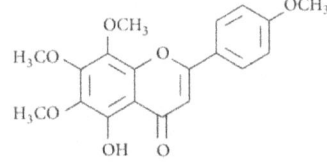

5-Demethyltangeretin

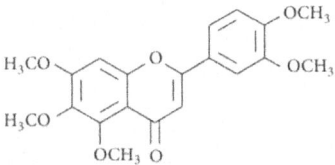

Sinensetin

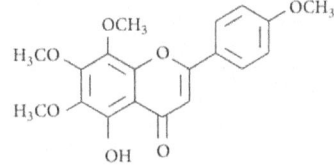

5 HTMF

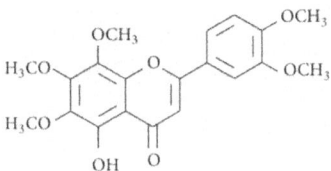

5HPMF

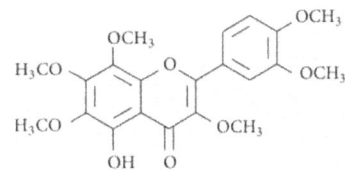

5HHMF

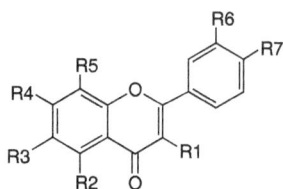

R^1 = H, R^2 = OCH$_3$, R^3 = OCH$_3$, R^4 = OCH$_3$, R^5 = H, R^6 = H, R^7 = OCH$_3$ – Sinsetin; R^1 = H, R^2 = OH, R^3 = H, R^4 = OCH$_3$, R^5 = H, R^6 = H, R^7 = OH – Sakuranetin; R^1 = H, R^2 = OH, R^3 = OCH$_3$, R^4 = OCH$_3$, R^5 = H, R^6 = H, R^7 = OCH$_3$ – Pedunculin

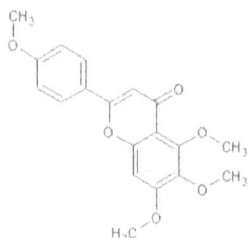

Tetra-*O*-methyl scutellarin

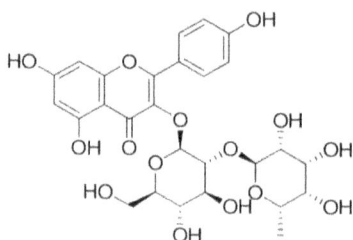

Kaempferol-3-*O*-neoheperidoside

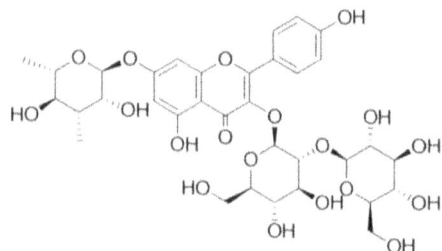

Kaempferol-3-*O*-hexosyl(1→2)hexoside-7-*O*-rhamnoside

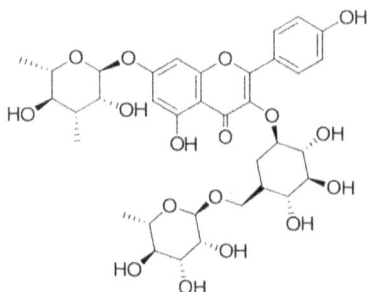

Kaempferol-3-*O*-rutinoside-7-*O*-rhamnoside

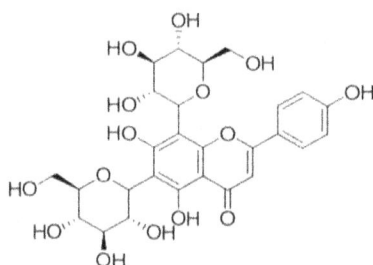

6,8-di-C-Glucosylapigenin

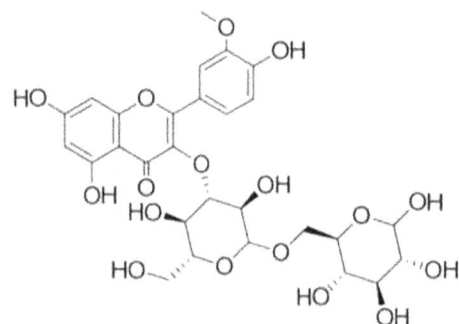

Isorhamnetin-3-*O*-hexosyl(1→6)hexoside

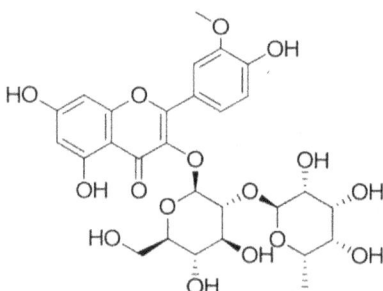

Isorhamnetin-3-*O*-neoheperidoside

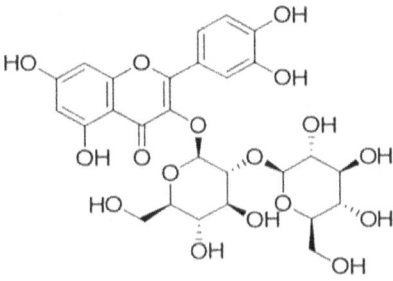
Quercetin-3-*O*-hexosyl(1→2)hexoside

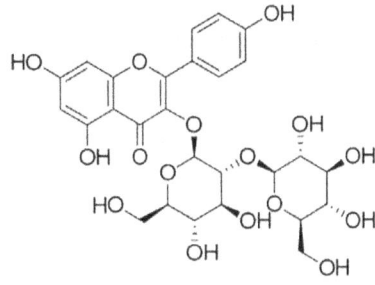
Kaempferol-3-*O*-hexosyl(1→2)hexoside

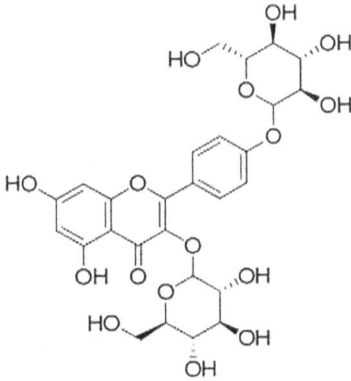
Kaempferol-3,4'-di-*O*-hexoside

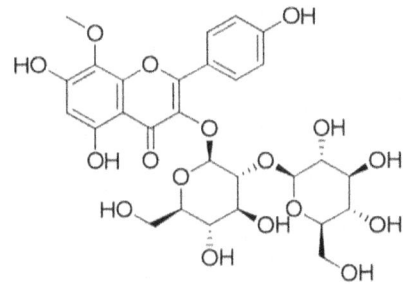
8-Methoxykaempferol-3-*O*-hexosyl(1→2)hexoside

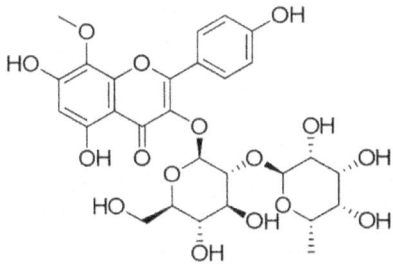
8-Methoxykaempferol-3-*O*-neoheperidoside

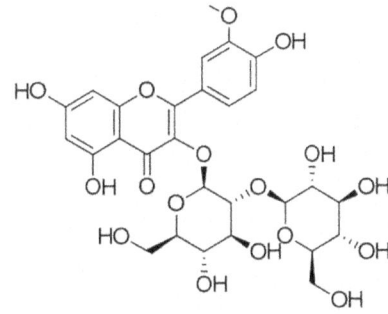
Isorhamnetin-3-*O*-hexosyl(1→2)hexoside

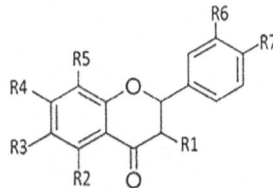

R^1 = H, R^2 = OH, R^3 = H, R^4 = OH, R^5 = H, R^6 = OH, R^7 = OH – Naringin; R^1 = H, R^2 = OCH$_3$, R^3 = OCH$_3$, R^4 = OH, R^5 = OCH$_3$, R^6 = H, R^7 = OCH$_3$ – Tangeretin; R^1 = H, R^2 = OCH$_3$, R^3 = OCH$_3$, R^4 = OH, R^5 = OCH$_3$, R^6 = OCH$_3$, R^7 = OCH$_3$ – Nobiletin; R^1 = OH, R^2 = OH, R^3 = H, R^4 = OH, R^5 = H, R^6 = H, R^7 = OCH$_3$ – Chrysoeriol; R^1 = OH, R^2 = OH, R^3 = H, R^4 = OH, R^5 = OCH$_3$, R^6 = OCH$_3$, R^7 = OH – Limocitrin; R^1 = OH, R^2 = OH, R^3 = OCH$_3$, R^4 = OH, R^5 = OCH$_3$, R^6 = OCH$_3$, R^7 = OH – Limocitrol; R^1 = OH, R^2 = OH, R^3 = OH, R^4 = OH, R^5 = H, R^6 = OH, R^7 = OH – Quercetagetin; R^1 = H, R^2 = H, R^3 = H, R^4 = OH, R^5 = H, R^6 = H, R^7 = OCH$_3$ – Isosakuranetin

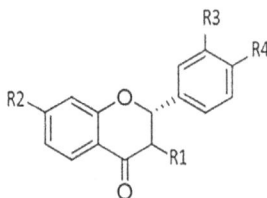

R^1 = H, R^2 = OCH$_3$, R^3 = H, R^4 = OH – Sakuratin; R^1 = O-Glc, R^2 = OCH$_3$, R^3 = OCH$_3$, R^4 = OH - 3,5,4'-Trihydroxy-7, 3'-dimethoxy-flavanone-glc; R^1 = H, R^2 = O-Glc, R^3 = OH, R^4 = OCH$_3$ – Hesperetin; R^1 = H, R^2 = O-Glc-rha, R^3 = OH, R^4 = O-Glc - Narirutin 4'-glucoside; R^1 = H, R^2 = O-Glc-rha, R^3 = OH, R^4 = OCH$_3$ – Hesperidin; R^1 = H, R^2 = O-Glc-rha, R^3 = H, R^4 = OH – Narirutin; R^1 = H, R^2 = O-Glc-rha, R^3 = H, R^4 = OCH$_3$ – Isosakuranetin

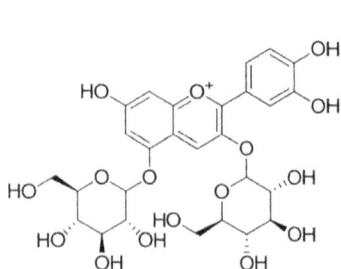

Cyanidin 3,5-diglucoside

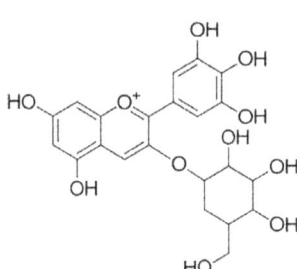

Delphinidin-3-glucoside

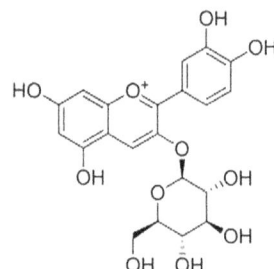

Cyanidin 3-glucoside

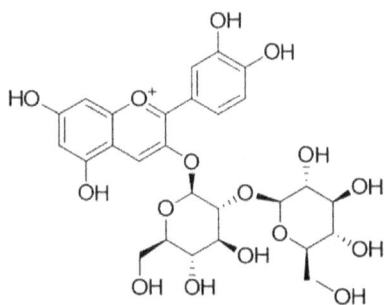

Cyanidin 3-sophoroside

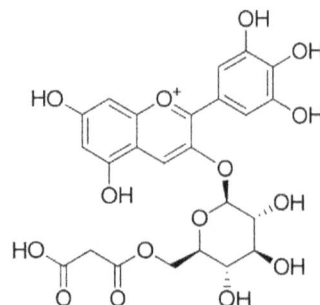

Delphinidin 3-(6″-malonylglucoside)

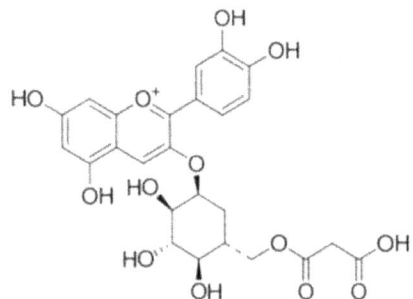

Cyanidin 3-(6″-malonylglucoside)

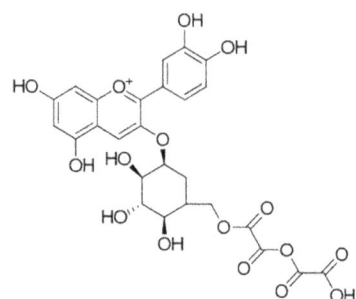

Cyanidin 3-(6″-dioxalylglucoside)

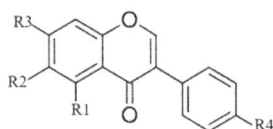

R^1 = H, R^2 = OH, R^3 = OH, R^4 = OH – Daidzein; R^1 = OH, R^2 = H, R^3 = OH, R^4 = OH – Genistein; R^1 = H, R^2 = H, R^3 = OH, R^4 = OCH$_3$ – Formononetin; R^1 = H, R^2 = H, R^3 = OCH$_3$, R^4 = OH

– Isoformononetin; R^1 = OH, R^2 = H, R^3 = OH, R^4 = OCH$_3$ – Biochanin A; R^1 = OH, R^2 = H, R^3 = OCH$_3$, R^4 = OH – Prunetin; R^1 = H, R^2 = OCH$_3$, R^3 = OH, R^4 = OH – Glycitein

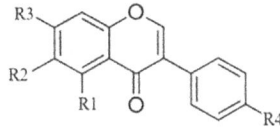

R^1 = H, R^2 = H, R^3 = O-Glu, R^4 = OH – Daidzin; R^1 = OH, R^2 = H, R^3 = O-Glu, R^4 = OCH$_3$ – Sissotrin; R^1 = H, R^2 = OCH$_3$, R^3 = O-Glu, R^4 = OH – Glycitin; R^1 = H, R^2 = H, R^3 = O-Glu, R^4 = OCH$_3$ – Ononin; R^1 = OH, R^2 = H, R^3 = O-Glu, R^4 = OH – Genistin

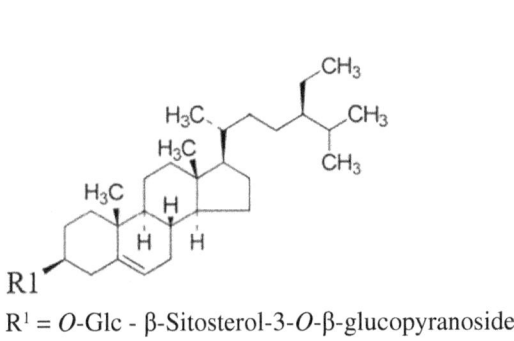

R^1 = *O*-Glc - β-Sitosterol-3-*O*-β-glucopyranoside

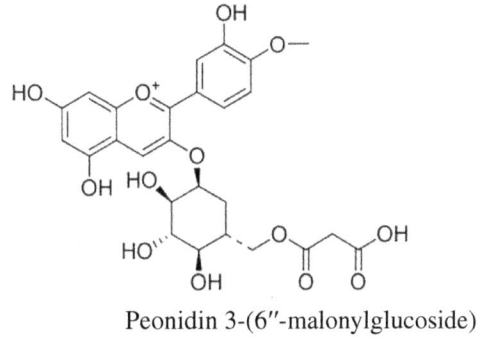

Peonidin 3-(6″-malonylglucoside)

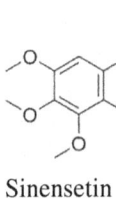

Sinensetin

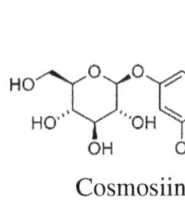

Cosmosiin

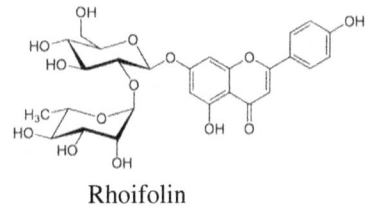

Rhoifolin

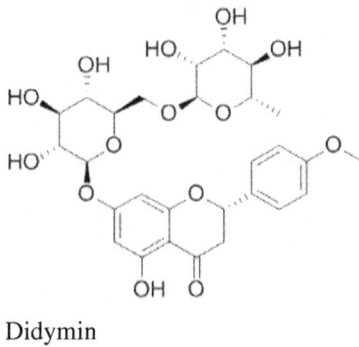

Didymin

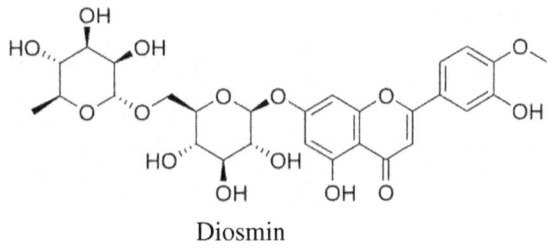

Diosmin

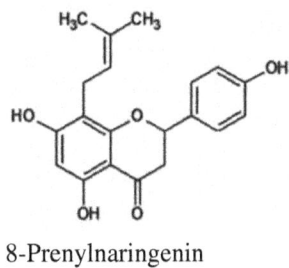

8-Prenylnaringenin

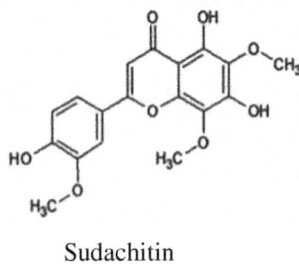

Sudachitin

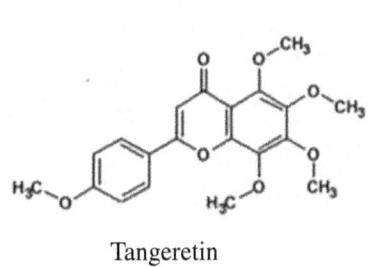

Tangeretin

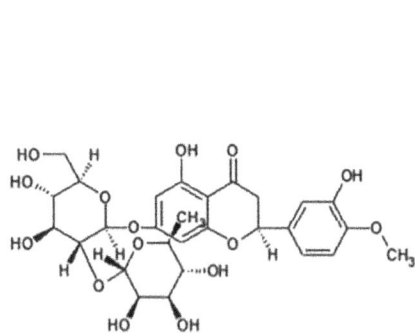

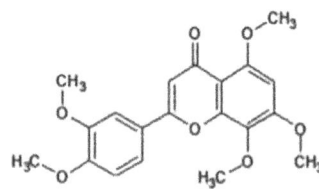

Neohesperidin   Xanthohumol   Isosinensetin

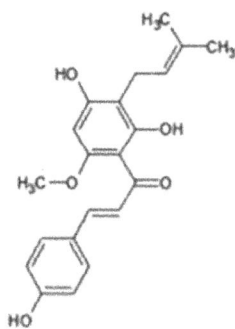

9.3　CULTURE CONDITIONS

For callus development, the *in vitro* regenerated plantlets were transferred to an MS (Murashige and Skoog 1962) medium. Maximum growth of callus was achieved in the MS medium containing 1-naphthaleneacetic acid (10 mg/l) and benzyl adenine (0.25 mg/l). The continuous callusing was maintained by periodic sub-culturing (Duran-Vila et al. 1989). Leaf explants of *C. sinensis* were cultured on MS media containing 2,4-dichlorophenoxyacetic acid alone or in combination with benzyl adenine and α-naphthalene acetic acid. The combination of 2,4-dichlorophenoxyacetic acid (4.53 μM) and 1-naphthaleneacetic acid (5.37 μM) in MS medium–induced callus formation (93.33%) and proliferation (86.67%; Pandey and Tamta 2016). Cotyledon explants of *C. sinensis* were cultured into an MS medium containing 2,4-D (4 mg/l). Non-embryogenic callus was developed in all the transferred explants after 20–24 days of culture (Kiong et al. 2008). Callus induction was maximum when segments of *C. sinensis* shoots were cultured on MS media containing 2, 4-D (2 mg/l) and coconut milk (20%; Mukhtar et al. 2005; Usman et al. 2005). Optimum induction and growth of callus were achieved in MS medium with supplementation of 2,4-D (1.5 mg/l) from all types of explants (stem, leaf and root of *C. jambhiri*). The stem explants showed the maximum response (92%) in the formation of callus (Ali and Mirza 2006). Shoot-tip explants from seedlings of *C. grandis* were cultured in MS medium with various concentrations of 6-benzylaminopurine and thidiazuron (Paudyal and Haq 2000).

REFERENCES

Addi M, Elbouzidi A, Abid M, Tungmunnithum D, Elamrani A, Hano C. 2022. An overview of bioactive flavonoids from *Citrus* fruits. Appl Sci 12, 29.

Adina AB, Goenadi FA, Handoko FF, Nawangsari DA, Hermawan A, Jenie RI. 2014. Combination of ethanolic extract of *Citrus aurantifolia* peels with doxorubicin modulate cell cycle and increase apoptosis induction on MCF-7 cells. Iran J Pharm Res 13, 919–926.

Aibinu I, Adenipekun T, Adelowotan T, Ogunsanya T, Odugbemi T. 2007. Evaluation of the antimicrobial properties of different parts of *Citrus aurantifolia* (lime fruit) as used locally. Afr J Tradit Complement Altern Med 4, 185–190.

Aktar K, Foyzun T. 2017. Phytochemistry and pharmacological studies of *Citrus macroptera*: A medicinal plant review. Evid Based Complement Altern Med 2017, Article ID 9789802.

Ali AM, Gabbar MA, Abdel-Twab SM, Fahmy EM, Ebaid H, Alhazza IM, Ahmed OM. 2020. Antidiabetic potency, antioxidant effects, and mode of actions of *Citrus reticulata* fruit peel hydroethanolic extract, hesperidin, and quercetin in nicotinamide/streptozotocin-induced Wistar diabetic rats. Oxid Med Cell Longev 2020, 1730492.

Ali S, Mirza B. 2006. Micropropagation of rough lemon (*Citrus jambhiri* Lush.): Effect of explant type and hormone concentration. Acta Bot Croat 65, 137–146.

Anonymous. 2000. The Wealth of India: A Dictionary of Indian Raw Materials and Industrial Products. National Institute of Science Communication and Information Resources, New Delhi.

Apraj VD, Pandita NS. 2016. Evaluation of skin anti-aging potential of *Citrus reticulata* Blanco peel. Pharmacogn Res 8, 160–168.

Assini JM, Mulvihill EE, Huff MW. 2013. *Citrus* flavonoids and lipid metabolism. Curr Opin Lipidol 24, 34–40.

Balogun FO, Ashafa AOT. 2019. A review of plants used in south African traditional medicine for the management and treatment of hypertension. Planta Med 85, 312–334.

Barreca D, Gattuso G, Bellocco E, Calderaro A, Trombetta D, Smeriglio A, Laganà G, Daglia M, Meneghini S, Nabavi SM. 2017. Flavanones: *Citrus* phytochemical with health-promoting properties. Biofactors 43, 495–506.

Beatriz AA, Luis R-L. 2005. Pharmacological properties of citrus and their ancient and medieval uses in the Mediterranean region. J Ethnopharmacol 97, 89–95.

Bhatia H, Pal Sharma Y, Manhas RK, Kumar K. 2015. Traditional phytoremedies for the treatment of menstrual disorders in district Udhampur, J&K, India. J Ethnopharmacol 160, 202–210.

Bhavsar SK, Joshi P, Shah MB, Santani DD. 2007. Investigation into hepatoprotective activity of *Citrus limon*. Pharm Biol 45, 303–311.

Bora H, Kamle M, Mahato DK, Tiwari P, Kumar P. 2020. *Citrus* essential oils (CEOs) and their applications in food: An overview. Plants (Basel) 9, 357.

Borgatti M, Mancini I, Bianchi N, Guerrini A, Lampronti I, Rossi D, Sacchetti G, Gambari R. 2011. Bergamot (*Citrus bergamia* Risso) fruit extracts and identified components alter expression of interleukin 8 gene in cystic fibrosis bronchial epithelial cell lines. BMC Biochem 12, 15.

Boughendjioua H, Boughendjioua Z. 2017. Chemical composition and biological activity of essential oil of mandarin (*Citrus reticulata*) cultivated in Algeria. Int J Pharm Sci Rev Res 44, 179–184.

Calomme M, Pieters L, Vlietinck A, Berghe DV. 1996. Inhibition of bacterial mutagenesis by *Citrus* flavonoids. Planta Med 62, 222–226.

Cappello AR, Dolce V, Iacopetta D, Martello M, Fiorillo M, Curcio R, Muto L, Dhanyalayam D. 2016. Bergamot (*Citrus bergamia* Risso) flavonoids and their potential benefits in human hyperlipidemia and atherosclerosis: An overview. Mini Rev Med Chem 16, 619–629.

Caputo L, Cornara L, Bazzicalupo M, Francesco CD, Feo VD, Trombetta D, Smeriglio A. 2020. Chemical composition and biological activities of essential oils from peels of three *Citrus* species. Molecules 25, 1890.

Chaudhari SY, Ruknuddin G, Prajapati P. 2016. Ethnomedicinal values of *Citrus* genus: A review. Med J DY Patil Univ 9, 560–565.

Cheong M-W, Loke X-Q, Liu S-Q, Pramudya K, Curran P, Yu B. 2011. Characterization of volatile compounds and aroma profiles of Malaysian pomelo (*Citrus grandis* (L.) Osbeck) blossom and peel. J Essent Oil Res 23, 34–44.

Chopra RN, Nayar SL, Chopra IC. 1986. Glossary of Indian Medicinal Plants (Including the Supplement). CSIR, New Delhi.

Chowdhury SA, Sohrab MH, Datta BK, Hasan CM. 2008. Chemical and antioxidant studies of *Citrus macroptera*. Bangladesh J Sci Ind Res 43, 449–454.

Cirmi S, Bisignano C, Mandalari G, Navarra M. 2016. Anti-infective potential of *Citrus bergamia* Risso et Poiteau (bergamot) derivatives: A systematic review. Phytother Res 30, 1404–1411.

Clement YN, Baksh-Comeau YS, Seaforth CE. 2015. An ethnobotanical survey of medicinal plants in Trinidad. J Ethnobiol Ethnomed 11, 1–28.

Clement YN, Morton-Gittens J, Basdeo L, Blades A, Francis MJ, Gomes N, Janjua M, Singh A. 2007. Perceived efficacy of herbal remedies by users accessing primary healthcare in Trinidad. BMC Complement Altern Med 7, 4.

Dassanayake MD. 1985. A Revised Handbook to the Flora of Ceylon, Vol V. Amerind Publishing Co Ltd, New Delhi.

Dasuki S. 2011. 202 Khasiat Herba (202 Benefits of Herbs). Grup Buku Karangkraf, Selangor.

Donna LD, Luca GD, Mazzotti F, Napoli A, Salerno R, Taverna D, Sindona G. 2009. Statin-like principles of bergamot fruit (*Citrus bergamia*): Isolation of 3-hydroxymethylglutaryl flavonoid glycosides. J Nat Prod 72, 1352–1354.

Duran-Vila N, Ortega V, Navarro L. 1989. Morphogenesis and tissue cultures of three *Citrus* species. Plant Cell Tissue Organ Cult 16, 123–133.

Effiom OE, Avoaja DA, Ohaeri CC. 2012. Mosquito repellent activity of phytochemical extracts from fruit peels of *Citrus* fruit species. Glob J Sci Front Res 12, 5–8.

Enejoh OS, Ogunyemi IO, Bala MS, Oruene IS, Suleiman MM, Ambali SF. 2015. Ethnomedical importance of *Citrus aurantifolia* (Christm) Swingle. Pharm Innov J 4, 1–6.

Escudero-López B, Cerrillo I, Herrero-Martín G, Hornero-Méndez D, Gil-Izquierdo A, Medina S, Ferreres F, Berná G, Martín F, Fernández-Pachón MS. 2013. Fermented orange juice: Source of higher carotenoid and flavanone contents. J Agric Food Chem 61, 8773–8782.

Etebu E, Nwauzoma AB. 2014. A review on sweet orange (*Citrus sinensis* Osbeck): Health, diseases, and management. Am J Res Commun 2, 33–70.

Fortin H, Vigora C, Lohezic-Le F, Robina V, le Bosse B, Boustiea J, Arnoros M. 2002. *In vitro* antiviral activity of thirty-six plants from La Reunion Island. Fitoterapia 73, 346–350.

Gandhi GR, Vasconcelos ABS, Wu D-T, Li H-B, Antony PJ, Li H, Geng F, Gurgel RQ, Narain N, Gan R-Y. 2020. *Citrus* flavonoids as promising phytochemicals targeting diabetes. and related complications: A systematic review of *in vitro* and *in vivo* studies. Nutrients 12, 2907.

García-Salas P, Gómez-Caravaca AM, Arráez-Román D, Segura-Carretero A, Guerra-Hernández E, García-Villanova B, Fernández-Gutiérrez A. 2013. Influence of technological processes on phenolic compounds, organic acids, furanic derivatives, and antioxidant activity of whole-lemon powder. Food Chem 141, 869–878.

Gil-Izquierdo A, Gil MI, Ferreres F, Tomas-Barberan FA. 2001. *In vitro* availability of flavonoids and other phenolics in orange juice. J Agric Food Chem 49, 1035–1041.

Gmitter FG, Hu X. 1990. The possible role of Yunnan, China, in the origin of contemporary 78 *Citrus* species (Rutaceae). Econ Bot 44, 267–277.

Goetz P. 2014. *Citrus limon* (L.) Burm. f. (Rutacées) Citronnier. Phytotherapie 12, 116–121.

Goldhamer DA, Intrigliolo DS, Castel JR, Fereres E. 2012. *Citrus*. In: Crop Yield Response to Water: FAO Irrigation and Drainage Paper 66, Vol 49, Pasquale Steduto P, Theodore C, Hsiao TC, Elias Fereres E, Dirk Raes D (eds). Food & Agriculture Organization, United Nations, Rome, pp 300–315.

González-Mas MC, Rambla JL, López-Gresa MP, Blázquez MA, Granell A. 2019. Volatile compounds in *Citrus* essential oils: A comprehensive review. Front Plant Sci 10, 12.

Goudeau D, Uratsu, SL, Inoue K, daSilva FG, Leslie A, Cook D, Reagan L, Dandekar AM. 2008. Tuning the orchestra: Selective gene regulation and orange fruit quality. Plant Sci 174, 310–320.

Gualdani R, Cavalluzzi MM, Lentini G, Habtemariam S. 2016. The chemistry and pharmacology of citrus limonoids. Molecules 21, 1530.

Hamdan DI, Mohamed ME, El-Shazly AM. 2016. *Citrus reticulata* Blanco cv. Santra leaf and fruit peel: A common waste products, volatile oils composition and biological activities. J Med Plant Res 10, 457–467.

Han ST. 2008. Medicinal Plants in the South Pacific. World Health Organization, Regional Publication, Western Pacific Series, Manila.

He D, Shan Y, Wu Y, Liu G, Chen B, Yao S. 2011. Simultaneous determination of flavanones, hydroxycinnamic acids and alkaloids in citrus fruits by HPLC-DAD-ESI/MS. Food Chem 127, 880–885.

He J-Z, Shao P, Liu J-H, Ru Q-M. 2012. Supercritical carbon dioxide extraction of flavonoids from pomelo (*Citrus grandis* (L.) Osbeck) peel and their antioxidant activity. Int J Mol Sci 13, 13065–13078.

He W, Li Y, Liu M, Yu H, Chen Q, Chen Y, Ruan J, Ding Z, Zhang Y, Wang T. 2018. *Citrus aurantium* L. and its flavonoids regulate TNBS-induced inflammatory bowel disease through anti-inflammation and suppressing isolated jejunum contraction. Int J Mol Sci 19, 3057.

He W, Liu M, Li Y, Yu H, Wang D, Chen Q, Chen Y, Zhang Y, Wang T. 2019. Flavonoids from *Citrus aurantium* ameliorate TNBS-induced ulcerative colitis through protecting colonic mucus layer integrity. Eur J Pharmacol 857, 172456.

Hillebrand S, Schwarz M, Winterhalter P. 2004. Characterization of anthocyanins and pyranoanthocyanins from blood orange (*Citrus sinensis* (L.) Osbeck) juice. J Agric Food Chem 52, 7331–7338.

Hooker J. 1872. The Flora of British India, Part I. L Reeve, London.

Hou H-S, Bonku EM, Zhai R, Zeng R, Hou Y-L, Yang Z-H, Quan C. 2019. Extraction of essential oil from *Citrus reticulata* Blanco peel and its antibacterial activity against *Cutibacterium acnes* (formerly *Propionibacterium acnes*). Heliyon 5, e02947.

Huang S, Liu X, Xiong B, Qiu X, Sun G, Wang X, Zhang X, Dong Z, Wang Z. 2019. Variation in limonin and nomilin content in citrus fruits of eight varieties determined by modified HPLC. Food Sci Biotechnol 28, 641–647.

Huong DTT, Takahashi Y, Ide T. 2006. Activity and mRNA levels of enzymes involved in hepatic fatty acid oxidation in mice fed citrus flavonoids. Nutrition 22, 546–552.

Intekhab J, Aslam M. 2009. Isolation of a flavonoid from the roots of *Citrus sinensis*. Malays J Pharm Sci 7, 1–8.

Jantan I, Ahmad AS, Ahmad AR, Ali NAM, Ayop N. 1996. Chemical composition of some *Citrus* oils from Malaysia. J Essent Oil Res 8, 627–632.

Jhade RK, Huchche AD, Dwivedi SK. 2018. Phenology of flowering in *Citrus*: Nagpur mandarin (*Citrus reticulata* Blanco) perspective. Int J Chem Stud 6, 1511–1517.

Jitin R. 2013. An ethnobotanical study of medicinal plants in Taindol village, district Jhansi, Region of Bundelkhand, Uttar Pradesh, India. J Med Plants Stud 1, 59–71.

Joy PP, Thomas J, Mathew S, Skaria BP. 1998. Medicinal Plants. Aromatic and Medicinal Plants Research Station, Kerala Agricultural University, Kerala.

Jung UJ, Lee M-K, Park YB, Kang MA, Choi M-S. 2006. Effect of citrus flavonoids on lipid metabolism and glucose-regulating enzyme mRNA levels in type-2 diabetic mice. Int J Biochem Cell Biol 38, 1134–1145.

Kelebek H, Selli S. 2011. Determination of volatile, phenolic, organic acid and sugar components in a Turkish cv. Dortyol (*Citrus sinensis* L. Osbeck) orange juice. J Sci Food Agric 91, 1855–1862.

Khalid H, Nisar M, Majeed A, Nawaz K, Bhatti K. 2010. Ethnomedicinal survey for important plants of Jalalpur Jattan, District Gujrat, Punjab, Pakistan. Ethnobot Leafl 14, 807–825.

Khan MA, Ali M, Alam P. 2010. Phytochemical investigation of the fruit peels of *Citrus reticulata* Blanco. Nat Prod Res 24, 610–620.

Khan RA, Riaz A. 2015. Behavioral effects of *Citrus limon* in rats. Metab Brain Dis 30, 589–596.

Khare CP. 2007. Indian Medicinal Plants: An Illustrated Dictionary. Springer, Berlin.

Kharjul A, Kharjul M, Vilegave K, Chandankar P, Gadiya M. 2012. Pharmacognostic investigation on leaves of *Citrus maxima* (Burm.) Merr. (Rutaceae). Int J Pharm Sci Res 3, 4913–4918.

Kim G-S, Park HJ, Woo J-H, Kim M-K, Koh P-O, Min W, Ko Y-G, Kim C-H, Won C-K, Cho J-H. 2012a. *Citrus aurantium* flavonoids inhibit adipogenesis through the Akt signaling pathway in 3T3-L1 cells. BMC Complement Altern Med 12, 31.

Kim J, Jayaprakasha GK, Uckoo RM, Patil BS. 2012c. Evaluation of chemopreventive and cytotoxic effect of lemon seed extracts on human breast cancer (MCF-7) cells. Food Chem Toxicol 50, 423–430.

Kim J-A, Park H-S, Kang S-R, Park K-I, Lee D-H, Nagappan A, Shin S-C, Lee W-S, Kim E-H, Kim G-S. 2012b. Suppressive effect of flavonoids from Korean *Citrus aurantium* L. on the expression of inflammatory mediators in L6 skeletal muscle cells. Phytother Res 26, 1904–1912.

Kimira Y, Taniuchi Y, Nakatani S, Sekiguchi Y, Kim HJ, Shimizu J, Ebata M, Wada M, MAtsumoto A, Mano H. 2015. *Citrus* limonoid nomilin inhibits osteoclastogenesis *in vitro* by suppression of NFATc1 and MAPK signaling pathways. Phytomedicine 22, 1120–1124.

Kiong ALP, Wan LS, Hussein S, Ibrahim R. 2008. Induction of somatic embryos from explants different of *Citrus sinensis*. J Plant Sci 3, 18–32.

Klimek-Szczykutowicz M, Szopa A, Ekiert H. 2020. *Citrus limon* (Lemon) phenomenon—a review of the chemistry, pharmacological properties, applications in the modern pharmaceutical, food, and cosmetics industries, and biotechnological studies. Plants 9, 119.

Koh D, Ong CN.1999. Phytophotodermatitis due to the application of *Citrus hystrix* as a folk remedy. Br J Dermatol 140, 737–738.

Kritikar KR, Basu BD, 2008. Indian Medicinal Plants, Vol I. Int Book Distributors, Dehradun.

Kunow MA. 2003. Maya Medicine: Traditional Healing in Yucatan. University of New Mexico Press, Albuquerque.

Lapcík O, Klejdus B, Davidová M, Kokoska L, Kubán V, Moravcová J. 2004. Isoflavonoids in the Rutaceae family: 1. *Fortunella obovata*, *Murraya paniculata* and four *Citrus* species. Phytochem Anal 15, 293–299.

Lee D-H, Park K-I, Park H-S, Kang S-R, Nagappan A, Kim J-A, Kim E-H, Lee W-S, Hah Y-S, Chung H-J, An S-J, Kim G-S. 2012. Flavonoids isolated from Korea *Citrus aurantium* L. induce G2/M phase arrest and apoptosis in human gastric cancer AGS cells. Evid Based Complement Altern Med 2012, Article ID 515901.

Lee HJ, Nagappan A, Park HS, Hong GE, Yumnam S, Raha S, Saralamma VVG, Lee WS, Kim EH, Kim GS. 2015. Flavonoids isolated from *Citrus platymamma* induce mitochondrial-dependent apoptosis in AGS cells by modulation of the PI3K/AKT and MAPK pathways. Oncol Rep 34, 1517–1525.

Leite MP, Fassin Jr. J, Baziloni EMF, Almeida RN, Mattei R, Leite JR. 2008. Behavioral effects of essential oil of *Citrus aurantium* L. inhalation in rats. Braz J Pharmacogn 18, 661–666.

Leuzzi U, Caristi C, Panzera V, Licandro G. 2000. Flavonoids in pigmented orange juice and second-pressure extracts. J Agric Food Chem 48, 5501–5506.

Li S, Lambros T, Wang Z, Goodnow R, Ho CT. 2007. Efficient and scalable method in isolation of polymethoxyflavones from orange peel extract by supercritical fluid chromatography. J Chromatogr B Anal Technol Biomed Life Sci 846, 291–297.

Li S, Lo C-Y, Ho C-T. 2006. Hydroxylated polymethoxyflavones and methylated flavonoids in sweet orange (*Citrus sinensis*) peel. J Agric Food Chem 54, 4176–4185.

Liu W, Zheng W, Cheng L, Li M, Huang J, Bao S, Xu Q, Ma Z. 2022. Citrus fruits are rich in flavonoids for immunoregulation and potential targeting ACE2. Nat Prod Bioprospect 12, 4.

Liu X, Lin C, Ma X, Tan Y, Wang J, Zeng M. 2018. Functional characterization of a flavonoid glycosyltransferase in sweet orange (*Citrus sinensis*). Front Plant Sci 9, 166.

Liu Y, Zhang X, Wang Y, Chen F, Yu Z, Wang L, Chen S, Guo M. 2013. Effect of *Citrus lemon* oil on growth and adherence of *Streptococcus mutans*. World J Microbiol Biotechnol 29, 1161–1167.

Liu YQ, Heying E, Tanumihardjo SA. 2012. History, global distribution, and nutritional importance of *Citrus* fruits. Compr Rev Food Sci Food Saf 11, 530–542.

Lv X, Zhao S, Ning Z, Zeng H, Shu Y, Tao O, Xiao C, Lu C, Liu Y. 2015. *Citrus* fruits as a treasure trove of active natural metabolites that potentially provide benefits for human health. Chem Cent J 9, 68.

Lyle S. 2006. How to use citrus fruit peels in the home and garden. In: Discovering Fruit and Nuts: A Comprehensive Guide to the Cultivation, Uses and Health Benefits of Over 300 Food-Producing Plants, Lyle S (ed.). Landlinks Press, Collingwood.

Mabberley DJ. 2004. *Citrus* (Rutaceae): A review of recent advances in etymology, systematics and medical applications. Blumea J Plant Taxon Plant Geogr 49, 481–498.

Mahato N, Sinha M, Sharma K. 2019. Modern extraction and purification techniques for obtaining high purity food-grade bioactive compounds and value-added co-products from *Citrus* wastes. Foods 8, 523.

Manconi M, Manca ML, Marongiu F, Caddeo C, Castangia I, Petretto GL, Pintore G, Sarais G, D'Hallewin G, Zaru M. 2016. Chemical characterization of *Citrus limon* var. pompia and incorporation in phospholipid vesicles for skin delivery. Int J Pharm 506, 449–457.

Mandalari G, Bennett RN, Bisignano G, Saija A, Dugo G, Lo Curto RB, Faulds CB, Waldron KW. 2006. Characterization of flavonoids and pectins from bergamot (*Citrus bergamia* Risso) peel, a major byproduct of essential oil extraction. J Agric Food Chem 54, 197–203.

Mandalari G, Bennett RN, Bisignano G, Trombetta D, Saija A, Faulds CB, Gasson MJ, Narbad A. 2007. Antimicrobial activity of flavonoids extracted from bergamot (*Citrus bergamia* Risso) peel, a byproduct of the essential oil industry. J Appl Microbiol 103, 2056–2064.

Manthey JA. 2004. Fractionation of orange peel phenols in ultrafiltered molasses and mass balance studies of their antioxidant levels. J Agric Food Chem 52, 7586–7592.

Matsubara Y, Yusa T, Sawabe A, Iizuka Y, Takekuma S, Yoshida Y. 1991. Structures of new cyclic peptides in young unshiu (*Citrus unshiu* Marcov.), orange (*Citrus sinensis* Osbeck.) and amanatsu (*Citrus natsudaidai*) peelings. Agric Biol Chem 55, 2923–2929.

Md Othman SNA, Hassan MA, Nahar L, Basar N, Jamil S, Sarker SD. 2016. Essential oils from the Malaysian *Citrus* (Rutaceae) medicinal plants. Medicines 3, 13.

Miceli N, Mondello MR, Monforte MT, Sdrafkakis V, Dugo P, Crupi ML, Taviano MF, Pasquale RD, Trovato A. 2007. Hypolipidemic effects of *Citrus bergamia* Risso et Poiteau juice in rats fed a hypercholesterolemic diet. J Agric Food Chem 55, 10671–10677.

Milind P, Dev C. 2012. Orange: Range of benefits. Int Res J Pharm 3, 59–63.

Miller EG, Peacock JJ, Bourland TC, Taylor SE, Wright JM, Patil BS, Miller EG. 2008. Inhibition of oral carcinogenesis by citrus flavonoids. Nutr Cancer 60, 69–74.

Mukhtar R, Khan MM, Fatima B, Abbas M, Shahid A. 2005. *In vitro* regeneration and somatic embryogenesis in (*Citrus aurantifolia* and *Citrus sinensis*). Int J Agric Biol 7, 414–416.

Murashige T, Skoog F. 1962. A revised medium for rapid growth and bioassays with tobacco tissue culture. Physiol Plant 15, 473–497.

Musara C, Aladejana EB, Mudyiwa SM. 2020. Review of the nutritional composition, medicinal, phytochemical and pharmacological properties of *Citrus reticulata* Blanco (Rutaceae). F1000Research 9, 1387.

Musumeci L, Maugeri A, Cirmi S, Lombardo GE, Russo C, Gangemi S, Calapai G, Navarra M. 2020. *Citrus* fruits and their flavonoids in inflammatory bowel disease: An overview. Nat Prod Res 34, 122–136.

Nadkarni AK. 1996. Indian Materia Medica, Vol I. Popular Prakashan, Bombay.

Nagappan A, Lee HJ, Saralamma VVG, Park HS, Hong GE, Yumnam S, Raha S, Charles SN, Shin SC, Kim EH, Lee WS, Kim GS. 2016. Flavonoids isolated from *Citrus platymamma* induced G2/M cell cycle arrest and apoptosis in A549 human lung cancer cells. Oncol Lett 12, 1394–1402.

Nagappan A, Park H-S, Hong G-E, Yumnam S, Lee H-J, Kim D-H, Kim E-H, Kim G-S. 2014. Anti-cancer and anti-inflammatory properties of korean citrus fruits (*Citrus aurantium* L.). J Korean Clin Health Sci 2, 73–78.

Nagy S. 1980. Vitamin C contents of citrus fruit and their products: A review. J Agric Food Chem 28, 8–18.

Nauman MC, Johnson JJ. 2019. Clinical application of bergamot (*Citrus bergamia*) for reducing high cholesterol and cardiovascular disease markers. Integr Food Nutr Metab 6, 1–7.

Nikolić MM, Jovanović KK, Marković TL, Marković DL, Gligorijević NN, Radulović SS, Kostić M, Glamočlija JM, Soković MD. 2017. Antimicrobial synergism and cytotoxic properties of *Citrus limon* L., *Piper nigrum* L. and *Melaleuca alternifolia* (Maiden and Betche) Cheel essential oils. J Pharm Pharmacol 69, 1606–1614.

Nishikawa F. 2013. Regulation of floral induction in citrus. J Jpn Soc Hortic Sci 82, 283–292.

Norkaew O, Pitija K, Pripdeevech P, Sookwong P, Wongpornchai S. 2013. Supercritical fluid extraction and gas chromatographic-mass spectrometric analysis of terpenoids in fresh Kaffir lime leaf oil. Chiang Mai J Sci 40, 240–247.

Orhan I, Aydin A, Çölkesen A, Sener B, Isimer AI. 2003. Free radical scavenging activities of some edible fruit seeds. Pharm Biol 41, 163–165.

Ortuño A, Báidez A, Gómez P, Arcas MC, Porras I, García-Lidón A, DelRío JA. 2006. *Citrus paradisi* and *Citrus sinensis* flavonoids: Their influence in the defence mechanism against *Penicillium digitatum*. Food Chem 98, 351–358.

Orwa C, Mutua A, Kindt R, Jamnadass R, Simons A. 2009. Agroforestree Database: A Tree Species Reference and Selection Guide Version 4.0. World Agroforestry Centre ICRAF, Nairobi.

Otang WM, Afolayan AJ. 2016. Antimicrobial and antioxidant efficacy of *Citrus limon* L. peel extracts used for skin diseases by Xhosa tribe of Amathole District, Eastern Cape, South Africa. S Afr J Bot 102, 46–49.

Pandey A, Tamta S. 2016. Efficient micropropagation of *Citrus sinensis* (L.) Osbeck from cotyledonary explants suitable for the development of commercial variety. African J Biotechnol 15, 1806–1812.

Papp N, Bartha S, Boris G, Balogh L. 2011. Traditional uses of medicinal plants for respiratory diseases in Transylvania. Nat Prod Commun 6, 1459–1460.

Paudyal KP, Haq N. 2000. *In vitro* propagation of pummelo (*Citrus grandis* L. Osbeck). *In Vitro* Cell Dev Biol Plant 36, 511–516.

Paul S, Islam MA, Tanvir EM, Ahmed R, Das S, Rumpa N-E-N, Hossen MS, Parvez M, Gan SH, Khalil MI. 2016. Satkara (*Citrus macroptera*) fruit protects against acetaminophen-induced hepatorenal toxicity in rats. Evid Based Complement Altern Med 2016, Article ID 9470954.

Peter E, Peter J, Nes B and Asukwo G. 2008. Physiochemical properties and fungi toxicity of the essential of *Citrus medica* L. against ground nut storage fungi. Turk J Bot 32, 161–164.

Peterson JJ, Beecher GR, Bhagwat SA, Dwyer JT, Gebhardt SE, Haytowitz DB, Holden JM. 2006. Flavanones in grapefruit, lemons, and limes: A compilation and review of the data from the analytical literature. J Food Compost Anal 19, S74–S80.

Putnik P, Barba FJ, Lorenzo JM, Gabrić D, Shpigelman A, Cravotto G, Kovačević DB. 2017. An integrated approach to Mandarin processing: Food safety and nutritional quality, consumer preference, and nutrient bioaccessibility. Compr Rev Food Sci Food Saf 16, 1345–1358.

Rani G, Yadav L, Kalidhar SB. 2009. Chemical examination of *Citrus sinensis* flavedo variety pineapple. Indian J Pharm Sci 71, 677–679.

Rao MN, Soneji JR, Sahijram L. 2011. Citrus: Genoics resources developed. In: Wild Crop Relatives: Genomic and Breeding Resources, Chittaranjan K (ed.). Springer, Berlin, Heidelberg, pp 43–59.

Riaz A, Khan RA, Algahtani HA. 2014. Memory boosting effect of *Citrus limon*, pomegranate and their combinations. Pak J Pharm Sci 27, 1837–1840.

Rong X, Xu J, Jiang Y, Li F, Chen Y, Dou QP, Li D. 2021. *Citrus* peel flavonoid nobiletin alleviates lipopolysaccharide-induced inflammation by activating IL-6/STAT3/FOXO3a-mediated autophagy. Food Funct 12, 1305–1317.

Russo M, Bonaccorsi I, Costa R, Trozzi A, Dugo P, Mondello L. 2015. Reduced time HPLC analyses for fast quality control of citrus essential oils. J Essent Oil Res 27, 307–315.

Saidan I. 2013. Dalam Dusun Melayu (in Malay Orchard). Dewan Bahasa dan Pustaka, Kuala Lumpur.

Saleem M, Farooq A, Ahmad S, Shafiq N, Riaz N, Jabbar A, Arshad M, Malik A. 2010. Chemical constituents of *Citrus sinensis* var. Shukri from Pakistan. J Asian Nat Prod Res 12, 702–706.

Samah, B. 2009. Serangan Jantung: Punca, Pencegahan & Kaedah Meredakannya (Heart Attack: The Cause, Prevention & Treatment). Selangor, Alaf.

Scora RW. 1975. On the history and origin of citrus. Bull Torrey Bot Club 102, 369–375.

Sergeev IN, Ho C-T, Li S, Colby J, Dushenkov S. 2007. Apoptosis-inducing activity of hydroxylated polymethoxyflavones and polymethoxyflavones from orange peel in human breast cancer cells. Mol Nutr Food Res 51, 1478–1484.

Shakour ZTA, Fayek NM, Farag MA. 2020. How do biocatalysis and biotransformation affect *Citrus* dietary flavonoids chemistry and bioactivity? A review. Crit Rev Biotechnol 40, 689–714.

Sharon-Asa, L, Shalit, M, Frydman, A, Bar, E, Holland, D, Or E, Lavi U, Lewinsohn E, Eyal Y. 2003. *Citrus* fruit flavor and aroma biosynthesis: Isolation, functional characterization, and developmental regulation of Cstps1, a key gene in the production of the sesquiterpene aroma compound valencene. Plant J 36, 664–674.

Simeone GDR, Matteo AD, Rao MA, Vaio CD. 2020. Variations of peel essential oils during fruit ripening in four lemons (*Citrus limon* (L.) Burm. F.) cultivars. J Sci Food Agric 100, 193–200.

Singh VK, Ali ZA. 1998. Herbal Drugs of Himalaya: Medicinal Plants of Garhwal and Kumaon Regions of India. Today & Tomorrow's Printers and Publishers, New Delhi.

Srinivasan D, Ramasamy S and Sengottuvelu S. 2008. Protective effect of polyherbal formulation on experi-
    mentally induced ulcer in rats. Pharmacologyonline 1, 331–350.
Sultana HS, Ali M, Panda BP. 2012. Influence of volatile constituents of fruit peels of *Citrus reticulata* Blanco on
    clinically isolated pathogenic microorganisms under *in-vitro*. Asian Pac J Trop Biomed 2, S1299–S1302.
Suryawanshi J. 2011. An overview of *Citrus aurantium* used in treatment of various diseases. Afr J Plant Sci
    5, 390–395.
Takemoto JK, Remsberg CM, Yáñez JA, Vega-Villa KR, Davies NM. 2008. Stereospecific analysis of
    sakuranetin by high-performance liquid chromatography: Pharmacokinetic and botanical applications. J
    Chromatogr B Anal Technol Biomed Life Sci 875, 136–141.
Thavanapong N, Wetwitayaklung P, Charoenteeraboon J. 2010. Comparison of essential oils compositions of
    Citrus maxima Merr. peel obtained by cold press and vacuum stream distillation methods and of its peel
    and flower extract obtained by supercritical carbon dioxide extraction method and their antimicrobial
    activity. J Essent Oil Res 22, 71–77.
Tranchida PQ, Presti ML, Costa R, Dugo P, Dugo G. 2006. High-throughput analysis of bergamot essen-
    tial oil by fast solid-phase microextraction-capillary gas chromatography-flame ionization detection. J
    Chromatogr A 1103, 162–165.
Tripoli E, Guardia ML, Giammanco S, DiMajo D, Giammanco M. 2007. *Citrus* flavonoids: Molecular struc-
    ture, biological activity and nutritional properties: A review. Food Chem 104, 466–479.
Trovato A, Taviano MF, Pergolizzi S, Campolo L, Pasquale RD, Miceli N. 2010. *Citrus bergamia* Risso &
    Poiteau juice protects against renal injury of diet-induced hypercholesterolemia in rats. Phytother Res
    24, 514–519.
Uddin U, Hasan MR, Hossain MM. 2014. Antioxidant, brine shrimp lethality and antimicrobial activities of
    methanol and ethyl-acetate extracts of *Citrus macroptera* Montr. fruit using in vitro assay models. Br J
    Pharm Res 4, 1725–1738.
Ulloa FV, García SS, Estrada MC, López DLP, Cruz MCR, García PS. 2012. Interpretation methods of nutrient
    diagnosis in orange cv. Valencia (*Citrus sinensis* L. Osbeck). Terra Latinoam 30, 139–145.
Usman M, Muhammad S, Fatima B. 2005. *In vitro* multiple shoot induction from nodal explants of *Citrus*
    cultivars. J Cent Eur Agric 6, 435–442.
Vijaylakshmi P, Radha R. 2015. An overview: *Citrus maxima*. J Phytopharm 4, 263–267.
Waikedre J, Dugay A, Barrachina I, Herrenknecht C, Cabalion P, Fournet A. 2010. Chemical composition and
    antimicrobial activity of the essential oils from New Caledonian *Citrus macroptera* and *Citrus hystrix*.
    Chem Biodivers 7, 871–877.
Wang L, Wang J, Fang L, Zheng Z, Zhi D, Wang S, Li S, Ho C-T, Zhao H. 2014. Anticancer activities of *Citrus*
    peel polymethoxyflavones related to angiogenesis and others. BioMed Res Int 2014, Article ID 453972.
Wang Y, Qian J, Cao J, Wang D, Liu C, Yang R, Li X, Sun C. 2017. Antioxidant capacity, anticancer ability and
    flavonoids composition of 35 citrus (*Citrus reticulata* Blanco) varieties. Molecules 22, 1114.
Xiao H, Yang CS, Li S, Jin H, Ho C-T, Patel T. 2009. Monodemethylated polymethoxyflavones from sweet
    orange (*Citrus sinensis*) peel inhibit growth of human lung cancer cells by apoptosis. Mol Nutr Food Res
    53, 398–406.
Xiong Y, Chen D, Yu C, Lv B, Peng J, Wang J, Lin Y. 2015. *Citrus* nobiletin ameliorates experimental colitis
    by reducing inflammation and restoring impaired intestinal barrier function. Mol Nutr Food Res 59,
    829–842.
Yabesh JE, Prabhu S, Vijayakumar S. 2014. An ethnobotanical study of medicinal plants used by traditional
    healers in silent valley of Kerala, India. J Ethnopharmacol 154, 774–789.
Zhu X, Luo F, Zheng Y, Zhang J, Huang J, Sun C, Li X, Chen K. 2013. Characterization, purification of ponci-
    rin from edible citrus ougan (*Citrus reticulata* cv. Suavissima) and its growth inhibitory effect on human
    gastric cancer cells SGC-7901. Int J Mol Sci 14, 8684–8697.
Zou W, Wang Y, Liu H, Luo Y, Chen S, Su W. 2013. Melitidin: A flavanone glycoside from *Citrus grandis*
    'Tomentosa'. Nat Prod Commun 8, 457–458.
Zouaghi G, Najarc A, Aydid A, Claumanne CA, Zibettif AW, Mahmoudc KB, Jemmalic A, Bletong J, Moussag
    F, Abderrabba M, Chammem N. 2019. Essential oil components of *Citrus* cultivar 'Maltaise Demi
    Sanguine' (*Citrus sinensis*) as affected by the effects of rootstocks and viroid infection. Int J Food Prop
    22, 438–448.

10 *Clerodendrum* Species

10.1 MORPHOLOGICAL FEATURES, DISTRIBUTION, ETHNOPHARMACOLOGICAL PROPERTIES, PHYTOCHEMISTRY, AND PHARMACOLOGICAL ACTIVITIES

Clerodendrum genus (Fam.—Verbenaceae) is widely distributed in tropical Africa, southern Asia, tropical Americas, northern Australasia, and temperate regions of eastern Asia (Harley et al. 2004; Muthu et al. 2013). Clerodendrum is a genus of about 450 species and several species are medicinally important viz., *C. serratum, C. inerme, C. bungei, C. phlomidis, C. serratum* var. amplexifolium, *C. infortunatum, C. trichotomum, C. chinense, C. petasites, C. grayi, C. indicum,* and *C. trichotomum* (Staples and Herbst 2005; Mabberley 2008; Yuan et al. 2010). In Indian, Chinese, Korean, Japanese, Thailand, and African medicine, *C. indicum, C. phlomidis, C. serratum* var. amplexifolium, *C. trichotomum, C. chinense,* and *C. petasites* have been used in the treatment of cold, hyperpyrexia, asthma, furunculosis, hypertension, rheumatism, dysentery, toothache, anorexia, leukoderma, leprosy, and other inflammatory diseases (Shrivastava and Patel 2007; Chethana et al. 2013; Baker et al. 1995; Hazekamp et al. 2001). Various species of the Clerodendrum genus have been used in the treatment of syphilis, typhoid, cancer, jaundice, and hypertension (Shrivastava and Patel 2007; Chander et al. 2015; Jadhav 2006; Rao et al. 2006; Jiji 2014; Singh et al. 2015; Park et al. 2018). As per the Indian, Chinese, and Japanese traditional medicine systems, *C. paniculatum* is prescribed as a medicine for treating rheumatism, jaundice, body ache, snake bite and giddiness, ulcer, neuralgia, inflammation, and wounds. In the Thai medicine system, it is used as an anti-inflammatory and antipyretic drug (Phuneerub et al. 2015). *C. indicum, C. phlomidis, C. serratum, C. trochotomum, C. chinense,* and *C. petasites* root and leaf extracts are used for the treatment of rheumatism, asthma, and other inflammatory diseases (Kang et al. 2003). *C. indicum* and *C. inerme* are useful in coughs, beriberi disorder, scrofulous infection, buboes complaints, venereal and skin infections, and used as a vermifuge and febrifuge (Hazekamp et al. 2001). The isolated compounds and crude extracts from the *Clerodendrum* species possess anti-inflammatory, antinociceptive, antioxidant, antihypertensive, anticancer, antimicrobial, antidiarrheal, hepatoprotective, hypoglycemic, and hypolipidemic properties (Wang et al. 2018; Youssef et al. 2022). The root and leaf extracts of *C. phlomidis, C. petasites, C. serratum, C. trichotomum, C. chinense* and *C. indicum* are used in the treatment of hypertension, diabetes, rheumatism, asthma, coughs, skin diseases, vermifuge, febrifuge, malaria, and inflammatory diseases (Siddik et al. 2021; Ugbaja et al. 2021).

C. serratum (L.) Moon is a perennial woody shrub with blunty, quadrangular stems and branches, growing up to 1–3 m high. The stem bark is thin and can be easily separated from a broad wood. Leaves are alternate, or sometimes opposite, oblong, or elliptic, and serrated. Flower is bisexual, zygomorphic, rarely sub-actinomorphic, blue, and contains epipetalous stamens. Fruit is purple drupe and seeds with or without endosperm. The shrub is found in Africa and southern Asia, tropical America, and northern Australia, Malaysia, India, and Sri Lanka as well as in temperate regions of eastern Asia (Mabberley 2008). As per Ayurvedic, Siddha, and Unani medicine, it is used for the treatment of syphilis, typhoid, cancer, jaundice, and hypertension. The roots possess antioxidant, antibacterial, and antifungal properties (Singh et al. 2012). The ethanol extract of *C. serratum* roots displayed significant antioxidant activity in DPPH radical scavenging, ferric-reducing antioxidant power, and hydrogen peroxide radical scavenging assays (Bhujbal et al. 2009). The roots and leaves of *C. serratum* are useful in asthma, allergy, fever, inflammation, and liver diseases. Icosahydropicenic acid and ursolic acid isolated from this species demonstrated anti-allergic and hepatoprotective activities (Patel et al 2014). Anticancer activity of aqueous and methanol extracts

DOI: 10.1201/9781003398035-10

of *C. serratum* roots (100 and 200 mg/kg/day/p.o., for 14 days) was evaluated in cancerous mice. The methanol extract demonstrated significant anticancer activity by increasing life span, body weight, and other hematological parameters viz., RBC, WBC, and hemoglobin (Zalke et al. 2010). The ethanol extract of *C. serratum* leaves (200 and 500 mg/kg) showed significant analgesic activity in tail flick and acetic acid-induced writhing tests in Wistar rats (Saha et al. 2012). The ethanol extract (50, 100, 200 mg/kg/p.o.) showed significant reduction in pyrexia. The results were compared with paracetamol (Narayanan et al. 1999). Ethanol extract (25 mg/ml) also demonstrated significant antibacterial activity against *S. aureus, E. coli, P. aeruginosa*, and *B. subtilis* (Gupta and Sharma 2008). Apigenin 7-glucoside, hispidulin, scutellarein-7-*O*-β-D-glucuronate, acteoside, and verbascoside were reported from *C. serratum*. Apigenin 7-glucoside and hispidulin possess maximum ability in binding interactions with 17 cancer drug targets in respect to docking weighted network pharmacological analysis (Gogoi et al. 2017; Kalonio et al. 2017).

C. trichotomum Thunb is a deciduous tree, and height reaches to 4–6 m. Leaves are dark green, ovate, lanceolate, simple, and opposite, petiole 4–10 cm long, and pubescent. Inflorescence is terminal, long (6–16 cm) and across (12–25 cm), pedicels 2–12 cm long, and softly pubescent. Flowers are long (2.5–3 cm), bisexual, and reddish. The fruit is drupe, blue-purple, and surrounded by red persistent calyx (Das et al. 2014). Aqueous extract of *C. trichotomum* significantly decreased respiratory syncytial virus replication, respiratory syncytial virus-induced cell death, respiratory syncytial virus gene transcription, respiratory syncytial virus protein synthesis, and prohibited formation of syncytia. Oral administration of extract significantly improved viral clearance in BALB/c mice lungs. Acteoside demonstrated similar antiviral activity as shown by aqueous extract against respiratory syncytial virus (*in vitro* and *in vivo*); therefore, *C. trichotomum* may be considered as a potent antiviral agent against respiratory syncytial virus infections (Chathuranga et al. 2019). Anti-inflammatory effects of methanol extract of *C. trichotomum* leaves were evaluated in rats, mice, and in Raw 264.7 cells. Extract (30%, and 60%; 1 mg/kg) and indomethacin (1 mg/kg) significantly suppressed the volume of carrageenan-induced hind paw edema (19.5%, 23.0%, and 20.5%). The methanol extract (60%) also suppressed Evans blue dye leakage (47.0%), which was higher (10%) than indomethacin. Methanol extract inhibited the formation of prostaglandin E2 in RAW 264.7 macrophage cells. The methanol (60%) extract showed significant suppression in the carrageenan-induced rat paw oedema, vascular permeability, and the formation of PGE2 (Choi et al. 2004).

C. petasites (Lour.) S.Moore is a shrub plant species, 3–5 m in height and is distributed in Thailand, India, Malaysia, Sri Lanka, Vietnam, and Southern China. It is used in the treatment of cough, and respiratory diseases (Deb 1983; Kottegoda 1994; Panthong et al. 1986; Mabberley 1997). As per Thai medicine, *C. petasites* is used in the treatment of asthma, inflammation, fever, cough, vomiting, and skin disorders. Hispidulin, vanillic acid, verbascoside, and apigenin have been reported as major compounds from this species. Hispidulin is useful in cancer, osteolytic bone disorders, and neurological complaints (Brimson et al. 2019). Ethanolic extract of *C. petasites* (2.25–9.0 mg/ml) produced significant concentration-dependent relaxation in tracheal smooth muscle (EC_{50} 4.8 mg/ml) in guinea pigs. Similarly, isolated hispidulin (flavonoid) also displayed significant results [EC_{50} $(3.0 \pm 0.8) \times 10^{-5}$ M]; therefore, hispidulin may be used in treating asthma (Hazekamp et al. 2001). The ethyl acetate fraction of *C. petasites* consisted of greater amounts of hesperidin and hesperetin flavonoids than ethanolic root extract. The findings reveal that ethyl acetate fraction (compared with ethanol extract) and hesperetin (compared with hesperidin) demonstrated stronger anti-inflammatory effects upon Spike S1 stimulation via a significant decrease in IL-6, IL-1β, and IL-18 cytokine liberates in A549 cells culture supernatant ($p < 0.05$). Moreover, ethyl acetate fraction and hesperetin significantly suppressed the Spike S1-stimulated inflammatory gene expressions (*NLRP3, IL-1β*, and *IL-18*, $p < 0.05$). Spontaneously, ethyl acetate fraction and hesperetin reduced inflammasome machinery protein expressions (NLRP3, ASC, and Caspase-1) and deactivated the Akt/MAPK/AP-1 pathway (Arjsri et al. 2022).

C. thomsoniae Balf.f. is a twining, vine-like shrub plant species and found in tropical West Africa (Hunt et al. 1978; Schmid and Riffle 1998). *C. thomsoniae* leaves and flowers are used in curing bruises, cuts, skin rashes, and sores (Ra et al. 2004). The ethyl acetate extract of *C. thomsoniae* aerial parts displayed significant anticancer effects against MCF-7, Hep-G2, A549, HT-29, MOLT-4, and HeLa (IC$_{50}$ 29.43 ± 1.44 µg/ml, 43.22 ± 1.02 µg/ml, 56.93 ± 1.41 µg/ml, 60.68 ± 1.05 µg/ml, 69.83 ± 1.33 µg/ml, and 40.02 ± 1.14 µg/ml) cell lines. The results suggest that *C. thomsoniae* ethyl acetate extract possess significant cytotoxic activity against breast cancer cell lines (Muhammed Ashraf et al. 2021).

C. infortunatum L. (syn *C. viscosum* Vent.) is a terrestrial and perennial shrub plant species with square, blackish stems. The leaves are simple, opposite, decussate, petiolate, exstipulate, coriaceous, and hairy (Nadkarni and Nadkarni 2002). In Indian traditional medicine, aerial parts and roots of this species are used in the treatment of colic, scorpion sting and snake bite, tumors, and skin infections (Nadkarni and Nadkarni 2002; Nandi and Lyndem 2016). Isolated flavonoids from *C. infortunatum* such as hispidulin, ursolic, cleroflavone, and a diterpenoid clerodin exhibit potent antioxidant, antimicrobial, antiasthmatic, antitumor, and CNS-binding properties (Sindhu et al. 2020). *C. infortunatum* has been prescribed as a remedy for the treatment of fever, pain, diabetes, inflammation, wounds, rheumatism, deworming, dysentery, diarrhea, snakebite, blood pressure, and cancers (Verma 2014; Singh 2015; Mitra and Mukherjee 2010; Dey et al. 2017; Bharadwaj and Seth 2017). Its sprout is used as an antidote in snakebites. The roots are useful in curing bronchitis and asthma (Shetty et al. 2002; Punekar and Lakshminarasimhan 2011). The leaves are slightly bitter and used in the treatment of inflammation, skin infections, and smallpox (Chopra et al. 1992). The plant species showed the presence of triterpenes, steroids, and flavonoids (Joshi et al. 1977; Akihisa et al. 1988; Sinha et al. 1981; Manzoor-Khuda and Sarela 1965). The species possess antioxidant and hepatoprotective (Gopal and Sengottuvelu 2008; Sannigrahi et al. 2009; Vidya et al. 2007; Chae et al. 2005), antimicrobial (Rajakaruna et al. 2002), antimalarial (Goswami et al. 1998), and analgesic activities (Pal et al. 2009). Ethanol extract of *C. infortunatum* leaves displayed significant inhibitory (maximum inhibition zone) activity against *B. megaterium*, *S. typhi*, *K. pneumoniae*, *A. niger*, and *C. albicans* (with respect to maximum inhibition zone). Ethanol extract also showed significant inhibition (with respect to minimum inhibitory concentration) to *B. megaterium*, *S. typhi*, and *K. pneumoniae* (MIC 64 µg/ml); *S. aureus*, *Streptococcus*-β-*hemolyticus*, and *E. coli* (128 µg/ml). The order of activity was recorded as leaf extract>root extract>stem extract against tested microbes (Waliullah et al. 2014; Prashith Kekuda et al. 2019). High performance thin layer chromatographic studies revealed that gallic acid (0.244 mg/g) was found as a major constituent in aqueous extract of crude powder of this species (Verma and Gupta 2014). The oleanolic acid and clerodinin A were identified from methanol extract of *C. infortunatum*. The methanol extract showed significant reduction in the tumor cell volume but increased the life span of treated animals. The median survival time of Ehrlich's ascites carcinoma was recorded as 19.42 ± 0.91 days (control group), while methanol extract treatment (200 mg/kg) increased the median survival time (27.57 ± 2.57 days) in treated animals. The hematological parameters, malonaldehyde content, and antioxidant enzymes' activity were normalized in treated animals. Methanol extract showed significant cytotoxicity (IC$_{50}$ 498.33 µg/ml). The anticancer activity may be attributed due to the presence of oleanolic acid and clerodinin A (Sannigrahi et al. 2012). The methanolic extract *C. infortunatum* showed maximum quantity of polyphenolics and flavonoids, which indicates potent antioxidant activity. Methanol extract revealed the presence of alkaloids, carbohydrates, glycosides, phytosterols, phenols, tannins, and flavonoids (Kokate et al. 2022).

Aqueous extract of *C. viscosum* roots demonstrated pro-apoptotic, antiproliferative, and antimigratory effect in a dose-dependent manner against cervical cancer cell lines. The extract did not show any effect on apoptosis of primary fibroblasts (control healthy cells). Glycoprotein was isolated as a major compound from aqueous extract (Sun et al. 2013). Chloroform and ethyl acetate fractions of methanol (70%) extract of *C. viscosum* leaves displayed significant free radical scavenging activities as well as suppressed the growth of human lung cancer (A459), breast (MCF-7), and brain (U87) cell lines. In addition, both fractions obstructed the cell cycle at the G2/M phase of breast and brain cancers. The presence of tannic acid, quercetin, ellagic caid, gallic acid, reserpine, and methyl gallate was confirmed in both fractions. Both fractions showed significant reactive oxygen species and reactive nitrogen species scavenging abilities. Both the fractions prevented the cell cycle at the G2/M phase in MCF-7 and U87 cell lines, which lead to stimulate apoptosis (Shendge et al. 2017).

Methanol extract (70%) of *C. viscosum* leaves revealed the presence of tannic acid, catechin, rutin, and reserpine. The extract showed potent cytotoxicity against the breast cancer cell line (MCF-7). Moreover, the extract confirmed an apoptosis-inducing effect by annexin staining (Shendge et al. 2021a). Apigenin, isolated from *C. viscosum* leaves, displayed cytotoxicity against MCF-7 cells (IC_{50} 56.72 ± 2.35 µM) but displayed negligible cytotoxicity against WI-38 cells. Apigenin treatment in the presence of Pifithrin-µ displayed reduced apoptotic population. The findings indicated the vital role of p53 in apigenin-stimulated apoptosis in MCF-7 cells. The apigenin-induced intracellular ROS in MCF-7 cells followed by stimulation of the G2/M phase cell cycle arrest and moreover apoptosis via the regulation of p53 and caspase-cascade signaling pathway (Shendge et al. 2021b).

C. phlomidis L.f. [syn. *C. multiflorum* (Burm.f.) Kuntze] is a perennial shrub and is found in the arid areas of tropical deserts (Maruga Raja and Mishra 2010). The anti-obesity effect of ethanol and methanol extracts *C. phlomidis* roots were evaluated on high fat diet-induced obesity in C57BL/6J female mice. Out of these two extracts, the methanol extract showed greater anti-obesity effect in treated animals (LD_{50} 2000 mg/kg; Chidrawar et al. 2012). Chloroform extract of *C. phlomidis* showed significantly lower concentrations (LC_{50} and LC_{90} 5.02, 61.63 ppm and 32.86, 73.62 ppm) against *Culex quinquefasciatus* and *Aedes aegypti*. Similarly, the isolated compound pectolinaringenin displayed the lowest values (LC_{50} and LC_{90} 0.62, 2.87 ppm and 0.79, 5.31 ppm) against *C. quinquefasciatus* and *A. aegypti* respectively. The study results show that the chloroform extract can be used as a strong source and pectolinaringenin as a novel mosquito larvicidal agent (Muthu et al. 2012). Ethyl acetate and hexane extracts of *C. phlomidis* leaves and stem were tested for their antifungal activity. Both extracts of stem and leaf demonstrated significant suppression against plant and human pathogenic fungi (*Epidermophyton floccosum*, *Trichophyton mentagrophytes*, *T. rubrum*, *T. tonsurans*, *Aspergillus flavus*, *A. niger*, *Botrytis cinerea*, *Curvularia lunata*, *F. oxysporum*). Hexane extract of leaves (1 mg/ml) showed better inhibition of plant pathogenic fungi than the human dermatophytes (Anitha and Kannan 2006). Ethanol extract (100, 200, and 400 mg/kg) of *C. phlomidis* leaves displayed maximum suppression against inflammation stimulated by carrageenan in rats (100 mg/kg–47.73%; 200 mg/kg–54.00% and 400 mg/kg–65.15%). Ethanol extract (100, 200, and 400 mg/kg) showed significant decrease in granuloma weight in cotton pellet–induced granuloma. Ethanol extract also displayed a significant decrease in paw thickness (100 mg/kg–51.71%; 200 mg/kg–57.58%, and 400 mg/kg–62.48%) in Freund's complete adjuvant-induced arthritis model. Ethanol extract significantly decreased the levels of pro-inflammatory cytokines in a dose-dependent manner in treated animals. Future studies will offer new insights into the anti-inflammatory effects of *C. phlomidis* and the identification of compounds, from which may lead to formation of novel anti-inflammatory drugs (Prakash Babu et al. 2011). The 3-hydroxy, 2-methoxy-sodium butanoate isolated from *C. phlomidis* leaves decreased the paw edema stimulated by carrageenan and Freund's complete adjuvant in a dose-dependent manner. The isolated compound decreased the levels of lysosomal enzymes, plasma acute phase protein, and protein-bound carbohydrates in treated rats. The extract also decreased the levels of protein and mRNA expression of pro-inflammatory cytokines TNF, IL-1, and IL-6 in the joints of treated animals. 3-hydroxy,2-methoxy sodium butanoate displayed significant anti-inflammatory activity and possesses a prominent anti-arthritic effect (Prakash Babu et al. 2014). The *n*-hexane extract of *C. volubile* leaves and aqueous extract of *C. phlomidis* leaves possess antioxidant activity on DPPH, hydroxyl radical scavenging, and nitric oxide radical scavenging assays (Salahudeen and Bolaji 2019; Krishnaveni and Rajan 2021).

C. inerme (L.) Gaertn is an evergreen shrub, with woody and smooth stem, growing up to 1–1.8 m high. Leaves are green, opposite, smooth, ovate to elliptical, long (5–10 cm), and with acute to acuminate tips. Inflorescence is of cyme type. Corolla are white, fused, with 4 stamens. Fruit is green, ovoid, long (1–1.5 cm), and turning black on maturity (Turner and Wasson 1997). It is naturalized in India, Nepal, Bangladesh, Sri Lanka, Southeast Asia, and the Mediterranean (Rabiul et al. 2011). It is used as a febrifugal and uterine stimulant, antiseptic, and in treating asthma, hepatitis, ringworm, and stomach pains (Muthu et al. 2006). In Indian tribal medicine, leaves are used for the treatment of fever, cough, skin rashes, chronic pyrexia, and boils (Nadkarni 1976). The roots are prescribed as medicine in the treatment of scrofulous and venereal infections (Al-Snaf 2016). Fresh leaf juice is used topically in treating skin infections. Its roots are boiled in oil and used in rheumatic affections (Sharaf et al. 1969; Chourasiya et al. 2010, 2011). Chloroform and ethanol extract of *C. inerme* leaves (200 mg/kg and 400 mg/kg) showed significant diuretic activity after 24 h in treated animals (Upmanyu et al. 2011). Cytotoxic and antiproliferative potential of hydroalcoholic (methanol and water, 70:30, v/v) extract was assessed on brine shrimp lethality assay, Daudi cell line culture, dye exclusion, and MTT assays, and animal model. The extract displayed cytotoxic effect below 100 ppm and (LD_{50}) in the brine shrimp lethality method. A significant reduction in cell viability was recorded at 213 µg/ml concentration and displayed antitumor effect against Burkitt's lymphoma cells as well the tumor model in female mice. The extract (200 and 400 mg/kg bw) increased the life span in treated animals at a total greater than control group. The reference compound (doxorubicin, 2.5 mg/kg bw) showed significant effect on tumor parameters (Chouhan et al. 2017). Aqueous extract of *C. inerme* leaves showed significant ($p < 0.001$) analgesic effect in hot plate, tail flick, and tail immersion tests in albino rats. Aqueous extract displayed a significant ($p < 0.001$) dose-dependent decrease of pyrexia in rabbits (Thirumal et al. 2013). *C. inerme* leaves possess an antimicrobial effect (Prasad et al. 1995) and stimulate uterine motility in rats as well as suppress intestinal motility (Husain et al. 1992). Chloroform-water and ethanol extracts showed significant antioxidant activity in nitric oxide scavenging activity and ferric chloride reductive ability assays (Sayyed et al. 2011). Aqueous extract of *C. inerme* leaves (500 mg/kg bw) administration to 7,12-dimethylbenz(a) anthracene painted animals stopped the incidence, volume, and burden of tumors. Aqueous extract also produced strong antilipidperoxidative activity and increased the antioxidant defense system in 7,12-dimethylbenz(a) anthracene-painted animals. The extract showed significant inhibition of tumor formation (80%) in 7,12-dimethylbenz(a) anthracene-painted animals. The chemopreventive efficacy of the extract is possibly due to its antilipidperoxidative activity or the presence of some strong bioactive chemopreventive compounds in *C. inerme* leaves (Manoharan et al. 2006). The ethanol extract of *C. inerme* leaves displayed significant antibacterial effect against *Mycobacterium tuberculosis*, whereas *C. splendens* presented moderate activity (Elaskary et al. 2020; Okaiyeto et al. 2021).

C. indicum (L.) Kuntze is an erect, subshrub, unbranched stem, and grows up to 2 m high. Leaves are straight, lanceolate, or elliptic-lanceolate, and sessile or subsessile. Inflorescence is terminal panicles. Roots are used in the treatment of asthma, cough, and scrofulous affections. Its resin possesses antirheumatic property. The whole plant is also useful in fever, atrophy, emaciation of cachexia, and consumption (Sidde et al. 2018). Antinociceptive, antidiarrheal and antimicrobial activities of methanol extract of *C. indicum* leaves were investigated in acetic acid-induced writhing test in mice, castor oil-induced diarrhea, and disc diffusion methods. Methanol extract (200 and 400 mg/kg) displayed a significant ($p < 0.001$) decrease in the number of writhes (62.57% and 70.76%) in acetic acid-induced writhing test, whereas the carbon tetrachloride fraction (200 and 400 mg/kg) demonstrated strong antinociceptive effect ($p < 0.001$; 73.09% inhibition) of writhing. The inhibition was higher than that of standard diclofenac sodium (55.56%). The methanol extract and chloroform fraction (400 mg/kg) showed significant (21.74% and 26.96%) inhibition of defecation and compared to the standard drug loperamide (37.39% inhibition, at 50 mg/kg concentration) with respect to diarrhea severity. The methanol extract, carbon tetrachloride, and chloroform fractions displayed moderate antimicrobial activity against the tested microorganisms (*B. sereus, B. megaterium, B. subtilis, S. aureus, Sarcina lutea, E. coli, P. aeruginosa, S. paratyphi, S. typhi, Shigella boydii, S. dysenteriae, V. mimicus, V. parahemolyticus, C. albicans, A. niger, and S. cerevisiae*) in respect to both inhibition zones (ranging from 9–13 mm, 10–13 mm, and 10–13 mm, at 400 µg/ disc concentration) and spectrum of activity. Methanol extract and its fractions (carbon tetrachloride and chloroform) of leaves of *C. indicum* possess significant antinociceptive, antidiarrheal, and antimicrobial activities (Pal et al. 2012). Phytochemicals (taraxerol, friedelin, and stigmasterol) from twelve *Clerodendrum* species were analyzed by GC-MS and investigated for their antiviral potential against novel coronavirus SARS-CoV-2. Taraxerol demonstrated higher binding energy scores with respect to viral proteins than the standard drugs. The present study offers a new avenue in further studies of taraxerol *in vitro* and *in vivo* toward its successful use as an anti-SARS-CoV-2 agent and in combating the catastrophic COVID-19 infections (Kar et al. 2021). Dichloromethane extracts of *C. indicum* and *C. villosum* demonstrated significant cytotoxicity and specificity against five (SW620, ChaGo-K-1, HepG2, KATO-III, and BT-474) cell lines. The oleanolic acid 3-acetate and betulinic acid showed significant cytotoxicity to all tested five cell lines (IC_{50} 1.66–20.49 µmol/l) while 3β-hydroxy-D:B-friedo-olean-5-ene and taraxerol were found cytotoxic to the SW620 cell line only (IC_{50} = 23.39 and 2.09 µmol/l). Lupeol displayed strong cytotoxicity against SW620 (IC_{50} = 1.99 µmol/l) and KATO-III cell lines (IC_{50} = 1.95 µmol/l), whereas pectolinarigenin showed moderate cytotoxicity against SW620 and KATO-III cell lines (IC_{50} = 13.05 and 24.31 µmol/l). The stigmasterol showed significant effectivity against the SW620 cell line (IC_{50} = 2.79 µmol/l) whereas β-sitosterol displayed activity against SW620 (IC_{50} = 11.26 µmol/l), BT-474 (IC_{50} = 14.11 µmol/l) and HepG2 cancer cell lines (IC_{50} = 20.47 µmol/l; Somwong and Suttisri 2018).

C. colebrookianum Walp (syn *C. glandulosum* Coleb.) is a perennial shrub, with quadrangular stem, robust branching, and sparsely pubescent with corky internodes, growing up to 1.5–3 m high. Leaves are opposite, broad-ovate, acute, entire, and petiolated. Inflorescence is terminal, compact, and corymbose cymes. Flowers are white, terminal compact, numerous, and pedicelated (Kalita et al. 2012). The shrub is distributed in Bangladesh, Bhutan, China, India, Indonesia, Malaysia, Myanmar, Nepal, Sri Lanka, and Vietnam (Hooker 1885; Jadeja et al. 2011; Yang et al. 2000). Its leaves are boiled in water, mixed with a ground piece (1–3) of *Allium sativum* and salt, taken (as soup or decoction, orally, twice a day for 90 days) for the treatment of hypertension (Nath and Bordoloi 1993; Nath and Choudhury 2010). Leaf extract (3 teaspoons) is mixed with common salt (small amount) and is taken thrice a daily for the treatment of abdominal pain (Kalita and Phukan 2010). Leaf juice and decoction are used in the treatment of diabetes, diarrhea, and dysentery (Saklani and Jain 1994; Sharma et al. 2001). Docking studies revealed that three chemical constituents (acteoside, martinoside, and osmanthuside β6 from *C. colebrookianum*) interacted with the antihypertensive drug targets with good glide score. Moreover, the compounds developed strong H-bond interactions with the key residues Met156/Met157 of ROCK I/ROCK II and Gln817 of PDE5. The study suggests that antihypertensive activity of the plant is due to acteoside and osmanthuside β6 interaction with ROCK and PDE5 drug targets (Arya et al. 2018). Aqueous extract and its aqueous, n-butanol, ethyl-acetate, and chloroform fractions of *C. colebrookianum* leaves (200 mg/kg/p.o.) showed significant suppression to carrageenan and histamine-stimualted inflammation and cotton pallet-induced granuloma formation in rats. Aqueous extract and ethyl-acetate and chloroform fractions demonstrated significant inhibition in COX-1 and COX-2 *in-vitro* assays. The aqueous extract, aqueous, n-butanol, ethyl-acetate, and chloroform fractions (100 µg concentration) demonstrated significant DPPH radical-scavenging activity (54.37%, 33.88%, 62.85%, 56.28%, and 57.48%). The study results established the anti-inflammatory activity of fractions in acute and chronic stages of inflammations (Deb et al. 2013). The aqueous extract and its aqueous, *n*-butanol, ethyl acetate, and chloroform fractions of *C. colebrookianum* leaves (100µg/ml) displayed calcium antagonism in rat ileum. The ethyl acetate fraction (75 µg/ml) demonstrated significant inhibition of rho-kinase-II (68.62%) and phosphodiesterase type 5 (52.28%) activities but did not show any effect on angiotension converting enzyme activity. The aqueous extract and its fractions also displayed negative inotropic and chronotropic actions on an isolated frog heart and significant ($p < 0.001$) decrease in systolic blood pressure and heart rate in hypertensive rats (Deb and Dutta 2012).

C. eriophyllum Gürke is a small shrub and grows up to 0.5–2 m high. It is distributed in dry bushlands of Eastern Kenya. In Kenyan medicine, it is used in the treatment of malaria (Beenje 1994). Methanol extract of *C. eriophyllum* root bark demonstrated weak *in vitro* antiprotozoal activity against *Plasmodium falciparum* D6 and W2 clones (IC_{50} 9.51–10.56 µg/ml). The methanol and aqueous extracts (100 mg/kg/bw) displayed significant chemoinhibition (*in vivo*, 90.1% and 61.5%) against *P. berghei* infected mice (Muthaura et al. 2007).

Methanol extract of *C. wallichii* leaves (800 mg/kg b.w.) displayed maximum decrease in body weight (4.76 ± 0.372 vs. 16.92 ± 0.846) and liver weight (3.06 ± 0.128 vs. 5.55 ± 0.311). Methanol extract (1000 mg/kg) significantly ($p < 0.01$) decreased the levels of serum glutamic pyruvic transaminase. Similarly, serum glutamic-oxaloacetic transaminase and alkaline phosphatase levels were significantly reduced by methanol extract (1000 mg/kg). Extract (400, 600, 800, and 1000 mg/kg) significantly enhanced antioxidant enzymes such as glutathione and superoxide dismutase. Methanol extract (600 and 800 mg/kg) significantly improved the levels of catalase in liver tissues, while it significantly minimized the levels of malondialdehyde (Tian et al. 2022).

10.2 PHYTOCHEMISTRY

The monosaccharide derivative (serratumin A) from aerial parts (Hui et al. 2000), apigenin-7-glucoside and [7-(β-D-glucopyranosyloxy)-5-hydroxy2-(4-hydroxyphenyl)-4H-1-benzopyran-4-one] from *C. serratum* roots (Bhujbal et al. 2010), vanillic acid, verbascoside, 4-coumaric acid, ferulic acid, nepetin, luteolin, apigenin, naringenin, hispidulin, hesperetin, and chrysin were reported from ethanol extract of *C. petasites* (Thitilertdecha et al. 2014, 2015).

β-(3′, 4′-dihydroxyphenyl)ethyl-*O*-α-L-rhamnopyranosyl (1→3)-β-D-(4-*O*-caffeoyl)-glucopyranoside, acteoside (verbascoside), β-(3′, 4′-dihydroxyphenyl)ethyl-*O*-α-L-rhamnopyranosyl (1→3)-β-D-(6-*O*-caffeoyl)-glucopyranoside, isoacteoside, β-(3′, 4′-dihydroxyphenyl) ethyl-*O*-α-L-rhamnopyranosyl (1→3)-β-D-glucopyranoside, and decaffeoylacteoside were obtained from methanol extract of *C. trichotomum* leaves. These isolated compounds showed antioxidant [DPPH reduction and TBARS assays on Cu $^{(2+)}$-induced oxidized LDL, PGE$_{(2)}$ assays] and anti-inflammatory (carrageenan-induced hind paw edema) activities (Kim et al. 2009). The caffeic acid and phenylpropanoid glycosides {1-*O*-caffeoyl glycoside and acteoside [β-(3′,4′-dihydroxyphenyl) ethyl-*O*-α-L-rhamnopyranosyl(1→3)-β-D-(4-*O*-caffeoyl)-glucopyranoside]} and ethyl acetate fraction evaluated on the aerosolized ovalbumin challenge in the ovalbumin-sensitized guinea pigs. Acteoside and 1-*O*-caffeoyl glycoside (25 mg/kg) significantly ($p < 0.05$) suppressed specific airway resistance (32.14 and 26.79%) in immediate-phase response and in late-phase response (55.88% and 52.94%). The caffeic acid (25 mg/kg) suppressed specific airway resistance (30.36%) in immediate-phase response and in late-phase response (44.12%) and compared to control. Moreover, 1-*O*-caffeoyl glycoside and acteoside [β-(3′,4′-dihydroxyphenyl) ethyl-*O*-α-L-rhamnopyranosyl(1→3)-β-D-(4-*O*-caffeoyl)-glucopyranoside] (25 mg/kg) significantly suppressed the generation of leukocytes (neutrophils and eosinophils) in the lungs of treated animals. In addition, 1-*O*-caffeoyl glycoside, acteoside, and caffeic acid significantly ($p < 0.05$) suppressed the levels of protein (at 25mg/kg dose) and histamine contents in bronchoalveolar lavage fluids. Acteoside showed more effectivity than caffeic acid and 1-*O*-caffeoyl glycoside. The study results revealed that caffeic acid and its glycosides (25 mg/kg) possess significant anti-asthmatic effect (Lee et al. 2011). Acteoside, leucosceptoside A, martynoside, acteoside isomer, and isomartynoside were obtained from *C. trichotomum* stems. The isolated compounds suppressed angiotensin converting enzyme activities in a dose-dependent manner (IC$_{50}$ 373 ± 9.3 µg/ml, 423 ± 18.8 µg/ml, 524 ± 28.1 µg/ml, 376 ± 15.6 µg/ml, 505 ± 26.7 µg/ml; Kang et al. 2003). Seven compounds [1-hydroxy-1-(8-palmitoyloxyethyl) cyclohexanone, 5-*O*-butyl cleroindin D, rengyolone, cleroindin C, cleroindin B, rengyol and isorengyol] were isolated and characterized from *C. trichotomum* leaves. 1-hydroxy-1-(8-palmitoyloxyethyl) cyclohexanone and 5-*O*-butyl cleroindin D showed cytotoxicity against A549 human tumor cell lines (Xu et al. 2014). Seven phenylpropanoid glycosides (acteoside, acteoside isomer, leucosceptoside A, plantainoside C, jionoside D, martynoside, and isomartynoside) were identified from *C. trichotomum*. Acteoside and acteoside isomer displayed strong inhibitory effects against HIV-1 integrase activity (IC$_{50}$ 7.8 ± 3.6 and 13.7 ± 6.0 µM; Kim et al. 2001). Nine abietane diterpenoids (Villosin C, Villosin B, cyrtophyllone B, uncinatone, teuvincenone B, sugiol, teuvincenone F, teuvincenone A, and teuvincenone H) have been obtained from *C. trichotomum* stems. The isolated compounds (villosin C, Villosin B, cyrtophyllone B, uncinatone) showed significant cytotoxicity (IC_{50} 8.79 to 35.46 μM) against tested cancer cell lines (A549, HepG-2, MCF-7 and 4T1; Li et al. 2014).

Pectolinaringenin, scutellarein, clerodin, clerodendrin, clerosterol, 4,2′,4′-trihydroxy-6′-methoxyxchalcone-4,4′α-D-diglucoside, clerodendrin, 24β-Ethylcholesta-5,22*E*,25-triene-3β-ol, lup-20(29)-en-3-triacontanoate,α-L-rhamnopyranosyl-(1→2)α-D-glucopyranosyl-7-*O*-naringin-4′-*O*-α-D-glucopyranoside-5-methyl ether, 7-hydroxyflavone, 7-hydroxyflavanone-7-*O*-glucoside) were obtained from *C. phlomidis*. The ethanol and aqueous extracts possess analgesic, antidiarrheal, antiplasmodial, hypoglycemic, minor tranquilizers, anti-asthmatic, antifungal, nematicidal, anti-amnestic, and anti-arthritic properties (Maruga Raja and Mishra 2010). Two new flavonoid glycosides {3,4,5-trihydroxy-6-[5-hydroxy-3-methoxy-2-(4- methoxy-phenyl)-4-oxo-4H-chromen-7-yloxy]-tetrahydro-pyran-2-carboxylic acid; 3,4,5-trihydroxy-6-[5-hydroxy-3-methoxy-2-(4-methoxy-phenyl)-4-oxo-4H-chromen-7-yloxy]-tetrahydro-pyran-2-carboxylic acid methyl ester} together with six other compounds (pectolinaringenin, pectolinaringenin-7-*O*-β-D-glucopyranoside, 24β-ethylcholesta-5,22*E*,25-triene-3β-ol, 24β-ethylcholesta-5,22*E*,25-triene-3β-*O*-β-D-glucopyranoside, (2*S*,3*S*,4*R*, 10*E*)-2-[(2′ *R*)-2′ -hydroxytetracosanoylamino]-10-octadecene-1,3,4-triol and andrographolide) have been identified from *C. phlomidis* (Bharitkar et al. 2015).

A phenylethanoid glycoside {2-(3-methoxy-4-hydroxylphenyl) ethyl-*O*-2″,3″-diacetyl-α-L-rhamnopyranosyl-(1→3)-4-*O*-(*E*)-feruloyl-β-D-glucopyranoside together with monomelittoside, melittoside, inerminoside A1, verbascoside, isoverbascoside, campneoside I from aerial parts (Nan et al. 2005), three iridoid biglycosides (inerminosides A1, C and D) from leaves (Caliş et al. 1994a), ridoid biglycosides {2′-*O*-[5″-*O*-(8-hydroxy-2,6-dimethyl-2(*E*)-octenoyl)-β-D-apiofuranosy l]-mussaenosidic acid (inerminoside A), and 2′-*O*-[5″-*O*-(8-hydroxy-2,6-dimethyl-2(*E*),6(*E*)-octadienoyl)-β-D- apiofuranosyl]-8-epi-loganicacid(inerminosideB)} fromleaves(Calişetal. 1994b), two chalcones (3-hydroxy-3′,4′-dimethoxychalcone and 3,2′-dihydroxy-3′,4′-dimethoxychalcone) together with two flavones (7-*O*-methylwogonin and eucalyptin) from flowers (Shahabuddin et al. 2013), two megastigmane glucosides (sammangaosides A and B), a iridoid glucoside (sammangaoside C) along with 15 other compounds {verbascoside, isoverbascoside, leucosceptoside A, decaffeoylverbascoside, darendoside B, monomelittoside, melittoside, (7*S*, 8*R*)-dehydrodiconiferyl alcohol 9-*O*-β-glucopyranoside, (7*S*, 8*R*)-dehydrodiconiferyl alcohol 4-*O*-β-glucopyranoside, benzyl alcohol β-glucopyranoside, benzyl alcohol β-(2′ -*O*-β-xylopyranosyl) glucopyranoside, salidroside, (Z)-3-hexenyl-β-glucopyranoside and 2,6-dimethoxy-*p*-hydroquinone 1-*O*-β-glucopyranoside, seguinoside K; Kanchanapoom et al. 2001}, nine other compounds (verbascoside, isoverbascoside, decaffeoylverbaseoside, campneoside 1, cistanoside E, markhamioside F, isocistanoside F, benzyl glucoside and purpureaside B) from aerial parts (Nan et al. 2008), two sterols (4α-methyl-24β-ethyl-5α-cholesta-14, 25-dien-3β-ol and 24β-ethylcholesta-5, 9(11), 22*E*-trien-3b-ol) and a aliphatic ketone (11-pentacosanone) together with aliphatic ketone (6-nonacosanone) and a diterpene (clerodermic acid, Δ^4 -cholesten-3-one) from aerial parts (Pandey et al. 2003), β-friedoolean-5-ene-3-β-ol, β-sitosterol, stigmasta-5,22,25-trien-3-β-ol, betulinic acid, and 5-hydroxy-6,7,4′-trimethoxyflavone have been obtained from methanol extract of *C. inerme* aerial parts. The methanol extract and 5-hydroxy-6,7,4′-trimethoxyflavone displayed scavenging activity in a DPPH assay (Ibrahim et al. 2014). Three neo-clerodane diterpenoids (inermes A, B, epimeric mixture of 14,15-dihydro-15-hydroxy-3-epicaryoptin, and 14,15-dihydro-15β-methoxy-3-epicaryoptin, 14,15-dihydro-15-hydroxy-3-epicaryoptin from the hexane extract leaves (Pandey et al. 2005), a neolignane (5,8-epoxy-6,7-dimethyl 2′,3′,2″,3″-dimethylene dioxy-4′,1″-dimethoxy-1,2:3,4-dibenzo-1,3-cyclooctadiene) was reported from the petroleum ether extract of *C. inerme* seeds (Spencer and Flippen-Anderson 1981).

A sterol glycoside clerosterol 3β-*O*-(β-D-glucoside) along with clerosterol, sitosterol, octacosanol, and fatty acids from leaves (Goswami et al. 1996), colebrin A–E have been isolated and characterized from the aerial parts of *C. colebrookianum* (Yang et al. 2000). Ten abietane diterpenoids (12-hydroxy-8,12-abietadiene-3,11,14- trione, royleanone, taxodione, 11- hydroxy-7,9(11),13-abietatrien-12-one, sugiol, ferruginol, 6-hydroxysalvinolone, 6,11,12,16-tetrahydroxy-5,8,11,13-abietatetra-en7-one, uncinatone, and 11-hydroxy8,11,13-abietatriene-12-*O*-β-xylopyranoside) have been separated and identified from *C. eriophyllum*. Taxodione and ferruginol demonstrated strong antifungal activity (IC$_{50}$/MIC 0.58/1.25 and 0.96/2.5 μg/ml) against *C. neoformans*, whereas taxodione, ferruginol, 6-hydroxysalvinolone displayed strong antibacterial effect against *S. aureus* and methicillin-resistant *S. aureus* (IC$_{50}$/MIC 1.33–1.75/2.5–5 and 0.96–1.56/2.5 μg/ml). Moreover, taxodione and uncinatone demonstrated strong antileishmanial effects (IC$_{50}$ 0.08 and 0.20 μg/ml) against *Leishmania donovani* (Machumi et al. 2010).

The abietane derivatives (bungnates A, B, 15-dehydrocyrtophyllone A and 15-dehydro-17-hydroxycyrtophyllone A), two phenylethanoid glycosides (bunginoside A and 3″,4″-di-O-acetylmartynoside) together with nine abietane derivatives and fourteen phenylethanoid glycosides [12,16-epoxy-1-1,14,17-trihydroxy6-methoxy-17(15→16)-abeo-abieta-5,8,11,13-tetraene-7-one, cyrtophyllone A, villosin C, trichotomoside, *O*-2-(3-hydroxy-4- methoxyphenyl)-ethyl *O*-2,3-di-*O*-acetyl-α-L-rhamnopyranosyl-(1→3)-(4-*O*-cis-feruloyl)-β-D-glucopyranoside, isomartynoside, isoacteoside, darendoside B, phlomisethanoside, darendoside A, 3″,4″-di-*O*-acetylmartynoside, acetylmartynoside B, verbascoside, acetylmartynoside A, 3″-*O*-acetylmartyonside, 2″-*O*-acetylmartynoside, martynoside, leucosceptoside A, 12,16-epoxy-11,14-dihydroxy-6-methoxy17(15→16)-abeo-abieta-5,8,11,13,15-pentaene-3,7-dione, 12-*O*-β-D-glucopyranosyl-3,11,16-trihydroxyabieta-8,11,13-triene, uncinatone, bunginoside A, teuvincenone F, 19-hydroxyteuvincenone F, mandarone E, bungnate A, bungnate B, 15-dehydrocyrtophyllone A, 15-dehydro-17-hydroxycyrtophyllone A, 12,16-epoxy-11,14,17-trihydroxy6-methoxy-17(15→16)-abeo-abieta-5,8,11,13-tetraene-7-one, cyrtophyllone A, villosin C] were identified from the ethanol extract of *C. bungei* Steud roots (Liu et al. 2014). Two royleanones (bungone A and B) from the stem (Tianpei et al. 1999) and five compounds (12-*O*-β-D-glucopyranosyl-3,11,16-trihydroxyabieta-8,11,13-triene, 3,12-*O*-β-D-diglucopyranosyl-11,16-dihydroxyabieta-8,11,13-triene, ajugaside A, uncinatone, and 19-hydroxyteuvincenone F) were separated and identified from acetone extract of *C. bungei* roots. The isolated compounds displayed suppressive effect against a complement system (IC$_{50}$ 24 μm, 138 μm, 116 μm, 87 μm and 232 μm). Among the isolated compounds, 12-*O*-β-D-glucopyranosyl-3,11,16-trihydroxyabieta-8,11,13-triene displayed the strongest anticomplement activity (IC$_{50}$ 24 μm; Kim et al. 2010). 2-phenylethyl 3-*O*-(6-deoxy-α-L-mannopyranosyl)-β-D-glucopyranoside, 6″-*O*-[(*E*)-caffeoyl] rengyoside B, clerodenone A, 2-({6-*O*-[(4-hydroxy-3-methoxyphenyl)carbonyl]-β-D-glucopyranosyl}oxy)-2-methylbutanoic acid, and 2-{(2*S*,5*R*)-5-[(1*E*)-4-hydroxy-4-methylhexa-1,5-dien-1-yl]-5-methyltetrahydrofuran-2-yl}propan-2-yl β-D-glucopyranoside were isolated and characterized from *C. bungei* roots. 2-phenylethyl 3-*O*-(6-deoxy-α-L-mannopyranosyl)-β-D-glucopyranoside displayed modest *in vitro* suppression of the growth of the HeLa human cervical carcinoma 14,15-dihydro-15-hydroxy-3-epicaryoptin (CCL-2; IC$_{50}$ 3.5–8.7 μM; Liu et al. 2009). The 5-*O*-ethylcleroindicin D and bungein A, acteoside, betulinic acid, cleroindicin A, cleroindicin C, cleroindicin E, cleroindicin F, clerosterol, clerosterol 3β-*O*-β-D-glucopyranoside, martinoside, octadecnoic acid, n-pentacosane, and 5,7,4′ -trihydroxyflavone were obtained from *C. bungei* aerial parts (Yang et al. 2002).

Three pheophorbide-related compounds and (10*S*)-hydroxypheophytin were isolated and characterized from *C. calamitosum* L. leaves and stems. Pheophorbide-related compounds demonstrated potent cytotoxic activity against human lung carcinoma (A549), ileocecal carcinoma (HCT-8), kidney carcinoma (CAKI-1), breast adenocarcinoma (MCF-7), malignant melanoma (SK-MEL-2), ovarian carcinoma (1A9), and epidermoid carcinoma of the nasopharynx (KB), and its etoposide-(KB-7d), vincristine- (KB-VCR), and camptothecin-resistant (KB−CPT) cell lines (Cheng et al. 2001). 3′,5′, 5-trihydroxy 4′- methoxy flavonone 7-*O*-β-D gluconopyranosyl methyl glucopyranose and 3′, 5′, 7-trihydroxy-4 methoxy flavonone have been obtained from ethanol extract of *C. splenden* G.Don aerial parts. The isolated compound 3′,5′, 5-trihydroxy 4′- methoxy flavonone 7-*O*-β-D gluconopyranosyl methyl glucopyranose possesses antioxidant, anti-inflammatory, antimalarial, and antimicrobial properties (Okwu and Iroabuchi 2008). Triancontanol, (22*E*, 24*S*)-stigmasta-5, 22, 25-trien-3β-ol, and 3-*O*-D-glucopyranoside of (22*E*, 24*S*)-stigmasta-5,22,25-trien-3β-ol have been reported from methanol extract of *C. splendens* leaves. The isolated compounds showed significant reduction effect on Fe^{2+}. In addition, triancontanol, and (22E, 24S)-stigmasta-5, 22, 25-trien-3β-ol displayed greater IC_{50} than that of vitamin C (5.613 ± 0.117). (Oscar et al. 2018).

Iridoid diglucoside, iridoid glucosides, and cyclohexylethanoids (monomelittoside, melittoside, harpagide, 5-O-β-glucopyranosyl-harpagide, racemic rengyolone, racemic dihydrorengyolone, rengyoxide, rengyoside B, cornoside, and dihydrocornoside) were obtained from methanol extract of aerial parts (Kanchanapoom et al. 2005); verbascoside, isoverbascoside, decaffeoylverbascoside, cornoside, rengyolone, hispidulin, lupeol, and icariside B5 were separated and identified from *C. chinense* (Osbeck) Mabb. leaves. The methanol extract of *C. chinense* leaves and verbascoside displayed significant analgesic, anti-inflammatory, and antipyretic activities (Wahba et al. 2011).

Two neoclerodane diterpenoids (clerodinins A and B) along with three other compounds (clerodin, stigmasta-5, 22, 25-trien-3β-ol and 3-epi-glutinol) have been isolated from the hexane extract of *C. brachyanthum* Schauer leaves (Lin et al. 1989). Six compounds (cleroindicins A–F) were isolated and characterized from the aerial parts of *C. indicum* (L.) Kuntze (Tian et al. 1997). β-sitosterol, lupeol, oleanolic aldehyde acetate, (22E)-stigmasta-4,22,25-trien-3-one, stigmasta-4,25-dien-3-one, and (3β)- stigmasta-4,22,25-trien-3-ol were reported from *C. paniculatum* L roots (Prashith Kekuda and Sudharshan 2018). Ten compounds viz., 3β-taraxerol, 3β-taraxerol acetate, (3β)-stigmasta-4,22,25-trien3ol, 1-monoacetin, seguinoside K, trichotomoside, 2",3"-O-acetylmartynoside, 2",4"-O-acetylmartynoside, martynoside, and acteoside were obtained from *C. longisepalum* Dop. Acteoside showed significant anti-α-glucosidase activity (Phaopongthai et al. 2017). Two abietane derivatives (cyrtophyllones A and B) together with six other compounds (teuvincenone F, unicinatone, sugiol, friedelin, clerodolone, stigmasta-5, 22, 25-trien-3β-ol and clerosterol) from the stem (Tian et al. 1993) and twelve phenolic components (jionoside C, jionoside D, luteolin, cirsilineol, cirsimartin, acteoside, martynoside, cirsilineol-4′-O-β-D-glucoside, cirsimarin, jaceosidin 7-O-β-D-glucoside) were separated and identified from the ethyl acetate fraction of *C. cyrtophyllum* Turcz leaves. The acteoside showed significant antioxidant activity in DPPH and ABTS radical scavenging assays (IC_{50} 79.65 ± 3.4 and 23.00 ± 1.5 µg/ml; Zhou et al. 2020).

A hydroquinone diterpenoid (7,11-dihydroxy-3,4,9,11b-tetramethyl-1,2,8,9,11β-pentahydrophenanthro [3,2-$β$]furan-6(2H)-one) has been obtained from *C. uncinatum* Schinz (syn—*Kalaharia uncinata* (Schinz) Moldenke). The isolated diterpene showed strong antifungal activity against *Cladosporium cucumerinum* (Dorsaz et al. 1985). An abietane diterpenoid [(3S,16R)-12,16-epoxy-3,6,11,14,17-pentahydroxy17(15→16)-abeo-5,8,11,13-abietatetraen-7-one] along with four diterpenoids (teuvincenone G, teuvincenone H, 14-deoxycoleon U, 12-methylcoleon U) were isolated and characterized from *C. kaichianum* stem. (3S,16R)-12,16-epoxy-3,6,11,14,17-pentahydroxy17(15→16)-abeo-5,8,11,13-abietatetraen-7-one displayed significant cytotoxicity against the HL-60 and A-549 tumor cell lines (Xu et al. 2011a). (16R)-12,16-epoxy-11,14,17-trihydroxy-17(15→16)-abeo-8,11,13-abietatrien-7-one, villosin A, salvinolone, 14-deoxyloleon U, 5,6-dehydrosugiol, and coleon U were reported from *C. kaichianum*. (16R)-12,16-epoxy-11,14,17-trihydroxy-17(15→16)-abeo-8,11,13-abietatrien-7-one, and villosin A demonstrated strong cytotoxic activities against the HL-60 tumor cell line (Xu et al. 2011b). An abeo-abietane diterpenoid (12-methoxy-6,11,14,16-tetrahydroxy-17(15Ñ16)-abeo5,8,11,13-abietatetraen-3,7-dione along with cryptojaponol and fortunin E were reported from the aqueous-ethanol extract of *C. kiangsiense*. The isolated compounds displayed significant cytotoxic activity against human cancer cells lines viz., HL-60, SMMC-7721, A-549 and MCF-7 (Xu et al. 2016).

6′-O-caffeoyl-12-glucopyranosyloxyjasmonic acid, isoacteoside, acteoside, 2"-O-acetylmartyonside, 3"-O-acetyl-martyonside, martynoside, brachynoside, leucosceptoside A, jionoside C, jionoside D, incanoside C, apigenin 7-O-glucuronide, and acacetin 7-O-glucuronide were obtained from *C. infortunatum* (Uddin et al. 2020). The ethanol extract of *C. fragrans* leaves showed the presence of significant levels of quercetin (13.47%; Sapiun et al. 2020). Pilocarpine, glyceric acid, pangamic acid, quercetin, and gallic acid were isolated and identified by HPTLC from ethanol extract of *C. paniculatum* flowers. The ethanol extract showed hepatoprotective activity in carbon tetrachloride-intoxicated model systems (Kopilakkal et al. 2021).

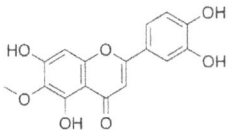

Nepetin

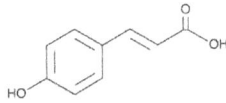

4-Coumaric acid

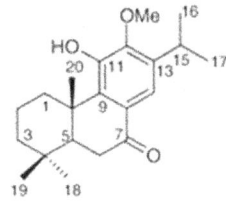

Cryptojaponol

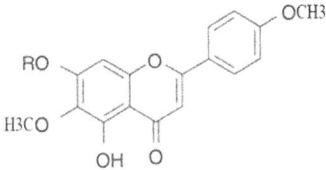

R = D-Glucose - Pectolinaringenin-7-*O*-β-D-glucopyranoside

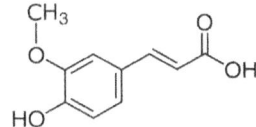

Ferulic acid

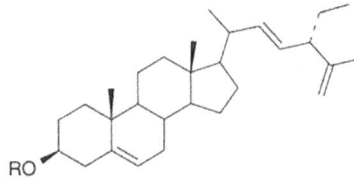

R = H - 24β-Ethylcholesta-5,22*E*,25-triene-3β-ol; R = D-Glucose - 24β-Ethylcholesta-5,22*E*,25-triene-3β-*O*-β-D-glucopyranoside

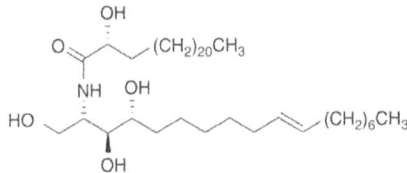

(2*S*,3*S*,4*R*,10*E*)-2- [(2′ *R*)-2′ -Hydroxytetracosanoylamino]-10-octadecene-1,3,4-triol

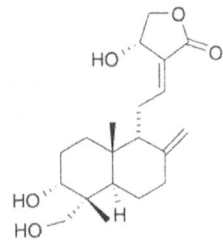

Andrographolide

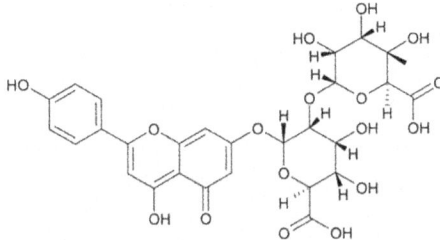

Clerodendrin

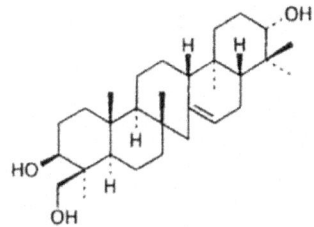

Serratumin A

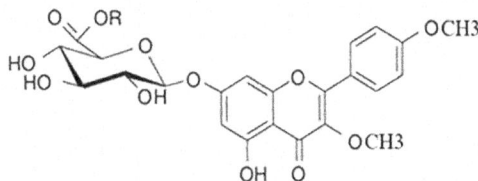

R = H - 3,4,5-Trihydroxy-6-[5-hydroxy-3-methoxy-2-(4- methoxy-phenyl)-4-oxo-4H-chromen-7-yloxy]-tetrahydro-pyran-2-carboxylic acid; R = CH₃ - 3,4,5-Trihydroxy-6-[5-hydroxy-3-methoxy-2-(4- methoxy-phenyl)-4-oxo-4H-chromen-7-yloxy]-tetrahydro-pyran-2-carboxylic acid methyl ester

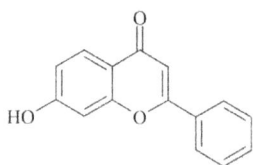

7-Hydroxyflavone

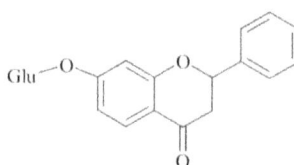

7-Hydroxyflavanone-7-*O*-glucoside

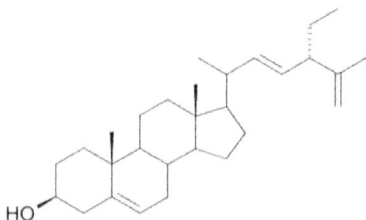

24β-Ethylcholesta-5,22E,25-triene-3β-ol

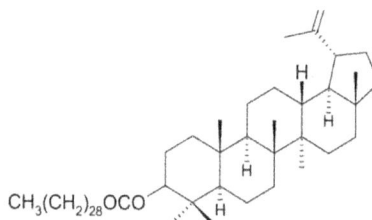

Lup-20(29)-en-3-triacontanoate

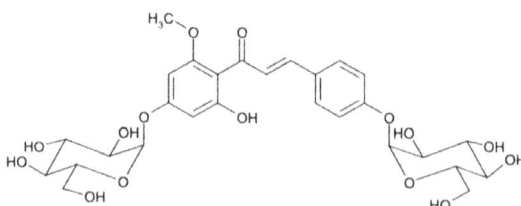

4,2′,4′-Trihydroxy-6′-methoxyxchalcone-4,4′α-D-diglucoside

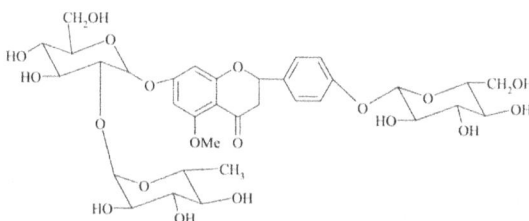

α-L-Rhamnopyranosyl-(1→2)α-D-glucopyranosyl-7-*O*-naringin-4′-*O*-α-D-glucopyranoside-5-methyl ether

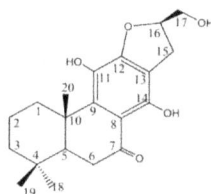

(16*R*)-12,16-Epoxy-11,14,17-trihydroxy-17(15→16)-abeo-8,11,13-abietatrien-7-one

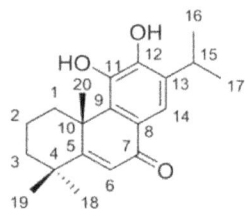

Salvinolone

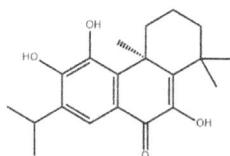

14-Deoxycoleon U

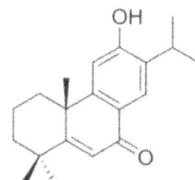

5,6-Dehydrosugiol

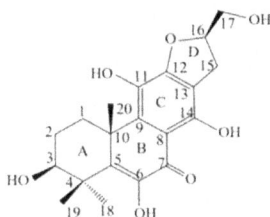

(3S,16R)-12,16-epoxy-3,6,11,14,17-pentahydroxy17(15→16)-abeo-5,8,11,13-abietatetraen-7-one

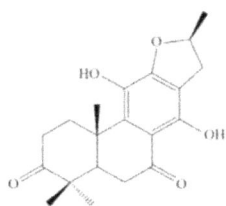

Teuvincenone G

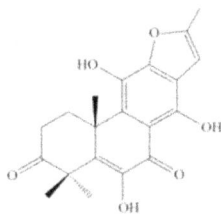

Teuvincenone H

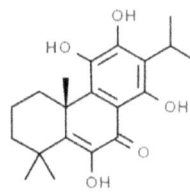

Coleon U

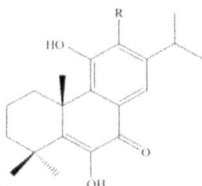

R = OH - 14-Deoxycoleon U; R = OCH$_3$ - 12-Methylcoleon U

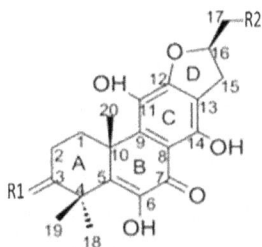

R^1 = H$_2$, R^2 = OH - Villosin C; R^1 = H$_2$, R^2 = H$_2$ - Teuvincenone B; R^1 = O, R^2 = H - Teuvincenone A

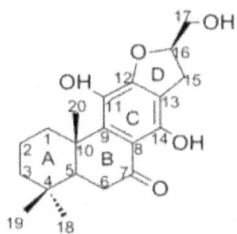

Villosin B

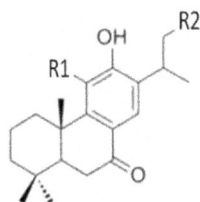

R^1 = OH, R^2 = OH - Cyrtophyllone B

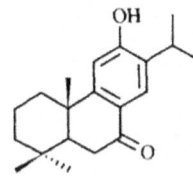

Sugiol

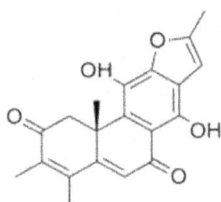

Teuvincenone F

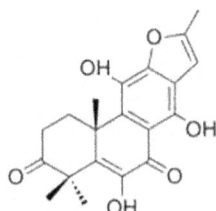

Teuvincenone H

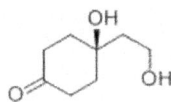

Cleroindicin B

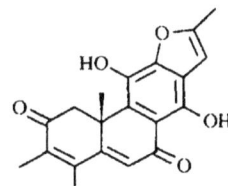

Teuvincenone F

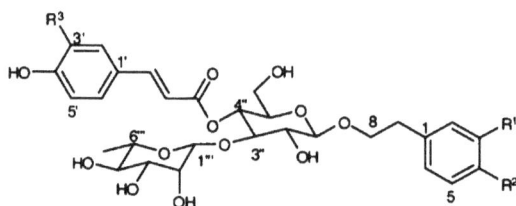

$R^1 = R^2 = H$, $R^3 = OH$ - Jionoside C; $R^1 = R^3 = OH$, $R^3 = OCH_3$ - Jionoside D

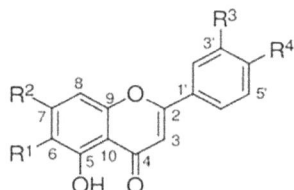

$R^1 = H$, $R^2 = R^3 = R^4 = OH$ - Luteolin; $R^1 = R^4 = OH$, $R^2 = R^3 = OCH_3$ - Cirsilineol; $R^1 = R^2 = R^3 = OCH_3$, $R^4 = OH$ - Cirsimartin; $R^1 = OH$, $R^2 = R^3 = OCH_3$, $R^4 = O$-Glc - Cirsilineol-4'-O-β-D-glucoside; $R^1 = R^2 = R^3 = OCH_3$, $R^4 = O$-Glc - Cirsimarin; $R^1 = R^3 = OCH_3$, $R^2 = O$-Glc, $R^4 = OH$ - Jaceosidin 7-O-β-D-glucoside

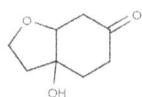

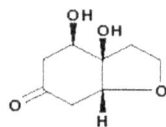

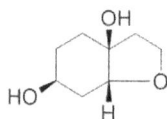

Cleroindicin C   Cleroindicin D   Cleroindicin E   Cleroindicin F

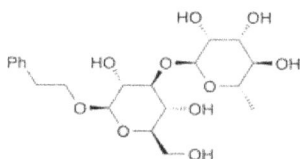

2-Phenylethyl 3-O-(6-deoxy-α-L-mannopyranosyl)-β-D-glucopyranoside

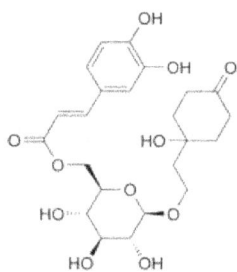

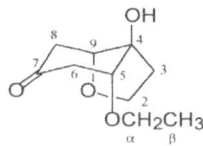

6''-O-[(E)-caffeoyl] rengyoside B   5-O-Ethylcleroindicin D

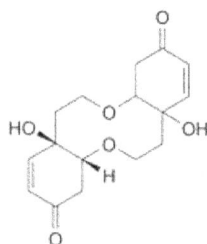

Clerodenone A

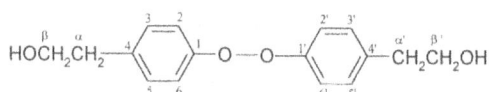

Bungein A

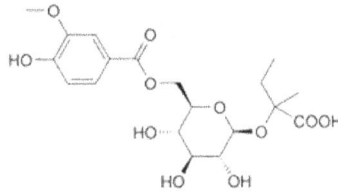

2-({6-*O*-[(4-Hydroxy-3-methoxyphenyl)carbonyl]-β-D-glucopyranosyl}oxy)-2-methylbutanoic acid

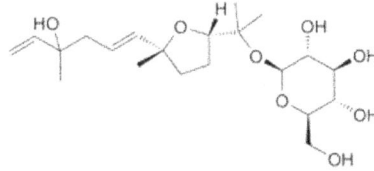

2-{(2*S*,5*R*)-5-[(1*E*)-4-Hydroxy-4-methylhexa-1,5-dien-1-yl]-5-methyltetrahydrofuran-2-yl} propan-2-yl β-D-glucopyranoside

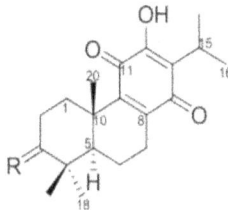

R = O - 12-Hydroxy-8,12-abietadiene-3,11,14-trione; R = H2 – Royleanone

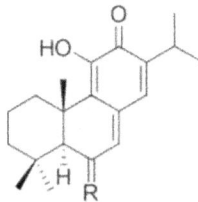

R = O - Taxodione; R = H$_2$ - 11- Hydroxy-7,9(11),13-abietatrien-12-one

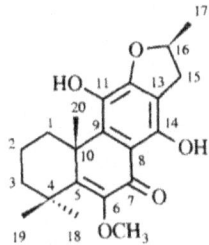

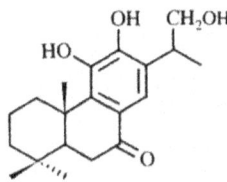

Cyrtophyllone A   Cyrtophyllone B   Clerodolone

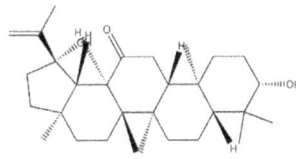

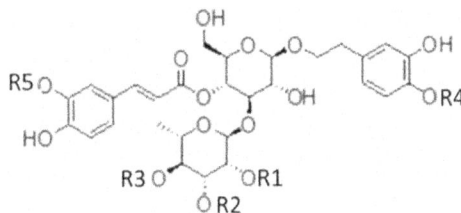

R^1 = R^2 = Ac, R^3 = H, R^4 = R^5 = CH$_3$ - 2",3"-*O*-Acetylmartynoside; R^1 = H, R^2 = R^3 = Ac, R^4 = R^5 = CH$_3$ - 2",4"-*O*-Acetylmartynoside

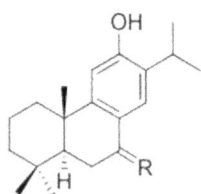

R = H$_2$ – Ferruginol

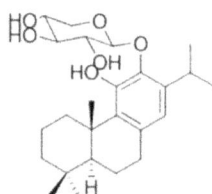

11-Hydroxy8,11,13-abietatriene-12-O-β-xylopyranoside

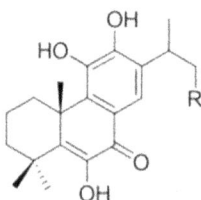

R = H - 6-Hydroxysalvinolone; R = OH - 6,11,12,16-Tetrahydroxy-5,8,11,13-abietatetra-en7-one

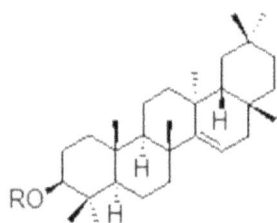

R = Ac - 3β-Taraxerol acetate

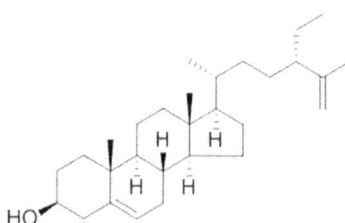

Clerosterol

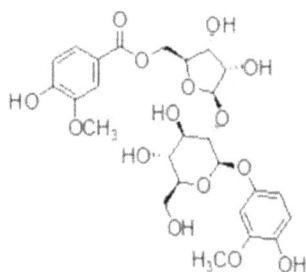

Seguinoside K

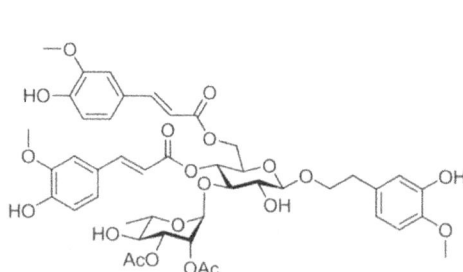

Trichotomoside

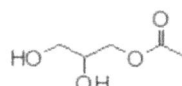

1-Monoacetin

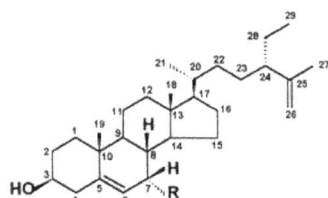

R = $\overset{O}{\underset{OCH}{\|}}$ - Colebrin A

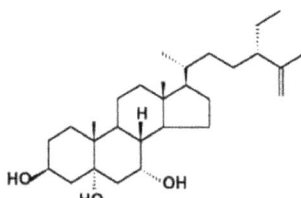

Colebrin B

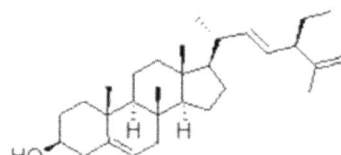

(3β)-Stigmasta-4,22,25-trien-3-ol

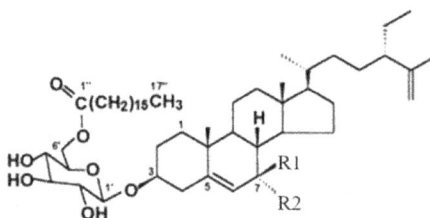

R^1 = H, R^2 = H - Colebrin C; R^1 = H, R^2 = OH - Colebrin D; R^1 = OH, R^2 = H - Colebrin E

O-2-(3-Hydroxy-4- methoxyphenyl)-ethyl *O*-2,3-di-*O*-acetyl-α-L-rhamnopyranosyl- (1→3)-(4-*O*-cis-feruloyl)-β-D-glucopyranoside

R^1 = CH₃, R^2 = Feruloyl - Isomartynoside; R^1 = H, R^2 = Caffeoyl - Isoacteoside; R^1 = CH₃, R^2 = H - Darendoside B

R = Vanilloyl - Phlomisethanoside; R = H - Darendoside A

R^1 = H, R^2 = R^3 = Ac, R^4 = R^5 = CH₃ - 3″,4″-Di-*O*-Acetylmartynoside; R^1 = H, R^3 = Ac, R^2 = H, R^4 = R^5 = CH₃ - Acetylmartynoside B; R^1 = R^2 = R^3 = R^4 = R^5 = H - Verbascoside; R^1 = R^2 = Ac, R^3 = H, R^4 = R^5 = CH₃ - Acetylmartynoside A; R^1 = R^3 = H, R^2 = Ac, R^4 = R^5 = CH₃ - 3″-O-Acetylmartyonside; R^2 = R^3 = H, R^1 = Ac, R^4 = R^5 = CH₃ - 2″-*O*-Acetylmartynoside; R^1 = R^2 = R^3 = H, R^4 = R^5 = CH₃ - Martynoside; R^1 = R^2 = R^3 = R^4 = H, R^5 = CH₃ - Leucosceptoside A

12,16-Epoxy-11,14-dihydroxy-6-methoxy17(15 →16)-abeo-abieta-5,8,11,13,15-pentaene-3,7-dione

12-O-β-D-Glucopyranosyl-3,11,16-trihydroxyabieta-8,11,13-triene

Uncinatone

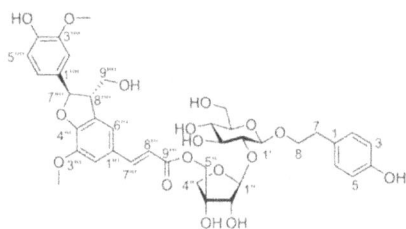

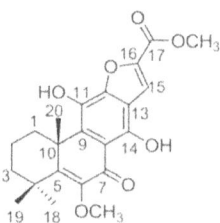

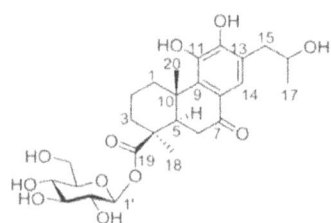

Bunginoside A   Bungnate A   Bungnate B

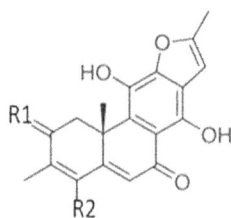

R^1 = O, R^2 = CH_3 - Teuvincenone F; R^1 = O, R^2 = CH_2OH - 19-Hydroxyteuvincenone F; R^1 = H_2, R^2 = CH_3 - Mandarone E

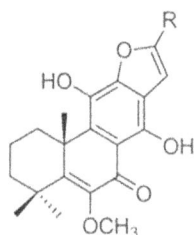

R = CH_3 - 15-Dehydrocyrtophyllone A, R = CH_2OH - 15-Dehydro-17-hydroxycyrtophyllone A

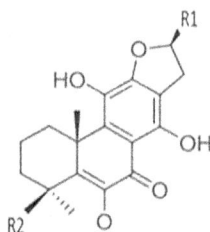

R^1 = CH_2OH, R^2 = CH_3 - 12,16-epoxy-11,14,17-trihydroxy6-methoxy-17(15→16)-abeo-abieta-5,8,11, 13-tetraene-7-one; R^1 = R^2 = CH_3 - Cyrtophyllone A; R^1 = CH_2OH, R^2 = H - Villosin C

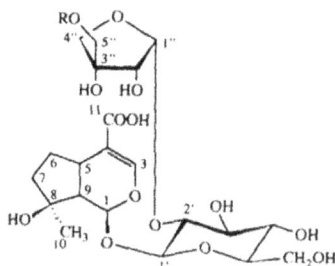

R = H – Inerminoside A1; R = p-Hydroxybenzoyl - Inerminoside D

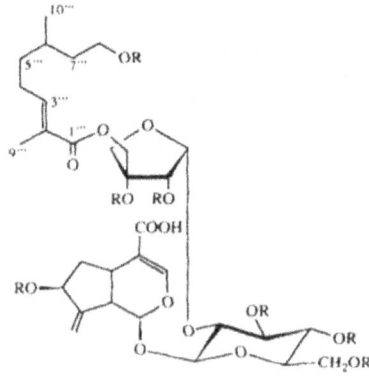

R = H - Inerminoside C; R = Ac - Inerminoside C heptaacetate

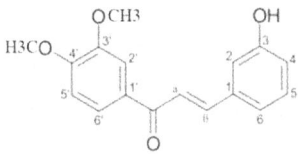

3-Hydroxy-3′,4′- dimethoxychalcone

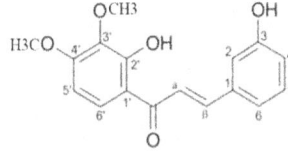

3, 2′-Dihydroxy-3′,4′-dimethoxychalcone

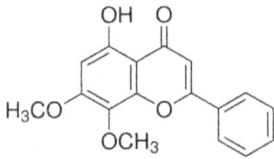

7-*O*-Methylwogonin

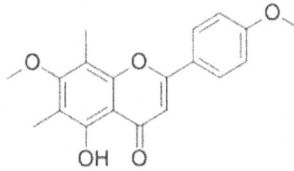

Eucalyptin

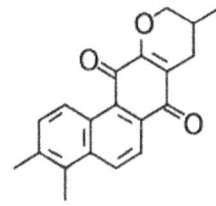

Bungone A

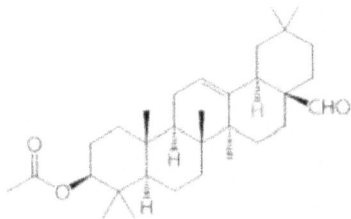

Oleanolic aldehyde acetate

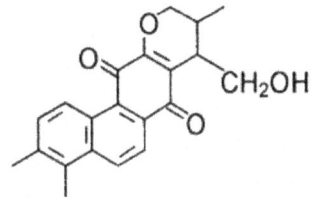

Bungone B

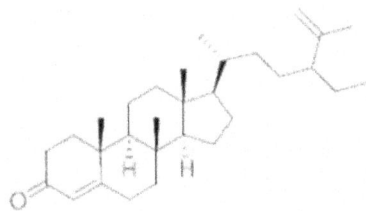

Stigmasta-4,25-dien-3-one

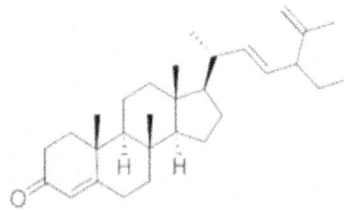

(22E)-Stigmasta-4,22,25-trien-3-one

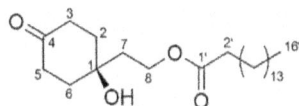

1-Hydroxy-1-(8-palmitoyloxyethyl) cyclohexanone

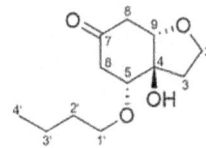

5-*O*-Butyl cleroindin D

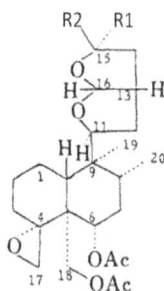

R1 = H, R² = OCH₃ – Clerodinin A; R1 = OCH₃, R² = H - Clerodinin B

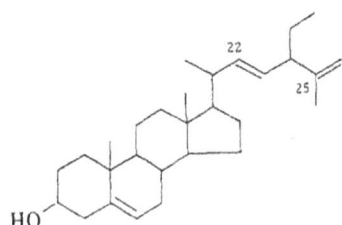

Stigmasta-5,22,25-trien-3β-ol

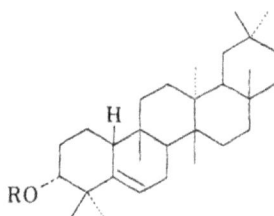

R = H – 3-*epi*-Glutinol

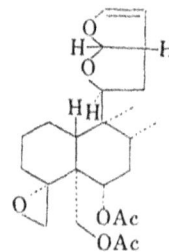

Clerodin

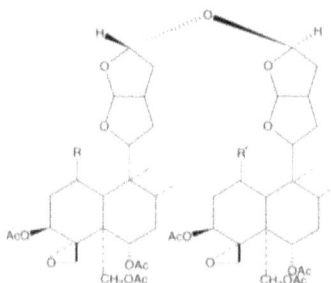

R, R' = H – Inermes A; R, R' = OCH₃,H/H,OCH₃ – Inermes B

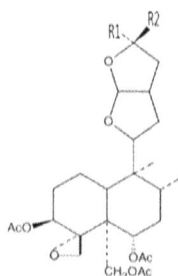

R¹ = H, R² = OCH₃ - 14,15-Dihydro-15β-methoxy-3-epicaryoptin; R¹, R² = OH, H/H, OH - 14,15-Dihydro-15-hydroxy-3-epicaryoptin

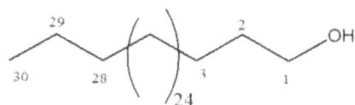

Triancontanol

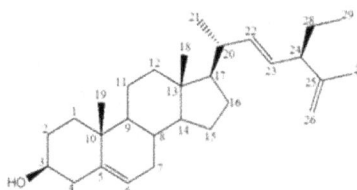

(22*E*, 24*S*)- Stigmasta–5,22,25-trien-3β–ol

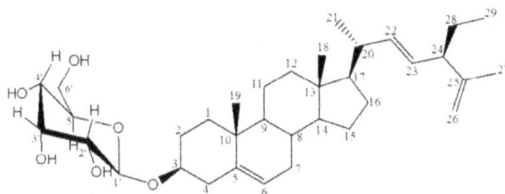

3-*O*-D-Glucopyranoside of (22*E*, 24*S*)-stigmasta – 5, 22, 2 -trien-3β–ol

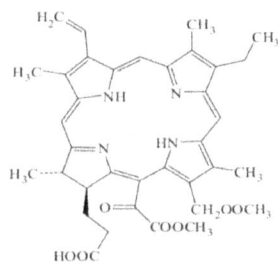

Pheophorbide related compound I

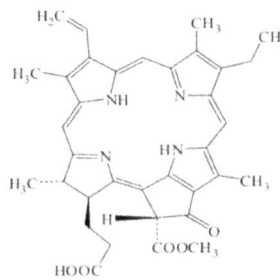

Pheophorbide related compound II

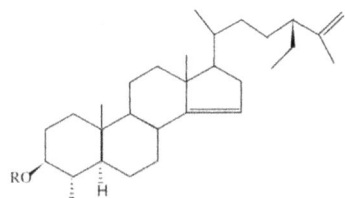

R = H - 4α-Methyl-24β-ethyl-5α-cholesta-14, 25-dien-3β-ol

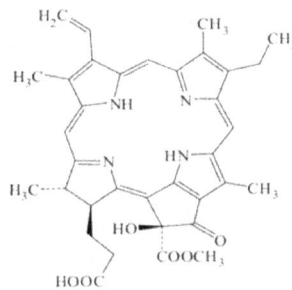

Pheophorbide related compound III

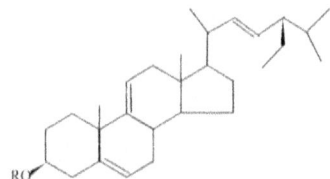

R = H - 24β-Ethylcholesta-5, 9 (11), 22E-trien-3β-ol

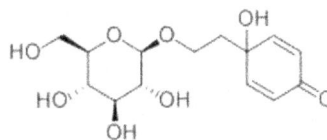

Cornoside

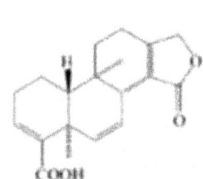

Clerodermic acid

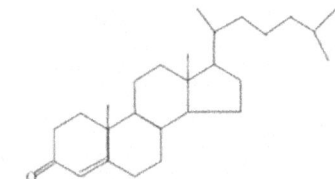

Δ^4 -Cholesten-3-one

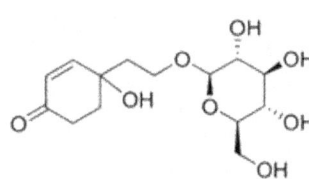

Dihydrocornoside

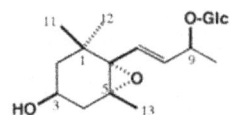

Sammangaoside A

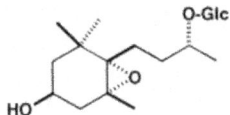

Sammangaoside B

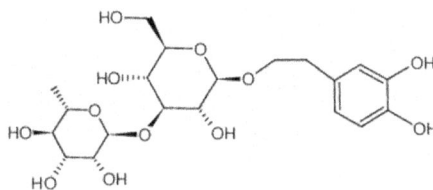

Decaffeoyl verbascoside

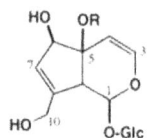

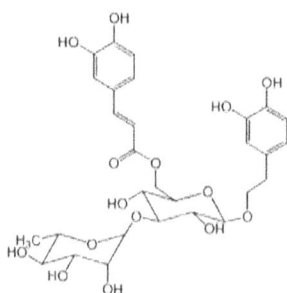

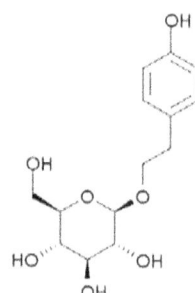

R = Glc(3′-1″)-glc - Sammangaoside C   Isoverbascoside   Salidroside

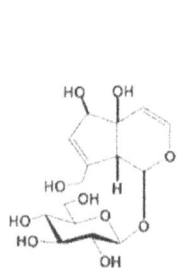

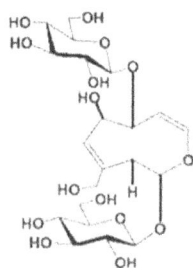

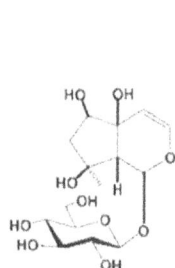

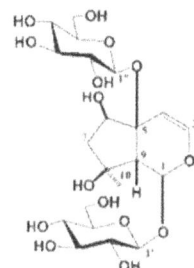

Monomelittoside   Melittoside   Harpagide   5-O-β-Glucopyranosyl-harpagide

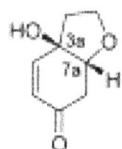

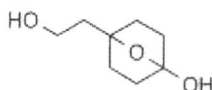

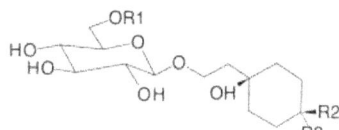

Racemic rengyolone   Rengyoxide $R^1 = R^2 = = O, R^3 = O$ - Rengyoside B

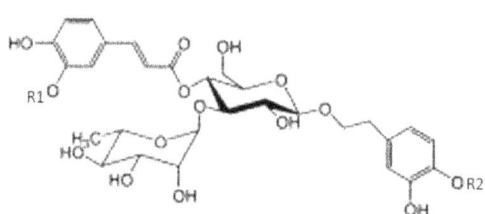

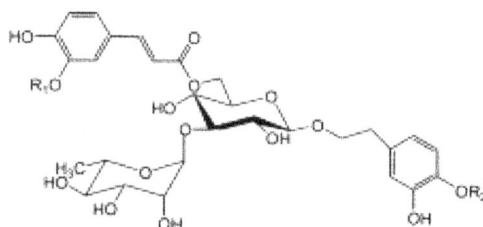

$R^1 = H, R^2 = H$ – Acteoside $R^1 = H, R^2 = H$ - Acteoside isomer

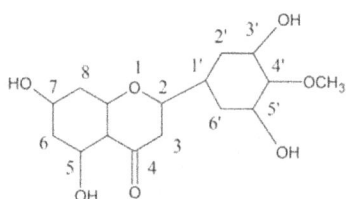

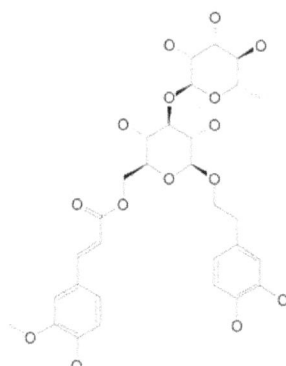

3′, 5′, 7-Trihydroxy - 4 methoxy flavonone   Plantainoside C

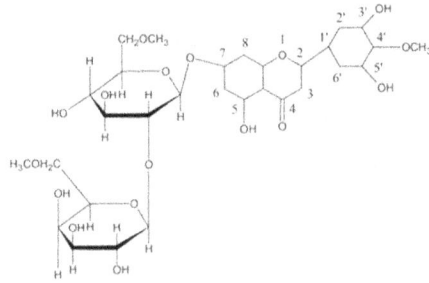

3', 5', 5- Trihydroxy -4' methoxy flavonone 7-O- β-D- glucucronopyranosyl methyl glucopyranose

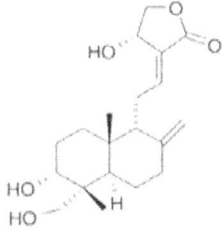

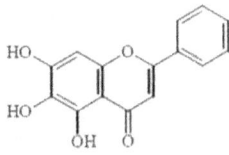

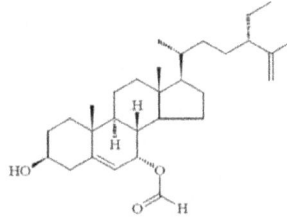

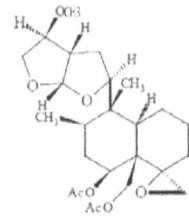

Andrographolide   Scutellarein   Colebrin A   Clerodinin A

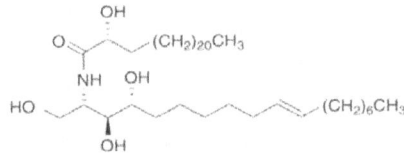

(2*S*,3*S*,4*R*,10*E*)-2- [(2' *R*)-2' -Hydroxytetracosanoylamino]-10-octadecene-1,3,4-triol

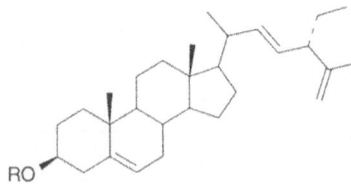

R = H - 24β-Ethylcholesta-5,22*E*,25-triene-3β-ol; R = D-glucose - 24β-Ethylcholesta-5,22*E*,25-triene-3β-*O*-β-D-glucopyranoside

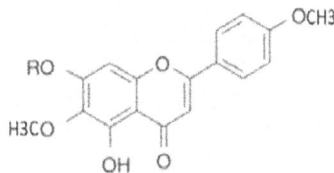

R = H - Pectolinaringenin; R = D-glucose - Pectolinaringenin-7-*O*-β-D-glucopyranoside

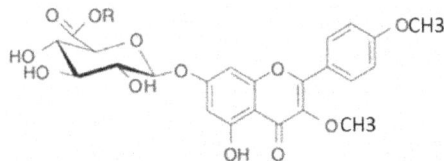

R = H - 3,4,5-Trihydroxy-6-[5-hydroxy-3-methoxy-2-(4- methoxy-phenyl)-4-oxo-4H-chromen-7-yloxy]-tetrahydro-pyran-2-carboxylic acid; R = CH$_3$ - 3,4,5-Trihydroxy-6-[5-hydroxy-3-methoxy-2-(4- methoxy-phenyl)-4-oxo-4H-chromen-7-yloxy]-tetrahydro-pyran-2-carboxylic acid methyl ester

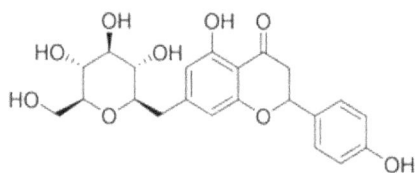

Apigenin-7-glucoside

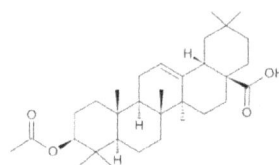

Oleanolic acid 3-acetate

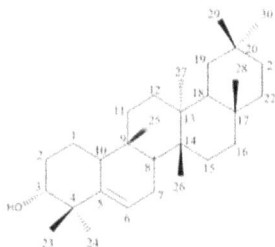

D:B-friedo-olean-5-en-3α-ol

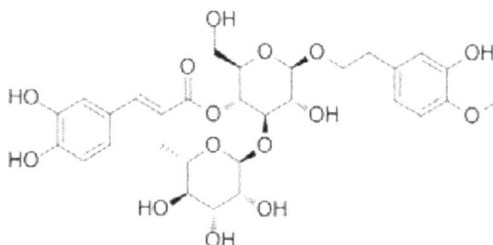

Jionoside D

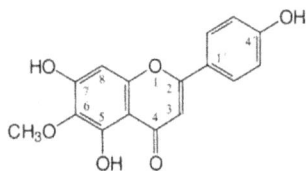

Hispidulin

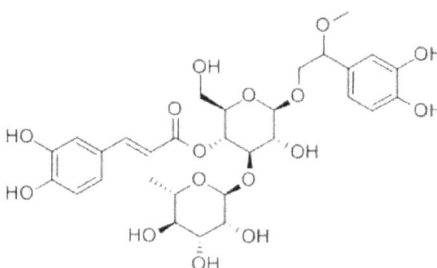

Campneoside I

10.3   CULTURE CONDITIONS

Leaf and nodal explants of *C. phlomidis* were cultured on an MS medium (Murashige and Skoog 1962) containing different concentration of growth hormones. Maximum callus formation was achieved with supplementation of 2,4-dichlorophenoxyacetic acid (2 mg/l) and 1-naphthaleneacetic acid (2 mg/; Devika and Kovilpillai 2012). Similarly, nodal explants of *C. phlomidis* were inoculated on an MS medium supplemented with 6-benzyladenine, kinetin, thidiazuron, N 6 -(2-isopentenyl) adenine, *trans*-zeatin, and meta-topolin. Callus growth occurred at all concentrations and developed from the base of nodal explants (Kher et al. 2016). Similarly, the nodal and leaf explants of *C. inerme* were surface sterilized and transferred to an MS medium containing indole-3-butyric acid and 2,4-dichlorophenoxyacetic acid. The combination of indole-3-butyric acid and 2,4-dichlorophenoxyacetic acid showed better response in the aformation of callus (Farooq et al. 2018). The nodal explants of *C. incisum* were inoculated on an MS medium with supplementation of benzyl adenine (5 μM) for callus formation (Goyal et al. 2010). The single shoots of *C. wallichii* and *C. colebrookianum* were cultured *in vitro* on an MS medium containing benzyl adenine (0.1, 0.5, 1.0 and 2.0 mg/l) for 3 weeks for obtaining callus formation. The growth hormone displayed better results in the formation of callus (Mao et al. 1995; Jirakiattikul and Boonha 2009). Apical leaves of *C. indicum* were transferred to an MS medium for the development of callogenesis. Iron is considered a significant nutritional constituent for the growth of plants. The iron uptake induced the formation and growth of callus (Mukherjee et al. 2013; Mukherjee and Bandyopadhyay 2014). Nodal explants of *C. serratum* were cultured on an MS liquid medium for callus regeneration. A liquid medium supplemented with 6-benzylaminopurine (0.5–1.5 mg/l) showed better response in formation and growth of callus (Vidya et al. 2005, 2012; Sharma et al. 2009).

REFERENCES

Akihisa T, Matsubara Y, Ghosh P, Thakur S, Shimizu N, Tamura T, Matsumoto T. 1988. The 24α- and 24β-epimers of 24-ethylcholesta5,22-dien-3β-ol in two *Clerodendrum* species. Phytochemistry 27, 1169–1172.

Al-Snaf AE. 2016. Chemical constituents and pharmacological effects of *Clerodendrum inerme*—a review. SMU Med J 3, 129–153.

Anitha R, Kannan P. 2006. Antifungal activity of *Clerodendrum inerme* (L). and *Clerodendrum phlomidis* (L). Turk J Biol 30, 139–142.

Arjsri P, Srisawad K, Mapoung S, Semmarath W, Thippraphan P, Umsumarng S, Yodkeeree S, Dejkriengkraikul P. 2022. Hesperetin from root extract of *Clerodendrum petasites* S. Moore inhibits SARS-CoV-2 spike protein S1 subunit-induced NLRP3 inflammasome in A549 lung cells via modulation of the Akt/MAPK/AP-1 pathway. Int J Mol Sci 23, 10346.

Arya H, Syed SB, Singh SS, Ampasala DR, Coumar MS. 2018. *In silico* investigations of chemical constituents of *Clerodendrum colebrookianum* in the anti-hypertensive drug targets: ROCK, ACE, and PDE5. Interdiscip Sci 10, 792–804.

Baker JT, Borris RP, Carte B. 1995. Natural product drug discovery and development; new perspective on international collaboration. J Nat Prod 58, 1325–1357.

Beenje HK. 1994. Kenyan Trees, Shrubs and Lianas. National Museums of Kenya. Nairobi.

Bharadwaj J, Seth MK. 2017. Medicinal plant resources of Bilaspur, Hamirpur and Una districts of Himachal Pradesh: An ethnobotanical enumeration. J Med Plants Stud 5, 99–110.

Bharitkar YP, Hazra A, Shah S, Saha S, Matoori AK, Mondal NB. 2015. New flavonoid glycosides and other chemical constituents from *Clerodendrum phlomidis* leaves: Isolation and characterization. Nat Prod Res 29, 1850–1856.

Bhujbal SS, Kewatkar SMK, More LS, Patil MJ. 2009. Antioxidant effects of roots of *Clerodendrum serratum* Linn. Pharmacogn Res 1, 294–298.

Bhujbal SS, Nanda RK, Deoda RS, Kumar D, Kewatkar SM, More LS, Patil MJ. 2010. Structure elucidation of a flavonoid glycoside from the roots of *Clerodendrum serratum* (L.) Moon, Lamiaceae. Braz J Pharmacogn 20, 1001–1002.

Brimson JM, Onlamoon N, Tencomnao T, Thitilertdecha P. 2019. *Clerodendrum petasites* S. Moore: The therapeutic potential of phytochemicals, hispidulin, vanillic acid, verbascoside, and apigenin. Biomed Pharmacother 118, 109319.

Caliş I, Hosny H, Yürüker A. 1994a. Inerminosides A1, C and D, three iridoid glycosides from *Clerodendrum inerme*. Phytochemistry 37, 1083–1085.

Caliş I, Hosny M, Yürüker A, Wright AD, Sticher O. 1994b. Inerminosides A and B, two novel complex iridoid glycosides from *Clerodendrum inerme*. J Nat Prod 57, 494–500.

Chae S, Kim JS, Kang KA, Bu HD, Lee Y, Seo YR, Hyun JW, Kang SS. 2005. Antioxidant activity of isoactoeoside from *Clerodendron trichotomum*. J Toxicol Environ Health A 68, 389–400.

Chander PM, Kartick C, Vijayachari P. 2015. Herbal medicine and healthcare practices among Nicobarese of Nancowry group of Islands – an indigenous tribe of Andaman and Nicobar Islands. Indian J Med Res 141, 720–744.

Chathuranga K, Kim MS, Lee H-C, Kim T-H, Kim J-H, Gayan Chathuranga WA, Ekanayaka P, Wijerathne HMSM, Cho W-K, Kim H-I, Ma JY, Lee J-S. 2019. Anti-respiratory syncytial virus activity of *Plantago asiatica* and *Clerodendrum trichotomum* extracts *in vitro* and *in vivo*. Viruses 11, 604.

Cheng HH, Wang HK, Ito J, Bastow KF, Tachibana Y, Nakanishi Y, Xu Z, Luo TY, Lee KH. 2001. Cytotoxic phenphorbide-related compounds from *Clerodendrum calamitosum* and *C. crytophyllum* J Nat Prod 64, 915–919.

Chethana GS, Hari VKR, Gopinath SM. 2013. Review on *Clerodendrum inerme*. J Pharm Sci Innov 2, 38–40.

Chidrawar VR, Patel KN, Chitme HR, Shiromwar SS. 2012. Pre-clinical evolutionary study of *Clerodendrum phlomidis* as an anti-obesity agent against high fat diet induced C57BL/6J mice. Asian Pac J Trop Biomed 2, S1509–S1519.

Choi J-H, Wang W-K, Kim H-J. 2004. Studies on the anti-inflammatory effects of *Clerodendron trichotomum* Thunberg leaves. Arch Pharm Res 27, 189–193.

Chopra RN, Nayer SL, Chopra IC. 1992. Glossay of Indian Medicinal Plant. PID, CSIR, New Delhi.

Chouhan MK, Hurkadale PJ, Hegde HV. 2017. Evaluation of *Clerodendrum inerme* (L.) Gaertn. on Burkitt's lymphoma cancer. Indian J Pharm Educ Res 52, 241–246.

Chourasiya RK, Jain PK, Jain SK, Nayak SS. Agrawal RK. 2010. In vitro antioxidant activity of *Clerodendron inerme* (L.) Gaertn leaves. Res J Pharm Biol Chem Sci 1, 119–123.

Chourasiya RK, Jain PK, Sharma S, Ganesh N, Nayak SS, Agrawal RK. 2011. Genomic stability and tissue protection of *Clerodendron inerme* (L.) Gaertn leaves. Med Chem Res 20, 1674–1679.

Das N, Sarma J, Bortahur SK. 2014. *Clerodendrum trichotomum* Thunberg (Lamiaceae): A new record to the flora of India from Assam. Pleione 8, 503–505.

Deb DB. 1983. The Flora of Tripura State, Vol II. Today & Tomorrow's Printers and Publishers, New Delhi.

Deb L, Dey A, Sakthivel G, Bhattamishra SK, Dutta A. 2013. Protective effect of *Clerodendrum colebrookianum* Walp., on acute and chronic inflammation in rats. Indian J Pharmacol 45, 376–380.

Deb L, Dutta A. 2012. Evaluation of mechanism for antihypertensive action of *Clerodendrum colebrookianum* Walp., used by folklore healers in north-east India. J Ethnopharmacol 143, 207–212.

Devika R, Kovilpillai J. 2012. Phytochemical and *in vitro* micropropogation studies of *Clerodendrum phlomidis* L. J Pharm Res 5, 4396–4398.

Dey A, Gorai P, Mukherjee A, Dhan R, Modak BK. 2017. Ethnobiological treatments of neurological conditions in the Chota Nagpur Plateau, India. J Ethnopharmacol 198, 33–44.

Dorsaz A-C, Marston A, Stoeckli-Evans H, Msonthi JD, Hostettmann K. 1985. Uncinatone, a new antifungal hydroquinone diterpenoid from *Clerodendrum uncinatum* SCHINZ. Helvetica 68, 1605–1610.

Elaskary HI, Sabry OM, Khalil AM, El Zalabani SM. 2020. UPLC-PDA-ESI-MS/MS profiling of *Clerodendrum inerme* and *Clerodendrum splendens* and significant activity against *Mycobacterium tuberculosis*. Pharmacogenomics J 12, 1518–1524.

Farooq SA, Farook TT, Al Rawahy SH. 2018. Adventitious and *de novo* somatic embryogenesis from nodal and leaf explants of *Clerodendrum inerme*. Am J Plant Physiol 13, 53–57.

Gogoi B, Gogoi D, Silla Y, Kakoti BB, Bhau BS. 2017. Network pharmacology-based virtual screening of natural products from *Clerodendrum* species for identification of novel anti-cancer therapeutics. Mol Biosyst 13, 406–416.

Gopal N, Sengottuvelu S. 2008. Hepatoprotective activity of *Clerodendrum inerme* against CCl4 induced hepatic injury in rats. Fitoterapia 79, 24–26.

Goswami A, Dixit VK, Srivastava BK. 1998. Anti-malarial activity of aqueous extract of *Clerodendrum infortunatum*. Bionature 48, 45–48.

Goswami P, Kotoky J, Chern Z, Lu Y. 1996. Asterol glycoside from leaves of *Clerodendron colebrookianum*. Phytochemistry 41, 279–281.

Goyal S, Shahzad A, Anis M, Khan S. 2010. Multiple shoot regeneration in *Clerodendrum incisum* L.,-an ornamental woody shrub. Pak J Bot 42, 873–878.

Gupta AK, Sharma M. 2008. Reviews on Indian Medicinal Plants, Vol 7. ICMR, New Delhi.

Harley RM, Atkins S, Budantsev AL. 2004. The Family and Genera of Vascular Plants, Vol VII. Springer-Verlag, Berlin, Heidelberg.

Hazekamp A, Verpoorte R, Panthong A, Hazekamp A, Verpoorte R, Panthong A. 2001. Isolation of a bronchodilator flavonoid from the Thai medicinal plant *Clerodendrum petasites*. J Ethnopharmacol 78, 45–49.

Hooker JD. 1885. Flora of British India, Vol IV. L. Reeve and Co. Ltd, Kent.

Hui Y, Aijun H, Bei J, Zhongwen L, Handong S. 2000. Serratumin A, a novel compound from *Clerodendrum serratum*. Acta Bot Yunnan 22, 75–80.

Hunt DR, Bailey LH, Bailey EZ. 1978. Hortus third. A concise dictionary of plants cultivated in the United States and Canada. Kew Bull 32, 801.

Husain A, Virmani OP, Popli SP, Misra LN, Gupta MM, Srivastava GN, Abraham Z, Singh AK. 1992. Dictionary of Indian Medicinal Plants. CIMAP, Lucknow.

Ibrahim SRM, Alshali KZ, Fouad MA, Elkhayat ES, Al Haidari RA, Mohamed GA. 2014. Chemical constituents and biological investigations of the aerial parts of Egyptian *Clerodendrum Inerme*. Bull Fac Pharm Cairo Univ 52, 165–170.

Jadeja RN, Thounaojam MC, Ramani UV, Devkar RV, Ramachandran A. 2011. Anti-obesity potential of *Clerodendron glandulosum* Coleb. leaf aqueous extract. J Ethnopharmacol 135, 338–343.

Jadhav D. 2006. Ethnomedicinal plants used by Bhil tribe of Bibdod, Madhya Pradesh. Indian J Tradit Knowl 5, 263–267.

Jiji P. 2014. Ethnomedicinal uses of wild vegetables used by TaiShyam people of Sivasagar district, Assam, India. Int Res J Biol Sci 3, 63–65.

Jirakiattikul Y, Boonha K. 2009. *In vitro* proliferation and root induction of *Clerodendrum wallichii*. Thai Sci Technol J 17, 61–67.

Joshi KC, Prakash L, Shah RK. 1977. Chemical constituents of *Clerodendrum infortunatum* and *Ficus racemosa*. J Indian Chem Soc 54, 1104–1106.

Kalita D, Phukan B. 2010. Some ethnomedicines used by the Tai Ahom of Dibrugarh district, Assam, India. Indian J Nat Prod Res 1, 507–511.

Kalita J, Singh SS, Khan ML. 2012. *Clerodendrum colebrookianum* Walp.: A potential folk medicinal plant of Northeast India. Asian J Pharm Biol Res 2, 256–261.

Kalonio DE, Hendriani R, Barung eEN. 2017. Anticancer activity of plant genus *Clerodendrum* (Lamiaceae): A review. Tradit Med J 22, 182–189.

Kanchanapoom T, Chumsri P, Kasai R, Otsuka H, Yamasaki K. 2005. A new iridoid diglycoside from *Clerodendrum chinense*. J Asian Nat Prod Res 7, 269–272.

Kanchanapoom T, Kasaia R, Chumsri P, Hiraga Y, Yamasaki K. 2001. Megastigmane and iridoid glucosides from *Clerodendrum inerme*. Phytochemistry 58, 333–336.

Kang DG, Lee YS, Kim HJ, Lee YM, Lee HS. 2003. Angiotensin converting enzyme inhibiting phenglypropanoid glycosides from *Clerondendron trichotomum*. J Ethnopharmacol 89, 151–154.

Kar P, Sharma NR, Singh B, Sen A, Roy A. 2021. Natural compounds from *Clerodendrum* spp. as possible therapeutic candidates against SARS-CoV-2: An in-silico investigation. J Biomol Struct Dyn 39, 1–12.

Kher MM, Soner D, Srivastava N, Nataraj M, Teixeira da silva JA. 2016. Micropropagation of *Clerodendrum phlomidis* L.F. J Hortic Res 24, 21–28.

Kim HJ, Woo ER, Shin CG, Hwang DJ, Park H, Lee YS. 2001. HIV-1 integrase inhibitory phenylpropanoid glycosides from *Clerodendron trichotomum*. Arch Pharm Res 24, 286–291.

Kim KH, Kim S, Jung MY, Ham IH, Wang WK. 2009. Anti-inflammatory phenylpropanoid glycosides from *Clerodendron trichotomum* leaves. Arch Pharm Res 32, 7–13.

Kim S-K, Cho S-B, Moon H-I. 2010. Anti-complement activity of isolated compounds from the roots of *Clerodendrum bungei* Steud. Phytother Res 24, 1720–1723.

Kokate A, Limsay RP, Dubey SA. 2022. The phytochemistry and antioxidant activity of methanolic extract of *Clerodendrum infortunatum* leaves. Pharm Innov J 11, 661–664.

Kopilakkal R, Chanda K, Balamurali MM. 2021. Hepatoprotective and antioxidant capacity of *Clerodendrum paniculatum* flower extracts against carbon tetrachloride-induced hepatotoxicity in rats. ACS Omega 6, 26489–26498.

Kottegoda SR. 1994. Flowers of Sri Lanka. Royal Asiatic Society of Sri Lanka, Colombo.

Krishnaveni R, Rajan S. 2021. *In vitro* antioxidant and anti-inflammatory activity of the aqueous and ethanol leaf extracts of *Clerodendrum phlomidis* Linn. F. Res J Pharm Technol 14, 5217–5221.

Lee JY, Lee JG, Sim SS, Wang W-K, Kim CJ. 2011. Anti-asthmatic effects of phenylpropanoid glycosides from *Clerodendron trichotomum* leaves and *Rumex gmelini* herbes in conscious guineapigs challenged with aerosolized ovalbumin. Phytomedicine 18, 134–142.

Li L, Wu L, Wang M, Sun J, Liang J. 2014. Abietane diterpenoids from *Clerodendrum trichotomum* and correction of NMR data of villosin C and B. Nat Prod Commun 9, 907–910.

Lin Y-L, Kuo Y-H, Chen Y-L. 1989. Two new clerodane-type diterpenoids, clerodinins A and B, from *Clerodendron brachyanthum* Schauer. Chem Pharm Bull 37, 2191–2193.

Liu Q, Hu H-J, Li P-F, Yang Y-B, Wu L-H, Chou G-X, Wang Z-T. 2014. Diterpenoids and phenylethanoid glycosides from the roots of *Clerodendrum bungei* and their inhibitory effects against angiotensin converting enzyme and a-glucosidase. Phytochemistry 103, 196–202.

Liu S-S, Zhou T, Zhang S-W, Xuan L-J. 2009. Chemical constituents from *Clerodendrum bungei* and their cytotoxic activities. Helvetica 92, 1023–1224.

Mabberley DJ. 1997. The Plant-Book, A Portable Dictionary of Vascular Plants. Cambridge University Press, Cambridge.

Mabberley DJ. 2008. Mabberley's Plant-Book, 3rd ed. Cambridge University Press, Cambridge.

Machumi F, Samoylenko V, Yenesew A, Derese S, Midiwo JO, Wiggers FT, Jacob MR, Tekwani BL, Khan SI, Walker LA, Muhammad I. 2010. Antimicrobial and antiparasitic abietane diterpenoids from the roots of *Clerodendrum eriophyllum*. Nat Prod Commun 5, 853–858.

Manoharan S, Kavitha K, Senthil N, Renju GL. 2006. Evaluation of anticarcinogenic effects of *Clerodendron inerme* on 7,12- dimethylbenz(a) anthracene-induced hamster buccal pouch carcinogenesis. Singapore Med J 47, 1038.

Manzoor-Khuda M, Sarela S. 1965. Constituents of *Clerodendron infortunatum* (bhat)-I. Isolation of clerodolone, clerodone, clerodol and clerosterol. Tetrahedron 21, 797–802.

Mao AA, Wetten A, Fay M, Caligari PD. 1995. *In vitro* propagation of *Clerodendrum colebrookianum* Walp., a potential natural anti-hypertension medicinal plant. Plant Cell Rep 14, 493–496.

Maruga Raja MKM, Mishra SH. 2010. Comprehensive review of *Clerodendrum phlomidis*: A traditionally used bitter. Zhong Xi Yi Jie He Xue Bao 8, 510–524.

Mitra S, Mukherjee SK. 2010. Ethnomedicinal usages of some wild plants of North Bengal plain for gastrointestinal problems. Indian J Tradit Knowl 9, 705–712.

Muhammed Ashraf VK, Kalaichelvan VK, Venkatachalam VV, Ragunathan R. 2021. Evaluation of *in vitro* cytotoxic activity of different solvent extracts of *Clerodendrum thomsoniae* Balf.f and its active fractions on different cancer cell lines. Future J Pharm Sci 7, 50.

Mukherjee A, Bandyopadhyay A. 2014. Inducing somatic embryogenesis by polyamines in medicinally important *Clerodendrum indicum* L. Int J Curr Microbiol Appl Sci 3, 12–26.

Mukherjee A, Bandyopadhyay A, Dutta S, Basu S. 2013. Phytoaccumulation of iron by callus tissue of *Clerodendrum indicum* (L). Chem Ecol 29, 564–571.

Murashige T, Skoog F. 1962. A revised medium for rapid growth and bioassays with tobacco tissue cultures. Physiol Plant 15, 473–497.

Muthaura CN, Rukunga GM, Chhabra SC, Omar SA, Guantai AN, Gathirwa JW, Tolo FM, Mwitari PG, Keter LK, Kirira PG, Kimani CW, Mungai GM, Njagi ENM. 2007. Antimalarial activity of some plants traditionally used in Meru district of Kenya. Phytother Res 21, 260–267.

Muthu C, Ayyanar M, Raja N, Ignacimuthu C. 2006. Medicinal plants used by traditional healers in Kancheepuram District of Tamil Nadu. India J Ethnobiol Ethnomed 2, 43.

Muthu C, Baskar K, Ignacimuthu S, Ai-Khaliel. 2013. Ovicidal and oviposition deterrent activities of the flavonoid pectolinaringenin from *Clerodendrum phlomidis* against Earias vittella. Phytoparasitica 41, 365–372.

Muthu C, Reegan AD, Kingsley S, Ignacimuthu S. 2012. Larvicidal activity of pectolinaringenin from *Clerodendrum phlomidis* L. against *Culex quinquefasciatus* Say and *Aedes aegypti* L. (Diptera: Culicidae). Parasitol Res 111, 1059–1065.

Nadkarni KM. 1976. Indian Materia Medica. Popular Prakashan, Bombay.

Nadkarni KM, Nadkarni AK. 2002. Indian Materia Medica. Popular Publishing, Mumbai.

Nan H-H, Wu J, Zhang S. 2005. A new phenylethanoid glycoside from *Clerodendrum inerme*. Pharmazie 60, 798–799.

Nan H-H, Yin H, Zhang S. 2008. Phenylethanoid glycosides from *Clerodendrum inerme*. Nat Prod Res Dev 20, 1008–1011.

Nandi S, Lyndem LM. 2016. *Clerodendrum viscosum*: Traditional uses, pharmacological activities and phytochemical constituents. Nat Prod Res 30, 497–506.

Narayanan N, Thirugnanasambantham P, Viswanathan S, Vijayasekaran V, Sukumar E. 1999. Antinociceptive, anti-inflammatory and antipyretic effects of ethanol extract of *Clerodendron serratum* roots in experimental animals. J Ethnopharmacol 65, 237–241.

Nath M, Choudhury MD. 2010. Ethno-medico-botanical aspects of Hmar tribe of Cachar district, Assam (Part-I). Indian J Tradit Knowl 9, 760–764.

Nath SC, Bordoloi DN. 1993. Diversity of economic flora in Arunachal Pradesh: Plant folk medicines among the Chakma, Singpho and Tangsa tribals. In: Himalayan Biodiversity Conservation Strategies, Dhar U (ed.). GB Pant National Institute of Himalayan Environment, Almora, pp 179–189.

Okaiyeto K, Falade AO, Oguntibeju OO. 2021. Traditional uses, nutritional and pharmacological potentials of *Clerodendrum volubile*. Plants 10, 1893.

Okwu TE, Iroabuchi F. 2008. Isolation of an antioxidant flavonone diglycoside from the Nigerian medicinal plant *Clerodendron splendens* A. Cheval. Int J Chem Sci 6, 631–636.

Oscar NDY, Joel TNS, Ange AANG, Desire S, Brice SNF, Barthelemy N. 2018. Chemical constituents of *Clerodendrum splendens* (Lamiaceae) and their antioxidant activities. J Dis Med Plants 4, 120–127.

Pal A, Mahmud ZA, Akter N, SaifulIslam M, Bachar SC. 2012. Evaluation of antinociceptive, antidiarrheal and antimicrobial activities of leaf extracts of *Clerodendrum indicum*. Pharmacogenomics J 4, 41–46.

Pal D, Sannigrahi S, Mazumder UK. 2009. Analgesic and anticonvulsant effects of saponin isolated from the leaves of *Clerodendrum infortunatum* Linn. in mice. Indian J Exp Biol 47, 743–747.

Pandey R, Verma RK, Gupta MM. 2005. Neo-clerodane diterpenoids from *Clerodendrum inerme*. Phytochemistry 66, 643–648.

Pandey R, Verma RK, Singh SC, Gupta MM. 2003. 4α-Methyl-24β-ethyl-5α-cholesta-14,25-dien-3β-ol and 24β-ethylcholesta-5, 9(11), 22E-trien-3β-ol, sterols from *Clerodendrum inerme*. Phytochemistry 63, 415–420.

Panthong A, Kanjanapothi D, Taylor WC. 1986. Ethnobotanical review of medicinal plants from Thai traditional books, Part I: Plants with anti-inflammatory, anti-asthmatic and antihypertensive properties. J Ethnopharmacol 18, 213–228.

Park JH, Kang HS, Bang M, Cheng HC, Jin HY, Ahn TH. 2018. Bounithiphonh C, Phongoudome C, Floristic inventory of vascular plant in Nam Ha National Biodiversity Conservation Area, Lao People's Democratic Republic. J Asia Pac Biodivers 11, 300–304.

Patel JJ, Acharya SR, Acharya NS. 2014. *Clerodendrum serratum* (L.) Moon.—a review on traditional uses, phytochemistry and pharmacological activities. J Ethnopharmacol 154, 268–285.

Phaopongthai S, Sichaem J, Phaopongthai J. 2017. *In vivo* and *in vitro* antidiabetic effects of *Clerodendrum longisepalum*. J Sci Technol 39, 317–323.

Phuneerub P, Limpanasithikul W, Palanuvej C, Ruangrungsi N. 2015. *In vitro* anti-inflammatory, mutagenic and antimutagenic activities of ethanolic extract of *Clerodendrum paniculatum* root. J Adv Pharm Technol Res 6, 48–52.

Prakash Babu N, Pandikumar P, Ignacimuthu S. 2011. Lysosomal membrane stabilization and anti-inflammatory activity of *Clerodendrum phlomidis* L.f., a traditional medicinal plant. J Ethnopharmacol 135, 779–785.

Prakash Babu N, Saravanan S, Pandikumar P, Bala Krishna K, Karunai Raj M, Ignacimuthu S. 2014. Anti-inflammatory and anti-arthritic effects of 3-hydroxy, 2-methoxy sodium butanoate from the leaves of *Clerodendrum phlomidis* L.f. Inflamm Res 63, 127–138.

Prasad V, Srivastava S, Varsha Verma HN. 1995. Two basic proteins isolated from *Clerodendrum inerme* Gaertn. are inducers of systemic antiviral resistance in susceptible plants. Plant Sci 110, 73–82.

Prashith Kekuda TR, Dhanya Shree VS, Saema Noorain GK, Sahana BK, Raghavendra HL. 2019. Ethnobotanical uses, phytochemistry and pharmacological activities of *Clerodendrum infortunatum* L. (Lamiaceae): A review. J Drug Deliv Ther 9, 547–559.

Prashith Kekuda TR, Sudharshan SJ. 2018. Ethnobotanical uses, phytochemistry and biological activities of *Clerodendrum paniculatum* L. (Lamiaceae): A comprehensive review. J Drug Deliv Ther 8, 28–34.

Punekar SA, Lakshminarasimhan P. 2011. Flora of Anshi National Park: Western Ghats-Karnataka. Biospheres Publication, Pune.

Ra DF, Sl M, Crepin J. 2004. Medicinal Plants of the Guianas. Department of Botany, National Museum of Natural History, Smithsonian Institution, Guyana.

Rabiul H, Subhasish M, Sinha S, Roy MG, Sinha D, Gupta S. 2011. Hepatoprotective activity of *Clerodendron inerme* against paracetamol induced hepatic injury in rats for pharmaceutical product. Int J Drug Dev Res 3, 118–126.

Rajakaruna N, Harris CS, Towers GHN. 2002. Antimicrobial activity of plants collected from serpentine outcrops in Sri Lanka. Pharm Biol 40, 235–244.

Rao DM, Rao UVUB, Sudharshanam G. 2006. Ethno-medicobotanical studies from Rayalaseema region of Southern Eastern Ghats, Andhra Pradesh, India. Ethnobot Leafl 10, 198–207.

Saha D, Talukdar A, Das T, Ghosh SK, Rahman H. 2012. Evaluation of analgesic activity of ethanolic extract of *Cleodendrum serratum* linn leaves in rats. Res J Pharm Appl Sci 2, 33–37.

Saklani A, Jain SK. 1994. Cross-cultural Ethnobotanical Studies in Northeast India. Deep Publications, Agra.

Salahudeen, Bolaji A. 2019. Antioxidant activity of the leaf extract of *Clerodendrum volubile* (Verbenaceae). Acta Sci Pharm Sci 3(6), 8–12.

Sannigrahi S, Mazumder UK, Pal D, Mishra SL. 2012. Terpenoids of methanol extract of *Clerodendrum infortunatum* exhibit anticancer activity against Ehrlich's ascites carcinoma (EAC) in mice. Pharm Biol 50, 304–309.

Sannigrahi S, Mazumder UK, Pal DK, Parida S. 2009. *In vitro* antioxidant activity of methanol extract of *Clerodendrum infortunatum* Linn. Orient Pharm Exp Med 9, 128–134.

Sapiun Z, Pangalo P, Imran AK, Wicita PS, Daud RPA. 2020. Determination of total flavonoid levels of ethanol extract sesewanua leaf (*Clerodendrum fragrans* Wild) with maceration method using UV-Vis spectrofotometry. Pharmacogn J 12, 356–360.

Sayyed HY, Patel MR, Patil JK, Suryawanshi HP, Ahirrao RA. 2011. *In vitro* antioxidant activity of leaves extracts of *Clerodendrum innerme* (L.) GAERTN. J Pharm Res 4, 2941–2942.

Schmid R, Riffle RL. 1998. The tropical look: An encyclopedia of dramatic landscape plants. Taxon 47, 985.

Shahabuddin SK, Munikishore R, Trimurtulu G, Gunasekar D, Deville A, Bodo B. 2013. Two new chalcones from the flowers of *Clerodendrum inerme*. Nat Prod Commun 8, 459–460.

Sharaf A, Aboulezz AF, Abdul-alim MA, Golviaa N 1969. Some pharmacological studies on the leaves of *Clerodendron inerme*. Qual Plant Mater XVII, 293–298.

Sharma HK, Chhangte L, Dolui AK. 2001. Traditional medicinal plants in Mizoram, India. Fitoterapia 72, 146–161.

Sharma M, Rai SK, Purshottam DK, Jain M, Chakrabarty D, Awasthi A, Nair KN, Sharma AK 2009. *In vitro* clonal propagation of *Clerodendrum serratum* (Linn.) Moon (barangi): A rare and threatened medicinal plant. Acta Physiol Plant 31, 379–383.

Shendge AK, Basu T, Chaudhuri D, Panja S, Mandal N. 2017. *In vitro* antioxidant and antiproliferative activities of various solvent fractions from *Clerodendrum viscosum* leaves. Pharmacogn Mag 13(Suppl 2), S344–S353.

Shendge AK, Basu T, Mandal N. 2021a. Evaluation of anticancer activity of *Clerodendrum viscosum* leaves against breast carcinoma. Indian J Pharmacol 53, 377–383.

Shendge AK, Chaudhuri D, Basu T, Mandal N. 2021b. A natural flavonoid, apigenin isolated from *Clerodendrum viscosum* leaves, induces G2/M phase cell cycle arrest and apoptosis in MCF-7 cells through the regulation of p53 and caspase-cascade pathway. Clin Transl Oncol 23, 718–730.

Shetty BV, Kaveriappa KM, Bhat GK. 2002. Plant resources of Western Ghats and lowlands of Dakshina Kannada and Udupi districts. Pilikula Nisarga Dhama Society, Mangalore.

Shrivastava N, Patel T. 2007. Clerodendrum and healthcare: An overview. Med Aromat Plant Sci Biot 1, 209–223.

Sidde L, Malathi S, Swethalatha S, Rajani K. 2018. A brief review on *Clerodendrum indicum*. Int J Indig Herb Drug 3, 1–4.

Siddik ASNU, Alam S, Borgohain R, Chutia P. 2021. Antioxidant properties of *Clerodendrum* species found in northeast India: A review. J Pharmacogn Phytochem 10, 390–394.

Sindhu TJ, Akhilesh KJ, Jose A, Binsiya KP, Thomas B, Wilson E. 2020. Antibacterial Screening of *Clerodendrum infortunatum* leaves: Experimental and molecular docking studies. Asian J Res Chem 13, 128–132.

Singh MK, Khare G, Iyer SK, Sharwan G, Tripathi DK. 2012. *Clerodendrum serratum*: A clinical approach. J Appl Pharm Sci 2, 11–15.

Singh PN, Gajurel PR, Rethy P. 2015. Ethnomedicinal value of traditional food plants used by the Zeliang tribe of Nagaland. Indian J Tradit Knowl 14, 298–305.

Singh R. 2015. Medicinal plants: An overview. J Plant Sci 3, 50–55.

Sinha NK, Seth KK, Pandey VB, Dasgupta B, Shah AH. 1981. Flavonoids from the flowers of *Clerodendron infortunatum*. Planta Med 42, 296–298.

Somwong P, Suttisri R. 2018. Cytotoxic activity of the chemical constituents of *Clerodendrum indicum* and *Clerodendrum villosum* roots. J Integr Med 16, 57–61.

Spencer GF, Flippen-Anderson JL. 1981. Isolation and X-ray structure determination of a neolignan from *Clerodendron inerme* seeds. Phytochemistry 20, 2757–2759.

Staples GW, Herbst DR. 2005. A Tropical Garden Flora. Bishop Museum Press, Honolulu.

Sun C, Nirmalananda S, Jenkins CE, Debnath S, Balambika R, Fata JE, Raja KS. 2013. First ayurvedic approach towards green drugs: Anti cervical cancer-cell properties of *Clerodendrum viscosum* root extract. Anticancer Agents Med Chem 13, 1469–1476.

Thirumal M, Srimanthula S, Kishore G, Vadivelan R, Anand Kumar AVS. 2013. Analgesic and antipyretic effects of aqueous extract from *Clerodendrum inerme* (L.) Gaertn. leaves in animal models. Der Pharm Lett 5, 315–323.

Thitilertdecha P, Guy RH, Rowan MG. 2014. Characterisation of polyphenolic compounds in *Clerodendrum petasites* S. Moore and their potential for topical delivery through the skin. J Ethnopharmacol 154, 400–407.

Thitilertdecha P, Rowan MG, Guy RH. 2015. Topical formulation and dermal delivery of active phenolic compounds in the Thai medicinal plant--*Clerodendrum petasites* S. Moore. Int J Pharm 478, 39–45.

Tian J, Zhao Q-S, Zhang H-J, Lin Z-W, Sun H-D. 1997. New cleroindicins from *Clerodendrum indicum*. J Nat Prod 60, 766–769.

Tian X-D, Min Z-D, Xie N, Lei Y, Tian Z-Y, Zheng Q-T, Xu R-N, Tanaka T, Iinuma M, Mizuno M. 1993. Abietane diterpenes from *Clerodendron cyrtophyllum*. Chem Pharm Bull 14, 1415–1417.

Tian Y, Liang N, Jing T, Yuan F, Sarker MMR, Maruf MRA, Chen S. 2022. Clerodendrum wallichii Merr methanol extract protected alcohol-induced liver injury in Sprague-Dawley rats by modulating antioxidant enzymes. Evid Based Complement Altern Med 2022, Article ID 5635048.

Tianpei FAN, Zhida MIN, Munekazu IINUMA. 1999. Two novel diterpenoids from *Clerodendrum Bungei*. Chem Pharm Bull 47, 1797–1798.

Turner RJ Jr, Wasson E. 1997. Botanica. Mynah Publishing, New South Wales.

Uddin MJ, Çiçek SS, Willer J, Shulha O, Abdalla MA, Sönnichsen F, Girreser U, Zidorn C. 2020. Phenylpropanoid and flavonoid glycosides from the leaves of *Clerodendrum infortunatum* (Lamiaceae). Biochem Syst Ecol 92, 104131.

Ugbaja RN, Akinhanmi TF, James AS, Ugwor EI, Babalola AA, Ezenandu EO, Ugbaja VC, Emmanuel EA. 2021. Flavonoid-rich fractions from *Clerodendrum volubile* and *Vernonia amygdalina* extenuates arsenic-invoked hepatorenal toxicity via augmentation of the antioxidant system in rats. Clin Nutr Open Sci 35, 12–25.

Upmanyu G, Tanu M, Gupta M, Gupta AK, Sushma A, Dhakar RC. 2011. Acute toxicity and diuretic studies of leaves of *Clerodendrum inerme*. J Pharm Res 4, 1431–1432.

Verma RK. 2014. An ethnobotanical study of plants used for the treatment of livestock diseases in Tikamgarh District of Bundelkhand, Central India. Asian Pac J Trop Biomed 4(Suppl 1), S460–S467.

Verma S, Gupta R. 2014. Pharmacognostical and high performance thin layer chromatography studies on leaves of *Clerodendrum infortunatum* L. Ayu 35, 416–422.

Vidya SM, Krishna V, Manjunatha BK. 2005. Micropropagation of *Clerodendrum serratum* L. from leaf explants. J Non Timber For Prod 12, 57–60.

Vidya SM, Krishna V, Manjunatha BK, Mankani KL, Ahmed M, Singh SD. 2007. Evaluation of hepatoprotective activity of *Clerodendrum serratum* L. Indian J Exp Biol 45, 538–542.

Vidya SM, Krishna V, Manjunatha BK, Pradeepa. 2012. Micropropagation of *Clerodendrum serratum* L. through direct and indirect organogenesis. Plant Tissue Cult Biotechnol 22, 179–185.

Wahba HM, AbouZid SF, Sleem AA, Apers S, Pieters L, Shahat AA. 2011. Chemical and biological investigation of some *Clerodendrum* species cultivated in Egypt. Pharm Biol 49, 66–72.

Waliullah TM, Yeasmin AM, Wahedul IM, Parvez H. 2014. Evaluation of antimicrobial study in in vitro application of *Clerodendrum infortunatum* Linn. Asian Pac J Trop Dis 4, 484–488.

Wang J-H, Luan F, He X-D, Wang Y, Li M-X. 2018. Traditional uses and pharmacological properties of *Clerodendrum* phytochemicals. J Tradit Complement Med 8, 24–38.

Xu M-F, Shen L-Q, Wang K-W, Du Q-Z, Wang N. 2011a. A new rearranged abietane diterpenoid from *Clerodendrum kaichianum* Hsu. J Asian Nat Prod Res 13, 260–264.

Xu M-F, Shen L-Q, Wang K-W, Du Q-Z, Wang N. 2011b. Bioactive diterpenes from *Clerodendrum kaichianum*. Nat Prod Commun 6, 3–5.

Xu M-F, Wang S, Jia O, Zhu Q, Shi L. 2016. Bioactive diterpenoids from *Clerodendrum kiangsiense*. Molecules 21, 86.

Xu R-L, Wang R, Ha W, Shi Y-P. 2014. New cyclohexylethanoids from the leaves of *Clerodendrum trichotomum*. Phytochem Lett 7, 111–113.

Yang H, Hou A-J, Mei S-X, Sun S-D, Che C-T. 2002. Constituents of *Clerodendrum bungei*. J Asian Nat Prod Res 4, 165–169.

Yang H, Wang J, Hou AJ, Gou YP, Lin ZW, Sun HD. 2000. New steroids from *Clerodendrum colebrookianum*. Fitoterapia 71, 641–648.

Youssef FS, Sobeh M, Dmirieh M, Bogari HA, Koshak AE, Wink M, Ashour ML, Elhady SS. 2022. Metabolomics-based profiling of *Clerodendrum speciosum* (Lamiaceae) leaves using LC/ESI/MS-MS and *in vivo* evaluation of its antioxidant activity using *Caenorhabditis elegans* model. Antioxidants (Basel) 11, 330.

Yuan YW, Mabberley DJ, Steane DA, Olmstead RG. 2010. Further disintegration and redefinition of *Clerodendrum* (Lamiaceae): Implications for the understanding of the evolution of an intriguing breeding strategy. Taxon 59, 125–133.

Zalke AS, Kulkarni AV, Shirode DS, Duraiswamy B. 2010. *In vivo* anticancer activity of *Clerodendrum serratum* (L) moon. Res J Pharm Biol Chem 1, 89–95.

Zhou J, Yang Q, Zhu X, Lin T, Hao D, Xu J. 2020. Antioxidant activities of *Clerodendrum cyrtophyllum* Turcz leaf extracts and their major components. PLoS One 15, e0234435.

11 *Convolvulus* Species

11.1 MORPHOLOGICAL FEATURES, DISTRIBUTION, ETHNOPHARMACOLOGICAL PROPERTIES, PHYTOCHEMISTRY, AND PHARMACOLOGICAL ACTIVITIES

Convolvulus arvensis L (Syn. *C. ambigens* House, *C. incanus* Vahl, syn. *Strophocaulos arvensis* L. Small, *C. chinensis* Ker Gawl.) is a perennial vine, 0.3–1.8 m high, hairless, often twined to form dense and tangled mats. Leaves are green with visible veins, round, ovate or oblong, some may even be linear, hastate or sagittate, extremely variable, flattened petiole, and upper-side grooved. Flowers are long (2–2.5 cm), funnel-like corolla, with white and pale pink petals. The fruit is wide (8 mm), rounded, and light brown. Seeds are dark brown to black, with rough surfaces (Arora and Malhotra 2011; Mehrafarin et al. 2009). The vine is distributed in North America and China, India, Sri Lanka, and other regions (Austin 2000). Its aerial parts are used in the healing of wounds, jaundice and possess antispasmodic antihemorrhagic, and anti-angiogenetic properties (Munz and Keck 1959; Desta 1993; Alkofahi et al. 1996; Austin 2000; Amin et al. 2014). Moreover, it is used as a diuretic agent and in the treatment of skin infections (Leporatti and Ivancheva 2003). Aqueous decoction is useful in cough and flu, and used to treat painful joints, inflammations, and swellings (Ali et al. 2013). The ethanol extract of *C. arvensis* roots demonstrated significant antioxidant activity in 2,2′-azino-bis-3-ethylbenzothiazoline-6-sulphonic acid radical cation, oxygen radical absorbance capacity, and th ferric-reducing scavenging (1.62 mmol Trolox equivalents/g dry weight, 1.71 mmol trolox equivalents/g dry weight and 2.11 mmol Trolox equivalents/g dry weight) assays (Mohd Azman et al. 2015). The aqueous and acetone extracts *C. arvensis* showed significant antibacterial activity against *S. aureus, S. pyogenes, E. coli*, and *K. pneumoniae* (different inhibitory concentrations viz., 500 mg/ml, 250 mg/ml, 125 mg/ml, 0.06 mg/ml and 0.03 mg/ml; Abu-Mejdad et al. 2010). Glycosides-rich fraction of *C. arvensis* leaves (1250, 2500, 5000, and 10000, μg/ml) demonstrated significant cell features (apoptosis) such as cell volume shrinking, chromatin condensation, and nuclear fragmentation of rhabdomyosarcoma tumor cell lines at 72h. Moreover, apoptotic property was confirmed by DNA fragmentation into low molecular weight fragments (180–200 pb) as well as high molecular weight fragments (AL-Asady et al. 2014). Methanol extract of *C. arvensis* stimulated a dose-dependent (0.4–2.8 mg/ml) relaxation of rabbit's duodenal smooth muscles (Atta and Mouneir 2004; Sethiya et al. 2009). *Convolvulus* species possess antiulcer activity (Awaad et al. 2014). Zinc oxide nanoparticles were formed using *C. arvensis* leaf extract (Al-Senani 2020). The scammonin resin, dihydroxy cinnamic acid, β-methylesculetin, ipuranol, surcose, reducing sugar, and starch were reported from *C. arvensis* and *C. scammonia*. Both species possess cytotoxic, antioxidant, vasorelaxant, immunostimulant, antibacterial, antidiarrheal, and diuretic properties (Al-Snafi 2016).

C. pluricaulis var. macra Clarke (syn *C. prostratus* Forssk., *C. pluricaulis* Choisy) is a perennial, hairy, procumbent, and diffuse herb. The leaves are alternate, elliptic to oblong, lanceolate, obtuse, and mucronate. Its branches spread horizontally and grow up to 30 cm length. It is distributed in India, China, Nepal, and Bangladesh (Chopra et al. 1969; Kumar 2007). It is used in the treatment of hypertension, neurodegenerative disorders, ulcers, high blood pressure, epilepsy, vomiting, diabetes, sun stroke, and bleeding. Moreover, it is useful in improving memory and decreasing the cholesterol levels (Dubey et al. 2007). *C. pluricaulis* is used in the treatment of liver, epileptic, cytotoxic, microbial, viral, and central nervous system diseases. (Agarwal et al. 2014). As per clinical studies, *C. pluricaulis* demonstrated beneficial effects on the patients with anxiety neurosis. It reduced the levels of stress, mental fatigue, anxiety, and neuroticism arising due to various levels of stresses (Bhowmik et al. 2012).

DOI: 10.1201/9781003398035-11

The antidepression effects of petroleum ether, chloroform, and ethyl acetate fractions of ethanol extract of *C. pluricaulis* were evaluated in the forced swim test and tail suspension tests. Chloroform fraction (50 and 100 mg/kg) significantly decreased the immobility time in both forced swim test and tail suspension test. Chloroform fraction did not affect locomotor activity. Chloroform fraction restored the reserpine-induced extension of immobility period in forced swim and tail suspension tests, displaying a significant antidepressant-like activity in mice by interaction with the adrenergic, dopaminergic, and serotonergic systems (Dhingra and Valecha 2007; Chen et al. 2018). Aqueous-methanol extract of *C. pluricaulis* (100 mg/kg, p.o.) displayed maximum nootropic and anxiolytic effects ($p < 0.001$) but did not demonstrate any antidepressant effect. The extract possesses nootropic, anxiolytic, and central nervous system–depressant activities; therefore, it may be used in increasing the memory-related actions (Malik et al. 2011). Methanol extract of *C. pluricaulis* aerial parts was tested on a chronic animal model of depression. The extract (50 and 100 mg/kg) significantly enhanced the sucrose preference index, decreased immobility time in the forced swimming test, and enhanced the number of squares crossed, the number of rearings in the open field test, and locomotion in the chronic unpredictable mild stress-exposed rats. In addition, the extract (50 and 100 mg/kg) reversed the increased levels of pro-inflammatory cytokines viz., IL-1β, IL-6, TNF-α and liver biomarkers (alanine aminotransferase, aspartate aminotransferase) to normal levels. In addition, the extract (50 and 100 mg/kg, for one week) normalized the levels of serotonin and noradrenaline in the hippocampus as well as in the prefrontal cortex of the chronic unpredictable mild stress-exposed rats. Therefore, the extract produced significant antidepressant-like effect in the stressed rats (Gupta and Fernandes 2019). Ethanol extract of *C. pluricaulis* flower petals (100, 200, and 400 mg/kg, p.o.) significantly increased the time spent on open arms after acute, sub-acute, and chronic extract treatment. Lower doses of extract did not affect locomotor effects. As per results, the petal extract of the plant produced anti-anxiety effects in mice on an elevated plus-maze (Sharma et al. 2009).

Chloroform and ethanol extracts of *C. pluricaulis* displayed significant neuroprotective effect by reducing the levels of lipid peroxidation ($p < 0.001$) and by enhancing the levels of superoxide dismutase ($p < 0.01$, $p < 0.001$), catalase ($p < 0.01$, $p < 0.001$), glutathione ($p < 0.001$), and total thiol ($p < 0.001$) in treated animals. The cerebral infarction area, blood brain barrier disruption, and microtubule-associated protein 2 immunohistochemical and histopathological parameters supported the protective effect of the methanol extract. *C. pluricaulis* possesses strong neuroprotection against bilateral common carotid artery occlusion stimulated cerebral ischemic reperfusion injury (Shalavadi et al. 2020). Ethanol extract of *C. pluricaulis* and its ethyl acetate and aqueous fractions were examined for their memory increasing property. Two doses (100 and 200 mg/kg, p.o.) of ethanol extract and ethyl acetate and aqueous fractions were given in separate groups of animals. Both the doses of extract and fractions showed significant improvement in learning and memory of treated rats. Both doses significantly changed the amnesia stimulated by scopolamine (0.3 mg/kg, i.p.). In addition, extract and its fractions demonstrated strong memory-increasing effects in the step-down and shuttle-box avoidance paradigms (Nahata et al. 2008). Ethanol extracts of *A. racemosus* and *C. pluricaulis* whole plants (200 mg/kg, po) displayed a greater percentage of retentions than piracetam. A lower percentage of retention in older mice (18–20 months) was compared with younger mice. Older mice displayed greater transfer latency values on the first and second day (after 24 h) as compared to younger mice, revealing impairment in learning and memory of treated mice. Ethanol extracts (both species; for 7 days) increased the memory in treated older mice. A greater retention percentage was recorded in extract (200 mg/kg, po)-treated mice than piracetam (10 mg/kg, po). In the case of acetylcholinesterase activity, the extract increased the hippocampal regions connected with learning and memory functions. The basic mechanism of these functions of ethanol extracts may be associated to their antioxidant, neuroprotective, and cholinergic activities (Sharma et al. 2010).

The neuroprotective effects of aqueous extract of *C. pluricaulis* were investigated against aluminum chloride stimulated neurotoxicity in the rat cerebral cortex. Daily intake of aqueous extract

(150 mg/kg, for 3 months) together with aluminum chloride (50 mg/kg) reduced the increased acetylcholine esterase activity and also suppressed the reduction in Na$^{(+)}$/K$^{(+)}$ATPase activity. The extract maintained the levels of mRNA of muscarinic receptor 1, choline acetyl transferase, and nerve growth factor-tyrosine kinase A receptor in treated animals. It also boosted the expression of upregulated protein of cyclin dependent kinase5 as stimulated by aluminum. The ability of aqueous extract to suppress aluminum-stimulated toxicity was compared with rivastigmine tartrate (standard drug, 1 mg/kg; Bihaqi et al. 2009). The neuroprotective activity of *C pluricaulis* aqueous extract was examined against scopolamine-induced neurotoxicity in the cerebral cortex of male Wistar rats. Aqueous extract (150 mg/kg bw) significantly decreased the scopolamine-stimulated enhancement in the transfer latency in an elevated plus-maze test, while in a Morris water maze test, the extract increased the impairment of spatial memory. The acetylcholinesterase activity was significantly suppressed by extract treatment within the cortex and hippocampus. Reduced levels of glutathione reductase, superoxide dismutase, and reduced glutathione within the cortex and hippocampus stimulated by scopolamine were increased by the extract. Results revealed that the extract may apply its potent-enhancing effect via both anti- acetylcholinesterase activity and antioxidant accomplishments (Bihaqi et al. 2011). Ethanol extract of *C. pluricaulis* roots (150 mg/kg) decreased the elevated levels of protein and mRNA of tau and AβPP followed by a decrease in the levels of Aβ in scopolamine treated rats. The ability of the extract to stop scopolamine neurotoxicity was observed in microscopic analysis. Extract attenuated neurotoxic effect of scopolamine revealed its potential to act as a potent neuroprotective agent (Bihaqi et al. 2012). Aqueous-methanol extract of *C. pluricaulis* and its ethyl acetate, butanol, and aqueous fractions were tested for their neuroprotective effects against 3-nitropropionic acid stimulated neurotoxicity in rats. The 3-nitropropionic acid induced Huntington's disease–like features (decreased body weight, locomotor activity, memory, grip strength, and oxidative defense). Aqueous-methanol extract (200 mg/kg), ethyl acetate (30 mg/kg), and butanol (50 mg/kg) fractions significantly ($p < 0.001$) reduced the 3-nitropropionic acid induced reduction in locomotor activity, grip strength, memory, body weight, and oxidative defense. The study results suggest that aqueous-methanol extract possesses a protective role against 3-nitropropionic acid-stimulated neurotoxicity and may be further explored for its activity against Huntington's disease (Malik et al. 2015). The methanol extract and its chloroform, ethyl acetate, n-butanol, and aqueous fractions were also tested for their neuroprotective actions in 3-nitropropionic acid–induced Huntington's disease symptoms in Wister rats. Systemic delivery of 3-nitropropionic acid displayed marked motor deficits and oxidative injury in rats. Scopoletin was isolated from ethyl acetate fraction and demonstrated significant antioxidant effects in treated animals. Study results displayed that ethyl acetate fraction (20 mg/kg) significantly reduced the loss in body weight, but increased the locomotor activity, and grip strength in treated rats. It reduced the elevated levels of malondialdehyde and nitrite and normalized the activities of superoxide dismutase and decreased glutathione in the striatum and cortex of 3-nitropropionic acid-treated rats. The study suggests that *C. pluricaulis* extract and its fractions possess neuroprotective activity by elevating the brain antioxidant defenses in 3-nitropropionic acid-treated animals (Kaur et al. 2016).

Neuroprotective effects of aqueous extract of *C. pluricaulis* were examined against human microtubule–associated protein tau–stimulated neurotoxicity in an Alzheimer's disease *Drosophila* model. Aqueous extract significantly neutralized the human microtubule-associated protein tau–stimulated early death and extended the life span and minimized the level of τ protein in tauopathy *Drosophila*. The extract also enhanced the activities of antioxidant enzymes and boosted the τ-stimulated oxidative stress, normalizing the exhausted acetylcholinesterase activity in *Drosophila* model. As per the study, *C. pluricaulis* extract together with the regular standard food improved the neurotoxic effect of human microtubule–associated protein tau in an Alzheimer's disease *Drosophila* model; therefore, it may be used as a potent neuroprotective agent (Anupama Kizhakke et al. 2019). Methanol extract of *C. pluricaulis* showed significant neuroprotective effects against β-amyloid stimulated neurotoxicity in Neuro 2A cells. The effect may be associated with the presence of steroids (stigmasterol and betulinic acid), coumarins (scopoletin), and flavonoids (β-carotene and chlorogenic acid) in methanol extract. Therefore, it may be used as a promising candidate in the management of neuronal disorders (Sethiya et al. 2019).

Ethanol extract of *C. pluricaulis* whole plant (administered twice a day for five consecutive days, 375 and 750 mg/kg bw) displayed significant antiulcerogenic activity in ethanol, aspirin, 2 hr cold restraint stress, and 4 hr pyloric ligation–induced gastric ulcer models. The results were compared with the reference drug sucralfate (250 mg/kg). The antiulcerogenic activity of extract was associated with enlargement of mucin secretion, life span of mucosal cells, and glycoproteins rather than acid-pepsin (Sairam et al. 2001). The scopoletin content was recorded as 0.1738% in *C. pluricaulis* and possessed antihypertensive, antiproliferative, anti-inflammatory, neurological, antidopaminergic, anti-adrenergic, and antidiabetic properties (Firmansyah et al. 2021). The hydroalcoholic extract of *C. pluricaulis* showed the presence of alkaloids, steroids, tannins, terpenoid, and phenol contents (Swamy et al. 2019). The total phenolic content of *C. pluricaulis* was 668.143 + 1.107 µg gallic acid equivalents/ml extract. The inflammation was significantly decreased at 50, 100, and 300 µg/ml in *C. pluricaulis* in heat hemolysis. The extract showed effective suppression of hemolysis in two *in vitro* models such as heat-stimulated and hypotonic solution–stimulated hemolysis (Jain and Patil 2020; Yin et al. 2022).

The methanol extract of *C. pluricaulis* significantly improved the lipid anomalies in high fat-diet–streptozotocin-treated experimental animals ($p < 0.001$) but failed to change the hyperglycaemia produced in diabetic rats ($p > 0.05$). Glipizide displayed a preventive effect on both lipid anomalies and hyperglycemia by changing lipid parameters and the levels of insulin. The study results confirm that methanol extract exhibits significant hypolipidemic effects (Garg et al. 2020). Similarly, the antihyperlipidemic effect of methanolic extract of *C. pluricaulis* (100, 200, and 400 mg/kg of extract, and 65 mg/kg of Fenofibrate, reference drug) was examined on triton WR-1339–induced hyperlipidemia in rats. Methanol extract significantly reduced the levels of total cholesterol, triglycerides, low-density-lipoprotein-cholesterol, malonaldehyde, and atherogenic index, whereas high-density-lipoproteins-cholesterol and glutathione levels were increased. Methanol extract (400 mg/kg) displayed a consistent effect on all parameters of lipid profiles. Lower doses (100 and 200 mg/kg) did not show a significant decrease in levels of low-density-lipoproteins-cholesterol (Garg et al. 2018).

The hepatoprotective effects of aqueous extracts of *C. pluricaulis* leaves were evaluated against thioacetamide–stimulated liver injury in rats. The aqueous extract of *C. pluricaulis* (200 mg/kg, 400 mg/kg, and 600 mg/kg) was administered to treated groups for 21 days. The different parameters such as alanine transferase, aspartate transferase, alkaline phosphatase, total bilirubin, direct bilirubin, albumin, total protein, ions, clotting time, and liver weight were analyzed in treated animals. The obtained results suggest that aqueous extract of leaves possess significant hepatoprotective effects (Ravichandra et al. 2013). Ethanol extract of *C. pluricaulis* roots reduced the serum T3 concentration and hepatic 5′-monodeiodinase and glucose-6-phosphatase activities without changing in hepatic lipid peroxidation, revealing the possible regulation of hyperthyroidism. It seems that the actions of ethanol extract on thyroid functions is basically mediated via the suppression of hepatic 5′-monodeiodinase activity (Panda and Kar 2001).

The methanol extract of *C. pluricaulis* whole plant demonstrated significant antibacterial activity against Gram-negative (*E. coli* ATCC 8739) and Gram-positive (*S. aureus* ATCC 6538) bacterial species. The results were compared with tetracycline (standard compound). It is concluded that *C. pluricaulis* methanol extract is more active against *E. coli* than *S. aureus* (Verma et al. 2011).

Ethanol extract of *C. pluricaulis* leaves demonstrated antioxidant ability in free radical scavenging assay. The extract suppressed H_2O_2 stimulated macromolecule injury [plasmid DNA damage and 2,2′-azobis (2-amidinopropane) dihydrochloride-stimulated oxidation] of bovine serum albumin and lipid peroxidation of hepatic tissues of rats. Ethanol extract (50 µg/ml) demonstrated 50% cell survival against a 100 µM H_2O_2 challenge for 24 h, and it also decreased the lactate dehydrogenase leakage. *C. pluricaulis* extract normalized and regulated the antioxidant and apoptosis markers (superoxide dismutase, catalase, p53, and caspase 3) as well as suppressed the formation of reactive oxygen species and depolarization of the mitochondrial membrane (Rachitha et al. 2018). Methanol extract of *C. pluricaulis* whole plant demonstrated significant antioxidant effects in 1, 1-diphenyl-2-picryl-hydrazyl free radical scavenging assay (IC_{50} for methanol extract 41 μg/ml; 2.03 μg/ml ascorbic acid). Methanol extract (500 and 1000 mg/kg) did not eradicate hind limb extension but decreased the mean recovery time from convulsion in treated animals (Verma et al. 2012).

The presence of kaempferol, ferulic acid, caffeic acid, quercetin umbelliferone, scopoletin, asculetin, and scopoline was confirmed in methanol extract of *C. fatmensis*. Methanol extract (400 mg/kg bw) significantly decreased the number of alcohol-induced gastric ulcers (curative ratio; 32.6%). Extract increased high curative ratio (75%), but reduced the number of gastric ulcers, total acidity, and total proteins in aspirin-stimulated gastric ulcers ($p < 0.05$). The extract (500 mg/kg bw, p.o.) did not affect urine output or sodium excretion, whereas potassium and chloride excretion were significantly enhanced in treated animals. The specific dose (1000 mg/kg bw) significantly enhanced the levels of excretion of sodium, potassium, and chloride but did not affect the volume of urine. In addition, it showed mild hepatoprotective effect against carbon tetrachloride-induced hepatotoxicity in treated animals. No indications of acute toxicity were recorded after oral administration (2.75 g/kg bw dose) of the extract in treated animals (Atta et al. 2007).

Ethanol extract of *C. pilosellifolius* (500 mg/kg, p.o.) displayed strong antiulcerogenic effect in ethanol-stimulated ulcer models in rats (protection 84.6%). Kaempferol, quercetin, amanitate, and asmatol have been isolated and characterized from ethanol extract. Two isolated compounds (amanitate and asmatol; 50 mg/kg) demonstrated potent antiulcerogenic effects (95.4% and 55.84% protection) in treated animals. Kaempferol and quercetin displayed significant antiulcerogenic effects (78.38% and 5.38% protection). Ethanol extract (LD_{50} 5000 mg/kg) was found to be more highly efficacious and safe than isolated compounds. The administration of ethanol extract (500 mg/kg, p.o.) to rats (for 35 consecutive days) displayed no changes in the liver and kidney roles (Awaad et al. 2016). The profiles of quercetin and kaempferol (1.20 and 0.79 μg/ml) were estimated by high performance liquid chromatography from methanol extract of *C. pilosellifolius* (Al-Rifai et al. 2015).

C. phrygius methanol extract exhibited maximum antioxidant activity in 2,2-azinobis-(3-ethylbenzothiazoline-6-sulfonate), nitric oxide, ferric reducing antioxidant power, phosphomolybdenum, and metal chelating assays; therefore, methanol extract may be used as a potent natural antioxidant agent (Ozay and Mammadov 2019).

Methanol extract of *C. microphyllus* caused changes in the general behavior pattern, decrease in activity of spontaneous motor, hypothermia, and consolidation of pentobarbitone sleeping time in exploratory behavioral patterns and inhibition of aggressive behavior. The extract also displayed a suppressive effect on conditioned avoidance response and antagonism to amphetamine toxicity. The study results suggest that the whole plant extract possesses a strong central nervous system-depressant effect (Pawar et al. 2001).

C. austroaegyptiacus Abdallah & Sa'ad is a creeping or twining herb or shrub. The leaves are lobed. The flowers are funnel-shaped, and with 5-angled or 5-plaited limbs (Arora and Malhotra 2011). In Saudi Arabian traditional medicine, it is used in the treatment of ulcers (Bajpai et al. 2010; Banerjee et al. 2008). The total flavonoid contents of this species showed significant antioxidant activity in DPPH radical scavenging assay (IC_{50} 17.62 μg/ml). The total flavonoid content also demonstrated antibacterial activity against *E. coli*, *P. aeruginosa*, and *B. subtilis* (Al-Rifai et al. 2017).

The *C. scammonia* resin is widely used in Unani medicine to treat skin infections, chronic headache, bilious fever, conjunctivitis, and jaundice (Ansari et al. 2022; Soriano et al. 2022). The ethanolic extract of *C. betonicifolia* showed DPPH scavenging, ABTS scavenging, ferric ions reduction, cupric ions reducing capacity, and ferrous ions binding assays. The extract exhibited activity against acetylcholinesterase (IC_{50} 1.946 μg/ml), α-glycosidase (IC_{50} 0.815 μg/ml), and α-amylase (IC_{50} 0.675 μg/ml) enzymes (Bingol et al. 2021).

11.2 PHYTOCHEMISTRY

The coumarins (umbelliferone, esculetin, scopoletin, and scopoletin 7-*O*-glucoside), flavonoids and phenolic acids [15-hydroxyisocostic acid, isocostic acid 4-carboxaldehyde, methyl 15-oxo-eudesome-4,

11(13)-diene-12-oate, 1α,9α-dihydroxy-α-cyclo-costunolide, isorhamnetin 3-sulphate, isorhamnetin 3-*O*-rutinoside, rhamnetin, epicatechin, 6,3'-dihydroxy-3, 5,7,4'-tetramethoxyflavone, 3,6,7,3',4'-pentamethoxy flavone, 6,4'-dihydroxy-3, 7-dimethoxyflavone, 6,4-dihydroxy-3,5,7-trimethoxyflavone, protocatechic, *p*-hydroxybenzoic, syringic, vanillin, benzoic, ferulic, caffeic, gentisic, *p*-coumaric, syringic, vanillic, *p*-hydroxyphenylacetic, and *p*- hydroxybenzoic acids), lipids (n-butyric, iso-butyric, palmitic, oleic, stearic, behenic, linolenic, linoleic, methyl- 7,10-octadecadienoate, and arachidic acids), α-amyrin, campesterol, stigmasterol, β-sitosterol, apigenin, chrysin, genistein, hesperidin, kaempferol, luteolin, myricetin, naringenin, quercetin, rutin, tricine, vitexin, cuscohygrine, dihydroquercetin (taxifolin), neophytadiene, hexadecanamide, 9-octadecanamide, 1,2-benzendicarboxylic acid, stigment-5-en-3-ol, 5-β-pregn-7-en-3,20-dione, quercetagetin 3,5,6,7,3',4'-hexamethyl ether, sesquiterpene (eudesm-4(15),11(13)- diene-12,5β-olide), 3,5-dicaffeoyl quinic acid, β-methylesculetin, calystegins, convovulin, umbelliferone, chlorogenic acid, esculetin, scopoletin, and scopoletin-7-*O*-glucoside, kaempferol sugar derivatives (3-*O*-β-D-glucoside, 7-*O*-β-D-glucoside, 3-*O*-α-L-rhamnosyl, 7-*O*-β-D-glucoside, 3-*O*-rutinoside, 7-*O*-rutinoside, 3-*O*-α-L-rhamnoside, and 3-*O*-β-D-galactorhamnoside), quercetin sugar derivatives (3-*O*-α-L-rhamnoside and 3-*O*-rutinoside), alkaloids (pseudotropine, tropine, tropinone, meso- cuscohygrine, hygrine, calystegines, and atropine have been reported from *C. arvensis* (Al-Enazi 2018; Bazzaz and Haririzadeh 2003; Borchardt et al. 2008; Raza et al. 2012; Edrah et al. 2013; Khan et al. 2015, 2017; Miri et al. 2013; Salehi et al. 2020). Tropane alkaloids (tropine, pseudotropine, and tropinone) and the pyrrolidine alkaloids (cuscohygrine and hygrine), polyhydroxytropanes (cuscohygrine and calystegines) have been reported from *C. arvensis* roots. Pseudotropine is known to influence motility and might represent a responsible agent for the observed cases of equine intestinal fibrosis (Todd et al. 1995).

The presence of stigmasterol and rutin was determined in aerial parts of *C. arvensis* (Alwan and Hamad 2020). Six glycosidic acids (arvensic acids E–J) have been isolated and identified from a glycosidic acid fraction *C. arvensis* whole plant (Fan et al. 2019). Similarly, four glycosidic acids (arvensic acids A-D) were reported from the alcoholic extract of *C. arvensis* whole plant (Fan et al. 2018). Two glycosidic acids (arvensic acids K and L) were obtained from the resin glycoside fraction of *C. arvensis* whole plant (Lu et al. 2019).

The presence of scopoletin was determined by high performance thin layer chromatography from whole plant (Kapadia et al. 2006), β-carotene, rutin, scopoletin, chlorogenic acid, and mangiferin was confirmed by thin layer chromatography (Sethiya et al. 2013), ursolic acid, betulinic acid, stigmasterol, and lupeol were estimated by high performance thin layer chromatography from *C. pluricaulis* (Sethiya and Mishra 2015).

A total of 24 compounds (*trans*-pinocarveol, *trans*-verbenol, verbenone, *trans*-carveol, methyl carvacrol, β-maaliene, α-copaene, β-caryophyllene, 2,5-dimethoxy-*p*-cymene, α-humulene, (*E*)-geranylacetone, germacrene D, β-selinene *cis*-β-guaiene, (*E,E*)-α-farnesene, δ-cadinene, germacrene B, caryophyllene oxide, *cis*-arteannuic alcohol, 1-*epi*-cubenol, t-cadinol, t-muurolol, α-cadinol pentadecanal) were determined by GC-MS analysis from *C. althaeoides* flowers. The essential oils showed significant cytotoxic activity against human breast cancer MCF-7 cells (IC_{50} ¼ 8.16 mg/ml; Hassine et al. 2014). An acylated anthocyanin trioside {cyanidin 3-*O*-[6-*O*-(4-*O*-(6-*O*-(*E*-caffeoyl)-β-D-glucopyranosyl)-α-L-rhamnopyranosyl)-β-D-glucopyranoside]-5-*O*-β-D-glucopyranoside} has been identified from *C. althaeoides* flowers (Cabrita 2015).

A glycosidic acid (scammonic acid A) together with isobutyric, 2S-methylbutyric and tiglic acid, two kinds of resin glycosides (scammonin I and II) from ether-soluble resin glycoside extract (Noda et al. 1990), four kinds of ether-soluble resin glycosides (scammonins III-VI) obtained from roots (Kogetsu et al. 1991), and two minor ether-soluble resin glycosides (scammonins VII and VIII) together with (2S)-2-methylbutyric acid and tiglic acid, orizabic acid A and glycosidic acids (scammonic acid and B) were isolated and characterized from the roots of *C. scammonia* (Noda et al. 1992).

2,4-*N*-methylpyrrolidinylhygrine, *N*-methylpyrrolidinylcuscohygrine A, hygrine, tropinone, 5-(2-oxopropyl)-hygrine, cuscohygrine, 5-(2-hydroxypropyl)-hygrine, 2-[2-hydroxy-3-(N-methyl-

2-pyrrolidinyl)-propanyl]-N-methylpyrrolidine, 2,4-*N*-methylpyrrolidinylhygrine,2,5-Di-(2-hydroxy-propyl)-*N*-methylpyrrolidine, and phygrine have been reported from *C. lanatus* aerial parts and roots (El-Shazly and Wink 2008). The β-sitosterol, stigmasterol, vanillin, vanillic acid, syringic acid, ferulic acid, isoferulic acid, isoscopoletin, and stilbene carboxylic acid were reported from Saudi Arabian *C. hystrix* (Dawidar et al. 2000). Quercetin, quercetin 3-*O*-rutinoside, quercetin-7-*O*-rhamnoside, and ferulic acid have been isolated and characterized from ethyl acetate and *n*-butanol extracts of *C. hystrix*. The ethyl acetate extract displayed significant antioxidant activity in 2–2-diphenyl-1-picrylhydrazyl assay (Mohammed Donia et al. 2011).

Phellandrenes (α and β), *p*-hydroxyphenylacetic acid, scopoletin, ferulic acid, syringic acid, pinosylvin, apigenin/galangin, naringenin, kaempferol/luteolin/fisetin, eriodictyol/aromadendrin, quercetin, taxifolin scopoline from *C. austroaegyptiacus* and phellandrenes, *p*-hydroxyphenylacetic acid, protocatechuic acid/gentisic acid, vanillic acid, scopoletin, ferulic acid, syringic acid, pinosylvin, apigenin/galangin, naringenin, kaempferol/luteolin/fisetin, eriodictyol/aromadendrin, quercetin, taxifolin, myricetin, and scopolin were reported from ethanol extract of *C. pilosellifolius* (Al-Rifai et al. 2017).

An alkaloid (convolamine), flavonoid (kaempferol), and phenolics (scopoletin, β-sitosterol and ceryl alcohol) have been separated from *C. prostratus* (Balkrishna et al. 2020). The levels of scopoletin were estimated by LC-MS from the leaf, stem, and root of *C. prostratus* at different growth stages of plant development (30, 45, 60, and 90 days after sowing). A maximum level of scopoletin was recorded in stem (732 μg/g dw) and leaf (650 μg/g dw), collected 90 days after sowing, while a minimum quantity was recorded at 45 days of sowing in leaf (90.00 μg/g dw) and stem (110 μg/g dw). The study revealed that the growth stage of the plant and expression of F6'H regulate the scopoletin accumulation in *C. prostratus* (Rafaliya Rutul et al. 2021). Apigenin, chrysin, genistein, hesperidin, isorhamnetin, kaempferol, luteolin, myricetin, naringenin, quercetin, rhamnetin, rutin, tricine, and vitexin were obtained from ethanol extract of *C. lineatus* flowers (Noori et al. 2017). Convolvine, convolinine, convolamine, convolidine, phyllalbine, and phyllalbine N-oxide and the aminoalcohol nortropine were isolated and characterized from *C. subhirsutus* aerial parts (Gapparov et al. 2007, 2010).

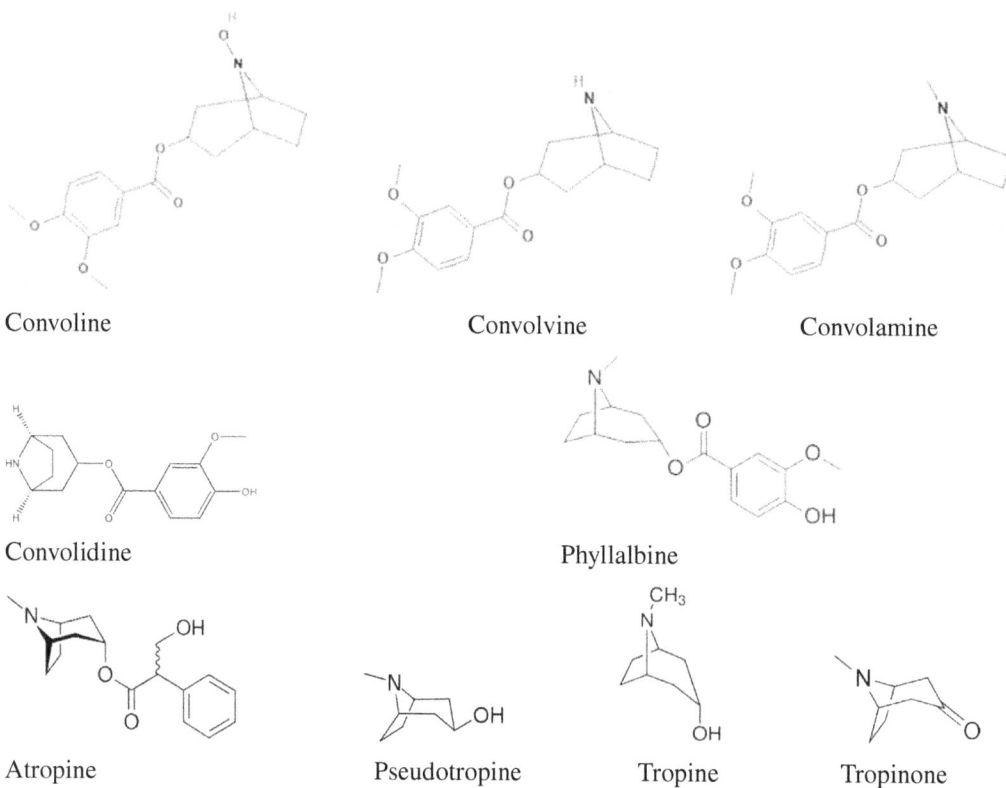

Convoline   Convolvine   Convolamine

Convolidine   Phyllalbine

Atropine   Pseudotropine   Tropine   Tropinone

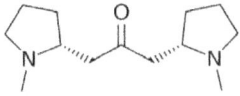

Meso- cuscohygrine

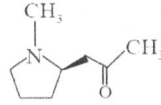

Hygrine

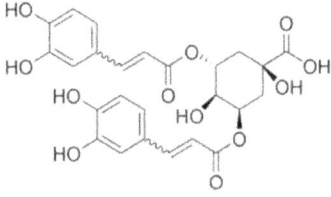

3,5-Dicaffeoyl quinic acid

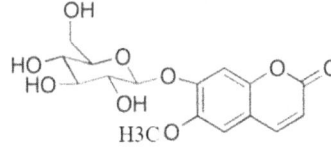

Scopoletin-7-*O*-glucoside

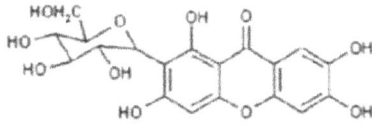

Mangiferin

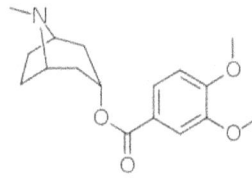

Convolamine

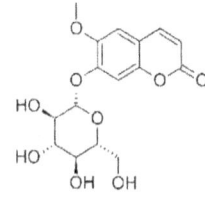

Scopolin

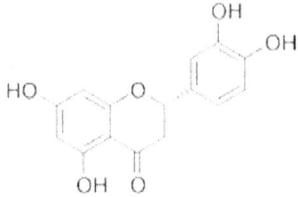

Eriodictyol

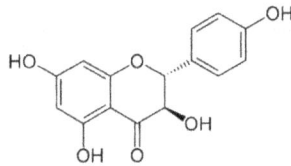

Aromadendrin

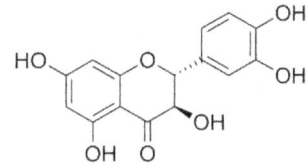

Taxifolin

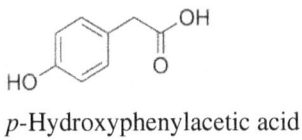

p-Hydroxyphenylacetic acid

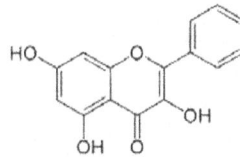

Galangin

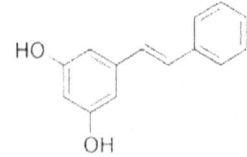

Pinosylvin

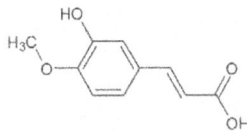

Isoferulic acid

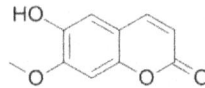

Isoscopoletin

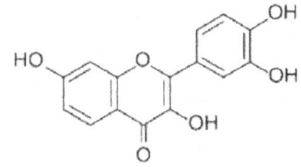

Fisetin

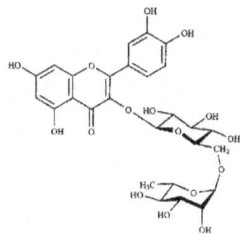

Quercetin 3-*O*-rutinoside

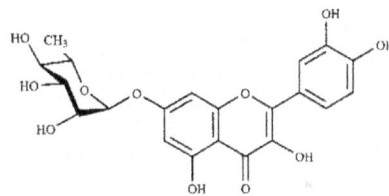

Quercetin-7-*O*-rhamnoside

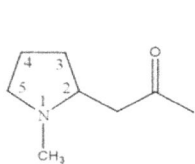
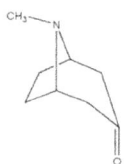
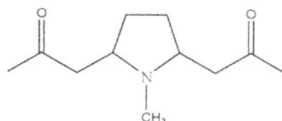
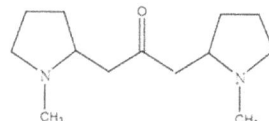

Hygrine   Tropinone   5-(2-Oxopropyl)-hygrine   Cuscohygrine

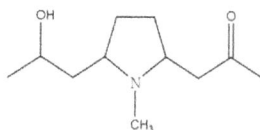

5-(2-Hydroxypropyl)-hygrine

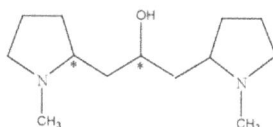

2-[2-Hydroxy-3-(N-methyl-2-pyrrolidinyl)-propanyl]-N-methylpyrrolidine

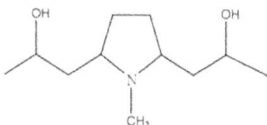
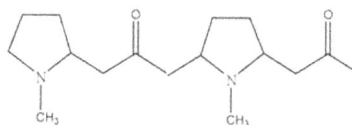

2,5-Di-(2-hydroxypropyl)-*N*-methylpyrrolidine   Phygrine

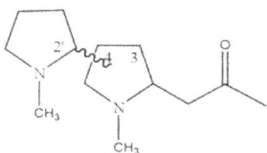
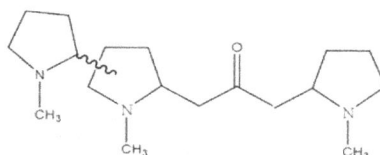

2,4-*N*-Methylpyrrolidinylhygrine   *N*-Methylpyrrolidinylcuscohygrine A

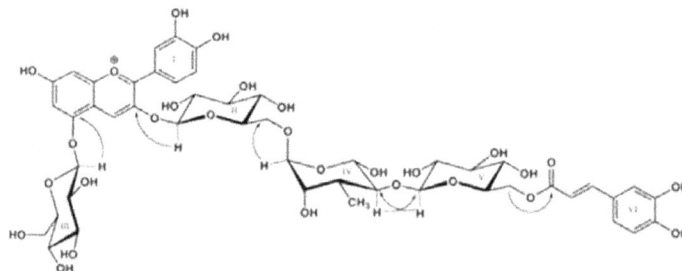

Cyanidin 3-*O*-[6-*O*-(4-*O*- (6-*O*-(*E*-caffeoyl)-β-D-glucopyranosyl)-α-L-rhamnopyranosyl)-β-D-glucopyranoside]-5-*O*-β-D-glucopyranoside

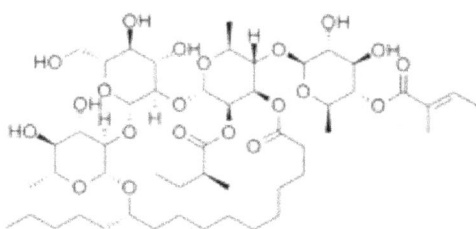
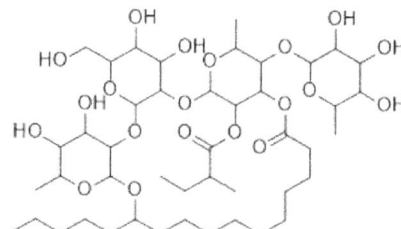

Scammonin I   Scammonin II

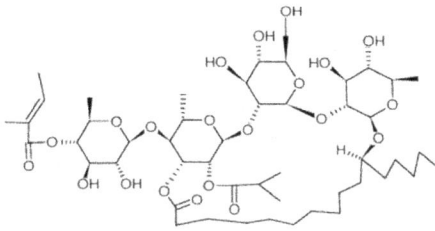

Scammonin III

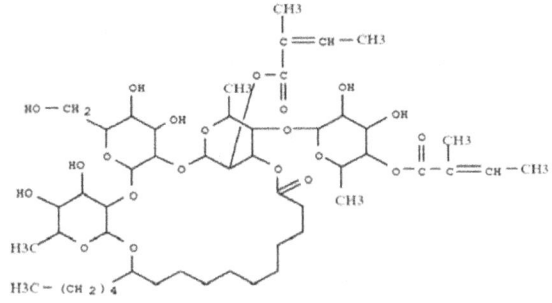

Scammonin IV

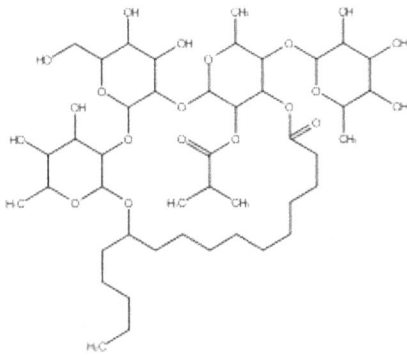

Sacmmonin V

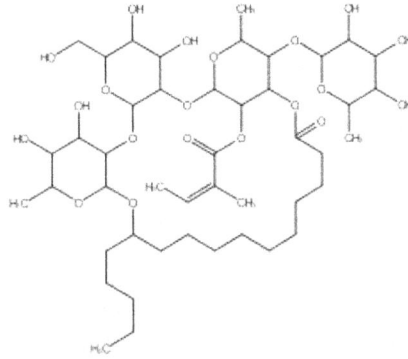

Scammonin VI

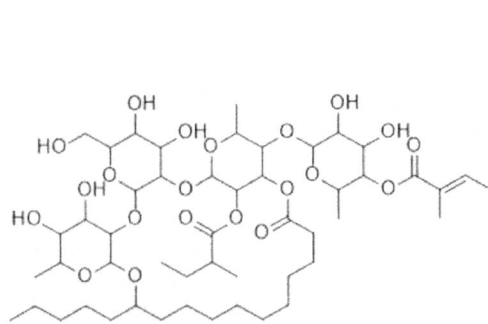

Scammonin VII

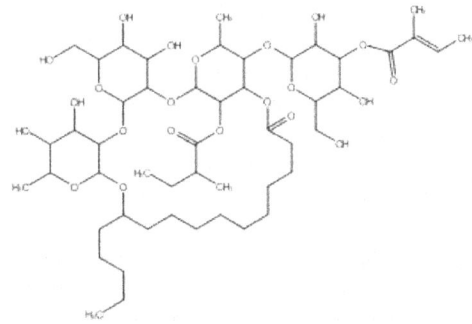

Scammonin VIII

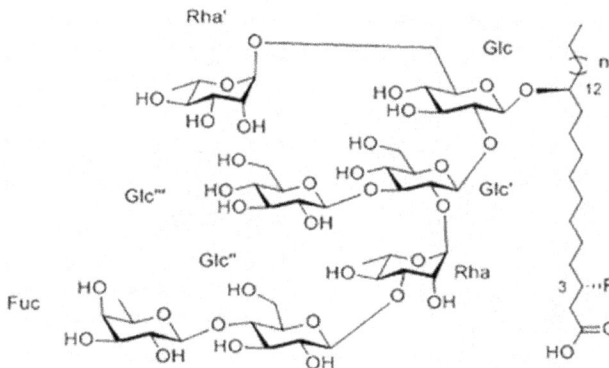

R = -H, n = 1 - Arvensic acid A; R = -H, n = 2 - Arvensic acid B; R = -OH, n = 1 - Arvensic acid C; R = -OH, n = 2 - Arvensic acid D

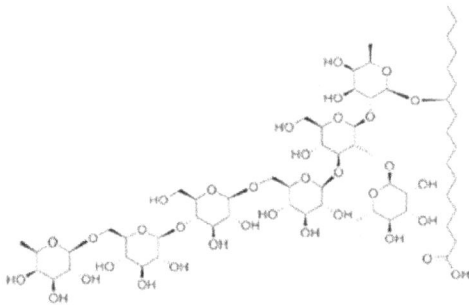

Arvensic acid E

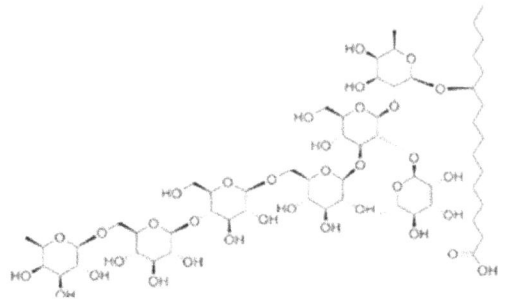

Arvensic acid F

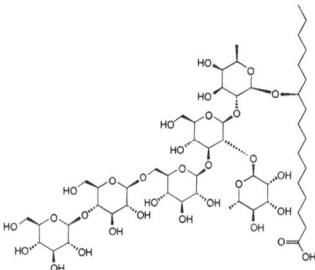

Arvensic acid G

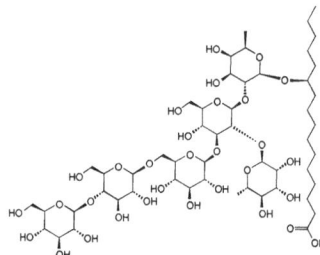

Arvensic acid H

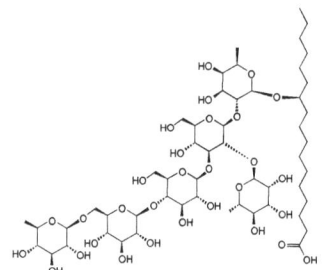

Arvensic acid I

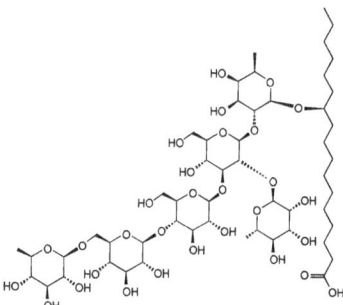

Arvensic acid J

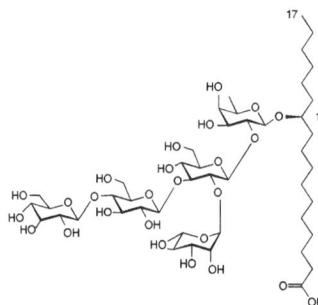

Arvensic acid K

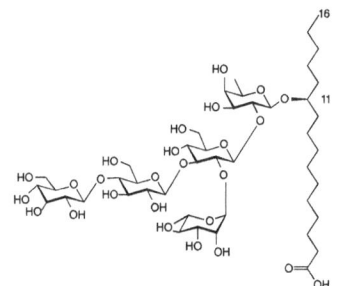

Arvensic acid L

11.3   CULTURE CONDITIONS

Pieces of roots, hypocotyls, stem, and leaves of *C. persicus* were transferred to an MS medium (with normal salts) for regeneration of callus. Maximum growth of callus was achieved with supplementation of 6-benzyl-amino-purine, 2,4-dichlorophenoxyacetic acid, and activated charcoal after 25–30 days of inoculation (Holobiuc et al. 2015). The terminal buds and flower explants of *C. alsinoides* were inoculated on an MS medium (Murashige and Skoog 1962) containing 2, 4-D (3 mg/l) and was found more effective in callus growth on terminal buds and flower explants. The developed calli were brownish white, friable, and found meristematic in nature. Kaempferol was isolated from the callus of *C. alsinoides* and showed potent effect in decreasing the levels of fat associated with Alzheimer's and Parkinson's diseases (Kaladhar 2012). Addition of mannitol, sorbitol, sucrose, or a mixture of calcium chloride and potassium chloride increased the growth of callus in *C. arvensis* suspension cultures (Ruesink 1978). Stem explants of *C. arvensis* were inoculated on an MS medium containing 2,4-dichlorophenoxy acetic acid (0.5 mgm/l), 15 per cent coconut milk, and kinetin (1 mg/l). Maximum callus growth was achieved with 2,4-dichlorophenoxy acetic acid (0.05 mgm/l) and 15 per cent coconut milk supplementation (Hill 1967). *C. pluricaulis* explants were inoculated on an MS medium supplemented with various concentrations of 2,4-dichlorophenoxy acetic acid and 6-benzylaminopurine and kinetin. The best callus growth was obtained on an MS medium with fortification of 2,4-dichlorophenoxy acetic acid (1.0 mg/l and 2.0 mg/l), 6-benzylaminopurine (1–3.0 mg/l) and kinetin (1.0 mg/l; Chandel and Kharoliwal 2017).

REFERENCES

Abu-Mejdad NMJ, Shaker HA and Al-Mazini MAA. 2010. The effect of aqueous and acetonic plant extracts of *Tagete patula* L, *Ammi visnaga* L and *Convolvulus arvensis* L in growth of some bacteria in vitro. J Basrah Res (Sci) 36, 23–32.

Agarwal P, Sharma B, Fatima A, Jain SK. 2014. An update on Ayurvedic herb *Convolvulus pluricaulis* Choisy. Asian Pac J Trop Biomed 4, 245–252.

Al-Asady AAB, Suker DK, Hassan KK. 2014. Apoptotic activity of glycoside extract (fraction I) from leaves of *Convolvulus arvensis* on rhabdomyosarcoma (RD) tumor cell line is associated with its induction of DNA damage. Int J Med Sci Clin Invent 1, 203–209.

Al-Enazi NM. 2018. Phytochemical screening and biological activities of some species of *Alpinia* and *Convolvulus* plants. Int J Pharmacol 14, 301–309.

Ali M, Qadir MI, Saleem M, Janbaz KH, Gul H, Hussain L, Ahmad B. 2013. Hepatoprotective potential of *Convolvulus arvensis* against paracetamol-induced hepatotoxicity. Bangladesh J Pharmacol 8, 300–304.

Alkofahi A, Batshoun R, Owais W, Najib N. 1996. Biological activity of some Jordanean medicinal plant extracts. Fitoterapia 67, 435–442.

Al-Rifai A, Aqel A, Al-Warhi T, Wabaidur SM, Al-Othman ZA, Yacine Badjah-Hadj-Ahmed A. 2017. Antibacterial, antioxidant activity of ethanolic plant extracts of some *Convolvulus* species and their DART-ToF-MS profiling. Evid Based Complement Altern Med 2017, Article ID 5694305.

Al-Rifai A, Aqel A, Awaad A, Al-Othman ZA. 2015. Analysis of quercetin and kaempferol in an alcoholic extract of *Convolvulus pilosellifolius* using HPLC. Commun Soil Sci Plant Anal 46, 1411–1418.

Al-Senani GM. 2020. Synthesis of ZnO-NPs using a *Convolvulus arvensis* leaf extract and proving its efficiency as an inhibitor of carbon steel corrosion. Materials (Basel) 13, 890.

Al-Snafi AE. 2016. The chemical constituents and pharmacological effects of *Convolvulus arvensis* and *Convolvulus scammonia*–a review. IOSR J Pharm 6, 64–75.

Alwan MH, Hamad MN. 2020. Phytochemical investigation of the aerial part of Iraqi *Convolvulus arvensis*. Iraqi J Pharm Sci 29, 62–69.

Amin, H., Sharma, R., Vyas, M., Prajapati, P., & Dhiman, K. (2014). Shankhapushpi (Convolvulus pluricaulis Choisy): Validation of the Ayurvedic therapeutic claims through contemporary studies. Int J Green Pharm 8, 193–200.

Ansari H, Ansari AP, Qayoom I, Reshi BM, Hasib A, Ahmed NZ, Anwar N. 2022. Saqmunia (*Convolvulus scammonia* L.), an important drug used in Unani system of medicine: A review. J Drug Deliv Ther 12, 231–238.

Anupama Kizhakke P, Shilpa O, Antony A, Siddanna Tilagul K, Gurushankara Hunasanahally P. 2019. *Convolvulus pluricaulis* (Shankhapushpi) ameliorates human microtubule-associated protein tau (hMAPτ) induced neurotoxicity in Alzheimer's disease Drosophila model. J Chem Neuroanat 95, 115–122.

Arora M, Malhotra M. 2011. A review on macroscopical, phytochemical and biological studies on *Convolvulus arvensis* (field bindweed). Pharmacologyonline 3, 1296–1305.

Atta AH, Mohamed NH, Nasr SM, Mouneir SM. 2007. Phytochemical and pharmacological studies on *Convolvulus fatmensis* Ktze. J Nat Remedies 7, 109–119.

Atta AH, Mouneir SM. 2004. Antidiarrhoeal activity of some Egyptian medicinal plant extracts. J Ethnopharmacol 92, 303–309.

Austin DF. 2000. Bindweed (*Convolvulus arvensis*, Convolvulaceae) in North America from medicine to menace. Bull Torrey Bot Club 127, 172–177.

Awaad AS, Al-Refaie A, El-Meligy R, Zain M, Soliman H, Marzoke MS, El-Sayed N. 2016. Novel compounds with new anti-ulcergenic activity from *Convolvulus pilosellifolius* using bio-guided fractionation. Phytother Res 30, 2060–2064.

Awaad AS, Al-Rifai AA, El-Meligy RM, Alafeefy AM, Alqasoumi SI. 2014. Antiulcerogenic activity of *Convolvulus* species. Austin Chromatogr 1, 9.

Bajpai V, Sharma D, Kumar B, Madhusudanan KP. 2010. Profiling of *Piper betle* Linn cultivars by direct analysis in real time mass spectrometric technique. Biomed Chromatogr 24, 1283–1286.

Balkrishna A, Thakur P, Varshney A. 2020. Phytochemical profile, pharmacological attributes and medicinal properties of *Convolvulus prostratus*—a cognitive enhancer herb for the management of neurodegenerative etiologies. Front Pharmacol 11, 171.

Banerjee S, Madhusudanan KP, Khanuja SPS, Chattopadhyay SK. 2008. Analysis of cell cultures of *Taxus wallichiana* using direct analysis in real-time mass spectrometric technique. Biomed Chromatogr 22, 250–253.

Bazzaz BSF, Haririzadeh G. 2003. Screening of Iranian plants for antimicrobial activity. Pharm Biol 41, 573–583.

Bhowmik D, Kumar KPS, Paswan S, Srivatava S, Yadav A, Dutta A. 2012. Traditional Indian herbs *Convolvulus pluricaulis* Choisy and its medicinal importance. J Pharmacogn Phytochem 1, 50–58.

Bihaqi SW, Sharma M, Singh AP, Tiwari M. 2009. Neuroprotective role of *Convolvulus pluricaulis* on aluminium induced neurotoxicity in rat brain. J Ethnopharmacol 124, 409–415.

Bihaqi SW, Singh AP, Tiwari M. 2011. *In vivo* investigation of the neuroprotective property of *Convolvulus pluricaulis* in scopolamine-induced cognitive impairments in Wistar rats. Indian J Pharmacol 43, 520–525.

Bihaqi SW, Singh AP, Tiwari M. 2012. Supplementation of *Convolvulus pluricaulis* attenuates scopolamine-induced increased tau and amyloid precursor protein (AβPP) expression in rat brain. Indian J Pharmacol 44, 593–598.

Bingol Z, Kızıltas H, Goren AC, Kose LP, Topal M, Durmaz L, Alwasel SH, Gulcin I. 2021. Antidiabetic, anticholinergic and antioxidant activities of aerial parts of shaggy bindweed (*Convulvulus betonicifolia* Miller subsp.)—profiling of phenolic compounds by LC-HRMS. Heliyon 7, e06986.

Borchardt JR, Wyse DL, Sheaffer CC, Kauppi KL, Fulcher RG, Ehlke NJ, Bey RF. 2008. Antioxidant and antimicrobial activity of seed from plants of the Mississippi river basin. J Med Plant Res 3, 707–718.

Cabrita L. 2015. A novel acylated anthocyanin with a linear trisaccharide from flowers of *Convolvulus althaeoides*. Nat Prod Commun 10, 1965–1968.

Chandel U, Kharoliwal S. 2017. An effective method for high frequency multiple shoots regeneration and callus induction of *Convolvulus pluricaulisc* Choisy: An important medicinal plant. Int J Pharm Bio Sci 8, 98–102.

Chen G-T, Lu Y, Yang M, Li J-L, Fan B-Y. 2018. Medicinal uses, pharmacology, and phytochemistry of Convolvulaceae plants with central nervous system efficacies: A systematic review. Phytother Res 32, 823–864.

Chopra RN, Chopra IC, Varma BS. 1969. Supplementary to Glossary of Indian Medicinal Plants. CSIR, New Delhi.

Dawidar AM, Ezmiriy ST, Abdel-Mogib M, el-Dessouki Y, Angawi RF. 2000. New stilbene carboxylic acid from *Convolvulus hystrix*. Pharmazie 55, 848–849.

Desta B. 1993. Ethiopian traditional herbal drugs. Part II. Antimicrobial activity of 63 medicinal plants. J Ethnopharmacol 39, 129–139.

Dhingra D, Valecha R. 2007. Evaluation of the antidepressant-like activity of *Convolvulus pluricaulis* choisy in the mouse forced swim and tail suspension tests. Med Sci Monit 13, 155–161.

Dubey NK, Kumar R, Tripathi P. 2007. Global promotion of herbal medicine: India's opportunity. Curr Sci 86, 37–41.

Edrah S, Osela E, Kumar A. 2013. Preliminary phytochemical and antibacterial studies of *Convolvulus arvensis* and *Thymus capitatus* plants extracts. Res J Pharmacogn Phytochem 5, 220–223.

El-Shazlya A, Wink M. 2008. Tropane and pyrrolidine alkaloids from *Convolvulus lanatus* Vahl. Z Naturforsch 63c, 321–325.

Fan B-Y, He Y, Lu Y, Yang M, Zhu Q, Chen G-T, Li J-L. 2019. Glycosidic acids with unusual aglycone units from *Convolvulus arvensis*. J Nat Prod 82, 1593–1598.

Fan B-Y, Lu Y, Yin H, He Y, Li J-L, Chen G-T. 2018. Arvensic acids A-D, novel heptasaccharide glycosidic acids as the alkaline hydrolysis products of crude resin glycosides from *Convolvulus arvensis*. Fitoterapia 131, 209–214.

Firmansyah A, Winingsih W, Manobi JDY. 2021. Review of scopoletin: Isolation, analysis process, and pharmacological activity. Biointerface Res Appl Chem 11, 12006–12019.

Gapparov AM, Aripova SF, Tashkhodzhaev B, Levkovich MG, Aripov O. 2010. Convolinine, a new alkaloid from *Convolvulus subhirsutus* of Uzbekistan flora. Chem Nat Comp 46, 590–592.

Gapparov AM, Razzakov NA, Aripova SF. 2007. Alkaloids of *Convolvulus subhirsutus* from Uzbekistan. Chem Nat Comp 43, 291–292.

Garg G, Patil A, Singh J, Kaushik N, Praksah A, Pal A, Chakrabarti A. 2018. Pharmacological evaluation of *Convolvulus pluricaulis* as hypolipidaemic agent in Triton WR-1339-induced hyperlipidaemia in rats. J Pharm Pharmacol 70, 1572–1580.

Garg G, Patil AN, Kumar R, Bhatia A, Kasudhan KS, Pattanaik S. 2020. Protective role of *Convolvulus pluricaulis* on lipid abnormalities in high-fat diet with low dose streptozotocin-induced experimental rat model. J Ayurveda Integr Med 11, 426–431.

Gupta GL, Fernandes J. 2019. Protective effect of *Convolvulus pluricaulis* against neuroinflammation associated depressive behavior induced by chronic unpredictable mild stress in rat. Biomed Pharmacother 109, 1698–1708.

Hassine M, Zardi-Berguaoui A, Znati M, Flamini G, Ben Jannet H, Hamza MA. 2014. Chemical composition, antibacterial and cytotoxic activities of the essential oil from the flowers of Tunisian *Convolvulus althaeoides* L. Nat Prod Res 28, 769–775.

Hill GP. 1967. Morphogenesis in stem-callus cultures of *Convolvulus arvensis* L. Ann Bot 31, 437–446.

Holobiuc I, Voichiță C, Cătană R. 2015. *In vitro* conservation of the critically endangered taxon *Convolvulus persicus* L. and regenerants evaluation. Oltenia J Stud Nat Sci 31, 51–59.

Jain G, Patil UK. 2020. Phytochemical investigations and in vitro anti-inflammatory activity of ethanolic extract of *Convolvulus pluricaulis* Choisy and *Evolvulus alsinoides* Linn.: A comparative study on shankhpushpi species. Ann Ayurvedic Med 9, 267–278.

Kaladhar DSVGK. 2012. An *in vitro* callus induction and isolation, identification, virtual screening and docking of drug from *Convolvulus alsinoides* Linn against aging diseases. Int J Life Sci Biotechnol Pharm Res 1, 93–103.

Kapadia NS, Acharya NS, Acharya SA, Shah MB. 2006. Use of HPTLC to establish a distinct chemical profile for Shankhpushpi and for quantification of scopoletin in *Convolvulus pluricaulis* Choisy and in commercial formulations of Shankhpushpi. J Planar Chromatogr 19, 195–199.

Kaur M, Prakash A, Kalia AN. 2016. Neuroprotective potential of antioxidant potent fractions from *Convolvulus pluricaulis* Chois in 3-nitropropionic acid challenged rats. Nutr Neurosci 19, 70–78.

Khan MU, Ghori N, Hayat MQ. 2015. Phytochemical analyses for antibacterial activity and therapeutic compounds of *Convolvulus arvensis* L, collected from the salt range of Pakistan. Adv Life Sci 2, 83–90.

Khan S, Ur-Rehman T, Mirza B, Ul-Haq I, Zia M. 2017. Antioxidant, antimicrobial, cytotoxic and protein kinase inhibition activities of fifteen traditional medicinal plants from Pakistan. Pharm Chem J 51, 391–398.

Kogetsu H, Noda N, Kawasaki T, Miyahara K. 1991. Scammonin III-VI, resin glycosides of *Convolvulus scammonia*. Phytochemistry 30, 957–963.

Kumar DC. 2007. Pharmacognosy can help minimize accidental misuse of herbal medicine. Curr Sci 93, 1356–1358.

Leporatti ML and Ivancheva S. 2003. Preliminary comparative analysis of medicinal plants used in the traditional medicine of Bulgaria and Italy. J Ethnopharmacol 87, 123–142.

Lu Y, He Y, Min Yang M, Fan B-Y. 2019. Arvensic acids K and L, components of resin glycoside fraction from *Convolvulus arvensis*. Nat Prod Res 35, 2303–2307.

Malik J, Choudhary S, Kumar P. 2015. Protective effect of *Convolvulus pluricaulis* standardized extract and its fractions against 3-nitropropionic acid-induced neurotoxicity in rats. Pharm Biol 53, 1448–1457.

Malik J, Karan M, Vasisht K. 2011. Nootropic, anxiolytic and CNS-depressant studies on different plant sources of shankhpushpi. Pharm Biol 49, 1234–1242.

Mehrafarin A, Meighani F, Baghestani MA, Labbafi MR, Mirhadi MJ. 2009. Investigation of morphophysiological variation in field bindweed (*Convolvulus arvensis* L.) populations of Karaj, Varamin, and Damavand in Iran. Afr J Plant Sci 3, 64–73.

Miri A, Sharifi Rad J, Mahsan Hoseini Alfatemi S, Sharifi Rad M. 2013. A study of antibacterial potentiality of some plants extracts against multi- drug resistant human pathogens. Ann Biol Res 4, 35–41.

Mohammed Donia AER, Alqasoumi SI, Awaad AS, Cracker L. 2011. Antioxidant activity of *Convolvulus hystrix* Vahl and its chemical constituents. Pak J Pharm Sci 24, 143–147.

Mohd Azman NA, Gallego MG, Juliá L, Fajari L, Almajano MP. 2015. The effect of *Convolvulus arvensis* dried extract as a potential antioxidant in food models. Antioxidants (Basel) 4, 170–184.

Munz PA, Keck DD. 1959. A California Flora. University of California Press, Berkeley.

Murashige T, Skoog F. 1962. A revised medium for rapid growth and bioassays with tobacco tissue culture. Physiol Plant 15, 473–497.

Nahata A, Patil UK, Dixit VK. 2008. Effect of *Convulvulus pluricaulis* Choisy. on learning behaviour and memory enhancement activity in rodents. Nat Prod Res 22, 1472–1482.

Noda N, Kogetsu H, Kawasaki T, Miyahara K. 1990. Scammonins I and II, the resin glycosides of radix scammoniae from Convolvulus scammonia. Phytochemistry 29, 3565–3569.

Noda N, Kogetsu H, Kawasaki T, Miyahara K. 1992. Scammonins VII and VIII, two resin glycosides from *Convolvulus scammonia*. Phytochemistry 31, 2761–2766.

Noori M, Bahrami B, Mousavi A, Khalighi A, Jafari A. 2017. Flower flavonoids of *Convolvulus* L. species in Markazi Province, Iran. Asian J Plant Sci 16, 45–51.

Ozay C, Mammadov R. 2019. Antioxidant activity, total phenolic, flavonoid and saponin contents of different solvent extracts of *Convolvulus phrygius* Bornm. Curr Pers MAPs 2, 23–28.

Panda S, Kar A. 2001. Inhibition of T3 production in levothyroxine-treated female mice by the root extract of *Convolvulus pluricaulis*. Horm Metab Res 33, 16–18.

Pawar SA, Dhuley JN, Naik SA. 2001. Neuropharmacology of an extract derived from *Convolvulus microphyllus*. Pharm Biol 39, 253–258.

Rachitha P, Krupashree K, Jayashree GV, Kandikattu HK, Amruta N, Gopalan N, Rao MK, Khanum F. 2018. Chemical composition, antioxidant potential, macromolecule damage and neuroprotective activity of *Convolvulus pluricaulis*. J Tradit Complement Med 8, 483–496.

Rafaliya Rutul V, Sakure Amar A, Parekh Mithil J, Sushil K, Amarjeet Singh ST, Desai Parth J, Patil Ghanshyam B, Mistri Jigar G, Subhash N. 2021. Study of dynamics of genes involved in biosynthesis and accumulation of scopoletin at different growth stages of *Convolvulus prostratus* Forssk. Phytochemistry 182, 112594.

Ravichandra VD, Ramesh C, Sridhar KA. 2013. Hepatoprotective potentials of aqueous extract of *Convolvulus pluricaulis* against thioacetamide induced liver damage in rats. Biomed Aging Pathol 3, 131–135.

Raza M, Fozia S, Rehman A, Wahab A, Iqbal H, Ullah H, Shah SM. 2012. Comparative antibacterial study of *Convolvulus arvensis* collected from different areas of Khyber Pakhtunkhwa, Pakistan. Int Res J Pharm 3, 220–222.

Ruesink AW. 1978. Leucine uptake and incorporation by *Convolvulus* tissue culture cells and protoplasts under severe osmotic stress. Physiol Plant 44, 48–56.

Sairam K, Rao CV, Goel RK. 2001. Effect of *Convolvulus pluricaulis* Chois on gastric ulceration and secretion in rats. Indian J Exp Biol 39, 350–354.

Salehi B, Krochmal-Marczak B, Skiba D, Patra JK, Das SK, Das G, Popović-Djordjević JB, Kostić AZ, Anil Kumar NV, Tripathi A, Al-Snafi AE, Arserim-Uçar DK, Konovalov DA, Csupor D, Shukla I, Azmi L, Mishra AP, Sharifi-Rad J, Sawicka B, Martins N, Taheri Y, Fokou PVT, Capasso R, Martorell M. 2020. *Convolvulus* plant—a comprehensive review from phytochemical composition to pharmacy. Phytother Res 34, 315–328.

Sethiya NK, Mishra S. 2015. Simultaneous HPTLC analysis of ursolic acid, betulinic acid, stigmasterol and lupeol for the identification of four medicinal plants commonly available in the Indian market as Shankhpushpi. J Chromatogr Sci 53, 816–823.

Sethiya NK, Mohan MK, Raja M, Mishra SH. 2013. Antioxidant markers-based TLC-DPPH differentiation on four commercialized botanical sources of Shankhpushpi (A Medhya Rasayana): A preliminary assessment. J Adv Pharm Technol Res 4, 25–30.

Sethiya NK, Nahata A, Mishra SH, Dixit VK. 2009. An update on Shankhpushpi, a cognition boosting Ayurvedic medicine. Zhong Xi Yi Jie He Xue Bao 7, 1001–1022.

Sethiya NK, Nahata A, Singh PK, Mishra SH. 2019. Neuropharmacological evaluation on four traditional herbs used as nervine tonic and commonly available as Shankhpushpi in India. J Ayurveda Integr Med 10, 25–31.

Shalavadi MH, Chandrashekhar VM, I S Muchchandi IS. 2020. Neuroprotective effect of *Convolvulus pluricaulis* Choisy in oxidative stress model of cerebral ischemia reperfusion injury and assessment of MAP2 in rats. J Ethnopharmacol 249, 112393.

Sharma K, Arora V, Rana AC, Bhatnagar M. 2009. Anxiolytic effect of *Convolvulus pluricaulis* choisy petals on elevated plus maze model of anxiety in mice. J Herb Med Toxicol 3, 41–46.

Sharma K, Bhatnagar M, Kulkarni SK. 2010. Effect of *Convolvulus pluricaulis* Choisy and *Asparagus racemosus* Willd on learning and memory in young and old mice: A comparative evaluation. Indian J Exp Biol 48, 479–485.

Soriano G, Fernández-Aparicio M, Masi M, Vilariño-Rodríguez S, Cimmino A. 2022. Complex mixture of arvensic acids isolated from *Convolvulus arvensis* roots identified as inhibitors of radicle growth of broomrape weeds. Agriculture 12, 585.

Swamy G, Holla R, Rao SR. 2019. Evaluation of hydroalcoholic extract of *Convolvulus pluriculis* (Shankapushpi) for standardization by colorimetric method: A preliminary report. J Health Allied Sci 9, 121–126.

Todd FG, Stermitz FR, Schultheis P, Knight AP, Traub-Dargatz J. 1995. Tropane alkaloids and toxicity of *Convolvulus arvensis*. Phytochemistry 39, 301–303.

Verma S, Sinha R, Kumar P, Amin F, Jain J, Tanwar J. 2012. Study of *Convolvulus pluricaulis* for antioxidant and anticonvulsant activity. Cent Nerv Syst Agents Med Chem 12, 55–59.

Verma S, Sinha R, Singh V, Tanwar S, Godara M. 2011. Antibacterial activity of methanolic extract of whole plant of *Convolvulus pluricaulis*. J Pharm Res 4, 4450–4452.

Yin Q, Abdulla R, Kahar G, Aisa HA, Li C, Xin X. 2022. Mass defect filtering-oriented identification of resin glycosides from root of *Convolvulus scammonia* based on quadrupole-orbitrap mass spectrometer. Molecules 27, 3638.

12 *Cordia* Species

12.1 MORPHOLOGICAL FEATURES, DISTRIBUTION, ETHNOPHARMACOLOGICAL PROPERTIES, PHYTOCHEMISTRY, AND PHARMACOLOGICAL ACTIVITIES

The Cordia genus (Fam.—Boraginaceae) consists of more than 300 species (Quattrocchi 2012) and are widely distributed in the tropical region of east Africa, Mexico, West Indies, Central America, South America, Pakistan, West Africa, Nigeria, Ghana, Sri Lanka, and India (Polhill 1991). Several Cordia species (*C. dichotoma, C. latifolia, C. macleodii, C. myxa, C. rothii* and *C. obliqua*) have been used in Ayurvedic, Unani, and Siddha medicines. *Cordia* species are used in the treatment of wounds, boils, tumors, gout, and ulcers. Leaf decoction is useful in flu, fever, cough, cold, asthma, menstrual cramps, dysentery, diarrhea, headache, and snakebites. It can be taken as tonic. Its bark is used as an astringent and liver stimulant. Decoction of roots is used to cure tuberculosis, bronchitis, and malaria. The poultice is prepared from leaves and is applied externally to treat migraine, inflammations, and wounds. Mucilaginous fruits are used as a demulcent, blood purifier, and in the treatment of complaints of spleen, kidneys, and lungs (Kirtikar and Basu 2005). Different parts (leaves, fruit, bark, and seed) of Cordia species (*C. obliqua* Willd., *C. dichotoma* G. Forst. and *C. collococca*) are used in the treatment of respiratory diseases, stomachache, wound, cough, dysentery, and diarrhea (Oza and Kulkarni 2017; Aimey et al. 2020). The different species of the Cordia genus (viz., *C. retusa* Vahl found in the Philippines, India, China, Japan, and different region of eastern and Southeast Asia). The leaves are used in the treatment of stomachache, cough, dysentery, and diarrhea (Pullaiah 2006). *C. rothii* Roem is found in India and Sri Lanka. The bark decoction is used as a gargle (Kirtikar and Basu 2005). *C. rufescens* A. DC Ramela de Velho is naturalized in areas of North-eastern Brazil. The plant species is used as abortive and anti-inflammatory agents and in the treatment of dysmenorrhea (de Oliveira et al. 2007). *C. salicifolia* Cham is native to Brazil and is found in tropical parts of Argentina and Paraguay. The plant is used as an appetite inhibitor, blood purifier, and diuretic agent (Caparroz-Assef et al. 2008). *C. sebestena* L is found in Brazil, Costa Rica, and Jamaica. Its leaves are used as an emollient and in the treatment of bronchitis, cough, fever, and influenza (Grandtner and Chevrette 2013). *C. spinescens* L is naturalized in North and South America including Mexico, Panama, Venezuela, and Peru. The whole plant is used in the treatment of skin infections and fever. Leaves are used externally on burns. Leaf decoction is used to sedating worms and to treat postpartum pain (Balick and Arvigo 2015; Lans 2007). *C. trichotoma* (Vell.) Arráb. ex Steud is found in Southern and northeastern parts Brazil, temperate regions of South America (Menezes et al. 2005; Roldao et al. 2008). *C. globifera* Smith Sak Hin is found in Thailand (Parks et al. 2010), and *C. globosa* (Jacq.) Kunth is distributed in the northern and southeastern United States. Leaf tea is useful in painful menstruation (Daniel 2004). *C. goetzei* Gurke is naturalized in Somalia, Kenya, and Tanzania. The water decoction of leaves and roots is used in the treatment of leprosy and malaria (Schmelzer and Gurib-Fakim 2008). *C. latifolia* Roxb is found in India and Pakistan. The fruits are used in the treatment of cough, chest pain, and for uterus and urethral complaints (Singh and Panda 2005). *C. leucocephala* Moric is found in northeastern Brazil. The leaves and roots are used to treat digestive complaints, dyspepsia, menstrual colic, and rheumatism (Diniz et al. 2008; Oliveira et al. 2012). *c. chacoensis* Chodat is naturalized in South America, Mexico, California, and the West Indies. Its leaves are used to treat anemia, stomachache, menstrual cramps, and respiratory diseases (Grandtner 2005). *C. obliqua* showed the presence of alkaloids, flavonoids, phenolics, and tannins. *C. obliqua* possesses antimicrobial, hypotensive, respiratory stimulant, diuretic, and anti-inflammatory properties (Gupta and Gupta 2014).

DOI: 10.1201/9781003398035-12

C. dichotoma G. Forst. is a medium-sized, deciduous tree, and consists of a short, crooked trunk. Leaves are simple, elliptical-lanceolate to broad ovate, entire, and slightly dentate, with a round and cordate base (Hussain and Kakoti 2013). Inflorescence is of corymb type. Flowers are stalked, hermaphroditic, and white to pink. Fruits are drupe and edible with gummy flesh mass (Patel et al. 2011a). It is distributed in Pakistan, Sri Lanka, Bangladesh, Nigeria, the Western Ghats in India, and tidal forests in Myanmar (Jamkhande et al. 2013), southern China, Malaysia, tropical Australia, and Polynesia (Lanting and Palaypayon 2002). Its leaves and stems are used in the healing of wounds and mouth sores. Its seeds are an aphrodisiac. The stem bark is useful in jaundice and ulcerative colitis (Kirtikar and Basu 1956; Nadkarni 1991; Chaubey et al. 2016). Different parts of this species are used to cure complaints of kidney, liver, spleen, heart, and blood. It also possesses antipyretic and anti-anemic properties (Nadkarni 1991; Anonymous 2004). Its fruits are used to treat colic pain, anxiety, and eczema, and to improve memory (Mishra and Garg 2011). Pulp and fruit fibers possess positive effects on leptin hormone and exhibit antioxidant activity (Patel et al. 2011b). The methanol fraction of methanol extract of *C. dichotoma* showed lower pathological scores and good healing. The fraction decreased the level of myeloperoxidase and malondialdehyde significantly in blood and tissues of treated animals. It displayed antioxidant effects in 1,1-diphenyl-2-picrylhydrazyl (IC_{50} 26.25 µg/ml trolox equivalent/g), 2,2'-azinobis[3-ethylbenzthiazoline]-6-sulfonic acid (2.03 µg/ml trolox equivalent/g) and ferric-reducing antioxidant potential (2.45 µg/ml trolox equivalent/g) assays. Therefore, methanol fraction of bark might be used for the treatment of ulcerative colitis (Ganjare et al. 2011).

Antidiabetic effects of the powder of pulp and peel of *C. dichotoma* fruits were investigated. *C. dichotoma* powder showed significant increase in body weight gain but decrease in the levels of serum blood glucose, aspartate aminotransferase, alanine aminotransferase, total bilirubin, and alkaline phosphatase in treated rats. It also increased the levels of serum creatinine. *C. dichotoma* powder reduced the levels of serum low density lipoproteins, triglycerides, total cholesterol, and malondialdehyde but enhanced the levels of high-density lipoproteins and activity of antioxidant enzymes. Therefore, *C. dichotoma* pulp + peel fiber powder may be used as a beneficial agent in the management of diabetic disease (Mohamed and Elkhamisy 2019). Methanolic extract of *C. dichotoma* bark (500 mg/kg/b.w.) showed significant suppression of paw edema (48.6%) when compared to standard indomethacin (5 mg/kg/b.w.; 56%). Methanol extract displayed *in vitro* antioxidant effects in DPPH radical scavenging assay (IC_{50} 62.46 µg/ml) when compared to the standard drug ascorbic acid (IC_{50} of 27.66 µg/ml). The findings suggest that the anti-inflammatory effect of *C. dichotoma* might be associated with the presence of phenolic compounds in the methanolic extract of stem bark (Hussain et al. 2020). *C. dichotoma* contains phenols, triterpenoids, pyrrolizidine, alkaloids, and coumarins, possessing germicidal, antifungal, hypoglycemic, anti-inflammatory, parasiticidal, injury, and anticancer effects (Vikas et al. 2022).

C. myxa L (Syn. *C. obliqua* Willd., *C. crenata* Delile) is a deciduous tree and grows up to 10.5 m high (Abdel El-Aleem et al. 2017). The leaves are broad, elliptic-lanceolate, and cordate at base. It is distributed in the regions of the Middle East to tropical Africa and Iran (Al-Ati 2011). Several essential oils (phytol and linalool), cordiaquinones, and pyrrolizidine alkaloids have been isolated from *C. myxa* (Yadav and Yadav 2013; Kendir et al. 2021). Its fruits and seeds are used as expectorants and in the treatment of lung diseases. The gum of raw fruits is useful in gonorrhea. In southern Iranian traditional medicine, the fruits are used in curing joint pain and throat burning. It is also useful in treating complaints of the spleen (Sharma et al. 2009). *C. myxa*-mediated zinc oxide micro-molecules demonstrated significant bactericidal effects against Gram-negative (*E. coli*) and Gram-positive (*S. aureus*) bacterial species (Saif et al. 2019). *C. myxa* fruit extract showed antiradical effect in α, α-diphenyl-β-picrylhydrazyl assay (10.0±1.24 ascorbic acid equivalent). A significant ($p = 0.05$) liver recovery was reported in treated animals (Afzal et al. 2009). *C. myxa* silver nanoparticles showed significant cytotoxicity against SW480 and HCT116 human colon cancer cell lines. In addition, the silver particles also displayed antioxidant activity in a ferric-reducing antioxidant power assay (Samari et al. 2019). Flavonoid-rich methanol extract of *C. myxa* fruit pulp displayed significant antioxidant effects in 1,1-diphenyl-2-picrylhydrazyl, total antioxidant activity, and reducing power assays (Murthy et al. 2019). The *n*-hexane fraction of ethanol extract of *C. myxa* leaves showed maximum inhibitory effects on α-glucosidase activity (IC_{50} 0.53 ppm; acarbose 6.85 ppm; Najib et al. 2019). The ethyl acetate fraction of ethanol extract of *C. myxa* displayed maximum antioxidant effect with higher levels of phenolic and flavonoid contents (31.03 ± 0.15 mg gallic acid equivalent/g dry weight and 811.91 ± 0.07 mg/rutin equivalent/g dry weight). Dichloromethane and ethyl acetate fractions demonstrated significant anti-inflammatory effects (with 45.16% and 40.26% inhibition), and the results were compared with indomethacin (51.61%). The petroleum ether and dichloromethane fractions displayed maximum analgesic activity (reaction time 289.00 ± 3.00 and 288.33 ± 20.82). The results of antipyretic activity revealed that the ethanol extract and different fractions displayed greater antipyretic effects after 2 h. The study results were compared with standard acetyl salicylic acid (30 min). The ethanol extract and the petroleum ether fraction showed significant hypoglycemic activity (with decrease in blood glucose level 95.67 ± 5.77 mg/dl and 87.67 ± 10.26 mg/dl). *C. myxa* extract and fractions demonstrated antioxidant, anti-inflammatory, analgesic, antipyretic, and antidiabetic effects, which may be associated with the presence of active phytoconstituents (Abdel-Aleem et al. 2019).

Hydro-alcoholic extract of *C. myxa* fruits was tested for their analgesic and anti-inflammatory activities in formalin and acetic acid-induced models. The fruit extract was found significantly effective in formalin and acetic acid test (Ranjbar et al. 2013). *C. myxa* fruits and leaves are useful in urinary and respiratory tract diseases, chronic fever, liver diseases, asthma, and used as anthelmintic, diuretic, expectorant, and purgative agents (Singh et al. 2022). The aqueous and alcoholic extracts (200 mg/ml) of *C. myxa* showed antibacterial activity against *Pseudomonas flurscence* (inhibition zone 17 mm), *Salmonella* and *E. coli* (both species inhibition zone 25 mm), *Shigella*, and *Staphylococcus aureus* (both IZ 19 mm; Hamdia and Al-Faraji 2017). The ethyl acetate and butanol fraction of *C. myxa* showed significant wound healing activity in Wistar albino rats using the excision wound model ($p < 0.001$; Aljeboury 2021). Ethanol extract of *C. myxa* leaves significantly reduced scopolamine-stimulated cognitive dysfunction in mice as determined by passive avoidance and novel object recognition assays. Extract administration improved MK-801-stimulated per pulse suppression deficits. Some behavioral outcomes were also recorded as enhanced phosphorylation phosphatidylinositol 3-kinase, protein kinase B, and glycogen synthase kinase 3β in the cortex and extracellular signal-regulated kinase and cAMP response element-binding protein levels in the hippocampus. The observed findings suggest that it might be used as a potential candidate for curing cognitive dysfunction and sensorimotor gating deficits detected in persons with neurodegenerative disorders (Kendir et al. 2022). Ethanol extract of *C. myxa* showed the presence of substantial phenol and flavonoid contents (113.71 ± 0.04 mg gallic acid/g dried extract and 68.9 ± 0.002 mg quercetin/g dried extract, respectively) which displayed the maximum DPPH suppression (86.45%). The extract also demonstrated potent antibacterial effect against *S. aureus, E. coli, S. enterica, B. subtilis*, and *P. aeruginosa* (inhibition zone 17.5 ± 1.0, 14.9 ± 1.0, 13.3 ± 1.5, 15.7 ± 1.0, and 13.8 ± 1.5 mm, respectively; Al-Musawi et al. 2022).

C. macleodii Hook.f. & Thomson is a perennial shrub or tree with gray bark and grows up to 8–10 m high. Leaves are ovate (broad), scabrous, with cordate base. Inflorescence is of corymb type. The flowers are white and polygamous. Fruits are drupes, 1.2 to 1.9 cm long, and ovoid (Chandrakar and Dixit 2017). The shrub is distributed in Brazil, tropical Africa, America, tropical Asia, Australia, Africa, India, Bangladesh, and Sri Lanka (Saxena 1995). Stem bark is used in wound healing and in the treatment of jaundice (Bhide et al. 2011). In Indian traditional medicine, its leaves and bark are used in the healing of wounds. The seeds are an aphrodisiac (Dubey et al. 2008; Rakesh et al. 2018). In Asian and Australian medicine, the macerated leaves are used to treat trypanosomiasis, and powdered bark is applied for the healing of broken bones (Anjaria 1998). Bark juice mixed with coconut oil is used in gripe (bowel pain) and to treat headache and ulcers (Parmar and Kaushal 1982). Fresh leaf paste is applied on forehead in the treatment of fever (Quattrocchi 2012). Leaf juice is useful in eye infections. The leaves and stem bark are used in the treatment of leprosy and chest pain. Stem bark decoction is used in bath form. Roots are recommended as a remedy for the treatment of vomiting and malaria (Quattrocchi 2012; Ruffo et al. 2002; Glover et al. 1966; Pradheeps and Poyyamoli 2013). In Brazil, it is used as an expectorant and in the treatment of contusion (Vieira et al. 1994). Its different parts possess anti-microbial, wound healing, antioxidant, hepatoprotective, antisnake venom, analgesic, and anti-inflammatory properties. Quercetin, p-hydroxyphenylacetic acid, β-sitosterol, stigmasterol, camphesterol, and Cholest-5-en-3ol (3β)-carbonyl chlorinated have been isolated from this plant species (Nayak and Kalidass 2016). Methanol extract of *C. macleodii* bark demonstrated significant inhibition of *E. coli, P. aeruginosa, S. pyogenes, S. aureus, A. niger,* and *C. albicans* (Nariya et al. 2010).

C. alliodora (Ruiz & Pav.) Oken is a perennial tree and grows up to 8–10 m high. It is naturalized in Spain, tropical America (from Mexico to Argentina), Caribbean islands, and Panama. Leaf decoction is used as a tonic to treat pulmonary diseases. Leaf paste is externally applied to bruises and swellings. As per Caribbean islands traditional medicine, an ointment is prepared using seeds and applied to skin infections (Faridah and Van der Maesen 1997; Morton 1981; Orwa et al. 2009).

C. americana (L.) Gottschling & J.S.Mill. is a perennial tree. Leaves are alternate, simple, elliptic-obovate, and slightly serrate. It is distributed in South America, Bolivia, Paraguay, Brazil, and North Argentina (Ló and Duarte 2011). Its leaf decoction is used in the washing of wounds and in inflammations. It is recommended as a remedy in the treatment of ulcers, diarrhea, liver diseases, and as an emollient in treating syphilis (Korbes 1995; Mentz et al. 1997; Gottschling 2003).

C. africana Lam is a small- to medium-sized tree. Its leaves are alternate and broad. It is found in Ethiopia. The root and root bark are useful in treating diarrhea. It is used in the treatment of hepatic complaints, amebiasis, stomachache, and diarrhea (Giday et al. 2007). Methanol extract of *C. africana* root bark (200 mg/kg, $p < 0.05$ and 400 mg/kg, $p < 0.01$) showed significant reduction in the number of diarrheic drops in treated animals. The extract (400 mg/kg) inhibited the accumulation of intestinal fluid (53.66% inhibition) also. The mean percentage of intestinal length moved by the charcoal meal was significantly reduced by the extract (400 mg/kg, 51.66% inhibition) in the extract treated group. Therefore, the root bark extract is effective in checking castor oil-induced diarrhea and intestinal motility in a dose-dependent manner (Asrie et al. 2016). Chloroform extract of *C. africana* leaf evaluated for antimalarial activity against *Plasmodium berghei.* Crude extract indicated 2000 mg/kg as a lethal dose against *P. berghei.* Chloroform extract (600 mg/kg) demonstrated significant parasitemia inhibition in the four-day inhibition (51.19%), curative (57.14%), and prophylactic (46.48%) tests. The n-butanol fraction (400 mg/kg) of chloroform extract showed the maximum chemosuppression (55.62%) in tested models (Wondafrash et al. 2019). The ethanolic extract of *C. africana* stem bark showed acute toxicity (LD50 ≥ 5 g/kg) for oral route and greater than 1 g/kg (LD50 ≥ 1 g/kg) for intraperitoneal route. The extract significantly ($p < 0.05$) decreased pain in formalin-induced pain model in the second phase (3.2 g/kg and 4.8 g/kg oral doses) while the extract (1.6 g/kg and 3.2 g/kg/b.w.) significantly ($p < 0.05$) decreased abdominal writhes and pain in acetic acid–induced abdominal writhings and hot plate–induced pain models (Tijjani et al. 2016). Antimicrobial effects, total phenolics/flavonoids, and cytotoxic effects of methanol extract of *C. Africana* were evaluated. Methanol extract presented potency vs. antimycobacterium species (MICs ranging from 32 to 1024 μg/ml). Methanol extract suppressed LOX enzyme activity (IC_{50} 55 ± 0.9 μg/ml). Methanol extract had significant ($p < 0.05$) free-radical scavenging activity (IC_{50} range 6.79 ± 0.07 to 331.98 ± 0.07 μg/ml) on a DPPH scavenging assay. Total flavonoid and total phenolic contents of *C. Africana* showed cytotoxic effect on Vero cells (LC_{50} 81.79 ± 13.31 and 99.67 ± 16.10 μg/ml; Isa et al. 2016). Different parts (leaves, stem, park, and fruit) of *C. africana* showed antioxidant against a DPPH assay (inhibition range 37 ± 0.10 to 95 ± 0.00% RSA; Alhadi et al. 2015; Sarah et al. 2013).

C. dentata Poir is found in Central America, Colombia, Venezuela, southern United states, Panama, and Costa Rica. Its flowers possess emollient, pectoral, sudorific, and diaphoretic properties. Its flower decoction is useful in induction of perspiration. The flowers and leaves are useful in chest complaints. The wood charcoal of this species is used in the treatment of stomach infection

(Richard et al. 2010; Standley 1926). Ethyl acetate extract of *C. sebestena* demonstrated significant antibacterial activity against *B. cereus, B. subtilis, S. aureus, E. coli,* and *P. aeruginosa*, therefore, it may be used in the treatment of bacterial infections (Osho et al. 2016).

C. verbenacea A.DC is a perennial shrub and grows up to 8–10 m high. It is distributed in Brazilian coastal regions. Aromatic leaves are used in folk medicine in healing wounds (Souza et al. 2004). Ethanol extract of *C. verbenacea* leaves demonstrated strong antiulcer effect in ethanol/HCl and absolute ethanol-stimulated gastric lesions. The extract showed significant inhibitory concentration (IC$_{50}$ 76.11mg/ml) on lipid peroxidation. *C. verbenacea* displayed a strong antiulcer effect (125mg/kg dose), and the activity may be attributed to an improvement in antioxidant mechanisms of the stomach (de Freitas Roldão et al. 2008). The petroleum ether and alcoholic extracts of *C. francisci, C. myxa,* and *C. serratifolia* leaves showed significant analgesic, anti-inflammatory, and anti-arthritic effects in treated rats. Four flavonoid glycosides (robinin, rutin, datiscoside, and hesperidin), one flavonoid aglycone (dihydrorobinetin), and two phenolic derivatives (chlorogenic and caffeic acid) were isolated and characterized (Ficarra et al. 1995). The methanol and hexane extracts of *C. verbenacea* leaves showed moderate antibacterial activity against *S. aureus, E. coli, P. aeruginosa* (MIC of 256 μg/ml), and *P. aeruginosa* (512 μg/ml; Matias et al. 2016). Gallic acid, chlorogenic acid, caffeic acid, glycoside phenol, rutin, and quercetin have been identified from the methanol extract of *C. verbenacea* leaves. The methanol extract did not show clinically relevant antibacterial activity but showed synergistic effect when mixed with antibiotic of aminoglycosides against *E. coli* (EC27), *S. aureus* 358 (SA358), and *P. aeruginosa* (PA03; Matias et al. 2013). Essential oils (β-caryophyllene, bicyclogermacrene, δ-cadinene, and α-pinene) were determined by GC-MS from *C. verbenacea* leaves. The essential oils showed antifungal (*C. albicans* and *C. krusei*) and antibacterial activities (*S. aureus, B. cereus,* and *E. coli* 27; MIC 64 μg/ml). Therefore, essential oils may be used as an adjuvant in antibiotic treatment against respiratory tract bacterial infections (Rodrigues et al. 2012). The essential oils of *C. verbenacea* suppressed the bone loss significantly when compared with animals in the non-treated group ($p < 0.05$). A significant reduction in the levels of IL-1α and enhancement in IL-10 level was recorded in the treated group ($p < 0.05$). Topical administration minimized the alveolar bone resorption and decreased the frequency of recognition of *Porphyromonas gingivalis* (Pimentel et al. 2012; Martim et al. 2021). Essential oils of *C. verbenacea* (300–600 mg/kg bw, p.o.) significantly decreased the carrageenan-induced rat paw oedema and myeloperoxidase activity. The essential oils blocked the carrageenan-induced exudation and neutrophil influx to the rat pleura and the migration of neutrophils into carrageenan-stimulated mouse air pouches. In addition, essential oils suppressed the oedema induced by *Apis mellifera* venom or ovalbumin in exposed rats and ovalbumin-evoked allergic pleurisy. Essential oil significantly reduced the formation of TNF-α, without influencing IL-1β, in carrageenan-induced rat paws. Neither the prostaglandin E$_2$ formation after intrapleural injection of carrageenan nor the cyclooxygenase-1or cyclooxygenase-2 activities in *in vitro* were affected by the essential oil. α-humulene and *trans*-caryophyllene (50 mg/kg, p.o.) significantly inhibited the volume of carrageenan-induced paw edema in mice. Therefore, *C. verbenacea* essential oils may be used as a new therapeutic option for the treatment of inflammations (Passos et al. 2007). The α-humulene and (-)-*trans*-caryophyllene were isolated and identified from *C. verbenacea*. Both isolated compounds showed significant inhibitory activity in different inflammatory models in mice and rats. Both compounds decreased the formation of prostaglandin E$_2$, as well as the expression of inducible nitric oxide synthase and cyclooxygenase. The results reveal that both compounds [α-humulene and (-)-*trans*-caryophyllene] may be used in the management of inflammatory diseases (Fernandes et al. 2007). α-humulene or (-)-*trans*-caryophyllene decreased the migration of neutrophils and activation of NF-κB as stimulated by lipopolysaccharide in the rat paw. α-humulene significantly decreased the increased levels of TNF-α and IL-1β, paw oedema, and the upregulation of B1 receptors in treated animals. Both compounds did not show any interference with the activation of the mitogen-activated protein kinases, p38 kinases, and c-Jun *N*-terminal kinases. Therefore, both compounds of *C. verbenacea* may be useful in the management of inflammatory diseases (Medeiros et al. 2007).

12.2 PHYTOCHEMISTRY

The thirty-one essential oils (*trans*-2-hexenal, tricyclene, α-pinene, camphene, sabinene, β-pinene, myrcene, limonene, 1,8-cineole, *trans*-β-ocimene, α-terpinolene, nonanal, fenchyl acetate, bornyl acetate, α-copaene, β-bourbonene, β-cubebene, italicene, β-caryophyllene, β-copaene, α-humulene, α-amorphene, γ-curcumene, germacrene D, bicyclogermacrene, α-muurolene, γ-cadinene, β-bisabolene, δ-cadinene, *trans*-γ-bisabolene, pimaradiene) have been determined by GC-MS from *C. verbenacea* aerial parts. The essential oils demonstrated significant antibacterial activity against *S. aureus* (ATCC 6538; MIC 170 μg/ml) and *E. faecalis* (ATCC 29212, MIC200 μg/ml; Meccia et al. 2009). Two dammarane-type triterpenes (cordialin A and B) were also reported from *C. verbenacea* aerial parts (Velde et al. 1982).

Meroterpene (globiferane) and glutarimide alkaloids (cordiarimide A and B) were identified from *C. globifera* roots. Cordiarimide B displayed radical scavenging activity by inhibiting superoxide anion radical formation in the xanthine/xanthine oxidase assay and also inhibited superoxide anion formation in differentiated HL-60 human promyelocytic leukemia cells (Parks et al. 2010). A ten-membered ring meroterpene (globiferin) and cordiachrome C were isolated and characterized from *C. globifera* roots. The globiferin and cordiachrome C demonstrated significant antimycobacterial activity (MIC 6.2 and 1.5 μg/ml; Dettrakul et al. 2009).

Eight compounds (cordiachrome A, cordiachrome B, cordiachrome C, cordiaquinol C, cordiaquinol I, cordiaquinol J, and cordiaquinol K) were separated and identified from methanol extract of *C. fragrantissima*. The isolated compounds showed antiprotozoal activity against *Leishmania major*, *L. panamensis*, and *L. guyanensis* (Mori et al. 2008).

Six geranyl-hydroquinone-derived compounds (alliodorol, allioquinol C, cordiachromen A, cordallinol, cordiaquinol C, and cordiol A) from the acetone extract of heartwood (Manners and Jurd 1977), phenylpropanoid derivative [1-(3′-methoxypropanoyl)-2,4,5-trimethoxybenzene] and a prenylated hydroquinone [2-(2Z)-(3-hydroxy-3,7-dimethylocta-2, 6-dienyl)-1,4-benzenediol] have been isolated and characterized from *C. alliodora* root bark. Both compounds [1-(3′-methoxypropanoyl)-2,4,5-trimethoxybenzene] and a prenylated hydroquinone [2-(2Z)-(3-hydroxy-3,7-dimethylocta-2, 6-dienyl)-1,4-benzenediol] displayed antifungal activity against *Cladosporium cucumerinum*. 1-(3′-methoxypropanoyl)-2,4,5-trimethoxybenzene showed significant activity against larvae of *Aedes aegypti* (Ioset et al. 2000a). 5-*O*-[β-D-apiofuranosyl-(1→6)-β-D-glucopyranosyl]-1-isoindolinone, *N*-(2E)-3-[(2S,3R)-2-(4-hydroxy-3-methoxyphenyl)-3-(hydroxymethyl)-7-methoxy-2, 3-dihydro-1-benzofuran-5-yl]acryloylglycine, allantoin, *N*-carbamoylputrescine, rosmarinic acid and canthoside C from dichloromethane fraction of whole plant, and six triterpenoids (3α-hydroxyolean-12-en-27-oic acid, 3-oxoolean-12-en-27-oic acid, 3,29-dioxoolean-12-en-27-oic acid, 3α-hydroxy-29-oxoolean-12-en-27-oic acid, 3α,29-dihydroxyolean-12-en-27-oic acid, and 3α-hydroxyolean-12-ene-27,29-oic acid) have been isolated and characterized from *C. alliodora* leaves (Chen et al. 1983).

In addition to cordiaquinones A and B, two meroterpenoid naphthoquinones (cordiaquinones J and K) have been isolated from *C. curassavica* roots. All the isolated naphthoquinones displayed significant antifungal effects against *Cladosporium cucumerinum*, *Candida albicans*, and toxic effects against *A. aegypti* (Ioset et al. 2000b). The presence of essential oils [α-phellandrene, α-pinene, camphene, sabinene, β-pinene, myrcene, p-cymene, limonene, γ-terpinene, linalol, camphor, borneol, 4-terpineol, α-copaene, β-bourbonene, β-elemene, longifolene, (*E*)-caryophyllene, α-bergamotene, γ-elemene, α-humulene, α-amorphene, germacrene D, bicyclogermacrene, cubebol, δ-cadinene, cadina-1,4-diene, germacrene B, spathulenol, caryophyllene oxide, globulol, viridiflorol, α-bisabolol] was determined by GC-MS and GC-FID from *C. curassavica* leaves (Santos et al. 2006). Essential oils (β-caryophyllene, α-humulene, α-pinene, bicyclogermacrene, and sabinene) were determined from *C. curassavica* leaves (Andrade et al. 2022).

Cordiaquinones E, F, G, H, cordiaquinone B and naphthoxirene have been isolated from *C. linnaei* roots. The isolated compounds showed antifungal activity against *Cladosporium cucumerinum*,

C. albicans, and *A. aegypti* (Ioset et al. 1998). Two meroterpenoid naphthoquinones (cordiaquinone C and D) have been identified from the dichloromethane extract of *C. linnaei* (Ioset et al. 1999). The cordiachrome (alliodorin) from *C. millenii* (Manners and Jurd 1973) and terpenoid benzoquinones (cordiachromes A–F) have been isolated and identified from the *C. millenii* heartwood (Moir and Thomson 1973).

Cordiaquinones B, L, E, N, and O were also isolated and determined by spectroscopic analysis from *C. polycephala* roots. All the isolated cordiaquinones showed significant anticancer activity against HCT-8 (colon), HL-60 (leukemia), MDA-MB-435 (melanoma), and SF295 (glioblastoma) human cancer cell lines (IC_{50} 1.2 to 11.1 mmol/l). Cordiaquinones N and O displayed potent activity against leukemia HL-60 cells (IC_{50} 2.2 µmol/l; Freitas et al. 2012). The cordiaquinone E has been identified from *C. polycephala* roots. The isolated compound showed significant inhibition of promastigote (IC_{50} 4.5 ± 0.3 µM) and axenic amastigote (IC_{50} 2.89 ± 0.11 µM). The concentration (CC_{50} 246.81 ± 14.5 µM) did not show any toxicity to the host cell (CC_{50} 246.81 ± 14.5 µM). The antiamastigote effect was attributed with enhanced levels of TNF-α, IL-12, NO, and ROS and reduced IL-10 levels. The reported findings offer the development of new leishmanicidal agents from *C. polycephala* (Rodrigues et al. 2021).

1,4-naphthoquinone (+)-cordiaquinone J was obtained from *C. leucocephala* roots. The isolated compound showed anticancer activity against (IC_{50} 2.7–6.6µM) HL-60 and SF-295 cell lines. The (+)-cordiaquinone J (1.5 and 3.0 µM) reduced the rate of cell viability of HL-60 leukemia cells. The isolated compound also displayed fast induction of apoptosis (caspase activation, DNA fragmentation, morphologic changes, and rapid induction of necrosis) in treated cells (Marinho-Filho et al. 2010). Two meroterpenoid naphthoquinones (cordiaquinone L and cordiaquinone M) were isolated from *C. leucocephala* roots (Diniz et al. 2009). The β-caryophyllene and bicyclogermacrene were determined by GC-MS from *C. leucocephala* leaves (Diniz et al. 2008).

Meroterpenoid benzoquinones {(1a S*,1b S*,7a S*,8a S*)-4,5-dimethoxy-1a,7a-dimethyl-1,1a,1b,2,7, 7a,8,8a-octahydrocyclopropa cyclopenta[1,2-b]naphthalene-3,6-dione and microphyllaquinone)} have been isolated from *C. globosa* roots. (1a S*,1b S*,7a S*,8a S*)-4,5-dimethoxy-1a,7a-dimethyl-1,1a,1b,2,7, 7a,8,8a-octahydrocyclopropa cyclopenta[1,2-b]naphthalene-3,6-dione showed potent cytotoxicity against leukemic cells (IC_{50} 1.2 to 5.0 µg/ml; de Menezes et al. 2005a). Narigenin-4′,7-dimethyl ether and eriodictyol were reported from *C. globosa* (da Silva et al. 2010).

Cordinoic acid and cordicilin along with other significant compounds (cordinoic acid, cordicilin, rosmarinic acid, tetraacetyl rosmarinic acid, rosmarinic acid 3-*O*-β-D-glucoside, latifolicinin A, latifolicinin B, latifolicinin C, latifolicinin D, cordicinol cordinol, cordioic acid, cordifolic acid) from fruits, bark, and leaves (Begum et al. 2011), and two abietane diterpenes (cordioic acid and cordifolic acid) were isolated and characterized from the methanol extract of *C. latifolia* stem bark (Siddiqui et al. 2006).

The methanol extract and isolated compounds {2-(3,4, 5-trihydroxyphenyl)-5,7-dihydroxy-3-[α-L-rhamnopyranosyl-(1→6)-β-D-glucopyranosyloxy]-4H-chromen-4-one} from *C. obliqua* seeds and leaves displayed significant antibacterial activity against Gram-positive (*S. mutans*, *S. mitis* and *S. sanguis*) and Gram-negative (*A. actinomycetemcomitans*, *P. gingivalis* and *B. forsythus*) bacterial species and fungal stain (*C. albicans*; Yadav et al. 2015). Hesperetin 7-rhamnoside (Chauhan et al. 1978), α-amyrin, betulin, octacosanol, lupeol-3-rhamnoside, β-sitosterol, β-sitosterol-3-glucoside, hentricontanol, hentricontane, taxifolin-3, 5-dirhamnoside and hesperetin-7-rhamnoside were obtainedfrom seeds (Agnihotri et al. 1987), flavonoids (7-methoxyflavone, and 5,7,3′,4′-tetrahydroxy-3-methoxyflavone) from *C. globosa* (da Silva et al. 2004), dimethyl-3,4′-kaempferol from *C. boissieri* (Dominguez 1973), and dimethoxytaxifolin-3-*O*-α-L-rhamnopyranoside and distylin-3-xyloside from the roots of *C. obliqua* (Chauhan and Srivastava 1977; Srivastava 1980).

Several compounds (lantanolic acid, 3-epipomolic acid, lantic acid, icterogenin, lantadene A, lantadene B, ursomic acid, pomonic acid, pachypodol, retusin cordiaketal A, cordiaketal B, cordianal A, cordianone, cordianal B, and cordianal C, lantanolic acid, 3-epipomolic acid, lantic acid, icterogenin, lantadene A, lantadene B, ursomic acid, pomonic acid, pachypodol and retusin) were obtained from ethyl acetate extract (Kuroyanagi et al. 2001), and nine dammarane-type triterpenes (cordianols A–I) along with cordialin A from ethyl acetate fraction (Kuroyanagi et al. 2003), *trans*-phytol, taraxerol, 3,7,4'-trimethoxyflavone, 5,3'-dihydroxy-3,7,4'-trimethoxyflavone, quercetin, tiliroside, and rutin were identified from *C. multispicata* leaves. Isolated compounds (quercetin, tiliroside, and rutin) showed antioxidant activity in DPPH quenching (IC_{50} 7.7 ±3.6 to 79.3 ±3.4 mg/l) and lipid peroxidation (IC_{50} 80.1 ±0.98 to 88.7 ±3.62 mg/l) assays (Correia Da Silva et al. 2010).

βsitosteryl-3β-glucopyranoside-6'-*O*-palmitate, nervonyl 4-hydroxy-trans-cinnamate ester, β-sitosterol, and chlorophyll were obtained from the leaves and 1,2-dilinoleoyl-3-linolenoylglycerol from dichloromethane extracts of *C. dichotoma* twigs (Ragasa et al. 2015). One triterpenoids (α-amyrin) and fatty acids methyl esters (pentene 2,2 di methyl, 2-methyl pentane, sec-butylcarbinol, bis-isopropyl, undecane, 3-methyl heptane, 1-nonene, 2-methyl, di isoamyl, 2,4-dimethyl-3-ethyl pentane, cicloesano, sextone, gem-di methyl cyclopentane, 1,2,3, tri methyl cyclopentane, butyric acid-2,2 di methyl vinyl ester, hexadecanoic acid, methyl ester or palmitic acid, methyl ester, heptadecanoic acid methyl ester or margaric, acid methyl ester, 9-octadeconoic acid(Z)-, methyl ester or oleic acid, methyl ester) were isolated and characterized from *C. dichotoma* (Nariya et al. 2018). Two bioactive flavonoids (3,5,7,3',4'-tetrahydroxy-4-methoxyflavone-3-*O*-L-rhamnopyranoside (MECD-1) and 5,7,3'-trihydroxy-4-methoxyflavone-7-*O*-L-rhamnopyranoside (MECD-2) were isolated and characterized from the methanolic extract *C. dichotoma* bark (Hussain et al. 2021).

Phenylpropanoid [3-(2',4',5'-trimethoxyphenyl)propanoic acid], (+)-1β,4β,6α-trihydroxyeudesmane, (-)-1β,4β,7α-trihydroxyeudesmane and (+)-1β,4β,11-trihydroxyoppositane from heartwood (de Menezes et al. 2004), essential oils [α-cadinol, α-muurolol, *epi*-α-muurolol, δ-cadinene and guaia-3, 10(4)-dien-11-ol] from heartwood and sapwood (de Menezes et al. 2005b), sesquiterpene (trichotomol) together with cordiachrome C, α-cadinol, oleanolic acid, oncocalyxone A, β-sitosterol, β-sitosterol-β-D-glucoside, allantoin. and sucrose were isolated and characterized from the ethanol extract of *C. trichotoma* heartwood (de Menezes et al. 2001).

3β-*O*-[α-L-rhamnopyranosyl-(1→2)-β-D-glucopyranosyl]ursolic acid 28-*O*-[β-D-glucopyranosyl-(1 → 6)-β-D-glucopyranosyl] ester from methanol extract of stem (Santos et al. 2003), two bidesmoside triterpenoid saponins [3-*O*-α-L-rhamnopyranosyl-(1→2)-β-D-glucopyranosyl pomolic acid 28-*O*-β-D-glucopyranosyl ester and 3-*O*-α-L-rhamnopyranosyl-(1→2)-β-D-glucopyranosyl oleanolic acid 28-*O*-β-D-glucopyranosyl-(1→6)-β-D-glucopyranosyl ester] from stem (Santos et al. 2007), two saponins {3β-*O*-[α-L-rhamnopyranosyl-(1→2)-β-D-glucopyranosyl]pomolic acid 28-*O*-[β-D-glucopyranosyl-(1→6)-β-D-glucopyranosyl] ester and 3β-*O*-[α-L-rhamnopyranosyl-(1→2)-β-D-glucopyranosyl]oleanolic acid 28-*O*-[β-D-xylopyranosyl-(1→2)-β-D-glucopyranosyl-(1→6)-β-D-glucopyranosyl] ester} from stem (Santos et al. 2005a), monodesmoside triterpenoid saponin (3β-*O*-α-L-rhamnopyranosyl- (1→2)-β-D-glucopyranosyl pomolic acid) together with quinovic acid, cincholic acid, cincholic 3β-*O*-6-deoxy-β-D-glucopyranoside acid, and β-sitosterol-β-D-glucoside and quinovic 3β-*O*-β-D-glucopyranoside acid were identified from *C. piauhiensis* (Santos et al. 2005b).

Nineteen compounds {8,8'dimethyl-3,4,3',4'-dimethylenedioxy-7-oxo-2,7'cyclolignan, 8,8'-dimethyl-4,5- dimethoxy-3',4'-methylenodioxy-7-oxo-2,7'cyclolignan, sitosterol, stigmasterol, sitosterol-3-*O*-β-D-glucopyranoside, stigmasterol-3-*O*-β-D-glucopyranoside, phaeophytin A, 13^2-hydroxyphaeophytin A, 17^3-ethoxypheophorbide A, 13^2-hydroxy-17^3-ethoxypheophorbide A, *m*-methoxy-*p*-hydroxybenzaldehyde, (*E*)-7-(3,4-dihydroxyphenyl)-7-propenoic acid, 1-benzopyran-2-one, 7-hydroxy-1-benzopyran2-one, 2,5-bis-(3',4'-methylenedioxiphenyl)-3,4-dimethyltetrahydrofuran, 3,4,5,3',5'- pentamethoxy-1'-allyl-8.O.4'-neolignan, 3,5,7,3',4'-pentahydroxyflavonol, 5,7-dihydroxy-4'-methoxyflavone, 5,8-dihydroxy-7,4'-dimethoxyflavone, kaempherol 3-*O*-β-D-glucosyl-6''-α-L-rhamnopyranoside, and kaempherol 3,7-di-*O*-α-L-rhamnopyranoside have been identified from *C. exaltata* (Nogueira et al. 2013).

Rufescenolide, β-sitosterol, stigmasterol, syringaldehyde, 3-β-*O*-D-glucopyranosyl-sitosterol, methyl caffeate, 4-methoxy-protocatechuic acid, and methyl rosmarinate from stem (do Vale et al. 2012), three compounds [stigmasterol, cholest-5-en-3ol(3β)-carbonyl chlorinated, camphesterol] were determined from *C. macleodii* bark (Nariya et al. 2014), two triterpenes [3α,6β,25-trihydroxy-20(*S*),24(*S*)-epoxydammarane, 3α-acetoxy-6β,25-dihydroxy-20(*S*),24(*S*)-epoxydammarane, cabra-leadiol and 3α-acetoxy-6β,25-dihydroxy-20(*S*),24(*S*)-epoxydammarane] were isolated from the methanol extract of *C. spinescens* leaves (Nakamura et al. 1997).

Merosesquiterpenoid quinones (cordiaquinones A and B) from roots (Bieber et al. 1990), cordia-quinones A and B were isolated and characterized from *C. corymbosa* roots (Silva Filho et al. 1993). The afzelin and quercitrin have been reported from methanol extract of *C. colloccoca* leaves. The afzelin showed significant cytotoxicity against two lung cancer cell lines, A549 and NSCLC-N6 (adenocarcinoma and epidermoid lung cancer; IC_{50} 6.7 and 12.9 μg/ml) followed by 5 (12.7 μg/ml; Fouseki et al. 2016). *C. lutea* contains substantial levels of rutin, quercitin, linolenic acid, hexadeca-noic acid, and hexadecanoic acid glyceryl ester (Mayevych et al. 2015). The phenolic compounds (*p*-hydroxy benzoic acid, caffeic acid, syringic acid, ferulic acid, rutin, *o*-coumaric acid, myric-etin, quercetin, rosmarinic acid, naringenin, and kaempferol) and others (α-amyrin, β-sitosterol, rosmarinic acid, and methyl rosmarinate) were isolated from ethanolic extract of *C. africana* stem bark (Sabry et al. 2022). Eleven compounds [allantoin, rosmarinic acid, caffeic acid, isoquercetin, rutin, quercetin-3-*O*-β-D-neohesperidoside, kaempferol 3-*O*-β-D-neohesperidoside, helichryso-side, kaempferol 3-*O*-(2″-*O*-α-L-rhamnosyl-6″ *trans-p*-coumaroyl)-β-D-glucoside, quercetin 3-*O*-(6″ *trans-p*-coumaroyl)-β-D-galactoside, and 4-hydroxyphenyl lactic acid] have been isolated and structurally elucidated from *C. bicolor*, *C. megalantha* and *C. dentata* (Marini et al. 2018).

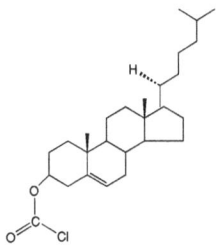

Cholest-5-en-3ol (3-β) -carbonyl chlorinated

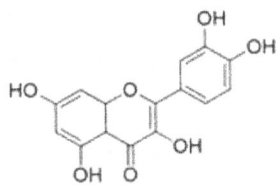

3,5,7,3′,4′-Pentahydroxyflavonol

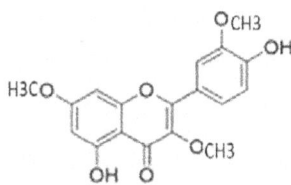

Pachypodol

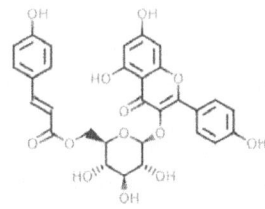

Tiliroside

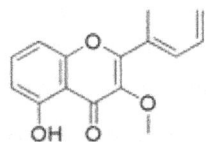

5-Hydroxy-3,7,4′-trimethoxyflavone

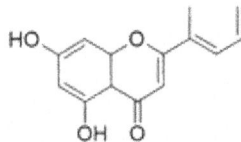

5,7-Dihydroxy-4′-methoxyflavone

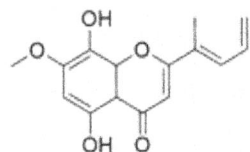

5,8-Dihydroxy-7,4′ -dimethoxyflavone

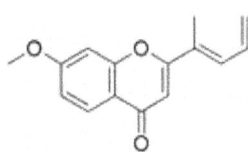

7-Methoxyflavone

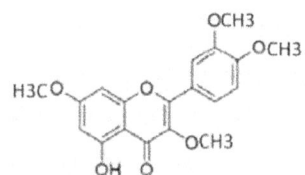

Retusin

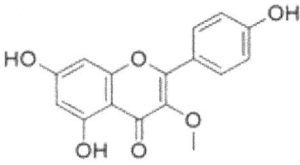

5,7,3',4'-Tetrahydroxy-3-methoxyflavone

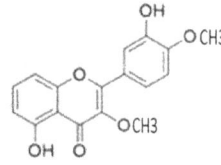

3,7,4'-Trimethoxyflavone

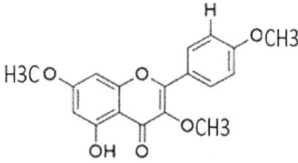

5,3'-Dihydroxy-3,7,4'-trimethoxyflavone

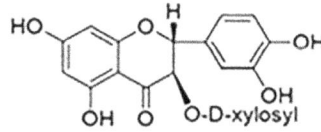

Distylin-3-xyloside

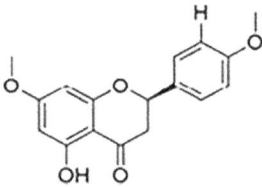

Narigenin-4',7-dimethyl ether

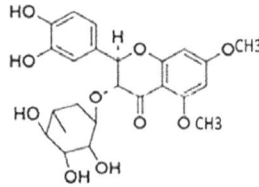

5,7-Dimethoxytaxifolin-3-O-α-L-rhamnopyranoside

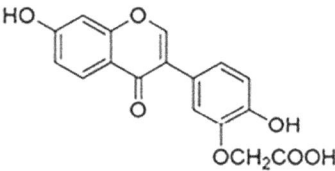

7,4'-Dihydroxy-5'-carboxymethoxy isoflavone

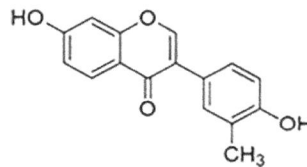

7,4'-Dihydroxy-5'-Me isoflavone

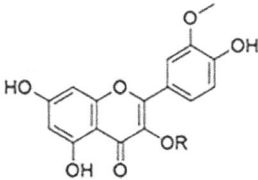

R = Glc-rha - Isorhamnetin-3-O-rutinoside

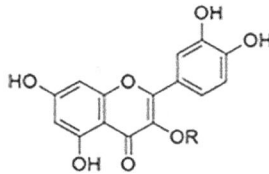

R = Glc-rha - Quercetin-3-O-rutinoside

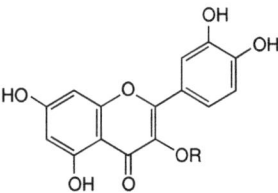

R = $\overset{\text{Glc-rha}}{\underset{\text{Rha}}{|}}$ - Quercetin-3-O-2G-rhamnosylrutinoside

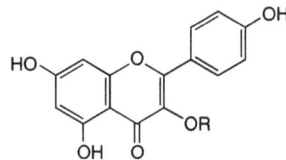

R = Glc-rha - Kaempferol-3-O-robinoside

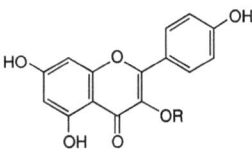

R = $\overset{\text{Glc-rha}}{\underset{\text{Rha}}{|}}$ - Kaempferol-3-O-2G-rhamnosylrutinoside

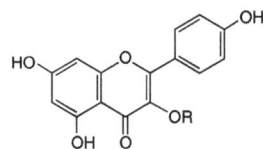

R = Glu-rha - Kaempferol-3-O-rutinoside

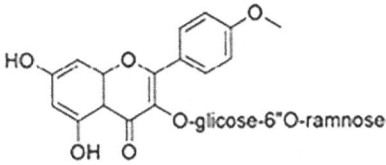

Kaempferol-3-*O*-β-D-glucosyl-6″-α-L-rhamnopyranoside

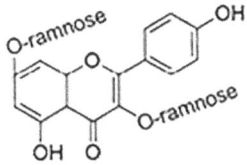

Kaempferol-3,7-di-*O*-α-L-rhamnopyranoside

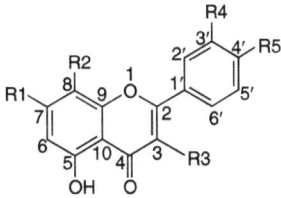

R^1 = OH, R^2 = H, R^3 = OH, R^4 = OH, R^5 = OH - 3,4,5,3′,5′- Pentamethoxy-1′-allyl-8.O.4′-neolignan (14), 3,5,7,3′,4′-pentahydroxyflavonol; R^1 = OH, R^2 = H, R^3 = H, R^4 = H, R^5 = OCH$_3$ - 5,7-Dihydroxy-4′-methoxyflavone; R^1 = OCH$_3$, R^2 = OH, R^3 = H, R^4 = H, R^5 = OCH$_3$ - 5,8-Dihydroxy-7,4′-dimethoxyflavone; R^1 = OH, R^2 = H, R^3 = -O-Glucose-6″-O-rhamnose, R^4 = H, R^5 = OH - Kaempferol 3-*O*-β-D-glucosyl-6″-α-L-rhamnopyranoside; R^1 = O-Rhamnose, R^2 = H, R^3 = -O-Rhamnose, R^4 = H, R^5 = OH - Kaempferol 3,7-di-*O*-α-L-rhamnopyranoside

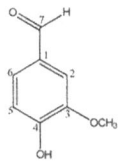

m-Methoxy-*p*-hydroxybenzaldehyde

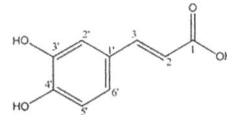

(*E*)-7-(3,4-Dihydroxyphenyl)-7-propenoic acid

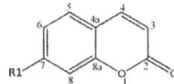

R^1 = H - 1-benzopyran-2- one; R = OH - 7-Hydroxy-1-benzopyran-2-one

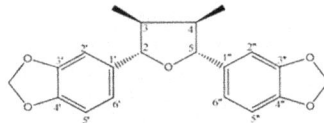

2,5-bis-(3′,4′-Methylenedioxiphenyl)-3,4-dimethyltetrahydrofuran

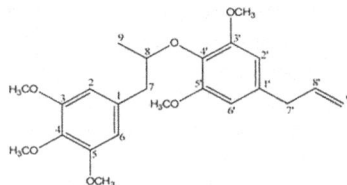

3,4,5,3′,5′- Pentamethoxy-1′-allyl-8.O.4′-neolignan

R^1 = Fitil, R^2 = H - Phaeophytin A; R^1 = Fitil, R^2 = OH - 13^2-Hydroxyphaeophytin A; R^1 = Etil, R^2 = H - 17^3-Ethoxypheophorbide A; R^1 = Etil, R^2 = OH -13^2- Hydroxy-17^3-ethoxypheophorbide A

R^1 = H, R^2 = R^3 = OCH$_3$O - 8,8′-Dimethyl-3,4,3′,4′-dimethylenedioxy-7-oxo-2,7′cyclolignan; R^1 = OCH$_3$, R^2=OCH$_3$, R^3=H- 8,8′-Dimethyl-4,5- dimethoxy-3′,4′-methylenodioxy-7-oxo-2,7′cyclolignan

Sitosterol-3-O-β-D-glucopyranoside

Stigmasterol-3-O-β-D-glucopyranoside

R = Glc –Stigmast-5-en-3-O-β-D-glucoside

3β-O-[α-L-rhamnopyranosyl-(1→2)-β-D-glucopyranosyl] pomolic acid 28-O-[β-D-glucopyranosyl-(1→6)-β-D-glucopyranosyl] ester

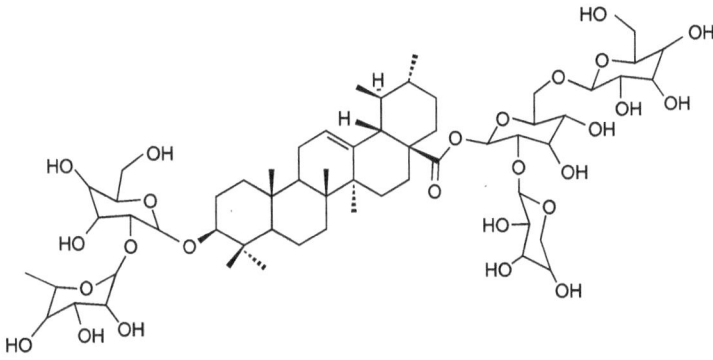

3β-*O*-[α-L-Rhamnopyranosyl-(1→2)-β-D-glucopyranosyl]oleanolic acid 28-*O*-[β-D-xylopyranosyl-(1→2)-β-D-glucopyranosyl-(1→6)-b-D-glucopyranosyl]

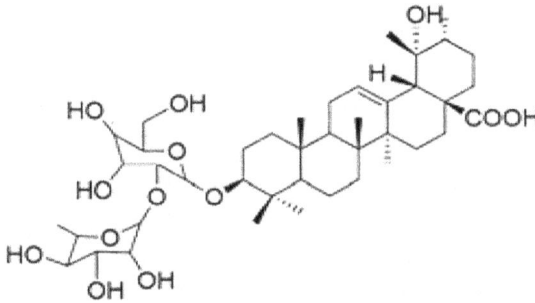

3β-*O*-α-L-Rhamnopyranosyl-(1→2)-β-D-glucopyranosyl pomolic acid

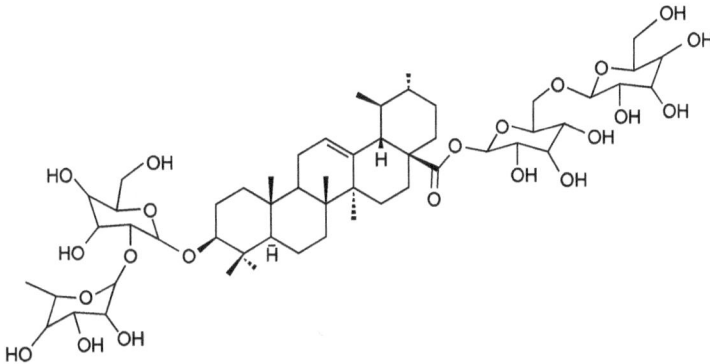

3β-*O*-[α-L-rhamnopyranosyl-(1→2)-β-D-glucopyranosyl]ursolic acid 28-*O*-[β-D-glucopyranosyl-(1→6)-β-D-glucopyranosyl]

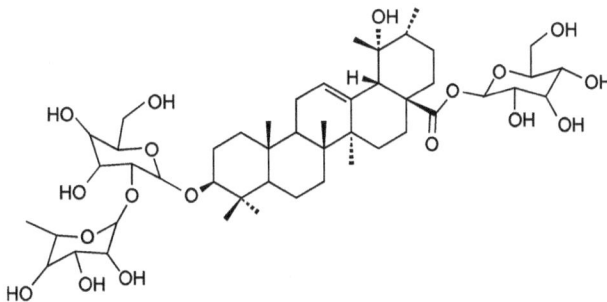

3-*O*-α-L-Rhamnopyranosyl-(1→2)-β-D-glucopyranosyl pomolic acid 28-*O*-β-D-glucopyranosyl ester

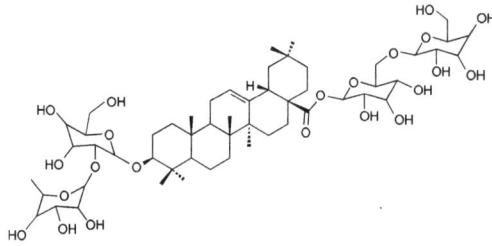

3-*O*-α-L-Rhamnopyranosyl-(1→2)-β-D-glucopyranosyl oleanolic acid 28-*O*-β-D-glucopyranosyl-(1→6)-β-D-glucopyranosyl ester

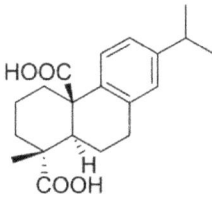

Cordioic acid

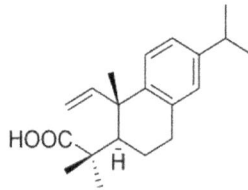

Cordifolic acid

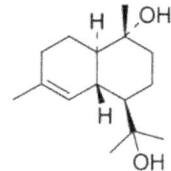

Trichotomol

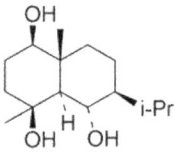

(+)-1β,4β,6α-Trihydroxyeudesmane

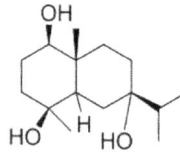

(-)-1β,4β,7α-Trihydroxyeudesmane

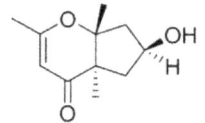

Cordinol

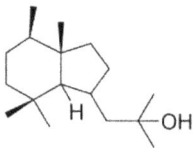

(+)-1β,4β,11-Trihydroxyoppositane

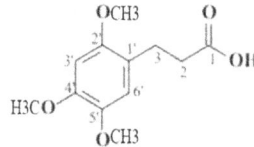

3-(2′,4′,5′-Trimethoxyphenyl) propanoic acid

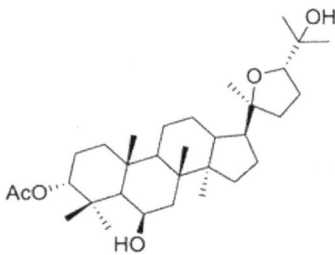

3α-Acetoxy-6β, 25-dihydroxy-20(*S*), 24(*S*)-epoxydammarane

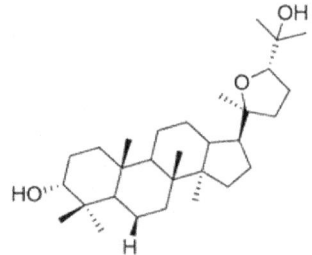

Cabraleadiol

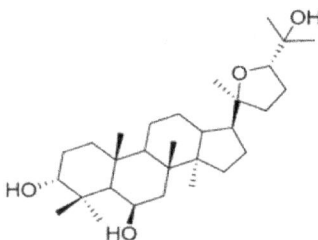

3α, 6β, 25-Trihydroxy-20(*S*), 24(*S*)-epoxydammarane

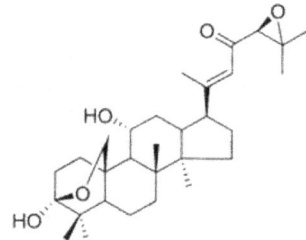

Cordialin A

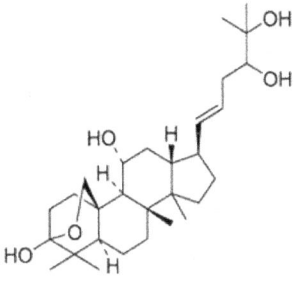

Cordialin B

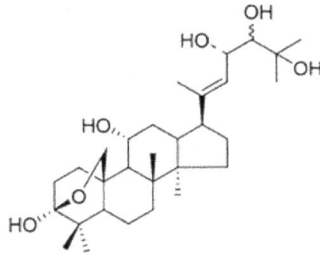

Cordianol-A

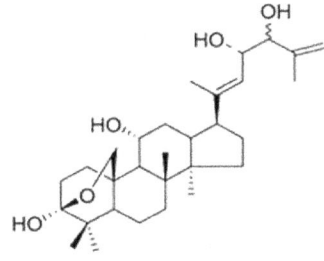

Cordianol-B

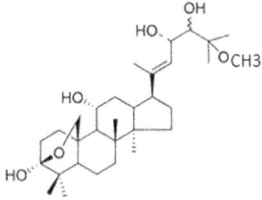

Cordianol-C

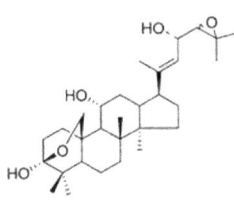

Cordianol-D

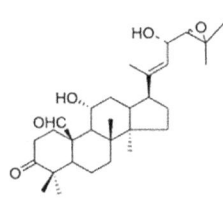

Cordianol-E

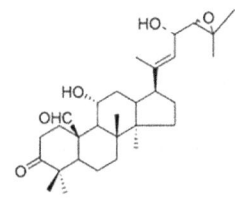

Cordianol-F

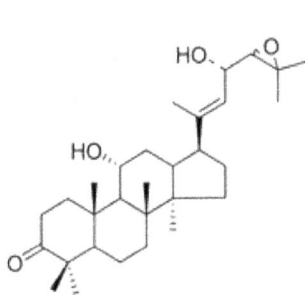

Cordianol-G

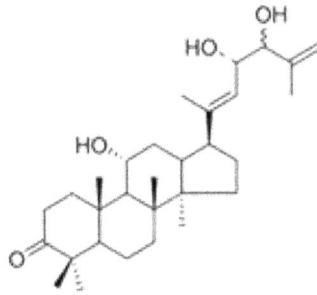

Cordianol-H

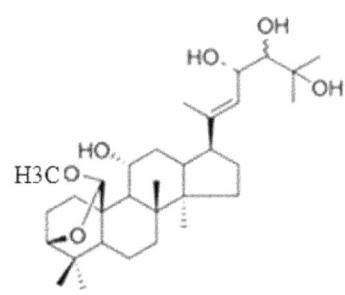

Cordianol-I

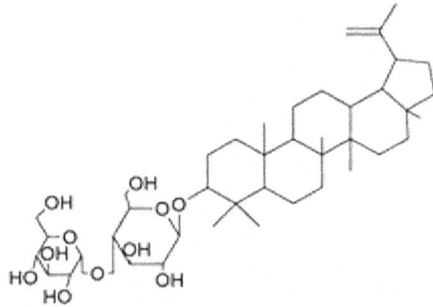

Lup-20(29)-ene-3-*O*-β-D-maltoside

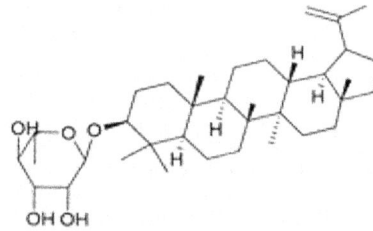

Lupa-20(29)-ene-3-*O*-α-L-rhamnopyranoside

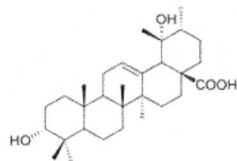

3-Epipomolic acid

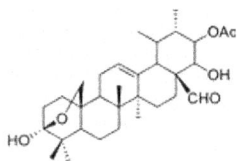

Cordiaketal A

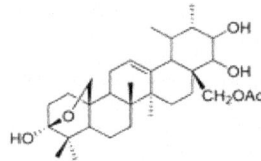

Cordiaketal B

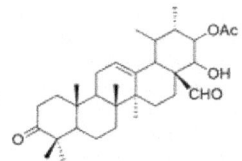

Cordianal A

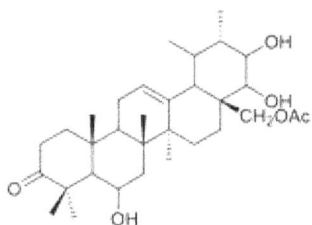

Cordianone

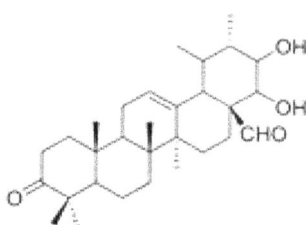

Cordianal B

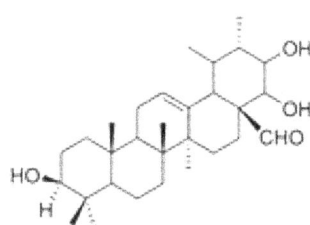

Cordianal C

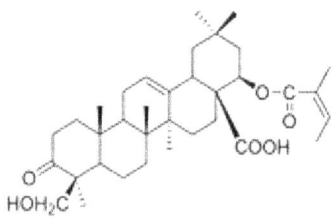

Icterogenin

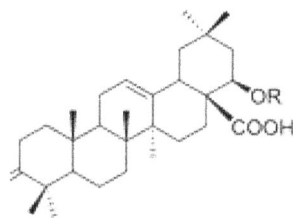

R = 2-Methyl-2-Z-butenoyl – Lantadene A

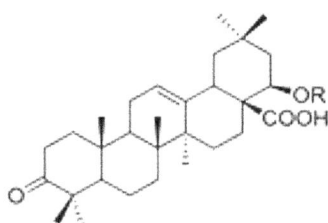

R = 3-Methylbutenoyl - Lantadene B

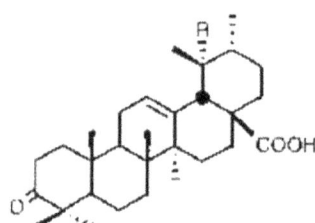

R = H - Ursomic acid; R = OH - Pomonic acid

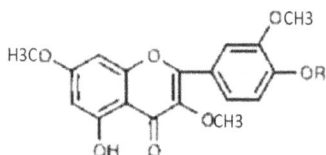

R = H - Pachypodol; R = CH$_3$ – Retusin

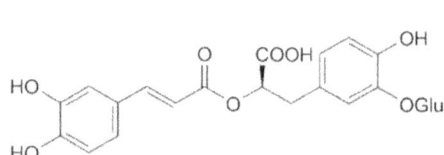

Rosmarinic acid 3-O-β-D-glucoside

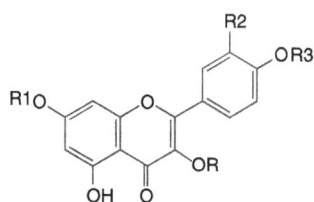

R = R^1 = R^3 = CH$_3$, R^2 = OH - 3,7,4'-Trimethoxyflavone; R = R^1 = R^3 = CH$_3$, R^2 = H - 5,3'-Dihydroxy-3,7,4'-trimethoxyflavone

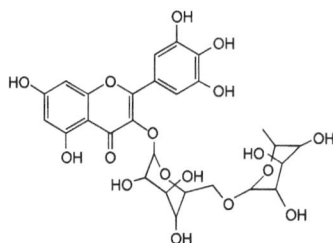

2-(3,4, 5-Trihydroxyphenyl)-5,7-dihydroxy-3-[α-L-rhamnopyranosyl-(1→6)-β-D-glucopyranosyloxy]-4H-chromen-4-one

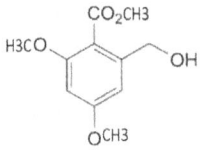

Cordicinol

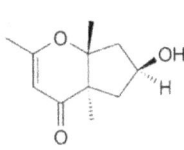

Cordinol

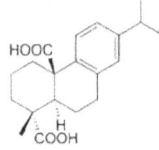

Cordioic acid

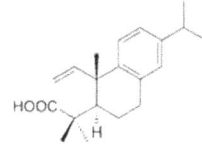

Cordifolic acid

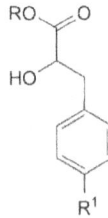

R = Bu, R^1 = OH – Latifolicinin A; R = Et, R^1 = OH - Latifolicinin B; R = CH$_3$, R^1 = OH - Latifolicinin C; R = H, R^1 = H$_3$CO - Latifolicinin D

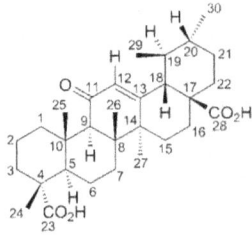

Cordinoic acid

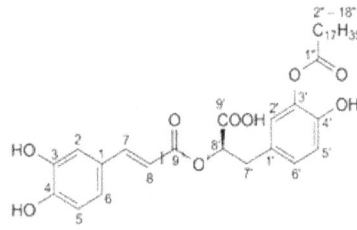

Cordicilin

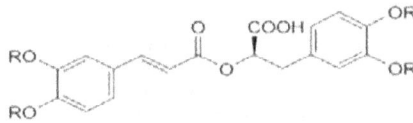

R = H - Rosmarinic acid; R = Ac - Tetraacetyl rosmarinic acid

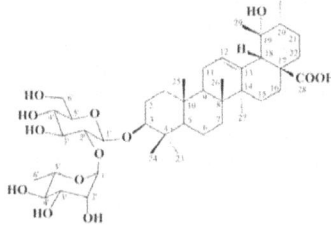

3β-*O*-α-L-rhamnopyranosyl- (1→2)-β-D-glucopyranosyl pomolic acid

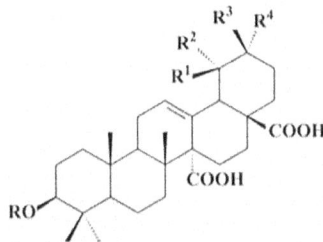

R = R^2 = R^3 = H, R^1 = R^4 = CH$_3$ - Quinovic acid; R = R^1 = R^2 = H, R^3 = R^4 = CH$_3$ - Cincholic acid; R = 6-Deoxy-β-D-glucosyl, R^2 = R^3 = H, R^1 = R^4 = CH$_3$ - Cincholic 3β-*O*-6-deoxy-β-D-glucopyranoside acid; R = β-D-glucosyl, R^2 = R^3 = H, R^1 = R^4 = CH$_3$ - quinovic 3β-*O*-β-D-gluycopyranoside acid

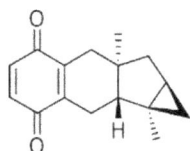

Globiferane

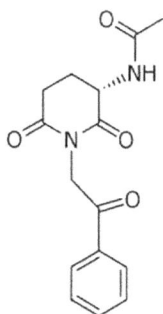

Cordiarimide A

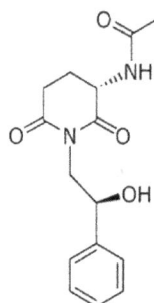

Cordiarimide B

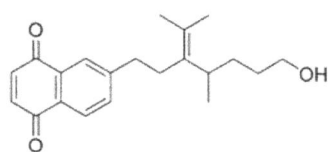

Cordiaquinone A

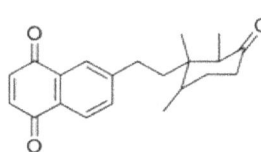

Cordiaquinone B

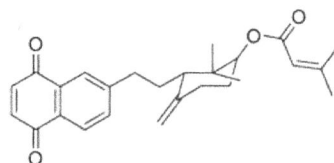

Cordiaquinone C

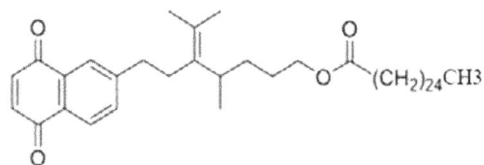

Cordiaquinone D

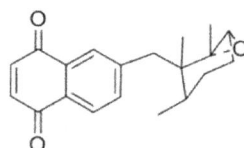

Cordiaquinone E

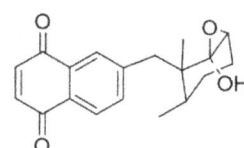

Cordiaquinone F

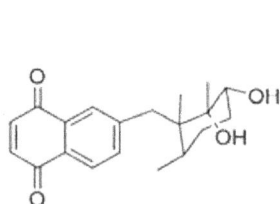

Cordiaquinone G

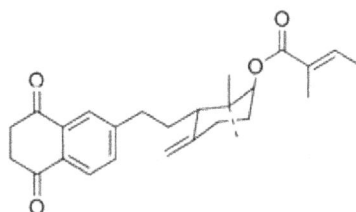

Cordiaquinone H

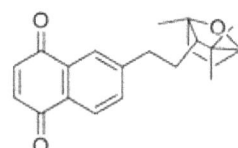

Cordiaquinone J

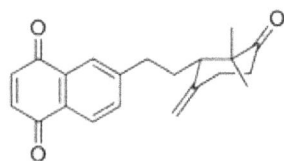

Cordiaquinone K

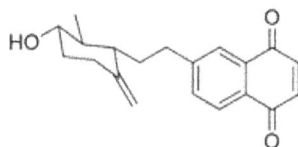

Cordiaquinone L

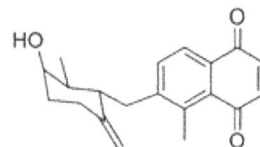

Cordiaquinone M

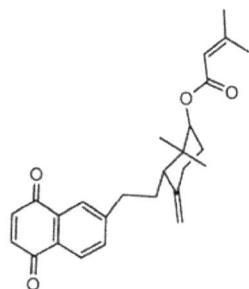

Cordiaquinone N

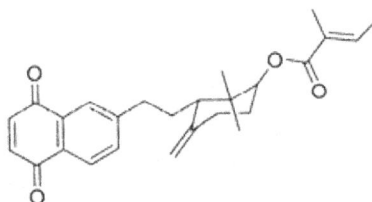

Cordiaquinone O

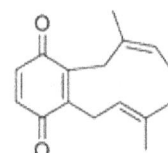

Globiferin

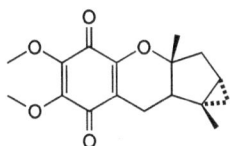

(1a*S**,1b*S**,7a*S**,8a*S**)-4,5-Dimethoxy-1a,7a-dimethyl-1,1a, 1b,2,7,7a,8,8a-octahydrocyclopropa[3,4] cyclopenta[1,2-b]naphthalene-3,6-dione

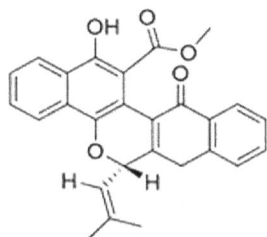

Microphyllaquinone

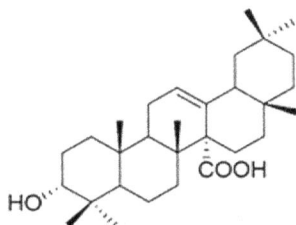

3α-Hydroxyolean-12-en-27-oic acid

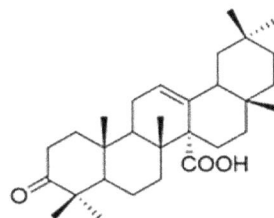

3-Oxoolean-12-en-27-oic acid

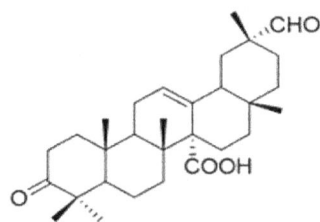

3,29-Dioxoolean-12-en-27-oic acid

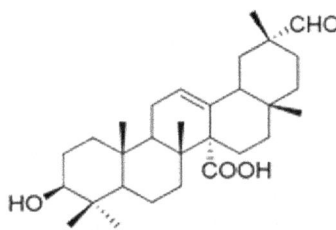

3α-Hydroxy-29-oxoolean-12-en-27-oic acid

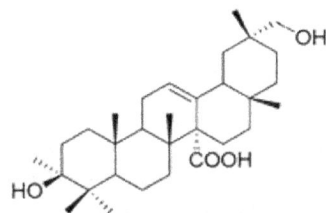

3α,29-Dihydroxyolean-12-en-27-oic acid

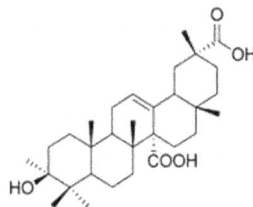

3α-Hydroxyolean-12-ene-27,29-oic acid

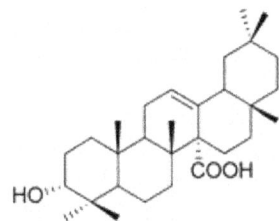

3α-Hydroxyolean-12-en-27-oic acid

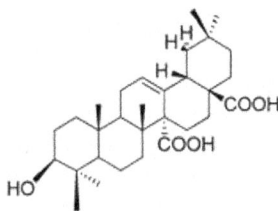

Cincholic acid

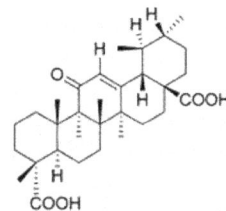

Cordinoic acid

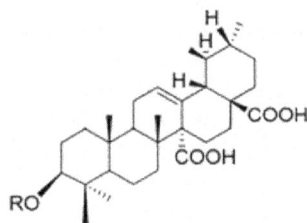

R = 6-Deoxy-β-D-glucosyl - Cincholic 3β-*O*-6-deoxy-β-D-glucopyranoside acid

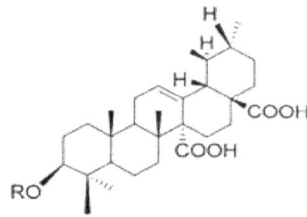

R = β-D-Glucosyl - Quinovic 3β-*O*-β-D-glucopyranoside acid

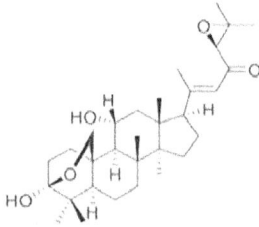

Cordialin A

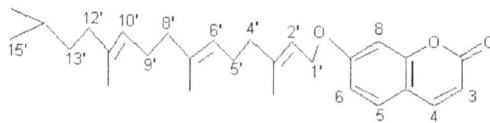

7(3′,7′,11′,14′-Tetramethy)pentadec-2′,6′,10′-trienyloxycoumarin

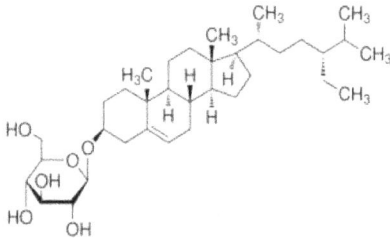

β-Sitosterol-β-D-glucoside

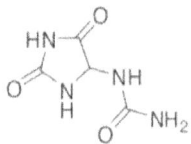

Allontoin

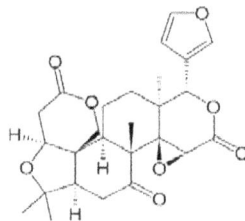

Limonin

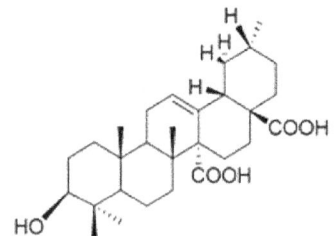

Quinovic acid

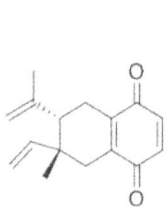

Cordiachrome C

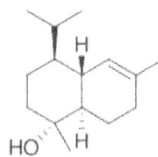

α-Cadinol

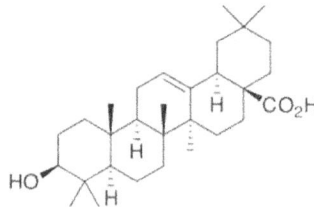

Oleanolic acid

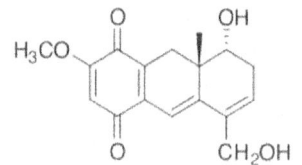

Oncocalyxone A

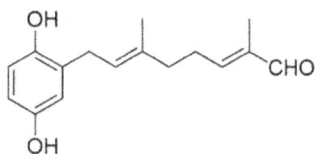

Alliodorin

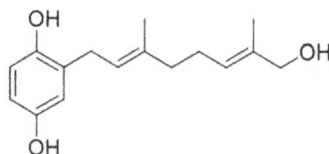

Alliodorol

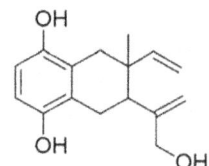

Allioquinol C

Cordiachromen A   Cordiachrome A   Cordiachrome B   Cordiachrome C

Cordiachrome G   Cordallinol   Cordiol A   Cordiaquinol C

Cordiaquinol I   Cordiaquinol J   Cordiaquinol K   Trichotomol

Leucocordiachrome H   2-(2Z)-(3-hydroxy-3,7-dimethylocta-2,6-dienyl)-1,4-benzenediol

MECD-1   MECD-2

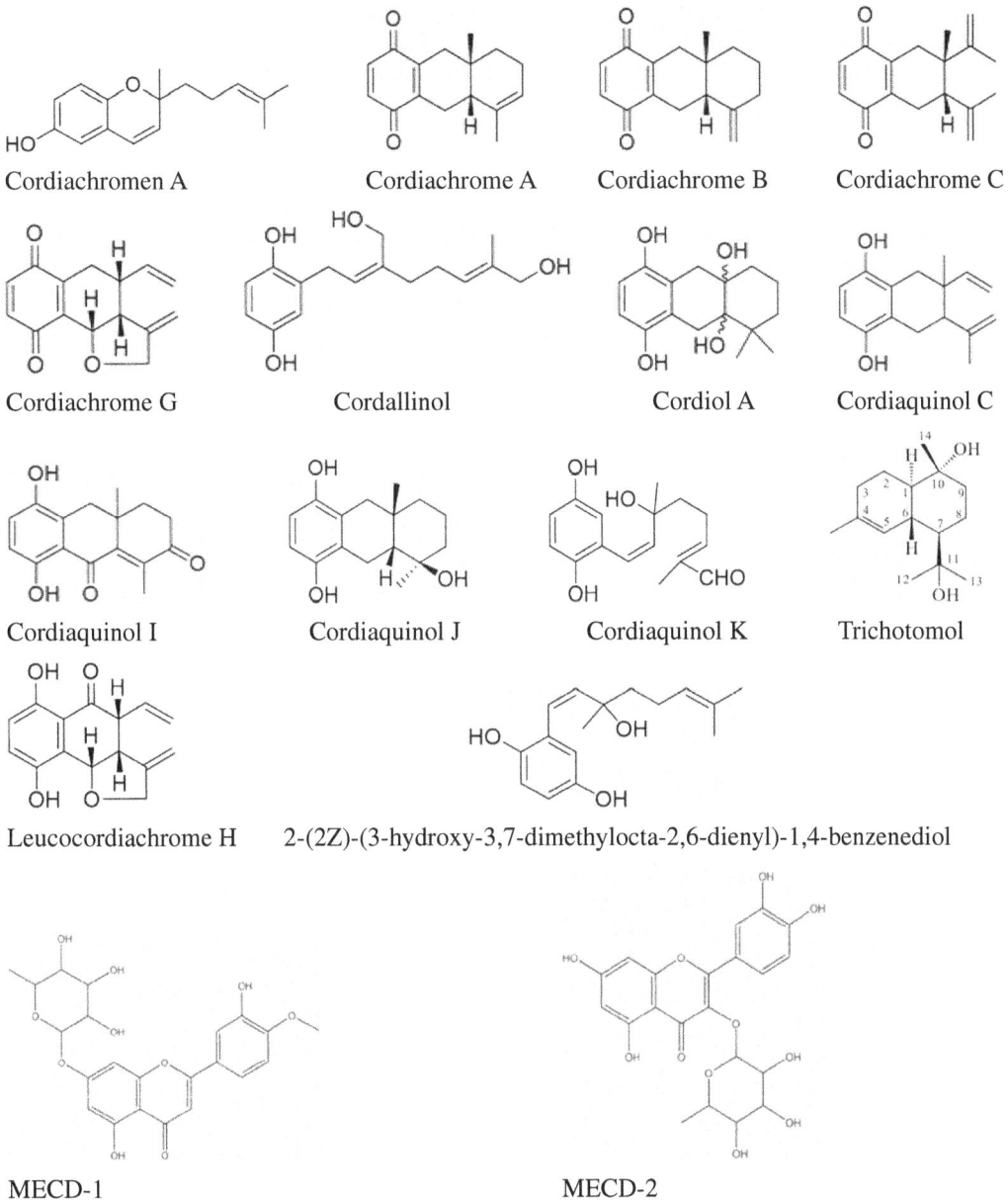

12.3 CULTURE CONDITIONS

The cell suspension cultures of *C. verbenacea* leaf explants were established for increasing the production of flavonoids. The explants were inoculated onto an MS (Murashige and Skoog 1962) medium containing kinetin (2.32 µM) + 1-naphthaleneacetic acid (10.74 µM). The maximum callus growth rate (37%) was recorded from the fourth to twelfth day. 7,4′-dihydroxy-5′-carboxymethoxy isoflavone and 7,4′- dihydroxy-5′-methyl isoflavone were isolated and identified from the callus. Higher accumulation of these compounds was higher in *in vitro* cultured cells than intact plant (Lameira et al. 2009). The leaves of *C. myxa* were cultured on an MS medium. Maximum callus growth was recorded in MS medium with supplementation of 1-naphthaleneacetic acid (2.0 mg/l) and benzyl adenine (2.0 mg/l; Taha 2016a). Single nodal segments of *C. myxa* were transferred to an MS medium containing kinetin (2.0, 4.0, and 6.0 mg/l) and 6-benzylaminopurine alone or in combination of 1-naphthaleneacetic acid (0.01 mg/l). The best callus growth response (93.6%) was reported in the presence of kinetin (4.0 mg/l; Krishna and Singh 2013). For increasing the production of secondary metabolites, biotic and abiotic elicitors were added to the cell cultures. The addition of elicitors increased the accumulation of phenolic compounds (rutin, rubinin, and chlorogenic acid). Addition of zinc sulphate (0.5 mg/l) significantly increased the profiles of dihydroxyrobmetin, rubinin, and chlorogenic acid in cell cultures (182.95, 218.44, 60.99 µg/ml). Similarly, zinc sulphate (1 mg/l) increased the levels of hesperdin and rutin concentration (100.65, 106.50 µg/ml). Cobalt chloride (1mg/l) addition led to enhanced production of robinin and hesperdin (239.43 and 55.99 µg/ml) cell cultures (Taha 2016b).

REFERENCES

Abdel El-Aleem ER, Seddik FE-Z, Samy MN, Desoukey SY. 2017. Botanical studies of the leaf of *Cordia myxa* L. J Pharmacogn Phytochem 6, 2086–2091.

Abdel-Aleem ER, Attia EZ, Farag FF, Samy MN, Desoukey SY. 2019. Total phenolic and flavonoid contents and antioxidant, anti-inflammatory, analgesic, antipyretic and antidiabetic activities of *Cordia myxa* L. leaves. Clin Phytoscience 5, 29.

Afzal M, Obuekwe C, Khan AR, Barakat, H. 2009. Influence of *Cordia myxa* on chemically induced oxidative stress. Nutr Food Sci 39, 6–15.

Agnihotri VK, Srivastava SD, Srivastava SK, Pitre S, Rusia K. 1987. Constituents from the seeds of *Cordia obliqua* as potential anti-inflammatory agents. Indian J Pharm Sci 49, 66–69.

Aimey Z, Goldson-Barnaby A, Bailey D. 2020. A review of Cordia species found in the Caribbean: *Cordia obliqua* Willd., *Cordia dichotoma* G. Forst. and *Cordia collococca* L. Int J Fruit Sci 20, S884–S893.

Al-Ati T. 2011. Assyrian plum (*Cordia myxa* L.). Postharvest Biol Technol Trop Subtrop Fruits 2011, 116–126.

Alhadi EA, Khalid HS, Alhassan MS, Ali AA, Babiker SG, Alabdeen EMZ, Kabbashi AS. 2015. Antioxidant and cytotoxicity activity of *Cordia africana* in Sudan. J Med Plant Res 3, 29–32.

Aljeboury GH. 2021. Anti-inflammatory effect of *Cordia myxa* extract on bacteria that infected wounds in rats as a model for human. Ann RSCB 25, 16040–16045.

Al-Musawi MH, Ibrahim KM, Albukhaty S. 2022. In vitro study of antioxidant, antibacterial, and cytotoxicity properties of *Cordia myxa* fruit extract. Iran J Microbiol 14, 97–103.

Andrade KCR, Martins DHN, Barros DA, Souza PM, Silveira D, Fonseca-Bazzo YM, Magalhães PO. 2022. Essential oils of Cordia species, compounds and applications: A systematic review. Bol Latinoam Caribe Plant Med Aromat 21, 156–175.

Anjaria J. 1998. Natural heals, a glossary of selected indigenous medicinal plants of India. Indian J Pharmacol 30, 126.

Anonymous. 2004. The Wealth of India, Vol IV. CSIR, New Delhi.

Asrie AB, Abdelwuhab M, Shewamene Z, Gelayee DA, Adinew GM, Birru EM. 2016. Antidiarrheal activity of methanolic extract of the root bark of *Cordia africana*. J Exp Pharmacol 8, 53–59.

Balick MJ, Arvigo R. 2015. Messages from the Gods: A Guide to the Useful Plants of Belize. Oxford University Press, New York.

Begum S, Perwaiz S, Siddiqui BS, Khan S, Fayyaz S, Ramzan M. 2011. Chemical constituents of *Cordia latifolia* and their nematicidal activity. Chem Biodivers 8, 850–861.

Bhide B, Acharya RN, Naria P, Pillai APG, Shukla VJ. 2011. Pharmacognostic evaluation of *Cordia macleodii* Hook. stem bark. Pharmacogenomics J 3, 49–53.

Bieber LW, Messana I, Lins SCN, da Silva Filho AA, Chiappeta AA, De Méllo JF. 1990. Meroterpenoid naphthoquinones from *Cordia corymbosa*. Phytochemistry 29, 1955–1959.

Caparroz-Assef SM, Grespan R, Batista RCF, Bersani-Amado FA. 2008. Toxicity studies of *Cordia salicifolia* extract. Acta Sci Health Sci 27, 41–44.

Chandrakar J, Dixit AK. 2017. *Cordia macleodii* Hook.f. & Thomson.—a potential medicinal plant. Int J Phytomed 9, 394–398.

Chaubey ON, Upadhyay R, Tripathi NK, Ranjan A. 2016. Isolations & characterization of compounds from *Cordia macleodii* Hook bark & leaves. IOSR J Appl Chem 9, 83–85.

Chauhan JS. Srivastava SK. 1977. 5, 7- Dimethoxytaxifolin-3-*O*-α-L-rhamnopyranoside from the roots of *Cordia obliqua* Linn. Indian J Chem Sect B 15, 760–761.

Chauhan JS, Srivastava SK, Sultan M. 1978. Hesperetin 7- rhamnoside from *Cordia obliqua*. Phytochemistry 17, 334.

Chen TK, Ales DC, Baenziger NC, Wiemer DF. 1983. Ant-repellent triterpenoids from *Cordia alliodora*. J Org Chem 48, 3525–3531.

Correia Da Silva TB, Souza VKT, Da Silva APF, Lemos RPL, Conserva LM. 2010. Determination of the phenolic content and antioxidant potential of crude extracts and isolated compounds from leaves of *Cordia multispicata* and *Tournefortia bicolor*. Pharm Biol 48, 63–69.

Daniel FA. 2004. Florida Ethnobotany. CRC Press, Boca Raton.

da Silva AS, de Fátima Agra, Tavares JF, da-Cunha EVL, Barbosa-Filho JM, da Silva MS. 2010. Flavanones from aerial parts of *Cordia globosa* (Jacq.) Kunth, Boraginaceae. Rev Bras Farmacogn 20, 682–685.

da Silva SAS, Rodrigues MSL, Agra MF, Leitãoda-Cunha EV, Barbosa-Filho JM, Silva MS. 2004. Flavonoids from *Cordia globosa*. Biochem Syst Ecol 32, 359–361.

de Freitas Roldão E, Witaicenis A, Seito LN, Hiruma-Lima CA, Di Stasi LC. 2008. Evaluation of the antiulcerogenic and analgesic activities of *Cordia verbenacea* DC. (Boraginaceae). J Ethnopharmacol 119, 94–98.

de Menezes JE, Lemos TL, Pessoa OD, Braz-Filho R, Montenegro RC, Wilke DV, Costa-Lotufo LV, Pessoa C, de Moraes MO, Silveira ER. 2005a. A cytotoxic meroterpenoid benzoquinone from roots of *Cordia globosa*. Planta Med 71, 54–58.

de Menezes JESA, Lemos TLG, Silveira E, Andrade-Neto M. 2005b. Volatile constituents of *Cordia trichotoma* Vell. from the northeast of Brazil. Flavour Fragr J 20, 149–151.

de Menezes JESA, Lemos TLG, Silveira ER, Braz-Filho R, Pessoa ODL. 2001. Trichotomol, a new cadinenediol from *Cordia trichotoma*. J Braz Chem Soc 12, 787–790.

de Menezes JESA, Machado FEA, Lemos TLG, Silveira ER, Filho RB, Pessoa ODL. 2004. Ssquiterpenes and a phenylpropanoid from *Cordia trichotoma*. Z Naturforsch 59c, 19–22.

de Oliveira JCS, da Camara CAG, Schwartz MOE. 2007. Volatile constituents of the stem and leaves of *Cordia* species from mountain forests of Pernambuco (north-eastern Brazil). J Essent Oil Res 19, 444–448.

Dettrakul S, Surerum S, Rajviroongit S, Kittakoop P. 2009. Biomimetic transformation and biological activities of globiferin, a terpenoid benzoquinone from *Cordia globifera*. J Nat Prod 72, 861–865.

Diniz JC, Viana FA, de Oliveira OF, Silveira ER, Pessoa ODL. 2008. Chemical composition of the leaf essential oil of *Cordia leucocephala* Moric from northeast of Brazil. J Essent Oil Res 20, 495–496.

Diniz JC, Viana FA, Oliveira OF, Maciel MAM, de Menezes Torres MdC, Braz-Filho R, Silveira ER, Pessoa ODL. 2009. ^{1}H- and ^{13}C-NMR assignments for two new cordiaquinones from roots of *Cordia leucocephala*. Magn Reson Chem 47, 190–193.

Dominguez XA. 1973. Dimethyl-3, 4'-kaempferol of *Cordia boissieri*. Phytochemistry 12, 724–725.

do Vale AE, David JM, dos Santos EO, David JP, e Silva LCRC, Bahia MV, Brandão HN. 2012. An unusual caffeic acid derived bicyclic [2.2.2] octane lignan and other constituents from *Cordia rufescens*. Phytochemistry 76, 158–161.

Dubey PC, Sikarwar RLS, Tiwari A. 2008. Ethobotany of *Cordia macleodii*. Shodha Samagya 2(1&2), 31.

Faridah H, van der Maesen LG. 1997. Plant Resources of South-East Asia 11, Auxiliary Plants. Prosea Foundation, Bogor.

Fernandes ES, Passos GF, Medeiros R, da Cunha FM, Ferreira J, Campos MM, Pianowski LF, Calixto JB. 2007. Anti-inflammatory effects of compounds α-humulene and (-)-*trans*-caryophyllene isolated from the essential oil of *Cordia verbenacea*. Eur J Pharmacol 569, 228–236.

Ficarra R, Ficarra P, Tommasini S, Calabrò ML, Ragusa S, Barbera R, Rapisarda A. 1995. Leaf extracts of some *Cordia* species: Analgesic and anti-inflammatory activities as well as their chromatographic analysis. Farmaco 50, 245–256.

Fouseki MM, Damianakos H, Karikas GA, Gupta M, Roussakis C, Chinou I. 2016. Chemical constituents from *Cordia alliodora* and *Cordia colloccoca* from Panama and their antimicrobial and cytotoxic activities. Planta Med 82, S1–S381.

Freitas HPS, Maia AIV, Silveira ER, Marinho Filho JDB, Moraes MO, Pessoa C, Lotufo LVC, Pessoa ODL. 2012. Cytotoxic cordiaquinones from the roots of *Cordia polycephala*. J Braz Chem Soc 23, 1558–1562.

Ganjare AB, Nirmal SA, Rub RA, Patil AN, Pattan SR. 2011. Use of *Cordia dichotoma* bark in the treatment of ulcerative colitis. Pharm Biol 49, 850–855.

Giday M, Teklehaymanot T, Animut A, Mekonnen Y. 2007. Medicinal plants of the Shinasha, Agew-awi and Amhara peoples in northwest Ethiopia. J Ethnopharmacol 110, 516–525.

Glover PE, Stewart J, Gwynne MD. 1966. Notes on East African plants, Part III—medicinal uses of plants. East Afr Agric For J 32, 200–207.

Gottschling M. 2003. Phylogenetic Analysis of Selected Boraginales. PhD Thesis, Freien Universitat Berlin, Berlin.

Grandtner MM. 2005. Elsevier's Dictionary of Trees, Vol 1: North America. Elsevier, Amsterdam.

Grandtner MM, Chevrette J. 2013. Dictionary of Trees, South America: Nomenclature, Taxonomy and Ecology. Academic Press, Elsevier, Amsterdam.

Gupta R, Gupta GD. 2014. A review on plant *Cordia obliqua* Willd. (Clammy cherry). Pharmacogn Rev 9, 127–131.

Hamdia MSAl-H, Al-Faraji ASA. 2017. Evaluation of inhibitory activity of *Cordia myxa* fruit extract on microorganisms that causes spoilage of food and its role in the treatment of certain disease states. J Biol Agric Healthcare 7, 43–49.

Hussain N, Kakoti BB. 2013. Review on ethnobotany and psychopharmacology of *Cordia dichotoma*. J Drug Deliv Ther 3, 110–113.

Hussain N, Kakoti BB, Rudrapal M, Rahman Z, Rahman M, Chutia D, Sarwa KK. 2020. Anti-inflammatory and antioxidant activities of *Cordia dichotoma* Forst. Biomed Pharmacol J 13, 2093–2099.

Hussain N, Kakoti BB, Rudrapal M, Sarwa KK, Celik I, Attah EI, Khairnar SJ, Bhattacharya S, Sahoo RK, Walode SG. 2021. Bioactive antidiabetic flavonoids from the stem bark of *Cordia dichotoma* Forst.: Identification, docking and ADMET studies. Molbank 2021, M1234.

Ioset JR, Marston A, Gupta MP, Hostettmann K. 1998. Antifungal and larvicidal meroterpenoid naphthoquinones and a naphthoxirene from the roots of *Cordia linnaei*. Phytochemistry 47, 729–734.

Ioset JR, Marston A, Gupta MP, Hostettmann K. 2000a. Antifungal and larvicidal compounds from the root bark of *Cordia alliodora*. J Nat Prod 63, 424–426.

Ioset JR, Marston A, Gupta MP, Hostettmann K. 2000b. Antifungal and larvicidal cordiaquinones from the roots of *Cordia curassavica*. Phytochemistry 53, 613–617.

Ioset JR, Wolfender J-L, Marston A, Gupta MP, Hostettmann K. 1999. Identification of two isomeric meroterpenoid naphthoquinones from *Cordia linnaei* by liquid chromatography-mass spectrometry and liquid chromatography-nuclear magnetic resonance spectroscopy. Phytochem Anal 10, 137–142.

Isa AI, Saleh MIA, Abubakar A, Dzoyem JP, Adebayo SA, Musa I, Sani UF, Daru PA. 2016. Evaluation of anti-inflammatory, antibacterial and cytotoxic activities of *Cordia africana* leaf and stem bark extracts. Bayero J Pure Appl Sci 9, 228.

Jamkhande PG, Barde SR, hailesh L. Patwekar SL, Tidke PS. 2013. Plant profile, phytochemistry and pharmacology of *Cordia dichotoma* (Indian cherry): A review. Asian Pac J Trop Biomed 3, 1009–1012.

Kendir G, Bae HJ, Kim J, Jeong Y, Bae HJ, Park K, Yang X, Cho Y-j, Kim J-Y, Jung SY, Köroğlu A, Jang DS, Ryu JH. 2022. The efects of the ethanol extract of *Cordia myxa* leaves on the cognitive function in mice. BMC Complement Med Ther 22, 215.

Kendir G, Özek G, Köroğlu A. 2021. Leaf essential oil analysis and anatomical study of *Cordia myxa* from Turkey. Plant Biosyst 155, 204–210.

Kirtikar KR, Basu BD. 1956. Indian Medicinal Plants, 2nd ed., Vol II. International Book Distributors, Dehradun.

Kirtikar KR, Basu RR. 2005. Indian Medicinal Plants. International Book Distributors, Dehradun.

Korbes CV. 1995. Manual de plantas medicinais. Assesoar Press de imprensa, Minas Gerais.

Krishna H, Singh D. 2013. Micropropagation of lasora (*Cordia myxa* Roxb.). Indian J Hortic 70, 323–327.

Kuroyanagi M, Kawahara N, Sekita S, Satake M, Hayashi T, Takase Y, Masuda K. 2003. Dammarane type triterpenes from the Brazilian medicinal plant *Cordia multispicata*. J Nat Prod 66, 1307–1312.

Kuroyanagi M, Seki T, Tatsuo Hayashi T, Nagashima Y, Kawahara N, Sekita S, Satake M. 2001. Anti-androgenic triterpenoids from the Brazilian medicinal plant, *Cordia multispicata*. Chem Pharm Bull 49, 954–957.

Lameira OA, Pinto JEB, Cardoso MG, Arrigoni-Blank MF. 2009. Establishment of cell suspension cultures and flavonoid identification in *Cordia verbenacea* DC. Rev Bras Plant Med 11, 7–11.

Lans C. 2007. Ethnomedicines used in Trinidad and Tobago for reproductive problems. J Ethnobiol Ethnomed 3, 13.

Lanting MV, Palaypayon CM. 2002. Forest tree species with medicinal uses. Depart Environ Nat Resour Recomm 11, 1–24.

Ló SMS, Duarte MR. 2011. Leaf and Stem morpho-anatomy of *Cordia americana* (L.) Gottschling & J.S. Mill., Boraginaceae. Lat Am J Pharm 30, 823–828.

Manners GD, Jurd L. 1973. Alliodorin, a phenolic terpenoid from *Cordia alliodora*. Tetrahedron Lett 31, 2955–2958.

Manners GD, Jurd L. 1977. The hydroquinone terpenoids of *Cordia alliodora*. J Chem Soc Perkin Trans 1, 405–410.

Marinho-Filho JD, Bezerra DP, Araújo AJ, Montenegro RC, Pessoa C, Diniz JC, Viana FA, Pessoa ODL, Silveira ER, de Moraes MO, Costa-Lotufo LV. 2010. Oxidative stress induction by (+)-cordiaquinone J triggers both mitochondria dependent apoptosis and necrosis in leukemia cells. Chem Biol Interact 183, 369–379.

Marini G, Graikou K, Zengin G, Karikas GA, Gupta MP, Chinou I. 2018. Phytochemical analysis and biological evaluation of three selected *Cordia* species from Panama. Ind Crops Prod 120, 84–89.

Martim JKP, Maranho LT, Thais A Costa-Casagrande TA. 2021. Review: Role of the chemical compounds present in the essential oil and in the extract of *Cordia verbenacea* DC as an anti-inflammatory, antimicrobial and healing product. J Ethnopharmacol 265, 113300.

Matias EFF, Alves EF, Santos BS, de Souza CES, Ferreira JVdA, de Lavor AKLS, Figueredo FG, de Lima LF, dos Santos FAV, Peixoto FSN, Colares AV, Boligon AA, Saraiva RdA Athayde ML, da Rocha JBT, Menezes IRA, Coutinho HDM, da Costa JGM. 2013. Biological activities and chemical characterization of *Cordia verbenacea* DC. as tool to validate the ethnobiological usage. Evid Based Complement Altern Med 2013, Article ID 164215.

Matias EFF, Alves EF, Silva MKN, Carvalho VRA, Medeiros CR, Santos FAV, Bitu VCN, Souza CES, Figueredo FG, Boligon AA, Athayde ML, Costa JGM, Coutinho HDM. 2016. Potentiation of antibiotic activity of aminoglycosides by natural products from *Cordia verbenacea* DC. Microb Pathog 95, 111–116.

Mayevych I, Manuel López-Romero J, Cabanillas J. 2015. Chemical composition of *Cordia lutea* L.: Absence of pyrrolizidine alkaloids. Nat Prod Chem Res 3, 6.

Meccia G, Rojas LB, Velasco J, Díaz T, Usubillaga A, Arzola JC, Ramos S. 2009. Chemical composition and antibacterial activity of the essential oil of *Cordia verbenacea* from the Venezuelan Andes. Nat Prod Commun 4, 1119–1122.

Medeiros R, Passos GF, Vitor CE, Koepp J, Mazzuco TL, Pianowski LF, Campos MM, Calixto JB. 2007. Effect of two active compounds obtained from the essential oil of *Cordia verbenacea* on the acute inflammatory responses elicited by LPS in the rat paw. Br J Pharmacol 151, 618–627.

Menezes JESA, Lemos TLG, Pessoa ODL, Braz-Filho R, Montenegro RC, Wilke DV, Costa Lotufo LV, Pessoa C, Moraes MO, Silveira ER. 2005. A cytotoxic meroterpenoid benzoquinone from roots of *Cordia globosa*. Planta Med 71, 54–58.

Mentz LA, Lutzemberger LC, Schenkel EP. 1997. Daflora medicinal do Rio Grande do Sul: Notas sobre a obra de D'avila. Caderno de Farmacia 13, 25–48.

Mishra A, Garg GP. 2011. Antidiabetic activity of fruit pulp of *Cordia dichotoma* in alloxan induced diabetic rats. Int J Pharm Sci Res 2, 2314–2319.

Mohamed SS, Elkhamisy AES. 2019. Anti-diabetic potential of *Cordia dichotoma* pulp and peel (functional fiber) in type II diabetic rats. Int J Pharmacol 15, 102–109.

Moir M, Thomson RH. 1973. Naturally occurring quinones. Part XXII. Terpenoid quinones in Cordia spp. J Chem Soc Perkin Trans 1, 1352–1357.

Mori K, Kawano M, Fuchino H, Ooi T, Satake M, Agatsuma Y, Kusumi T, Sekita S. 2008. Antileishmanial compounds from *Cordia fragrantissima* collected in Burma (Myanmar). J Nat Prod 71, 18–21.

Morton JF. 1981. Atlas of Medicinal Plants of Middle America. Charles C Thomas, Springfield.

Murashige T, Skoog F. 1962. A revised medium for rapid growth and bioassay with tobacco tissue culture. Physiol Plant 15, 473–497.

Murthy HN, Joseph KS, Gaonkar AA, Payamalle S. 2019. Evaluation of chemical composition and antioxidant activity of *Cordia myxa* fruit pulp. J Herbs Spices Med Plants 25, 192–201.

Nadkarni AK. 1991. Indian Materia Medica, 3rd ed., Vol I. Bombay Popular Prakashan, Bombay.

Najib A, Ahmad AR, Handayani V. 2019. ELISA test on *Cordia myxa* L leaf extract for α-glucosidase inhibitor. Pharmacogenomics J 11, 358–361.

Nakamura N, Kojima S, Lim YA, Meselhy MR, Hattori M, Gupta MP, Correa M. 1997. Dammarane-type triterpenes from *Cordia spinescens*. Phytochemistry 46, 1139–1141.

Nariya PB, Bhalodia NR, Shukla VJ, Nariya MB. 2010. *In vitro* evaluation of antimicrobial and antifungal activity of *Cordia macleodii* bark. (Hook.F. & Thomson). Int J Pharmtech Res 2, 2522–2526.

Nariya PB, Shukla VJ, Acharya RN, Nariya MB, Bhatt PV, Pandit CM, Tada R. 2014. Isolation and characterization of phytosterols from *Cordia macleodii* (Hook F. and Thomson) bark by chromatographic and spectroscopic method. Asian J Pharm Clin Res 7, 86–88.

Nariya PB, Shukla VJ, Acharya RN, Nariya MB, Dhalani JM, Patel AS, Ambasana PA. 2018. Triterpenoid and fatty acid contents from the stem bark of *Cordia dichotoma* (Forst f.). Folia Med (Plovdiv) 60, 594–600.

Nayak P, Kalidass C. 2016. Ethnobotany, phytochemistry, pharmacognostic and pharmacological aspects of *Cordia macleodii* Hook.f. & Thomson—a review. J Non Timber For Prod 23, 67–71.

Nogueira TB, de Sá de Sousa Nogueira RB, e Silva DA, Tavares JF, de Oliveira Lima E, de Oliveira Pereira F, de Souza Fernandes MMM, de Medeiros FA, Sarquis RdFSR, Braz Filho R, da Silva Maciel JK, de Souza MdFV. 2013. First chemical constituents from *Cordia exaltata* Lam and antimicrobial activity of two neolignans. Molecules 18, 11086–11099.

Oliveira FM, Diniz JC, Viana FA, Rocha SAS, da Silva Junior HM. 2012. Quantificac̦áo por CLAE de nafto-quinonas do extrato das raızes de Cordia leucocephala Moric. Holos 1, 41–48.

Orwa C, Mutua A, Kindt R, Simons A, Jamnadass RH. 2009. Agroforestree Database: A Tree Reference and Selection Guide Version 4.0. World Agroforestry Centre ICRAF, Nairobi.

Osho A, Otuechere CA, Adeosun CB, Oluwagbemi T, Atolani O. 2016. Phytochemical, sub-acute toxicity, and antibacterial evaluation of Cordia sebestena leaf extracts. J Basic Clin Physiol Pharmacol 27, 163–170.

Oza MJ, Kulkarni YA. 2017. Traditional uses, phytochemistry and pharmacology of the medicinal species of the genus Cordia (Boraginaceae). J Pharm Pharmacol 69, 755–789.

Parks J, Gyeltshen T, Prachyawarakorn V, Mahidol C, Ruchirawat S, Kittakoop P. 2010. Glutarimide alkaloids and a terpenoid benzoquinone from Cordia globifera. J Nat Prod 73, 992–994.

Parmar C, Kaushal MK. 1982. Wild Fruits of the Sub-Himalayan Region. Kalyani Publishers, New Delhi.

Passos GF, Fernandes ES, da Cunha FM, Ferreira J, Pianowski LF, Campos MM, Calixto JB. 2007. Anti-inflammatory and anti-allergic properties of the essential oil and active compounds from Cordia verbenacea. J Ethnopharmacol 110, 323–333.

Patel AK, Pathak N, Trivedi H, Gavania M, Patel M, Panchal N. 2011a. Phytopharmacological properties of Cordia dichotoma as a potential medicinal tree: An overview. Int J Inst Pharm Life Sci 1, 40–51.

Patel AK, Pathak NL, Trivedi HD, Patel LD, Gavania MG, Trivedi H. 2011b. Role of Cordia dichotoma on behavioral changes by using long-term hypoperfusion in rats. Int J Pharm Res Dev 3, 6–17.

Pimentel SP, Barrella GE, Casarin RCV, Cirano FR, Casati MZ, Foglio MA, Figueira GM, Ribeiro FV. 2012. Protective effect of topical Cordia verbenacea in a rat periodontitis model: Immune-inflammatory, anti-bacterial and morphometric assays. BMC Complement Altern Med 12, 224.

Polhill RM. 1991. Flora of Tropical East Africa—Boraginaceae. CRC Press, Rotterdam.

Pradheeps M, Poyyamoli G. 2013. Ethnobotany and utilization of plant resources in Irula villages (Sigur plateau, Nilgiri Biosphere Reserve, India). J Med Plant Res 7, 267–276.

Pullaiah T. 2006. Encyclopaedia of World Medicinal Plants, Vol 1. Regency Publications, New Delhi.

Quattrocchi U. 2012. CRC World Dictionary of Medicinal and Poisonous Plants: Common Names, Scientific Names, Eponyms, Synonyms, and Etymology. CRC Press, Boca Raton.

Ragasa CY, Ebajo Jr. V, De Los Reyes MM, Mandia EH, Tan MCS, Brkljača R, Urban S. 2015. Chemical constituents of Cordia dichotoma G. Forst. J Appl Pharm Sci 5(Suppl 2), 16–21.

Rakesh G, Patel AG, Shukla VJ, Nariya MB, Acharya RN. 2018. Phytochemical analysis of successive extracts of the Cordia macleodii leaves Hook.: A folklore medicinal plant. J Ayurvedic Herb Med 4, 14–17.

Ranjbar M, Varzi HN, Sabbagh A, Bolooki A, Sazmand A. 2013. Study on analgesic and anti-inflammatory properties of Cordia myxa fruit hydro-alcoholic extract. Pak J Biol Sci 16, 2066–2069.

Richard C, Rolando P, Nefertaris D. 2010. Trees of Panama and Costa Rica. Princeton University, Princeton.

Rodrigues FFG, Oliveira LGS, Rodrigues FFG, Saraiva ME, Almeida SCX, Cabral MES, Campos AR, Costa JGM. 2012. Chemical composition, antibacterial and antifungal activities of essential oil from Cordia verbenacea DC leaves. Pharmacogn Res 4, 161–165.

Rodrigues RRL, Nunes LTA, de Araújo AR, Filho JDBM, da Silva MV, de Amorim Carvalho FA, Pessoa ODL, Freitas HPS, da Franc Rodrigues KA, Araújo AJ. 2021. Antileishmanial activity of cordiaquinone E towards Leishmania (Leishmania) amazonensis. Int Immunopharmacol 90, 107124.

Ruffo CK, Birnie A, Tengnas B. 2002. Edible Wild Plants of Tanzania. Regional Land Management Unit, Nairobi.

Sabry MM, El-Fishawy AM, El-Rashedy AA, El Gedaily RA. 2022. Phytochemical investigation of Cordia Africana Lam. stem bark: Molecular simulation approach. Molecules 27, 4039.

Saif S, Tahir A, Asim T, Chen Y, Khan M, Adil SF. 2019. Green synthesis of ZnO hierarchical microstructures by Cordia myxa and their antibacterial activity. Saudi J Biol Sci 26, 1364–1371.

Samari F, Parkharia P, Eftekhar E, Mohseni F, Yousefinejad S. 2019. Antioxidant, cytotoxic and catalytic degradation efficiency of controllable phyto-synthesised silver nanoparticles with high stability using Cordia myxa extract. J Exp Nanosci 14, 141–159.

Santos RP, Lemos TLG, Pessoa ODL, Braz-Filho R, Rodrigues-Filho E, Viana FA, Silveira ER. 2005b. Chemical constituents of Cordia piauhiensis—Boraginaceae. J Braz Chem Soc 16, 662–665.

Santos RP, Nunes EP, Nascimento RF, Santiago GMP, Menezes GHA, Silveira ER, Pessoa ODL. 2006. Chemical composition and larvicidal activity of the essential oils of Cordia leucomalloides and Cordia curassavica from the Northeast of Brazil. J Braz Chem Soc 17, 1027–1030.

Santos RP, Silveira ER, de A Uchôa DE, Pessoa ODL, Viana FA, Braz-Filho R. 2007. ^{1}H and ^{13}C NMR spectral data of new saponins from Cordia piauhiensis. Magn Reson Chem 45, 692–694.

Santos RP, Silveira ER, Lemos TLG, Viana FA, Braz-Filho R, Pessoa ODL. 2005a. Characterization of two minor saponins from Cordia piauhiensis by 1H and 13C NMR spectroscopy. Magn Reson Chem 43, 494–496.

Santos RP, Viana FA, Lemos TLG, Silveira ER, Braz-Filho R, Pessoa ODL. 2003. Structure elucidation and total assignment of 1H and 13C NMR data for a new bisdesmoside saponin from *Cordia piauhiensis*. Magn Reson Chem 41, 735–738.

Sarah T-B, Fagertun RS, Kebede A, Judith N, Fetien A, Trude W. 2013. Ferric reducing antioxidant power and total phenols in *Cordia Africana* fruit. Afr J Biochem Res 7, 215–224.

Saxena HO. 1995. The Flora of Orissa. Regional Research Laboratory, Bhubaneshwar.

Schmelzer GH, Gurib-Fakim A. 2008. PROTA Medicinal Plants, Vol 1. Backhuys Publisher, Leiden, The Netherlands.

Sharma RA, Singh B, Singh D, Chandrawat P. 2009. Ethnomedicinal, pharmacological properties and chemistry of some medicinal plants of Boraginaceae in India. J Med Plant Res 3, 1153–1175.

Siddiqui BS, Perwaiz S, Begum S. 2006. Two new abietane diterpenes from *Cordia latifolia*. Tetrahedron 62, 10087–10090.

Silva Filho AAda, Lima RMOC, Nascimento SCdo, Silva EC, Andrade MSAS, Lins SCN, Bieber LW. 1993. Biological activity of cordiaquinones A and B, isolated from *Cordia corymbosa*. Fitoterapia 64, 78–80.

Singh MP, Panda H. 2005. Medicinal Herbs with Their Formulations. Daya Books, New Delhi.

Singh V, Sood A, Pruthi S, Singh M, Saini B, Singh M, Thakur G, Kumar A. 2022. Chemical constituents and biological activities of *Cordia myxa* L.: A review. Nat Prod J 12, 30–41.

Souza GC, Haas AP, Von Poser GL, Schapoval EE, Elisabetsky E. 2004. Ethnopharmacological studies of antimicrobial remedies in the south of Brazil. J Ethnopharmacol 90, 135–143.

Srivastava SK. 1980. Distylin-3-xyloside from *Cordia oblique*. Indian J Pharm Sci 42, 95–96.

Standley PC. 1926. Trees and Shrubs of Mexico. Contributions from the United States National Herbarium, Vol 23. Smithsonian Institution, Washington.

Taha AJ. 2016a. Callus induction and plant regeneration of *Cordia myxa* L via tissue culture system. Int J Pharm Integr Life Sci 4, 18–29.

Taha AJ. 2016b. Effect of some chemical elicitors on some secondary metabolite induction of *Cordia myxa* L. *in vitro*. IOSR J Pharm 6, 15–20.

Tijjani RG, Umar ML, Hussaini IM, Shafiu R. 2016. Anti-nociceptive activities of the ethanolic stem bark extract of *Cordia africana* (Boraginaceae) in rats and mice. Ann Biol Sci 4, 6–12.

Velde V, Lavie D, Zelnik R, Matida AK, Panizza S. 1982. Cordialin A and B, two new triterpenes from *Cordia verbenacea* DC. J Chem Soc Perkin Trans 1, 2697–2700.

Vieira IC, Uhl C, Nepstad DC. 1994. The role of the shrub *Cordia multispicata* Cham. as a 'succession facilitator' in an abandoned pasture, Paragominas, Amazonia. Vegetation 115, 91–99.

Vikas, Anisha, Kumar M. 2022. A critical review on *Cordia dichotoma*: Its therapeutic value. Pharm Innov J 11, 668–677.

Wondafrash DZ, Bhoumik D, Altaye BM, Tareke HB, Assefa BT. 2019. Antimalarial activity of *Cordia africana* (Lam.) (Boraginaceae) leaf extracts and solvent fractions in *Plasmodium berghei*-infected mice. Evid Based Complement Altern Med 2019, Article ID 8324596.

Yadav R, Mohan G, Choubey A, Soni UN, Patel JR. 2015. Isolation, spectroscopic characterization and screening of antimicrobial activity of isolated compounds from leaves and seeds of *Cordia obliqua* against some oral pathogens. Indo Am J Pharm Res 5, 3921–3933.

Yadav R, Yadav SK. 2013. Evaluation of antimicrobial activity of seeds and leaves of *Cordia obliqua* wild against some oral pathogens. Indo Am J Pharm Res 3, 6035–6043.

13 *Crassocephalum* Species

13.1 MORPHOLOGICAL FEATURES, DISTRIBUTION, ETHNOPHARMACOLOGICAL PROPERTIES, PHYTOCHEMISTRY, AND PHARMACOLOGICAL ACTIVITIES

Crassocephalum crepidioides (Benth.) S. Moore (Fam.—Asteraceae) is an annual herb, with erect stem, straight, soft-ribbed, branches densely pubescent, and grows up to 40–120 cm high. Leaves are alternate, elliptic, or oblong-elliptic, petiolated (2–2.5 cm), base often long, both surfaces glabrous or sub-glabrous, uppermost leaves smaller, sessile, and apex acuminate. Inflorescence is of capitulum type. Flowers are numerous in terminal corymbiform cymes, homogamous, and shortly pedunculate. Fruits are achenes, narrowly oblong (1.8–2.3 mm), ribbed, cylindric-linear, and thinly pubescent (Belcher 1955; Dairo and Adanlawo 2007; Adjatin et al. 2012a; Vu and Nguyen 2017; Mishra et al. 2018). It is distributed in tropical Africa, Ethiopia, South Africa, Madagascar, Mauritius, China, Vietnam, Japan, tropical and subtropical Asia, Australia, the New Hebrides, Fiji, Tonga and Samoa, and the Americas (Aniya et al. 2005; Vanijajiva and Kadereit 2009; Che et al. 2009; Tan et al. 2012). *C. crepidioides* (Benth.) possess anti-inflammatory, anti-diabetic, antimalarial, and blood regulation activities and are used in the treatment of indigestion, liver complaints, colds, intestinal worms, and hepatic insufficiency (Dansi et al. 2008, 2012).

Hydroethanolic extract of *C. crepidioides* (50 mg/kg/day) decreased wound closure time by about 3.5 days, compared to vehicle treatment. The extract displayed a significant reduction (2.8-fold) on day 7 in inflammatory cells density, significant enhancement (1.9-fold) in the fibroblast density, and a higher number of blood vessels. The levels of mRNA expression of NF-κB1 and TNF-α mRNA were decreased in extract-treated wounds (4.6 and 3.3 times) but showed significant increase in levels of transforming growth factor-β-1 and vascular endothelial growth factors (3.3 and 2.4 times). Wound-healing effects of hydroalcoholic extract of *C. crepidioides* leaves is associated with its antioxidant, anti-inflammation, fibroblast proliferation, wound contraction, and angiogenesis effects (Can and Thao 2020).

The blood coagulation effects of methanol extract of *C. crepidioides* leaves were evaluated in healthy human volunteers. Methanol extract and fractions (hexane, ethyl acetate, and butanol) significantly ($p < 0.05$) increased the clotting time, prothrombin, and activated partial thromboplastin times in blood of volunteers. Hexane fraction (10 mg/ml) showed maximum prolongation effect in treated volunteers. The activity of hexane fraction may be associated with the presence of unsaturated fatty acids and esters, phenolic compounds, flavonoids, and coumarins. As per study results, *C. crepidioides*, contains bioactive compounds and may be utilized in the treatment of blood coagulation diseases (Ayodele et al. 2019). Methanol extract of *C. crepidioides* leaves was examined on blood coagulation effect in diabetic Wistar rats. Methanol extract and hexane fraction (50, 100, and 200 mg/kg) showed significant increase in times of bleeding (58–200%), clotting (65–133%), prothrombin (176–441%), and activated partial thromboplastin (209–518%) in diabetic rats ($LD_{50} \geq$ 5000 mg/kg). Hexane fraction (100 mg/kg, bw) showed maximum prolongation effects in treated diabetic group. The levels of plasma calcium and platelet counts in fraction treated diabetic rats were significantly ($p < 0.05$) decreased. The fraction increased the red blood cells counts, hemoglobin concentration, and packed cell volume in treated rats. Therefore, the fraction of leaves may be used as a potential source of novel anticoagulant and nutraceutical for treatment of thrombotic diseases in diabetic patients (Ayodele et al. 2020a).

Potential toxicities of extract of *C. crepidioides* leaves were assessed in brine shrimp lethality bioassay. The extracts consist of tannins, coumarins, combined anthracene derivatives C-heterosides, flavonoids, mucilage, and steroids. Extracts of leaves showed different lethal doses (LC_{50} 0.901 mg/ml

DOI: 10.1201/9781003398035-13

for *C. crepidioides* and 0.374 mg/ml for *C. rubens*) against brine shrimp model. Therefore, the potent doses indicate the nontoxicity of both plant species (Adjatin et al. 2012b, 2013). The different parts are commonly eaten and possess medicinal properties (Grubben 2004). It is used in the treatment of indigestion, stomachache, epilepsy, sleeping sickness, and swollen lips (Bahar et al. 2016). It possesses hepatoprotective, antihyperlipidemic, and antioxidative activity (Aniya et al. 2005; Adeyemi et al. 2021). Aqueous extract of *C. crepidioides* aerial parts was evaluated for their antitumor effects in S-180-cell-bearing mice. The extract slowed down the tumor growth in S-180-bearing mice but did not show any suppressive effect in growth of S-180 cells *in vitro*. Supernatant of cultured extract–induced RAW264.7 macrophages showed cytotoxic effects to S-180 cells. Cytotoxic effect was attributed to the production of nitric oxide. Isochlorogenic acid (*C. crepidioides* extract) stimulated the nuclear factor-κB activation and expression of inducible nitric oxide synthase. The study results showed oncolytic and immunopotentiation activities of aqueous extract-mediated via nuclear factor-κB-induced formation of nitric oxide from macrophages (Tomimori et al. 2012).

Methanol (80%) extract of *C. crepidioides* aerial parts was investigated for their antioxidant activity against oxidative stress-induced degenerative diseases such as diabetes. Methanol extract displayed significant ($p < 0.05$, $p < 0.01$) activity on a hyperglycemia model. The results were compared with a standard drug (gliclazide) in an oral glucose tolerance test. The extract demonstrated the protective effect of pancreatic β-cell from cell death in an INS-1 cell line by reducing ($p < 0.05$, $p < 0.01$) alloxan-induced apoptosis. Moreover, the extract demonstrated significant ($p < 0.05$, $p < 0.01$) effect on hyperglycemia by enhancing β-cells (percentage) in each islet (45%60%) when compared with the diabetic group. The study results demonstrate that *C. crepidioides* possesses β-cell protection and antidiabetic effects in pancreatic β-cell culture and Wistar albino rats (Bahar et al. 2017). Methanol extract of *C. crepidioides* (500 mg/kg bw; p.o.; twice a daily) significantly reduced the levels of blood and serum glucose but increased the sperm quality and levels of testosterone, follicle stimulating hormone, and luteinizing hormone in diabetic rats. The extract also improved the degeneration and inflammatory responses in the testicular cells of the diabetic rats. As per this study, the extract enhanced the quality, viability, and motility of semen and levels of reproductive hormones; therefore, the extract may be useful in maintaining testicular roles and male genital organs in diabetic rats (Adelakun and Ogunlade 2018).

Hexane, ethyl acetate, ethanol, and water extracts of *C. crepidioides* aerial parts were evaluated for their antioxidant, anti-inflammatory, cytotoxicity, and cytoprotective effects. The ethanol extract showed significant antioxidant activity and suppressed 5-lipoxygenase activity. The ethanol extract (100 µg/ml) did not show any toxicity against MRC-5 and HepG2 cell lines. Ethanol extract (25 µg/ml) displayed strong cytoprotection (69.47%). The presence of phenols and flavonoids with 422.22 gallic acid equivalent (mg/g) and 3.46 quercetin equivalent (mg/g) has been determined in ethanol extract (Wijaya et al. 2011). The distilled water, hot water, and ethanol extract of *C. crepidioides* leaves were examined for their antioxidant activity in 2,2-diphenyl-1-picrylhydrazyl and 2,2-azinobis-(3-ethylbenzothiazoline-6-sulfonate) scavenging assays. The ethanol extract showed maximum levels of total phenolic content (175.06±0.574 µg/ml) and total flavonoid content (139.72±0.923 µg/ml), followed by hot aqueous extract (total phenolic content 54.45 ± 0.818 µg/ml and total flavonoid content 25.07±0.156 µg/ml). Ethanol extract showed strongest (%) antioxidant activity in 2,2-diphenyl-1-picrylhydrazyl (85.4±1.64%) and 2,2-azinobis-(3-ethylbenzothiazoline-6-sulfonate) (85.2±0.57%) assays followed by hot aqueous extract with 2,2-diphenyl-1-picrylhydrazyl (65.4±3.87%) and 2,2-azinobis-(3-ethylbenzothiazoline-6-sulfonate; 79.4±3.2%) assays. Distilled water extract displayed minimum antioxidant activities in 2,2-diphenyl-1-picrylhydrazyl (55.0±0.7 2%) and 2,2-azinobis-(3-ethylbenzothiazoline-6-sulfonate (71.35±2.61%) assays (Awang-Kanak et al. 2019). Ethanol extract of *C. crepidioides* leaves (37.5 mg/kg bw) reduced the levels of malonaldehyde (from 21.24 nmol/l to 16.33 nmol/l). As per Tukey's Honestly Significant Difference test, effects of extract were not significantly different from that of the control group (Widayanti et al. 2020).

Water extract of *C. crepidioides* leaves (300 mg/kg bw; p.o.; 30 days) showed no significant abnormal changes in the frontal cortex, liver, and testes to one side from changes recorded in the kidneys of 4 rats (glomerular distortion, vacuolations, and tubular necrotic bodies). Water extract showed histotoxic effects on the histological profile of the kidneys of the rats (Musa et al. 2011). Ethanol extract of *C. crepidioides* leaves (14 g/kg) displayed significant analgesic activity. The results were compared with tramadol (6.5 mg/kg) in treated animals. It is suggested that leaf extract may be used as an analgesic agent (Yumniati et al. 2016). Hot water extract of *C. crepidioides* displayed significant antibacterial activity against *S. aureus, K. pneumonia, E. coli* (MIC 15 mg/ml) and *S. aureus* (MIC of 45 mg/ml; Omotayo et al. 2015).

The ethanolic leaf extract of *C. crepidioides* leaves stimulated a significant ($p < 0.001$) enhancement in the mobility time in tail suspension. The extract significantly ($p < 0.01$) enhanced superoxide dismutase activity but significantly ($p < 0.01$) reduced the levels of malondialdehyde. These findings reveal that the ethanolic extract of *C. crepidioides* possesses a neuroprotective effect, which might be attributed to its antioxidant effects (Jean et al. 2022). *C. crepidioides* is used in the treatment of stomach ulcers, indigestion, wounds, ulcers, burns, treatment of wounds, and gastric ulcers (Silalahi 2021; Schramm et al. 2021). Total phenolic contents were obtained as 637.22 mg/g gallic acid from methanol extract of *C. crepidioides*. The methanolic extract showed antioxidant activity on DPPH assay (IC_{50} 136.016 µg/ml) when compared to ascorbic acid (IC_{50} 94.12 µg/ml). The lethal concentrations of vincristine sulphate and the plant extract were 3.064 µg/ml and 69.245 µg/ml respectively; the plant demonstrated significant antioxidant, cytotoxic, and thrombolytic effects (Kabir et al. 2021).

C. bauchiense (Hutch.) Milne-Redh. is an erect bushy annual herb and grows up to 1 ft high. It is distributed in rough open ground in Nigeria and Cameroon (Hutchinson and Dalziel 1963). In traditional medicine of Cameroon, it is used in the treatment of gastrointestinal infections, liver disorders, epilepsy, pain, arthritis, colics, behavioral disturbances in mentally retarded children, and neuropathic pain (Sofowara 1996; Arbonnier 2000). The water extract of *C. bauchiense* leaves (100, 200, and 400 mg/ml/kg bw) significantly ($p < 0.05$) increased the level of follicle-stimulating hormone, but the enhancement in luteinizing hormone was insignificant ($p > 0.05$). The aqueous extract significantly ($p < 0.05$) enhanced the levels of catalase and peroxidase activities (Ngoula et al. 2019).

Ethyl acetate extract of *C. bauchiense* leaves showed the presence of alkaloids, phenols, tannins, and sterols. Extract showed significant antibacterial activity against *S. aureus* (ATCC 25922), *E. faecalis* (ATCC 10541), *E. coli* (ATCC 11775), *P. aeruginosa* (ATCC 27853), *S. typhi* (ATCC 6539) and three clinical isolates (*E. coli, P. aeruginosa, S. aureus*; MIC 0.04–6.25 mg/ml). A formulated gel [Shea butter oil (30 g) thawed by heating; bee wax (7.5 g) added and homogenized to develop the vehicle; ethyl acetate extract mixed to form extract-gel concentrations of 0.5, 1.0 and 2.0% w/v); 32 g/kg bw] did not show any visible symptom of toxicity. Daily dermal use of extract gel formulation (for 28 days) showed positive effect on alanine aminotransferase, low density lipoprotein, high density lipoprotein, and triglycerides significantly ($p < 0.05$). The study shows that ethyl acetate extract of *C. bauchiense* leaves may be used safely for the treatment of some bacterial diseases (Mouokeu et al. 2011). Ethyl acetate extract of *C. bauchiense* leaves showed significant activity against dermatophytes and yeast cells (MIC 0.125–4 mg/ml). The ethyl acetate extracts also displayed potent scavenging activity in 2,2-diphenyl-1-picrylhydrazyl (CI_{50} 28.57–389.38 µg/ml) assay. *C. bauchiense* ethyl acetate extract possesses significant antifungal and antioxidant activity; therefore, it may be considered as a source of active compounds that might be used as an antifungal and antioxidant agent (Mouokeu et al. 2014).

Aqueous extract and alkaloid-rich fraction *C. bauchiense* leaves produced dose-dependent suppression of novelty-stimulated rearing behavior, reduced the apomorphine-stimulated stereotypy and fighting, and showed significant fall in body temperature. The aqueous extract increased the sodium pentobarbital sleeping time. This prolongation time was not normalized by bicuculline. Although, the activity of water extract on sodium pentobarbital-stimulated sleeping time was stopped by *N*-methyl-β-carboline-3-carboxamide and flumazenil. As per this study, the antipsychotic and sedative activities of *C. bauchiense* are possibly mediated through the obstruction of dopamine D-2 receptors and γ-aminobutyric acid activation (Taïwe et al. 2012a). Aqueous extract and alkaloid-rich fraction showed significant antinociceptive activity in the acetic acid, formalin, glutamate, capsaicin, and hot plate tests. The extract and fraction did not change the animal movement in the open-field or rotarod assays, which suggest a lack of a central depressant effect. Aqueous extract and alkaloid-rich fraction caused dose-dependent antinociception in chemical and thermal nociception models via changing opioidergic pathway (Taïwe et al. 2012b). Catalase and total peroxidase activities were reduced significantly, whereas the levels of malondialdehyde and superoxide dismutase enhanced significantly by ethanol extract (400 mg/kg) of *C. bauchiense* in potassium dichromate-induced oxidative stress. Leaf ethanol extract did not show any toxicity against the toxic effects of potassium dichromate; therefore, the ethanol extract might be used to improve female reproduction (Momo et al. 2021).

Ethanol extract of *C. macropappum* (Sch.Bip. ex. A.Rich.) S. Moore aerial parts displayed the maximum levels of total phenolic (101.48 mg gallic acid equivalents/g) and flavonoid (293.25 mg quercetin equivalent/g) contents. Phytochemical analysis showed the presence of saponins, tannins, anthraquinones, steroids, terpenoids, and flavonoids in ethanol extract. The extract showed strong antioxidant activity in 2,2-diphenyl-1-picryl hydrazyl scavenging assay ($IC_{50} \leq 100$ µg/ml) and displayed a significant protective effect against oxidative DNA damage (Robi et al. 2019). 1 M hydrochloric acid and methanol (1:1, v/v) extract of *C. crepidioides* leaves showed the presence of four phenolic acids (gallic, chlorogenic, caffeic, and ellagic acids) and three flavonoids (catechin, rutin and quercetin). The extract showed significant antioxidant, acetylcholinesterase, and butyrylcholinesterase suppressive activities. The significant activities may be associated with the presence of higher levels of the phenolic compounds (Adedayo et al. 2015).

C. rubens (Juss. ex Jacq.) S. Moore is an erect herb, about 80 cm in height, and is naturalized in southwestern part of Nigeria, Yemen, South Africa, and the islands of the Indian Ocean. In Nigeria, its leaves and stems are used in soups and stews (Ojo and Adenegan-Alakinde 2017). It is also used in the treatment of liver dysfunction, stomach inflammation, ocular and earaches, burns, leprosy, and breast cancer (Ojo and Adenegan-Alakinde 2017). *C. rubens* showed maximum levels of phenolic contents and free radical scavenging activity (Ademoyegun et al. 2013). The methanol extract of *C. rubens* leaves significantly ($p < 0.05$) decreased the levels of total cholesterol (75.45 ± 2.44–96.09 ± 1.65 mg/dL), total glycerides (84.04 ± 1.79–127.00 ± 0.47 mg/dL), very-low-density lipoprotein-C (16.81 ± 0.36–25.42 ± 0.10 mg/dL), low density lipoproteins-cholesterol (03.49 ± 1.03–26.04 ± 1.64 mg/dL), and aspartate aminotransferase of the pretreated myocardial infarcted rats versus isoproterenol-induced myocardial infarction control (168.70 ± 9.85 mg/dL, 146.60 ± 1.74 mg/dL, 29.31 ± 0.34 mg/dL, 77.73 ± 9.67 mg/dL, respectively), but enhanced the level of plasma high density lipoproteins-C. *C. rubens* extract displayed significant cardioprotective effects on the integrity of the heart tissues of the experimental animals. The results indicate that antihyperlipidemic and cardioprotective potentials can offer a possible lead for purification of secondary products from *C. rubens* in the management of hyperlipidaemia and its related disorders (Ayodele et al. 2022).

C. vitellinum ethanol extract administration significantly stopped the levels of rifampicin-induced enhancement in serum levels of liver biomarker enzymes and reduced the hepatocellular necrosis and inflammatory cells infiltration. The ethanolic extract reduces the liver biomarker enzymes (alanine transaminase, alkaline phosphatase, and aspartate transaminase) and protects the histomorphology of the hepatocytes, suggesting that plant extract possesses hepatoprotective properties (Kiyimba et al. 2022).

13.2   PHYTOCHEMISTRY

Sabinene, β-pinene, myrcene, α-phellandrene, p-cymene, β-phellandrene, (*Z*)-β-ocimene, (*E*)-β-ocimene, γ-terpinene, terpinolene, linalool, borneol, terpin-4-ol, α-terpineol, thymol, methyl ether, carvacrol, methyl ether, bornyl acetate, cyclosativene, α-copaene, isocomene, β-elemene, cyperene, α-gurjunene, β-caryophyllene, 2,5-dimethoxy-p-cymene, trans-α-bergamotene, α-humulene, (*E*)-β-farnesene, β-chamigrene, *cis*-β-guaiene, α-selinene, α-muurolene, α-bulnesene, cubebol, δ-cadinene, occidentalol, germacrene B, ledol, caryophyllene oxide, guaiol, hinesol, valerianol, occidentalol acetate, cedr-8-en-13-ol, 14-oxy-α-muurolene, cedrene-8,13-ol acetate, khusinol acetate were obtained from roots (Joshi 2014); essential oils (α-pinene, camphene, sabinene, β-pinene, myrcene, p-cymene, limonene, β-phellandrene, (Z)-β-ocimene, (E)-β-ocimene, 2,6-dimethyl-2,6-octadiene, perillene, linalool, trans-verbenol, terpinen-4-ol, cryptone, trans-4-caranone, cis-4-caranone, cuminal, neryl formate, bornyl acetate, p-cymen-7-ol, 3-hydroxycineole, 4-hydroxycryptone, 3-oxo-p-menth-1-en-7-al, neryl acetate, α-copaene, β-elemene, cyperene, β-caryophyllene, (3E)-4,8-dimethyl-3,7-nonadien2-ol, trans-α-bergamotene, (E)-β-farnesene, α-humulene, germacrene D, β-selinene, α-selinene, α-muurolene, neryl isobutanoate, δ-cadinene, (E)-nerolidol, spathulenol, caryophyllene oxide, cis-bisabol-11-ol, humulene epoxide II, cyperotundone, τ-cadinol, τ-muurolol, α-cadinol, selin-11-en-4α-ol, cis-calamenen-10-ol, phytone, phytol) have been separated from C. crepidioides stem, leaf and flowers. The identified essential oils demonstrated antilarvicidal activity against wild-caught Aedes albopictus (IC$_{50}$ 14.3 µg/ml), A. aegypti (IC$_{50}$ 4.95 µg/ml), and Culex quinquefasciatus (IC$_{50}$ 18.4 µg/ml; Hung et al. 2019).

α-pinene, sabinene, β-pinene, β-myrcene, cyclooctyne, p-cymene, limonene, β-phellandrene, β-ocimene, perillone, mentha-2-en-1-ol'cis-para', cryptone, cumic aldehyde, bornyl acetate, terpinen-7-al, α-copaene, β-elemene, trans-caryophyllene, α-bergamotene, α-humulene, germacrene D, α-selinene, apofarnesol, δ-cadinene, caryophyllene oxide, humulene epoxide II, and cubenol were identified from C. crepidioides. The isolated essential oils (100 µg/ml) showed anticancer activity against human cervical cancer SiHa, human oral epidermal carcinoma KB, and human adenocarcinoma Colo-205 (59.8±3.7, 67.9±0.5 and 84.5±3.6 µg/ml) cell lines at 48 h (Thakur et al. 2019). The essential oils viz., m-tert-butyl phenol, p-myrecene, 1-methyl-2-(1-methylethyl)- 3-(1-methylethylidene)cyclopropane, myrtenol, 2,6-dimethyl-3,5,7-octatriene-2-ol, 2-isopropylidene-3-methylhexa-3,5dienal, Z-1-[1-butenyl]aziridine, thymol, 3-methylene-p-menth-8-ene, artemisia triene, copaene, 2-dodecanone, β-elemene, caryophyllene, α-caryophyllene, α+β-caryophyllene, cis-β-farnesene, β-cubebene, 2-tridecanone, β-elemene, α-fernesene, δ-cadinene, γ-elemene, 3,7,11-trimethyl-1,6,10-dodecatrien-3-ol, caryophyllene oxide, humulene epoxide ii, α-santalol, Z-ocimene, 2-methylene-6,8,8-trimethyl tricycle[5.2.2.0(1,6)]undecan-3-ol, 2,6,8-trimethyl decanoic acid methylester, 6-phenyl dodecane, 1,19-eicosadione, phytol, n-hexadecanoic acid, and 4-cyclohexylbutyramide have been determined from leaves and stems (Owokotomo et al. 2012); 7-butyl-6,8-dihydroxy-3(R)-pent-11-enylisochroman-1-one, 7-but-15-enyl-6,8-dihydroxy-3(R)-pent-11-enylisochroman-1-one, and 7-butyl-6,8-dihydroxy-3(R)-pentylisochroman-1-one have been obtained from C. crepidioides (Kongsaeree et al. 2003).

The presence of butyrolactone, benzene acetaldehyde, 1-methyl, 2-pyrrolidinone, erythritol, dl-threitol, glycerin, phenylethyl alcohol, phthalic acid, benzofuran, indole, N-hydroxylamine, glutaric acid, mequinol, thujone, phytol, phenol, eugenol, 1,7-nonadiene,4,8-dimethyl-, isocyclocitral, 2,4 dimethylanisole, 2,4,6-trimethyl-2-(4-methyl-pent-3-enyl) 2H-pyran, 2-tridecanone, benzofuranone, dodecanoic acid, vanillin/propyl ester, 3-cyclohexen-1-carboxaldehyde, butyrophenone, 1,9 octadecadiene, citronellol, aromadendrene oxide, longipinocarveol, N-acetyl-d-serine, 2-Aminoresorcinol, α-guaiene, caryophyllene, 3-buten-2-one, 4-hydroxy trimethyl-7-oxabicyclo-heptyl-, 1-acetyl-3methylurea, ledol/cedvanoxide, 2-Benzothiozolamine, 2,5-octadiene-tetramethyl-, Spiro[2.3] hexan-4-one, 5,5-dichloro, Cyclodecanone, Semicarbazone, Orcinol, 2-(1-Hydroxycyclohexyl)-furan, paradrine, alloaromadendrene oxide, 1,1,4,7-tetramethyldecahydro-1H-cyclopropa[e]azulene-4,7-diol, 2-acetylbenzoic acid, thumbergol, hexadecanoic acid, methyl ester, n-hexadecanoic acid, metanephrine, methoxamine, 7,10,13-hexadecatrienoic acid, methyl ester, 2-furanmethanol, tetrahydro-acetate, 9,12,15-octadecatrienoic acid (α-linolenic acid), ethyl 9,12-hexadecadinoate, bicyclo heptane, 7,7-dimethyl 1–2-methylene, doconexent/methyl parinarate 9-octadecenamide, tocainide, hydroquinone, catechol, glucopyranuronamide, phenylephrine/adrenalone was determined by GC-MS analysis from hexane extract of C. crepidioides leaves (Ayodele et al. 2020b)

A clerodane diterpenoid (ent-2β,18,19-trihydroxycleroda-3,13-dien-16,15-olide) along with two flavonoids (3',5-dihydroxy-4',5',6,7,8-pentamethoxyflavone and 4',5-dihydroxy-3',5',6,7,8-pentamethoxyflavone) were isolated and characterized from C. bauchiense whole plant. The isolated compound (3′,5-dihydroxy-4′,5′,6,7,8-pentamethoxyflavone) displayed weak antiprotozoal activity against chloroquine-sensitive 3D7 strain of *Plasmodium falciparum* (IC_{50} = 10.1 g/ml) (Tchinda et al. 2015). The presence of α-phellandrene, *p*-cymene, pinenes, germacrene D, myrcene, limonene, and (*E*)-β-ocimene have been identified by GC-MS from leaves of seven *Crassocephalum* species (Zollo et al. 2000).

Three dihydroisocoumarins (biafraecoumarins A, B, and C), two triterpenes (fernenol and sorghumol acetate) and a ceramide [(2*S*,3*R*,4*E*)-2-*N*-(2′-hydroxytetracosanoyl)heptadecasphinga-4-ene)] have been isolated and characterized from chloroform extract of *C. biafrae* stem bark. Isolated compounds (biafraecoumarins A, B, and C) demonstrated significant antimicrobial activities against *Escherichia coli, Bacillus subtilis, Staphylococcus aureus, Pseudomonas picketti, Trichphyton longifusus, Aspergillus flavus, Microsporum canis, Fusarium solani, Candida albicans*, and *Candida glabrata* (Tabopda et al. 2009).

Essential oils {methoxy-phenyl-oxime, p-methoxybenzaldehyde, 2-methyl-3,5-diethylpyrazine, phthalic acid, cyclobutyl tridecyl ester, 2,2-dimethyl-propyl 2,2-dimethyl-propanesulfinyl sulfone, 1-iodohexadecane, 1,2,3-trimethyldiaziridine, 2-bromotetradecane, 2-ethyl-2-methyl-tridecanol, 1-iodo-decane, 2,6,10,14,18-pentamethyl-2,6,10,14,18-eicosapentaene from aqueous extract; 3-hepten-2-one, 3-ethoxypentane, 1,1-dimethylcyclopentane, 2-ethyl-2-hexenal, 2-oxopentanedioic acid, 2-trimethyl-silyloxy-1,3-butadiene, butanal, 1-(3,4-methylenedioxybenzylidene) semicarbazide, (*Z*)-2-Buten-1-ol, ([(*E*)-2-cyclopropylethenyl]oxy)(trimethyl)silane, 4,4-dimethyl-8-methylene-2-propyl-1-oxaspiro[2.5] octane, 1,1-dimethylethylamine, 1,1-dimethyl-2-propynyl ethyl ether, 1,1-dimethylethylamine, 1,1-dimethyl-2-propynyl ethyl ether, 2,3-hexanediol, 2,3-tetramethyleneaziridine, 4-ethylformanilide, methyl butyrate, 2-acetylisoxazolidine, 4-ethylformanilide, 2,3-hexanediol from ethyl acetate extract; 3-hepten-2-one, 3-ethoxypentane, 3,4-dimethy-1-pentanol, 2-ethyl-2-hexenal, 1,1-dibutylhydrazine, 2-trimethylsilyloxy-1,3-butadiene, butanal, ([(*E*)-2-cyclopropylethenyl]oxy)(trimethyl)silane, 2-methyl-2-propanamine, 6-octen-1-ol, 3,7-dimethyl-, propanoate, 4-ethylformanilide, methyl butyrate, methyl 2-methylundecanoate, phytol, 4-methyl-1,4-heptadiene, diethyl(decyloxy) borane, oxalic acid, allyl butyl ester, oxalic acid, allyl butyl ester} were reported from ethanol extract of *C. rubens* leaves. Gold nanoparticles were developed and findings displayed the production of green gold nanoparticles (wavelength 538 nm) and mostly spherical gold nanoparticle with 20 ± 5 nm size. The gold nanoparticles demonstrated anticancer activity against MCF-7 and Caco-2 cells (125 and 250 mg/ml; Adewale et al. 2020). A new 14,15-dinor-labdane glucoside (crassoside A) was isolated and characterized from *C. mannii* aerial parts (Hegazy et al. 2008).

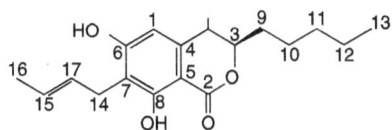

Biafraecoumarin A

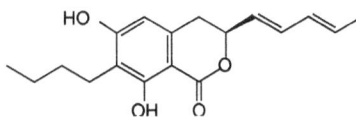

Biafraecoumarin B

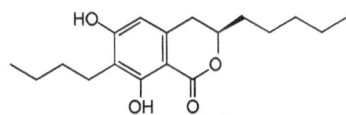

Biafraecoumarin C

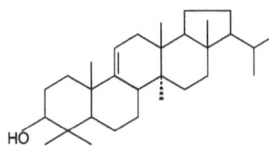

Fernenol

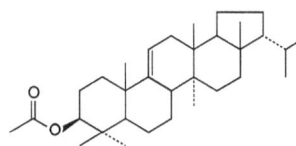

Sorghumol acetate

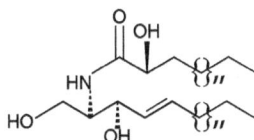

(2*S*,3*R*,47)-2-*N*-(2′-Hydroxytetracosanoyl)heptadecasphinga-4-ene

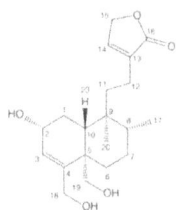

ent-2β,18,19-Trihydroxycleroda-3,13-dien-16,15-olide

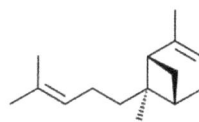

trans-α-Bergamotene

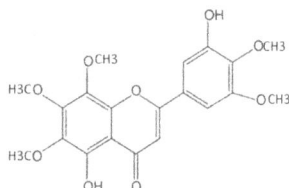

3′,5-Dihydroxy-4′,5′,6,7,8-pentamethoxyflavone

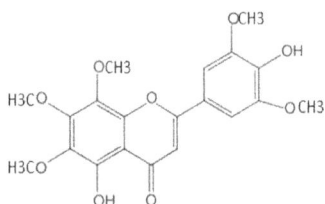

4′,5-Dihydroxy-3′,5′,6,7,8-pentamethoxyflavone

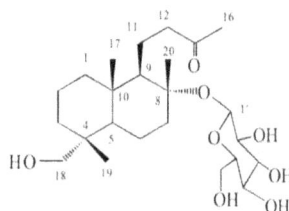

Crassoside A

13.3 CULTURE CONDITIONS

The shoots and roots of *Crassocephalum crepidioides* were inoculated on an MS (Murashige and Skoog 1962) medium containing 6-benzylaminopurine alone and in combination with zeatin and 1-naphthalene acetic acid. Maximum growth of callus was achieved with supplementation of 6-benzylaminopurine (10 mg/l), 1-naphthalene acetic acid (15 mg/l), and zeatin (1.5 mg/l) therefore, the callus may be used for increasing the production of secondary metabolites (Opabode et al. 2016).

REFERENCES

Adedayo BC, Oboh G, Oyeleye SI, Ejakpovi II, Boligon AA, Athayde ML. 2015. Blanching alters the phenolic constituents and *in vitro* antioxidant and anticholinesterases properties of fireweed (*Crassocephalum crepidioides*). J Taibah Univ Med Sci 10, 419–426.

Adelakun SA, Ogunlade B. 2018. Responses to the bioactive component of *Crassocephalum crepidioides* on histomorphology, spermatogenesis and steroidogenesis in streptozotocin-induced diabetic male rats. J Reprod Endocrinol Infertil 3, 6.

Ademoyegun OT, Akin-Idowu PE, Ibitoye DO, Adewuyi GO. 2013. Phenolic contents and free radical scavenging activity in some leafy vegetables. Int J Veg Sci 19, 126–137.

Adewale OB, Anadozie SO, Potts-Johnson SS, Onwuelu JO, Obafemi TO, Osukoya OA, Fadaka AO, Davids H, Roux S. 2020. Investigation of bioactive compounds in *Crassocephalum rubens* leaf and *in vitro* anti-cancer activity of its biosynthesized gold nanoparticles. Biotechnol Rep 28, e00560.

Adeyemi KD, Sola-Ojo FE, Ajayi DO, Banni F, Isamot HO, Lawal MO. 2021. Influence of dietary supplementation of *Crassocephalum crepidioides* leaf on growth, immune status, caecal microbiota, and meat quality in broiler chickens. Trop Anim Health Prod 53, 125.

Adjatin A, Dansi A, Badoussi E, Loko YL, Dansi M, Azokpota P, Gbaguidi F, Ahissou H, Akoègninou A, Akpagana K, Sanni A. 2013. Phytochemical screening and toxicity studies of *Crassocephalum rubens* (Juss. ex Jacq.) S. Moore and *Crassocephalum crepidioides* (Benth.) S. Moore consumed as vegetable in Benin. Int J Curr Microbiol Appl Sci 2, 1–13.

Adjatin A, Dansi A, Eze CS. 2012b. Ethnobotanical investigation and diversity of Gbolo (*Crassocephalum rubens* (Juss. ex Jacq.) S. Moore and *Crassocephalum crepidioides* (Benth.) S. Moore), a traditional leafy vegetable under domestication in Benin. Genet Resour Crop Evol 59, 1867–1881.

Adjatin A, Dansi A, Eze CS, Assogba P, Dossou-Aminon I, Akpagana K, Akoégninou A, Sanni A. 2012a. Ethnobotanical investigation and diversity of Gbolo (*Crassocephalum rubens* (Juss. ex Jacq.) S. Moore and *Crassocephalum crepidioides* (Benth.) S. Moore), a traditional leafy vegetable under domestication in Benin. Genet Resour Crop Evol 59, 1867–1881.

Aniya Y, Koyama T, Miyagi C, Miyahira M, Inomata C, Kinoshita S, Ichiba T. 2005. Free radical scavenging and hepatoprotective actions of the medicinal herb, *Crassocephalum crepidioides* from the Okinawa Islands. Biol Pharm Bull 28, 19–23.

Arbonnier M. 2000. Arbres, arbustes et lianes des zones se'ches d'Afrique de l'Ouest (Trees, shrubs and liana of dried zones of west Africa). CIRAD, Montpellier.

Awang-Kanak F, Bakar MFA, Mohamed M. 2019. Ethnobotanical note, total phenolic content, total flavonoid content, and antioxidative activities of wild edible vegetable, *Crassocephalum crepidioides* from Kota Belud, Sabah. IOP Conf Ser Earth Environ Sci 269, 012012.

Ayodele OO, Banigo TC, Okoro EE, Ojo OO. 2022. *Crassocephalum rubens* extract and fractions: Hypolipidemic and cardioprotective activities. GSC Biol Pharm Sci 20, 253–261.

Ayodele OO, Onajobi FD, Omolaja R. Osoniyi OR. 2020b. Phytochemical profiling of the hexane fraction of *Crassocephalum crepidioides* Benth S. Moore leaves by GC-MS. Afr J Pure Appl Chem 14, 1–8.

Ayodele OO, Onajobi FD, Osoniyi O. 2019. *In vitro* anticoagulant effect of *Crassocephalum crepidioides* leaf methanol extract and fractions on human blood. J Exp Pharmacol 11, 99–107.

Ayodele OO, Onajobi FD, Osoniyi OR. 2020a. Modulation of blood coagulation and hematological parameters by *Crassocephalum crepidioides* leaf methanol extract and fractions in STZ-induced diabetes in the rat. Sci World J 2020, 1036364.

Bahar E, Akter K-M, Lee G-H, Lee H-Y, Rashid H-O, Choi M-K, Bhattarai KR, Hossain MMM, Ara J, Mazumder K, Raihan O, Chae H-J, Yoon H. 2017. β-Cell protection and antidiabetic activities of *Crassocephalum crepidioides* (Asteraceae) Benth. S. Moore extract against alloxaninduced oxidative stress via regulation of apoptosis and reactive oxygen species (ROS). BMC Complement Altern Med 17, 179.

Bahar E, Siddika MS, Nath B, Yoon H. 2016. Evaluation of *in vitro* antioxidant and *in vivo* antihyperlipidemic activities of methanol extract of aerial part of *Crassocephalum crepidioides* (Asteraceae) Benth S Moore. Trop J Pharm Res 15, 481–488.

Belcher RO. 1955. The typification of *Crassocephalum* Moench and Gynura CASS. Kew Bull 10, 455–465.

Can NM, Thao DTP. 2020. Wound healing activity of *Crassocephalum crepidioides* (Benth.) S. Moore. leaf hydroethanolic extract. Oxid Med Cell Longev 2020, 2483187.

Che GQ, Guo SL, Huang QS. 2009. Invasiveness evaluation of fireweed (*Crassocephalum crepidioides*) based on its seed germination features. Weed Biol Manag 9, 123–128.

Dairo FAS, Adanlawo IG. 2007. Nutritional quality of *Crassocephalum crepidioides* and *Senecio biafrae*. Pak J Nutr 6, 35–39.

Dansi A, Adjatin A, Adoukonou-Sagbadja H, Faladé V, Yedomonhan H, Odou D, Dossou B. 2008. Traditional leafy vegetables and their use in the Benin Republic. Genet Resour Crop Evol 55, 1239–1256.

Dansi A, Vodouhé R, Azokpota P, Yedomonhan H, Assogba P, Adjatin A, Loko LY, Dossou-Aminon I, Akpagana K. 2012. Diversity of the neglected and underutilized crop species of importance in Benin. Sci World J 2012, 932947.

Grubben GJH. 2004. Plant Resources of Tropical Africa: Vegetables. Backhuys Publishers, Prota Foundation, Leiden.

Hegazy M-FF, Aly AA, Ahmed AA, Pierre DC, Tane P, Ahmed MM. 2008. A new 14,15-dinor-labdane glucoside from *Crassocephalum mannii*. Nat Prod Commun 3, 869–872.

Hung NH, Satyal P, Dai DN, Tai TA, Huong LT, Chuong NTH, Hieu HV, Tuan PA, Vuong PV, William N. Setzer WN. 2019. Chemical compositions of *Crassocephalum crepidioides* essential oils and larvicidal activities against *Aedes aegypti, Aedes albopictus*, and *Culex quinquefasciatus*. Nat Prod Commun 14, 1–5.

Hutchinson JJ, Dalziel JM. 1963. Flora of West Tropical Africa. Crown Agents, London.

Jean BG, Irène FA, Donatien AA, Hervé-Hervé NA, Bertrand BM, Gael A-DN, Merline NY, Franklik ZG, Bertrand DA, Théophile D. 2022. Neuroprotective effects of the ethanolic leaf extract of *Crassocephalum crepidioides* (Asteracaeae) on diazepam-induced amnesia in mice. Adv Pharmacol Pharm Sci 2022: 1919469.

Joshi RK. 2014. Study on essential oil composition of the roots of *Crassocephalum crepidioides* (Benth.) S. Moore. J Chil Chem Soc 59, 2363–2365.

Kabir MT, Samiha M, Yasmin H, Rahman MS, Rahman MS, Ashraf GM, Akter R. 2021. Free radical scaveng-ing, thrombolytic and cytotoxic effects of the medicinal herb, *Crassocephalum crepidioides*. Res J Pharm Technol 14, 2205–2210.

Kiyimba K, Ayikobua ET, Mwandah DC, Obakiro SB. 2022. Assessing the protective effect of *Crassocephalum vitellinum* against rifampicin- induced hepatotoxicity in Wistar rats. Afr Health Sci 22, 352–360.

Kongsaeree P, Prabpai S, Sriubolmas N, Vongvein C, Wiyakrutta S. 2003. Antimalarial dihydroisocoumarins produced by *Geotrichum* sp., an endophytic fungus of *Crassocephalum crepidioides*. J Nat Prod 66, 709–711.

Mishra R, Jena GSJP, Satapathy KB. 2018. *Crassocephalum* Moench (Asteraceae) an invasive alien genus: A new record for the state of Odisha, India. Int J Curr Adv Res 7, 13859–13861.

Momo CMM, Pascal MTF, Pasima N, Simplice MR, Narcisse VB, Ferdinand N, Joseph T. 2021. Effects of *Crassocephalum bauchiense* ethanolic extract on reproductive parameters in rabbit does exposed to potassium dichromate. Am J Anim Vet Sci 16, 151–161.

Mouokeu RS, Ngane RA, Njateng GS, Kamtcheung MO, Kuiate JR. 2014. Antifungal and antioxidant activity of *Crassocephalum bauchiense* (Hutch.) Milne-Redh ethyl acetate extract and fractions (Asteraceae). BMC Res Notes 7, 244.

Mouokeu RS, Ngono RAN, Lunga PK, Koanga MM, Tiabou AT, Njateng GSS, Tamokou JDD, Kuiate J-R. 2011. Antibacterial and dermal toxicological profiles of ethyl acetate extract from *Crassocephalum bauchiense* (Hutch.) Milne-Redh (Asteraceae). BMC Complement Altern Med 11, 43.

Murashige T, Skoog F. 1962. A revised medium for rapid growth and bioassays with tobacco tissue cultures. Physiol Plant 15, 473–497.

Musa AA, Adekomi DA, Tijani AA, Muhammed OA. 2011. Some of the effect of *Crassocephalum crepidioides* on the frontal cortex, kidney, liver and testis of adult male sprague dawley rats: Microanatomical study. Eur J Exp Biol 1, 228–235.

Ngoula F, Chongsi MMM, Ngouateu OBK, Makona AMN, Kenfack A, Vemo BN, Tchoumboue J. 2019. Effects of *Crassocephalum bauchiense* (Hutch) leaf aqueous extract on toxicity indicators and reproductive char-acteristics in Oryctolagus cuniculus exposed to potassium dichromate. Asian Pac J Reprod 8, 276–282.

Ojo FM, Adenegan-Alakinde TA. 2017. Phytochemical studies of four indigenous vegetables commonly con-sumed in ile-ife, South-West Nigeria. Int J Curr Sci 21, E6–13.

Omotayo MA, Avungbeto O, Sokefun OO, Eleyowo OO. 2015. Antibacterial activity of *Crassocephalum crepidioides* (fireweed) and *Chromolaena odorata* (siam weed) hot aqueous leaf extract. Int J Pharm Biol Sci 5, 114–122.

Opabode JT, Ajibola OV, Lamidi T. 2016. *In vitro* propagation of *Crassocephalum crepidioides*—An endan-gered African traditional leaf vegetable and molecular analysis of micropropagated plants. Int J Veg Sci 23, 18–30.

Owokotomo IA, Ekundayo O, Oladosu IA, Aboaba SA. 2012. Analysis of the essential oils of leaves and stems of *Crassocephalum crepidioides* growing in south western Nigeria. Int J Chem 4, 34–37.

Robi AG, Megersa N, Mehari T, Muleta D, Kim Y-M. 2019. Phytochemical composition and antioxidant property of mandillo, *Crassocephalum macropappum* (Sch.Bip. ex. A.Rich.) S. Moore. Prev Nutr Food Sci 24, 197–201.

Schramm S, Rozhon W, Adedeji-Badmus AN, Liang Y, Nayem S, Winkelmann T, Poppenberger B. 2021. The orphan crop *Crassocephalum crepidioides* accumulates the pyrrolizidine alkaloid jacobine in response to nitrogen starvation. Front Plant Sci 12, 702985.

Silalahi M. 2021. *Crassocephalum crepidioides* (bioactivity and utilization). ICES, Jakarta.

Sofowara EA. 1996. Plantes M'edicinales et M'edecine Tropicale d'Afrique. Karthala 1, 1–256.

Tabopda TK, Fotso GW, Ngoupayo J, Mitaine-Offer A-C, Ngadjui BT, Lacaille-Dubois M-A. 2009. Antimicrobial dihydroisocoumarins from *Crassocephalum biafrae*. Planta Med 75, 1258–1261.

Taïwe GS, Bum EN, Talla E, Dawe A, Moto FCO, Ngoupaye GT, Sidiki N, Dabole B, Dzeufiet PDD, Dimo T, Waard MD. 2012a. Antipsychotic and sedative effects of the leaf extract of *Crassocephalum bauchiense* (Hutch.) Milne-Redh (Asteraceae) in rodents. J Ethnopharmacol 143, 213–220.

Taïwe GS, Bum EN, Talla E, Dimo T, Sidiki N, Dawe A, Nguimbou RM, Dzeufiet PDD, Waard MD. 2012b. Evaluation of antinociceptive effects of *Crassocephalum bauchiense* Hutch (Asteraceae) leaf extract in rodents. J Ethnopharmacol 141, 234–241.

Tan D, Thu P, Dell B. 2012. Invasive plant species in the national parks of Vietnam. Forests 3, 997–1016.

Tchinda AT, Mouokeu SR, Ngono RAN, Ebelle MRE, Mokale ALK, Nono DK, Frédérich M. 2015. A new ent-clerodane diterpenoid from *Crassocephalum bauchiense* Huch. (Asteraceae). Nat Prod Res 29, 1990–1994.

Thakur S, Koundal R, D. Kumar D, Maurya AK, Padwad YS, Lal B, Agnihotri VK. 2019. Volatile composi-tion and cytotoxic activity of aerial parts of *Crassocephalum crepidioides* growing in Western Himalaya, India. Indian J Pharm Sci 81, 167–172.

Tomimori K, Nakama S, Kimura R, Tamaki K, Ishikawa C, Mori N. 2012. Antitumor activity and macrophage nitric oxide producing action of medicinal herb, *Crassocephalum crepidioides*. BMC Complement Altern Med 12, 78.

Vanijajiva O, Kadereit JW. 2009. Morphological and molecular evidence for interspecific hybridisation in the introduced African genus *Crassocephalum* (Asteraceae: Senecioneae) in Asia. Syst Biodivers 7, 269–276.

Vu DT, Nguyen TA. 2017. The neglected and underutilized species in the Northern mountainous provinces of Vietnam. Genet Resour Crop Evol 64, 1115–1124.

Widayanti NP, Ayu SLW, Apriyanthi DPRV, Nyoman Arijana IGK. 2020. Lipid peroxidation inhibition activity of sintrong (*Crassocephalum crepidioides*) leaf extract in rats consuming arak jembrana. Makara J Sci 24, 228–232.

Wijaya S, Nee TK, Jin KT, Din WM, Wiart C. 2011. Antioxidant, anti-inflammatory, cytotoxicity and cytoprotection activities of *Crassocephalum crepidioides* (Benth.) S. Moore. extracts and its phytochemical composition. Eur J Sci Res 67, 1–10.

Yumniati I, Yuniarni U, Hazar S. 2016. Uji aktivitas analgetika ekstrak etanol daun sintrong (*Crassocephalum crepidioides* (Benth.) S. Moore) terhadap mencit jantan galur ddy. Farmasi Gelombang 2, 413–419.

Zollo PHA, kuiate JR, Menut C, Bessiere JM. 2000. Aromatic plants of tropical central Africa. XXXVI. Chemical composition of essential oils from seven Cameroonian *Crassocephalum* species. J Essent Oil Res 12, 533–536.

14 *Cyclopia* Species

14.1 MORPHOLOGICAL FEATURES, DISTRIBUTION, ETHNOPHARMACOLOGICAL PROPERTIES, PHYTOCHEMISTRY, AND PHARMACOLOGICAL ACTIVITIES

Cyclopia genus (Fam.—Fabaceae) consists of about 23 species and is distributed in different regions of South Africa (Schutte 1997; Schutte 1997; Joubert et al. 2009, 2011). *C. genistoides* D.Don is a woody shrub with yellowish twigs and 1.5 to 3 m in height. The leaves are trifoliolate, narrow, and pointed; the shrub produces attractive yellow flowers and brown pods bearing arillate seeds (Schutte 1997; van Wyk and Gericke 2000; van Wyk et al. 1997, 2009; Kokotkiewicz and Luczkiewicz 2009). Due to its unique aroma and taste, *C. intermedia* is used in preparation of herbal tea (Kamara et al. 2003; Joubert et al. 2008). It possesses antidiabetic activity also (Ajuwon et al. 2018). Polyphenol compounds (mangiferin, isomangiferin, hesperidin, eriocitrin, eriodictyol glucoside, iriflophenone-3-C-β-glucoside, 3-hydroxyphloretin-3,5-di-C-hexoside phloretin-3,5-di-C-glucoside, scolymoside, luteolin) were obtained from *C. genistoides, C. sessiliflora, C. intermedia*, and *C. subternata* (Louw et al. 2013; van Wyk 2008). Hot water and ethanol-water extracts (40%) of *C. genistoides, C. subternata*, and *C. maculata* showed immune-regulating effects against murine splenocytes and mesenteric lymph node cells (*in vitro*). Both extracts enhanced the ratio of $CD4^+CD25^+Foxp3^+$ Treg cells to total $CD4^+$ cells and revealed stimulation of $Foxp3^+$ cells when mesenteric lymph node cells were grown in the presence of both extracts (Murakami et al. 2018; van Dyk et al. 2022). Honeybush extracts are used in functional foods and the formation of cosmetics products (Niemandt et al. 2018; Slabbert et al. 2019). The mangiferin showed a modulatory effect on the apoptotic activity of hesperidin. Both compounds have been identified from Cyclopia species (Bartoszewski et al. 2014).

C. intermedia E.Mey. is an erect and much-branched shrub. It is one of the major sources of honeybush tea. It is collected in large quantities from the wild and is also sometimes cultivated. The tea is commonly drunk in South Africa and is also becoming more popular on a global basis. The fermented extract (12.5 µg/ml) improved sperm motility and kinetic parameters, maintained plasma membrane integrity, and decreased lipid peroxidation in the samples exposed to Fe^{2+}/ascorbate ($p < 0.05$). In the preserved samples, positive effects of honeybush on sperm parameters (motility, kinetics, acrosome, and mitochondria) were recorded from 48 h until 120 h of semen storage ($p < 0.05$). Study results show the protective effects of fermented extract on sperm motility, thus improving its use as a natural source of antioxidants for boar semen (Ros-Santaella et al. 2020). Methanol extract of *C. intermedia* showed significant antimicrobial activity against *S. aureus* and *C. albicans*, whereas chloroform extract was found most effective against *S. pyogenes* (Dube et al. 2017).

Polyphenol-rich organic fraction (iriflophenone-3-C-β-D-glucoside-4-O-β-D-glucoside, hesperidin, mangiferin, and neoponcirin) of *C. intermedia* suppressed lipid accumulation in 3T3-L1 pre-adipocytes and reduced lipid content in mature 3T3-L1 adipocytes. Iriflophenone-3-C-β-D-glucoside-4-O-β-D-glucoside, hesperidin, and mangiferin induced lipolysis in mature adipocytes. Iriflophenone-3-C-β-D-glucoside-4-O-β-D-glucoside and hesperidin increased the expression of messenger RNA of hormone sensitive lipase of mature adipocytes (Jack et al. 2018, 2019). Unfermented honeybush tea significantly ($p < 0.05$) reduced fumonisin B1-stimulated lipid peroxidation in the liver. The tea demonstrated varying effects on fumonisin B1-stimulated alterations in catalase, glutathione peroxidase and glutathione reductase activities as well as the glutathione status. Unfermented honeybush significantly ($p < 0.05$) decreased the total number of foci (>10µM)

DOI: 10.1201/9781003398035-14

also (Marnewick et al. 2009). Polyphenols (mangiferin and the flavonones hesperitin and isokuranetin) of *C. intermedia* showed significant antioxidant and antimutagenic activities *in vitro*. Several studies revealed that herbal tea (*C. intermedia*) possess strong antioxidant, immune-modulating, and chemopreventive activities (McKay and Blumberg 2007).

The *in vitro* effect of water extract of fermented honeybush (*C. intermedia*) was evaluated on osteoclast formation and bone resorption in RAW264.7 murine macrophages. Aqueous extract suppressed osteoclast proliferation and tartrate-resistant acid phosphatase activity, which was followed by decreased bone resorption and interruption of prominent cytoskeletal elements of mature osteoclasts without cytotoxic effects. Moreover, aqueous extract reduced the expression of key osteoclast specific genes, matrix metalloproteinase-9, tartrate resistant acid phosphatase, and cathepsin K. The study suggests that aqueous extract have potential anti-osteoclastogenic effects; therefore, the extract should be further explored for its useful effects on bone (Visagie et al. 2015).

The ethanol/acetone fraction of unprocessed *C. intermedia* inhibited skin tumorigenesis significantly (90%, $p < 0.001$). The availability of flavanol/proanthocyanidin and flavonol/flavone composition and/or nonpolyphenolic chemicals in unprocessed *C. intermedia* are likely an important determining factor in the suppression of tumor formation in mouse skin (Marnewick et al. 2005). Antimutagenic effect of aqueous extracts of *C. subternata, C. genistoides and C. sessiliflora* in aflatoxin B and 2-acetylaminofluorene-induced *Salmonella* mutagenicity assay. *C. intermedia* and *C. subternata* demonstrated significant protection against aflatoxin B-induced mutagenicity (Marnewick et al. 2004; van der Merwe et al. 2006). Honeybush tea significantly ($p < 0.05$) increased cytosolic glutathione S-transferase α activity. The levels of oxidized glutathione were significantly ($p < 0.05$) decreased in the liver of all tea-treated rats whereas reduced glutathione level was remarkably enhanced in the liver of the herbal tea–treated rats. Transformation of drug metabolizing enzymes in the liver may be the focal point in the protection against negative effects associated with mutagenesis and oxidative injury (Marnewick et al. 2003).

The antiwrinkle effect of fermented *C. intermedia* extract (HU-018) was examined in 120 Korean subjects with crow's feet. The subjects were categorized in three groups (low-dose extract 400 mg/day; high-dose extract 800 mg/day; placebo negative control, only dextran) with extract received for 12 weeks. Both doses significantly improved global skin wrinkle grade, skin hydration, and elasticity in treated subjects when compared with the placebo group. A significant decrease in transepidermal water loss in treated groups was recorded and results compared with placebo. During this study, no negative effects were observed in treated subjects. Study results show that fermented extract (HU-018) is effective in ameliorating skin wrinkles, elasticity, and hydration. So, daily consumption of the extract as a food supplement may be useful in skin protection against skin aging (Choi et al. 2018). Similarly, the anti-aging effect of herbal tea of fermented leaves and stems of *C. intermedia* was also studied in African subjects. The fermented herbal tea decreased the length and depth of skin winkles produced by UV irradiation and suppressed the thickening of the epidermal layer. The inhibition of collagen tissue breakdown also was recorded, showing its possible use as an anti-aging agent. The study reveals that fermented extract causes significant antiwrinkle effects and is therefore of interest in antiwrinkle skin care agents (Im et al. 2014). The protective effects of fermented honeybush extract (methanol) and scale-up fermented honeybush extract were evaluated against ultraviolet B-stimulated injury in HaCaT keratinocytes. Ultraviolet B significantly reduced the viability of HaCaT cells, while fermented honeybush extract and scale-up fermented honeybush extract did not show cytotoxic effects and enhanced the viability of the HaCaT cells. Ultraviolet B-stimulated treatment decreased the actions of antioxidant enzymes and skin barrier roles, while fermented honeybush extract or scale-up fermented honeybush extract enhanced their actions. The fermented honeybush extract shows cytoprotective effect against Ultraviolet B-stimulated oxidative stress in HaCaT cells. Moreover, both extracts inhibited the ultraviolet B-stimulated expression of inflammatory mediators, such as IL-1β, IL-6, and IL-8. Furthermore, both extracts also changed the phosphorylation of mitogen-activated protein kinase stimulated by ultraviolet B-irradiation. Interestingly, both extracts remarkably suppressed ultraviolet B-stimulated activation of extracellular signal-regulated kinase, p38, and c-JUN N-terminal kinase. The study suggests that both extracts may be used as potent skin antiphotoaging agents (Im et al. 2016). The skin hydration effects of scaled-up fermented honeybush extract (HU-018) were evaluated against ultraviolet B irradiation in HaCaT immortalized human keratinocytes and hairless mice. HU-018 minimized the reduced levels of hyaluronic acid and mRNA expression of genes encoding involucrin, filaggrin, and loricrin in pretreated HaCaT cells. HU-018 treatment also improved the reduced stratum corneum hydration and the enhanced levels of transepidermal water loss and erythema index in hairless mice. HU-018 also induced alterations in patterns of gene expression of hyaluronan synthase 2, transforming growth factor-β 3, and elastin in ultraviolet B exposed mice. HU-018 reduced the expression of matrix metalloproteinase-1 but enhanced the expression of procollagen type-1, elastin, and TGF-β1. HU-018 elevated skin hydration processes in ultraviolet B-exposed keratinocytes and hairless mice by regulating the expression of involucrin, filaggrin, loricrin, and hyaluronic acid that reduced the visible signs of photoaging. Therefore, HU-018 extract may be used as a potent skin hydration agent for skin care (Im et al. 2020).

The unfermented *C. intermedia* extract inhibited tumor multiplicity significantly ($p < 0.05$) in male F344 rats. A daily intake of unfermented extract (7 mg/100 g bw) showed significant effect in papilloma development (Sissing et al. 2011). The aqueous extract of *Cyclopia intermedia* was found more effective in reducing cell viability in premalignant, normal, and malignant skin cells (Magcwebeba et al. 2016).

C. maculata (Andrews) Kies is an erect, much branched shrub, and indigenous to the southeast and southwest coastal areas of South Africa. The anti-obesity effects of hot water extracts of *C. maculata* and *C. subternata* were assessed in *in vitro* models. The different levels of total polyphenol content of unfermented *C. subternata* (25.6 gallic acid equivalent/100g extract), unfermented *C. maculata* (22.4 gallic acid equivalent/100g extract) and fermented *C. maculata* (10.8g gallic acid equivalent/100g extract) were recorded in significant amounts during this study. The phloretin-3′,5′-di-C-glucoside in *C. subternata*, the mangiferin in unfermented *C. maculata* and the hesperidin in fermented C. maculata were reported as major compounds. All extracts suppressed the accumulation of intracellular triglycerides and fats and reduced the expression of PPARγ2 in treated *in vitro* models. The unfermented *C. maculata* (800 and 1600 μg/ml) and *C. subternata* (1600 μg/ml) extracts were found cytotoxic in reference to reduced mitochondrial dehydrogenase activity. Both fermented and unfermented *C. maculata* extract (more than 100 μg/ml) reduced the cellular ATP contents. Similarly, *C. maculata* and *C. subternata* suppressed adipogenesis *in vitro*, showing their effectivity as anti-obesity agents (Dudhia et al. 2013). Hot water extract of *C. maculata* (80 μg/ml) induced maximum lipolysis (1.8-fold, $p < 0.001$). The enhanced lipolysis was followed by an enhancement in the expression of lipase (1.6-fold, $p < 0.05$) and perilipin (1.6-fold, $p < 0.05$). The extract (0–100 μg/ml) did not show any cytotoxicity to mitochondrial dehydrogenase and adenosine-5′-triphosphate activities. The study results suggest that *C. maculata* induces lipolysis in mature 3T3-L1 adipocytes, offering applications as an anti-obesity agent (Pheiffer et al. 2013).

Methanol extract of *C. subternata* leaves was evaluated for its anticancer activity against orthotopic model of LA7 cell-stimulated mammary tumors. The extract inhibited tumor growth to the same levels as displayed by tamoxifen in treated models. The study results suggest chemosuppressive effect of methanol extract on mammary tumor growth, which was comparable to that of tamoxifen, without inducing adverse side effects. Therefore, this model may be used in examining potential endocrine therapies for hormone responsive breast cancer (Oyenihi et al. 2018). Similarly, methanol extract stimulated a significant cell cycle G0/G1 phase arrest as with the tamoxifen. The extract increased the tumor latency by 7 days and median tumor free survival by 42 days but reduced the frequency of palpable tumor (32%), tumor mass (40%), and tumor volume (53%). Therefore, methanol extract may be used as a chemosuppressive agent against the development and progression of breast cancers (Visser et al. 2016).

Methanol extract of *C. genistoides* and *C. subternata* were examined for estrogen receptor subtypes specific to agonism and antagonism (in transactivation and transrepression). In case of transactivation, the extracts of both species showed estrogen receptor α antagonism and estrogen receptor β agonism. In case of transrepression, methanol extract *C. genistoides* demonstrated uniform agonism, whereas methanol extract of *C. subternata* showed antagonism. Furthermore, both extracts antagonize cell proliferation in the presence of estrogen at lower concentrations in breast cancer cell proliferation assays. In addition, the absence of uterine growth and slowed vaginal opening in an immature rat uterotrophic model were also observed. The studies show that extracts may be used in hormone replacement therapy associated disorders (Visser et al. 2013). Methanol extract of unfermented *C. genistoides* was examined for phytoestrogenic activity in terms of transactivation of an estrogen responsive elements-containing promoter reporter, proliferation of MCF-7-BUS and MDA-MB-231 breast cancer cells. Methanol extract transactivated estrogen responsive element-containing promoter reporters via estrogen receptor β. Methanol extract produced proliferation of the estrogen-sensitive MCF-7-BUS (Michigan Cancer Foundation-7) cells. Growth of MCF-7-BUS cells was estrogen receptor-dependent as ICI 182,780 reversed proliferation. Methanol extract stimulated the formation of estrogen-insensitive MDA-MB-231 cells, revealing that the extract can stimulate estrogen receptor-dependent and estrogen receptor-independent cell formation. Methanol extracts show phytoestrogenic effect and act predominantly through estrogen receptorβ (Verhoog et al. 2007). Luteolin-rich ethyl acetate fraction of methanol extract of *C. subternata* also showed stronger estrogenic activity (Mfenyana et al. 2008).

Hot water and ethanol extracts (0.1–1 ng/ml) of *C. subternata*, *C. genistoides*, and *C. longifolia* showed beneficial effect on bioenergetics by enhancing ATP production, respiration, and mitochondrial membrane potential in human neuroblastoma SH-SY5Y cells. The water extracts of *C. subternata* and *C. genistoides* displayed a protective effect by rescuing the bioenergetic and mitochondrial deficits under oxidative stress conditions (400 µM H_2O_2 for 3 h). The results reveal that extracts of *Cyclopia* species may be used as potent subjects in the prevention of oxidative stress and age-related neurodegenerative diseases (Agapouda et al. 2020).

Dichloromethane, ethyl acetate, ethanol, and aqueous extracts of *C. genistoides* demonstrated significant scavenging and reducing-power effects. The aqueous and ethyl acetate extracts displayed strong antioxidant activity. Both extracts (aqueous and ethyl acetate) also mediate antihyperglycemia effects by suppressing lipid and carbohydrate digestion. The aqueous and ethyl acetate extracts suppressed lipid and carbohydrate digestive enzymes linked to type 2 diabetes, as well as modulate oxidative pancreatic damage. The study promotes its utilization as a potential nutraceutical agent in the cure of diabetes and its associated disorders (Xiao et al. 2020). The presence of genistein, naringenin, isoliquiritigenin, luteolin, helichrysin B, and 5,7,3′,5′-tetrahydroxyflavanone was determined by HPLC from fermented and unfermented extract of *C. genistoides*. The fermented and unfermented extracts suppressed the proliferation of human cancer cell lines A2780 and T47D (Roza et al. 2017). *C. genistoides* extracts displayed antioxidant activity at above 0.6250 mg/ml concentration when compared to the toxin, while mangiferin and hesperidin did not display any antioxidant effect on their own. Both extracts presented the potential to release their own antioxidant effects. Both extracts ameliorated skin smoothness but did not ameliorate skin hydration when compared to the placebos (Gerber et al. 2015).

14.2 PHYTOCHEMISTRY

Mangiferin, isomangiferin, vicenin-2, scolymoside, luteolin, iriflophenone-3-C-β-D-glucoside-4-*O*-β-D-glucoside, iriflophenone-3-C-β-D-glucoside, 3-hydroxy-phloretin-3′,5′-di-C-hexoside, phloretin-3′,5′-di-C-β-D-glucoside, eriocitrin, hesperidin, *p*-coumaric acid, and protocatechuic acid were obtained from methanol extract of *C. subternata* aerial parts. Methanol fraction stimulated antagonism of E2-induced breast cancer cell proliferation and robust ERβ agonist activity (Mortimer et al. 2015). Pinitol, shikimic acid, *p*-coumaric acid, 4-glucosyltyrosol, epigallocatechin gallate, orobol, hesperidin, narirutin, and eriocitrin, luteolin, 5-deoxyluteolin, scolymoside, mangiferin, and C-6-glucosylkaempferol were reported from unfermented leaves of *C. subternata* (Irene

Kamara et al. 2004). The effect of predrying treatments and storage temperatures were evaluated on the phenolic composition of green *C. subternata*. The comminution + drying decreased the content of mangiferin, isomangiferin, and eriocitrin. It is also suggested that oxidation of eriocitrin to scolymoside occurred with comminution and steaming (Joubert et al. 2010). The xanthones (mangiferin, isomangiferin), flavanones (hesperidin, eriocitrin), a flavone (scolymoside), a benzophenone (iriflophenone-3-C-β-glucoside), and dihydrochalcones (phloretin-3',5'-di-C-β-glucoside, 3-hydroxyphloretin-3',5'-di-C-hexoside), 3-hydroxyphloretin-glycoside, iriflophenone-di-*O*,C-hexoside, eriodictyol-di-C-hexoside, (*R*)-eriodictyol-di-*C*-hexoside, iriflophenone-3-*C*-β-glucoside, apigenin-6,8-di-*C*-glucoside, vicenin-2, eriodictyol-*O*-glucoside, 3-hydroxyphloretin-3',5'-di-C-hexoside, eriocitrin, luteolin-7-*O*-rutinoside, scolymoside, phloretin-3',5'-di-*C*-β-glucoside, hesperetin-7-*O*-rutinoside, naringenin-di-*C*-hexoside, naringenin-*O*-*C*-dihexoside, aspalathin, hesperidin, luteolin, iriflophenone-3-*C*-β-glucoside, 3-hydroxyphloretin-3',5'-di-*C*-hexoside, scolymoside, phloretin-3',5'-di-*C*-β-glucoside) have been reported from aqueous extract of *C. subternata* shoots and leaves (de Beer et al. 2012). Mangiferin, isomangiferin, iriflophenone-3-C-β-D-glucoside-4-*O*-β-D-glucoside, iriflophenone-3-C-β-D-glucoside, scolymoside, and phloretin-3',5'-di-C-β-D-glucoside from aqueous extracts (Schulze et al. 2016), mangiferin and isomangiferin from methanol extract of leaves (de Beer et al. 2009), mangiferin, scolymoside, hesperidin, narirutin, iriflophenone 3-C-β-glucoside, phloretin 3',5'-di-C-β-glucoside, and isorhoifolin, calycosin 7-*O*-β-glucoside, rothindin, and ononin were obtained from *C. subternata* (Kokotkiewicz et al. 2012). The isolated compound (scolymoside; isolated from *C. subternata*) elicited anticoagulant activity in mice by significantly reducing the PAI-1 to t-PA ratio. The results reveal that scolymoside exhibits anticoagulant effects and might be developed as a novel anticoagulant agent (Yoon et al. 2015). 3-β-D-Glucopyranosyl-4- *O*-β-D-glucopyranosyliriflophenone, 3-β-D-glucopyranosyliriflophenone, mangiferin, isomangiferin, 3',5'-di-β-D-glucopyranosyl-3-hydroxyphloretin, vicenin-2,3',5'-di-β-D-glucopyranosylphloretin, eriocitrin, scolymoside, hesperidin, and *p*-coumaric acid were isolated and identified from the hot water extract of *C. subternata* (Dippenaar et al. 2022).

Diosmetin, liquiritigenin, hesperetin, iriflophenone 2-*O*-β-glucopyranoside, iriflophenone 3-C-β-glucopyranoside, piceol, 4-hydroxybenzaldehyde, (6a*R*, 11a*R*)-(-)-2-methoxymaackiain, (6a*R*, 11a*R*)-(-)-maackiain, afrormosin, luteolin, and formononetin have been identified from methanol extract of fermented and nonfermented *C. genistoides* leaves and stems. Luteolin and diosmetin significantly inhibited the xanthine oxidase activity *in vitro* (Roza et al. 2016). Maclurin-di-*O*, C-hexoside, iriflophenone-di-*O*, C-hexoside isomer, maclurin-3-C-glucoside, iriflophenone-di-*O*,C-hexoside isomer, iriflophenone-di-*O*,C-hexoside isomer, iriflophenone-3-C-glucoside, iriflophenone-di-C-hexoside, tetrahydroxyxanthone-C-hexoside dimer, tetrahydroxyxanthone-C-hexoside dimer, tetrahydroxyxanthone-di-*O*,C-hexoside, aspalathin derivative of (iso)mangiferin, nothofagin derivative of (iso)mangiferin, mangiferin (2-C-β-D-glucopyranosyl-1,3,6,7- tetrahydroxyxanthone), isomangiferin (4-C-β-Dglucopyranosyl-1,3,6,7- tetrahydroxyxanthone), tetrahydroxyxanthone-C-hexoside isomer, tetrahydroxyxanthone-C-hexoside isomer, schoepfin A derivative of (iso)mangiferin, tyrosine, phenylalanine, glycosylated phenolic acids, dihydroxybenzoic acid-*O*-pentoside, dihydroxybenzoic acid-*O*-dipentoside, phenyllactic acid 2-*O*- hexoside, coumaric acid-*O*- (pentosyl)hexoside, caffeic acid-*O*- (pentosyl)hexoside, apigenin-6,8-di-C-glucoside (vicenin-2), diosmetin-7-*O*-rutinoside (diosmin), dihydrochalcones x 33.54 3-hydroxyphloretin-3',5'-diC-hexoside, phloretin-3',5'-di-C-glucoside, eriodictyol-*O*-hexose-*O*-pentose, eriodictyol-*O*-hexose-*O*-pentose, eriodictyol-*O*-hexose-*O*-deoxyhexose, eriodictyol-*O*-hexose-*O*-deoxyhexose, naringenin derivative, naringenin derivative, naringenin-*O*-hexose-*O*-deoxyhexose, naringenin-*O*-hexose-*O*-deoxyhexose, eriocitrin, narirutin, hesperidin, naringenin-*O*-deoxyhexose(1→2) hexose, maclurin, mangiferin, vicenin-2, aspalathin, maclurin-di-*O*,C-hexoside, iriflophenone-di-*O*,C-hexoside, maclurin-3-C-glucoside, iriflophenone-3-C-glucoside, tetrahydroxyxanthone-C-hexoside dimer, tetrahydroxyxanthone di-*O*,C-hexoside, and eriodictyol-*O*-hexose-*O*-deoxyhexose have been determined by HPLC-DAD method from *C. genistoides* hot water extract (Beelders et al. 2014a). 3-C-β-D-glucopyranosyl-4-*O*-β-D-glucopyranosyliriflophenone together with 3-C-β-D-glucopyranosylmaclurin were identified from *C. genistoides*. Both compounds displayed α-glucosidase suppressive effect against an enzyme mixture separated from rat intestinal acetone powder. The 3-C-β-D-glucopyranosylmaclurin demonstrated significant ($p < 0.05$) inhibitory effect (54%) at 200 μM concentration (Beelders et al. 2014b). Mangiferin, isomangiferin, iriflophenone-3-C-β-glucoside, and hesperidin were identified from *C. genistoides* leaves (Joubert et al. 2014).

4-hydroxycinnamic acid, formononetin, afrormosin, calycosin, pseudobaptigenin, fujikinetin, naringenin, eriodictyol, hesperitin, hesperidin, medicagol, flemichapparin, sophoracoumestan B, mangiferin, isomangiferin, luteolin, and (+)-pinitol were identified from *C. intermedia* stem and leaves (Ferreira et al. 1998). Mangiferin and 3-β-D-glucopyranosyl-4-β-D-glucopyranosyloxyiriflophenone were reported from *C. pubescens* leaves (Walters et al. 2019). Hesperidin, 3-β-D-glucopyranosyl-4-*O*-β-D-glucopyranosyliriflophenone and mangiferin from fermented extract (mean > 5 mg/l) and isomangiferin, vicenin-2, 3-β-D-glucopyranosyliriflophenone, neoponcirin, hesperetin and protocatechuic acid (mean > 1 mg/L) were reported from infusion of *C. intermedia* (de Beer et al. 2021).

Citric acid, ferulic acid pentoside isomer, ferulic acid rhamnose isomer, piscidic acid, iriflophenone-di-*O*,C-hexoside, (3-β-D-glucopyranosyl-4-β-D-glucopyranosyloxyiriflophenone), protocatechuic acid (dihydroxybenzoic acid), maclurin-3-C-glucoside (3-β-D-glucopyranosylmaclurin), dihydroxybenzoic acid-*O*-pentoside, (iso)mangiferin-*O*-hexoside (tetrahydroxyxanthone-di-O,C-hexose), piceol-hexoside-pentoside isomer, piceol-hexoside-rhamnoside (sibiricaphenone), iriflophenone-3-C-glucoside (3-β-D-glucopyranosyliriflophenone), dihydroxybenzoic acid-*O*-dipentoside, *p*-coumaric acid hexoside, *p*-coumaric acid-*O*-pentose-*O*-hexoside, eriodictyol-*O*-hexose-*O*-rhamnose isomer, mangiferin, eriodictyol-O-hexose-O-rhamnose isomer, isomangiferin, naringenin-O-hexoside-O-rhamnose isomer, orobol/luteolin-*O*-hexoside, 3-hydroxyphloretin-3′,5′-di-C-hexoside, eriodictyol-*O*-hexose-*O*-rhamnose isomer, naringenin-*O*-hexoside isomer, luteolin-*O*-rutinoside (scolymoside), phloretin-3′,5′-di-C-glucoside, naringenin-*O*-hexoside isomer, orobol/kaempferol/luteolin-*O*-hexoside, eriodictyol-*O*-hexose-*O*-rhamnose isomer, naringenin-*O*-hexoside-*O*-rhamnose isomer/narirutin, olmelin-*O*-hexoside, hesperidin (hesperetin-*O*-rutinoside), naringenin-*O*-hexoside-*O*-rhamnose isomer, butein/butin, butein-hexoside isomer1, isomer, butein-hexoside isomer, orobol, eriodictyol, didymin/neoponcirin (isosakuranetin-7-*O*-rutinoside), luteolin, naringenin, hesperetin, and (iso)sakuranetin were isolated and identified from Cyclopia extracts (Stander et al. 2019).

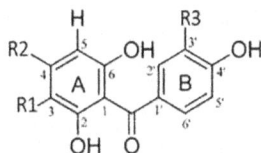

R^1 = β-D-Glucopyranosyl, R^2 = *O*-β-D-Glucopyranosyl, R^3 = H - 3-C-β-D-glucopyranosyl-4-*O*-β-D-glucopyranosyliriflophenone; R^1=β-D-Glucopyranosyl, R^2=R^3=OH-3-C-β-D-Glucopyranosylmaclurin; R^1 = β-D-Glucopyranosyl, R^2 = OH, R^3 = H – 3-C-β-D-Glucopyranosyliriflophenone; R^1 = H, R^2 = R^3 = OH – Maclurin

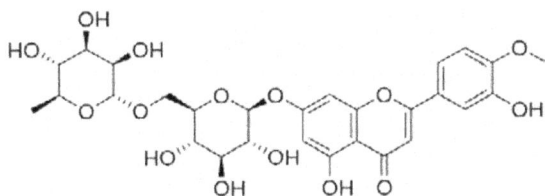

Diosmin

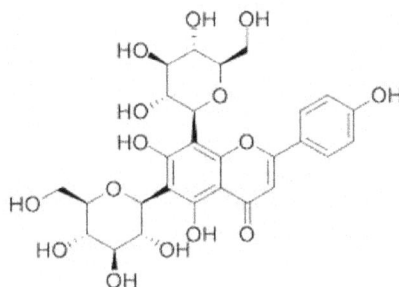

Vicenin-2

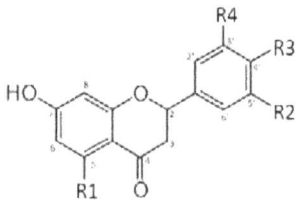

$R^1 = R^2 = H$, $R^3 = OH$, $R^4 = H$ - Liquiritigenin

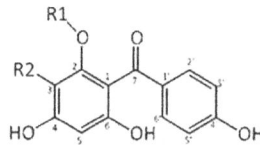

$R^1 = O$-β-Glucosyl, $R^2 = H$ – Iriflophenone 2-O-β-glucopyranoside; $R^1 = H$, $R^2 = C$-β-D-Glucosyl – Iriflophenone 3-C-β-glucopyranoside

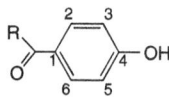

$R = CH_3$ - Piceol; $R = H$ – 4-Hydroxybenzaldehyde

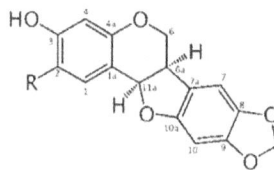

$R = OCH_3$ – (6aR, 11aR)-(-)-2-Methoxymaackiain; $R = H$ -(6aR, 11aR)-(-)-Maackiain

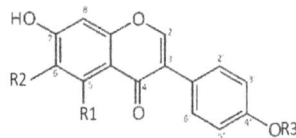

$R^1 = H$, $R^2 = OCH_3$, $R^3 = CH_3$ - Afrormosin; $R^1 = R^2 = H$, $R^3 = CH_3$ - Formononetin

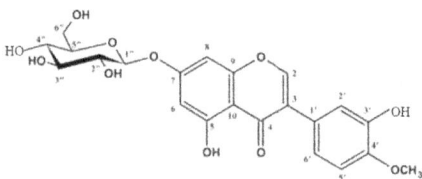

Calycosin 7-O-β-glucoside

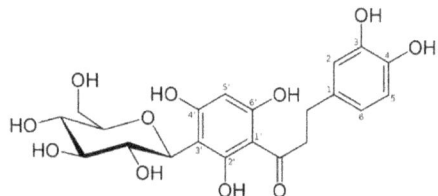

Aspalathin

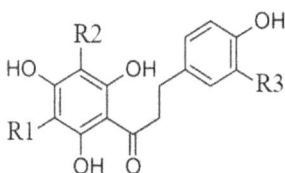

$R^1 = R^2 = $ Glucosyl, $R^3 = H$ - Phloretin 3′,5′-di-C-β-glucoside

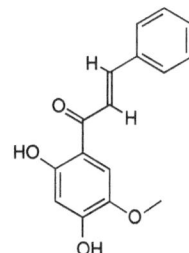

Flemichapparin

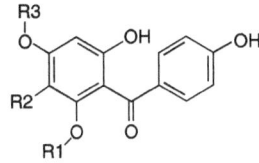

$R^1 = R^3 = H$, $R^3 = $ Glucosyl - Iriflophenone 3-C-β-glucoside

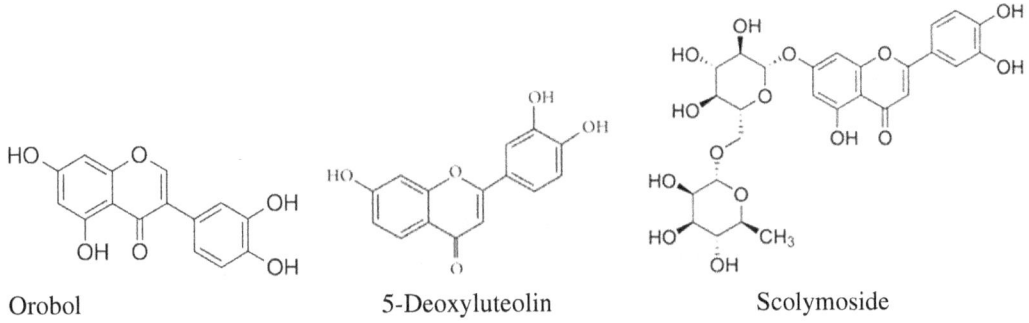

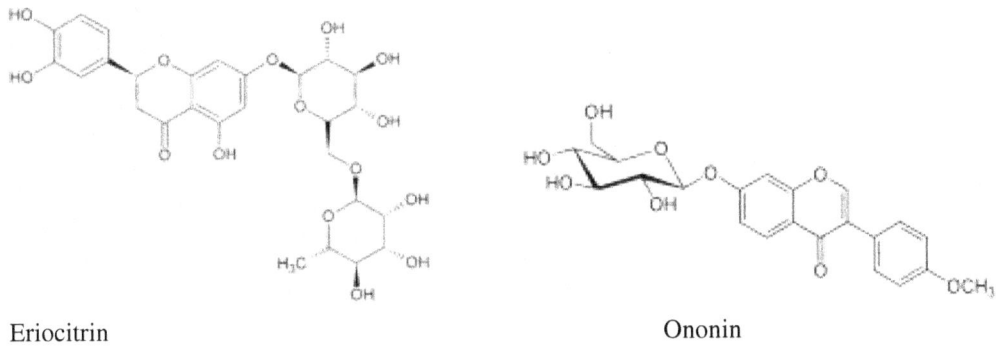

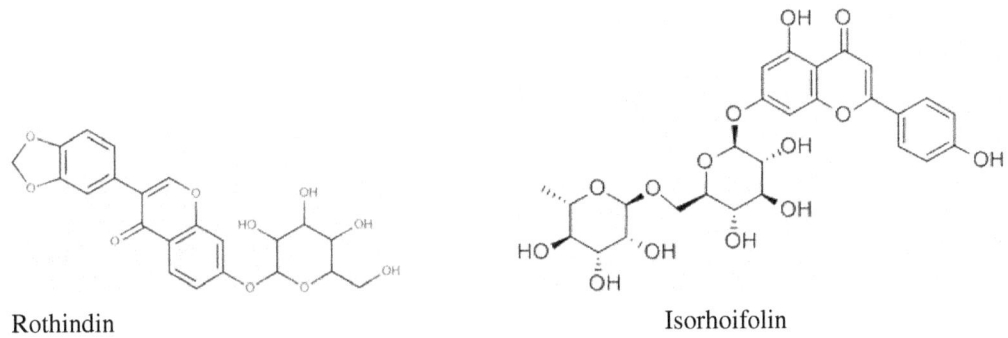

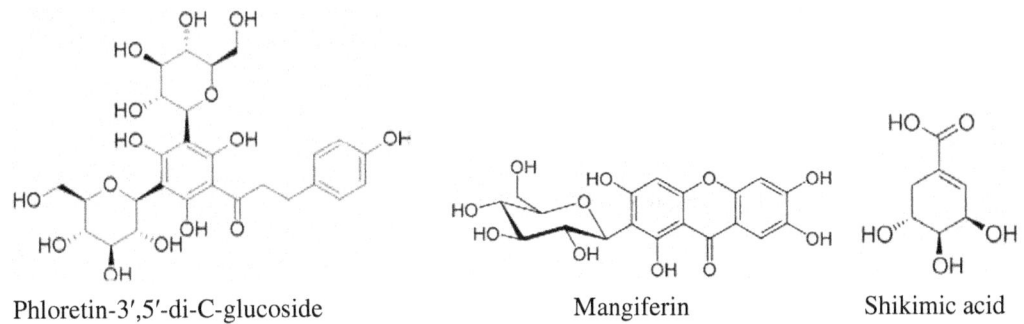

Orobol

5-Deoxyluteolin

Scolymoside

Eriocitrin

Ononin

Rothindin

Isorhoifolin

Phloretin-3′,5′-di-C-glucoside

Mangiferin

Shikimic acid

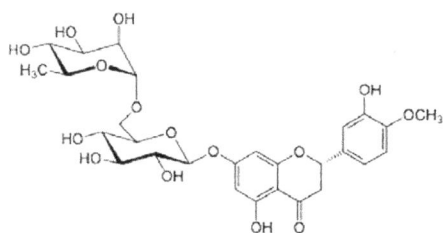

Hesperidin

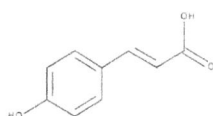

4-Hydroxycinnamic acid

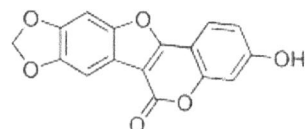

Medicagol

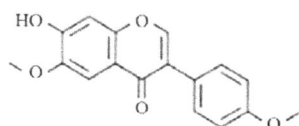

Afrormosin

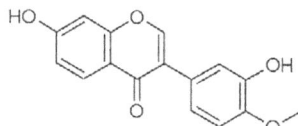

Calycosin

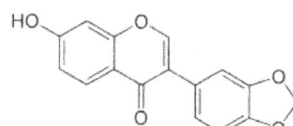

Pseudobaptigenin

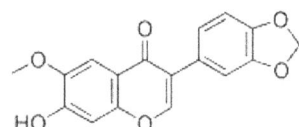

Fujikinetin

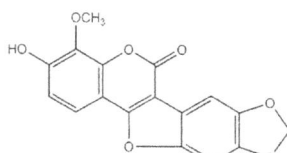

Sophoracoumestan B

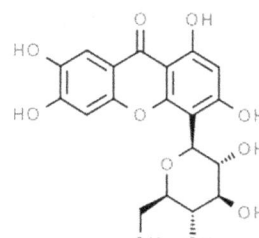

Isomangiferin

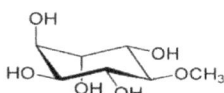

(+)-Pinitol

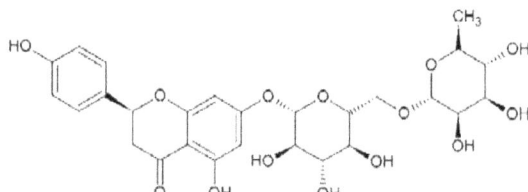

Narirutin

14.3   CULTURE CONDITIONS

In vitro shoots of *C. genistoides* were cultivated in membrane rafts and a temporary immersion biore-actor for increasing the production of phenolic compounds. The maximum levels of mangiferin, iso-mangiferin, and iriflophenone 3-C-β-glucoside were recorded as 1,843.59, 712.02, and 594.29 mg/100 g dw in membrane rafts grown *in vitro* shoots. Similarly, higher levels of mangiferin (2,622.70 mg/100 g dw), isomangiferin (757.40 mg/100 g dw), and iriflophenone 3-C-β-glucoside (648.30 mg/100 g dw) were achieved in temporary immersion bioreactor grown shoots. Therefore, these cultures of *C. genistoides* may be used as an alternative source for increasing the production of phenolic compounds (Kokotkiewicz et al. 2015). The maximum amount of mangiferin (1.55%) and isomangiferin (0.56%) were reported in calli of *C. subternata* microshoots (Kokotkiewicz et al. 2009).

REFERENCES

Agapouda A, Butterweck V, Hamburger M, de Beer D, Joubert E, Eckert A. 2020. Honeybush extracts (*Cyclopia* spp.) rescue mitochondrial functions and bioenergetics against oxidative injury. Oxid Med Cell Longev 2020, 1948602.

Ajuwon OR, Ayeleso AO, Adefolaju GA. 2018. The potential of South African herbal tisanes, rooibos and honeybush in the management of type 2 diabetes mellitus. Molecules 23, 3207.

Bartoszewski R, Hering A, Marszałł M, Stefanowicz Hajduk J, Bartoszewska S. 2014. Mangiferin has an additive effect on the apoptotic properties of hesperidin in Cyclopia sp. tea extracts. PLoS One 9, e92128.

Beelders T, Brand DJ, de Beer D, Malherbe CJ, Mazibuko SE, Muller CJF, Joubert E. 2014b. Benzophenone C- and O-glucosides from *Cyclopia genistoides* (Honeybush) inhibit mammalian α-glucosidase. J Nat Prod 77, 2694–2699.

Beelders T, de Beer D, Stander MA, Jouber E. 2014a. Comprehensive phenolic profiling of *Cyclopia genistoides* (L.) Vent. by LC-DAD-MS and -MS/MS reveals novel xanthone and benzophenone constituents. Molecules 19, 11760–11790.

Choi SY, Hong JY, Ko EJ, Kim BJ, Hong S-W, Lim MH, Yeon SH, Son RH. 2018. Protective effects of fermented honeybush (*Cyclopia intermedia*) extract (HU-018) against skin aging: A randomized, double-blinded, placebo-controlled study. J Cosmet Laser Ther 20, 313–318.

de Beer D, du Preez BVP, Joubert E. 2021. Development of HPLC method for quantification of phenolic compounds in *Cyclopia intermedia* (honeybush) herbal tea infusions. J Food Comp Anal 104, 104154.

de Beer D, Jerz G, Joubert E, Wray V, Winterhalter P. 2009. Isolation of isomangiferin from honeybush (*Cyclopia subternata*) using high-speed counter-current chromatography and high-performance liquid chromatography. J Chromatogr A 1216, 4282–4289.

de Beer D, Schulze AE, Joubert E, de Villiers A, Malherbe CJ, Stander MA. 2012. Food ingredient extracts of *Cyclopia subternata* (Honeybush): Variation in phenolic composition and antioxidant capacity. Molecules 17, 14602–14624.

Dippenaar C, Shimbo H, Okon K, Miller N, Joubert E, Yoshida T, de Beer D. 2022. Anti-allergic and antioxidant potential of polyphenol-enriched fractions from *Cyclopia subternata* (Honeybush) produced by a scalable process. Separations 9, 278.

Dube P, Meyer S, Marnewick JL. 2017. Antimicrobial and antioxidant activities of different solvent extracts from fermented and green honeybush (*Cyclopia intermedia*) plant material. S Afr J Bot 110, 184–193.

Dudhia Z, Louw J, Muller C, Joubert E, de Beer D, Kinnear C, Pheiffer C. 2013. *Cyclopia maculata* and *Cyclopia subternata* (honeybush tea) inhibits adipogenesis in 3T3-L1 pre-adipocytes. Phytomedicine 20, 401–408.

Ferreira D, Kamara BI, Brandt EV, Joubert E. 1998. Phenolic compounds from *Cyclopia intermedia* (Honeybush Tea). 1. J Agric Food Chem 46, 3406–3410.

Gerber GSFW, Fox LT, Gerber M, du Preez JL, van Zyl S, Boneschans B, du Plessis J. 2015. Stability, clinical efficacy, and antioxidant properties of Honeybush extracts in semi-solid formulations. Pharmacogn Mag 11, S337–S351.

Im A-R, Song JH, Lee MY, Yeon SH, Um KA, Chae S. 2014. Anti-wrinkle effects of fermented and non-fermented *Cyclopia intermedia* in hairless mice. BMC Complement Altern Med 14, 424.

Im A-R, Yeon SH, Ji K-Y, Son RH, Um KA, Chae S. 2020. Skin hydration effects of scale-up fermented *Cyclopia intermedia* against ultraviolet b-induced damage in keratinocyte cells and hairless mice. Evid Based Complement Altern Med 2020, 3121936.

Im A-R, Yeon SH, Lee JS, Um KA, Ahn Y-J, Chae S. 2016. Protective effect of fermented *Cyclopia intermedia* against UVB-induced damage in HaCaT human keratinocytes. BMC Complement Altern Med 16, 261.

Irene Kamara B, Jacobus Brand D, Vincent Brandt E, Joubert E. 2004. Phenolic metabolites from honeybush tea (*Cyclopia subternata*). J Agric Food Chem 52, 5391–5395.

Jack BU, Malherbe CJ, Mamushi M, Muller CJF, Joubert E, Louw J, Pheiffer C. 2019. Adipose tissue as a possible therapeutic target for polyphenols: A case for *Cyclopia* extracts as anti-obesity nutraceuticals. Biomed Pharmacother 120,109439.

Jack BU, Malherbe CJ, Willenburg EL, de Beer D, Huisamen B, Joubert E, Muller CJF, Louw J, Pheiffer C. 2018. Polyphenol-enriched fractions of *Cyclopia intermedia* selectively affect lipogenesis and lipolysis in 3T3-L1 adipocytes. Planta Med 84, 100–110.

Joubert E, de Beer D, Hernández I, Munné-Bosch S. 2014. Accummulation of mangiferin, isomangiferin, iriflophenone-3-C-β-glucoside and hesperidin in honeybush leaves (*Cyclopia genistoides* Vent.) in response to harvest time, harvest interval and seed source. Ind Crops Prod 56, 74–82.

Joubert E, Gelderblom W, De Beer D. 2009. Phenolic contribution of South African herbal teas to a healthy diet. Nat Prod Commun 4, 701–718.

Joubert E, Joubert ME, Bester C, De Beer D, De Lange JH. 2011. Honeybush (*Cyclopia* spp.): From local cottage industry to global markets—the catalytic and supporting role of research. S Afr J Bot 77, 887–907.

Joubert E, Manley M, Maicu C, de Beer D. 2010. Effect of pre-drying treatments and storage on color and phenolic composition of green honeybush (*Cyclopia subternata*) herbal tea. J Agric Food Chem 58, 338–344.

Joubert E, Richards ES, Merwe JDVD, De Beer D, Manley M, Gelderblom WC. 2008. Effect of species variation and processing on phenolic composition and *in vitro* antioxidant activity of aqueous extracts of
Cyclopia spp.(honeybush tea). J Agric Food Chem 56, 954–963.

Kamara BI, Brandt EV, Ferreira D, Joubert E. 2003. Polyphenols from honeybush tea (*Cyclopia intermedia*). J
Agric Food Chem 51, 3874–3879.

Kokotkiewicz A, Bucinski A, Luczkiewicz M. 2015. Xanthone, benzophenone and bioflavonoid accumulation
in *Cyclopia genistoides* (L.) Vent. (honeybush) shoot cultures grown on membrane rafts and in a temporary immersion system. Plant Cell Tissue Organ Cult 120, 373–378.

Kokotkiewicz A, Luczkiewicz M. 2009. Honeybush (*Cyclopia* sp.)—a rich source of compounds with high
antimutagenic properties. Fitoterapia 80, 3–11.

Kokotkiewicz A, Łuczkiewicz M, Sowinski P, Glod D, Gorynski K, Bucinski A. 2012. Isolation and structure
elucidation of phenolic compounds from *Cyclopia subternata* Vogel (Honeybush) intact plant and *in vitro*
cultures. Food Chem 133, 1373–1382.

Kokotkiewicz A, Malgorzata Wnuk M, Bucinski A, Luczkiewicz M. 2009. *In vitro* cultures of Cyclopia plants
(Honeybush) as a source of bioactive xanthones and flavanones. Z Naturforsch 64c, 533–540.

Louw A, Joubert E, Visser K. 2013. Phytoestrogenic potential of *Cyclopia* extracts and polyphenols. Planta
Med 79, 580–590.

Magcwebeba TU, Riedel S, Swanevelder S, Swart P, de Beer D, Joubert E, Christoffel W, Gelderblom A. 2016.
The potential role of polyphenols in the modulation of skin cell viability by *Aspalathus linearis* and
Cyclopia spp. herbal tea extracts *in vitro*. J Pharm Pharmacol 68, 1440–1453.

Marnewick J, Joubert E, Joseph S, Swanevelder S, Swart P, Gelderblom WCA. 2005. Inhibition of tumour promotion in mouse skin by extracts of rooibos (*Aspalathus linearis*) and honeybush (*Cyclopia intermedia*),
unique South African herbal teas. Cancer Lett 224, 193–202.

Marnewick JL, Batenburg W, Swart P, Joubert E, Swanevelder S, Gelderblom WCA. 2004. *Ex vivo* modulation
of chemical-induced mutagenesis by subcellular liver fractions of rats treated with rooibos (*Aspalathus
linearis*) tea, honeybush (*Cyclopia intermedia*) tea, as well as green and black (*Camellia sinensis*) teas.
Mutat Res 558, 145–154.

Marnewick JL, Joubert E, Swart P, van der Westhuizen F, Gelderblom WCA. 2003. Modulation of hepatic
drug metabolizing enzymes and oxidative status by rooibos (*Aspalathus linearis*) and honeybush
(*Cyclopia intermedia*), green and black (*Camellia sinensis*) teas in rats. J Agric Food Chem 51,
8113–8119.

Marnewick JL, van der Westhuizen FH, Joubert E, Swanevelder S, Swart P, Gelderblom WCA. 2009.
Chemoprotective properties of rooibos (*Aspalathus linearis*), honeybush (*Cyclopia intermedia*) herbal
and green and black (*Camellia sinensis*) teas against cancer promotion induced by fumonisin B1 in rat
liver. Food Chem Toxicol 47, 220–229.

McKay DL, Blumberg JB. 2007. A review of the bioactivity of South African herbal teas: Rooibos (*Aspalathus
linearis*) and honeybush (*Cyclopia intermedia*). Phytother Res 21, 1–16.

Mfenyana C, DeBeer D, Joubert E, Louw A. 2008. Selective extraction of *Cyclopia* for enhanced in vitro phy-
toestrogenicity and benchmarking against commercial phytoestrogen extracts. J Steroid Biochem Mol
Biol 112, 74–86.

Mortimer M, Visser K, de Beer D, Joubert E, Louw A. 2015. Divide and conquer may not be the optimal
approach to retain the desirable estrogenic attributes of the *Cyclopia* nutraceutical extract, SM6Met.
PLoS One 10, e0132950.

Murakami S, Miura Y, Hattori M, Matsuda H, Malherbe CJ, Muller CJF, Joubert E, Yoshida T. 2018.
Cyclopia extracts enhance Th1-, Th2-, and Th17-type T cell responses and induce Foxp3+ cells in murine
cell culture. Planta Med 84, 311–319.

Niemandt M, Roodt-Wilding R, Tobutt KR, Bester C. 2018. Microsatellite marker applications in *Cyclopia*
(Fabaceae) species. S Afr J Bot 116, 52–60.

Oyenihi OR, Krygsman A, Verhoog N, de Beer D, Saayman MJ, Mouton TM, Louw A. 2018. Chemoprevention
of LA7-induced mammary tumor growth by SM6Met, a well-characterized *Cyclopia* extract. Front
Pharmacol 9, 650.

Pheiffer C, Dudhia Z, Louw J, Muller C, Joubert E. 2013. *Cyclopia maculata* (honeybush tea) stimulates lipolysis in 3T3-L1 adipocytes. Phytomedicine 20, 1168–1171.

Ros-Santaella JL, Kadlec M, Pintus E. 2020. Pharmacological activity of honeybush (*Cyclopia intermedia*) in
boar spermatozoa during semen storage and under oxidative stress. Animals 10, 463.

Roza O, Lai W-C, Zupkó I, Hohmann J, Jedlinszki N, Chang F-R, Csupor D, Eloff JN. 2017. Bioactivity guided
isolation of phytoestrogenic compounds from *Cyclopia genistoides* by the pER8:GUS reporter system.
S Afr J Bot 110, 201–207.

Roza O, Martins A, Hohmann J, Lai W-C, Eloff J, Chang F-R, Csupor D. 2016. Flavonoids from *Cyclopia genistoides* and their xanthine oxidase inhibitory activity. Planta Med 82, 1274–1278.

Schulze AE, de Beer D, Mazibuko SE, Muller CJF, Roux C, Willenburg EL, Nyunaï N, Louw J, Manley M, Joubert E. 2016. Assessing similarity analysis of chromatographic fingerprints of *Cyclopia subternata* extracts as potential screening tool for *in vitro* glucose utilization. Anal Bioanal Chem 408, 639–649.

Schutte AL. 1997. Systematics of the genus *Cyclopia* Vent. (Fabaceae, Podalyrieae). Edinb J Bot 54, 125–170.

Sissing L, Marnewick J, de Kock M, Swanevelder S, Joubert E, Gelderblom WCA. 2011. Modulating effects of rooibos and honeybush herbal teas on the development of esophageal papillomas in rats. Nutr Cancer 63, 600–610.

Slabbert EL, Malgas RR, Veldtman R, Addison P. 2019. Honeybush (*Cyclopia* spp.) phenology and associated arthropod diversity in the Overberg region, South Africa. Bothalia 49, a2430.

Stander MA, Redelinghuys H, Masike K, Long H, van Wyk B-E. 2019. Patterns of variation and chemosystematic significance of phenolic compounds in the genus Cyclopia (Fabaceae, Podalyrieae). Molecules 24, 2352.

van der Merwe JD, Joubert E, Richards ES, Manley M, Snijman PW, Marnewick JL, Gelderblom WCA. 2006. A comparative study on the antimutagenic properties of aqueous extracts of *Aspalathus linearis* (rooibos), different *Cyclopia* spp. (honeybush) and *Camellia sinensis* teas. Mutat Res 611, 42–53.

van Dyk L, Verhoog NJD, Louw A. 2022. Combinatorial treatments of tamoxifen and SM6Met, an extract from *Cyclopia subternata* Vogel, are superior to either treatment alone in MCF-7 cells. Front Pharmacol 13, 1017690.

van Wyk B-E, Gericke N. 2000. People's Plants: A Guide to Useful Plants of Southern Africa. Briza Publications, Pretoria.

van Wyk B-E, van Oudtshoorn B, Gericke N. 1997. Medicinal Plants of South Africa. Briza Publications, Pretoria.

vanWyk B-E, van Oudtshoorn B, Gericke N. 2009. Medicinal Plants of South Africa. Revised and Expanded Edition. Briza Publications, Pretoria.

van Wyk BE. 2008. A broad review of commercially important southern African medicinal plants. J Ethnopharmacol 119, 342–355.

Verhoog NJD, Joubert E, Louw A. 2007. Evaluation of the phytoestrogenic activity of *Cyclopia genistoides* (honeybush) methanol extracts and relevant polyphenols. J Agric Food Chem 55, 4371–4381.

Visagie A, Kasonga A, Deepak V, Moosa S, Marais S, Kruger MC, Coetzee M. 2015. Commercial honeybush (*Cyclopia* spp.) tea extract inhibits osteoclast formation and bone resorption in RAW264.7 murine macrophages-an *in vitro* study. Int J Environ Res Public Health 12, 13779–13793.

Visser K, Mortimer M, Louw A. 2013. *Cyclopia* extracts act as ERα antagonists and ERβ agonists, *in vitro* and *in vivo*. PLoS One 8, e79223.

Visser K, Zierau O, Macejová D, Goerl F, Muders M, Baretton GB, Vollmer G, Louw A. 2016. The phytoestrogenic *Cyclopia* extract, SM6Met, increases median tumor free survival and reduces tumor mass and volume in chemically induced rat mammary gland carcinogenesis. J Steroid Biochem Mol Biol 163, 129–135.

Walters NA, de Beer D, De Villiers A, Walczak B, Joubert E. 2019. Genotypic variation in phenolic composition of *Cyclopia pubescens* (honeybush tea) seedling plants. J Food Compost Anal 78, 129–137.

Xiao X, Erukainure OL, Beseni B, Koorbanally NA, Islam MS. 2020. Sequential extracts of red honeybush (*Cyclopia genistoides*) tea: Chemical characterization, antioxidant potentials, and anti-hyperglycemic activities. J Food Biochem 44, e13478.

Yoon E-K, Ku S-K, Lee W, Kwak S, Kang H, Jung B, Bae J-S. 2015. Antitcoagulant and antiplatelet activities of scolymoside. BMB Rep 48, 577–582.

15 *Cymbopogon* Species

15.1 MORPHOLOGICAL FEATURES, DISTRIBUTION, ETHNOPHARMACOLOGICAL PROPERTIES, PHYTOCHEMISTRY, AND PHARMACOLOGICAL ACTIVITIES

Cymbopogon genus (Fam.—Poaceae) contains about 144 species and is widely distributed in the tropical and subtropical areas of Africa, Asia, and America (Avoseh et al. 2015). The essential oils from different species of *Cymbopogon* have been used in the cosmetics, pharmaceuticals, and perfume industries (Khanuja et al. 2004, 2005). *Cymbopogon citratus* (DC.) Stapf is a perennial herb, and grows up to 1.2–1.8 m. The leaves are small, needle-like, wide (1.3–2.5 cm), and long (0.9 cm) with loose tips (Tajidin et al. 2012). The leaf blade is long (18–36 cm) with parallel venation. The inflorescence is a raceme of spikelets (30–60 cm; Carlson et al. 2001; Paviani et al. 2006; Chanthal and Ruangviriyachai 2012). The glumes are equal to subequal; the lower glume is lance-shaped while the upper glume is lanceolate (Negrelle and Gomes 2007). It is naturalized in India, Thailand, Bangladesh, Madagascar, China, America, and Africa. It is used in the treatment of digestive complaints and fevers (Jeong et al. 2009; Desai and Parikh 2012b). Its leaves possess anti-inflammatory, antiseptic, antidyspeptic, antifever, antispasmodic, analgesic, antipyretic, tranquilizer, antihermetic, and diuretic activities (Ademuyiwa et al. 2017; Tajidin et al. 2012; Ademuyiwa et al. 2017; Tajidin et al. 2012). In Asia, South America, and Africa, it is used as a deodorant in perfumes, local soaps, and candles (Shah et al. 2011; Dutta et al. 2016). Lemongrass is also used as a potent food preservative (Majewska et al. 2019). It is useful in fever, cough, elephantiasis flu, leprosy, malaria, and other digestive complaints (Gupta et al. 2019). In Argentina, the leaf decoction is taken for the treatment of sore throat, empacho, and used as an emetic (Filipoy 1994). In Brazil, leaf tea is used as an antispasmodic, analgesic anti-inflammatory, antipyretic, diuretic, and sedative agent (Leite et al. 1986; Souza-Formigoni et al. 1986). Hot water decoction of leaves is taken by Cuban people (orally) for the treatment of catarrh and rheumatism (Carbajal et al. 1989). Hot aqueous extract of dried leaves and stem is taken as a remedy in renal antispasmodic and diuretic in Egypt (Locksley et al. 1982). In India, its essential oil (two drops; mixed in hot water) is useful in gastric complaints; a few drops of essential oil mixed with lemon juice is useful in cholera. The leaf tea is used as a sedative for the treatment of disorders of the central nervous system (Rao and Jamir 1982; John 1984; Nair 1977). In Indonesian and Malaysian medicine systems, hot water extract of whole plant is used as an emmenagogue (Quisumbing 1951; Burkhill 1966). In Thailand, the whole plant is used as a condiment (Praditvarn and Sambhandharaksa 1950); hot water extract of the whole plant is stomachic (Wasuwat 1967). The hot water extract of roots is taken in management of diabetes (Ngamwathana and Kanchanapee 1971). Nowadays, it is cultivated in the tropical, subtropical, Savannah regions, and Sri Lanka (Negrelle and Gomes 2007). In Tunisia, it is cultivated for decoration, and for medicinal purpose. In gardens, it is grown as an insect repellant agent. The essential oil is also used against cockroaches, flies, and mosquitoes (Dhaou et al. 2010). In South Africa, it is used in the treatment of gastrointestinal complaints, wound management, headache, back pain, and sprains. As per Brazilian system of medicine, it is useful in colds, coughs, heart problems, and urinary tract inflammations and allergies (Negrelle and Gomes 2007; Francisco et al. 2013). In Egyptian medicine, its leaves and essential oil are useful in cough, fever, vomiting, headache, insomnia, and depression (Oloyede 2009; Tarkang et al. 2012). In the US state of Minnesota, boiled suspension is used as a topical lotion for the healing of wounds and bone fractures (Spring 1989). In folklore systems of medicine, aqueous extract of lemongrass leaves

DOI: 10.1201/9781003398035-15

is used in inflammation-associated diseases (Shah et al. 2011), cardiovascular disorders (Runnie et al. 2004), digestive complaints (Bergonzelli et al. 2003), and nervous system related diseases (Barbosa et al. 2008; Linck et al. 2010; Ekpenyong et al. 2015).

The inhibitory effects of lemongrass oil (288 μg/ml) and lemongrass oil vapor (32.7 μg/ml) were investigated against *C. albicans*. The root mean square value was significantly greater in *C. albicans* cells (control, 211.97 nm) than lemongrass oil (143 nm) and lemongrass oil vapor (5.981 nm)–treated cells. The results demonstrate lemongrass oil vapor showed greater inhibition and cellular damage in *C. albicans* cells than lemongrass oil in liquid phase; therefore, lemongrass oil vapor may be used as a potent antifungal agent against *C. albicans* (Tyagi and Malik 2010a). Similarly, lemongrass oil (0.288 mg/ml) and lemongrass oil vapor (0.567 mg/ml) were evaluated for their antibacterial activity against *E. coli*. As per scanning electron microscopic studies, lemongrass oil–treated cells were aggregated and partially deformed, whereas lemongrass oil vapor–treated cells lost their turgidity, with the cytoplasmic material completely released from the cells. In transmission electron microscopic observations, intracytoplasmic alterations and various irregularities were noticed in lemongrass oil vapor–treated cells. The study reveals that lemongrass oil vapor is more effective than lemongrass oil against *E. coli* (Tyagi and Malik 2012). Ethanol extract of *C. citratus* leaves showed significant antibacterial activity against *S. typhi* (MIC 50 mg/ml) but found ineffective against *E. coli, L. monocytogenes* and *S. aureus* (Asaolu et al. 2009). *C. citratus* leaf extract (5.0, 7.5 and 10 mg/g) showed significant antibacterial activity against *Aeromonas veronii*. Phytochemical analysis showed the presence of alkaloids (6.78 ± 1.00 mg/g), flavonoids (3.22 ± 1.00 mg/g), saponin (0.42 ± 1.00), tannins (1.68 ± 1.00), and phenols (4.29 ± 1.00 mg/g) in the leaf extract (Awe et al. 2019). Antimicrobial effect of lemongrass oil was examined against *Aeromonas hydrophila, A. caviae, C. freundii, S. enterica, Edwardsiella tarda, P. aeruginosa*, and *P. mirabilis*. Lemongrass oil showed significant inhibition against *Aeromonas hydrophila, A. caviae, C. freundii, S. enterica, Edwardsiella tarda*, and *P. mirabilis*. A significant minimum inhibitory concentration was observed against *E. tarda, A. hydrophilla, C. freundii, P. mirabilis*, and *S. enterica* (de Silva et al. 2017).

Ethanol extract of the *C. citratus* stem was evaluated for its hepatocarcinogenic effects. The extract was administered as a dietary supplement (0.2, 0.6 or 1.8% concentration) from the end of week 4 for 10 weeks. The extract decreased the number of putatively preneoplastic, glutathione S-transferase placental form-positive lesions and oxidative hepatocyte nuclear DNA injury levels. Conversely, the extract did not show any effect on the size of the preneoplastic lesions, hepatocyte proliferation, phase II enzyme activity, or hepatocyte extranuclear oxidative damages. The study results show inhibitory effects of extract on the early phase hepatocarcinogenesis in animals (Puatanachokchai et al. 2002). Aqueous extract (200 mg/kg b. w., p.o.) demonstrated significant reduction in biochemical parameters (alkaline phosphate, serum glutamic oxaloacetic transaminases, serum glutamic pyruvic transaminase, gamma transaminase, and total bilirubin). The study results reveal that, due to presence of bioactive compounds, the extract possesses significant hepatoprotective effect (Eraj et al. 2016). Protective effect of ethanol extract of *C. citratus* was evaluated in a doxorubicin-stimulated male testicular injury and infertility model. Ethanol extract (300 mg/kg b.w.) was administered for 10 days to the animals. Pretreatment with extract reduced the levels of injury. Extract showed higher testicular activity than normal, which reflected the potential of extract in improvement of reproductive health performance. Similarly, apigenin and quercetin were isolated from ethanol extract. Both compounds (apigenin and quercetin) demonstrated more significant anticancer activity against PC-3 cells than HCT-116 cells (Ahmed et al. 2018).

The hypoglycemic and hypolipidemic effects of aqueous extract of *C. citratus* leaves were evaluated in normal male Wistar rats for 42 days. The extract did not show any toxicity at higher doses (5000 mg/kg bw/p.o.) in treated animals. Aqueous extract (500 mg/kg b.w.) reduced the levels of fasting plasma glucose ($p < 0.05$) and lipid parameters but increased the levels of plasma high density lipoproteins-cholesterol significantly ($p < 0.05$). The study results confirm its folkloric use and safety in the management of type 2 diabetes (Adeneye and Agbaje 2007).

The aqueous extract of *C. citratus* was evaluated for their antioxidant activity in DPPH assay. The activity was also assessed in male Sprague-Dawley rats. The extract (250, 500, and 1,000 mg/kg/day b. w.) was orally administered for one month. The extract consists of flavonoids (496.17 mg gallic acid/g extract) and phenolic constituents (4020.18 mg catechin/g extract). The extract showed potent scavenging effect in DPPH assay (EC_{50} 917.76 $\pm$ 86.89 µg/ml). The extract (1000 mg/kg/day b.w.) significantly increased the expression of γ-glutamylcysteine ligase and heme oxygenase-1 in treated animals. In this study, water extract demonstrated significant antioxidant activity and the stimulation of antioxidant enzymes (Somparn et al. 2014).

Aqueous and ethanolic extracts of *C. citratus* leaves were evaluated against $AlCl_3$-induced Alzheimer's disease in rats. Both extracts stopped pathological alterations and maintained the levels of oxidative stress and inflammatory markers. Moreover, ethanol extract significantly suppressed the acetylcholinesterase activity (2.11 $\pm$ 0.11 mg/ml) and demonstrated a strong antioxidant activity (24.99 $\pm$ 0.00 µg/ml). A total of 28 essential oils and three flavonoids (isoorientin, isoschaftoside, and luteolin-7-*O*-neohesperidoside) were isolated from ethanol extract of *C. citratus* leaves. The powdered leaves contain significant amounts of caffeic acid (3.49 mg/g dry wt) and isoorientin (7.37 mg/g dry w. t.). Ethanol extract reduced the levels of $AlCl_3$-induced neurotoxicity in rats via inhibiting oxidative stress and inflammatory markers. The activity is attributed with the presence of phenolic compounds and flavonoids in crude extract (Madi et al. 2020; Zahra et al. 2021).

C. citratus aqueous extract significantly suppressed the formation of lipopolysaccharide-induced nitric oxide and the expression of inducible nitric oxide synthase. The flavonoid-, tannin-, and phenolic acid-rich fractions decreased the levels of inducible nitric oxide synthase and the formation of nitric oxide in skin-derived dendritic cell lines. It is suggested that flavonoid-rich fraction, due to the presence of luteolin glycosides, showed anti-inflammatory effect. *C. citratus* possesses nitric oxide scavenging effect and suppresses the expression of inducible nitric oxide synthase; therefore, it should be used in the treatment of infections of the gastrointestinal tract (Figueirinha et al. 2010). Aqueous extracts of *C. citratus* leaves suppressed the expression of inducible nitric oxide synthase and the formation of nitric oxide and lipopolysaccharides-stimulated pathways (p38 mitogen-activated protein kinase, c-jun NH_2-terminal kinase 1/2 and the transcription nuclear factor-κB). The extract did not affect extracellular signal-regulated kinase 1/2 and the activity of phosphatidylinositol-3-kinase/Akt. Both (phenolic acid-and tannin-rich) fractions significantly suppressed the activation of transcription nuclear factor-κB, expression of inducible nitric oxide synthase, and the formation of nitric oxide. The aqueous extract and polyphenol-rich fraction did not affect the expression of lipopolysaccharide-stimulated cyclooxygenase-2 but suppressed the formation of lipopolysaccharide-stimulated prostaglandin E_2. Therefore, aqueous extract and polyphenolic acid-rich fraction may be used as a natural source for the development of anti-inflammatory drugs (Francisco et al. 2011). *C. citratus* leaf infusion and its polyphenol-rich fraction suppressed the formation of cytokine on human macrophages. The study provides anti-inflammatory effect of leaf infusion and polyphenol-rich fraction in physiologically relevant cells. Both (leaf infusion and polyphenol-rich fraction) extracts suppressed the activation of lipopolysaccharide-stimulated transcription nuclear factor-κB. Chlorogenic acid was identified from the leaf infusion and polyphenol-rich fraction; therefore, it is assumed that it might be responsible for anti-inflammatory effect. Moreover, leaf infusion and polyphenol-rich fraction also suppressed the activity of proteasome in human and mouse macrophages. Leaf infusion and polyphenol-rich fraction showed strong anti-inflammatory activity by inhibiting the proteosome activity and transcription nuclear factor-κB pathway and the expression of cytokine in lipopolysaccharide-induced human macrophages (Francisco et al. 2013).

C. citratus silver nanoparticles were developed for increasing the antibacterial efficacy. Antibacterial activity of nanoparticles (C_{25}-C_{150} µg/ml) was examined through minimum inhibitory concentration and minimum bactericidal concentration. The C_{25} (µg/ml) and C_{50} (µg/ml) concentration of nanoparticles were measured as the minimum inhibitory concentration and minimum bactericidal concentration for *P. aeruginosa*, *E. coli*, *B. cereus*, and *B. licheniformis*. The nanoparticles showed maximum inhibition [inhibition zone C_{150} (µg/ml) concentration] to *B. cereus* (20.12 ± 0.42), *B. licheniformis* (22.34 ± 0.4), *P. aeruginosa* (35.23 ± 0.46), and *E. coli* (31.87 ± 0.24; Basera et al. 2019).

C. winterianus Jowitt ex Bor is a tufted perennial herb, growing up to 90–95 cm high with 90–95 tillers. The leaf length is 65–70 cm, leaf sheath smooth, linear, acuminate, glaucous, with fibrous roots, and bearing a large inflorescence. Panicles are large, grayish, or grayish green with raceme pairs in dense masses (Singh and Kumar 2017). It is native to Sri Lanka but distributed in India, Indonesia, Brazil, Southeast Asia, Java, and South and Central America (Shasany et al. 2000; Burdock 2002; Avoseh et al. 2015). In Brazil, fresh leaf infusion is used in the treatment of pain and anxiety (Quintans-Júnior et al. 2008). *C. winterianus* essential oil possesses antifungal, antiparasitic, mosquito repellent, and antibacterial (Wany et al. 2013), antimycotic, and acaricidal properties (Tawatsin et al. 2001; Mendonça et al. 2005; Cassel and Vargas 2006). It is also used as an air freshener (Blank et al. 2007).

Essential oil of *C. winterianus* leaves (200 mg/kg, p.o.) significantly decreased ($p < 0.05$) the number of writhings and paw licking times in the first (0–5 min) and second (15–30 min) phases of the acetic acid-stimulated writhing and formalin tests. Essential oil suppressed the carrageenan-stimulated neutrophil migration to the peritoneal cavity (66.1%; 200 mg/kg, $p < 0.001$). In addition, essential oil demonstrated greater scavenging effect toward 1,1-diphenyl-2-picrylhydrazyl radical (IC_{50} 12.66 ± 0.56 µg/ml; Leite et al. 2010). Nondistilled plant materials and their solid residues of *C. winterianus* were Soxhlet extracted with petroleum ether, chloroform, ethyl acetate, acetone, ethanol, methanol, water, and combination of (50% and 75%) of methanol, ethanol, and acetone in water. The different extracts were screened for their antioxidant activity in 2,2-diphenyl-1-picrylhydrazyl, 2,2′-azino-bis(3-ethylbenzothiazoline-6-sulphonic acid), superoxide anion radical scavenging, ferric-reducing power, and iron chelating ability assays. The total phenol and flavonoid contents were also evaluated in different extracts. Compared to distilled materials, the nondistilled plant materials showed greater total phenol/total flavonoids content and demonstrated greater antioxidant effects. The aqueous-methanolic (50% or 75%) extract also displayed maximum DPPH, ABTS and SO scavenging effects and ferric-reducing power ability. The ethyl acetate and aqueous acetone extract (75%) of nondistilled and distilled plant materials displayed maximum iron chelating activity. The IC_{50} values for DPPH, ABTS, SO and metal chelating ability in nondistilled plant extract ranged from 64–387, 92–761, 285–870, and 164–924 µg/ml, while values of distilled materials ranged from 144–865, 239–792, 361–833, and 374–867 µg/ml. The EC_{50} value for FRAP assay was ranged from 118–840 and 151–952 µg/ml for nondistilled and distilled plant materials, respectively. The study results reveal the potential use of these by-products as a natural antioxidant (Saha et al. 2021).

The orofacial antinociceptive effect of *C. winterianus* essential oil complexed with β-cyclodextrin was evaluated in formalin-, capsaicin-, and glutamate-induced orofacial nociception in male Swiss mice. The orofacial nociceptive behavior was significantly ($p < 0.05$) decreased. The number of Fos-positive cells was significantly altered in the dorsal raphe nucleus ($p < 0.01$), locus coeruleus ($p < 0.001$), trigeminal nucleus ($p < 0.05$), and trigeminal thalamic tract ($p < 0.05$) of treated animals. The essential oil complexed with β-cyclodextrin did not produce any alteration in motor coordination in the rota-rod test. Therefore, study results suggest that essential oil complexed with β-cyclodextrin possesses an orofacial antinociceptive effect via the activation of the central nervous system without making any alteration in motor coordination (Santos et al. 2015). The antinociceptive effect of citronellal complexed with β-cyclodextrin and noncomplexed was evaluated in a chronic muscle pain model in mice. Citronellal-stimulated anti-hyperalgesic effect continued until 6 h ($p < 0.001$) whereas citronellal complexed with β-cyclodextrin continued until 8 h ($p < 0.001$ vs. vehicle and $p < 0.001$vs. citronellal from the 6th h). Citronellal complexed with β-cyclodextrin decreased the mechanical hyperalgesia on all days of treatment ($p < 0.05$), without altering muscle strength. Periaqueductal gray ($p < 0.01$) and rostroventromedular area ($p < 0.05$) displayed significant enhancement in the Fos protein expression, whereas decrease was recorded in the spinal cord ($p < 0.001$). Citronellal displayed favorable energy binding (−5.6 and −6.1) to GluR2-S1S2J protein based in the docking score role. In this study, β-cyclodextrin increased the antihyperalgesic effect of citronellal, and it appears to associate with descending pain-inhibitory mechanisms, which are treated as promising molecules for the treatment of chronic muscle pain (Santos et al. 2016).

C. proximus Stapf is a weed, ascending densely, tufted perennial grass, and found in the Egyptian desert. In Egyptian traditional medicine, it is used as a renal antispasmodic and diuretic agent (Taeckholm 1974; Boulos 1983). Its water decoction is useful in diuresis, colicky pains, removal of small stones from the urinary tracts, and used as an antipyretic in fevers (El-Askary et al. 2003).

C. flexuosus (Nees ex Steud.) W.Watson is a tall, tender, perennial, and evergreen grass. It is distributed in Sri Lanka, India, Thailand, and Burma. Its essential oil possesses antimicrobial, insecticidal, analgesic, and anticancer properties (Mukarram et al. 2021). *C. flexuosus* essential oil showed higher cytotoxicity (IC_{50} 69.33 µg/ml) than citral (IC_{50} 140.7 µg/ml) and geraniol (IC_{50} 117 µg/ml) against MCF-7 cells. HSP90 is the most important chaperone responsible for cancer protein–folding. Essential oils also inhibited the expression of HSP90 protein and HSP90-ATPase. The anticancer action, demonstrated by the essential oil of *C. flexuosus*, occurs importantly in animals due to chaperone-protein-HSP90 folding (Gaonkar et al. 2018). *C. flexuosus* essential oil displayed dose-dependent activity against various human cancer cell lines (IC_{50} 4.2 to 79 g/ml). Essential oil showed strongest cytotoxic activity against 502713 (colon; IC_{50} 4.2 g/ml) and IMR-32 (neuroblastoma) cell lines (IC_{50} 4.7 g/ml). Essential oil (200 mg/kg, i.p.) showed significant inhibition of both ascitic and solid tumor forms of Ehrlich Ascites carcinoma (97.34 and 57.83%). Essential oil (200 mg/kg, i.p.) also suppressed the growth of Sarcoma-180 (94.07 and 36.97%) in both ascitic and solid forms. Essential oil also caused the loss of surface projections, chromatin condensation and apoptosis of HL-60 cells. The condensation and fragmentation of nuclei of sarcoma-180 solid tumor cells were also changed by the essential oil (Sharma et al. 2009).

Sixteen essential oils [heptanal, α-thujene, α-pinene, camphene, sabinene, 6-methylhepta-5-en-2-one, myrcene, 1,8-cineole, *cis*-β-ocimene, linalool, citronellal, terpinen-4-ol, neral (citral B), geraniol, geranial (citral A), geranyl acetate] were identified from *C. citratus* leaves. Among identified essential oils, geranial, neral, and myrcene were found as major constituents. *C. citratus* essential oil (3,000 mg/kg b.w) reduced the edema over time in a dose-dependent manner. Essential oil increased longer tolerance of tail in a hot water bath (50 °C) than untreated animals, showing the analgesic activity. Moreover, essential oil reduced the number of acetic acid-induced abdominal cramps in treated animals (Gbenou et al. 2013). The sun-drying, shade-drying, and oven-drying parts of *C. citratus* were evaluated for the presence of essential oils. The maximum amount of essential oil was retrieved from the parts of the oven-drying method. The minimal saponification value (142.59 mgKOH/g) was measured in the sun-drying method, while the minimum acid (4.14 mgKOH/g) and iodine values (114.31gI 2/100g) were observed in the shade-drying method. Among the isolated compounds (myrcene, limonene, citronellal, *cis*-carveol, nerol, neral, geraniol, geranial, carveol, geranyl acetate, caryophellene), geranial, neral, caryophellene, and limonene were found as major constituents in *C. citratus* (Dutta et al. 2014). Volatile oils (β-myrcene, limonene, α-ocimene, β-ocimene, terpinolene, citronellal, neral, geranial, caryophellene, *trans*-bergamotene, α-humulene, α-farnesene, β-farnesene, β-bisabolene, δ-cadinene, carvacrol, caryophellene oxide, isopulegol, 2-undecanone, geranyl formate, neryl acetate, geranyl acetate, geranyl *N*-butyrate) have been identified from *C. citratus*. Essential oil demonstrated promising antifungal activity against *C. albicans*, *C. tropicalis*, and *A. niger* (with different inhibition zone diameters 35–90 mm). Inhibition zone diameters enhanced with enhancing oil volume. Significantly, greater anti-Candida effect was recorded in the vapor phase. Essential oil (10 mg/kg, p.o.) significantly decreased the volume of carrageenan-stimulated paw edema. The effects were compared with diclofenac (50 mg/kg, p.o.). Moreover, topical use of essential oil (*in vivo*) produced a potent anti-inflammatory activity in the mouse model of croton oil-stimulated ear edema. The present study reveals that essential oil possesses a promising potential for the formulation of medicines in the treatment of fungal diseases and skin inflammations (Boukhatem et al. 2014).

15.2 PHYTOCHEMISTRY

Several essential oils (β-myrcene, (Z)-β-ocimene, linalool, (S)-*cis*-verbenol, 4,5-epoxycarene, neral, geranial, undecan-2-one) have been determined from *C. citratus*. Essential oil (2000 mg/kg b. w.) was administered to Swiss albino mice for 21 days. 10% ointment formulation of *C. citratus* essential oil was topically painted on the skin of rabbit for the determination of skin irritation activity. Essential oil did not show any significant changes ($p > 0.05$) in the body weights, gross abnormalities of the organs, and biochemical parameters in treated mice compared to control. The use of 10% ointment of essential oil did not produce any skin irritation. Therefore, *C. citratus* essential oil may be used as a safe and nontoxic subject in the treatment of skin diseases (Lulekal et al. 2019). The essential oils (tricyclene, pinene, camphene, 3-carene, β-myrcene, limonene, β-ocimene, cineole, n-octanal, 6-methyl-hepten-2-one, myrtanal, β-citronellal, linalool, β-caryophyllene, β-citral, sabinol, α-cyclocitral, borneol, neryl acetate, germacrene-D, zingiberene, α-citral, verbenone, nerol, γ-cadinene, Z-carveol, geranial butyrate, caryophyllene oxide, epi-cubenol, isoeugenol, nerolic acid) were identified from *C. citratus*. *C. citratus* essential oil demonstrated significant antifungal effect against *C. albicans*. The minimum inhibitory concentration of essential oil in liquid phase was significantly greater than that in the vapor phase (32.7 mg/l). The 4 h cell exposure was adequate to produce 100% mortality of *C. albicans* cells. Scanning electron microscopy/atomic force microscopic analysis of *C. albicans* cells treated with essential oil in liquid and vapor phase displayed significant shrinkage and partial degeneration. Lemongrass essential oil is very efficacious in vapor phase against *C. albicans*, leading to adverse morphological alterations in structure and surfaces of cells (Tyagi and Malik 2010b). The presence of essential oils (neral, geranial, geraniol, limonene, citronellal, and β-myrcene) was determined by GC-MS analysis in *C. citratus*. For determination of essential oils, four methods of extraction were compared (solvent extraction, steam distillation extraction, accelerated solvent extraction, and supercritical fluid extraction). Sonication with nonpolar solvent extraction showed better results than the steam distillation method in identification of essential oils (Schaneberg and Khan 2002).

Essential oils viz., cyclohexane, 1,3,5-trimethyl-, (1α,3α,5α), 2-acetylcyclopentanone, 1,6-octadien-3-ol, 3,7-dimethyl-, linalool, 4H-pyran-4-one, 2,3-dihydro-3,5-dihydroxy-6-methyl-, cyclooctane, ethenyl-, furan-2-carbohydrazide, N2-(1-methylhexylidene)-, 7-oxabicyclo[4.1.0]heptane, 1-methyl-4-(1-methylethenyl)-, oxiranecarboxaldehyde, 3-methyl-3-(4-methyl-3-pentenyl)-, benzofuran, 2,3-dihydro-, 2,6-octadienal, 3,7-dimethyl-, (Z)-, geraniol, citral, epoxy-linalooloxide, cyclopentane, (1-methylethyl)-, 2-methoxy-4-vinylphenol, bicyclo[2.2.2]octan-1-amine, 3-cyclopropylcarbonyloxydodecane, triallylsilane, 3-heptanol, 2-methyl-, 1,5-heptadiene, 3,3-dimethyl-, (E)-, geranylacetate, cyclopropanemethanol, α, 2-dimethyl-2-(4-methyl-3-pentenyl)-, [1α(R*),2α]-, vanillin, 3,5-heptadienal, 2-ethylidene-6-methyl-, adamantane, 3-cyclopentylpropionic acid, but-3-yn-2-yl ester, 2-propanol, 1,1,1-trichloro-2-methyl-, 2,6-octadienal, 3,7-dimethyl-, (Z)-, 3-cyclohexene-1-acetaldehyde, α,4-dimethyl-, 3-n-propyl-2-pyrazolin-5-one, 4-methyl-5h-furan-2-one, dodecanoic acid, 1-methyl-3-n-propyl-2-pyrazolin-5-one, selina-6-en-4-ol, 2-(2-hydroxyethylthio)propionic acid, phenylacetylformic acid, 4-hydroxy-3-methoxy, tetradecanoic acid, benzene, 1,1′-ethylidenebis-, pyridine, 4-[(1,1-dimethylethyl)thio]-, *p*-hydroxycinnamic acid, ethyl ester, 2-propenoic acid, 3-(4-hydroxy-3-methoxyphenyl)-, *p*-fluoroethylbenzene, n-hexadecanoic acid, hexadecanoic acid, ethyl ester, heptadecanoic acid, 3-methyl-2-butenoic acid, 2-tridecyl ester, phytol, diboroxane, triethyl[(4-methyl-2-pyridyl)amino]-, 9,12-octadecadienoic acid (Z,Z)-, 9,12,15-octadecatrienoic acid, (Z,Z,Z)-, cyclooctene, 3-ethenyl, linoleic acid ethyl ester, ethyl 9,12,15-octadecatrienoate, p-menth-2-en-9-ol, trans27.25 0.3594 octadecanoic acid, ethyl ester, 5,9-undecadien-2-one, 6,10-dimethyl-, (E)-, naphtho[2,1-b:3,4-b′]difuran, 2,3,8,9-tetrahydro-2,9-dimethyl, cyclohexanol, 5-methyl-2-(1-methylethenyl)-, 1,6,10,14-Hexadecatetraen-3-ol, 3,7,11,15-tetramethyl-, (E,E)-, eicosanoic acid, methyl 19-methyl-eicosanoate, 9-tricosene, (Z)-, heptadecane, hexadecanoic acid, 2-hydroxy-1-(hydroxymethyl)ethyl ester, dichloroacetic acid, heptadecyl ester, hexacosane, cyclohexane, 1,1′-[4-(3-cyclohexylpropyl)-1,7-heptanediyl]bis-, 1-nonadecene, tetracosane, butane, 2,2-bis(5-acetyl-2-thienyl)-, squalene, nonacosane, nonadecyl heptafluorobutyrate, heptacosyl acetate, triacontane, triacontyl acetate, dl-α-tocopherol, benzene, 1-nitro-4-(phenylthio)-, campesterol, stigmasterol, 1,2,3,4–4h-isoquinolin-1,3-dione, 4,4,5,6,8-pentamethyl, γ-sitosterol, 2-furancarbox-amide, n-[3-methyl-1-(phenylmethyl)-1h-pyrazol-5-yl]-, tetratriacontane, 9,19-cyclolanost-24-en-3-ol, (3β)-, 4-[5-(3,4-diethoxy-benzyl)-[1,2,4]oxadiazol-3-yl]-furazan-3-ylamine, cannabidiol, eicosane, cyclopropane-1-carboxamide, 2-butyl-N-(5,6,7,8-tetrahydro-7,7-dimethyl-5- oxoquin-azolin-2-yl)-, 3-Methoxy-17β-(O-nitrobenzoyloxy)-estra-1,3,5(10)-triene, and 2-(Acetoxymethyl)-3-(methoxycarbonyl)biphenylene have been reported from ethanol extract of *C. citratus* leaves (Bolade et al. 2018).

The presence of caffeic acid, 3-feroylquinic acid, neochlorogenic acid, chlorogenic acid, *p*-coumaric acid, carlinoside, isoschaftoside, isoorientin, cynaroside, veronicastroside, luteo-lin 7-*O*-neohesperidoside, kurilensin A, cassiaoccidentalin B and isovitexin was determined by HPLC-DAD from *C. citratus* dried leaves. Phenolic acids and flavonoids [luteolin 6-C-glucosyl-8-Carabinoside, luteolin 6-C-glucoside, luteolin 7-*O*-glucoside, luteolin 7-*O*-neohesperidoside, luteolin 6-C-arabinosyl-2"-*O*-rhamnoside and luteolin 2"-*O*-rhamnosyl-C-(6-deoxy-ribo-hexos-3-ulosyl)] were identified from leaf infusion of *C. citratus* (Costa et al. 2015). The effect of arbuscu-lar mycorrhizal fungi associations and heavy metals (lead) was investigated on the production of *C. citratus* essential oils. The (500 and 1000 mg Pb/kg soil) lead together with arbuscular mycor-rhizal fungi association enhanced the production of essential oil (0.69%). A total of 21 essential oils [myrcene, menth 1.(7). 8.diene, linalool, citronellal, trans-carveol, citronellol, neral, geraniol, geranial, linalool oxide, citral, geranic acid, bergamotene, isoledene, farnesene (Z)-(β), germacrene, δ-selinene, γ-gurjunene, δ-cadinene, aromadendrene, muurolol] were identified and increasing lev-els of lead, causing an increase in levels of essential oils (Lermen et al. 2015). Thirty three essential oils [α-pinene, camphene, 6-methyl-5-hepten-2-one, myrcene, menthatriene-1,3,8-para, *p*-cymene, limonene, eucalyptol, *p*-cymenene, perillene, linalool, mentha-2,8-diene-1-ol *trans-para*, *cis*-limonene oxide, mentha-2,8-diene-1-ol *cis-para*, camphor, citronellal, *trans*-limonene oxide, *trans*-verbenol, isogeranial, *cis*-mentha-1(7),8-dien-2-ol, *cis*-dihydrocarvone, *trans*-dihydrocarvone, *cis*-4-caranone, citronellol, mentha-1(7),8-dien-2-ol *trans*, neral (or citral B), carvone, geraniol, geranial (or citral B), bornyl acetate, geranyl formate, neric acid, geranyl acetate, β-caryophyllene, *trans* α-bergamotene, α-cubebene, aromadendrene, α-farnesene, viridiflorene, bicyclogermacrene, spathulenol, caryophyllene oxide, *epi*-globulol] were identified from *C. citratus*. *C. citratus* essential oil displayed maximum ability to scavenge DPPH+ radicals (68% at 8 mg/ml), whereas *C. giganteus* essential oil demonstrated maximum ability to reduce ABTS+ (0.59 μmolET/g). *C. citratus* essential oil showed significant cytotoxicity to prostate cell line LNCaP (IC$_{50}$ = 6.36 μg/ml) and PC-3 (IC$_{50}$ = 32.1 μg/ml) and glioblastoma cell lines (SF-767, IC$_{50}$ = 45.13 μg/ml) and SF-763 (IC$_{50}$ = 172.05 μg/ml; Bayala et al. 2018). The presence of myrcene, 1,3,8-*p*-menthatriene, limonene, linalool, *trans*-*p*-mentha-2,8-dienol, *cis*-*p*-menth-2,8-dienol, *trans*-*p*-mentha-1(7),8-dien-2-ol, *trans*-carveol, *cis*-*p*-mentha-1(7),8-dien-2-ol, neral, carvone, geraniol, geranial was determined by GC-MS analysis from *C. citratus* and *C. giganteus*. The *C. giganteus* essential oil showed significant antibacte-rial activity against *E. coli* (CIP 105182), *E. aerogenes* (CIP 104725), *E. faecalis* (CIP 103907), *L. monocytogenes* (CRBIP 13.134), *P. aeruginosa* (CRBIP 19.249), *S. enterica* (CIP 105150), *S. typhimurium* (ATCC 13311), *S. dysenteriae* (CIP 54.51) and *S. aureus* (ATCC 9144), whereas *C. citratus* essential oil failed to suppress the growth of *P. aeruginosa* (Bassolé et al. 2018). The major components from *C. citratus* were reported as geranial (37.7–41.3%), neral, (30.0–33–4%) and myr-cene (5.6–18.6%; Chagonda et al. 2000).

A total of 94 essential oils [α-pinene, camphene, β-myrcene, β-pinene, (±)-2-carene, α-phellandrene, *p*-cymene, limonene, (Z)-β-ocimene, (E)-β-ocimene, melonal, α-terpinolene, thujol, fenchone-D, myrcenol, β-linalool, *trans*-3(10)-caren-2-ol, *p*-mentha-1,3,8-trienol, *trans-p*-mentha-2,8-dienol, *cis-p*-mentha-2,8-dienol, isopulegol, 4-isopropylidene-cyclohexanol, α-phellandren-8-ol, *trans*-2-caren-4-ol, *cis*-α-terpineol, *trans-p*-mentha-1(7),8-dien-2-ol, β-citronellal, *cis*-verbenol, *trans*-carane, 4,5-epoxy-, *cis-p*-mentha-1(7),8-dien-2-ol, *trans*-piperitol, *p*-menth-1-en-9-al, *cis*-carveol, 7-methyl-3-methylene-6-octen-1-ol, β-citronellol, *trans*-carveol, *cis*-carvone, neral, isoamyl hexanoate, *trans*-3-caren-2-ol, *cis*-geraniol, perillal, nerol, piperitone, *p*-mentha-1(7),8(10)-dien-9-ol, 1-methyl-2-decalone, limonene dioxide, 2-caren-10-al, geranial, piperitone oxide, exo-2-hydroxycineole acetate, nopol, β-bourbonene, geranyl acetate, β-elemene, 2-undecanone, 3-oxo-α-ionol, β-caryophyllene, neric acid, isoamyl caprylate, α-humulene, β-cubebene, geranic acid, α-himachalene, germacrene-D, β-eudesmene, τ-gurjunene, α-muurolene, seychellene, τ-muurolene, α-bergamotene, δ-cadinene, elemol, geranyl butyrate, cubenol, β-caryophyllene oxide, hedycaryol, ledol, eudesm-7(11)-en-4-ol, phenylethyl caproate, guaiol, τ-eudesmol, τ-cadinol, β-eudesmol, α-cadinol, isoaromadendrene epoxide, (Z,E)-farnesol, geranyl caproate, (-)-spathulenol, ledene alcohol, phenylethyl octanoate, 4,4-dimethylandrost-5-en-3-one, bolasterone, norethindrone] were determined by GC-MS from *Cymbopogon* species fresh leaves. *C. giganteus* essential oil showed antitrypanosomal activity against *Trypanosoma brucei*. *C. citratus* essential oil was found toxic against Chinese hamster ovary cells and moderately toxic to human noncancer fibroblast cell lines (WI38 cells; Kpoviessi et al. 2014). α-pinene, camphene, 6-methylhept-5-en-2-one, myrcene, dehydro-1,8-cineole, limonene, (Z)-β-ocimene, (E)-β-ocimene, c-terpinene, linalool, lavandulol, citronellal, *trans*-chrysanthemol, isoneral, p-mentha-1,5-dien-8-ol, borneol, isogeranial, α-terpineol, citronellol, nerol, neral, geraniol, geranial, 2-undecanone, dimethoxy-(Z)-citral, methyl geranate, dimethoxy-(E)-citral, geranic acid, eugenol, geranyl acetate, α-cedrene, (E)-caryophyllene, α-*trans*-bergamotene, α-humulene, germacrene D, β-chamigrene, δ-selinene, α-muurolene, *cis*-dihydroagarofuran, γ-cadinene, δ-cadinene, (E)-γ-bisabolene, caryophyllene oxide, 5-*epi*-7-*epi*-α-eudesmol, cedrol, valerianol, *epi*-α-muurolol, hinesol, *epi*-α-cadinol, α-cadinol, neo-intermedeol, intermedeol, (Z, Z)-farnesol, (E, E)-farnesol, and (E, Z)-farnesol were identified from the leaf and culm of *C. citratus*. *C. citratus* leaf essential oil demonstrated the most potent effects against A549 and H1975 cells (IC$_{50}$ 1.73 ± 0.37 and 4.01 ± 0.30 μg/ml). Essential oil produced the strongest cytotoxic effects against H1650 cells (IC$_{50}$ 4.86 ± 0.29 μg/ml; Trang et al. 2020). 6-methylhept-5-en-2-one, camphene, limonene, nonan-4-ol, citronellal, citronellol, neral, geraniol, citral, geranyl acetate, β-caryophyllene, γ-muurolene, and caryophyllene oxide have been identified from *C. citratus* leaves (Brügger et al. 2019). Thirteen essential oils (6-methylhept-5-en-2-one, camphene, limonene, nonan-4-ol, citronellal, citronellol, neral, geranial, citral, geranyl acetate, β-caryophyllene, γ-muurolene, caryophyllene oxide) were reported from *C. citratus* fresh leaves (Plata-Rueda et al. 2020). Twenty-five essential oils [myrcene, limonene, 1,8-cineole, (Z)-β-ocimene, (E)-β-ocimene,7-epoxymyrcene, linalool, exo-isocitral, trans-α-necrodol, citronellal, (Z)-isocitral, (E)-isocitral, nerol, citronellol, neral, geraniol, geranial, 2-undecanone, ethyl nerolate, (E)-caryophyllene, 2-tridecanone, myristicin] have been determined by GC-MS analysis from *C. citratus* leaves. Essential oil displayed a significant reduction in tumors as well as necrosis and mitosis of 7,12-dimethylbenz [a] anthracene-stimulated breast cancer in female rats. Carvacrol (100 mg/kg/day b. w.) showed a significant reduction in the cumulative tumor volume of treated cells {0.11 ± 0.05 cm3 down compared to 0.38 ± 0.04 cm3 of the 7,12-dimethylbenz [a] anthracene group ($p < 0.01$)}. It is suggested that essential oil and carvacrol possess an antitumor effect on 7,12-dimethylbenz [a] anthracene-stimulated breast cancer in female rats (Rojas-Armas et al. 2020). Essential oil of *C. citratus* demonstrated significant antibacterial activity (Ngan et al. 2020). Eight compounds [3β-methoxy lanosta-9(11)-en-27-ol, 3β-hydroxylanosta-9 (11)-en, (24S) -3β-methoxylanosta-9(11), 25-dien-24-ol, 8-hydroxyl-neo-menthol, (2E)-3,7-dimethyl-2,7-octadiene-1, 6-diol, (+)-citronellol, 7-hydroxymenthol and ethyl nonadecanoate] have been isolated and characterized from *C. citratus* (Zhang et al. 2014). Cymbopogone and cymbopogonol have been identified from leaf wax of *C. citratus* (Crawford et al. 1975). Citral α, citral β, nerol, geraniol, citronellal, terpinolene, geranyl acetate, myrecene, terpinol, methylheptenone, borneol, linalyl acetate, α-pinene, and β-pinene were determined from *C. citratus* leaves. Citral possesses hypoglycemic and hypolipidemic, hypocholesterolemic, free radical scavengers, and antioxidative activities (Kumar et al. 2010). Cymbopogonol, neral, geranial, and citral were isolated from *C. citratus*. Cymbopogonol and neral showed greater antimicrobial activity against *C. albicans* but low activity against *P. aeruginosa*, *E. coli*, *S. aureus*, and *Trichophyton mentagrophytes*. Citral showed cytotoxic activity to colon adenocarcinoma (HCT 116; IC$_{50}$ 10.35 μg/ml) and to human lung adenocarcinoma (A549; IC_{50} 17.74 µg/ml; Ragasa et al. 2008). The maximum amount of total phenol (50.017 gallic acid equivalent mg/g) content was recorded in 30% ethanol extract of *C. citratus* leaves. Ethanol extract of lemongrass showed significant antioxidant activity in DPPH (2,2-diphenyl-1-picrylhydrazyl) scavenging assay (IC_{50} 79.444 mg/l; Hasim et al. 2015).

Gallic acid, isoquercetin, quercetin, rutin, catechin, and tannic acid were isolated and identified from *C. citratus*. Water extract significantly reduced the levels of total cholesterol, low-density lipoprotein and atherogenic index in treated rats ($p < 0.05$). Expression of sterol regulatory element binding protein-1c and HMG-CoA reductase was also reduced significantly in treated animals ($p < 0.05$). In addition, serum antioxidant potential was enhanced in treated rats significantly ($p < 0.05$). The activity was attributed to reduced level of serum lipid peroxidation (Somparn et al. 2018). Ethanolic extract of *C. citratus* produced a significant antioxidant activity ($p < 0.05$) in nitric oxide, H_2O_2 scavenging, and reducing power assays but did not show any effect in ferric antioxidant power assay (Soares et al. 2013).

1,1-dimethyl diborane-D6, 1-p-mentha-1,8-diene, α-terpinolene, (-)-isopulegol, oxirane, octyl-, isopulegol, Z-citral, citronellyl acetate, geranyl acetate, cyclohexane, 1-ethenyl-1-methyl2,4-bis (1-methylethenyl)-, germacrene D, α-muurolene, spathulenol, elemol, torreyol, farnesol, 7-oxabicyclo[4.1.0]heptane, 1- methyl-4-(2-methyloxiranyl)-, (-)-spathulenol, palmitic acid, cyclopropanemethanol, 2-methyl-2- (4-methyl-3-pentenyl)-, tricosane, docosane, eicosane, *M-N*-undecyl phenol, pentatriacontane, and hexatriacontane from stem and cyclohexene, 1-methyl-4-(1- methylethenyl)-, α-terpinolene, (-)-isopulegol, oxirane, octyl-, cyclohexanol, 5-methyl-2-(1- methylethenyl)-, Z-citral, citronellyl acetate, geranyl acetate, eugenol, 2,4-diisopropenyl-1-methyl-1-vinylcyclohexane, germacrene D, δ-cadinene, elemol, "KW3 aus epiglobulol", kauran-18-al, 17-(acetyloxy)-, (4β)-, α-cadinol, nerolidol Z and E, tricosane, pentatriacontane, and heptacosane were identified from the leaves of *C. winterianus* (Andila et al. 2018). Maximum level of essential oil was reported as monoterpene (38.01% and 45.78% in stem and leaf oil) and sesquiterpene (27.67% and 17.78 % in stem and Leaf oil) in different parts of plant species. The essential oil showed significant antifungal activity against *Fusarium solani*, *Aspergillus niger*, and *Cladosporium* species (Andila et al. 2018). Neril acetate, β-elemene, γ -murolene, germacrene D, α-murolene, γ -cadinene, δ-cadinene, elemol, γ -eudesmol, τ -cadinol, β-cudesmol, α-cadinol, citronellyl acetate, eugenol, *trans*-geraniol, geranial, neral, citronellol, *cis*-dihydrocarvone, isopulegol, citronellal, limonene, and linalool have been identified from *C. winterianus* stems and leaves. Essential oil demonstrated significant antibacterial activity against *B. cereus*, *M. luteus*, and *S. aureus* but did not show any effectiveness against *E. coli*, *P. mirabilis*, and *P. tolasii* at tested concentrations. *C. winterianus* essential oil displayed stronger antifungal activity (MICs and MFCs 1–20 µl/ml) against *Phomopsis helianthi*, *Phoma macdonaldii*, *Cladosporium fulvum*, *Cladosporium cladosporioides*, *Aureobasidium pullulans*, and *Trichoderma viride* (Simic et al. 2008). Essential oil of *C. winterianus* showed significant antifungal activity against *Trichophyton mentagrophytes* (MIC 312 µg/ml and MFC was 2500 µg/ml; Pereira et al. 2011). Bergamal, terpineol, linalool, citronellol isobutanoate, lavandulyl acetate, citronellal, mentha-2,8-dien-1-ol, limonene, decenal, decenal-1-ol, citronellol, neral, geraniol, geranial, neryl acetate, elemene, dodecanal, caryophyllene, germacrene, geranyl acetate, bergamotene, isoeugenol, germacrene, elemol, elemol acetate, limonene, γ-cadinene, methyl linolate, α-damascene, γ-eudesmol, eremoligenol, α-bisabolene, α-damascene, farnesol, methyl isoeugenol, isophorene, myrtenol, linalyl acetate, α-pinene, camphene, β-pinene, sabinene, β-caryophyllene, 4-terpineol, *cis*-ocimene, *trans*-ocimene, *p*-cymene, terpinolene, 1- hexanol, and 1-borneol were identified from *C. winterianus* (Singh and Kumar 2017). Tricyclene, α-pinene, camphene, β-pinene, sabinene, myrcene, car-3-ene, α-phellandrene, α-terpineol, limonene, *cis*-ocimene, *trans*-ocimene, *p*-cymene, terpinolene, 1-hexanol, methyl heptanone, citronellal, camphor, bourbonene, linalool, linalyl acetate, α- erpineol, β-caryophyllene, 4-terpineol, menthol, citronellyl acetate, 1-borneol, geranyl formate, citronellol, nerol, geraniol, citronellol butyrate, geranyl butyrate, nerolidol, methyl eugenol, elemol, methyl isoeugenol, and farnesol have been determined from Java type and Ceylon type varieties of *C. winterianus* (Wijesekara 1973). Steam-distilled oil from *C. winterianus* and *C. citratus* was examined by GC-MS. The major components were reported as citronellal (33.4–41.6%), geraniol (23.4–25.1%), and citronellol (9.1–12.8%). The geraniol, citronellol, and citronellal were identified from *C. winterinus* (Akhila 1986). Eleven essential oils viz., limonene, citronellal, isopulegon, β-elemene, citronellyl acetate, α-amorphene, δ-cadinene, citronellol, geraniol, α-cubebene, and elemol were determined from *C. winterianus* oil (Chooluck et al. 2019).

Nineteen essential oils [limonene, α-pinene, piperitone, δ-3-carene, α-terpinene, p-cymene, 1,8-cineole, (Z)-β-ocimene, (E)-β-ocimene, γ-terpinene, α-terpinolene, linalool, linalyl acetate, β-bourbobene, β-elemene, elemol, γ-eudesmol, β-eudesmol, α-eudesmol] were reported from *C. proximus*. Essential oil showed strong inhibition of *Bacillus cereus* and *Salmonella choleraesuis* while methanol extract displayed moderate activity. Methanol extract showed significant scavenging effect in DPPH assay (IC$_{50}$ 48.66±3.1 µg/ml). So, *C. proximus* may be used as a natural food preservative (Selim 2011). The α-terpinene, carene, limonene, piperitone, *p*-menth-2-en-1-ol, elemol, bulnesol, and eudesmol were determined by GC-MS from aqueous extract of *C. proximus* whole plant. Volatile oil (0.2 and 1.2 ml/kg, i.p.) did not show any change in the electrocardiogram waves of animals (El Tahir and Abdel Kader 2008). The presence of elemol, piperitone, β-eudesmol, α-eudesmol, β-elemene, τ-cadinol, terpinolene, β-selinenol, 3-cyclo-hexen-1-one, 2-isopropyl-5-methyl, 4-carene, shyobunol, α-terpineol, cadina-1(10),4-diene, (–)-guaia-6,9-diene, limonene, terpinolene, β-caryophyllane, 4,8-epoxy, *cis*-calamenene, *trans*-geranylgeraniol, *epi*-cubenolespatulenol, 2-carene, cuparene, thymol, (Z)-β-ocimene, germacrene B, α-dihydroagarofuran, γ-muurolene, caryophyllene oxide, shyobunol, α-selinene, espatulenol, *p*-mentha-1,5-dien-8-ol, anethole, cadinene, aromandendrene, δ-elemene, isocaryophyllene, allo-ocimene, and α-amorphene was determined GC-MS from *C. Proximus*. The essential oil showed cardioprotection against isoproterenol- induced cardiac hypertrophy and fibrosis (Althurwi et al. 2020). Proximadiol, 5α-hydroxy-β-eudesmol, 1β-hydroxy-β-eudesmol, 1β-hydroxy-α-eudesmol, elemol, β-eudesmol, 5α-hydroperoxy-β-eudesmol, and 7α,11-dihydroxycadin-10(14)-ene were isolated and identified from the unsaponifiable fraction of the petroleum ether extract of *C. proximus* (Elgamal and Wolff 1987; El-Askary et al. 2003).

The presence of essential oils [α-thujene, α-pinene, sabinene, β-pinene, 3-octanone, myrcene, δ-3-carene, α-terpinene, *p*-cymene, limonene, β-phellandrene, eucalyptol, γ-terpinene, terpinolene, linalol, isopulegol, citronellal, δ-terpineol, α-terpineol, nerol, neral, geraniol, *cis*-acetate de pino-carveyle, acetate de citronellyle, eugenol, acetate de geranyle, α-copaene, β-elemene, β-bourbonene, δ-elemene, β-caryophyllene, β-copaene, (Z)-β-farnesene, α-humulene, γ-muurolene, germacrene-D, iso-butanoate de néryle, α-muurolene, γ-cadinene, δ-cadinene] were determined by GC-MS in *C. nardus* leaves. *C. nardus* essential oil demonstrated significant antiproliferative effect on the pros-tate cancer cell line LNCaP (IC$_{50}$ 58.0 ± 7.9 µg/ml; Bayala et al. 2020). Ctral, α-terpinol, β-*o*-cimene, α-pinene oxide, t-muurolol, 1-Octyn-3-ol, neral, geranial, citronellol, γ-muurolene, α-farnesene, δ-cadinene, β-myrcene, linalool, citronellal, nerol, geranyl-acetate, and t-cadinolwas were reported from *C. citratus* and camphene, dipentene, citronellal, geraniol, geranyl acetate, nerol, citronellol, farnesol, linalool, borneol and methyl eugenol, limonene, citronellal, citral, geraniol, citronellol, citronellal, eugenol, chavicol, elemol, citronellyl oxide, δ-cadinene, γcadinene, methyl eugenol, and vanillin were determined from *C. nardus* (Muttalib et al. 2018).

A new method was developed to optimize the extraction of citral from *C. flexuosus* leaves. By adopting the Taguchi method, the level of extraction was increased, and both hydrotrope solution (sodium salicylate and sodium cumene sulfonate) produced the maximum yield of citral at (con-centration 1.75 M, 5% solid loading) 30 °C temperature (Desai and Parikh 2012a). 4-chlorometh ylene-2-phenyl,-4,5-Dihydrooxazole-5-one, neohexene, ethyl butanoate, acetaldehyde ethyl propyl acetal, butyraldehyde diethyl acetal, 1-*O*-methyl fructose, isoamyl acetate, isovaleraldehyde diethyl acetal, benzyl chloride, benzyl alcohol, trimethylsilyl *p*-(trimethylsilyloxy), silyl ester benzoate, 4-ethoxy-2-butanone, *O*-tolualdehyde, 5-hydroxymethylfurfural, *p*-hydroxybenzaldehyde, 9-octa-decen-12-ynoic, 3,5-dichloro-2,4-dimethylphenol, dihydroactinolide, nonadecene, nonadecane, *N*-methylsaccharin, 2-hexylcinnamaldehyde, 2-myristynoyl pantetheine, eicosyne, phytol, cetyl alcohol, geraniol, dibutyl phthalate, 6-Pentadecen-1-ol, dihomo γ linolenic acid, ethyl linolenate, heptacosane, cholesterol benzoate, ethyl 4-morpholinyl, isooctyl phthalate, ethyl cholate, dioctyl isophthalate, campesterol, stigmasterol, β-sitosterol, and β-amyrin have been identified from *C. flexuosus* ethanol extract. Antidepressant and anxiolytic effects of *C. flexuosus* ethanol extract was evaluated against chronic mild stress model stimulated (activity box, open field activity, light and dark box, and elevated plus-maze) in rats. The ethanolic extract (50, 100, 200 mg/kg) significantly improved all the behavioral deficiencies that developed because of chronic mild stress. *C. flexuosus* ethanol extract (100 mg/kg) displayed the best effect among all other treated groups. The study results suggest that ethanol extract possesses a potential antidepressant and anxiolytic activity against chronic mild stress induced in rats (Nomier et al. 2021). Isointermedeol, a new sesquiterpene alcohol, has been isolated and characterized from *C. flexuosus* (Thappa et al. 1979).

Trans-geraniol, (*R*)-citronellal, β-eudesmol, (+)-limonene, (+)-citronellol, and α-elemol were identified from *C. distans* aerial parts (Zhang et al. 2011). Geranial, geranyl acetate, neral, geraniol, limonene, linalool, and citronellal have been reported as major compounds from *C. distans* leaves and inflorescence (Verma et al. 2013). *C. distans* essential oil showed maximum antifungal activity (MIC 0.625 mg/ml) against *A. flavus* and *Pythium* species (Naik et al. 2017). Essential oils viz., camphene, myrcene, limonene, α-pinene, β-pinene, *trans*-farnesene, linalool, borneol, isoborneol, citral (A and B), geraniol, geranyl acetate, β-bisabolol, farnesal, nerolidol, and farnesol were identified from *C. distans* (Melkani et al. 1985).

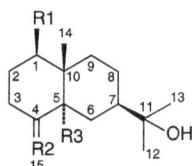

R^1 = H, R^2 = α-OH, CH_3, R^3 = H - Proximadiol; R^1 = H, R^2 = CH_2, R^3 = OH - 5α-Hydroxy-β-eudesmol; R^1 = H, R^2 = CH_2, R^3 = OOH - 5α-Hydroperoxy-β-eudesmol; R^1 = OH, R^2 = CH_2, R^3 = H - 1β-Hydroxy-β-eudesmol

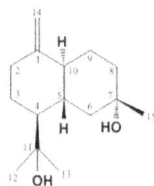

7α,11-Dihydroxycadin-10(14)-ene

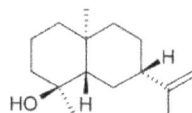

Isointermedeol

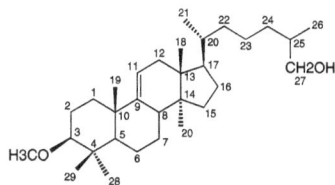

3β-methoxy lanosta-9(11)-en-27-ol

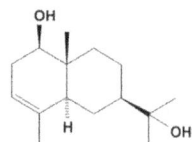

1β-Hydroxy-α-eudesmol

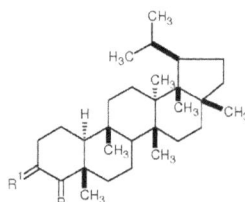

R = β-CH_3,H, R^1 = O – Cymbopogone; R = CH_2, R^1 = β-HO,H - Cymbopogonol

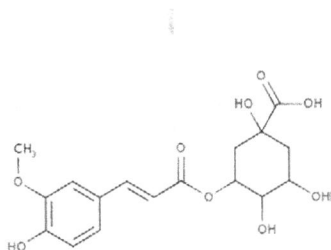
3-Feroylquinic acid

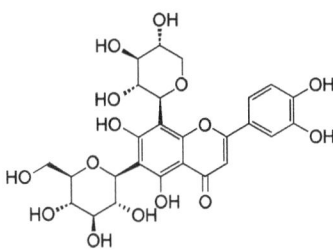
Carlinoside

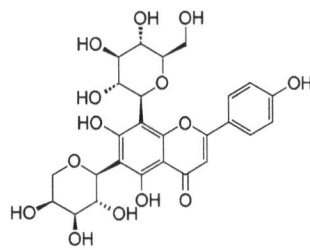
Isoschaftoside

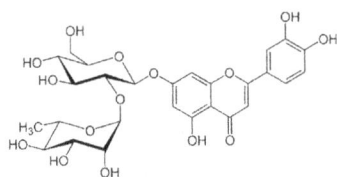
Veronicastroside

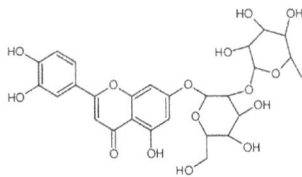
Luteolin 7-*O*-neohesperidoside

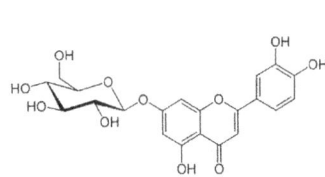
Cynaroside

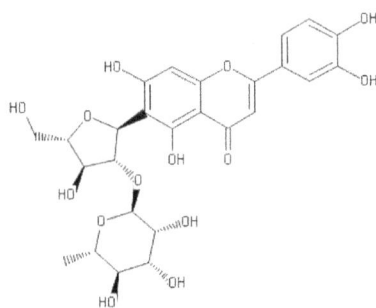
Kurilensin A

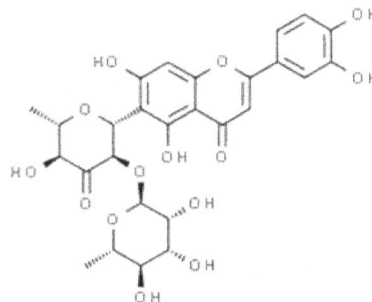
Cassiaoccidentalin B

15.3 CULTURE CONDITIONS

C. nardus nodal parts were cultured onto an MS (Murashige and Skoog 1962) medium containing 2,4-D (3 mg/l) and kinetin (0.5 mg/l). The supplementation showed maximum growth of callus (Nayak 1996). The callus formation of *C. flexuosus* explants was also induced on MS medium with supplementation of 2,4-D (5 mg/l), NAA (0.1 mg/l) and kinetin (0.5 mg/l; Nayak et al. 1996). Callus culture of *C. schoenanthus* subsp. proximus seeds was obtained on an MS medium containing 2,4-D (4.0 mg/l). Suspension cultures were also established in an MS medium with supplementation of 2,4-D (0.5, 1.0 and 2.0 mg/l) as well in different combinations of 2,4-D and 6-benzyl adenine (El-Bakry and Abdel-Salam 2012). The maximum fresh weight (FW; 62.2 and 66.2 g) of *C. citratus* callus was achieved with a frequency of four and six immersions per day, respectively. The dry weight value (6.4 g) and height (8.97 cm) were higher in the treated cells (Quiala et al. 2006). Four essential oils viz., geraniol, geranyl acetate, geranyl formate, and linalool were identified from suspension cultures of *C. martini*. Eight somaclones were selected on the basis of higher yield of oil over the donor line and high geraniol content in the oil was reported (Patnaik et al. 1999). The positive effect of N^6-benzylaminopurine along with light intensity was investigated on terpene production in cell culture of *C. citratus*. The N^6-benzylaminopurine (60, 120, and 180 mg/l) did not show any positive effect on the production of essential oil (1.08% on average). Contrary to this, the citral percentage was negatively affected by N^6-benzylaminopurine, while light intensity showed no effects. The N^6-benzylaminopurine (60 mg/l) increased the yield of citral (72% higher) in treated cells (Prins et al. 2013; Quiala et al. 2016).

REFERENCES

Ademuyiwa AJ, Elliot S, Olamide OY, Johnson OO. 2017. Studies on the nephroprotective and nephrotoxicity effects of aqueous extract of *Cymbopogon citratus* (lemon grass) on Wistar albino rats. Int J Contemp Appl Res 4, 81–95.

Adeneye AA, Agbaje EO. 2007. Hypoglycemic and hypolipidemic effects of fresh leaf aqueous extract of *Cymbopogon citratus* Stapf. in rats. J Ethnopharmacol 112, 440–444.

Ahmed NZ, Ibrahim SR, Ahmed- Farid OA. 2018. Quercetin and apigenin of *Cymbopogon citratus* mediate inhibition of HCT-116 and PC-3 cell cycle progression and ameliorate doxorubicin-induced testicular dysfunction in male rats. Biomed Res Ther 5, 2466–2479.

Akhila A. 1986. Biosynthesis of monoterpenes in *Cymbopogon winterianus*. Phytochemistry 25, 421–424.

Althurwi HN, Abdel-Kader MS, Alharthy KM, Salkini MA, Albaqami FF. 2020. *Cymbopogon proximus* essential oil protects rats against isoproterenol-induced cardiac hypertrophy and fibrosis. Molecules 25, 1786.

Andila PS, Hendra PA, Wardani PK, Tirta IG, Sutomo, Fardenan D. 2018. The phytochemistry of *Cymbopogon winterianus* essential oil from Lombok Island, Indonesia and its antifungal activity against phytopathogenic fungi. Nusantara Biosci 10, 232–239.

Asaolu MF, Oyeyemi OA, Olanlokun O. 2009. Chemical compositions, phytochemical constituents and *in vitro* biological activity of various extracts of *Cymbopogon citratus*. Pak J Nutr 8, 1920–1922.

Avoseh O, Oyedeji O, Rungqu P, Nkeh-Chungag B, Oyedeji A. 2015. Cymbopogon species; Ethnopharmacology, phytochemistry and the pharmacological importance. Molecules 20, 7438–7453.

Awe FA, Hammed AM, Olanloye OA. 2019. Effects of whole lemon grass (*Cymbopogon citratus*) extract on bacteria (*Aeromonas veronii*) infected sub-adult *Clarias gariepinus*. Asian J Appl Sci 12, 99–107.

Barbosa LCA, Pereira UA, Martinazzo AP, Maltha CRA, Teixeira RR, Melo EDC. 2008. Evaluation of the chemical composition of Brazilian commercial *Cymbopogon citratus* (D.C.) stapf samples. Molecules 13, 1864–1874.

Basera P, Lavania M, Agnihotri A, Lal B. 2019. Analytical investigation of *Cymbopogon citratus* and exploiting the potential of developed silver nanoparticle against the dominating species of pathogenic bacteria. Front Microbiol 10, 282.

Bassolé IHN, Lamien-Meda A, Bayala B, Obame LC, Ilboudo AJ, Franz C, Novak J, Nebié RC, Dicko MH. 2018. Chemical composition and antimicrobial activity of *Cymbopogon citratus* and *Cymbopogon giganteus* essential oils alone and in combination. Phytomedicine 18, 1070–1074.

Bayala B, Bassole IHN, Maqdasy S, Baron S, Simpore J, Lobaccaro J-MA. 2018. *Cymbopogon citratus* and *Cymbopogon giganteus* essential oils have cytotoxic effects on tumor cell cultures. Identification of citral as a new putative anti-proliferative molecule. Biochimie 153, 162–170.

Bayala B, Coulibaly AY, Djigma FW, Nagalo BM, Baron S, Figueredo G, Lobaccaro J-MA, Simpore J. 2020. Chemical composition, antioxidant, antiinflammatory and antiproliferative activities of the essential oil of *Cymbopogon nardus*, a plant used in traditional medicine. Biomol Concepts 11, 86–96.

Bergonzelli GE, Donnicola D, Porta N, Corthesy-Theulaz IE. 2003. Essential oils as components of a diet-based approach to management of Helicobacter infection. Antimicrob Agents Chemother 47, 3240–3246.

Blank AF, Costa AG, Arrigoni-Blank MF, Sócrates Cavalcanti SCH, Alves PB, Innecco R, Ehlert PAD, Sousa IF. 2007. Influence of season, harvest time and drying on Java citronella (*Cymbopogon winterianus* Jowitt) volatile oil. Rev Bras Farmacogn 17, 557–564.

Bolade OP, Akinsiku AA, Adeyemi AO, Williams AB, Benson NU. 2018. Dataset on phytochemical screening, FTIR and GC-MS characterisation of *Azadirachta indica* and *Cymbopogon citratus* as reducing and stabilizing agents for nanoparticles synthesis. Data Brief 20, 917–926.

Boukhatem MN, Ferhat MA, Kameli A, Saidi F, Kebir HT. 2014. Lemon grass (*Cymbopogon citratus*) essential oil as a potent anti-inflammatory and antifungal drugs. Libyan J Med 9, 25431.

Boulos L. 1983. Medicinal Plants of North Africa. Reference Publication Inc, Algonac.

Brügger BP, Martínez LC, Plata-Rueda A, Castro BMCE, Soares MA, Wilcken CF, Carvalho AG, Serrão JE, Zanuncio JC. 2019. Bioactivity of the *Cymbopogon citratus* (Poaceae) essential oil and its terpenoid constituents on the predatory bug, *Podisus nigrispinus* (Heteroptera: Pentatomidae). Sci Rep 9, 8358.

Burdock GA. 2002. Fanarali's Handbook of Flavor Ingredients. CRC Press, Boca Raton.

Burkhill IH. 1966. Dictionary of the Economic Products of the Malay Peninsula. Ministry of Agriculture and Co-operatives, Kuala Lumpur.

Carbajal D, Casaco A, Arruzazabala L, Gonzalez R, Tolon Z. 1989. Pharmacological study of *Cymbopogon citratus* leaves. J Ethnopharmacol 25, 103–107.

Carlson LH, Machad CB, Pereira LK, Bolzan A. 2001. Extraction of lemongrass essential oil with dense carbon dioxide. J Supercrit Fluids 21, 33–39.

Cassel E, Vargas RMF. 2006. Experiments and modeling of the *Cymbopogon winterianus* essential oil extraction by steam distillation. J Mex Chem Soc 50, 126–129.

Chagonda LS, Makanda C, Chalchat J-C. 2000. Essential oils of cultivated *Cymbopogon winterianus* (Jowitt) and of *C. citratus* (DC) (Stapf) from Zimbabwe. J Essent Oil Res 12, 478–480.

Chanthal SP, Ruangviriyachai C. 2012. Influence of extraction methodologies on the analysis of five major volatile aromatic compounds of citronella grass and lemongrass grown in Thailand. J AOAC Int 9, 763–772.

Chooluck K, Teeranachaideekul V, Jintapattanakit A, Lomarat P, Phechkrajang C. 2019. Repellency effects of essential oils of *Cymbopogon winterianus, Eucalyptus globulus, Citrus hystrix* and their major constituents against adult German cockroach (*Blattella germanica* Linnaeus (Blattaria: Blattellidae). Jordan J Biol Sci 12, 519–523.

Costa G, Nunes F, Vitorino C, Sousa JJ, Figueiredo IV, Batista MT. 2015. Validation of a RP-HPLC method for quantitation of phenolic compounds in three different extracts from *Cymbopogon citratus*. Res J Med Plant 9, 331–339.

Crawford M, Hanson SW, Koker MES. 1975. The structure of cymbopogone, a novel triterpenoid from lemongrass. Tetrahedron Lett 16, 3099–3102.

Desai MA, Parikh J. 2012a. Hydrotropic extraction of Citral from *Cymbopogon flexuosus* (Steud.) Wats. Ind Eng Chem Res 51, 3750–3757.

Desai MA, Parikh J. 2012b. Microwave assisted extraction of essential oil from *Cymbopogon flexuosus* (Steud.) wats: A parametric and comparative study. Sep Sci Technol 47, 1963–1970.

de Silva BCJ, Jung W-G, Hossain S, Wimalasena SHMP, Pathirana HNKS, Heo G-J. 2017. Antimicrobial property of lemongrass (*Cymbopogon citratus*) oil against pathogenic bacteria isolated from pet turtles. Lab Anim Res 33, 84–91.

Dhaou SO, Jeddi K, Chaieb M. 2010. Les Poaceae en Tunisie: systématique et utilité thérapeutique. Phytotherapie 8, 145–152.

Dutta D, Kumar P, Nath A, Verma N, Gangwar B. 2014. Qualities of lemongrass (*Cymbopogan citratus*) essential oil at different drying conditions. Int J Agric Environ Biotechnol 7, 301–309.

Dutta S, Munda S, Lal M, Bhattacharyya PR. 2016. A short review on chemical composition therapeutic use and enzyme inhibition activities of *Cymbopogon* species. Indian J Sci Technol 9, 1–9.

Ekpenyong CE, Akpan E, Nyoh A. 2015. Ethnopharmacology, phytochemistry, and biological activities of *Cymbopogon citratus* (DC.) Stapf extracts. Chin J Nat Med 13, 321–337.

El-Askary HI, Meselhy MR, Galal AM. 2003. Sesquiterpenes from *Cymbopogon proximus*. Molecules 8, 670–677.

El-Bakry AA, Abdel-Salam AM. 2012. Regeneration from embryogenic callus and suspension cultures of the wild medicinal plant *Cymbopogon schoenanthus*. Afr J Biotechnol 11, 10098–10107.

Elgamal MH, Wolff P. 1987. A further contribution to the sesquiterpenoid constituents of *Cymbopogon proximus*. Planta Med 53, 293–294.

El Tahir KEH, Abdel Kader MS. 2008. Chemical and pharmacological study of *Cymbopogon proximus* volatile oil. Res J Med Plants 2, 53–60.

Eraj A, Sarfaraz S, Usmanghani K. 2016. Hepatoprotective potential and phytochemical screening of *Cymbopogon citratus*. J Anal Pharm Res 3, 00074.

Figueirinha A, Cruz MT, Francisco V, Lopes MC, Batista MT. 2010. Anti-inflammatory activity of *Cymbopogon citratus* leaf infusion in lipopolysaccharide-stimulated dendritic cells: Contribution of the polyphenols. J Med Food 13, 681–690.

Filipoy A. 1994. Medicinal plants of the Pilaga of Central Chaco. J Ethnopharmacol 44, 181–193.

Francisco V, Costa G, Figueirinha A, Marques C, Pereira P, Neves BM, Lopes MC, García-Rodríguez C, Cruz MT, Batista MT. 2013. Anti-inflammatory activity of *Cymbopogon citratus* leaves infusion via proteasome and nuclear factor-κB pathway inhibition: Contribution of chlorogenic acid. J Ethnopharmacol 148, 126–134.

Francisco V, Figueirinha A, Neves BM, García-Rodríguez C, Lopes MC, Cruz MT, Batista MT. 2011. *Cymbopogon citratus* as source of new and safe anti-inflammatory drugs: Bio-guided assay using lipopolysaccharide-stimulated macrophages. J Ethnopharmacol 133, 818–827.

Gaonkar R, Shiralgi Y, Lakkappa DB, Hegde G. 2018. Essential oil from *Cymbopogon flexuosus* as the potential inhibitor for HSP90. Toxicol Rep 5, 489–496.

Gbenou JD, Ahounou JF, Akakpo HB, Laleye A, Yayi E, Gbaguidi F, Baba-Moussa L, Darboux R, Dansou P, Moudachirou M, Kotchoni SO. 2013. Phytochemical composition of *Cymbopogon citratus* and *Eucalyptus citriodora* essential oils and their anti-inflammatory and analgesic properties on Wistar rats. Mol Biol Rep 40, 1127–1134.

Gupta PK, Rithu BS, Shruthi A, Lokur AV, Raksha M. 2019. Phytochemical screening and qualitative analysis of *Cymbopogon citratus*. J Pharmacogn Phytochem 8, 3338–3343.

Hasim, Falah S, Ayunda RD, Faridah DN. 2015. Potential of lemongrass leaves extract (*Cymbopogon citratus*) as prevention for oil oxidation. J Chem Pharm Res 7, 55–60.

Jeong M-R, Park PB, Kim D-H, Jang Y-S, Jeong HS, Choi S-H. 2009. Essential oil prepared from *Cymbopogon citrates* exerted an antimicrobial activity against plant pathogenic and medical microorganisms. Mycobiol 37, 48–52.

John D. 1984. One hundred useful raw drugs of the Kani Tribes of Trivandrum Forest Division, Kerala, India. Int J Crude Drug Res 22, 17–39.

Khanuja SPS, Shasany AK, Pawar A, Lal RK, Darokar MP, Naqvi AA, Rajkumar S, Sundaresan V, Lal N, Kumar S. 2004. Essential oil constituents and RAPD markers to establish species relationship in *Cymbopogon* Spreng. (Poaceae). Biochem Syst Ecol 33, 171–186.

Khanuja SPS, Shasany AK, Pawar A, Lal RK, Darokar MP, Naqvi AA, Rajkumar S, Sundaresan V, Lal N, Kumar S. 2005. Essential oil constituents and RAPD markers to establish species relationship in *Cymbopogon* Spreng. (Poaceae). Biochem Syst Ecol 33, 171–186.

Kpoviessi S, Bero J, Agbani P, Gbaguidi F, Kpadonou-Kpoviessi B, Sinsin B, Accrombessi G, Frédérich M, Moudachirou M, Quetin-Leclercq J. 2014. Chemical composition, cytotoxicity and *in vitro* antitrypanosomal and antiplasmodial activity of the essential oils of four *Cymbopogon* species from Benin. J Ethnopharmacol 151, 652–659.

Kumar R, Krishan P, Swami G, Kaur P, Shah G, Kaur A. 2010. Pharmacognostical investigation of *Cymbopogon citratus* (DC) Stapf. Der Pharm Lett 2, 181–189.

Leite BLS, Bonfim RR, Antoniolli AR, Thomazzi SM, Araújo AAS, Blank AF, Estevam CS, Cambui EVF, Bonjardim LR, Albuquerque Júnior RLC, Quintans-Júnior LJ. 2010. Assessment of antinociceptive, anti-inflammatory and antioxidant properties of *Cymbopogon winterianus* leaf essential oil. Pharm Biol 48, 1164–1169.

Leite JR, Seabra ML, Maluf E, Assolant K, Suchecki D, Tufi KS. 1986. Pharmacology of lemongrass (*Cymbopogon citratus* Stapf). Assessment of eventual toxic, hypnotic and anxiolytic effects on humans. J Ethnopharmacol 17, 75–83.

Lermen C, Morelli F, Gazim ZC, daSilva AP, Gonçalves JE, Dragunski CD, Alberton O. 2015. Essential oil content and chemical composition of *Cymbopogon citratus* inoculated with arbuscular mycorrhizal fungi under different levels of lead. Ind Crops Prod 76, 734–738.

Linck VM, da Silva AL, Figueiro M, Caramao EB, Moreno PRH, Elisabetsky E. 2010. Effects of inhaled linalool in anxiety, social interaction and aggressive behavior in mice. Phytomedicine 17, 679–683.

Locksley HD, Fayez MB, Radwan AS, Chari VM, Cordell GA, Wagner H. 1982. Constituents of local plants XXV, Constituents of the antispasmodic principle of *Cymbopogon proximus*. Planta Med 45, 20–22.

Lulekal E, Tesfaye S, Gebrechristos S, Dires K, Zenebe T, Zegeye N, Feleke G, Kassahun A, Shiferaw Y, Mekonnen A. 2019. Phytochemical analysis and evaluation of skin irritation, acute and sub-acute toxicity of *Cymbopogon citratus* essential oil in mice and rabbits. Toxicol Rep 6, 1289–1294.

Madi YF, Choucry MA, El-Marasy SA, Meselhy MR, El-Kashoury A-SA. 2020. UPLC-Orbitrap HRMS metabolic profiling of *Cymbopogon citratus* cultivated in Egypt; neuroprotective effect against AlCl 3-induced neurotoxicity in rats. J Ethnopharmacol 259, 112930.

Majewska E, Kozłowska M, Gruczyńska-Sękowska E, Kowalska D, Tarnowska K. 2019. Lemongrass (*Cymbopogon citratus*) essential oil: Extraction, composition, bioactivity and uses for food preservation—a review. Pol J Food Nutr Sci 69, 327–341.

Melkani AB, Joshi P, Pant AK, Mathel CS. 1985. Constituents of the essential oils from two varieties of *Cymbopogon distans*. J Nat Prod 48, 995–997.

Mendonça FA, Silva KF, Santos KK, Ribeiro Júnior KA, Sant'Ana AE. 2005. Activities of some Brazilian plants against larvae of the mosquito *Aedes aegypti*. Fitoterapia 76, 629–636.

Mukarram M, Khan MMA, ZehrA A, Choudhary S, AfTab T, Naeem M. 2021. Biosynthesis of lemongrass essential oil and the underlying mechanism for its insecticidal activity. In: Medicinal and Aromatic Plants, Aftab A, Hakeem KR (eds.). Springer Nature, Berlin.

Murashige T, Skoog F. 1962. A revised medium for rapid growth and bio assays with tobacco tissue cultures. Physiol Plant 15, 473–497.

Muttalib SA, Edros R, Nor Azah MA, Kutty RV. 2018. A review: The extraction of active compound from *Cymbopogon* sp. and its potential for medicinal applications. Int J Eng Technol Sci 5, 82–98.

Naik G, Bhandari U, Gwari G, Lohani H. 2017. Evaluation of essential oil of *Cymbopogon distans* and *Cinnamomum tamala* against plant pathogenic fungi. Indian J Agric Res 51, 191–193.

Nair EV. 1977. Essential oil of east Indian lemon grass present position in India and scope of its development. In: Cultivation and Utilization of Medicinal and Aromatic Plants, Atal CK, Kapur BM (eds.). Regional Research Laboratory, Jammu-Tawi, pp. 204–206.

Nayak S. 1996. Plant regeneration from callus culture of *Cymbopogon* (Jamrosa). J Herbs Spices Med Plants 4, 39–46.

Nayak S, Debata BK, Sahoo S. 1996. Rapid propagation of lemongrass (*Cymbopogon flexuosus* (Nees) Wats.) through somatic embryogenesis *in vitro*. Plant Cell Rep 15, 367–370.

Negrelle RRB, Gomes EC. 2007. *Cymbopogon citratus* (DC.) Stapf: Chemical composition and biological activities. Rev Bras Plant Med Botucatu 9, 80–92.

Ngamwathana MO, Kanchanapee P. 1971. Investigation into Thai medicinal plants said to cure diabetes. J Med Assoc Thai 54, 105–11.

Ngan TTK, Hien TT, Danh PH, Nhan LTH, Tien LX. 2020. Formulation of the Lemongrass (*Cymbopogon citratus*) essential oil-based eco-friendly diffuse solution. IOP Conf Ser Mater Sci Eng 959, 012024.

Nomier Y, Asaad GF, Alshahrani S, Safhi S, Medrba L, Alharthi N, Rehman Z, Alhazmi H, Sanobar S. 2021. Antidepressant and anxiolytic profiles of *Cymbopogon flexuosus* ethanolic extract in chronic unpredictable mild stress induced in rats. Biomed Pharmacol J 14, 175–185.

Oloyede IO. 2009. Chemical profile and antimicrobial activity of *Cymbopogon citratus* leaves. J Nat Prod 72, 98–103.

Patnaik J, Sahoo S, Debata BK. 1999. Somaclonal variation in cell suspension culture-derived regenerants of *Cymbopogon martinii* (Roxb.) Wats var. motia. Plant Breeding 118, 351–354.

Paviani L, Pergher SB, Dariva C. 2006. Application of molecular sieves in the fractionation of lemongrass oil from high-pressure carbon dioxide extraction. Br J Chem Eng 23, 219–222.

Pereira FO, Wanderley PA, Viana FAC, de Lima RB, de Sousa FB, dos Santos SG, Lima EO. 2011. Effects of *Cymbopogon winterianus* Jowitt ex Bor essential oil on the growth and morphogenesis of *Trichophyton mentagrophytes*. Braz J Pharm Sci 47, 145–153.

Plata-Rueda A, Rolim GDS, Wilcken CF, Zanuncio JC, Serrão JE, Martínez LC. 2020. Acute toxicity and sublethal effects of lemongrass essential oil and their components against the Granary Weevil, *Sitophilus granaries*. Insects 11, 379.

Praditvarn L, Sambhandharaksa C. 1950. A study of the volatile oil from Siam lemongrass. J Pharm Assoc Siam 3, 87–92.

Prins CL, de Paiva Freitas S, de Assis Gomes MdM, Vieira IJC, Gravina GdA. 2013. Citral accumulation in *Cymbopogon citratus* plant as influenced by N6-benzylaminopurine and light intensity. Theor Exp Plant Physiol 25, 159–165.

Puatanachokchai R, Kishida H, Denda A, Murata N, Konishi Y, Vinitketkumnuen U, Nakae D. 2002. Inhibitory effects of lemon grass (*Cymbopogon citratus*, Stapf.) extract on the early phase of hepatocarcinogenesis after initiation with diethylnitrosamine in male 344 rats. Cancer Lett 183, 9–15.

Quiala E, Barbón R, Capote A, Pérez N, Jiménez E. 2016. *In vitro* mass propagation of *Cymbopogon citratus* Stapf., a medicinal Gramineae. Methods Mol Biol 1391, 445–457.

Quiala E, Barbón R, Jiménez E, De Feria M, Chávez M, Capote A, Pérez N. 2006. Biomass production of *Cymbopogon citratus* (D.C.) Stapf., a medicinal plant, in temporary immersion systems. *In Vitro* Cell Dev Biol Plant 42, 298–300.

Quintans-Júnior LJ, Souza TT, Leite BS, Lessa NMN, Bonjardim LR, Santos MRV, Alves PB, Blank AF, Antoniolli AR. 2008. Phythochemical screening and anticonvulsant activity of *Cymbopogon winterianus* Jowitt (Poaceae) leaf essential oil in rodents. Phytomedicine 15, 619–624.

Quisumbing E. 1951. Medicinal Plants of the Philippines. Tech Bull 16. Rep Philippines. Thomas Publisher, Department of Agriculture and Natural Resources, Manilla.

Ragasa CY, Phuong Ha HK, Hasika M, Maridable JB, Gaspillo PD, Rideout JA. 2008. Antimicrobial and cytotoxic terpenoids from *Cymbopogon citratus* Stapf. Philipp J Sci 45, 111–122.

Rao RR, Jamir NS. 1982. Ethnobotanical studies in Nagaland. Indian medicinal plants. Soc Econ Bot 36, 176–181.

Rojas-Armas JP, Arroyo-Acevedo JL, Palomino-Pacheco M, Herrera-Calderón O, Ortiz-Sánchez JM, Rojas-Armas A, Calva J, Castro-Luna A, Hilario-Vargas J. 2020. The essential oil of *Cymbopogon citratus* Stapt and carvacrol: An approach of the antitumor effect on 7,12-dimethylbenz-[α]-anthracene (DMBA)-induced breast cancer in female rats. Molecules 25, 3284.

Runnie I, Salleh MN, Mohamed S, Head RJ, Abeywardena MY. 2004. Vasorelaxation induced by common edible tropical plant extracts in isolated rat aorta and mesenteric vascular bed. J Ethnopharmacol 92, 311–316.

Saha A, Basak BB, Manivel P, Kumar J. 2021. Valorization of Java citronella (*Cymbopogon winterianus* Jowitt) distillation waste as a potential source of phenolics/antioxidant: Influence of extraction solvents. J Food Sci Technol 58, 255–266.

Santos PL, Araújo AAS, Quintans JSS, Oliveira MGB, Brito RG, Serafini MR, Menezes PP, Santos MRV, Alves PB, de Lucca Júnior W, Blank AF, Rocca VL, Almeida RN, Quintans-Júnior LJ. 2015. Preparation, characterization, and pharmacological activity of *Cymbopogon winterianus* Jowitt ex Bor (Poaceae) leaf essential oil of β-cyclodextrin inclusion complexes. Evid Based Complement Altern Med 2015, 502454.

Santos PL, Brito RG, Oliveira MA, Quintans JSS, Guimarães AG, Santos MRV, Menezes PP, Serafini MR, Menezes IRA, Coutinho HDM, Araújo AAS, Quintans-Júnior LJ. 2016. Docking, characterization and investigation of β-cyclodextrin complexed with citronellal, a monoterpene present in the essential oil of *Cymbopogon* species, as an anti-hyperalgesic agent in chronic muscle pain model. Phytomedicine 23, 948–957.

Schaneberg BT, Khan IA. 2002. Comparison of extraction methods for marker compounds in the essential oil of lemon grass by GC. J Agric Food Chem 50, 6, 1345–1349.

Selim SA. 2011. Chemical composition, antioxidant and antimicrobial activity of the essential oil and methanol extract of the Egyptian lemongrass *Cymbopogon proximus* Stapf. Grasas y Aceites 62, 55–61.

Shah G, Shri R, Panchal V, Sharma N, Singh B, Mann AS. 2011. Scientific basis for the therapeutic use of *Cymbopogon citratus*, stapf (Lemon grass). J Adv Pharm Technol Res 2, 3–8.

Sharma PR, Mondhe DM, Muthiah S, Pal HC, Shahi AK, Saxena AK, Qazi GN. 2009. Anticancer activity of an essential oil from *Cymbopogon flexuosus*. Chem-Biol Interact 179, 160–168.

Shasany AK, Lal RK, Darokar MP, Patra NK, Garg A, Kumar S, Khanuja SPS. 2000. Phenotypic and RAPD diversity among *Cymbopogon winterianus* Jowitt accessions in relation to *Cymbopogon nardus* Rendle. Genet Resour Crop Evol 47, 553–559.

Simic A, Rancic A, Sokovic MD, Ristic M, Grujic-Jovanovic S, Vukojevic J, Marin PD. 2008. Essential oil composition of *Cymbopogon winterianus* and *Carum carvi* and their antimicrobial activities. Pharm Biol 46, 437–441.

Singh A, Kumar A. 2017. Cultivation of Citronella (*Cymbopogon winterianus*) and evaluation of its essential oil, yield and chemical composition in Kannauj region. Int J Biotechnol Biochem 13, 139–146.

Soares MO, Alves RC, Pires PC, Beatriz M, Oliveira PP, Vinha AF. 2013. Angolan *Cymbopogon citratus* used for therapeutic benefits: Nutritional composition and influence of solvents in phytochemicals content and antioxidant activity of leaf extracts. Food Chem Toxicol 60, 413–418.

Somparn N, Saenthaweesuk S, Naowaboot J, Thaeomor A. 2014. Effects of *Cymbopogon citratus* Stapf water extract on rat antioxidant defense system. J Med Assoc Thai 97, S57–S63.

Somparn N, Saenthaweeuk S, Naowaboot J, Thaeomor A, Kukongviriyapan V. 2018. Effect of lemongrass water extract supplementation on atherogenic index and antioxidant status in rats. Acta Pharm 68, 185–197.

Souza-Formigoni ML, Lodder HM, Gianoĵ i FO, Ferreira TM, Carlini EA. 1986. Pharmacology of lemongrass (*Cymbopogon citratus* Stapf). Effects of daily two-month administration in male and female rats and in off spring exposed "in utero". J Ethnopharmacol 17, 65–74.

Spring MA. 1989. Ethanopharmacological analysis of medicinal plants used by Laotian Hmong refugees in Minnesota. J Ethnopharmacol 26, 65–91.

Taeckholm V. 1974. Students Flora of Egypt, 2nd ed. Cairo University Press, Cairo.

Tajidin NE, Ahmad SH, Rosenani AB, Azimah H, Munirah M. 2012. Chemical composition and citral content in lemongrass (*Cymbopogon citratus*) essential oil at three maturity stages, Afr J Biotechnol 11, 2685–2693.

Tarkang PA, Agbor GA, Tsabang N. 2012. Effect of long-term oral administration of the aqueous and ethanol leaf extract of *Cymbopogon citratus* (DC. Ex Ness) Stapf. Ann Biol Res 3, 5561–5570.

Tawatsin A, Wratten SD, Scott RR, Thavara U, Techadamrongsin Y. 2001. Repellency of volatile oils from plants against three mosquito vectors. J Vector Ecol 26, 76–82.

Thappa RK, Dhar KL, Atal CK.1979. Isointermedeol, a new sesquiterpene alcohol from *Cymbopogon flexuosus*. Phytochemistry 18, 671–672.

Trang DT, Hoang TKV, Nguyen TTM, Cuong PV, Dang NH, Dang HD, Quang TN, Dat NT. 2020. Essential oils of lemongrass (*Cymbopogon citratus* Stapf) induces apoptosis and cell cycle arrest in A549 lung cancer cells. BioMed Res Int 2020, Article ID 5924856.

Tyagi AK, Malik A. 2010a. *In situ* SEM, TEM and AFM studies of the antimicrobial activity of lemon grass oil in liquid and vapour phase against *Candida albicans*. Micron 41, 797–805.

Tyagi AK, Malik A. 2010b. Liquid and vapour-phase antifungal activities of selected essential oils against candida albicans: Microscopic observations and chemical characterization of *Cymbopogon citratus*. BMC Complement Altern Med 10, 65.

Tyagi AK, Malik A. 2012. Morphostructural damage in food-spoiling bacteria due to the lemon grass oil and its vapour: SEM, TEM, and AFM investigations. Evid Based Complement Altern Med 2012, 692625.

Verma RS, Padalia RC, Chauhan A. 2013. Compositional variation in leaves and inflorescence essential oils of *Cymbopogon distans* (Steud.) Wats. from India. Nat Acad Sci Lett 36, 615–619.

Wany A, Jha S, Nigam VK, Pandey DM. 2013. Chemical analysis and therapeutic uses of citronella oil from *Cymbopogon winterianus*: A short review. Int J Adv Res 1, 504–521.

Wasuwat SA. 1967. List of Thai Medicinal Plants. Report No 1 on Res Project 17. ASRCT, Bangkok.

Wijesekara ROB. 1973. The chemical composition and analysis of *Citronella* oil. J Nat Sci Council Sri Lanka 1, 67–81.

Zahra AA, Hartati R, Fidrianny I. 2021. Review of the chemical properties, pharmacological properties, and development studies of *Cymbopogon* sp. Biointerface Res Appl Chem 11, 10341–10350.

Zhang JS, Zhao NN, Liu QZ, Liu ZL, Du SS, Zhou L, Deng ZW. 2011. Repellent constituents of essential oil of *Cymbopogon distans* aerial parts against two stored-product insects. J Agric Food Chem 59, 9910–9915.

Zhang M-M, Sun L-L, Li C, Gao W, Yang J-B, Wang A-G, Su Y-L, Ji T-F. 2014. A new lanostane-type triterpenoid from *Cymbopogon citratus*. Zhongguo Zhong Yao Za Zhi 39, 1834–1837.

16 *Cyperus* Species

16.1 MORPHOLOGICAL FEATURES, DISTRIBUTION, ETHNOPHARMACOLOGICAL PROPERTIES, PHYTOCHEMISTRY, AND PHARMACOLOGICAL ACTIVITIES

Cyperus rotundus L. (Fam.—Cyperaceae) is a perennial and monocot erect herb, with a slender, scaly, and creepy rhizome. The leaves are oblong-ovate, smooth, shiny, and dark green, and three-angled. Its stem is bulbous at the bottom and arises separately from the tuber. The plant height is approximately 1.4 m. Inflorescence is spikelet; each spikelet 3.5 cm in length and consists of 10 to 40 flowers. The color of unripe tuber is yellow and black on ripening (Wills 1987; Galinato et al. 1999; Hall et al. 2009; Himaja et al. 2014). It is naturalized in Algeria, Egypt, Morocco, Tunisia, Ethiopia, Somalia, Sudan, Kenya, Tanzania, Uganda, Rwanda, Niger, Nigeria, Angola, Malawi, Mozambique, Zambia, Zimbabwe, Madagascar, Mauritius, Afghanistan, Iran, Iraq, Saudi Arabia, Turkey, Kyrgyzstan, Turkmenistan, Uzbekistan, China, Japan, Korea, Taiwan, India, Pakistan, Sri Lanka, Myanmar, Malaysia, Switzerland, Albania, Bulgaria, Greece, France, Portugal, Spain, the United States, Mexico, Brazil, and Argentina (Al-Snafi 2016). It is used in the treatment of stomach complaints, nausea, vomiting, food poisoning, indigestion, and bowel irritation (Talukdar et al. 2011; Yeung 1985; Duke and Ayensu 1985). Its tubers are useful in fever, malaria, cough, bronchitis, amenorrhea, dysmenorrhea, deficient lactation, insect bites, bronchitis, infertility, and menstrual diseases (Bown 1995; Chopra et al. 1956; Bajpay et al. 2018). In Ayurvedic medicine, its rhizome is used as an astringent, diuretic, analgesic, antispasmodic, aromatic, carminative, antitussive, emmenagogue, sedative, stimulant, and stomachic agent (Jiangsu New Medical College 1971; Sivapalan and Jeyadevan 2012; Sivapalan 2013; Kamala et al. 2018b).

Ethanol extract of *C. rotundus* rhizome (500 mg/kg) showed maximum inhibition (36%) to carrageenan-induced paw edema. The inhibition was compared with aspirin (300 mg/kg). Similarly on the same dose, it displayed significant antiulcer effect (41.2%) and results compared with cimetidine (positive control). The ethanol extract displayed mild reduction in open field, head dip, rearing traction, and forced swimming tests, demonstrating slight muscle relaxant activity (Ahmad et al. 2014; Pirzada et al. 2015). The antiulcer activity of methanol extract (70%) of *C. rotundus* rhizome was evaluated in aspirin-induced (400 mg/kg) gastric ulcers. The extract (500 mg/kg) significantly suppressed the aspirin-stimulated gastric ulceration in treated animals (53.15%). The activity was compared with the standard gastric ulcer drug ranitidine. The extract significantly enhanced the activity of superoxide dismutase, cellular glutathione, and glutathione peroxidase but suppressed the lipid peroxidation in the gastric mucosa of ulcerated rats. *C. rotundus* methanol extract possess the ability to suppress aspirin-stimulated gastric ulcers via activating an antioxidant defense system; therefore, the extract may be used as gastroprotective agent (Thomas et al. 2015).

The aqueous extract of *C rotundus* rhizome was examined for their lactogenic activity. The aqueous extract (600 mg/kg, p.o.) stimulated more milk (40%) in treated animals than animals of the control group. The extract also increased the weight in treated pups and mothers. In addition, extract also increased the protein and carbohydrate contents of mammary gland tissues in treated animals significantly. The extract also showed significant stimulatory effect on the formation of prolactin. Moreover, extract displayed significant effect on obvious lobulo-alveolar development with milk secretion in the mammary gland tissues. The extract did not show any toxic effect in treated animals. The study reveals that the aqueous extract induced milk formation in the female animals, therefore, may be effective in enhancing the lactation of humans (Badgujar and Bandivdekar 2015).

DOI: 10.1201/9781003398035-16

The total contents of flavonoids and phenolics were estimated in methanol and ethanol extracts of *C. rotundus* rhizome. The methanol extract contains higher levels of flavonoid (8.15–18.25 mg catechin equivalent/g/dry matter) and phenolic contents (27.40–37.85 mg gallic acid equivalent/g/ dry matter) than ethanol extract (6.44–13.77 mg catechin equivalent/g/dry matter for flavonoids and 25.21–30.23 mg gallic acid equivalent/g/dry matter for phenolic contents). Methanol extract also showed higher suppression of linoleic acid system (32.50%–48.17%) than ethanol extract (25.20%–45.53%). Methanol extract (51.50%–61.73%) also displayed greater DPPH free radical scavenging ability than ethanol extract (38.37%–47.86%). Similarly, methanol extract also showed higher potential (0.754–1.112) in reducing power assay than ethanol extract (0.711–0.837); therefore, it may be used as a potent antioxidant agent (Bashir et al. 2012).

Acetone extract (70%) of *C. rotundus* rhizome was analyzed for the presence of phenolic, flavonoid, and proanthocyanidin contents. The different levels of total flavonoid (0.036 ± 0.002 to 118.924 ± 5.946 µg/mg extract) and total phenolic (7.196 ± 0.359 to 200.654 ± 10.032 µg/mg extract), total proanthocyanidin (13.115 ± 0.656 to 45.901 ± 2.295 µg/mg extract) contents were estimated in acetone extract. The levels of total flavonoid content, total phenolic content, and total proanthocyanidin content were found to be the maximum in 70% acetone extract. Acetone (70%) and methanol (70%) extracts showed maximum radical scavenging effect. The presence of (2)-acetyl-3 (5)-styryl-5 (3)-methylthiopyrazole was confirmed by GC-MS in acetone (7%) extract. The study reveals that both extracts (acetone and methanol) possess antioxidant potential and therefore may be used as potent antioxidant agents (Kamala et al. 2018a). The aqueous extract of *C. rotundus* rhizome powder showed the presence of flavonoid (24.30 mg catechin equivalent/g) and polyphenol (353.10 mg gallic acid equivalent/g) contents. The aqueous extract also displayed significant inhibitory effect against *E. coli* (inhibition zone 16.3 mm) and *S. aureus*, (inhibition zone 11.7 mm; Eltilib et al. 2016).

The hyperlipidemic effect of ethanol extract of *C. rotundus* rhizome was evaluated on hyperlipidemia stimulated by carbimazole and cholesterol in male Wistar rats. Ethanol extract (500 mg/kg) and simvastatin (5 mg/kg) significantly ($p = 0.05$) decreased total cholesterol, triglycerides, low density lipoproteins, low density lipoproteins/high-density lipoproteins ratio, total non-high-density lipoproteins, and cholesterol but significantly ($p = 0.05$) enhanced the level of high-density lipoproteins-cholesterol. The study results reveal that *C. rotundus* rhizome contains bioactive compounds that have hypolipidemic potentials and hence, may be used as a hypolipidemic agent (Okwu et al. 2015).

Antiplatelet effects of ethanol extract and (+)-nootkatone of *C. rotundus* rhizome were evaluated on rat platelet aggregations *in vitro* and *ex vivo*, and on mice-tail bleeding times. Ethanol extract displayed significant and concentration-dependent suppressive effects on collagen-, thrombin-, and/ or arachidonic acid-stimulated platelet aggregation. The identified compound (+)-nootkatone, displayed significant inhibitory effect on collagen-, thrombin-, and arachidonic acid-stimulated platelet aggregation. Moreover, ethanol extract and (+)-nootkatone significantly extended the bleeding times in treated rats. In addition, (+)-nootkatone showed significant suppressive effect on rat platelet aggregation *ex vivo*. The study exhibits antiplatelet effects of ethanol extract and (+)-nootkatone; therefore, extract and isolated compound might be used as a therapeutic agent for the treatment of platelet-associated cardiovascular disorders (Seo et al. 2011).

The inhibitory effect of methanol extract of *C. rotundus* rhizomes was evaluated against α-glucosidase and α-amylase activities. Four compounds [(2RS,3SR)-3,4′,5,6,7,8-hexahydroxyflavane, cassigarol E, scirpusin A and B] were identified from methanol extract. The cassigarol E suppressed both α-glucosidase and α-amylase activities, whereas (2RS,3SR)-3,4′,5,6,7,8-hexahydroxyflavane only displayed inhibitory effect on α-amylase. Scirpusin A and B were found effective on α-glucosidase activity. The identified compounds may be considered as potent suppressors of α-glucosidase activity (Tran et al. 2014). Hydro-alcoholic extract of *C. rotundus* (500 mg/kg b.w.; once a day for seven days) significantly reduced the levels of blood glucose in alloxan-stimulated hyperglycemia in rats. This antihyperglycemic effect can be associated with its antioxidant effect as it displayed strong DPPH radical scavenging activity *in vitro* (Raut and Gaikwad 2006). Six compounds (cyperusphenol A, mesocyperusphenol A, cyperusphenol D, scirpusins B, scirpusins A, sugetriol) have been identified from aqueous-methanol extract of *C. rotundus* rhizome. Among isolated compounds, cyperusphenol A (IC_{50} 1.43 ± 0.11 µM), mesocyperusphenol A (IC_{50} 1.44 ± 0.17 µM), and cyperusphenol D (IC_{50}1.18 ± 0.11 µM) showed potent α-glucosidase inhibition activity (Deng et al. 2019).

Ethyl acetate extract and its crude fractions (ether and ethyl acetate) of *C. rotundus* rhizomes were examined for their hepatoprotective activity in carbon tetrachloride-damaged liver of rats. The ethyl acetate extract (100 mg/kg) demonstrated significant protective activity by reducing serum levels of glutamic oxaloacetic transaminase, glutamic pyruvic transaminase, alkaline phosphatase, and total bilirubin. The results were compared with silymarin (Suresh Kumar and Mishra 2005).

The anti-inflammatory effect of ethanol extract of *C. rotundus* leaves was evaluated in carrageenan-induced (acute model) paw edema and implantation of cotton pellets in the subplantar region. The extract (300 mg/kg) demonstrated significant anti-inflammatory effect by suppressing the volume of paw edema in treated animals. The extract (300 mg/kg) decreased the wet weight of the implanted granuloma (37.5%) significantly in treated rats. As per this study, ethanol extract possesses significant anti-inflammatory effect (Sundaram et al. 2008). Similarly, anti-inflammatory effect of three extracts (ethanol, ether and aqueous) of *C. rotundus* rhizome (200, 400 mg/kg b.w.) was also evaluated in carrageenan-induced paw edema in rats. Ethanol extract (400 mg/kg b.w.) displayed maximum suppression (%) to volume of paw edema. The study exhibits anti-inflammatory activity of ethanolic extract and displays equal effect to standard anti-inflammatory drug (Chithran et al. 2012). The analgesic, anti-inflammatory, and genotoxic effects of aqueous, ethyl acetate, methanol, and total oligomer flavonoids-enriched extracts *C. rotundus* aerial parts were evaluated in different models. All the tested extracts were able to decrease the mouse ear oedema as induced by xylene. The aqueous, ethyl acetate, methanol, and total oligomer flavonoids-enriched extracts (300 mg/kg b.w.) displayed significant suppressive effect against ear oedema (inhibition percentages 74.38%, 62.73%, 44.6%, and 77.25%). The results were compared with the positive control, dexamethasone (300 mg/kg, b.w.; inhibition percentage 68.81%). Similarly, aqueous, ethyl acetate, and methanol extracts (300 mg/kg, b.w.) showed different percentages of inhibition (65.7%, 63% and 43) to number of writhing as induced by acetic acid. The total oligomer flavonoids-enriched extract (300 mg/kg b. w.) showed maximum inhibitory effect (92.8%). The results were compared with diclofenac sodium (100 mg/kg, b.w.) which showed 100% reduction. The aqueous, ethyl acetate, methanol, and total oligomer flavonoids-enriched extracts, with the different extract concentrations, suppressed iron-induced lipid peroxidation in rat liver homogenate. The different doses (300, 150, and 50 µg/ml) of *C. rotundus* extracts did not show any toxicity in animals. The study results show that *C. rotundus* extracts consist of bioactive compounds, such as flavonoids, that may potentially be useful in modulating the immune cell functions, exerting analgesic, anti-inflammatory, and antioxidant activities (Soumaya et al. 2013). The topical anti-inflammatory effect of ethanolic extract of *C. rotundus* rhizome was evaluated in acute and chronic dermatitis models. The chlorogenic acid (45 µg/g) was isolated and characterized from ethanol extract. Ethanol extract, topically, decreased ear edema and cellular infiltrate in acute and chronic skin inflammation models (arachidonic acid and 12-*O*-tetradecanoylphorbol-13-acetate-stimulated skin inflammation). In addition, the extract also showed significant decrease in 12-*O*-tetradecanoylphorbol-13-acetate-stimulated keratinocyte hyperproliferation in topically treated animals. During topical treatment, extract did not produce any skin atrophy or alteration in lymphoid organ weight. Based on results, it is suggested that the extract may be used as a potent therapeutic agent for the treatment of inflammatory skin diseases (Rocha et al. 2020).

The methanol and aqueous extracts of *C. rotundus* aerial part were examined for their *in vitro* antioxidant and apoptotic activities. Both (methanol and aqueous) extracts (300, 150, and 50 µg/ml) were examined for their antioxidant effect in xanthine/xanthine oxidase assay systems. Aqueous (800, 400, and 200 µg/ml) and methanol extracts (350, 175, and 88 µg/ml) were screened against lipid peroxidation (stimulated by 75 μM H_2O_2) assay also. The cytotoxic effect of both extracts was tested against K562 and L1210 cell lines. The methanol and aqueous extracts displayed significant inhibitory effect (88% and 19%) in xanthine oxidase and lipid peroxidation (61.5% and 42.0%) assays. Both extracts (16 mg/ml) also suppressed the formation of hydroxyl radicals (27.1% and 25.3%) significantly. Only methanol extract stimulated the degradation of DNA. Orientin was reported as the major constituent identified from the butanol fraction of methanol extract. The results reveal that *C. rotundus* extracts demonstrated a potential use as a natural antioxidant and an apoptosis-inducing agents (Soumaya et al. 2014). The total of oligomers flavonoid-rich and ethyl acetate extracts showed significant inhibition in xanthine oxidase (IC_{50} 240 and 185 μg/ml and superoxide anion (IC_{50} 150 and 215 μg/ml) assays. Both extracts were found effective in decreasing the formation of thiobarbituric acid reactive substances and showed significant protection against H_2O_2/UV-photolysis–stimulated DNA injury. Ethyl acetate extract showed the strongest activity (equivalents of malonaldehyde concentration = 2.04 nM). Moreover, the total oligomer's flavonoid-rich extract displayed significant inhibition on K562 cells via induction of apoptosis. Three major constituents (catechin, afzelechin, and galloyl quinic acid) were identified from the total oligomer's flavonoid-rich extract while five major constituents (luteolin, ferulic acid, quercetin, 3-hydroxy, 4-methoxy-benzoic acid, and 6,7-dimethoxycoumarin) were isolated and characterized from ethyl acetate extract. Luteolin showed significant reduction in the formation of thiobarbituric acid reactive substances (malonaldehyde equivalent = 1.5 nM), significant suppression in the proliferation of K562 cells (IC_{50} = 25 μg/ml) and significant protection against H_2O_2/UV-photolysis–stimulated DNA injury (Soumaya et al. 2009). The total oligomers flavonoids and ethyl acetate extracts of *C. rotundus* tubers showed significant antibacterial activity against *S. enteritidis*, *S. aureus*, and *E. faecalis*. Moreover, both extracts displayed a significant potential to suppress the reduction of nitro-blue tetrazolium by the superoxide radical in a nonenzymatic superoxide generating assay. Both extracts inhibited the growth and proliferation of L1210 cells derived from murine lymphoblastic leukaemia (Kilani et al. 2009).

Four biflavone constituents (amentoflavone, ginkgetin, isoginkgetin, and sciadopitysin) were isolated and characterized from *C. rotundus* rhizome and their antitumor effect against uterine tumors were investigated. Amentoflavone significantly reduced the uterine coefficient and the levels of serum estrogen in rats with uterine fibroids, but enhanced the pathological conditions of uterine tissues of treated rats. The compound also decreased the number of Bcl-2- and Bax-positive dots in smooth muscles and significantly suppressed the tumor-like proliferation in model rats ($p < 0.01$). Amentoflavone also reduced the concentration of serum estradiol and serum progesterone and uterine homogenate NOS activity in treated animals significantly ($p < 0.05$). Amentoflavone possessed significant suppressive effect on uterine tumors in rats by increasing the expression of Bax protein, downregulating Bcl-2, forming homodimers Bax/Bax, and decreasing plasma estradiol and progesterone to promote apoptosis of uterine fibroid cells (Ying and Bing 2016). Patchoulenone, caryophyllene α-oxide, 10,12-peroxycalamenene, and 4,7-dimethyl-1-tetralone were isolated from *C. rotundus* tubers. 10,12-peroxycalamenene demonstrated significant antimalarial activity EC_{50} 2.33×10^{-6} M (Thebtaranonth et al. 1995).

C. scariosus R. Br is a glabrous herb with angular soft and slender stem and underground rhizomatous tubers. The stem is long (40–90 cm), slender, and triquetrous. The leaves are usually short, narrow, and weak. The inflorescence is umbel and slender (3 inch long). The rhizome is very short, woody, and stolons (Kasana et al. 2013). It is distributed in different regions of Bangladesh, eastern and southern parts of the Indo-Pak subcontinent, South Africa, China, and Pacific islands (Watt 1972; Srivastava et al. 2014). Its roots are used as a tonic, desiccant, emmenagogue, diaphoretic, and vermifuge (Chopra et al. 1956; Said 1982). It is also useful in diarrhea, epilepsy, fever, gonorrhea, liver damage, and syphilis (Kritikar and Basu 1918).

The aqueous-methanol extract of *C. scariosus* (3–10 mg/kg) exerted significant hypotensive and bradycardiac effects. The extract inhibited the spontaneous contractions of guinea-pig paired atria, rat uterus, and rabbit jejunum in a concentration-dependent (0.1–1 mg/ml) manner. It also suppressed histamine or acetylcholine-stimulated contractions of guinea pig ileum showing nonspecific spas-molytic activity. The extract suppressed norepinephrine (10 μM) as well as K⁺ (80 mM)-stimulated contractions in rabbit aorta (concentrations 0.1–1 mg/ml). The study results suggest that C. *scariosus* extract consists of Ca^{2+} channel blocker–like compounds, which may be associated with general spasmolytic effects (Gilani et al. 1994).

The hepatoprotective effect of water-methanolic extract of C. *scariosus* was evaluated against acetaminophen and carbon tetrachloride–stimulated liver damage. The extract (500 mg/kg) significantly reduced (30%) the mortality rate in acetaminophen-induced rats. Acetaminophen (640 mg/kg) caused liver injury in rats as displayed by an increase in the levels of serum alkaline phosphatase (430 ± 68 IU/l), glutamate oxaloacetate transaminase (867 ± 305 IU/l), and glutamate pyruvate transaminase (732 ± 212 IU/l; $n = 10$), respectively, which were0020compared with the respective control (202 ± 36, 59 ± 14, and 38 ± 7 IU/l). The extract (500 mg/kg) significantly reduced the levels of serum alkaline phosphatase, glutamate oxaloacetate transaminase, and glutamate pyruvate trans-aminase ($p < 0.05$). The extract (500 mg/kg) significantly blocked ($p < 0.05$) the carbon tetrachloride-induced increase in serum enzymes. The extract also blocked carbon tetrachloride-induced extension in pentobarbital sleeping time showing hepatoprotectivity. The study results reveal that C. *scariosus* extract has hepatoprotective activity and thus may be used in the treatment of hepatobiliary diseases (Gilani and Janbaz 1995).

Twelve compounds (stigmasterol, β-sitosterol, lupeol, gallic acid, quercetin, β- amyrin, oleanolic acid, β-amyrin acetate, 4-hydroxyl butyl cinnamate, 4-hydroxyl cinnamic acid, caffeic acid, and kaempferol) have been identified from C. *scariosus*. The gallic acid and quercetin showed strong radical scavenging effect (IC_{50} 0.43 and 0.067 μg/ml) in DPPH and ABTS assays. Gallic acid, quercetin, and 4-hydroxyl cinnamic acid displayed significant antidiabetic effects, whereas lupeol displayed strong IL-1 β activity suppression in THP-1 monocytic cells and also showed significant ($p < 0.0025$) *in vivo* anti-inflammatory effect (Kakarla et al. 2016).

Methanolic extract of C. *scariosus* rhizome was tested for its antioxidant and anti-inflammatory activities against different models. The antioxidant activity of methanolic extract (100 μg/ml) was like that of ascorbic acid (50 μg/ml). It indicates that methanol extract possesses potent antioxidant effects, associated with the presence of phenolic constituents. Anti-inflammatory activity of methanol extract was determined by recording suppression (percentage) of denaturation of bovine serum albumin. The extract suppressed bovine serum albumin denaturation in a dose-dependent manner (50–5000 μg/ml). A total of nine compounds [1,5-diphenyl-2H-1,2,4-triazoline, 2-propene-1- one,3-(4-nitrophenyl), phenylacetamide N-ethyl-N-(3-methyl), 6-(2-aminophenyl)- 1, 2, 4-triazine, benzene-1, 2-diol, 4-(4-bromo-3-chloro), thiazolidine-4-one, 2-(4-bromophenyl), (E)-2- bromobutyloxychalcone, N-methyl-1-adamantaneacetamide, 2- ethylacridine] were determined by GC-MS analysis from methanol extract. N-methyl-1-adamantaneacetamide and 1,5, diphenyl-2H-1,2,4- triazine make a hydrogen bond interaction with Ser-530 and Tyr-385, respectively, and showed similar interactions with the crystal structure of diclofenac bound COX-2 protein. Benzene-1, 2-diol, 4-(4-bromo-3 chlorophenyl iminomethyl makes hydrogen bond interactions with Thr-199 and Thr-200, indistinguishable from crystalized COX-2 protein with valdecoxib. The study results confirm that methanol extract consists of medicinally important anti-inflammatory constituents; hence, this advocates the use of this plant in the treatment of inflammation-associated diseases (Kakarla et al. 2014).

C. *esculentus* L is an erect herb with triangular stem, growing up to 20 to 60 cm high. The plant's superficial rhizomes contain a stored form of proteins and starches. The leaves are yellowish green. The plant has a complex, shallow underground system consisting of thin fibrous roots and scaly rhizome (Mulligan and Junkins 1976; AbdelKader et al. 2017). It is naturalized in the Mediterranean region, Africa, India, North America, Mexico, Peru (Holm et al. 1977; Schippers et al. 1995; CABI 2020), Nigeria, Senegal, Ghana, Spain, and other regions of Europe (Guillerm 1987; Ezeh et al. 2014; Castro et al. 2015), the northwest region of Turkey, and northeast parts of Anatolia (Defelice 2002; Bazine and Arslanoğlu 2020). Its tubers are used in the treatment of bone fractures and tissue injuries. Due to the presence of higher levels of phosphorus, potassium, calcium, magnesium, and iron, the plant species is useful in body development (Mohdaly 2019). The plant species possesses a protective effect against cardiovascular disorders and cancers (Gambo and Da'u 2014). The tuber contains dietary fibers; therefore, it is effective in colon cancer, coronary health disorders, obesity, and diabetes (Achoribo and Ong 2017). Its rhizome is known as an aphrodisiac, carminative, diuretic, emmenagogue, stimulant, and tonic agent. It is used in the treatment of flatulence, indigestion, diarrhea, dysentery, and excessive thirst (Adejuyitan 2011; Yang et al. 2016).

The inhibitory effect of aqueous extract of *C. esculentus* was evaluated on α-amylase and α-glucosidase activities. The extract strongly suppressed the *α*-amylase and *α*-glucosidase activities, respectively. The suppression was concentration-dependent with respect to its inhibitory concentrations (IC_{50} 5.19 and 0.78 mg/ml). The results were compared with the control (3.72 and 3.55 mg/ml). Aqueous extract showed strong free radical scavenging effects in a DPPH assay. The antioxidant effect is attributed with the presence of phytoconstituents (Sabiu et al. 2017). Ethanol extract of roasted *C. esculentus* tubers was evaluated for their antioxidant potential in the 2,2-diphenyl-1-picrylhydrazyl and 2,2'-azino-bis-3-ethylbenzothiazoline-6-sulphonic acid assays. The extract consists of a significant amount of total phenolic contents (45.67 gallic acid equivalents mg/100 ml). The extract showed significant inhibitory effect in both assays (Badejo et al. 2014). The water extracts of *C esculentus* tubers and *Adansonia digitata* fruit powder were mixed together and evaluated for their antioxidant activity in 2,2'-azino-bis(3-ethylbenzothiazoline-6-sulfonic acid) and 1,1-diphenyl-2-picrylhydrazyl assays. The addition of *A. digitata* fruit extract significantly ($p < 0.05$) enhanced the total phenolic (18%; from 31.06 to 36.83 mg gallic acid equivalent/100 ml) and the flavonoid contents (~15% to 20%). The aqueous extract possesses potential for use as a natural antioxidant and might be suggested for consumers with diets deficient in Ca and K (Badejo et al. 2020). The maximum level of δ-tocopherol content was recorded in *C esculentus* oil extracted with petroleum ether (54.91 mg/100g), whereas the minimum level was determined in roasted tubers of *C esculentus* oil extracted with *n*-hexane (50.77 mg/100 g; Aljuhaimi et al. 2018; Yang et al. 2018).

Blood glucose reducing potential of defatted soybean and tigernut flour was examined in streptozotocin-induced diabetic rats. Plantain-based dough meals (60.5–71.9%) increased the blood-glucose-reducing potential in streptozotocin-induced diabetic rats that were compared with acarbose (69%). Therefore, dough-meals may be used in the management of diabetes (Oluwajuyitan and Ijarotimi 2019).

C. alternifolius L is a monoecious, perennial herb, and grows up to 1.5 m high. Stems are cylindrical, green, and smooth. The leaves are linear, and long (20 cm). Flowers are small, green, and clustered. It is distributed in the tropical region of Africa, South Africa, and Madagascar (Kern 1974; Townsend and Guest 1985). Its roots are used in the treatment of gastric complaints as well as in healing wounds (Burkill 1985).

Methanol and ethyl acetate extracts of *C. alternifolius* tubers and aerial parts were evaluated for their antiulcer effects. Both extracts (100 mg/kg b.w.) significantly decreased ulcer number, total ulcer score, and TNF-α content in the stomach of treated animals. Methanol or ethyl acetate extracts of tubers were more effective than ranitidine (antiulcer drug). Both methanol extracts of tubers and aerial parts contain higher levels of phenolic acids. Ethyl acetate extract of the aerial part was rich in aldehydes; therefore, extracts might be useful in the treatment of gastric ulcers (Farrag et al. 2019).

C. articulatus L is perennial herb and naturalized in north, northeast, and southeast of Brazil (Goetghebeur 1998). Its stalks form small tubers, exuding a fresh, woody, and spicy scent, traditionally used in scented baths and in the manufacture of artisanal colonies. It is distributed in northern Brazil, Africa, Latin America, Asia and Oceania, Cameroon, Central Africa Republic, Gabon, and Senegal (Nicoli et al. 2006; Zoghbi et al. 2006). The rhizome decoction is used for the treatment of headaches and migraines (Abubakar et al. 2000). Its essential oil is used in the cosmetic industry (Zoghbi et al. 2008). The rhizomes (Rukunga et al. 2008) and leaves (Akendengué 1992) of *C. articulatus* are used in the treatment of malaria (Ghafari et al. 2013). In the Indian medicine system, it has been used in the treatment of headaches, migraine, and epilepsy (Bum et al. 2003). Its rhizome decoction possesses sedative property in mice (Rakotonirina et al. 2001). In African and American medicine systems, its rhizomes are useful in fevers, pain, seizures, urinary disorders, bleeding, irregular menstruation, cancer, and inducing abortion (Bum et al. 1996; Milliken and Albert 1996; Nguta et al. 2015).

The hepatoprotective effect of methanol extract of *C. articulatus* was evaluated against paracetamol-induced liver damage in rats. Methanol extract (400 mg/kg. p.o., daily for 16 days) reduced the toxic effect of paracetamol and controlled it significantly by restoring the levels of biochemical parameters (serum glutamic pyruvic transaminase, serum glutamic-oxaloacetic transaminase, alkaline phosphatase), total protein, and total bilirubin. The extract showed enhancement in the antioxidant status to or toward near normal values. The study results reveal that the methanol extract of *C. articulatus* has hepatoprotective effect against paracetamol-stimulated hepatotoxicity in rats (Datta et al. 2013).

The anti-epileptic effect of *C. articulatus* was examined on excitatory amino acid receptors. Aqueous extract of *C. articulatus* rhizome decreased the spontaneous epileptiform discharges and N-methyl-D-aspartate-induced depolarizations in the rat cortical wedge preparation. The study supports the beneficial effects of *C. articulatus* in treating epilepsy by inhibiting the N-methyl-D-aspartate-mediated neurotransmission (Bum et al. 1996). Aqueous extract suppressed spontaneous epileptiform discharges induced in Mg^{2+} artificial cerebrospinal fluid. The extract (0.5 mg/ml) decreased the frequency of spontaneous epileptiform discharges while the same extract (2.2 mg/ml) produced a complete block of spontaneous events within 6 min. The effect was concentration dependent. Epileptiform discharges were decreased up to 70% at a concentration of 0.1 M and were completely prevented at a 0.3 M concentration (Bum et al. 2003). The anticonvulsant effect of methanol extract of *C. articulates* rhizome was evaluated against maximal electroshock- and pentylenetetrazol-induced seizures in rats. The extract showed significant protection against maximal electroshock [ED_{50} 1005 (797–1200) mg/kg i.p.}- and pentylenetetrazol-induced {ED_{50} 306 (154–541) mg/kg i.p.] seizures. It also slowed the onset of seizures stimulated by isonicotinic acid hydrazide and strongly antagonized N-methyl-D-aspartate-stimulated turning behavior. The methanolic extract displayed protection (54%) from seizures stimulated by strychnine (1000 mg/kg i.p. dose) but shows moderate effect against picrotoxin- or bicuculline-induced seizures. Therefore, *C. articulatus* rhizome may be used in traditional medicine for the treatment of epilepsy (Bum et al. 2001).

The decoction of *C. articulatus* rhizome did not show any anesthetic or paralyzing effects. Decoction significantly decreased spontaneous motor activity. Though, in comparison to diazepam, decoction did not show any muscle relaxant activity. The decoction showed sleep induction and enhanced the total sleep time without any concomitant analgesic effect. The study suggests that the decoction of *C. articulatus* rhizome possesses sedative properties (Rakotonirina et al. 2001).

Ethanol extract of *C. articulatus* was tested on pentylenetetrazol-induced seizures in mice and on a DPPH assay. The pentylenetetrazol-extract (150 mg/kg b. w.) reduced the seizure scores ($p < 0.01$), latency ($p < 0.01$), frequency ($p < 0.01$), and duration ($p < 0.01$), which compared with control group. The extract showed significant antioxidant effect in a DPPH radical ($IC_{50} = 16.9 \pm 0.1$ µg/ml) scavenging assay and trolox-equivalent antioxidant capacity (2.28 ± 0.08, mmol trolox/g). The extract (150 mg/kg) significantly increased the content of gamma amino butyric acid and malondialdehyde significantly ($p < 0.01$). The study suggests that extract possesses a strong effect on pentylenetetrazol-induced seizures, and in addition, it showed antioxidant effect by increasing levels of gamma amino butyric acid and malondialdehyde (Herrera-Calderon et al. 2017).

C. articulatus essential oil loaded with chitosan nanoparticles showed significant antimicrobial activity against *S. aureus* (ATCC6538), and *E. coli* (ATCC 8739; Kavaz et al. 2019). Essential oil of *C. articulatus* rhizome showed significant antimalarial effect against the two *P. falciparum* strains ($IC_{50} = 1.21$ µg/ml for W2 and 2.30 µg/ml for 3D7). Essential oil significantly decreased the parasitemia and anemia as stimulated by *P. berghei* in mice (da Silva et al. 2019).

C. conglomeratus Rottb is naturalized in coastal sand dunes of Egypt and the southern coast of Iran. It is used as a pectoral, anthelmintic, antidiarrheal, emollient, stimulant, carminative, tonic, diuretic, and analgesic agent (Abdel-Mogib et al. 2000). Similarly, *C. distans* Lf is an annual herb growing up to 0.5–1.4 m high and found commonly in damp locations, along with rivers, roadside ditches, and in coastal and midland areas of Kwa Zulu Natal (Morimoto et al. 1999).

16.2 PHYTOCHEMISTRY

Essential oils viz., α-pinene, camphene, β-thujene, β-pinene, *p*-cymene, δ-limonene, 1,8-cineole, linalool, pinocarveol, terpinenol, *trans*-pinocarveol, pinocarvone, terpinen-4-ol, *allo*-aromadendrene, terpineol, *p*-cymen-8-ol, myrtenol, verbenone, *trans*-carveol, carvone, bornyl acetate, *trans*-anethole, α-copaene, β-elemene, cyperene, gurjunene, γ-muurolene, α-selinene, β-selinene α-muurolene, γ-cadinene, isolongifolen-5-one, α-cadinene, α-calacorene, spathulenol, β-caryophyllene oxide, α-cyperone, aristolone, nootkanone, α-cubebene, caryophyllane-2–6-β-oxide, α-humulene, vulgarol B, *tau*-calamenene, caryophyllenol, isorotundene, isocyperol, cyperol, *tau*-cadinol, muurolol, α-cadinol, mustakone, cyperotundone, calamanene, nardol, humulene epoxide, eudesma 5-en-11-α-ol, *epi*-cubenol, and intermediol were identified from *C. rotundus* rhizomes. These essential oils showed significant antioxidant activities, protective effects against DNA injury, and cytotoxicity against the human neuroblastoma SH-SY5Y cell (Hu et al. 2017). Sixty essential oils viz., α-pinene, verbenene, β-pinene, *o*-cymene, limonene, 1,8-cineole, *p*-cymene, α-fenchol, *trans*-pinocarveol, *cis*-verbenol, pinocarvone, isopinocamphon, *p*-cymen-8-ol, α-terpineol, myrtenol, verbenone, *trans*-carveol, *trans*-myrtenyl acetate, cuminic aldehyde, carvone, cinnamaldehyde, *trans*-anethole, thymol, 2,4-decadienal, carvacrol, eugenol, α-ylangene, α-copaene, β-elemene, cyperene, β-caryophyllene, β-gurjunene, α-guaiene, aromadendrene, isoaromadendrene, α-humulene, α-caryophyllene, rotundene, γ-gurjunene, γ-muurolene, n-dodecanol, β-selinene, α-selinene, α-longipinane, α-farnesene, *cis*-γ-bisabolene, *trans*-calamenene, *trans*-γ-bisabolene, α-calacorene, β-calacorene, γ-elemene, spathulenol, caryophyllene oxide, humulene epoxide II, γ-gurjunene epoxide, aristolone, α-cyperone, n-hexadecanoic acid, phytol, and methyl linoleate were identified from tubers (Ghannadi et al. 2012); α-pinene, β-pinene, *cis*-1-isopropenyl-4-methylcyclohexane, 4-methyl-decane, 2-ethyl-hexanol, δ-limonene, eucalyptol, 3,7-dimethyl-undecane, 4,6-dimethyl-dodecane, tridecane, 26-dimethyl-undecane, *trans*-pinocarveol, (-)-verbenol, β-terpineol, pinocarvone, *cis*-pinocamphone, terpinen-4-ol, 3*E*-dodecene, myrtenol, hexadecane, 2,3,8-trimethyldecane, 2,7,10-trimethyl-dodecane, naphthalenone3,4,4a,5,6,7-hexahydro-1,1,4a-trimethyl-2(1H), β-muurolane, β-vatirenene, α-copaene, cyperene, *cis*-α-bisabolene, γ-muurolene, eicosane, α-selinene, 1,2, 3,3a,4,5,6,7–1, 4-azulene,1,2,3,3a, 4,5,6,7-octahydro-1,4-dimethyl-7-(,1-methylethanyl)-, 4a,5,6,7,8,8a-4, 8a-6-(1-)-2 (1H)-naphthalenone, 4a5,6,7,8,8a-hexahydro-4, 8a-dimethyl-6-(1-methylethenyl)-2(1H)-, 1,1,6–1,2-naphthalene,1,1,6-trimethyl-1,2-dihydro-, δ-cadinene, calamenene, (+)-spathulenol, 15-copaenol, 4Z,6Z,9Z-Z,Z,Z-4,6,9-nonadecatriene (-) α- (-) α-gurjunene, patchoulane, (+)-calarene, β-caryophyllene oxide, dehydro-aromadendrene, naphthalene, 6-isoproenyl-4, 8a-dimethyl-1,2,3,5,6,7,8,8a-octahydro-, 1-dodecen-1-ol, acetate, *cis*-α-bisabolene epoxide, 1,3,4-trimethyl-3-cyclohexene carboxaldehyde, α-farnesene, 2–6–1,7–3–2-methyl-6-methylene-octa-1, 7-dien-3-ol, alloaromadendrene, α-cyperone, isolongifolenone, and valencene were identified from *C. rotundus* rhizome (Chen et al. 2011; Tam et al. 2007).

Undecane, 3,7-dimethyl, 2-cyclohexen-1-one, 3,5,5-tri methyl, 1H-pyrazole, 4,5-dihydro-5,5-dimethyl-4-isopropylidene, 1-dodecene, 3-tetradecene, tridecanol, hexadecane, heptadecane, octadecane, phenol, 3,5-bis (1,1dimethyl ethyl), 1-hexadecene, cinnamyl tiglate, propanedinitrile, dicyclohexyl, cyclopropane carboxylic acid, 1-methyl-,2,6-bis (1,1-dimethyl ethyl)-4-methylphenyl ester, octadecanophenone, hexadecanophenone, tetradecanophenone, docosanoic acid, 1-docosene, isopropyl myristate, 1,2-benzenedicarboxylic acid, butyl octyl ester, 1,2-benzenedicarboxylic acid, butyl-8-methylnonyl ester, 1,2-benzenedicarboxylic acid, bis (2-methylpropyl) ester, pentadecanoic acid, 9-octadecenal, 9,12-octadecadienoicacid (Z, Z), 3-tetradecenal from hexane extract and heptane, 2,5,5-tri methyl, 1-octanol, 1-heptanol, cyclopropane, pentyl, 1-tridecene, 1-dodecanol, 3-hexadecene, (Z), heptadecane, 1-heptadecene, 3,4-hexanediol, 2,5-dimethyl, nonadecane, phosphonic acid, di octa decyl ester, 1-nonadecene, 1-pentadecanol were identified from chloroform extract of *C. scariosus*. Spiro[2.4] heptane, 1,2, 4,5- tetramethyl- 6-methylene, tricyclo [3.3.0.0 (2, 8)]octan-3-one, 5,8-dimethyl, naphthalene, 1,2,3,5,6,7,8,8α- octahydro-1,8α-dimethyl-7- (1-methyl ethenyl)-, [1s-(1α,7α, 8α), 1-heptatriacotanol, cyclo hexane, 1-methyl-2,4-bis (1-methyl ethenyl), kauren-18-ol, acetate, (4β), 5,9-undecadien-1-yne, 6,10-dimethyl, 1-eicosanol, 1-heptacosanol, dodecanoic acid, 3-hydroxy, 12-methyl-*E,E*-2,13-octadeca- dien-1-ol, Pregna-1,4,7,16-tetraene-3, 20-dione, H-cyclopropa[3,4]benzyl[1,2-e]azulene 4α,5,7β,9,9α(1αH)-pentol, 3-[(acetyloxy)methyl]-1β,4,5,7α, 8, 9-hexahydro-1,1,6,8 tetra methyl-, 5,9,9α, hexacosane, 5,8-dimethyl-1-naphtaline-dicarbonic acid, and 1,4-dimethyl ester were identified from hexane extract, and 1,4-methanocyclo octa[d] pyridazine, 1,4,4α, 5,6,9,10,10α-octa hydro-11,11-dimethyl-, (1α, 4α, 4α,10α), cyclobutene, 4,4-dimethyl-1- (2,7-octa di enyl), 2-methyl-4-(2,6,6-tri methylcyclohex-1-enyl) but-2-en-1-ol, 2-hydroxy-2,4,4-trimethyl-3- (3-methyl buta-1,3-dienyl) cyclo- hexanone, androstan-17-one, 3-ethyl-3-hydroxy-, (5α), limonene diepoxide and 1H-cyclopropa[3,4] benz[1,2-e] azulene-4α, 5,7β, 9,9α (1αH)-pentol, 3-[(acetyloxy) methyl]-1β, 4,5,7α, 8,9-hexahydro-1,1,6,8-tetramethyl-, 5,9,9α-tr, 1H-cyclopropa[3,4] benz[1,2-e] azulene-4α, 5, 7β, 9,9α (1αH)-pentol, 3-[(acetyloxy) methyl]-1β, 4,5,7α, 8,9-hexahydro-1,1, 6,8-tetramethyl-, 9,9α-dia, 5,8-dimethyl-1,4,6,7-tetrahydronaphtalin dicarbonic acid, 1,4-di methyl ester, 1-tridecanol, β-stigmasterol, stigmasterol methyl ether, tetratriacontane were determined by GC-MS analysis from chloroform extract of *C. rotundus* (Kakarla et al. 2015). n-tricont-1-ol-21-one, 18-epi-α-amyrin glucuronoside, oleanolic acid arabinoside, along with α-amyrin glucopyranoside and β -amyrin glucopyranoside were isolated and characterized from methanol extract of *C. rotundus* tubers (Alam et al. 2012).

Sixty-four essential oils viz., α-pinene, benzaldehyde, cyclopentene-3-ethylidene-1-methyl -, sabinene, β-pinene, *p*-cymene, 1-limonene, 1–8-cineole, *p*-cymenene, *trans*-pinocarveol, terpinen-4-ol, citronellal, 4,4-dimethyl-tricyclo-(3,2,1) octan-6-on, *p*-cymen-8-ol, 1-α-terpineol, myrtenal, *cis*-dihydrocarvone, myrtenol, verbenone, 1-β-4,4-trimethyl-bicyclo (3,2) hept-6-en-2-ol, *trans*-carveol, isocitronellol, carvone, carvenone, α-cubebene, dihydro-carvylacetate, α-copaene, isolongifoline, cyperene, *trans*-caryophyllene, α-humulene, patchoulane, dihydro-aromadendrene, aromadendrene-epoxide, iso-aromadendrene-epoxide, naphthalene, 1,6-dimethyl-4-(1-methyl ethyl), β-silenene, α-silenene, γ-cadenine, *cis*-calamenene, *trans*-calamenene, elema-1,3,11 (13)-trien-12-ol, β-calacorene, ledane, caryophyllene-oxide, *cis*-12-caryophyll-5-en-2-one, caryophylla-2(12), 6(13) dien-5-one, cyclohexane, 1,1,2-trimethyl,3,5 bis- (1-methyl ethyl), cyclo-hexenone, 2,3,3-trimethyl (3-methyl- 1,3 butadienyl), 2(H)-naphthalenone, 4a,5,6,7,8,8a hexahydro-7-isopropyl,4a β, 8a β-dimethyl, longiverbenone, 10-*epi*-α-cyperone, (+) oxo-α-ylangene, (+) α−cyperone, caryophyllenol, vulgarol, vellerdiol, aristolone, vulgarol B, ledenoxide, 2,10, dimethyl-7-isopropenyl-bicyclo- (4–4) dec-1-en-3-one, *cis*-tetradecenylacetate, longifolinaldehyde, and longipynocarvone were determined by GC-MS analysis from *C. rotuntdus* and *C. alopecuroides* Rottb tubers (El-Gohary 2004).

Eleven compounds (12-methyl cyprot-3-en-2-one-13-oic acid, *n*-dotriacontan-15-one, *n*-tetracontan-7-one, *n*-pentadecanyl linoleate, *n*-hexadecanyl linoleate, *n*-hexadecanyl oleate, *n*-pentacos-13′-enyl oleate, stigmasterol laurate, stigmasterol myristate, β-sitosterol-3β-*O*-glucoside and lupenyl 3β-*O*-arabinpyranosyl 2′-oleate) have been identified from methanol extract of *C. rotundus* rhizome (Sultana et al. 2017). Four compounds (methyl 3,4-dihydroxy benzoate, ipolamiide, 6β-hydroxyipolamiide, and rutin) were isolated and identified from methanol extract of *C. rotundus* rhizomes. Methanol extract and ethyl acetate fraction showed significant protection against carbon tetrachloride-injured liver in rats (reference drug; Mohamed 2015). The essential oils viz., α-pinene, camphene, β-pinene, myrcene, α-phellandrene, bicyclo [3.2.0] hept-6-ene, *p*-cymene, limonene, 1,8-cineole, terpinolene, perillene, 3,3,5-trimethyl cyclohexene, fenchol, *trans*-pinocarveol, camphene hydrate, pinocarvone, *p*-mentha-1, 5-diene-8-ol, borneol, terpinen-4-ol, myrtenol, verbenone, *trans*-carveol, cuminaldehyde, carvone, α-copaene, β-elemene, cyperene, β-caryophyllene, α-gurjunene, α-humulene, *allo*-aromadendrene, eudesma-2,4,11-triene, β-selinene, α-selinene, germacrene B, spathulenol, caryophyllene oxide, (2*R*,5*E*)-caryophyll-5-en-12-al, humulene epoxide II, oplopenone, globulol, patchenol, 2-cyclopropylthiophene, caryophylla-3,8(13)-dien-5-β-ol, vulgarol B, caryophylla-3,8(13)-dien-5-α-ol, caryophyllenol, aromadendrene epoxide, aristolone, α-cyperone, oxo-α-ylangene, solavetivone, nootkatone, hexadecanoic acid, phytol from rhizomes (Lawal and Oyedeji 2009a), sesquiterpenes viz., sugebiol 6-acetate, sugetriol 6,9-diacetate, α-rotunol, β-rotunol, (–)-eudesma-3,11-diene-5-ol, ligucyperonol, 14-hydroxy-α-cyperone, and britanlin E and 1β,4β-dihydroxyeudesma-11-ene have been identified from ethanol extract of *C. rotundus* rhizome (Kim et al. 2013).

A total of thirty-seven sesquiterpenes [cyperene-3, 8-dione, 14-hydroxy cyperotundone, 14-acetoxy cyperotundone, 3β-hydroxycyperenoic acid, sugetriol-3, 9-diacetate, cyperenol; cyperenoic acid, cyperotundone, sugeonol, scariodione, sugetriol triacetate, α-cyperone, (4a*S*, 7*S*),-7- hydroxy-1, 4a-dimethyl-7-(prop-1-en-2-yl),-4, 4a, 5, 6, 7, 8- hexahydronaphthalen-2 (3H),-one, (4a*S*, 7*S*, 8*R*),-8-hydroxy-1, 4a-dimethyl-7-(prop-1-en-2-yl), -4, 4a, 5, 6, 7, 8-hexahydronaphtha-len-2 (3H), -one, cyperol, 1β-hydroxy-α-cyperone, 10-epieudesm-11-ene-3β, 5α-diol, 3β-hydroxyilicic alcohol (11 (13),-eudesmene-3, 4, 12-triol), cyperusol C, α-corymbolol, 3β, 4α-dihydroxy-7-epieudesm-11 (13), -ene, 2-oxo-α-cyperone, rhombitriol, 7α (H), 10β-eudesm-4-en-3-one-11,12-diol, α-rotunol, 7-epiteucrenone, oxyphyllol C, nootkatone, 12-hydroxynootkatone, 5-hydroxylucinone, oplopanone, 10-hydroxyamorph-4-en-3- one, cyperusol D, 2- hydroxy-14-calamenenone, 1-isopropyl2, 7 dimethylnaphthalene, argutosine D, 4, 5-seco-guaia-1 (10), 11-diene-4, 5-dioxo] were identified from the rhizomes of *C. rotundus*. Nine eudesmane-type sesquiterpenoids (cyperol, 1β-hydroxy-α-cyperone, 10-epieudesm-11-ene-3β, 5α-diol, 3β-hydroxyilicic alcohol (11 (13),-eudesmene-3, 4, 12-triol), cyperusol C, α-corymbolol, 3β, 4α-dihydroxy-7-epieudesm-11 (13), -ene, 7α (H), 10β-eudesm-4-en-3-one-11,12-diol(23), rhombitriol) significantly suppressed HBV DNA replication (IC$_{50}$ 42.7±5.9, 22.5±1.9, 13.2±1.2, 10.1±0.7, 14.1±1.1, 15.3±2.7, 13.8±0.9, 19.7±2.1 and 11.9±0.6 µM). 10-epieudesm-11-ene-3β, 5α-diol, 3β, 4α-dihydroxy-7-epieudesm-11 (13), -ene, 7α (H), 10β-eudesm-4-en-3-one-11,12-diol, rhombitriol demonstrated high SI values (250.4, 125.5, >259.6 and 127.5). *C. rotundus* ethanol extract contains anti-HBV compounds (Xu et al. 2015). The presence of essential oils viz., α-pinene, δ-Cadinene, β-Pinene, isorotundene, ocimenone, isocyperol, borneol, cyperol, cyprotene, *tau*-Cadinol, α-cubebene, α-muurolol, cyperene, *tau*-muurolol, caryophyllene-2,6-β-oxide, α-cadinol, α-humulene, mustakone, β-selinene, cyperotundone, *tau*-calamenene, and α-cyperone was determined by GC-MS analysis from *C. rotundus* aerial parts (Aghassi et al. 2013).

Two sesquiterpenes (corymbolone and mustakone) were isolated and characterized from the chloroform extract *C. articulatus* rhizome. Mustakone demonstrated significant antimalarial activity against the sensitive strains of the *Plasmodium falciparum* (Rukunga et al. 2008). α-bulnesene, cadalene, cyperotundone, *cis*-thujopsenal, cyclocolorenone, corymbolone, hexadecanoic acid ethyl ester, 9,12-octadecadienic acid ethyl ester, 9-octadecenoic acid ethyl ester, and cholesta-3,5-diene were identified from ethanol extract of *C. articulatus* rhizome. The extract demonstrated moderate antimalarial activity against the two strains of *P. falciparum* (IC$_{50}$ 1.21 ± 0.01 µg/ml against the W2 strain and IC$_{50}$ 1.10 ± 0.06 µg/ml against the 3D7 strain; de Assis et al. 2020). Twenty-seven essential oils [α-pinene, verbenene, β-pinene, p-cymene, limonene, isopinocarveol, β-phellandren-8-ol, α-phellandren-8-ol, terpinen-4-ol, α-terpineol, myrtenol, verbenone, α-copaene, β-elemene, α-gurjunene, β-caryophyllene, β-copaene, caryophyllene oxide, β-copaen-4-α-ol, spathulenol, globulol, muskatone, cyperol, pogostol, (*E,E*)-farnesol, cyclocolorenone, (*E*)-isogeraniol] were identified from *C. articulatus* rhizome. These essential oil displayed significant cytotoxicity (IC$_{50}$ 28.5 µg/ml for HepG2; IC$_{50}$ > 50 µg/ml for HCT116, and an IC$_{50}$ 46.0 µg/ml for MRC-5) against tested cancer lines. Essential oil enhanced the percentage of apoptotic-like cells; *in vivo* tumor mass suppression rates were 46.5%–50.0% (Nogueira et al. 2020). α-pinene, camphene, thuja-2,4(10)-diene, sabinene, *p*-pinene, *p*-cymene, limonene, 1,8-cineole, *p*-cymenene, α-campholenal, *trans*-pinocarveol, *trans*-verbenol, *cis*-verbenol, pinocarvone, *p*-mentha-1,5-dien-8-ol, terpinen-4-ol, *p*-cymen-8-ol, myrtenal + myrtenol, verbenone, *trans*-carveol, *cis*-carveol, carvone, *p*-cymen-7-ol, cypera-2,4-diene, α-copaene, *p*-elemene, cyperene, β-caryophyllene, α-guaiene, α-humulene, rotundene, germacrene D, eudesma-2,4, 11-triene 3-selinene, α-selinene, α-bulnesene, δ-cadinene, *trans*-calamenene, α-calacorene, ledol, caryophyllene oxide, humulene epoxide II, β-copaen-4α-ol, dillapiole, patchoulenone, caryophylla-4(14),8(15)-dien-5α-ol, M 218, eudesma-3,1 1-dien-5-ol, mustakone, cyperotundone M, α-cyperone, and aristolone have been reported from *C. articulatus* stem and rhizome (Zoghbi et al. 2006). 1,3,5-trioxepane, 3-furaldehyde, 5-methyl-2-furancarboxaldehyde, bicyclo(2.2.1)hepta-2,5-dien-7ol, 2-ethylfuran, α-thujene, α-pinene, camphene, sabinene, β-pinene, myrcene, *o*-cymene, *p*-cymene, limonene, *m*-cymene, eucalyptol, γ-terpinene, 5-ethylidene-1-methyl- cycloheptene, *cis*-limonene oxide, *trans*-linalool oxide, α-linalool, α-thujone, α-campholenal, *p*-cymenene, *trans*-pinocarveol, *cis*-verbenol, camphor, isoborneol, borneol, pinocarvone, terpinen-4-ol, α-terpineol, β-cyclocitral, dihydrocarveol, myrtenol, neral, *cis*-carveol, geraniol, piperitone, perillaldehyde, carvone, geraniol formate, 4-hydroxy-2-methyl-, benzaldehyde, neryl acetate, copaene, β-elemen, cyperene, β-caryophyllene, β-gurjunene, aromadendrene, α-cubebene, α-carophyllene, rotundene, γ-muurolene, germacrene D, *trans*-calamenene, γ-cadinene, *cis*-calamenane, δ-cadinene, ledene oxide, γ-patchoulene, amorphene, caryophyllene oxide, longipinocarvone, ledol, β-selinene, α-selinene, cedra-8-en-15-ol, globulol, cedrol, 1-epi-cubenol, α-bulnesene, dillapiol, α-cadinol, germacrone, β-bisabolol, cyperotundone, α-cyperone, cyclocolorenone, aristolone, ethylpalmitate, ethyl-(*E*)-9-octadecenoate, and ethyleicosanoate were identified from *C articulatus*, *C esculentus* and *C. papyrus* leaves (Hassanein et al. 2014).

Monoterpenes {camphenol, 6-, bicyclo[3.1.1]heptan-3-ol, 6,6-dimethyl-2-methylene-, [1S-(1 à,3à,5à)]-, bicyclo[3.1.1]hept-3-en-2-ol, 4,6,6-trimethyl-, [1*S*-(1 à,2a,5à)]-, bicyclo[3.1.0] hexan-3-ol, 4-methylene-1-(1-methylethyl)-, [1.S-(1 à,3a,5à)], (−)-myrtenol, bicyclo[3.1.0]hex-3-en-2-one, 4-methyl-1-(1-methylethyl)-, bicyclo[3.1.1]hept-3-en-2-one, 4,6,6-trimethyl-,(1S)-, 3,5-Heptadienal, 2-ethylidene-6-methyl-, bicyclo[4.4.0]dec-2-ene-4–01, 2-methyl-9-(prop-1-en-3 –01–2-yl)-}; sesquiterpenes {isolongifolene, 9,10-dehydro-, copaene, α-cubebene, cadina-1 (10),6,8-triene, 7-tetracyclo[6.2.1.0(3.8)O(3.9)]undecanol, 4,4,11,11-tetramethyl-, naphthalene, 1,2,3,4,4a,5,6,8a-octahydro-7-methyl-4-methylene-1-(1-methylethyl)-, (1à,4aà,8aà)-, naphthalene, 1,2,3,4-tetrahydro-1,6-dimethyl-4-(1-methylethyl)-,(1S-cis)-,(+)-Epi-bicyclosesquipheliandrene, 2,3,4-trifluorobenzoic acid, 4-nitrophenyl ester, 7-tetracyclo[6.2.1.0(3.8)0(3.9)]undecanol, 4,4,11,11-tetramethyl-, 2-naphthalenemethanol, 1,2,3,4,4a,8a-hexahydro-à,à,4a,8-tetramethyl-, [2*R*-(2à,4aà,8aà)]-, cycloisolongifolene, 8,9-dehydro-, caryophyllene oxide, longipinocarvone, 3-isopropyl-6, 7-dimethyltricyio[4.4.0.0(2,8)]decane-9, 1O-diol, *cis*-Z-α-bisabolene epoxide, longiverbenone, 2,2,7,7-tetramethyltricyclo[6.2.1.0(1,6)]undec-4-en-3-one, acetic acid, 3-hydroxy-6-isopropenyl-4,8a-dimethyl-1,2,3,5,6,7,8, 8aoctahydronaphthalen-2-yl ester, 5(1H)-Azulenone, 2,4,6,7,8, 8a-hexahydro-3, 8-dimethyl-4-(1-methylethylidene)-, (8S-cis)-, perhydrocyclopropa[e]azulene-4,5,6-triol, 1,1,4,6-tetramethyl, 1H-cycloprop[e]azulen-7–01, decahydro-1, 1,7 -trimethyl-4-methylene-, [1ar-(1 aà,4aà,7a,7aa,7bà)]-, (−)-spathulenol, corymbolone, spiro [4.5]decan-7-one, 1,8-dimethyl-8,9-epoxy-4-isopropyl-, 9H-cycioisolongifolene, 8-oxo-, 2(1H) naphthalenone, 3,5,6,7,8, 8a-hexahyd ro-4, 8adimethyl-6-(1-methylethenyl)-, *E*-1S-heptadecenal, 6-isopropenyl-4,8a-dimethyl-1,2,3,5,6,7,8,8aoctahydronaphthalene-2,3-diol, 2(1H)-naphthalenone, 4a,5,6,7,8,8a-hexahydro-6-[1-(hydroxymethyl)ethenyl]-4,8adimethyl-, [4ar-(4aà,6à,8aà)]-,1-naphthalenol,decahydro-1,4a-dimethyl-7-(1-methyleth-ylidene)-,[1R-(1à,4aa,8aà)]-}; triterpenes (cyclodecasiloxane, eicosamethyl-); and polyterpene (tetracosamethyl-cyclododecasiloxane) were identified from hexane extract of *C. articulatus* root/rhizome (Metuge et al. 2014).

Palmitic, oleic, heptadecanoic, linoleic, and minor; arachidonic, lignoceric, stearic, myristic acid, α-amyrin, and β-sitosterol were isolated from unsaponifiable matter of *C. conglomeratus*. The ethanol extract did not show any toxicity greater than 4000 mg/kg. In addition, the extract demonstrated antifungal activity against *Candida albicans* (MIC 23.1 ± 2.1, 0.98 µg/ml; Al-Hazmi et al. 2018). 7,3′-dihydroxy-5,5′-dimethoxy-8-prenylflavan and 5,7,3′-trihydroxy-5′-methoxy-8-prenylflavan, luteolin, and its 7-methyl ether were isolated and identified from dichloromethane-methanol (1:1) extract of *C. conglomeratus* (Abdel-Razik et al. 2006). The organic acids (quinic acid, malic acid, tetrahydroxypentanoic acid, citric acid/isocitric acid, malic acid, fumaric acid, leucine-hexose, homocitric acid, dihydroxybenzoic acid), phenolic compounds [dihydroxybenzoic acid methyl ester, gallocatechin, dihydroxy benzoic acid, dihydroxy benzoic acid methyl ester hexoside, *O*-hexosyl-*O*-methyl-myoinositol-dihydroxy benzoic acid, salicylic acid, piscidic acid, procyanidin B dimer, hexahydroxyflavan, C-hexosylprocyanidin B dimer, dimethoxyhomophthalic acid, caffeic acid, hydroxymethoxycinnamaldehyde, *O*-caffeoylquinic acid, *O*-syringoylquinic acid, caffeoquinone, procyanidin B dimer, syringoylmalic acid, syringic acid, dihydroxyhomophthalic acid dimethyl ester, hydroxycinnamic acid, (*epi*)-catechin, eriodictyol, scopoletin, hydroxydimethoxycinnamic acid, ferulic acid, dihydrocyperaquinone, caffeoquinone isomer, trihydroxycoumestan, trihydroxyflavanone, tetrahydroxyflavanone, longusol C, hydroxy-methoxycoumarin, trihydroxycinnamic acid dimethyl ether, luteolin, 3′,4′-dimethoxy luteolin, hesperitin, tetrahydroxyflavanone, tetrahydroxymethylaurone, trihydroxymethoxyprenyl isoflavone, tetrahydroxyflavanone methyl ether, trihydroxy-prenylflavan, trihydroxymethoxy prenylflavan] were identified from ethanol extract of *C. conglomeratus* (Elshamy et al. 2020). α-pinene, α-sabinene, β-pinene, sabinene hydrate-*trans*, camphor, borneol, cyprotene, cypera-2,4-diene, α-cubebene, β-cubebene, α-copaene, cyperene, β-damascone, β-caryophyllene, caryophyllane-2–6-β-oxide, α-humulene, rotundene, β-selinene, α-selinene, α-calamenene, α-muurolene, *tau*-calamenene, β-calamenene, δ-cadinene, α-calacorene, isorotundene, caryophyllene oxide, isocyperol, cyperol, *tau*-cadinol, cubenol-1-epi, α-muurolol, *tau*-muurolol, cubenol, α-cadinol, caryophyllene epoxide, mustakone, cyperotundone, and α-cyperone were identified from *C. conglomeratus* aerial parts (Feizbakhsh and Naeemy 2011). Three flavan derivatives {7,3′-dihydroxy-8,4′-dimethoxyflavan, 7,4′-dihydroxy-5,3′-dimethoxy-8-methylflavan and 7,4′-dihydroxy-5,3′-dimethoxy-8-prenylflavan}, two stilbene derivatives {4-hydroxy-5′- methoxy-6″,6″-dimethylpyran[2″,3″: 3′, 2′]stilbene and 4′-hydroxy-3,5-dimethoxy-2-prenylstilbene}, and four other compounds {5,4′-dihydroxy-7,3′-dimethoxyflavan, 3′,4′-dimethoxyluteolin, 3′,4-dihydroxy-5′-methoxy-2′-prenylstilbene, and 4,4′-dihydroxy-3,3′-dimethoxy-2′-prenylstilbene} were isolated and identified from *C. conglomeratus* (Zaki et al. 2018).

Flavonoids [chrysoeriol 7-*O*-β-(6′′′-*O*-acetyl-β-D-glucopyranosyl)-(1→4) glucopyranoside, chrysoeriol 7-*O*-β- (6′′′-O-acetyl-β-D-glucopyranosyl)-(1→4) glucopyranoside, apigenin, apigenin 7-*O*-β-glucopyranoside, luteolin, luteolin 7-*O*-β-glucopyranoside, chrysoeriol, chrysoeriol 7-*O*-β-glucopyranoside, and tricin] were isolated and identified from *C. laevigatus* aerial parts. The methanol and ethyl acetate fractions demonstrated significant anti-inflammatory effects in lipopolysaccharide-induced RAW 264.7 macrophages model by reducing nitric oxide formation (Elshamy et al. 2017).

A total of seven compounds (thunbergin A-B, *trans*-resveratrol, *trans*-scirpusin A, *trans*-cyperusphenol A, aureusidin and luteolin) have been identified from methanol extract of *C. thunbergii* aerial parts. Thunbergin A and *trans*-scirpusin A displayed significant suppression to arginase (IC_{50} 17.6 and 60.6 μM) activity. Moreover, methanol extract of *C. thunbergii* aerial parts demonstrated an endothelium and nitric oxide-dependent vasorelaxant effect on thoracic aortic rings from rats and improved endothelial dysfunction in an adjuvant-induced arthritis of a rat model (Arraki et al. 2021). α-pinene, β-pinene, limonene, 1,8-cineole, cyperene, β-cadinene, caryophyllene oxide, and humulene epoxide II were identified from *C. distans* fresh rhizome (Lawal and Oyedeji 2009b). Leptosidin 6-*O*-β-D-glucopyranosyl-*O*-α-l- rhamnopyranoside has been identified from *C. scariosus* leaves (Bhatt et al. 1981).

Tricin 5-glucoside, tricin 7-diglucoside, tricin 7-glucoside, sulphuretin, and aureusidin have been identified from *C. alopecuroides*, *C. difformis*, *C. digitatus*, *C. fenzelianus*, *C. maculatus*, and *C. rotundus* leaves; tricin 5-diglucoside and tricin 7,4′-diglucoside have been identified from *C. laevigatus*; tricin 7-glucuronide have been identified from *C. conglomeratus*, leaves (El-Habashy et al. 1989); quercetin 3-methyl ether, quercetin 3,7,3′-trimethyl ether, quercetin 3,7-dimethyl ether, quercetin 3,7,3′-trimethyl ether, kaempferol, kaempferol 3-methyl ether, and kaempferol 3,7-dimethylether were identified from *C. cunninghamii*, *C. rigidellus*, *C. dactylotes*, *C. rigidellus*, *C. tetraphyllus* leaves, tubers, and aerial parts (Harborne et al. 1982), quercetin 3,3'-dimethyl ether, quercetin 3,4'-dimethyl ether have been identified from *C. alopecuroides* inflorescence; kaempferol 3-*O*-β-D-(2^G-glucosylrutinoside) and kaempferol 3-*O*-β-D-(2^G-xylosylrutinoside) have been identified from *C. alopecuroides* aerial parts (Sayed et al. 2006). Trimethoxy-6-prenyl flavan, 7,3'-dihydroxy-5,5'-dimethoxy-8- prenylflavan and 5,7,3'-trihydroxy-5'- methoxy-8-prenylflavan from *C. conglomeratus* whole plant (Nassar et al. 1998; Abdel-Razik et al. 2006); 5,7-dihydroxy-3',5'- dimethoxy-6-prenyl-flavan, and 5-hydroxy-7, 3', 5'- trimethoxyflavan have been identified from *C. conglomeratus* tubers (Abdel-Mogib et al. 2000). Leptosidin-6-*O*-β-D- glucopyranosyl-*O*-α-L- rhamnopyranoside have been identified from *C. scariosus* leaves; 6,3',4'-trihydroxy-4-methoxy-5-methylaurone, 6,3'-dihydroxy-4, 4'- dimethoxy-5-methylaurone, 4,6,3',4'-tetramethoxy aurone, 4,6,3',4'-tetrahydroxy -5-methylaurone, 4,6,3',4'-tetrahydroxy-7-methylaurone, 5,3'-dihydroxy-6,4'- dimethoxyflavan and 5,7,4',5'-tetrahydroxy6,3'-diprenylflavanone, and 6, 3', 4'-trihydroxy-4- methoxy-7-methyl aurone were reported from *C. capitatus* rhizome and roots (Seabra et al. 1995, 1997, 1998; Abdel-Mogib and Serag 2001)

Rotundusides A, B, 6''-*O*-*p*-coumaroylgenipin gentiobioside, 1-[2,3-dihydro-6- hydroxy-4,7-dimethoxy2S-(prop-1-en-2- yl)benzofuran-5- yl]ethenone, 2S-isopropenyl-4,8- dimethoxy-5-methyl-2,3- dihydrobenzo-[1,2-b;5,4- b']difuran, 2S-isopropenyl-4,8- dimethoxy-5-hydroxy-6-methyl-2,3- dihydrobenzo[1,2-b;5,4- b']difuran, 1α-methoxy-3β-hydroxy-4α-(3',4'- dihydroxyphe-nyl)-1, and 2,3,4- tetrahydronaphthalin were reported from *C. rotundus* and *C. teneriffae* rhizome (Zhou and Zhang 2013; Amesty et al. 2011; Zhou and Yin 2012). Rotundine A-C, octopamine, 6, and 7-dihydro-2, 3- dimethyl-5- cyclopentapyrazine have been identified from *C. rotundus* and *C. esculentus* rhizome (Smith 1977; Cantalejo 1997; Jeong et al. 2000). *Trans-* scirpusin A, B, cassigarol E and G, and pallidol have been identified from *C. longus* whole plant (Morikawa et al. 2002).

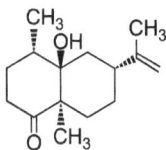 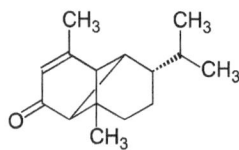 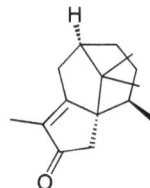 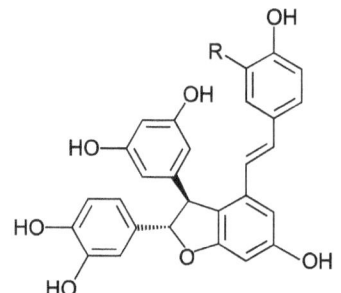

Corymbolone   Mustakone   Cyperotundone   R = H – Scirpusin A;
                                                                                        R = OH - Scirpusin B

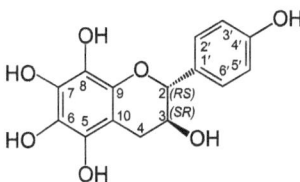

(2*RS*,3*SR*)-3,4' ,5,6,7,8-hexahydroxyflavane

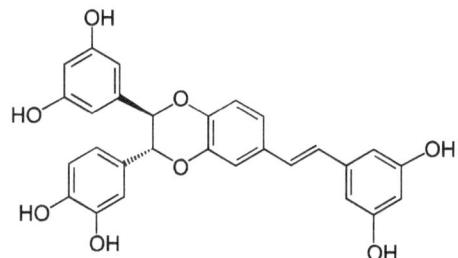

Cassigarol E

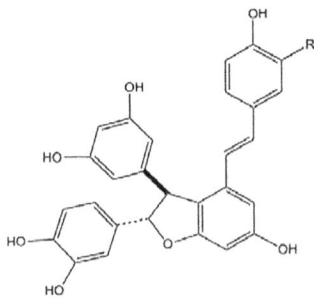

R = H – *trans*- Scirpusin A; R = OH - *trans*-Scirpusin B

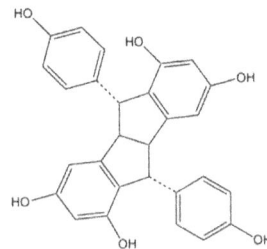

Pallidol

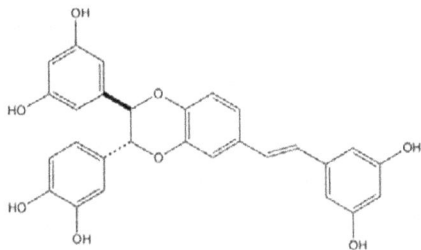

Cassigarol E

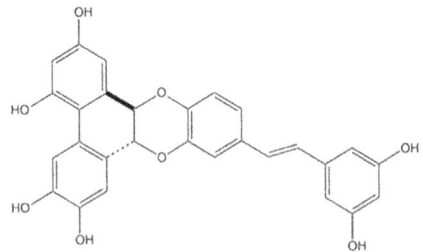

Cassigarol G

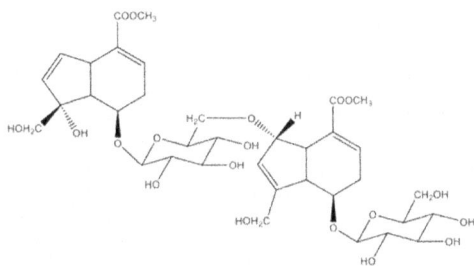

Rotunduside A

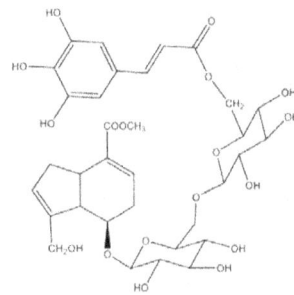

Rotunduside B

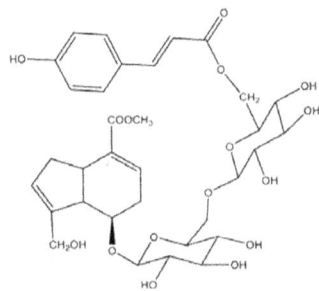

6″-*O*-*p*-Coumaroylgenipin gentiobioside

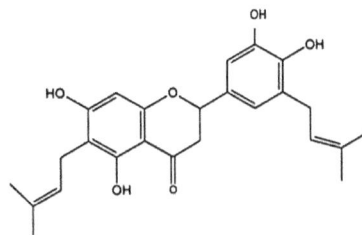

5,7,4′,5′-Tetrahydroxy6,3′-diprenylflavanone

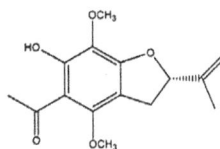

1-[2,3-Dihydro-6- hydroxy-4,7-dimethoxy2S-(prop-1-en-2- yl)benzofuran-5- yl]ethanone

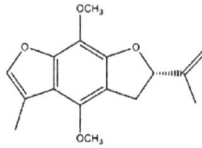

2S-Isopropenyl-4,8- dimethoxy-5-methyl-2,3- dihydrobenzo-[1,2-b;5,4-b′]difuran

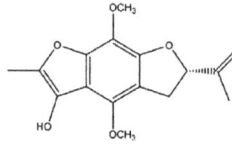

2S-Isopropenyl-4,8- dimethoxy-5-hydroxy-6- methyl-2,3- dihydrobenzo[1,2-b;5,4- b′]difuran

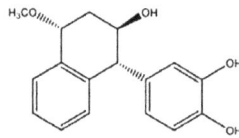

1α-Methoxy-3β-hydroxy-4α-(3′,4′- dihydroxyphenyl)-1, 2,3,4- tetrahydronaphthalin

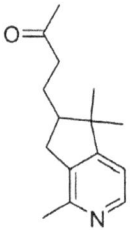      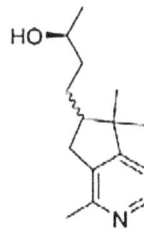   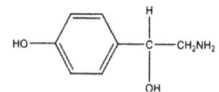

Rotundine A   Rotundine B   Rotundine C   Octopamine

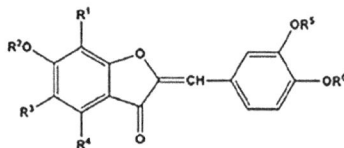

R^1 = OCH$_3$, R^2 = Glc (1→4) rham, R^3 = R^4 = R^5 = R^6 = H - Leptosidin–6–O–β-D– glucopyranosyl–O–α–L– rhamnopyranoside; R^1 = R^2 = R^5 = R^6 = H R^3 = CH$_3$, R^4 = OCH$_3$ - 6,3′,4′–Trihydroxy–4–me thoxy–5–methylaurone; R^1 = R^2 = R^5 = H R^3 = R^6 = CH$_3$, R^4 = OCH$_3$ - 6,3′–Dihydroxy–4, 4′–dimethoxy–5–methylaurone; R^1 = R^3 = H, R^2 = R^5 = R^6 = CH$_3$, R^4 = OCH$_3$ - 4,6,3′,4′–Tetramethoxy aurone; R^1 = R^2 = R^5 = R^6 = H, R^4 = OH, R^3 = CH$_3$ - 4,6,3′,4′–Tetrahydroxy –5–methylaurone; R^1 = CH$_3$, R^2 = R^3 = R^5 = R^6 = H, R^4 = OH - 4,6,3′,4′–Tetrahydroxy–7 –methylaurone; R^1 = CH$_3$, R^2 = R^3 = R^5 = R^6 = H, R^4 = OCH$_3$ - 6, 3′, 4′–Trihydroxy–4– methoxy–7–methyl aurone

6, 7-Dihydro-2, 3- dimethyl-5- cyclopentapyrazine

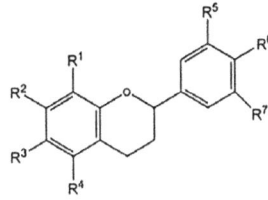

$R^1 = R^5 = R^7 = H$, $R^2 = R^4 = R^6 = OCH_3$, $R^3 = CH_2CH=C(CH_3)_2$ - 5,7,4′– Trimethoxy–6–prenyl flavan; $R^1 = R^6 = H$, $R^2 = R^4 = OH$, $R^3 = CH_2CH=C(CH_3)_2$, $R^5 = R^7 = OCH_3$ - 5,7-Dihydroxy–3′,5′-dimethoxy–6– prenylflavan; $R^1 = R^3 = R^6 = H$, $R^2 = R^5 = R^7 = OCH_3$, $R^4 = OH$ - 5-Hydroxy-7, 3′, 5′-trimethoxyflavan; $R^1 = CH_2CH=C(CH_3)_2$, $R^2 = R^5 = OH$, $R^3 = R^6 = H$, $R^4 = R^7 = OCH_3$ - 7,3′-Dihydroxy-5,5′-dimethoxy-8-prenylflavan; $R^1 = CH_2CH=C(CH_3)_2$, $R^2 = R^4 = R^5 = OH$, $R^3 = H$, $R^6 = H$, $R^7 = OCH_3$ - 5,7,3′-Trihydroxy-5′-methoxy-8-prenylflavan; $R^1 = R^2 = R^7 = H$, $R^3 = R^6 = OCH_3$, $R^4 = R^5 = OH$ - 5,3′-Dihydroxy-6,4′-dimethoxyflavan

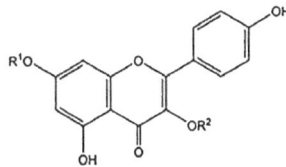

$R^1 = H$, $R^2 = O\text{-}\beta\text{–}D\text{–}(2^G$ - glucosylrutinoside) - Kaempferol 3–O-β–D– (2^G-glucosylrutinoside); $R^1 = H$, $R^2 = $ - O-β–D–(2^G - xylosylrutinoside) - Kaempferol 3-O-β–D– (2^G-xylosylrutinoside); $R^1 = H$, $R^2 = H$ – Kaempferol; $R^1 = H$, $R^2 = CH_3$ - Kaempferol 3–methyl ether; $R^1 = R^2 = CH_3$ - Kaempferol 3,7–dimethylether

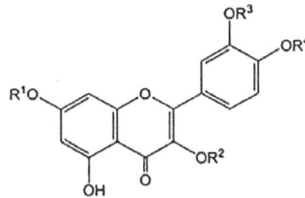

$R^1 = R^2 = R^3 = R^4 = H$ – Quercetin; $R^1 = R^3 = R^4 = H$, $R^2 = CH_3$ – Quercetin 3–methyl ether; $R^1 = R^2 = CH_3$, $R^3 = R^4 = H$ – Quercetin 3,7–dimethyl ether; $R^1 = R^2 = R^3 = CH_3$, $R^4 = H$ - Quercetin 3,7,3′–trimethyl ether; $R^1 = R^4 = H$, $R^2 = R^3 = CH_3$ - Quercetin 3,3′–dimethyl ether; $R^1 = R^3 = H$, $R^2 = R^4 = CH_3$ - Quercetin 3,4′–dimethyl ether

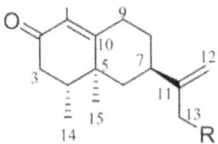

R = H - Nootkatone; R = OH - 12-Hydroxynootkatone

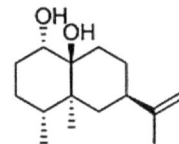

Oxyphyllol C

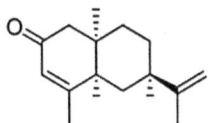

$R^1 = H$, $R^2 = OH$ - 7-*epi*-Teucrenone; $R^1 = OH$, $R^2 = H$ - α-Rotunol

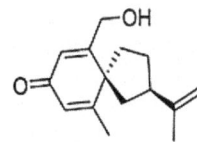

Argutosine D

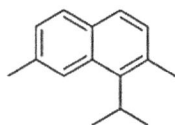

1-Isopropyl2, 7-dimethylnaphthalene

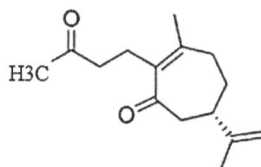

4, 5-Seco-guaia-1 (10), 11-diene-4, 5-dioxo

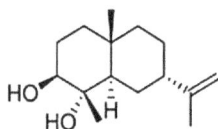

3β, 4α-Dihydroxy-7-epieudesm-11 (13), -ene

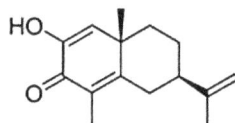

2-Oxo-α-cyperone

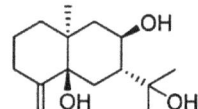

Rhombitriol

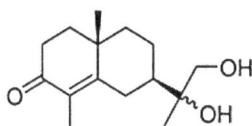

7α (H), 10β-Eudesm-4-en-3-one-11,12-diol

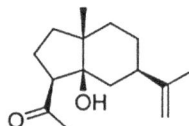

5-Hydroxylucinone

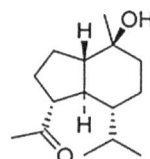

Oplopanone

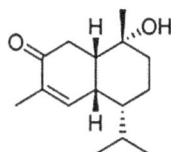

10-Hydroxyamorph-4-en-3- one

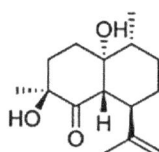

Cyperusol D

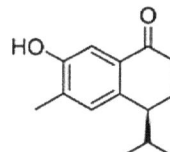

2- Hydroxy-14-calamenenone

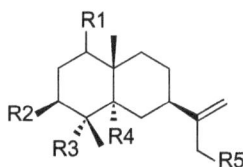

R^1 = H, R^2 = OH, R^3 = H, R^4 = OH, R^5 = H - 10-Epieudesm-11-ene-3β, 5α-diol; R^1 = H, R^2 = OH, R^3 = OH, R^4 = H, R^5 = OH - 3β-Hydroxyilicic alcohol (11 (13)-eudesmene-3, 4, 12-triol); R^1 = β-OH, R^2 = H, R^3 = OH, R^4 = H, R^5 = H - Cyperusol C; R^1 = α-OH, R^2 = H, R^3 = H, R^4 = OH, R^5 = H - α-Corymbolol

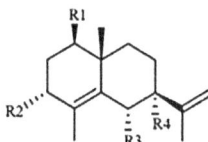

R^1 = H, R^2 = O, R^3 = H, R^4 = H - α-Cyperone; R^1 = H, R^2 = O, R^3 = H, R^4 = OH - (4aS, 7S),-7-Hydroxy-1, 4a-dimethyl-7-(prop-1-en-2-yl),-4, 4a, 5, 6, 7, 8- hexahydronaphthalen-2 (3H),-one; R^1 = H, R^2 = O, R^3 = OH, R^4 = H - (4aS, 7S, 8R),-8-Hydroxy-1, 4a-dimethyl-7-(prop-1-en-2-yl), -4, 4a, 5, 6, 7, 8-hexahydronaphtha-len-2 (3H), -one ; R^1 = H, R^2 = OH, R^3 = H, R^4 = H - Cyperol; R^1 = OH, R^2 = O, R^3 = H, R^4 = H - 1β-Hydroxy-α-cyperone

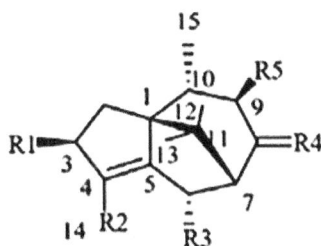

$R^1 = O$, $R^2 = CH_3$, $R^3 = H$, $R^4 = O$, $R^5 = H$ - Cyperene-3, 8-dione; $R^1 = O$, $R^2 = CH_2OH$, $R^3 = H$, $R^4 = H$, $R^5 = H$ - 14-Hydroxy cyperotundone; $R^1 = O$, $R^2 = CH_2OAc$, $R^3 = H$, $R^4 = H$, $R^5 = H$ - 14-Acetoxy cyperotundone; $R^1 = OH$, $R^2 = COOH$, $R^3 = H$, $R^4 = H$, $R^5 = H$ - 3β-Hydroxycyperenoic acid; $R^1 = OAc$, $R^2 = CH_3$, $R^3 = OH$, $R^4 = H$, $R^5 = OAc$ - Sugetriol-3, 9-diacetate; $R^1 = H$, $R^2 = CH_2OH$, $R^3 = H$, $R^4 = H$, $R^5 = H$ - Cyperenol; $R^1 = H$, $R^2 = COOH$, $R^3 = H$, $R^4 = H$, $R^5 = H$ - Cyperenoic acid; $R^1 = O$, $R^2 = CH_3$, $R^3 = H$, $R^4 = H$, $R^5 = H$ - Cyperotundone; $R^1 = O$, $R^2 = CH_3$, $R^3 = OH$, $R^4 = H$, $R^5 = H$ - Sugeonol; $R^1 = O$, $R^2 = CH_3$, $R^3 = O$, $R^4 = H$, $R^5 = H$ - Scariodione; $R^1 = OAc$, $R^2 = CH_3$, $R^3 = OAc$, $R^4 = H$, $R^5 = OAc$ - Sugetriol triacetate

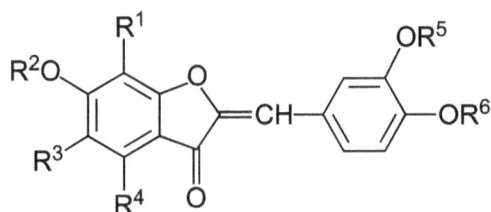

$R^1 = R^2 = R^3 = R^4 = R^5 = R^6 = H$ – Sulphuretin; $R^1 = R^2 = R^3 = R^5 = R^6 = H$, $R^4 = OH$ – Aureusidin

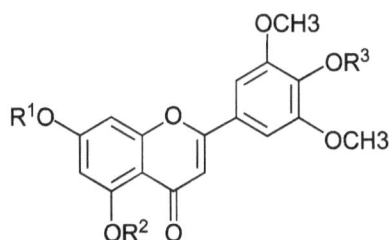

$R^1 = H$, $R^2 = Glc$, $R^3 = CH_3$ - Tricin 5–glucoside; $R^1 = H$, $R^2 = Glc$ - Glc, $R^3 = CH_3$ - Tricin 5–diglucoside; $R^1 = Glc$, $R^2 = H$, $R^3 = CH_3$ - Tricin 7–glucoside; $R^1 = Glucuronic acid$, $R^2 = H$, $R^3 = CH_3$ - Tricin 7–glucuronide; $R^1 = Glc-Glc$, $R^2 = H$, $R^3 = CH_3$ - Tricin 7–diglucoside; $R^1 = R^3 = Glc$, $R^2 = H$ - Tricin 7,4'-diglucoside

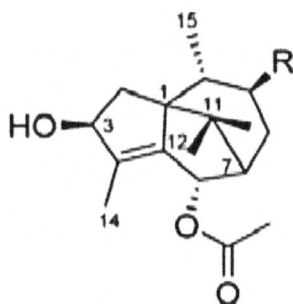

R = OAc - Sugetriol 6,9-diacetate; R = H - Sugebiol 6-acetate

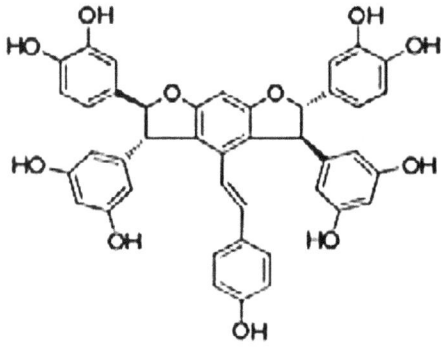

Cyperusphenol A

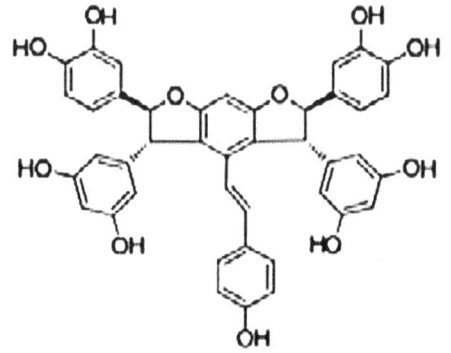

Mesocyperusphenol A

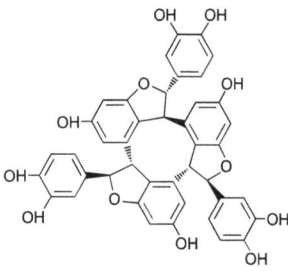

Cyperusphenol D

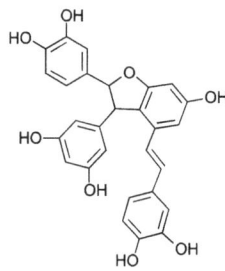

Scirpusins B

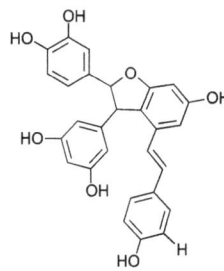

Scirpusins A

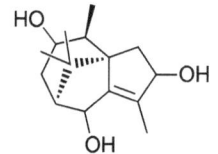

Sugetriol

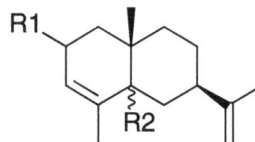

R^1 = O, R^2 = α-OH - α-Rotunol; R^1 = O, R^2 = β-OH - β-Rotunol; R^1 = H, R^2 = β-OH – (–)-Eudesma-3,11-diene-5-ol

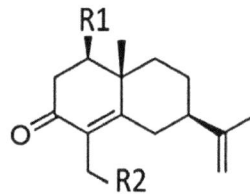

R^1 = OH, R^2 = H - Ligucyperonol; R^1 = H, R^2 = OH - 14-Hydroxy-α-cyperone

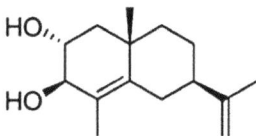

Britanlin E

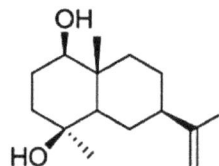

1β,4β-Dihydroxyeudesma-11-ene

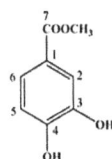

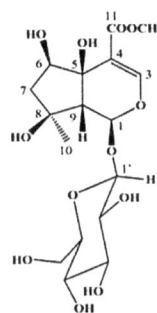

Methyl 3,4-dihydroxy benzoate   Ipolamiide   6β-Hydroxyipolamiide

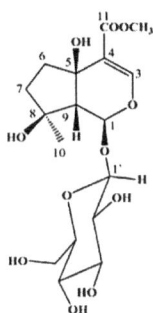

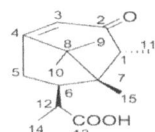

12-Methyl cyprot-3-en-2-one-13-oic acid

$$CH_3(CH_2)_{13}\overset{15}{C}O(CH_2)_{16}CH_3$$

n-Dotriacontan-16-one

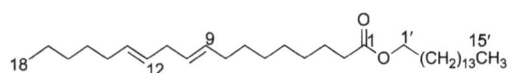

n-Pentadecanyl linoleate   n-Hexadecanyl linoleate

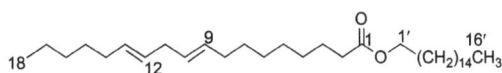

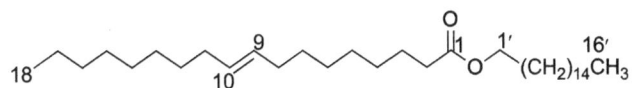

n-Hexadecanyl oleate

$$\overset{40}{CH_3}(CH_2)_{32}\overset{7}{C}O(CH_2)_5\overset{1}{CH_3}$$

n-Tetracontan-7-one

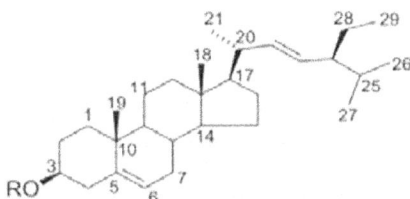

R = CO(CH$_2$)$_{10}$CH$_3$ - Stigmasteryl laurate; R = CO(CH$_2$)$_{12}$CH$_3$ - Stigmasteryl myristate

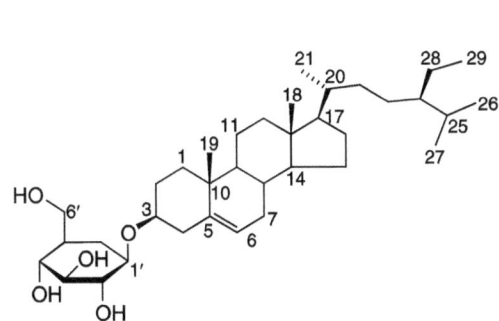

β-Sitosterol glucoside   Lupenyl arabinopyranosyl oleate

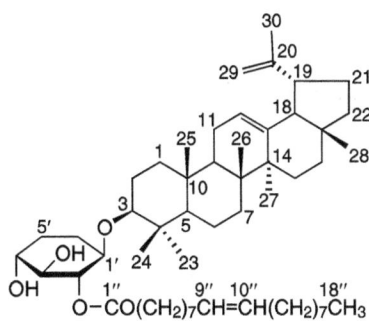

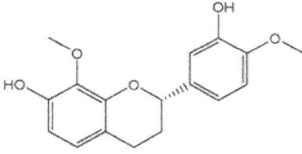

7,3′-Dihydroxy-8,4′-dimethoxyflavan

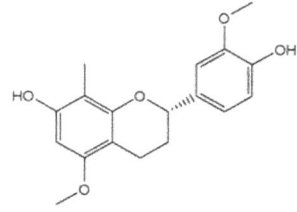

7,4′-Dihydroxy-5,3′-dimethoxy-8-methylflavan

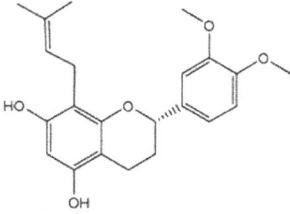

7,4′-Dihydroxy-5,3′-dimethoxy-8-prenylflavan

n-Pentacos-13-enyl oleate

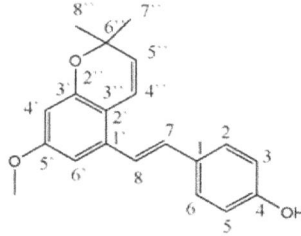

4-Hydroxy-5′- methoxy-6″,6″-dimethylpyran[2″,3″: 3′, 2′]stilbene

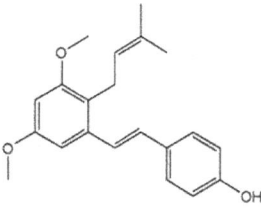

4′-Hydroxy-3,5-dimethoxy-2-prenylstilbene

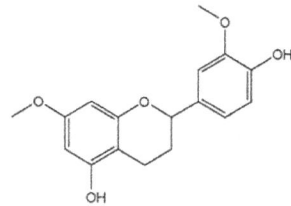

5,4′-Dihydroxy-7,3′-dimethoxyflavan

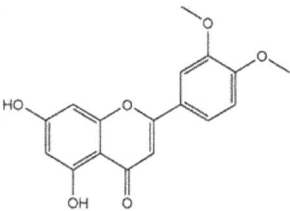

3′,4′-Dimethoxyluteolin

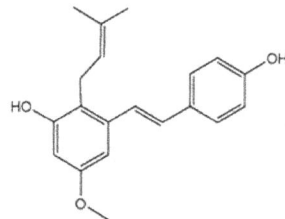

3′,4-Dihydroxy-5′-methoxy-2′-prenylstilbene

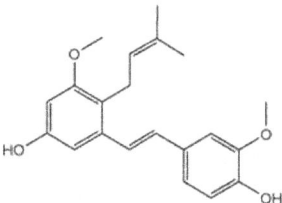

4,4′-Dihydroxy-3,3′-dimethoxy-2′-prenylstilbene

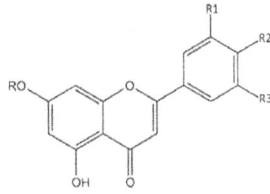

R = 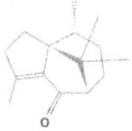, R¹ = OCH3, R² = OH, R³ = H - Chrysoeriol 7-O-β- (6‴-O-acetyl-β-D-glucopyranosyl)-(1→4) glucopyranoside; R = H, R¹ = H, R² = OH, R³ = H - Apigenin; R = β-Glu, R¹ = H, R² = OH, R³ = H - Apigenin 7-*O*-β-glucopyranoside; R = H, R¹ = OH, R² = OH, R³ = H - Luteolin; R = β-Glu, R¹ = OH, R² = OH, R³ = H - Luteolin 7-*O*-β-glucopyranoside; R = H, R¹ = OCH₃, R² = OH, R³ = H - Chrysoeriol; R = β-Glu, R¹ = OCH₃, R² = OH, R³ = H - Chrysoeriol 7-*O*-β-glucopyranoside; R = H, R¹ = OCH₃, R² = H, R³ = OCH₃ - Tricin

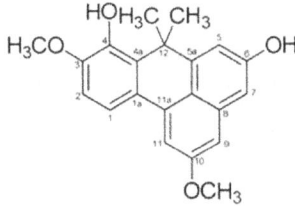

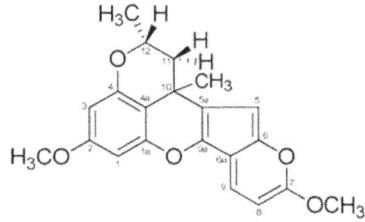

Patchoulenone   Thunbergin A   Thunbergin B

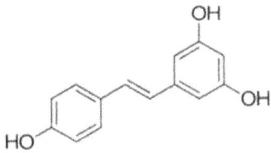

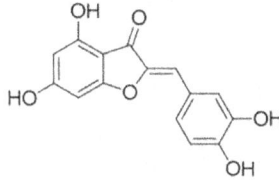

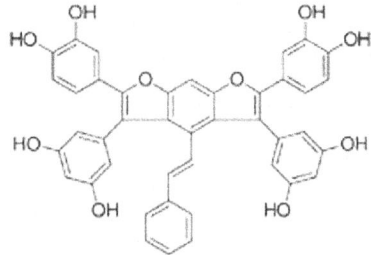

trans-Resveratrol   Aureusidin   *trans*-Cyperusphenol A

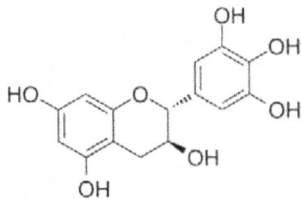

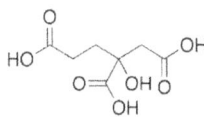

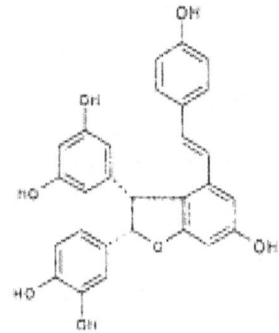

Gallocatechin   Homocitric acid   Longusol C

$$CH_3-(CH_2)_8 \overset{\overset{O}{\|}}{\underset{21}{C}} -(CH_2)_{19}-CH_2OH$$

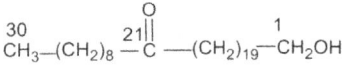

n-Triacont-1-ol-21-one

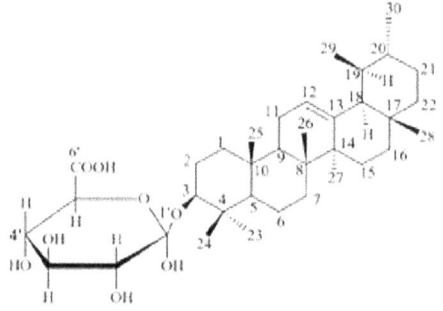

18-*epi*-α- Amyrin glucuronoside

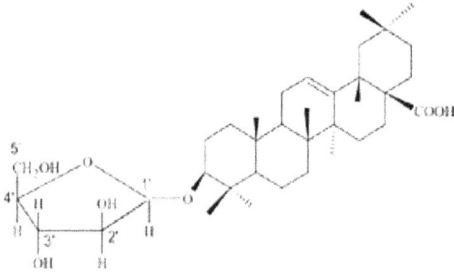

Oleanolic acid arabinoside

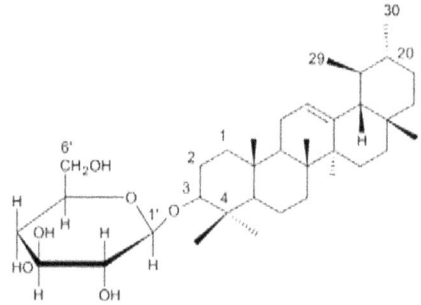

α-Amyrin glucopyranoside

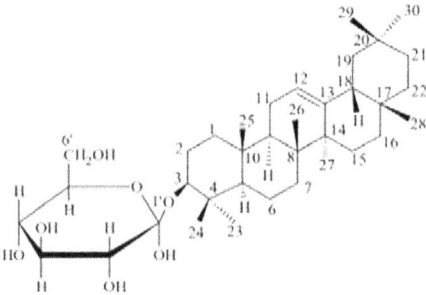

β-Amyrin glucopyranoside

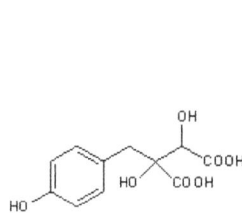

Piscidic acid

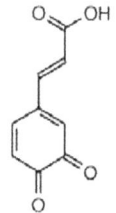

Caffeoquinone

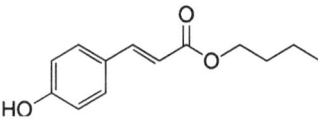

4-Hydroxyl butyl cinnamate

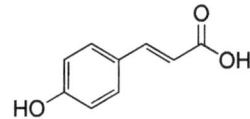

4-Hydroxyl cinnamic acid

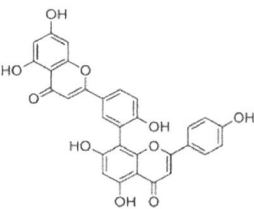

Amentoflavone

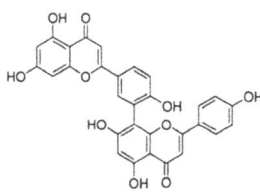

Ginkgetin

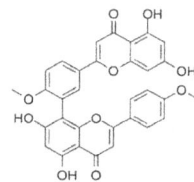

Isoginkgetin

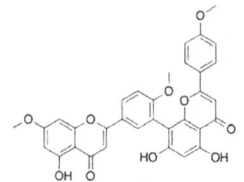

Sciadopitysin

16.3 CULTURE CONDITIONS

Tubers of *C. rotundus* were cultured on an MS (Murashige and Skoog 1962) medium containing 2,4-dichlorophenoxyaceticacid and kinetin. Maximum callus growth was achieved with supplementation of 2,4-dichlorophenoxyacetic acid (2.5 mg/l) and kinetin (0.5 mg/l). The highest amount of total flavonoid was recorded in six-week-old callus (1.96 mg/g dry weight) while a minimum amount (0.28 mg/g dry weight) was recorded in two-week-old callus. Quercetin, kaempferol, catechin, and myricetin were isolated and characterized from leaf and root and *in vitro* callus of this species (Samariya and Sarin 2013).

REFERENCES

AbdelKader H, Ibrahim F, Ahmed M, El-Ghadban E. 2017. Effect of some soil additives and mineral nitrogen fertilizer at different rates on vegetative growth, tuber yield and fixed oil of tiger nut (*Cyperus esculentus* L.) plants. J Plant Prod 8, 39–48.

Abdel-Mogib M, Basaif SA, Ezmirly ST. 2000. Two novel flavans from *Cyperus conglomeratus*. Pharmazie 55, 693–695.

Abdel-Mogib M, Serag MS. 2001. Prenylflavanone from *C. capitatus* Alex. J Pharm Sci 15, 129–131.

Abdel-Razik AF, Nassar MI, El-Khrisy E-DA, Dawidar AM. 2006. New prenylflavans from *Cyperus conglomeratus*. Fitoterapia 76, 762–764.

Abubakar MS, Abdurahman EM, Haruna AK. 2000. The repellant and antifeedant properties of *Cyperus articulatus* against *Tribolium casteneum* Hbst. Phytother Res 14, 281–283.

Achoribo ES, Ong MT. 2017. Tiger nut (*Cyperus esculentus*): Source of natural anticancer drug? Brief review of existing literature. Euro Medit Biomed J 12, 91–94.

Adejuyitan JA. 2011. Tigernut processing: Its food uses and health benefits. Am J Food Technol 6, 197–201.

Aghassi A, Naeemy A, Feizbakhsh A. 2013. Chemical composition of the essential oil of *Cyperus rotundus* L. from Iran. J Essent Oil-Bear Plants 16, 382–386.

Ahmad M, Rookh M, Rehman AB, Muhammad N, Amber, Younus M, Wazir A. 2014. Assessment of antiinflammatory, anti-ulcer and neuro-pharmacological activities of *Cyperus rotundus* Linn. Pak J Pharm Sci 27, 2241–2246.

Akendengué B. 1992. Medicinal plants used by the Fang traditional healers in Equatorial Guinea. J Ethnopharmacol 37, 165–173.

Alam P, Ali M, Aeri V. 2012. Isolation of keto alcohol and triterpenes from tubers of *Cyperus rotundus* Linn. J Nat Prod Plant Resour 2, 272–280.

Al-Hazmi GH, Awaad AS, Alothman MR, Alqasoumi SI. 2018. Anticandidal activity of the extract and compounds isolated from *Cyperus conglomertus* Rottb. Saudi Pharm J 26, 891–895.

Aljuhaimi F, Ghafoor K, Özcan MM, Miseckaite O, Babiker EE, Hussain S. 2018. The effect of solvent type and roasting processes on physico-chemical properties of tigernut (*Cyperus esculentus* L.) tuber oil. J Oleo Sci 67, 823–828.

Al-Snafi AE. 2016. A review on *Cyperus rotundus*: A potential medicinal plant. IOSR J Pharm 6, 32–48.

Amesty A, Burgueñ-Tapia E, Joseph-Nathan P, Ravelo ÁG, Estévez-Braun A. 2011. Benzodihydrofurans from *Cyperus teneriffae*. J Nat Prod 74, 1061–1065.

Arraki K, Totoson P, Decendit A, Zedet A, Maroilley J, Badoc A, Demougeot C, Girard C. 2021. Mammalian arginase inhibitory activity of methanolic extracts and isolated compounds from *Cyperus* species. Molecules 26, 1694.

Badejo AA, Damilare A, Ojuade TD. 2014. Processing effects on the antioxidant activities of beverage blends developed from *Cyperus esculentus, Hibiscus sabdariffa*, and *Moringa oleifera* extracts. Prev Nutr Food Sci 19, 227–233.

Badejo AA, Duyilemi TI, Falarunu AJ, Akande OA. 2020. Inclusion of baobab (*Adansonia digitata* L.) fruit powder enhances the mineral composition and antioxidative potential of processed tigernut (*Cyperus esculentus*) beverages. Prev Nutr Food Sci 25, 400–407.

Badgujar SB, Bandivdekar AH. 2015. Evaluation of a lactogenic activity of an aqueous extract of *Cyperus rotundus* Linn. J Ethnopharmacol 163, 39–42.

Bajpay A, Nainwal RC, Singh D, Tewari SK. 2018. Medicinal value of *Cyperus rotundus* Linn: An updated review. Med Plants 10, 165–170.

Bashir A, Sultana B, Akhtar FH, Munir A, Amjad M, Hassan Q. 2012. Investigation on the antioxidant activity of dheela grass (*Cyperus rotundus*). Afr J Basic Appl Sci 4, 1–6.

Bazine T, Arslanoğlu ŞF. 2020. Tiger nut (*Cyperus esculentus*): Morphology, products, uses and health benefits. Black Sea J Agricult 3, 324–328.

Bhatt SK, Saxena VK, Singh KV. 1981. A leptosidin glycoside from leaves of *Cyperus scariosus*. Phytochemistry 20, 2605.

Bown D. 1995. Encyclopaedia of Herbs and Their Uses. Dorling Kindersley, London.

Bum EN, Meier CL, Urwyler S, Wang Y, Herrling PL. 1996. Extracts from rhizomes of *Cyperus articulatus* (Cyperaceae) displace [3H]CGP39653 and [3H]glycine binding from cortical membranes and selectively inhibit NMDA receptor-mediated neurotransmission. J Ethnopharmacol 54, 103–111.

Bum EN, Rakotonirina A, Rakotonirina SV, Herrling P. 2003. Effects of *Cyperus articulatus* compared to effects of anticonvulsant compounds on the cortical wedge. J Ethnopharmacol 87, 27–34.

Bum EN, Schmutz M, Meyer C, Rakotonirina A, Bopelet M, Portet C, Jeker A, Rakotonirina SV, Olpe HR, Herrling P. 2001. Anticonvulsant properties of the methanolic extract of *Cyperus articulatus* (Cyperaceae). J Ethnopharmacol 76, 145–150.

Burkill H. M. 1985. The Useful Plants of West Tropical Africa, Vol 1. Royal Botanic Gardens, Kew.

CABI. 2020. Invasive Species Compendium. CAB International, Wallingford.

Cantalejo MJ. 1997. Analysis of volatile components derived from raw and roasted earth-almond (*Cyperus esculentus* L.). J Agric Food Chem 45, 1853–1860.

Castro OD, Gargiulo R, Guacchio ED, Caputo P, Luca PD. 2015. A molecular survey concerning the origin of *Cyperus esculentus* (Cyperaceae, Poales): Two sides of the same coin (weed vs. crop). Ann Bot 115, 733–745.

Chen Y, Zhao Y-Y, Wang X-Y, Liu J-T, Huang L-Q, Peng C-S. 2011. GC-MS analysis and analgesic activity of essential oil from fresh rhizome of *Cyperus rotundus*. Zhong Yao Cai 34, 1225–1229.

Chithran A, Ramesh babu T, Himaja N. 2012. Comparative study on anti-inflammatory activity of *Cyperus roduntus* (L.) using different solvent system in carrageenan induced paw edema in albino Wistar rats. Int J Phytopharm 3, 130–134.

Chopra RN, Nayar SL, Chopra IC. 1956. Glossary of Indian medicinal Plants. CSIR, New Delhi.

da Silva NC, Gonçalves SF, de Araújo LS, Kasper AAM, da Fonseca AL, Sartoratto A, Castro KCF, Moraes TMP, Baratto LC, de Pilla Varotti F, Barata LES, Moraes WP. 2019. *In vitro* and *in vivo* antimalarial activity of the volatile oil of *Cyperus articulatus* (Cyperaceae). Acta Amazon 49, 334–342.

Datta S, Dhar S, Nayak SS, Dinda SC. 2013. Hepatoprotective activity of *Cyperus articulatus* Linn against paracetamol induced hepatotoxicity in rats. J Chem Pharm Res 5, 314–319.

de Assis FFV, da Silva NC, Moraes WP, Barata LES, Minervino AHH. 2020. Chemical composition and *in vitro* antiplasmodial activity of the ethanolic extract of *Cyperus articulatus* var. nodosus residue. Pathogens 9, 889.

Defelice MS. 2002. Yellow nutsedge *Cyperus esculentus* L.: Snack food of the Gods. Weed Technol 16, 901–907.

Deng S, Xia L, Zhu X, Zhu J, Cai M, Wang X. 2019. Natural alpha-glucosidase inhibitors rapid fishing from *Cyperus rotundus* using immobilized enzyme affinity screening combined with UHPLC-QTOF MS. Iran J Pharm Res Summer 18, 1508–1515.

Duke JA, Ayensu ES. 1985. Medicinal Plants of China. Reference Publ Inc, Algonac.

El-Gohary HMA. 2004. Study of essential oils of the tubers of *Cyperus rotundus* L. and *Cyperus alopecuroides* Rottb. Bull Fac Pharm Cairo Univ 42, 157–163.

El-Habashy I, Mansour RMA, Zahran MA, El-Hadidi MN, Saleh NAM. 1989. Leaf flavonoids of *Cyperus* species in Egypt. Biochem Syst Ecol 17, 191–195.

Elshamy AI, El-Shazly M, Yassine YM, El-Bana MA, Farrag AR, Nassar MI, Singab AN, Noji M, Umeyama A. 2017. Phenolic constituents, anti-inflammatory and antidiabetic activities of *Cyperus laevigatus* L. Pharmacogenomics J 9, 828–833.

Elshamy AI, Farrag ARH, Ayoub IM, Mahdy KA, Taher RF, EI Gendy A-EG, Mohamed TA, Al-Rejaie SS, EI-Amier YA, Abd-EIGawad AM, Farag MA. 2020. UPLC-qTOF-MS phytochemical profile and antiulcer potential of *Cyperus conglomeratus* Rottb. alcoholic extract. Molecules 25, 4234.

Eltilib HHAB, Elgasim EA, Mohamed Ahmed IA. 2016. Effect of incorporation of *Cyperus rotundus* L. rhizome powder on quality attributes of minced beef meat. J Food Sci Technol 53, 3446–3454.

Ezeh O, Gordon MH, Niranjan K. 2014. Tiger nut oil (*Cyperus esculentus* L.): A review of its composition and physicochemical properties. Eur J Lipid Sci Technol 116, 783–794.

Farrag ARH, Abdallah HMI, Khattab AR, Elshamy AI, El Gendy AE-NG, Mohamed TA, Farag MA, Efferth T, Hegazy M-EF. 2019. Antiulcer activity of *Cyperus alternifolius* in relation to its UPLC-MS metabolite fingerprint: A mechanistic study. Phytomedicine 62, 152970.

Feizbakhsh A, Naeemy A. 2011. Chemical composition of the essential oil of *Cyperus conglomeratus* Rottb. from Iran. E-J Chem 8, S293–S296.

Galinato MI, Moody K, Piggin CM. 1999. Upland Rice Weeds of South and Southeast Asia. International Rice Research Institute, Makati City, Philippines.

Gambo A, Da'u A. 2014. Tiger nut (*Cyperus esculentus*): Composition, products, uses and health benefits—a review. Bayero J Pure App Sci 7, 56–61.

Ghafari S, Esmaeili S, Naghibi F, Mosaddegh M. 2013. Plants used to treat "tabe rebá" (malaria like fever) in Iranian traditional medicine. Int J Trad Herbal Med 1, 168–176.

Ghannadi A, Rabbani M, Ghaemmaghami L, Malekian N. 2012. Phytochemical screening and essential oil analysis of one of the Persian Sedges; *Cyperus rotundus* L. Int J Pharm Sci Res 3, 424–427.

Gilani AH, Ianbaz KH, Zaman M, Lateef A, Tariq SR, Ahmad HR. 1994. Hypotensive and spasmolytic activities of crude extract of *Cyperus scariosus*. Arch Pharm Res 17, 145–149.

Gilani A-UH, Janbazkh. 1995. Studies on protective effect of *Cyperus scariosus* extract on acetaminophen and CCl4-induced hepatotoxicity. General Pharmacol 26, 627–631.

Goetghebeur P. 1998. Cyperaceae. In: The Families and Genera of Vascular Plants. Flowering Plants. Monocotyledons, Vol IV, Kubitzki K (ed.). Springer Verlag, Heidelberg, pp 141–190.

Guillerm H. 1987. Le Souchet Comestible, un-Problem déSormais Present en France. Lyceé Agricole 'Charlemagne'. Station de Boigneville, Maisse; AGPM, Carcassonne.

Hall DW, Vandiver VV, Ferrell JA. 2009. Purple Nutsedge, *Cyperus rotundus* L. Institute of Food and Agricultural Sciences, University of Florida, Gainesville, Florida.

Harborne JB, Williams CA, Wilson KL. 1982. Flavonoids in leaves and inflorescences of Australian *Cyperus* species. Phytochemistry 21, 2491–2507.

Hassanein HD, Nazif NM, Shahat AA, Hammouda FM, Aboutable EA, Saleh MA. 2014. Chemical diversity of essential oils from *Cyperus articulatus, Cyperus esculentus* and *Cyperus papyrus*. J Essent Oil-Bear Plants 17, 251–264.

Herrera-Calderon O, Santiváñez-Acosta R, Pari-Olarte B, Enciso-Roca E, Montes VMC, Acevedo JLA. 2017. Anticonvulsant effect of ethanolic extract of *Cyperus articulatus* L. leaves on pentylenetetrazol induced seizure in mice. J Tradit Complement Med 8, 95–99.

Himaja N, Anitha K, Joshna A, Pooja M. 2014. Review article on health benefits of *Cyperus routundus*. Indian J Drugs 2, 136–141.

Holm LG, Plucknett DL, Pancho JV, Herberg JP. 1977. The World's Worst Weeds: Distribution and Biology. University of Hawai'i Press, Hawaii.

Hu Q-P, Cao X-M, Hao D-L, Zhang L-L. 2017. Chemical composition, antioxidant, DNA damage protective, cytotoxic and antibacterial activities of *Cyperus rotundus* rhizomes essential oil against foodborne pathogens. Sci Rep 7, 45231.

Jeong S, Miyamoto T, Inagaki M, Kim Y, Higuchi R. 2000. Rotundines A-C, three novel sesquiterpene alkaloids from *Cyperus rotundus*. J Nat Prod 65, 673–675.

Jiangsu New Medical College. 1971. Dictionary of Chinese Materia Medica. Shanghai People's Publishing House, Shanghai.

Kakarla L, Katragadda SB, Botlagunta M. 2015. Morphological and chemoprofile (liquid chromatography-mass spectroscopy and gas chromatography-mass spectroscopy) comparisons of *Cyperus scariosus* R. Br and *Cyperus rotundus* L. Pharmacogn Mag 11, 439–447.

Kakarla L, Katragadda SB, Tiwari AK, Kotamraju KS, Madhusudana K, Kumar DA, Botlagunta M. 2016. Free radical scavenging, α-glucosidase inhibitory and anti-inflammatory constituents from Indian sedges, *Cyperus scariosus* R.Br and *Cyperus rotundus* L. Pharmacogn Mag 12, S488–S496.

Kakarla L, Mathi P, Allu PR, Rama C, Botlagunta M. 2014. Identification of human cyclooxegenase-2 inhibitors from *Cyperus scariosus* (R.Br) rhizomes. Bioinformation 10, 637–646.

Kamala A, Middha SK, Gopinath C, Sindhura HS, Karigar CS. 2018a. *In vitro* antioxidant potentials of *Cyperus rotundus* L. rhizome extracts and their phytochemical analysis. Pharmacogn Mag 14, 261–267.

Kamala A, Middha SK, Karigar CS. 2018b. Plants in traditional medicine with special reference to *Cyperus rotundus* L.: A review. 3 Biotech 8, 309.

Kasana B, Sharma SK, Singh L, Mohapatra S, Singh T. 2013. *Cyperus scariosus*: A potential medicinal herb. Int Res J Pharm 4, 17–20.

Kavaz D, Idris M, Onyebuchi C. 2019. Physiochemical characterization, antioxidative, anticancer cells proliferation and food pathogens antibacterial activity of chitosan nanoparticles loaded with *Cyperus articulatus* rhizome essential oils. Int J Biol Macromol 123, 837–845.

Kern JH. 1974. Cyperaceae. Flora Malesiana Ser 1(7), 435–753.

Kilani S, Sghaier MB, Limem I, Bouhlel I, Boubaker J, Bhouri W, Skandrani I, Neffatti A, Ammar RB, Dijoux-Franca MG, Ghedira K, Chekir-Ghedira L. 2009. *In vitro* evaluation of antibacterial, antioxidant, cytotoxic and apoptotic activities of the tuber's infusion and extracts of *Cyperus rotundus*. Bioresour Technol 99, 9004–9008.

Kim SJ, Ryu B, Kim H-Y, Yang Y-I, Ham J, Choi J-H, Jang DS. 2013. Sesquiterpenes from the rhizomes of *Cyperus rotundus* and their potential to inhibit LPS-induced nitric oxide production. Bull Korean Chem Soc 34, 2207–2210.

Kritikar KR, Basu BD. 1918. Indian Medicinal Plants. Indian Press, Allahabad.

Lawal OA, Oyedeji AO. 2009a. Chemical composition of the essential oils of *Cyperus rotundus* L. from South Africa. Molecules 14, 2909–2917.

Lawal OA, Oyedeji AO. 2009b. The composition of the essential oil from *Cyperus distans* rhizome. Nat Prod Commun 4, 1099–1102.

Metuge JA, Nyongbela KD, Mbah JA, Samje M, Fotso G, Babiaka SB, Cho-Ngwa F. 2014. Anti-onchocerca activity and phytochemical analysis of an essential oil from *Cyperus articulatus* L. BMC Complement Altern Med 14, 223.

Milliken W, Albert B. 1996. The use of medicinal plants by the Yanomami Indians of Brazil. Econ Bot 50, 10–25.

Mohamed GA. 2015. Iridoids and other constituents from *Cyperus rotundus* L. rhizomes. Bull Fac Pharm Cairo Univ 53, 5–9.

Mohdaly AARAA. 2019. Tiger nut (*Cyperus esculentus* L.) oil. In: Fruit oils: Chemistry and functionality, Ramadan MF (ed.). Springer Nature, Dordrecht, pp. 243–269.

Morikawa T, Xu F, Matsuda H, Yashikawa M. 2002. Structures and radical scavanging activities of novel nor-stilbene dimer, longusone A, and new stilbene dimer, longusone A, B, and C, from Egyptian herbal medicine *Cyprus longus*. Heterocycles 57, 1983–1988.

Morimoto M, Fujii Y, Komai K. 1999. Antifeedants in Cyperaceae: Coumaran and quinones from *Cyperus* species. Phytochemistry 51, 605–608.

Mulligan GA, Junkins BE. 1976. The biology of Canadian weeds. 17. *Cyperus esculentus* L. Can J Plant Sci 56, 339–350.

Murashige T, Skoog F. 1962. A revised medium for rapid growth and bioassays with tobacco tissue cultures. Physiol Plant 15, 473–497.

Nassar MI, Abu-Mustafa EA, Abdel-Razik AF, Dawidar AM. 1998. A new flavanan isolated from *Cyperus conglomeratus*. Pharmazie 53, 806–807.

Nguta JM, Appiah-Opong R, Nyarko AK, Yeboah-Manu D, Addo PG. 2015. Medicinal plants used to treat TB in Ghana. Int J Mycobacteriol 4, 116–123.

Nicoli CM, Homma AOK, Matos GB, Menezes AJEA. 2006. Aproveitamento de biodiversidade amazônica: o caso da priprioca. Embrapa Amazônia Oriental, Belém, Pará.

Nogueira ML, de Lima EJSP, Adrião AAX, Fontes SS, Silva VR, de S Santos L, Soares MBP, Dias RB, Rocha CAG, Costa EV, da Silva FMA, Vannier-Santos MA, Cardozo NMD, Koolen HHF, Bezerra DP. 2020. *Cyperus articulatus* L. (Cyperaceae) rhizome essential oil causes cell cycle arrest in the G 2/M phase and cell death in HepG2 cells and inhibits the development of tumors in a xenograft model. Molecules 25, 2687.

Okwu GN, Abanobi SE, Nnadi UV, Ujowundu CO, Ene AC. 2015. Hypolipidemic properties of ethanol extract of *Cyperus rotundus* rhizome. Int J Biochem Res Rev 7, 132–138.

Oluwajuyitan TD, Ijarotimi OS. 2019. Nutritional, antioxidant, glycaemic index and antihyperglycaemic properties of improved traditional plantain-based (Musa AAB) dough meal enriched with tigernut (*Cyperus esculentus*) and defatted soybean (*Glycine max*) flour for diabetic patients. Heliyon 5, e01504.

Pirzada AM, Ali HH, Naeem M, Latif M, Bukhari AH, Tanveer A. 2015. *Cyperus rotundus* L.: Traditional uses, phytochemistry, and pharmacological activities. J Ethnopharmacol 174, 540–560.

Rakotonirina VS, Bum EN, Rakotonirina A, Bopelet M. 2001. Sedative properties of the decoction of the rhizome of *Cyperus articulatus*. Fitoterapia 72, 22–29.

Raut NA, Gaikwad NJ. 2006. Antidiabetic activity of hydro-ethanolic extract of *Cyperus rotundus* in alloxan induced diabetes in rats. Fitoterapia 77, 585–588.

Rocha FG, de Mello Brandenburg M, Pawloski PL, da Silva Soley B, Costa SCA, Meinerz CC, Baretta IP, Otuki MF, Cabrini DA. 2020. Preclinical study of the topical anti-inflammatory activity of *Cyperus rotundus* L. extract (Cyperaceae) in models of skin inflammation. J Ethnopharmacol 254, 112709.

Rukunga GM, Muregi FW, Omar SA, Gathirwa JW, Muthaura CN, Peter MG. 2008. Anti-plasmodial activity of the extracts and two sesquiterpenes from *Cyperus articulatus*. Fitoterapia 79, 188–190.

Sabiu S, Ajani EO, Sunmonu TO, Ashafa AOT. 2017. Kinetics of modulatory role of *Cyperus esculentus* L. on the specific activity of key carbohydrate metabolizing enzymes. Afr J Tradit Complement Altern Med 14, 46–53.

Said HM. 1982. Disease of the Liver: Greco Arab Concepts. Hamdard Foundation Press, Karachi.

Samariya K, Sarin R. 2013. Isolation and identification of flavonoids from *Cyperus rotundus* Linn. *in vivo* and *in vitro*. J Drug Deliv Ther 3, 109–113.

Sayed HM, Mohamed MH, Farag SF, Mohamed GA, Ebel R, Omobuwajo ORM, Proksch P. 2006. Phenolics of *Cyperus alopecuroides* Rottb. inflorescences and their biological activities. Bull Pharm Sci Assiut Univ 29, 9–32.

Schippers P, ter Borg SJ, Bos JJ. 1995. A revision of the infraspecific taxonomy of *Cyperus esculentus* (yellow nutsedge) with an experimentally evaluated character set. Syst Bot 20, 461–481.

Seabra RM, Andrade PB, Ferreres F, Moreira MM. 1997. Methoxylated aurones from *Cyperus capitatus*. Phytochemistry 45, 839–840.

Seabra RM, Moreira MM, Costa MAC, Paul MI. 1995. 6,3′,4′- trihydroxy-4-methoxy-5-methylaurone from *Cyperus capitatus*. Phytochemistry 40, 1579–1580.

Seabra RM, Silva AMS, Andrade PB, Moreira MM. 1998. Methylaurones from *Cyperus capitatus*. Phytochemistry 48, 1429–1432.

Seo EJ, Lee D-U, Kwak JH, Lee S-M, Kim YS, Jung Y-S. 2011. Antiplatelet effects of *Cyperus rotundus* and its component (+)-nootkatone. J Ethnopharmacol 135, 48–54.

Sivapalan SR. 2013. Medicinal uses and pharmacological activities of *Cyperus rotundus* Linn—a review. Int J Sci Res Publ 3, 1–8.

Sivapalan SR, Jeyadevan P. 2012. Physico-chemical and phyto-chemical study of rhizome of *Cyperus rotundus* Linn. Int J Pharmacol Technol 1, 2277–3436.

Smith TA. 1977. Phenethylamine and related compounds in plants. Phytochemistry 16, 9–18.

Soumaya K-J, Aicha N, Limem I, Boubaker J, Skandrani I, Sghair MB, Bouhlel I, Bhouri W, Mariotte AM, Ghedira K, Franca M-GD, Chekir-Ghedira L. 2009. Relationship correlation of antioxidant and anti-proliferative capacity of *Cyperus rotundus* products towards K562 erythroleukemia cells. Chem Biol Interact 181, 85–94.

Soumaya KJ, Dhekra M, Fadwa C, Zied G, Ilef L, Kamel G, Leila CG. 2013. Pharmacological, antioxidant, genotoxic studies and modulation of rat splenocyte functions by *Cyperus rotundus* extracts. BMC Complement Altern Med 13, 28.

Soumaya K-J, Zied G, Nouha N, Mounira K, Kamel G, Genviève FDM, Leila GC. 2014. Evaluation of *in vitro* antioxidant and apoptotic activities of *Cyperus rotundus*. Asian Pac J Trop Med 7, 105–112.

Srivastava RK, Singh A, Srivastava GP, Lehri A, Niranjan A, Tewari SK, Kumar K, Kumar S. 2014. Chemical constituents and biological activities of promising aromatic plant nagarmotha (*Cyperus scariosus* R.Br.): A review. Proc Indian Natl Sci Acad 80, 525–536.

Sultana S, Ali M, Mir SR. 2017. Chemical constituents from the rhizomes of *Cyperus rotundus* L. Open Plant Sci J 10, 82–91.

Sundaram MS, Sivakumar T, Balamurugan G. 2008. Anti-inflammatory effect of *Cyperus rotundus* Linn. leaves on acute and sub-acute inflammation in experimental rat models. Biomedicine 28, 302–304.

Suresh Kumar SV, Mishra SH. 2005. Hepatoprotective activity of rhizomes of *Cyperus rotundus* Linn against carbon tetrachloride-induced hepatotoxicity. Indian J Pharm Sci 67, 84–88.

Talukdar AD, Tarafdar RG, Choudhury MD, Nath D, Choudhury S. 2011. A review on pteridophyte antioxidants and their potential role in discovery of new drugs. Assam Univ J Sci Technol Biol Environ Sci 7, 151–155.

Tam CU, Yang FQ, Zhang QW, Guan J, Li SP. 2007. Optimization and comparison of three methods for extraction of volatile compounds from *Cyperus rotundus* evaluated by gas chromatography-mass spectrometry. J Pharm Biomed Anal 44, 444–449.

Thebtaranonth C, Thebtaranonth Y, Wanauppathamkul S, Yuthavong Y. 1995. Antimalarial sesquiterpenes from tubers of *Cyperus rotundus*: Structure of 10,12-peroxycalamenene, a sesquiterpene endoperoxide. Phytochemistry 40, 125–128.

Thomas D, Govindhan S, Baiju EC, Padmavathi G, Kunnumakkara AB, Padikkala J. 2015. *Cyperus rotundus* L. prevents non-steroidal anti-inflammatory drug-induced gastric mucosal damage by inhibiting oxidative stress. J Basic Clin Physiol Pharmacol 26, 485–490.

Townsend CC, Guest E. 1985. Flora of Iraq, Monocotyledones, Excluding Gramineae, Vol 8, Minister of Agriculture and Agrarian Reform, Erbil.

Tran HHT, Nguyen MC, Le HT, Nguyen TL, Pham TB, Chau VM, Nguyen HN, Nguyen TD. 2014. Inhibitors of α-glucosidase and α-amylase from *Cyperus rotundus*. Pharm Biol 52, 74–77.

Watt G. 1972. A Dictionary of the Economic Products of India. Vol II, Cosmo Publications, Delhi.

Wills GD. 1987. Description of purple and yellow nutsedge (*Cyperus rotundus* and *C. esculentus*). Weed Technol 1, 2–9.

Xu H-B, Ma Y-B, Huang X-Y, Geng C-A, Wang H, Zhao Y, Yang T-H, Chen X-L, Yang C-Y, Zhang X-M, Chen J-J. 2015. Bioactivity-guided isolation of anti-hepatitis B virus active sesquiterpenoids from the traditional Chinese medicine: Rhizomes of *Cyperus rotundus*. J Ethnopharmacol 171, 131–140.

Yang Z, Ji H, Liu D. 2016. Oil biosynthesis in underground oil-rich storage vegetative tissue: Comparison of *Cyperus esculentus* tuber with oil seeds and fruits. Plant Cell Physiol 57, 2519–2540.

Yang Z, Liu D, Ji H. 2018. Sucrose metabolism in developing oil-rich tubers of *Cyperus esculentus*: Comparative transcriptome analysis. BMC Plant Biol 18, 151.

Yeung H-C. 1985. Handbook of Chinese Herbs and Formulas. Institute of Chinese Medicine, Los Angeles.

Ying J, Bing X. 2016. Chemical constituents of *Cyperus rotundus* L. and their inhibitory effects on uterine fibroids. Afr Health Sci 16, 1000–1006.

Zaki AA, Ross SA, El-Amier YA, Khan IA. 2018. New flavans and stilbenes from *Cyperus conglomeratus*. Phytochem Lett 26, 159–163.

Zhou Z, Yin W. 2012. Two novel phenolic compounds from the rhizomes of *Cyperus rotundus* L. Molecules 17, 12636–12641.

Zhou Z, Zhang H. 2013. Phenolic and iridoid glycosides from the rhizomes of *Cyperus rotundus* L. Med Chem Res 22, 4830–4835.

Zoghbi MDG, Andrade EH, Carreira LM, Rocha EA. 2008. Comparison of the main components of the essential oils of "priprioca": *Cyperus articulatus* var. articulatus L., *C. articulatus* var. nodosus L., *C. prolixus* Kunth and *C. rotundus* L. J Essent Oil Res 20, 42–45.

Zoghbi MGB, Andrade EHA, Oliveira J, Carreira LMM, Guilhon GMSP. 2006. Yield and chemical composition of the essential oil of the stems and rhizomes of *Cyperus articulatus* L. cultivated in the state of Pará, Brazil. J Essent Oil Res 18, 10–12.

Index

Note: numbers in *italics* indicate a figure.
Note: for species, the ordering of entries is as follows: first letter [i.e., C. (as in *Callistemon*)] followed by all members of subspecies (i.e., *C. citrinus*); followed by formulas starting with numbers; followed by formulas starting with symbols. All other entries under species name are then alphabetized normally.

A

abortifacients
    C. ambrosioides, 63
    C. articulates, 282
    C. rufescens, 209
acacetin, 65
    acacetin 7-*O*-glucuronide, 172
acetone, 122, 140
    C. viminalis, 1
    Z, Z-farnesyl, 100, 101
acetone extract
    C. album, 61, 64
    C. arvensis, 194
    C. bungei, 171
    C. iners, 84, 86, 87
    C. limon, 140
    C. tenuiflorus, 38
    C. verum, 93
Achyranthes aspera, 124
Aedes spp.
    A. aegypti, 93, 94, 166, 214, 240
    A. albopictus, 240
allergic pleurisy, 213
allergy, 163, 234, 258
alkaline phosphatase, 61, 88, 122, 124–127, 140, 144, 169
alkaloids
    Callistemon, 2, 7
    Cinnamomum, 84
    Cissus, 129
    Clerodedrum, 165
    convolamine, 200
    Convolvulus, 197, 199, 220
    Cordia, 209, 201
    glutaramide, 214
    isoquinoline, 37
Allium sativa, 168
amentoflavone, 279, *298*
analgesics or analgesic effect
    Callistemon, 3, 8
    Carduus, 36
    Cinnamomum, 87, 89
    Cissus, 121, 127, 128
    Citrus, 139, 143
    Clerodedrum, 164, 165, 167, 170, 172
    Cordia, 211, 212, 213
anemia, 47, 88, 121, 209, 282
anthocyanidin
    proanthocyanidin, 84–85, 90, 102
anthocyanin, 7, 39, 102
    acylated anthocyanin trioside, 199
    monomeric, 49

anthraquinones, 239
aphrodisiac, 121, 143, 210, 211, 281
apigenin, 37, 144, 166
    apigenin-6,7-dimethyl ether, 8
    apigenin 6,8-di-C-glucoside, 145, 147
    apigenin-7-glucoside, 39, 42, *42*, 164, 169
    apigenin 7-*O*-glucuronide, 42, 172
    apigenin-8-C-glucoside, 146
    derivatives, 37
arthritis
    adjuvant, 48
    adjuvant-induced, 61, 166
    antarthritic, 135, 170
    rheumatic, 142
    rheumatoid, 129
Aspergillus spp.
    A. flavus, 92, 166
    A. fumigatus, 103
    A. niger, 92, 166, 241, 266
Ayurvedic system
    Carduus used in, 37
    Chenopodium used in, 60
    Cinnamomum used in, 84, 88, 91
    Cissus used in, 121
    Citrus used in, 139
    Clerodendrum used in, 163
    Cordia used in, 209

B

bergenin, 129, 130, *131*
bilirubin, 39, 61, 140, 144, 197, 210, 259, 278, 282
blumenol A, 6, *14*, 64, *71*
bronchitis
    Callistemon used to treat, 4
    Chenopodium used to treat, 63
    Citrus used to treat, 142
    Clerodendrum used to cure, 165
    Cyperus used to treat, 276

C

caffeic acid, 280, 287
    Carduus, 38, 39
    Chenopodium, 66, 73
    Chimaphila, 79
    Cinnamomum, 92
    Clerodendrum, 169
    Convolvulus, 198, 199
    Cordia, 213, 217
    Crassocephalum, 239
    Cyclopia, 250

Cymbopogon, 260, 264
Cyperus, 280, 287
caffeoylquinic acid
    1-*O*-caffeoylquinic acid, 38, *42*
    5-*O*-caffeoylquinic acid, 38, *42*
    O-caffeoylquinic acid, 38, 287
Callistemon species, 1–18
    C. citrinus, 2–4, 6–8, 18
    C. comboynensis, 4–5, 18
    C. lanceolatus, 2–6, 18
    C. rigidus, 4, 6, 8
    C. viminalis, 1–2, 5–6, 17
    C. viridiflorous, 8, 18
    3β-acetylmorolic acid, 7, *17*
    3β-Hydroxy-urs-11-en-13(28)-olide, *16*
    3-Methyltetradec-2-en-7-ol, 6, *16*
    5,7-Dihydroxy-6,8-dimethyl-4'-methoxy flavone, *16*
    5-Hydroxy-7,4'-dimethoxy-6,8-di, *16*
    5-Hydroxy-7,4'-dimethoxy-6-methylflavone, 6, *16*
    8-(2-hydroxypropan-2-yl)-5-hydroxy-7-methoxy-6-
        methyl-4'-methoxy flavone, *16*
    (–)-Myrtucommulone A and B, 9, *16, 17*
    blumenol A, 6, *14*
    callisalignenes, 8, *10, 12*
    callisalignones A–C, 8, *10*
    callisretones A, B, 8, *12*
    callistemenonone A, 5, 9, *16*
    callistenones A–P, 5–6, 8, *9, 11, 14–15*
    callistine, 7, *17*
    callistiviminenes A–O, 5, 8, *9–11*
    callistrilones A–Q, 6–8, *13–14, 17–18*
    callviminols, 6, *11–12*
    culture conditions, 17–18
    cyanidin 3,5-diglucoside, 7, *9*
    cyanin, 7, *9*
    dimethyleucaliptin, 7, 9, *17*
    epicallistrilone Q, 7, *17*
    eucaliptin, 7, 9, *14*
    eucaliptine, 6
    euglobal B, G, 8, *12*
    isoguaiacin, 9
    isomyrtucommulone B, 8, *11*
    morphological features, distribution,
        ethnopharmacological properties, phytochemistry,
        and pharmacological activities, 1–4
    myrtucommulone, A–D, 8, 9, *9, 10, 16, 17*
    nilocitin, 8, *16*
    peonidin, 7, *9*
    phytochemistry, 4–17
    piceatannol, 8, *17*
    quercetin-3-*O*-α-L-glucuronopyranoside, 8, *16*
    R = H - 8-Demethyleucalyptin, *17*
    R = H - Callistine A; R = CH3–6,8-Dimethyl-5,7-
        dihydroxy-4'-methoxy flavone, *17*
    scirpusin B, 8, *17*
    spathulenol, 4, 5, 6
    subulatone A, 9, *17*
    tormentic acid, 1–2, *15*
    ursenolide, 6, 7, *12*
    ursolic acid 6–7, *15*
    uvaol, 9
    viminalin B–L, 8, *11–12*
    watsonianone A, 5, *15*

Camellia sinensis, 50, 142, 147–148, 156
Canada, 36, 79
Candida spp.,
    C. albicans, 1,3 4, 5, 8, 26, 29, 63, 65, 65, 79, 86, 94,
        101, 165, 167, 212, 213, 214, 215, 246, 259, 262, 263
    C. glabrata, 26, 101, 241
    C. krusei, 26, 86
    C. parapsilosis, 26
    C. tropicalis, 26
carcinoma
    breast, 86, 103, 140
    colon adenocarcinoma, 265
    Ehrlich's ascites carcinoma (EAC), 165, 262
    hepatocellular, 6
    human cervical adenocarcinoma, 3, 171
    human esophageal adenocarcinoma, 96
    human gastric adenocarcinoma, 145
    human kidney, 171
    human lung, 171
    human oral epidermal adenocarcinoma, 145
    ileocecal, 171
    non-small-cell lung, 130
    urothelial, 91
Cardamomum species (*Elettaria cardamomum*), 23–32
    C.officianale, 23–32
    cardamonin, 29, *31*
    chemoprotective effects of, 28
    culture conditions, 31–32
    E. cardamomum, 23–32
    E. repens, 24
    green cardamom, 20, 33
    morphological features, distribution,
        ethnopharmacological properties, phytochemistry,
        and pharmacological activities, 23–29
    phytochemistry, 29–31
Carduus species, 36–44
    C. benedictus, 36, 44
    C. candicans 36, 39
    C. crispus, 36, 37, 39
    C. lanuginosus, 37
    C. nutans, 36, 39
    C. thoermeri, 36, 39
    1-*O*-caffeoylquinic acid, 38, *42*
    3,3' -biisofraxidin 38, *41*
    3,5-dihydroxyphenethyl alcohol 3-*O*-glucoside, 38, *41*
    3α, 24-Dihydroxyolean-12-en-28, 30-dioic acid
        dimethyl ester, 37, *43*
    3-*O*-acetyl-ursolic acid-28-ethyl ester, 37, *43*
    4-*O*-*p*-coumaroylquinic acid, 38, *41*
    5-*O*-caffeoylquinic acid, 38, *42*
    acanthoidine, 38, *40*
    acanthoine, 38, *40*
    antihypercholesterolemic effects of, 26
    apigenin-7-glucoside, 38, 39, *42*
    atherosclerosis treated with, 36
    bis (2-ethylhexyl) benzene-1,2-dicarboxylate, 37, *43*
    cnicin, *40*
    crispine, A–D, 37, *40, 41*
    culture conditions, 44
    cymaroside, 38, *42*
    cryptochlorogenic acid, 38, *42*
    erythrodiol 3-acetate, 38, *42*
    ethanol extract of, 23

diosmetin-7-O-α-L-arabinopyransyl (1‴→4″)-β-D-glucopyranoside, *43*

eurycarpin, 38, *41*

hispidulin, 39, *40*

isorhamnetin-3-glucoside, 39, *42*

kaempferol 3-glucoside, 38, *42*

kaempferol-3-O-α-L-rhamnoside, 37, 38, *43*

licochalcone, 38, *41*

luteolin, 37–39, *40, 42*

morphological features, distribution, ethnopharmacological properties, phytochemistry, and pharmacological activities, 36–37

onopordopicrin, *40*

phytochemistry, 37–44

R = Glc - Eleutheroside B1; R = H - Isofraxidin, 38, *41*

R = Glu - β-sitosterol-3-O-β-D-glucoside, 37, *44*

R = H - 5-O-p-coumaroylquinic acid, 38, *41*

salidroside, 38, *41*

scolymoside, 38, *42*

tachioside, 38, *41*

taraxasterol, 38, *42*

taraxasterol acetate, 38, *43*

tricin, 37, *42*

tricin-O-malonyl hexoside, 38

vanillic acid 4-O-β-D-glucoside, 38, *42*

carotene

β-carotene, 3, 49, 186, 199

β-carotene lineolic acid, 128

carotenoids

apocarotenoids, 64

Centaurea spp.

C. benedicta, 37

Centaurium species, 47–55

C. erythraea, 47–51, 55

C. littorale, 54, 55

1,3,6-trihydroxy-2,5-dimethoxyxanthone 1,6-Dihydroxy-3,5,7,8-tetramethoxyxanthone, *53*

1,3,8-trihydroxy-5,6-dimethoxy-xanthone, 51, *52*

1,6,8-trihydroxy-3,5,7-trimethoxyxanthone, *53*

1,6-dihydroxy-3,5-dimethoxyxanthone, *53*

1-hydroxy-2,3,4,5- tetramethoxyxanthone, *52*

1-hydroxy-2,3,4,7-tetramethoxyxanthone, *53*

1-hydroxy-2,3,5-trimethoxyxanthone, *52*

5-formyl-2,3-dihydroisocoumarin, *53*

amarogentin, 50, *52*

bellidifolin, *53*

culture conditions, 53–55

decussatin, 50, 51, *52*

demethyleustomin, 51, *52*

erythrocentaurin, 51, *52*

gentiopicrin, 51, *52*

gentiopicroside, 51, *52*

gentisin, *52*

isogentisin 1-Hydroxy-2,3,5-trimethoxyxanthone, 51, *52*

loganic acid, 50, 51, *52*

methylbellidifolin, 50, *52, 53,* 54

methylswertianin, 51, *53*

mesuaxanthone A, 51, *53*

morphological features, distribution, ethnopharmacological properties, phytochemistry, and pharmacological activities, 47–50

phytochemistry, 50–53

R^1 = H, R^2 = - amarogentin, 50, *52*

R = OH - swertiamarin; R = H – sweroside, 51, *52*

secologanin, *52*

secologanol, 51

secoxyloganin, 51, *52*

secoxyloganin, 51, *52*

swerchirin, *53*

sweroside, 51, *52*

tovopyrifolin C, *53*

chelation, 121

chemopreventative agents, 4, 121, 140, 148, 167

chemotherapeutics, 25, 147

Chenopodium, 60–73

C. ambrosioides, 63

C.album, 60–64

C.hircinum, 64

C. quinoa, 60–66, 73

2-(3, 4-Dihydroxyphenyl)-3, 5, 7- trihydroxy-4H-chromen-4-one trifolin, 64, *72*

3-O-(β-D-Glucopyranosyl)-oleanolic acid, *67*

(3R,6R,7E,9E,11E)-3-Hydroxy-13-apo-α-caroten-13-one (+)-abscisic alcohol, *70*

(6R,9R)-9-Hydroxy-4-megastigmen-3-one, *71*

(6S,7E,9E,11E)-3-Oxo-13-apo-α-caroten-13-one 3,6,9-Trihydroxy-4-megastigmene, 64, *71*

(6Z,9S)-9-Hydroxy-4,6-megastigmadien-3-one C-13 nor-terpenes, *70*

betanin, 66, *68*

blumenol A, 64, *71*

canthoside C, 65, *66*

chikusetsusaponin Iva, 64, 65, *70*

coniferyl alcohol, 63, *72*

culture conditions, 73

dihydroactinidiolide, 63, *72*

galactopyranosides, *67*

hexose-hexose-pentose, 73

isobetanin, 66, *68*

methyl spergulagenate, 65, *69, 70*

morphological features, distribution, ethnopharmacological properties, phytochemistry, and pharmacological activities, 60–63

neophytadiene, 50, 51, 63, *72*

oleanolic acid 3-O-β -D-glucopyranoside, *70*

oleanolic acid 3-O-β-D-glucuropyranoside, *70*

penstebioside, 65, *66*

phytochemistry, 63–73

quinoa protein isolate, 62

quinoa-saponin, 64, 65, 66, *70*

quinoa seed, 61, 63

R^1 = Glc(1–3)ara, R^2 = Glc - 3,23,30-trihydroxyolean-12-en-28-oic acid 3-O-β-D-Glucopyranosyl-(1,3)-α-L-arabinopyranosyl-28 -O-β-D-glucopyranoside, *68*

R^1 = Glc(1–3)ara, R^2 = Glc - 3β-O-β-D-glucopyranosyl-(1,3)- α -L-arabinopyranosyl-oxy-23-oxo- olean-12-en-28-oic acid β-D-glucopyranoside, *69*

R^1 = Glc(1–3)ara, R^2 = Glc - 3β-O-β-D-glucopyranosyl-(1,3)-α- L -arabinopyranosyl-oxy-27-oxo- olean-12-en-28-oic acid β-D-glucopyranoside, *68*

R^1 = H, R^2 = OH - Blumenol A; R^1 = R^2 = O - (+) dehydrovomifoliol, 64, *71*

R^1 = OH, R^2 = 2,6-Di-*O*-α-rhamnopyranosyl-
  β-galactopyranosyl - Quercetin
  3-*O*-(2″,6″-di- *O*-α-rhamnopyranosyl)-β-
R^1 = OH, R^2 = CH₃ - Grasshopper ketone; R^1 = CH₃,
  R^2 = OH – Zeaxantine, 64, *72*
R = OC₂H₅ - ethyl-*m*-digallate, *67*
R = α-L-Ara - 3-*O*-α-L-Arabinopyranosyl serjanic acid
  28-*O*-β-D-glucopyranosyl ester
R = β-D-Glc(1→3)-α-L-ara - 3β-[(O-β-D-
  glucopyranosyl-(1→3)-α-L-arabinopyranosyl)
  oxy]-23- oxo-olean-12-en-28-oic acid β-D-
  glucopyranoside, *67*
R = β-D-Glc(1→3)-α-L-ara - 3β-[(O-β-D-
  glucopyranosyl-(1→3)-α-L-arabinopyranosyl)
  oxy]-27- oxo-olean-12-en-28-oic acid β-D-
  glucopyranoside, *67*
R = β-D- GlcA - 3-*O*-β-D-Glucuronopyranosyl
  serjanic acid 28-*O*-β-D-glucopyranosyl ester,
  67
rhamnopyranose, *73*
S-(+)-3-Hydroxy-β-ionone, *71*
sinapinic acid, 39, 50, 62, *68*
trifolin, 64, *72*
zeaxantine, 64, *72*
Chimaphila species, 79–82
  C.umbellata, 79–80, 82
  4-Hydroxy-2,7-dimethylnaphthylene-1-*O*-β-D-
    glucopyranoside, *81*
  arbutin, 80, *80*
  avicularin, 80, *80*
  chimaphilin, 79, 80, *81*
  culture conditions, 82
  hyperoside, *80*
  isoarbutin, 80, *80*
  isofraxitin, 80, *81*
  methylsalicyclate, *80*
  morphological features, distribution,
    ethnopharmacological properties, phytochemistry,
    and pharmacological activities, 79
  phytochemistry, 79–82
  R = H - 1,3-Dihydroxy-2,7-dimethylnaphthyl 4-*O*-α-L-
    rhamnopyranoside, *81*
  R = H - 2,7-Dimethoxy-1,4,8-trihydroxynaphthalene, 8
  renifolin, 80, *80*
  taraxerol, *81*
  toluquinol, 80, *81*
  triacontane, *81*
cholesterol, 24–29, 61
  cholesterol esterase, 85, 86
  LDL, 122
Cinnamomum species, 84–111
  C. bejolghota, 87, 88, 101
  C. galucescens, 89, 102, 103
  C. perrottetti, 100, 103
  C. riparium, 95, 103
  C. sulphuratum, 96, 103
  C. tamala, 88, 99, 102
  C. verum, 84, 86, 78, 91, 92, 93, 103
  C. zeylanicum, 84–87, 91–92, 98, 104
  1, 4, cineole, 94, 95
  1, 8, cineole, 93, 94, 95, 96, 98, 99, 100, 101, 102, 103,
    104, *111*
  2-acetyl-5-dodecylfuran, 97, 98, *108*

4-methoxy stilbene, 100, 101, *109*
8S,14-cedranediol, 100, 101, *109*
α-copaene, 92, 93, 94, 95, 96, 97, 98, 99, 100, 101, 102,
  103, 104, *109*
α-guaiol, 93, 95
α-phellandrene, 91, 93, 94, 95, 96, 97, 98, 99, 100, 101,
  102, *110*
α-pinene, 91, 93, 94, 95, 96, 97, 98, 99, 100, 101, 102,
  110
α-terpineol, 91, 92, 93, 94, 95, 96, 97, 98, 99, 100, 101,
  102, 103, 104, *110*
α-terpinolene, 93, 94, 97, 100, 102, *110*
α-thujene, 91, 93, 94, 95, 96, 97, 98, 99, 100, 101, 102,
  103, *110*
α-*trans*-Bergamotene, *109*
α-ylangene, 93, 94, 96, 97, 99, 100, 101, 102, 104, *109*
α-zingiberene 100, 101, *109*
β-caryophyllene, 91, 92, 93, 94, 95, 96, 99, 100, 101,
  102, 103, 104, *111*
γ-terpinene, 91, 93, 94, 95, 96, 97, 99, 100, 101, 102,
  103, 104, *110*
(+)-diesamin, 102, *108*
(-)-sesamin, 97, 102, *105*
abienol, 100, 101, *109*
acorone 100, 101, *109*
actinodaphnine, 89, *104*
bergaptene, 100, 101, *109*
bicycloelemene, 93, 100, *110*
burmanol, 98, *105*
camphene, 89, 91, 92, 93, 94, 95, 96, 96, 98, 99, 100,
  101, 102, 103, 104, *110*
carisssone, 100, 101, *108*
caryolane-1,9β-diol, 98, *104*
cembrene A, 101, *108*
cineole, 97
cinnamaldehyde, 84, 86, 88, 90, 92, 93, 94, 95, 96, 96,
  98, 99, 100, 101, 102, 103, 104, *109*, *110*
cinnamtannin B1, 98, *104*
columellarin, 101, *108*
culture conditions, 111
E-2-hexyl cinnamaldehyde, 101, *109*
E-isoeugenyl benzyl ether, 101, *109*
epizonarene, 97
eugenol, 91, 92, 93, 94, 95, 96, 96, 98, 99, 100, 101,
  102, 103, 104, *109*, *111*
eumesm-7(11)-en-4-ol, *109*
erythro-guaiacylglycerol-β-*O*-4′-(5′)-
  methoxylariciresinol, 98, *104*
farnesol, 93, 95, 98, 100, 101, 102, *111*
feruloyl, 64, 72, 98, *107–108*
geraniol, 92, 93, 94, 95, 96, 98, 99, 100, 101, 102, 103,
  110
geranyl acetate 94, 96, 102, *109*
gernacrene A, 100, 104, *109*
guaiol, 94, 96, 100, 101, 102, *111*
isohibaene, 100, 101, *108*
isokotomolide A, 97, *108*
isolinderanolide B, 97, 98, *105*
isoobtusilactone A, 97, 98, *106*
isophilippinolide A, 98, *105*
juvibione, 100, 101, *109*
kotodiol, 97, 98, *108*
kotolactone A and B, 97, 98, *108*

limonene, 92, 93, 94, 95, 96, 97, 98, 99, 100, 101, 102, 103, 104, *110*

linalool/ linalool oxide, 92, 93, 94, 95, 96, 97, 98, 99, 100, 101, 102, 103, 104, *110*

lincolomide A and B, 98, *105*

morphological features, distribution, ethnopharmacological properties, phytochemistry, and pharmacological activities, 84–91

neryl isovalerate, 100, 101, *109*

obtusilactone A, 97, 98, *106*

phenethyl cinnamate, 100, 101, *109*

phytochemistry, 91–111

prezizaene, 100, 101, *109*

reticuol, 98, *105, 107*

sabinene, 91, 93, *110*

secosubamolide, 98, 99, *105, 107*

secosubamolide A, *107*

sclareolide, 100, 101, *109*

spathulenol, 92–103

subamolides A–E, 99, *106–107*

terpinen-4-ol, 91, 92, 94, 95, 96, 98, 99 100, 101, 102, 103, 104, *110*

trans-carveol, 93, 100, 101, 102, 103, *110*

trans-pinocarveol, 101, 102, *110*

trans-piperitol, 101, 102, *110*

trans-sabinene hydrate, 95, 101, 102, *110*

tricycline, *110*

validinol, 98, *105*

validinolide, 98, *105*

verbenone, 93, 94, 95, 99, 100, 101, 102, *110*

zonarene, 100, 101, *109*

(Z)-2-decenal, *110*

Z-α-*trans*-bergamotol, 101, *109*

Z-α-trans-bergamotol acetate, 101, *109*

(Z)-β-ocimene, *110*

(Z)-citral, *110*

(Z, Z)-farnesol, *111*

Cissus species, 121–133

   C. assamica, 129, 130

   C. javana, 129

   C. quadrangularis, 121–130, 133

   1,2-bis-(5-γ-Tocopheryl)ethane, 130, *132*

   3,3′,4,4′-tetrahydroxybiphenyl, *132*

   3β-Hydroxyl stigmast-5-en-7-one, 130, *131*

   7-oxoonocer-8-ene-3β,21α-diol, 130, *132, 133*

   β-sitostenone, 130, *131*

   β-sitosteryl glucoside, 130, *131*

   δ-amyrin, 130

   δ-amyrin acetate, *133*

   δ-Amyrone, 130, *133*

   bergenin, 129, 130, *131*

   betulinic acid, 130, *131*

   cissusic acid, 130, *131*

   cissuside, 130, *131*

   cissusol, 130, *131*

   culture conditions, 133

   diadzein, *133*

   epi-friedelinol, 130, *131*

   epi-Glut-5(6)-en-ol, 130, *132*

   ergosterol peroxide, 130, *131*

   estradiol, *133*

   friedelin, 126, 130, *131*

morphological features, distribution, ethnopharmacological properties, phytochemistry, and pharmacological activities, 121–129

pheophytin-a, 130, *131*

phytochemistry, 129–13

resveratrol, *133*

stigmasta-4,22-dien-3-one, 130, *131*

trans-resveratrol-3-*O*-glucoside, *133*

Citrus species, 139–156

   C. aurantium var bergamia, 139, 141–142, 144–145, 147

   C. bergamia, 139–141, 144–145

   C. limon, 139–147

   C. medica var. limetta, 139

   C. medica var. limon, 139

   C. natsudaidai, 148

   C. unshiu, 148

   Camellia sinensis, 142, 147–148, 156

   2′-hydroxy-3,4,3′,4′,5′,6′-hexamethoxychalcone, 147, *151*

   2′-hydroxy-3,4,4′,5′,6′-pentamethoxychalcone, 147, *151*

   3,5,6,7,3′,4′-hexamethoxyflavone, 147, *149*

   3,5,6,7,4′-pentamethoxyflavone, 147, *149*

   3,5,6,7,8,3′4′-heptamethoxyflavone, 147, *149*

   3,5,7,8,3′,4′-hexamethoxyflavone, 147, *149*

   3-hydroxy-5,6,7,4′-tetramethoxyflavone, 147, *150*

   3-hydroxy-5,6,7,8,4′-pentamethoxyflavone, *150*

   5,6,7,4′-tetramethoxyflavanone, *150*

   5,6,7,4′-tetramethoxyflavone, 147, *149*

   5,7,3′,4′-tetramethoxyflavone, 147, *149*

   5,7,4′-trimethoxyflavone, 147, *149*

   5,7,8,3′,4′-pentamethoxyflavone, 147, *149*

   5,7,8,4′-tetramethoxyflavone, 147, *149*

   5-demethyltangeretin 147, *151*

   5HHMF, 145, *151*

   5HPMF, 145, *151*

   5HTMF, 145, *151*

   5-hydroxy-3,6,7,8,3′,4′-hexamethoxyflavone, 147, *150*

   5-hydroxy-3,7,3′,4′-tetramethoxyflavone, *150*

   5-hydroxy-3,7,8,3′,4′-pentamethoxyflavone, 147, 148, *150*

   5-hydroxy-6,7,3′,4′-tetramethoxyflavone, 147, *150*

   5-hydroxy-6,7,4′-trimethoxyflavone, 147, *150*

   5-hydroxy-6,7,8,3′,4′-pentamethoxyflavanone, 147, *151*

   5-hydroxy-6,7,8,3′,4′-pentamethoxyflavone, 147, *151*

   5-hydroxy-6,7,8,4′-tetramethoxyflavone, *150*

   5-hydroxy-7,8,3′,4′-tetramethoxyflavone, 147, *149*

   6,8-di-C-glucosylapigenin, *152*

   7-hydroxy-3,5,6,3′,4′-pentamethoxyflavone, 147, *150*

   7-hydroxy-3,5,6,8,3′,4′-hexamethoxyflavone, 147, *150*

   8-methoxykaempferol-3-*O*-neoheperidoside, *153*

   8-prenylnaringenin, 145, *155*

   biochanin A, 147, *155*

   brutieridin, 145, *151*

   citrusin I–IV, 147, *148*

   cosmosiin, 145, *155*

   culture conditions, 156

   cyanidin 3-(6″-dioxalylglucoside), 147, *154*

   cyanidin 3-(6″-malonylglucoside), 147, *154*

   cyanidin 3-sophoroside, 147, *154*

   daidzein, 147, *154, 155*

   daidzin, 147, *155*

   delphinidin 3-(6″-malonylglucoside), 147, *154*

delphinidin-3-glucoside Cyanidin 3-glucoside, 147, *154*

didymin, 145, *155*

diosmin, 145, *155*

eriodictyol, 140, 144, 147

formononetin, 147, *154*, *155*

genistein, 147, *154*

genistin, 147, *155*

hesperidin, 140, 141, 143, 144, 145, 146, 147, 148, *154*

isoformononetin, *155*

isorhamnetin-3-*O*-hexosyl(1→6)hexoside, *152*

isorhamnetin-3-*O*-neoheperidoside, *152*

isorhamnetin-3-*O*-hexosyl(1→2)hexoside, *153*

isosakuranetin, 147, *153*, *154*

isosinensetin, 145, *156*

kaempferol-3,4'-di-*O*-hexoside 8-methoxykaempferol-
    3-*O*-hexosyl(1→2)hexoside, *153*

kaempferol-3-*O*-hexosyl(1→2)hexoside, *153*

kaempferol-3-*O*-hexosyl(1→2)hexoside-7-*O*-
    rhamnoside, *152*

kaempferol-3-*O*-neoheperidoside, *152*

kaempferol-3-*O*-rutinoside-7-*O*-rhamnoside, *152*

limocitrin, 147, *153*

limocitrol, 147, *153*

limolin, 144, *148*

luteolin, 144

melitidin, 145, 146, *151*

morphological features, distribution,
    ethnopharmacological properties, phytochemistry,
    and pharmacological activities, 139–144

naringin, 140, 141, 143, 144, 145, 146, 147, 148, *153*

narirutin, 144, 145, 146, 147, *154*

neoeriocitrin, 140, 141, 144, 145, 146

neohesperidin, 140, 141, 144, 145, 146, 147, *156*

neohesperidose, 147

nobiletin, 141, 142, 144, 145, 146, 147, 18, *151*

nomilin, 144, 145, *149*

pedunculin, *152*

peonidin 3-(6''-malonylglucoside), 147, *155*

phytochemistry, 144–156

poncirin, 141, 43, 144, 145, 146, *149*

prunetin, 147, *155*

quercetin-3-*O*-hexosyl(1→2)hexoside, *153*

rhoifolin, 145, 146, *155*

sinensetin, 144, 145, 146, 147, *151*, *155*

sissotrin, *155*

spathulenol, 145, 146

sudachitin, 145, *155*

tangeretin, 141, 144, 145, 146, 147, 148, *151*, *155*

tetra-*O*-methyl scutellarin, *152*

xanthohumol, 145, *156*

Cladosporium spp., 266

   C. cucumeriunum, 172, 214

Clerodendrum species, 163–187

   C. bungei, 163, 171

   C. chinense, 163, 172

   C. grayi, 163

   C. indicum, 163, 167, 168, 172, 186

   C. inerme, 163, 166, 167, 170, 186

   C. infortunatum, 163, 165, 172

   C. petasites, 163, 164, 169

   C. phlomidis, 163, 166, 170

   C. serratum, 163–164, 186

   C. serratum var. amplexifolium, 163

C. trichotomum, 163–164, 169

1-hydroxy-1-(8-palmitoyloxyethyl) cyclohexanone,
    169, *181*

1-monoacetin, 172, *178*

2-({6-*O*-[(4-Hydroxy-3-methoxyphenyl)carbonyl]-β-D-
    glucopyranosyl}oxy)-2-methylbutanoic acid, *177*

2-{(2*S*,5*R*)-5-[(1*E*)-4-Hydroxy-4-methylhexa-1,5-dien-
    1-yl]-5-methyltetrahydrofuran-2-yl} propan-2-yl
    β-D-glucopyranoside, *177*

(2*S*,3*S*,4*R*,10*E*)-2- [(2' *R*)-2'
    -hydroxytetracosanoylamino]-10-octadecene-1,3,4-
    triol *173*

3β-taraxerol acetate, 172, *178*

3-hydroxy-3',4'- dimethoxychalcone, 170, *181*

(3S,16R)-12,16-epoxy-3,6,11,14,17-
    pentahydroxy17(15→16)-abeo-5,8,11,13-
    abietatetraen-7-one, *175*

4,2',4'-Trihydroxy-6'-methoxyxchalcone-4,4'α-D-
    diglucoside, *174*

4-coumaric acid, 169, *173*

5-*O*-Ethylcleroindicin D, 171, *176*

6''-*O*-[(*E*)-caffeoyl] rengyoside B, *176*

7-hydroxyflavanone-7-*O*-glucoside, 170, *174*

7-hydroxyflavone, 170, *174*

7-*O*-methylwogonin, 170, *181*

(16*R*)-12,16-Epoxy-11,14,17-trihydroxy-17(15→16)-
    abeo-8,11,13-abietatrien-7-one, *174*

24β-ethylcholesta-5,22E,25-triene-3β-ol, 170, *174*

α-L-Rhamnopyranosyl-(1→2)α-D-glucopyranosyl-7-
    O-naringin-4'-*O*-α-D-glucopyranoside-5- methyl
    ether, *174*

acetylmartynoside, 171, 172, *177*, *179*

andrographolide, 170, *173*, *185*

apigenin-7-glucoside, *186*

bungein A, 171, *176*

bunginoside A, 171, *180*

bungnate A and B, 171, *180*

bungone A and B, 171, *181*

cirsilineol, 172, *176*

clerodendrin, 170, *173*

clerodenone A, 171, *176*

clerodinin A and B, 165, 172, *182*, *185*

clerodolone, 172, *177*

cleroindicin A–F, 169, 171, 172, *175–176*

clerosterol, 170, 171, 172, *178*

coleon U 172, *174*, *175*

cornoside, 172, *183*

culture conditions, 186–187

cryptojaponol, 172, *173*

cyrtophyllone A and B, 170, 171, 172, *175*, *177*, *180*

darendoside, 170, 171, *179*

decaffeoyl, 172, *183*

dihydrocornoside, 172, *183*

eucalyptin, 170, *181*

ferruginol, 170, 171, *178*

ferulic acid, 169, *173*

hispidulin, 169, 172, *186*

inerminoside, 170, *180*, *181*

isoverbascoside, 170, 172, *184*

jionoside C–D, 169, 172, *176*, *186*

lup-20(29)-en-3-triacontanoate, *174*

melittoside, 170, 172, *184*

monomelittoside, 170, 172, *184*

morphological features, distribution,
    ethnopharmacological properties, phytochemistry,
    and pharmacological activities, 163–169
nepetin, 169, *173*
oleanic acid, 165
oleanolic acid 3-acetate, 168, *186*
oleanolic aldehyde acetate, 172, *181*
pheophorbide-related compounds I–III, 171, *183*
phytochemistry, 169–186
plantainoside C, 169, *184*
R = D-Glucose - Pectolinaringenin-7-*O*-β-D-
    glucopyranoside, *173*
R = H - 3,4,5-Trihydroxy-6-[5-hydroxy-3-
    methoxy-2-(4- methoxy-phenyl)-4-oxo-4H-
    chromen-7- yloxy]-tetrahydro-pyran-2-carboxylic
    acid; R = CH₃ 3,4,5-Trihydroxy-6-[5-hydroxy-
    3-methoxy-2- (4- methoxy-phenyl)-4-oxo-4H-
    chromen-7-yloxy]-tetrahydro-pyran-2-carboxylic
    acid methyl ester, *173*
R = H - 24β-Ethylcholesta-5,22*E*,25-triene-3β-ol;
    R = D-Glucose - 24β-Ethylcholesta-5,22*E*,25-
    triene-3β-*O*-β-D-glucopyranoside, *173*
R = O - 12-Hydroxy-8,12-abietadiene-3,11,14-trione;
    R = H2 – royleanone, *177*
R = O - Taxodione; R = H₂ 11- Hydroxy-7,9(11),13-
    abietatrien-12-one, *177*
racemic rengyolone salidroside, 172, *184*
rengyoxide, 172, *184*
salvinolone 14-Deoxycoleon U 5,6-Dehydrosugiol, *174*
sammangaoside A–C, 170, *183, 184*
seguinoside K, 170, 172, *178*
serratumin A, 169, *173*
stigmasta-4,25-dien-3-one, 172, *181*
stigmasta-5, 170, 171, 172, *182*
sugiol, 169, 170, 172, *175*
teuvincenone A–H, 169, 171–172, *175*
trichotomoside, 171, 172, *178*
triancontanol, 171, 172, *182*
uncinatone, 169, 170, 171, *179*
verbascoside, 171, 172, *179, 183*
villosin B and C, 169–172, *175, 180*
colic, 84, 121, 125, 142, 165, 238, 262
    menstrual, 210
    renal, 50
Convolvulus species, 194–205
    C. alsinoids, 204
    C. arvensis, 194, 199, 204
    C. atheoides, 199
    C. austroaegyptiacus, 200
    C. hystrix, 200
    C. lineatus, 200
    C. persicus, 204
    C. pilosellifolius, 200
    C. pluricaulis, 194–197, 199, 204
    C. prostratus, 200
    C. scammonia, 194, 198, 199
    C. subhirsutus, 200
    2-[2-hydroxy-3-(N-methyl-2-pyrrolidinyl)-propanyl]-
        N-methylpyrrolidine, *202*
    2,4-*N*-methylpyrrolidinylhygrine A, 199, *202*
    2,5-di-(2-hydroxypropyl)-*N*-methylpyrrolidine, 200, *202*
    3,5-dicaffeoyl quinic acid, 199, *201*
    5-(2-hydroxypropyl)-hygrine, 199, *202*

5-(2-oxopropyl)-hygrine, 199, *202*
aromadendrin, 200, *201*
arvensic acids, E–J, 199, *203–204*
atropine, 199, *200*
convolamine, 200, *200*, 201
convolidine, 200, *200*
convoline, 200, *200*
convolvine, 200, *200*
culture conditions, 204–205
cuscohygrine, 199, *201*, 202
cyanidin 3-*O*-[6-*O*-(4-*O*- (6-*O*-(*E*-caffeoyl)-β-D-
    glucopyranosyl)-α-L-rhamnopyranosyl)-β-D-
    glucopyranoside]-5-*O*-β-D-glucopyranoside, 199,
    202
eriodictyol, 200, *201*
fisetin, 200, *201*
galangin, 200, *201*
hygrine, 199, 200, *201*, 202
isoferulic acid, 200, *201*
isoscopoletin, 200, *201*
luteolin, 199, *200*
mangiferin, *201*
meso-cuscohygrine, *201*
morphological features, distribution,
    ethnopharmacological properties, phytochemistry,
    and pharmacological activities, 194–198
N-methylpyrrolidinylcuscohygrine, 199, *202*
p-hydroxyphenylacetic acid, 199, 200, *201*
phygrine, 200, *202*
phyllalbine, 200, *200*
phytochemistry, 198–204
pinosylvin, 200, *201*
pseudotropine, 199, *200*
quercetin 3-*O*-rutinoside, 200, *201*
quercetin-7-*O*-rhamnoside, 200, *201*
scammonic acid A, 199
scammonin I–VIII, 199, *202–203*
scammonin resin, 194, 198
scopoletin, 196, 197, 198, 200
scopoletin-7-*O*-glucoside, 198, 199, *201*
scopolin, 200, *201*
scopoline, 198
taxifolin, 199, *201*
taxifolin scopoline, 200
tropine, 199, *200*
tropinone, 199, *200*, 202
Cordia species, 209–230
    C. alliodora, 212, 214,
    C. chacoensis 209
    C. collococca, 209, 217
    C. dichotoma, 209, 210, 216
    C. francisci, 213
    C. globifera, 209
    C. globosa, 209
    C. goetzei, 209
    C. latifolia, 209, 215
    C. leucocephala, 209, 215
    C. millenii, 215
    C. macleodii, 209, 211–212, 217
    C. myxa, 209, 210–211, 213, 229–230
    C. obliqua, 209, 210, 215
    C. rothii, 209
    C. rufescens, 209

C. salicifolia, 209

C. sebestena, 209

C. serratifolia, 213

C. spinescens, 209

C. trichotoma, 209

(1a*S**,1b*S**,7a*S**,8a*S**)-4,5-Dimethoxy-1a,7a-dimethyl-1,1a, 1b,2,7,7a,8,8a-octahydrocyclopropa [3,4] cyclopenta[1,2-b]naphthalene-3,6-dione, *227*

2-(2Z)-(3-hydroxy-3,7-dimethylocta-2,6-dienyl)-1,4-benzenediol, *229*

2,5-bis-(3′,4′-methylenedioxiphenyl)-3,4-dimethyltetrahydrofuran, 216, *219*

3-(2′,4′,5′-trimethoxyphenyl) propanoic acid, 216, *222*

3,4,5,3′,5′- pentamethoxy-1′-allyl-8.O.4′-neolignan, *219*

3,5,7,3′,4′-pentahydroxyflavonol, 216, *217, 219*

3,7,4′ -trimethoxyflavone, *218*

3,29-dioxoolean-12-en-27-oic acid

3α,6β, 25-trihydroxy-20(*S*), 24(*S*)-epoxydammarane, *222*

3α, 29-dihydroxyolean-12-en-27-oic acid, 214, *227*

3α-acetoxy-6β, 25-dihydroxy-20(*S*), 24(*S*)-epoxydammarane, *222*

3α-hydroxy-29-oxoolean-12-en-27-oic acid, 214, *227*

3α-hydroxyolean-12-ene-27,29-oic acid, 214, *227*

3α-hydroxyolean-12-en-27-oic acid, 214, *227*

3β-*O*-[α-L-rhamnopyranosyl-(1→2)-β-D glucopyranosyl] oleanolicacid 28-*O*-[β-D-xylopyranosyl- (1→2)-β-D-glucopyranosyl-(1→6)-b-D-glucopyranosyl], 216, *221*

3β-O-α-L-rhamnopyranosyl-(1→2)-β-D-glucopyranosyl pomolic acid, 216, *221*

3β-O-[α-L-rhamnopyranosyl-(1→2)-β-D-glucopyranosyl] pomolic acid 28-O-[β-D-glucopyranosyl- (1→6)-β-D-glucopyranosyl] ester, *220*

3β-*O*-[α-L-rhamnopyranosyl-(1→2)-β-D-glucopyranosyl]ursolic acid 28-*O*-[β-D-glucopyranosyl- (1→6)-β-D-glucopyranosyl], *221*

3-epipomolic acid, 215, *223*

3-*O*-α-L-rhamnopyranosyl-(1→2)-β-D-glucopyranosyl oleanolic acid 28-*O*-β-D-glucopyranosyl-(1→6)-β-D-glucopyranosyl ester, *222*

3-*O*-α-L-rhamnopyranosyl-(1→2)-β-D-glucopyranosyl pomolic acid 28-*O*-β-D-glucopyranosyl ester, *221*

3-oxoolean-12-en-27-oic acid, *227*

5,3′ -dihydroxy-3,7,4′-trimethoxyflavone, *217*

5,7,3′,4′ -tetrahydroxy-3-methoxyflavone, 215, *217*

5,7-dihydroxy-4′-methoxyflavone, *217*

5,7-dimethoxytaxifolin-3-*O*-α-L-rhamnopyranoside, *218*

5,8-dihydroxy-7,4′ -dimethoxyflavone, *217*

5-hydroxy-3,7,4′-trimethoxyflavone, *217*

7(3′,7′,11′,14′-Tetramethy)pentadec-2′,6′,10′-trienyloxycoumarin, *228*

7,4′-dihydroxy-5′-carboxymethoxy isoflavone, *218*

7,4′ -dihydroxy-5′-Me isoflavone, *218*

7-methoxyflavone, 215, *217*

α-cadinol, 216, *228*

β-sitosterol-β-D-glucoside, 216, *228*

(+)-1β,4β,6α-trihydroxyeudesmane, *222*

(+)-1β,4β,11-trihydroxyoppositane, *222*

(-)-1β,4β,7α-trihydroxyeudesmane, 216, *222*

alliodorin, 215, *228*

alliodorol, 214, *228*

allioquinol C, 214, *228*

allontoin, *228*

cabraleadiol, *222*

cholest-5-en-3ol (3-β) -carbonyl chlorinated, 212, 216, *217*

cincholic acid, 216, *227*

cordallinol, *229*

cordiachrome A–K, 214, 215, 216, *228–229*

cordiachromen A, 214, *229*

cordiaketal A and B, 215, *223*

cordialin A and B, 214, 216, *222–223, 228*

cordianal A–C, 215, *223–224*

cordianone, 215, *224*

cordiaquinol A–K, 214, *229*

cordiaquinones A–K, 210, 214–215, 217, *226*

cordiarimide A and B, 214, *226*

cordicilin, 215, *225*

cordifolic acid, 215, *222*

cordinoic acid, 215, *227*

cordinol, 215, 222, *225*

cordioic acid, 215, *222*

cordiol A, *229*

culture conditions, 229–230

distylin-3-xyloside, *218*

globiferane, 214, *226*

globiferin, 214, *226*

icterogenin, 215, *224*

kaempferol-3-*O*-β-D-glucosyl-6″-α-L-rhamnopyranoside, *219*

kaempferol-3,7-di-*O*-α-L-rhamnopyranoside, *219*

lantadene A and B, 215, *224*

latifolicinin A–D, 215, *225*

leucocordiachrome H, *229*

limonin, *228*

lup-20(29)-ene-3-*O*-β-D-maltoside, *223*

lupa-20(29)-ene-3-*O*-α-L-rhamnopyranoside, *223*

MECD-1, 216, *229*

MECD-2, 216, *229*

microphyllaquinone, 215, *227*

m-Methoxy-*p*-hydroxybenzaldehyde (*E*)-7-(3,4-Dihydroxyphenyl)-7-propenoic acid, *219*

morphological features, distribution, ethnopharmacological properties, phytochemistry, and pharmacological activities, 209–214

narigenin-4′,7-dimethyl ether, 215, *218*

oleanolic acid, 216, 221, 222, *228*

oncocalyxone A, 216, *228*

pachypodol, 215, *217*

phytochemistry, 214–229

quinovic acid, 216, 225, *228*

retusin, 215, *217*

R^1 = Fitil, R^2 = H - Phaeophytin A; R^1 = Fitil, R^2 = OH - 13^2 hydroxyphaeophytin A; R^1 = Etil, R^2 = H - 17^3 -ethoxypheophorbide A; R^1 = Etil, R^2 = OH -13^2 -Hydroxy-17^3 -ethoxypheophorbide A, *220*

R^1 = H - 1-benzopyran-2- one; R = OH - 7-Hydroxy-1-benzopyran-2-one, *219*

R^1 = H, R^2 = R^3 = OCH_3O - 8,8′-Dimethyl-3,4,3′,4′-dimethylenedioxy-7-oxo-2,7′cyclolignan; R^1 = OCH_3, R^2 = OCH_3, R^3 = H - 8,8′-Dimethyl-4,5- dimethoxy-3′,4′-methylenodioxy-7-oxo-2,7′cyclolignan, *220*

R = 2-methyl-2-Z-butenoyl – lantadene A, *224*

R = 3-methylbutenoyl - lantadene B, *224*

R = Glc-rha - isorhamnetin-3-*O*-rutinoside R = Glc-rha - Quercetin-3-*O*-rutinoside, *218*

R = Glc –Stigmast-5-en-3-*O*-β-D-glucoside, *220*

R = H - pachypodol; R = CH₃ – retusin, *224*

R = H - ursomic acid, *224*

R = - kaempferol-3-*O*-2G-rhamnosylrutinoside R = Glu-rha - kaempferol-3-*O*-rutinoside, *218*

R = OH - pomonic acid, *224*

R = - quercetin-3-*O*-2G-rhamnosylrutinoside R = Glc-rha - kaempferol-3-*O*-robinoside, *218*

rosmarinic acid, 214, 215, 217, *225*

rosmarinic acid 3-*O*-β-D-glucoside, *224*

sitosterol-3-*O*-β-D-glucopyranoside, 216, *220*

spathulenol, 214

stigmasterol-3-*O*-β-D-glucopyranoside, 216, *220*

tiliroside, 216, *217*

trichotomol, *222, 229*

ursomic acid, 215, *224*

Crassocephalum species, 236–242
  C. bauchiense, 238–240
  C. crepidioides, 236–240, 242
  C. macropappum, 239
  C. rubens, 237, 239, 241
  C. vitellinum, 239
  (2*S*,3*R*,47)-2-*N*-(2′-Hydroxytetracosanoyl) heptadecasphinga-4-ene, 241, *241*
  3′,5-Dihydroxy-4′,5′,6,7,8-pentamethoxyflavone, 240, 241, *242*
  4′,5-Dihydroxy-3′,5′,6,7,8-pentamethoxyflavone, *242*
  biafraecoumarins A–C, 241, *241*
  crassoside A, 241, *242*
  culture conditions, 242
  ent-2β,18,19-Trihydroxycleroda-3,13-dien-16,15-olide *trans*-α-bergamotene, *242*
  fernenol, 241, *241*
  morphological features, distribution, ethnopharmacological properties, phytochemistry, and pharmacological activities, 236–239
  phytochemistry, 239–242
  sorghumol acetate, 241, *241*
  spathulenol, 240

cryptochlorogenic acid, 38, 42

cryptone, 94, 99, 100, 101, 102, 103, 240

Culex quinquefasciatus, 94, 166, 240

cyanidin
  cyanidin 3,5-diglucoside, 7, 9, 147, *154*
  cyanidin-3-glucoside, 147, *154*
  cyanidin-3-*O*-glucoside, 7, 9
  cyanidin 3-*O*-[6-*O*-(4-*O*- (6-*O*-(*E*-caffeoyl)-β-D-glucopyranosyl)-α-L-rhamnopyranosyl)-β-D-glucopyranoside]-5-*O*-β-D- glucopyranoside, 199, *202*
  proanthocyanidin, 84, 85, 90, 102, 247
  procyanidin-B2, 86, 92

cyanin, 7, *9*
  anthocyanins, 7, 39, 49, 102, 199
  betacyanins, 66

Cyclopia species, 246–254
  C. genistoides, 246–251, 254
  C. intermedia, 246–248, 251
  C. longifolia, 249

C. subternata, 246–250, 254

afrormosin, 250, 251, *252, 254*

calycosin, 250, 251, *252, 254*

culture conditions, 254

diosmin, 250, *251*

eriocitrin, 246, 249, 250, *253*

eriodictyol, 246, 250–251

eriodictyol glucoside, 246

flemichapparin, 251, *252*

formononetin, 147, 250, 251, *252*

fujikinetin, 251, *254*

iriflophenone, 246, 249–251, *252–253*, 254

isomangiferin, 246, 249, 250, 251, *254*

isorhoifolin, 250, *253*

liquiritigenin, 250, *252*

luteolin, 246, 249, 250, 251

mangiferin, 246, 247, 248, 249, 250, 251, *253*

medicalgol, 251, *254*

morphological features, distribution, ethnopharmacological properties, phytochemistry, and pharmacological activities, 246–249

narirutin, 249, 250, 251, *254*

ononin, 250, *253*

orobol, 249, 251, *253*

phloretin, 249, 250, *252, 253*

phytochemistry, 249–254

rothindin, 250, *253*

scolymoside, 246, 249–251, *253*

shikimic acid, 1, 249, *253*

sophoracoumestan B, 251, *254*

vicenin-2, 249, 250, 251, *251*

Cymbopogon species, 258–269
  C. citratus, 258–260, 262–269
  C. flexuosus, 262, 267–269
  C. proximus, 262, 267, 269
  C. winterianus, 261, 266, 267
  1β-Hydroxy-α-eudesmol, *268*
  3β-methoxy lanosta-9(11)-en-27-ol, *268*
  3-feroylquinic acid, 264, *269*
  7α,11-Dihydroxycadin-10(14)-ene, *268*
  carlinoside, 264, *269*
  cassiaoccidentalin B, 264, *269*
  culture conditions, 269–270
  cymaroside, 264, *269*
  cymbopogonol, *268*
  isointermedeol, 268, *268*
  isoschaftoside, 260, 264, *269*
  kurilensin A, 264, *269*
  luteolin, 260, 264, *269*
  luteolin 7-*O*-neohesperidoside, 260, 264, *269*
  morphological features, distribution, ethnopharmacological properties, phytochemistry, and pharmacological activities, 258–263
  phytochemistry, 263–269
  spathulenol, 264, 265, 266
  veronicastroside, 264, *269*

cynaroside, 38, *42*, 264, *269*

Cyperus species, 276–299
  C. alopecuroides, 284, 287, 288
  C. esculentus, 280–281, 288
  C. longus, 288
  C. rotundus, 276–279, 283–285, 287–288, 299
  C. scariosus, 279–280, 284, 288

1-[2,3-Dihydro-6- hydroxy-4,7-dimethoxy2S-(prop-1-en-2- yl)benzofuran-5- yl]ethanone, *289*

1α-methoxy-3β-hydroxy-4α-(3′,4′- dihydroxyphenyl)-1, 2,3,4- tetrahydronaphthalin, 288, *290*

1-isopropyl2, 7-dimethylnaphthalene, *292*

2-hydroxy-14-calamenenone, *292*

2-oxo-α-cyperone rhombitriol, 285, *292*

2S-isopropenyl-4,8- dimethoxy-5-hydroxy-6- methyl-2,3- dihydrobenzo[1,2-b;5,4- b′]difuran, 288, *290*

2S-isopropenyl-4,8- dimethoxy-5-methyl-2,3- dihydrobenzo-[1,2-b;5,4-b′]difuran, 288, *290*

3β, 4α-Dihydroxy-7-epieudesm-11 (13), -ene, *292*

4, 5-seco-guaia-1 (10), 11-diene-4, 5-dioxo, 285, *292*

5,7,4′,5′-tetrahydroxy6,3′-diprenylflavanone, *289*

5-hydroxylucinone, 285, *292*

6″-O-p-coumaroylgenipin gentiobioside, *288*

6β-hydroxyipolamiide, 284, *295*

7α (H), 10β-E=eudesm-4-en-3-one-11,12-diol, *292*

10-hydroxyamorph-4-en-3-one, 285, *292* 12-methyl cyprot-3-en-2-one-13-oic acid, *295*

18-*epi*-α-amyrin glucuronoside, 284, *298*

α-amyrin glucopyranoside, 284, *298*

argutosine D, 285, *291*

aureusidin, 287, *293*, *297*

britanlin E, 285, *294*

cassigarol E and G, 277, 288, *288*, *289*

corymbolone, 285, 286, *288*

culture conditions, 299

cyperotundone, 283, 285, 286, 287, *288*

cyperusol C and D, 285, *292*

cyperusphenol A and D, 278, *294*

gallocatechin, 287, *297*

homocitric acid, 287, *297*

hydroxynootkatone 285, *291*

ipolamiide, 284, *295*

longusol C, 287, *297*

lupenyl arabinopyranosyl oleate, 284, *295*

luteolin, 279, 286, 287, *296*, *297*

mesocyperusphenol A, 278, *294*

methyl 3,4-dihydroxy benzoate, *295*

morphological features, distribution, ethnopharmacological properties, phytochemistry, and pharmacological activities, 276–283

mustakone, 283, 285, 286, 287, *288*

n-dotriacontan-16-one, 284, *295*

n-hexadecanyl linoleate, 284, *295*

n-hexadecanyl oleate, 284, *295*

nootkatone 283, 285, *291*

n-pentadecanyl linoleate, 284, *295*

n-tetracontan-7-one, 284, *295*

n-triacont-1-ol-21-one, 298

oplopanone, 285, *292*

oxyphyllol C, 285, *291*

pallidol, 288, *289*

patchoulenone, 279, 286, *297*

phytochemistry, 283–299

rotundine A–C, 288, *290*

rotunduside A and B, 288, *289*

scirpusin A and B, 277, 278, *288*, *294*

spathulenol, 283, 285, 286

sugebiol, 285

sugebiol 6-acetate, *293*

sugetriol, 278, 285, *293*, *294*

thunbergin A and B, 278, *297*

trans-cyperusphenol A, *297*

trans-resveratrol, 287, *297*

trans-scirpusin A and B, 287, *288*

cytokines, 24, 146, 166, 195

cytotoxicity
    Callistemon, 2, 4, 5, 6, 7, 8, 9
    Carduus, 36
    Chenopodium, 60, 65
    Chimaphilia, 80
    Cinnamomum, 87, 103
    Cissus, 124, 126, 130
    Citrus, 145
    Clerodendrum, 165, 166, 168, 169, 170, 172
    Cordia, 210, 215, 217
    Cyclopia, 248
    Cymbopogon, 262, 264
    Cyperus, 283, 285

D

daidzein, 63, 147, *154*, *155*

dehydrovomifoliol, 63, 71

dextran 247

diabetes
    Cadamomon, 24, 27, 29
    Centurium, 47, 48, 49
    Cinnamomum, 85, 91
    Cissus, 124, 136
    Clerodendrum, 163, 165, 168
    Crassocephalum, 237
    Cyclopia, 249
    Cymbopogon, 258
    Cyperus, 281
    type 2, 27, 249, 259

diarrhea
    antidiarrheals, 88, 163, 167, 170
    Callistemon, 1, 8
    Cadamomon, 23
    Carduus, 37
    Cinnamomum, 84, 98, 88, 89, 91
    Citrus, 139, 142
    Clerodendrum, 163, 165, 167, 168, 170
    Convolvulus, 194
    Cordia, 209, 212
    Cyperus, 279, 281

dysmenorrhea, 139, 209, 276

dyspepsia, 47, 60, 121, 125, 209

E

eczema, 36, 47, 210

Egypt, 267, 282
    traditional medicine of, 50, 258, 262

Ehrlich's ascites carcinoma (EAC), 165, 262

Elettaria cardamomum, 23–32
    E. cardamomum, 23–32
    E.repens, 24

ellagic acid, 5, 8, 130, 165, 239

Enterobacter spp., 1

eriodictyol

Citrus, 140, 144, 147
Convolvulus, 200, *201*
Cyclopia, 246, 250–251
eriodictyol glucoside, 246
eriodictyol-*O*-glucoside, 250
Erythrea centaurum, 47–48
Escherichia, 33
    E. coli, 86, 103, 241
    E. faecalis, 5, 6, 26, 37, 63, 86, 94, 140, 214, 238, 264,
        279
    E. faecium, 1, 37
estrogen
    phytoestrogen, 125
    receptors, 127, 248–249
    serum levels, in rats, 279

F

ferulic acid
    Carduus, 38, 39
    Centaurium, 50, 54
    Chenopodium, 60, 62, 63, 73
    Cinnamomum, 92
    Clerodendrum, 169, *173*
    Convolvulus, 198, 200
    Cordia, 217
    Cyclopia, 251
    Cyperus, 279, 287
    isoferulic acid, *201*
    trans-ferulic acid, 60
flavanones, 144, 147, 250
flavones, 6, 38, 141
    polymethoxylated, 147
flavonoids
    Callistemon, 3, 6, 7
    Carduus, 37, 39, 44
    Centaurium, 49, 51
    Chenopodium, 61, 62, 64, 66
    Chimaphilia, 79, 80
    Cinnamomum, 85, 87, 91, 92
    Cissus, 122, 128, 129, 130
    Citrus, 140, 141, 142, 144, 145, 146
    Clerodendrum, 164, 165, 170
    Convolvulus, 196, 198, 200
    Cordia, 210, 211, 212, 213, 215, 216, 229
    Crassocephalum, 236, 237, 239, 240
    Cymbopogon, 259, 260, 261, 264
    Cyperus, 277, 278, 279, 281, 287, 299
    isoflavanoids, 147
    oligomer, 278, 279
    polymethoxyflavanoids, 146
flavonol
    3,5,7,3′,4′-Pentahydroxyflavonol, 216, *217*
    acetylated flavonol glycosides, 51
    flavonol glycosides, 66
follicle stimulating hormone, 237, 238
formononetin, 147, *154*, *155*, 250, 251, *252*
Fusarium solani, 241, 266

G

gallic acid, 5, 6, 8, 24, 38, 39, 49, 51, 61, 79, 85, 88
gallocatechin, 287, *297*

epigallocatechin, 65, 249,
gastric disorders and diseases
    cancer cells, 143, 145
    lesions, in rats, 26
    ulcer, 48, 89, 197, 198, 238, 276
genistein, 63, 147, *154*, 199, 200, 249
genistin, 147, *155*
genotoxicity
    studies, 128
gonorrhea, 79, 210, 279
green tea, *see Camellia sinensis*

H

headache, 87, 89, 139, 209, 211, 258, 281
HeLa cancer cell line, 2, 3, 31, 65, 165, 171
Hemophilius influenzae, 102
hepatitis, 47, 167
hepatocarcinogenesis, 259
hepatotoxicity, 198, 282
herpes
    herpes simplex, 8
    treating, 142
hispidulin
    asthma treated with, 164
    Clerodendrum, 165, 169, 172, *186*
    hispidulin 7-glucoside, 39, *40*
    isolated, 164
hyperglycemia, 25
    antihyperglycemic activity and effects, 2, 47, 88, 90,
        143, 249
    in diabetic rats, 197
    in rats, 277
hyperlipidemia, 28
    antihyperlipidemic activity and effects, 86, 143, 237,
        239
    in rats, 141, 197
hyperoside, 6, 39, *80*
hypertension, 26, 47, 139, 142, 163, 188, 194
hypothalamus, 61
hypothermia, 198

I

insulin
    glipizide's impact on, 197
    insulin-like growth factor I and II, 125
    resistance, 24, 25, 29
    sensitivity, 25, 27, 141, 143
Iran, 60, 210, 276
isoquercitrin, 6, 8, 37, 217, 264, 266

J

Japan, 79, 91, 142, 143, 209, 236
    quinoa seeds, 66
    traditional medicine of, 163
jaundice, 37, 47, 139, 163, 194, 198, 210, 211

K

kaempferol
    8-methoxykaempferol-3-*O*-neoheperidoside, *153*

kaempferol 3-glucoside, 38, *42*
kaempferol-3,4′-di-*O*-hexoside 8-methoxykaempferol-
    3-*O*-hexosyl(1→2)hexoside, *153*
kaempferol-3,7-di-*O*-α-L-rhamnopyranoside, *219*
kaempferol-3-*O*-α-L-rhamnoside, 37, 38, *43*
kaempferol-3-*O*-β-D-glucosyl-6″-α-L-
    rhamnopyranoside, *219*
kaempferol-3-*O*-hexosyl(1→2)hexoside, *153*
kaempferol-3-*O*-hexosyl(1→2)hexoside-7-*O*-
    rhamnoside, *152*
kaempferol-3-*O*-neoheperidoside, *152*
kaempferol-3-*O*-robinoside, *218*
kaempferol-3-*O*-rutinoside-7-*O*-rhamnoside, *152*
Korea, 36, 79, 91, 163, 276
    C. intermedia extract's antiwrinkle effects on women
        from, 247
    C.platymamma, 146

L

Laos, 87
lemon juice, 84, 139
lemongrass, 116, 258, 259, 263
leptin, 123, 210
leucorrhea, 121
leukemia, 279
ligand expression, 125, 127
Listeria monocytegenes, 2, 120
liver
    cancer, 9
    damage, treating, 122
    disease, 211, 212
    disorders, 238
    enzymes, 25
    glycogen, 143
    homogenate, 121
    steatosis, 48
lupeol, 37, 64, 129, 168, 162, 199, 208, 215
lupenyl
    lupenyl arabinopyranosyl oleate, 284, *295*
luteinizing hormone, 237–238
luteolin
    Carduus, 37–39, *40*, *42*
    Citrus, 144
    Clerodendrum, 169, 172
    Convolvulus, 199, 200
    Cyclopia, 246, 249, 250, 251
    Cymbopogon, 260, 264, *269*
    Cyperus, 279, 286, 287, *296*, *297*
    3′,4′-dimethoxyluteolin, *296*
    5-deoxyluteolin, *253*

M

malaria
    antimalarial, 3, 165, 17, 279, 282
    curing, 84
    Kenyan traditional medicine to treat, 168
    treating, 142, 209, 281
Malassezia globosa, 80
mangiferin
    Convolulus, *201*

Cyclopia, 246, 247, 248, 249, 250, 251, *253*
    isomangiferin, 246, 249, 250, 251, *254*
menstruation
    dysmenorrhea, 139, 209, 276
    irregular, 121, 139, 282
    painful, 209
molluscum contagiosum, 66
monocytes, 87, 98
monoterpene, 8, 18, 25, 29, 30, 31, 286
    aldehyde, 146
    hydrocarbons, 94, 99, 144
mRNA, 80, 124–126, 141, 196
    glucokinase, 148
    NF-κB1, 236
    NFATc1, 145
    TNF-α, 236
    TRAP, 145
    UCP1, 124

N

neochlorogenic acid, 36, 38, 39, 264
nicotinamide, 143
nicotinic acid, 17
    isonicotinic acid, 282
Nigeria, 142, 209, 210, 238, 239, 276, 280

O

oleanolic acid, 6, 64–66, *67*, *71*, 130, 165, *228*
    oleanolic acid 3-acetate, 168, *186*
    oleanolic aldehyde acetate, 172, *181*
    oleanolic acid arbinoside, 284, *298*
oligomer flavonoids, 278, 279
oligomerization, 23, 24

P

Pakistan, 209, 210, 276
pancreatic α-amylase activities, 85
pancreatic β-cell, 237
pancreatic cancer, 6, 7, 9
pancreatic islets, 48
peonidin, 7, *9*, 147, *155*
phenolic acids, 39, 34, 54, 239, 250, 264, 281, 287
phenolic compounds, 39, 50, 54, 60, 66, 73, 74, 230, 236,
    254
    polyphenolic compounds, 5, 8
phenols, 165, 237, 238
Philippines, 89, 121, 209
phytosterol, 122, 165
pregnancy, 27
    in rats, 125
procyanidin
    B2, 86, 92
    B dimer, 287
pseudobaptigenin, 251, *254*
Pseudomas spp., 4
    P. aeruginosa, 103
    P. flursecence, 211
    P. picketti, 241
pseudoprotropine, 199, *200*

Q

quinoa plant, 61–66, 73

R

rheumatic pain, 50
    arthritis, 142
    antirheumatic properties, 60, 89, 178
rheumatism, 36, 47, 63, 84, 89, 01, 139, 163
Russia, 36, 79

S

salicylaldehyde, 91, 94, 101
salicylic acid, 211, 287
salicylic acid methyl ester, 80, 287
salidroside
    Carduus, 38, *41*
    Clerodendrum, 170, *184*
Salmonella spp., 211, 247
    S.choleraesuis, 267
    S. typhi, 3
Sarcina lutea, 167
scopolamine, 195–196, 211
scopoletin, 196–200, 287
    scopoletin-7-*O*-glucoside,199, *201*
scopolin, *201*
scopoline, 198, 200
scurvy, 121, 139
sesquisabinene hydrate, 145
sesquiterpenes, 31, 146, 285, 286
    oxygenated, 94, 99
Shigella spp., 211
    S.boydii, 167
    S.dysenteriae, 167
Siddha medicine, 121, 163, 209
silymarin, 278
snake bite, 63, 163, 165, 209
snake venom, 212
sore throat, 139, 258
spathulenol
    Callistemon, 4, 5, 6
    Cinnamomum, 92–103
    Citrus, 145, 146
    Cordia, 214
    Crassocephalum, 240
    Cymbopogon, 264, 265, 266
    Cyperus, 283, 285, 286
Sri Lanka, 23, 84, 87, 121, 123, 164, 166,
    168
Staphylococcus spp.
    S. aureus, 89, 102, 103, 165, 211, 241
stomachache, 36, 79, 209, 212, 237
Streptococcus spp., 128
    S. pneumoniae, 102
    Streptococcus-β-hemolyticus, 165
streptozotocin-induced diabetes in rats, 6, 27, 47–48, 86,
    88, 143, 156, 197, 206
superoxide
    CuZu superoxide dismuates, 48
    superoxide anion generation, 98, 130, 214

superoxide dismutase, 23, 24, 27, 28, 48, 122, 140, 143,
    169, 195, 196
superoxide radicals, 99
superoxide radical scavenging activity, 50
syphilis, 121, 163, 212, 279

T

tannic acid, 3, 92, 165, 166, 266
taraxerol, *81*, 130, 168, 172, 216
taraxerol acetate, 172, *178*
taraxsterol, 38, *42*
taraxsterol acetate, 38, *43*
terpenoids
    diterpenoids, 169, 170, 172
    meroterpenoids, 6, 7
    monoterpenoids, 94
    sesquiterpenoids, 285
    triterpenoids, 80, 129, 130, 210, 214, 216
testosterone, 24
Thailand, 87, 121, 129, 163, 164, 209, 258, 262
thrombolysis, 2
thrombosis, 33
thrombotic diseases 236
toothache, 87, 89, 91, 163
toothpowder, 139
traditional medicine
    Asia, 211
    Australia, 211
    Cameroon, 238
    Caribbean Islands, 212
    China, 1, 36, 163
    cinnamon used in, 84
    Egypt, 262
    Europe, 37, 47
    India, 163, 165, 210
    Iran, 211
    Japan, 163
    Nepal, 88
    Saudi Arabia, 198

U

ulcerative colitis, 142, 210
ulcer model
    aspirin-induced acute gastric, 28
    ethanol-induced, 26, 198
ulcers
    chronic, 121
    gastric, 89, 187, 238, 281
    peptic, 60
    stomach, 238
Unani system of medicine, 121, 163, 198, 209
urethral complaints, 209
urinary problems and diseases, 61, 211
    stones, 82, 87
    urinary tract infections, 142, 258
uterus, 209
    bleeding, 142
    developmental stage, 125
    fibroids, 279
    muscles, 49

rat, 279, 249, 279
stimulant, 167
tumors, 279

V

vagina
    slowing of opening, in rats, 249
venereal disease, 121
venereal infections, 167
venom
    Apis mellifera, 213
    antisnake, 212
vermifuge, 163, 279
Vietnam
    Cinnamomum in, 91, 94, 100

Cissus naturalized in, 129
Citrus in, 143
Clerodendrum in, 164, 168
Crassocephalum in, 236

X

xanthine/xanthine oxidase, 50, 140, 214, 250,
    278–279
xanthones, 49–55, 250–251

Y

yeast, 140, 238
    extract, 43, 44
Yunani medicine system, 91

For Product Safety Concerns and Information please contact our EU
representative GPSR@taylorandfrancis.com
Taylor & Francis Verlag GmbH, Kaufingerstraße 24, 80331 München, Germany

www.ingramcontent.com/pod-product-compliance
Lightning Source LLC
Chambersburg PA
CBHW080925220326
41598CB00034B/5680